Solutions Manual

to accompany Petrucci's
General Chemistry

Fourth Edition

Ralph H. Petrucci

California State University
San Bernardino

Macmillan Publishing Company
New York
Collier Macmillan Publishers
London

Printed in the United States of America

Macmillan Publishing Company
866 Third Avenue, New York, New York 10022

Collier Macmillan Canada, Inc.

ISBN: 0-02-394540-0

Printing: 5 6 7 8 Year: 8 9 0 1 2 3

TO THE STUDENT

The fundamental principles of chemistry are few in number yet powerful and wide ranging in their application. However, these principles are usually not mastered simply by reading about chemistry. They must be reinforced, through laboratory practice and through problem solving.

This supplement is designed to assist you in developing insight into chemical principles by offering brief solutions to the Review Problems, the Exercises, and the Self-Test Questions in the companion textbook: Petrucci, R. H., *General Chemistry, Principles and Modern Applications*, 4th ed., Macmillan, New York, 1985. We offer just a few suggestions for its most effective use.

The Review Problems and Exercises should be approached only after the text of the chapters has been studied carefully. They should be approached with pencil and paper in hand (an electronic calculator will also prove helpful); and a serious individual effort should be made to complete all of the Review Problems and at least a selective sampling of the Exercises. You may find it tempting to turn directly from the problems and exercises in the textbook to the solutions offered here, but if you succumb to this temptation, you will lose out on at least two counts. You will have reduced the problem solving aspect of chemistry just to some additional reading about chemistry, and you will have missed opportunities to discover, by yourself, alternative, and perhaps more original, solutions. Remember, it is not essential that you always "get the right answer". The right answers to these problems are known. What *is* most important is developing your ability to solve problems so that later you can successfully solve problems whose answers *are not* known.

For the most part, the methods used in this supplement are the same as in the textbook, although in some instances an alternative approach is suggested. Any logical method based on chemical principles that produces a correct answer (and sometimes even an incorrect answer) is an acceptable solution to a problem.

Problem solving means more than just successful manipulation of mathematical and chemical symbols in quantitative calculations. It requires also an ability to define terms clearly; to explain chemical phenomena; to represent chemical entities--atoms, ions, molecules, crystals--through names, formulas, and geometric sketches; to graph data and interpret these graphs; and so on. Try to develop your problem solving skills in this broadest sense.

Opportunities for making errors in problem solving are, unfortunately, all too numerous. You should not become discouraged as you make errors (take comfort in the fact that we all make them), but try to learn from your mistakes. First, if your answer disagrees with the one that is given only in the final digit, there probably is no error involved. The difference may stem from the method used in rounding off numbers. The method used in this supplement is to round off each intermediate result to the appropriate number of significant figures if the calculation is done in more than one step. If the calculation is done in a single step, only the final result is rounded off. Quite often an error is in a decimal point, that is, in a power of ten. In such cases you should expect a slip-up in exponential arithmetic; your chemical reasoning may be flawless. Other times answers may disagree by a simple numerical factor, that is, the answer may be twice what it should be, or only one half, and so on. In these cases you should suspect that a simple factor was omitted or used incorrectly. In any case, if you take the time to write down all the steps in the solution and to check your work carefully, you will find yourself less likely to make errors. Finally, you should be aware that often one learns best and most lastingly by making and then correcting errors.

Preparation of the pages of this book was a formidable task. It is a task that Jennifer Fisher and Linda Sheets undertook and saw through to the end, maintaining their good cheer no matter how complex some of the typing became. We are most grateful for their skill and assistance.

San Bernardino, California — Ralph H. Petrucci

Millersville, Pennsylvania — Robert K. Wismer

CONTENTS

Chapter 1

Matter--Its Properties and Measurement

Review Problems

1-1. (a) no. g = $1.17 \text{ kg} \times \frac{1000 \text{ g}}{1 \text{ kg}} = 1.17\times10^3 \text{ g}$

(b) no. m = $7115 \text{ mm} \times \frac{1 \text{ m}}{1000 \text{ mm}} = 7.115 \text{ m}$

(c) no. kg = $621 \text{ mg} \times \frac{1 \text{ g}}{1000 \text{ mg}} \times \frac{1 \text{ kg}}{1000 \text{ g}} = 0.000621 \text{ kg}$

(d) no. L = $673 \text{ mL} \times \frac{1 \text{ L}}{1000 \text{ mL}} = 0.673 \text{ L}$

(e) no. mm = $17.3 \text{ cm} \times \frac{10 \text{ mm}}{1 \text{ cm}} = 173 \text{ mm}$

(f) no. $\text{cm}^3 = 0.482 \text{ L} \times \frac{1000 \text{ mL}}{1 \text{ L}} \times \frac{1 \text{ cm}^3}{1 \text{ mL}} = 482 \text{ cm}^3$

(g) no. mg = $2.07 \text{ g} \times \frac{1000 \text{ mg}}{1 \text{ g}} = 2.07\times10^3 \text{ mg}$

(h) no. m = $0.481 \text{ km} \times \frac{1000 \text{ m}}{1 \text{ km}} = 481 \text{ m}$

1-2. (a) no. in = $21.18 \text{ ft} \times \frac{12 \text{ in}}{1 \text{ ft}} = 254.2 \text{ in}$

(b) no. lb = $416 \text{ oz} \times \frac{1 \text{ lb}}{16 \text{ oz}} = 26.0 \text{ lb}$

(c) no. in = $14.0 \text{ yd} \times \frac{36 \text{ in}}{1 \text{ yd}} = 504 \text{ in}$

(d) no. s = $2.5 \text{ h} \times \frac{60 \text{ min}}{1 \text{ h}} \times \frac{60 \text{ s}}{1 \text{ min}} = 9.0 \times 10^3 \text{ s}$

(e) no. yd = $2721 \text{ ft} \times \frac{1 \text{ yd}}{3 \text{ ft}} = 907.0 \text{ yd}$

(f) no. ft = $37.0 \text{ mi} \times \frac{5280 \text{ ft}}{1 \text{ mi}} = 1.95 \times 10^4 \text{ ft}$

1-3. (a) no. cm = $21 \text{ in} \times \frac{2.54 \text{ cm}}{1 \text{ in}} = 53 \text{ cm}$

(b) no. m = $22 \text{ ft} \times \frac{12 \text{ in}}{1 \text{ ft}} \times \frac{1 \text{ m}}{39.37 \text{ in}} = 6.7 \text{ m}$

(c) no. g = $11 \text{ oz} \times \frac{1 \text{ lb}}{16 \text{ oz}} \times \frac{454 \text{ g}}{1 \text{ lb}} = 3.1 \times 10^2 \text{ g}$

(d) no. lb = $63 \text{ kg} \times \frac{2.205 \text{ lb}}{1 \text{ kg}} = 1.4 \times 10^2 \text{ lb}$

(e) no. ft = $31.3 \text{ m} \times \frac{39.37 \text{ in}}{1 \text{ m}} \times \frac{1 \text{ ft}}{12 \text{ in}} = 103 \text{ ft}$

(f) no. oz = $5500 \text{ mg} \times \frac{1 \text{ g}}{1000 \text{ mg}} \times \frac{1 \text{ lb}}{454 \text{ g}} \times \frac{16 \text{ oz}}{1 \text{ lb}} = 0.19 \text{ oz}$

1-4. (a) no. $\text{m}^2 = 1 \text{ km}^2 \times \frac{(1000 \text{ m})^2}{(1 \text{ km})^2} = 1 \times 10^6 \text{ m}^2$

(b) no. $\text{ft}^2 = 1 \text{ mi}^2 \times \frac{(5280 \text{ ft})^2}{(1 \text{ mi})^2} = 2.79 \times 10^7 \text{ ft}^2$

(c) $\text{no. m}^2 = 1\text{ mi}^2 = \frac{(5280\text{ ft})^2}{(1\text{ mi})^2} \times \frac{(12\text{ in})^2}{(1\text{ ft})^2} \times \frac{(1\text{ m})^2}{(39.37\text{ in})^2} = 2.59\times10^6$

(d) $\text{no. in}^3 = 1\text{ ft}^3 \times \frac{(12\text{ in})^3}{(1\text{ ft})^3} = 1728\text{ in}^3$

(e) $\text{no. cm}^3 = 1\text{ ft}^3 \times \frac{(12\text{ in})^3}{(1\text{ ft})^3} \times \frac{(2.54\text{ cm})^3}{(1\text{ in})^3} = 2.83\times10^4\text{ cm}^3$

(f) $\text{no. nm}^3 = 1\text{ cm}^3 \times \frac{1\text{ m}^3}{(100\text{ cm})^3} \times \frac{(109\text{ nm})^3}{(1\text{ m})^3} = 1\times10^{21}\text{ nm}^3$

1-5. (a) $°F = (\frac{9}{5})°C + 32 = (\frac{9}{5} \times 35) + 32 = 63 + 32 = 95°F$

(b) $°C = \frac{5}{9}(°F - 32) = \frac{5}{9}(86 - 32) = \frac{5}{9} \times 54 = 30°C$

(c) $°C = \frac{5}{9}(°F - 32) = \frac{5}{9}(832 - 32) = 444°C$

(d) $°F = (\frac{9}{5})°C + 32 = (\frac{9}{5} \times -163) + 32 = -261°F$

1-6. Density = mass/volume = $3153\text{ g}/2500\text{ cm}^3 = 1.26\text{ g/cm}^3$

1-7. (a) $\text{no. g} = 3.50 \times 10^2\text{cm}^2 \times \frac{1.11\text{ g}}{\text{cm}^2} = 3.88\times10^2\text{ g}$

(b) $\text{no. L} = 2.50\text{ kg} \times \frac{1000\text{ g}}{1\text{ kg}} \times \frac{1\text{ cm}^3}{1.11\text{ g}} \times \frac{1\text{ L}}{1000\text{ cm}^3} = 2.25\text{ L}$

(c) $\text{no. lb} = 4.00\text{ gal} \times \frac{3.785\text{ L}}{1\text{ gal}} \times \frac{1000\text{ cm}^3}{1\text{ L}} \times \frac{1.11\text{ g}}{1\text{ cm}^3} \times \frac{1\text{ lb}}{454\text{ g}} = 37.0\text{ lb}$

1-8. no. g phosphorus = 25.0 lb fertilizer $\times \frac{454\text{ g fert.}}{1\text{ lb fert.}} \times \frac{6.6\text{ g phosphorus}}{100\text{ g fert.}} = 7.5\times10^2$ g phosphorus

1-9. no. g sample = 135.0 g sodium chloride $\times \frac{100.0\text{ g sample}}{99.2\text{ g sodium chloride}}$ = 136 g sample

1-10. In each case the decimal point is moved to the left or to the right to produce a coefficient with value between 1 and 10. The exponent (power of ten) indicates the number of places that the decimal point is moved. The exponent is positive if the decimal point is shifted to the left and negative if to the right.

(a) $3800 = 3.8\times10^3$

(b) $482000 = 4.82\times10^5$

(c) $0.100 = 1.00\times10^{-1}$

(d) $6,212 = 6.212\times10^3$

(e) $211,100 = 2.111\times10^5$

(f) $0.0000065 = 6.5\times10^{-6}$

(g) $0.0087 = 8.7\times10^{-3}$

(h) $0.0600 = 6.00\times10^{-2}$

(i) $22 = 2.2\times10^1$

1-11. The decimal point is moved to the left or to the right by the number shown in the exponent (power of ten). If the exponent is positive, the decimal point is moved to the right, and if negative, to the left.

(a) $6.18\times10^3 = 6180 = 6180$

(b) $8.12\times10^{-1} = 8.12 = 0.812$

(c) $4.613\times10^{-4} = 0004613 = 0.0004613$

(d) $43\times10^{-3} = 043 = 0.043$

(e) $3.47\times10^{-6} = 0.00000347$

(f) $670\times10^{-2} = 6.70$

(g) $6.2\times10^0 = 6.2$

(h) $2.98\times10^{10} = 29,800,000,000$

(i) $0.00168\times10^{-4} = 0.000000168$

1-12. (a) Three

(b) Two or three, it is not clear whether the terminal zero is significant.

(c) Two, the zero to the right of the decimal point is not significant.

(d) Five, both zeros are significant.

(e) Five, again, both zeros are significant.

(f) Five, because a zero is written to the right of the decimal point at the end of the number, both this zero and the one that precedes it are significant.

(g) Two to five, there is no indication how many, if any, of the zeros are significant.

(h) Four, the first of the zeros is not significant but the other two are.

(i) One, the zeros simply locate the decimal point.

1-13. The rules for "rounding off" numbers must be used here (see footnote on page 8 of the text).

(a) $3162.3 = 3162$ (d) $0.065045 = 0.06504$ (g) $186{,}000 = 1.860\times10^5$

(b) $3.2\times10^3 = 3.200\times10^3$ (e) $60\times10^{-5} = 6.000\times10^{-4}$ (h) $22\times10^4 = 2.200\times10^5$

(c) $218.51 = 218.5$ (f) $327.251 = 327.3$ (i) $14.7050 = 14.70$

1-14. (a) $200 \times 4000 = 2\times10^2 \times 4\times10^3 = 8\times10^5$

(b) $32 \times 30 \times 6100 = 3.2\times10^1 \times 3.0\times10^1 \times 6.1\times10^3 = 5.856\times10^6 = 5.9\times10^6$

(c) $0.087 \times 0.0040 = 8.7\times10^{-2} \times 4.0\times10^{-3} = 3.48\times10^{-4} = 3.5\times10^{-4}$

(d) $0.0070 \times 612 = 7.0\times10^{-3} \times 6.12\times10^2 = 42.84\times10^{-1} = 4.284 = 4.3$

(e) $\dfrac{5300}{0.0070} = \dfrac{5.3\times10^3}{7.0\times10^{-3}} = 0.757\times10^6 = 7.6\times10^5$

(f) $\dfrac{140 \times 600 \times 0.10}{0.030 \times 3.3} = \dfrac{1.40\times10^2 \times 6\times10^2 \times 1.0\times10^{-1}}{3.0\times10^{-2} \times 3.3} = 0.848\times10^5 = 8\times10^4$

1-15. (a) $$d = \frac{0.99984 + (5\times1.6945\times10^{-2}) - (5^2\times7.987\times10^{-6})}{1 + (5\times1.6880\times10^{-2})}$$

$$= \frac{0.99984 + 0.08472 - 0.00020}{1 + 0.08440} = \frac{1.08436}{1.08440} = 1.000\ \text{g/cm}^3$$

(b) $$d = \frac{0.99984 + (10\times1.6945\times10^{-2}) - (10^2\times7.987\times10^{-6})}{1 + (10\times1.6880\times10^{-2})}$$

$$= \frac{0.99984 + 0.16945 - 0.00080}{1 + 0.16880} = 0.9997\ \text{g/cm}^3$$

Exercises

Properties and classification of matter

1-1. An object displaying a physical property retains its basic chemical identity. Display of a chemical property is accompanied by a change in composition.

(a) Physical: The iron nail is not changed in any significant way when it is attracted to a magnet. Its basic chemical identity is unchanged.

(b) Chemical: The dark deposit on the surface of the silver indicates that the pure lustrous metal has been converted into a chemical compound, such as silver sulfide.

(c) Physical: The floating of ice on liquid water simply demonstrates the difference in a particular property of the two--density.

(d) Chemical: Large rubber molecules are broken down into smaller fragments by combining with ozone in the smog.

1-2. (a) Substance: Cane sugar is the chemical compound, sucrose, in a very high state of purity.

(b) Heterogeneous mixture: The chowder is visibly heterogeneous.

(c) Homogeneous mixture: Unleaded gasoline consists of a large number of hydrocarbon liquids and special additives.

(d) Heterogeneous mixture: Although mayonnaise has a smooth uniform appearance, suspended oil droplets can be detected through a microscope.

(e) Heterogeneous mixture: Pure salt is 100% sodium chloride (NaCl). Iodized salt contains small amounts of an iodide compound as well (usually postassium iodide).

(f) Homogeneous mixture: Tap water generally contains from a few to several hundred parts per million of dissolved mineral substances.

(g) Substance: Ice is simply solid water.

(h) Heterogeneous mixture: The paint has several components, and through the microscope, particles of the white pigment can be seen in the water- or oil-base medium.

1-3. (a) To obtain the elements, hydrogen and oxygen, from the compound water requires a chemical change.

(b) Seawater is a water solution containing a large quantity of dissolved substances (principally sodium chloride). Any process that separates the water from the dissolved salts is a physical change, e.g., distillation (boiling followed by condensation of the steam) or freezing (formation of ice crystals in the solution).

(c) Air is a simple physical mixture of gases--oxygen, nitrogen and traces of others. The individual gases can be separated by the physical process of liquefaction followed by distillation.

1-4. (a) Salt can be dissolved in water and the undissolved sand filtered off. The salt can be recovered by evaporating the solution to dryness.

(b) Draw a magnet through the mixture. Iron fillings are attracted to the magnet and the wood is left behind.

(d) Mineral oil is both insoluble in and less dense than water. The oil can be skimmed off the water or, better still, the water can be drawn off from the bottom of the mixture through a special funnel (called a separatory funnel).

1-5. (a) Extensive: The mass of air in a balloon depends on the volume of the balloon.

(b) Intensive: Ice melts at 0°C regardless of the mass or form of the ice involved--a block of ice, a glass of crushed ice, or a field of snow.

(c) Extensive: The melting point is an intensive property, but the length of time required to melt a sample of ice depends on the mass of the sample.

(d) Intensive: The color of the emitted light is very distinctive. It does not depend on the size or shape of the glass tube or the amount of neon present.

Scientific method

1-6. No. The greater the number of experiments that conform to predictions based on the law, the greater the confidence in the law. However, there is no point at which the law is ever verified with certainty.

1-7. The fewer assumptions involved in formulating a theory, and the greater the range of phenomena it explains, the more acceptable the theory becomes in comparison with alternative theories.

1-8. A given set of conditions, a cause, is expected to produce a certain result, an effect. Although cause-and-effect relationships may be very difficult to establish at times ("God is subtle"), they do exist nevertheless ("but He is not malicious").

Exponential arithmetic

1-9. Think first of how the number would be written out in decimal form. Then convert the number to the exponential form.

(a) 186 thousand = 186,000 = 1.86×10^5 mi/s

(b) 5 to 6 quadrillion = 5,000,000,000,000,000 to 6,000,000,000,000,000 = 5×10^{15} to 6×10^{15} t

(c) 173 thousand trillion = 173,000,000,000,000,000 = 1.73×10^{17} W

(d) one millionth = 0.000001 = 1×10^{-6} m = 1 μm

(e) ten millionths = 10 x 0.000001 = 1×10^{-5} m = 10 μm

1-10. (a) $0.048 + (62 \times 7.0\times10^{-4}) = 0.048 + 0.0434 = 0.091$

(b) $\dfrac{31 + 283 + (1.60\times10^{2})}{2.3\times10^{-1}} = \dfrac{474}{2.3\times10^{-1}} = 2061 = 2.1\times10^{3}$

(c) $\dfrac{(1.7\times10^{-3})^2}{0.060 + (2.0\times10^{-2})} = \dfrac{2.9\times10^{-6}}{0.080} = 3.6\times10^{-5}$

(d) $\dfrac{[(6.0\times10^{3}) + (4.2\times10^{4})]^2}{(2.2\times10^{3})^2 + 180{,}000} = \dfrac{(4.8\times10^{4})^2}{(4.8\times10^{6}) + (0.18\times10^{6})} = \dfrac{2.3\times10^{9}}{5.0\times10^{6}} = 4.6\times10^{2}$

Significant figures

1-11. (a) An exact number--12.

(b) The capacity of an irregularly shaped tank can only be determined by experiment and is thus subject to error.

(c) The distance between any pair of planetary bodies can only be determined through certain measured quantities. These measurements are subject to error.

(d) An exact number. Although the number of days may vary from one month to another (say from January to February), the month of January always has 31 days.

(e) The area is determined by calculation from measured dimensions; these are subject to error.

1-12. (a) $411 \times 183 = 75213 = 7.52\times10^{4}$

(b) $12.30\times10^{4} \times 3.5\times10^{5} = 43.05\times10^{9} = 4.3\times10^{10}$

(c) $\dfrac{5.11\times10^{2}}{1.7\times10^{5}} = 3.0059\times10^{-3} = 3.0\times10^{-3}$

(d) $35.24 + 36.3 + 1.08 = 72.62 = 72.6$

(e) $(1.561\times10^{3}) - (1.80\times10^{2}) + (2.02\times10^{4}) = (0.16\times10^{4}) - (0.02\times10^{4}) + (2.02\times10^{4}) = 2.16\times10^{-4}$

1-13. Expressed to the nearest milligram, each of the masses and their sum are

20.000 g + 1.000 g + 0.500 g + 0.200 g + 0.100 g + 0.050 g + 0.020 g + 0.005 g + 0.002 g = 21.877 g

1-14. The number 99.9 is expressed to 1 part in 999, and the number 1.008, to 1 part in 1008. The two numbers are about equally precise, i.e., known to 1 part per 1000, even though one number consists of three significant figures and the other, four. Their product, 100.7, would also be known to about 1 part per 1000. To round off this product to 101 would be reducing its precision to only 1 part per 100. Note also the products: 99.8 x 1.008 = 100.6, 99.9 x 1.008 = 100.7, and 100 x 1.008 = 100.8. These products are identical in the first three digits and differ in the fourth, four significant digits can be carried.

1-15. $\text{no. m} = 16.5\text{ ft} \times \dfrac{12\text{ in}}{1\text{ ft}} \times \dfrac{1\text{ m}}{39.37\text{ in}} = 5.03\text{ m}$

1-16. The prices must be expressed in the same units and then compared. For example, the cost of the 3 lb. can in \$/kg is

$$\frac{\text{no. \$}}{\text{kg}} = \frac{7.26}{3\text{ lb}} \times \frac{2.205\text{ lb}}{1\text{ kg}} = \$5.34/\text{kg}$$

This is cheaper than the other coffee, which is \$5.42/kg.

1-17. The quantity we are seeking has the units: s/100 m.

$$\frac{\text{no. s}}{100\text{ m}} = \frac{9.3\text{ s}}{100\text{ yd}} \times \frac{1\text{ yd}}{36\text{ in}} \times \frac{39.37\text{ in}}{1\text{ m}} = 10.2\text{ s}/100\text{ m}$$

1-18. The quantity sought is no. in.; the starting point is 1 link.

$$\text{no. in.} = 1\ \text{link} \times \frac{1\ \text{chain}}{100\ \text{link}} \times \frac{1\ \text{furlong}}{10\ \text{chain}} \times \frac{1\ \text{mi}}{8\ \text{furlong}} \times \frac{5280\ \text{ft}}{1\ \text{mi}} \times \frac{12\ \text{in}}{1\ \text{ft}} = 7.92\ \text{in.}$$

1-19. (a) $$\text{no. mg} = 2\ \text{tablets} \times \frac{5.0\ \text{gr}}{1\ \text{tablet}} \times \frac{1.0\ \text{g}}{15\ \text{gr}} \times \frac{1000\ \text{mg}}{1\ \text{g}} = 6.7\text{x}10^2\ \text{mg}$$

(b) $$\text{no.}\ \frac{\text{mg}}{\text{kg}} = \frac{6.7\text{x}10^2\ \text{mg}}{145\ \text{lb}} \times \frac{2.20\ \text{lb}}{1\ \text{kg}} = 10\ \frac{\text{mg}}{\text{kg}}$$

1-20. (a) $$\text{Volume} = (24 \times 36 \times 18)\ \text{in}^3 \times \frac{1\ \text{m}^3}{(39.37\ \text{in})^3} = 0.25\ \text{m}^3$$

(b) $$\text{surface area} = (2 \times 24 \times 36)\ \text{in}^2 + (2 \times 24 \times 18)\ \text{in}^2 + (2 \times 36 \times 18)\ \text{in}^2 = 3888\ \text{in}^2 = 3.9\text{x}10^3\ \text{in}^2$$

$$\text{no. cm}^2 = 3.9\text{x}10^3\ \text{in}^2 \times \frac{(2.54\ \text{cm})^2}{1\ \text{in}^2} = 2.5\text{x}10^4\ \text{cm}^2$$

1-21. (a) $$\text{no. ft}^2 = 1\ \text{acre} \times \frac{1\ \text{mi}^2}{640\ \text{acre}} \times \frac{(5280\ \text{ft})^2}{(1\ \text{mi})^2} = 43{,}560\ \text{ft}^2$$

(b) $$\text{no. hm}^2 = 1\ \text{acre} \times \frac{1\ \text{mi}^2}{640\ \text{acre}} \times \frac{(5280\ \text{ft})^2}{1\ \text{mi}^2} \times \frac{(12\ \text{in})^2}{1\ \text{ft}^2} \times \frac{1\ \text{m}^2}{(39.37\ \text{in})^2} \times \frac{1\ \text{hm}^2}{(100\ \text{m})^2} = 0.405\ \text{hm}^2$$

1-22. $$\text{no.}\ \frac{\text{km}}{\text{h}} = 1.38\ \text{Mach} \times \frac{1130\ \text{ft/s}}{1\ \text{Mach}} \times \frac{12\ \text{in}}{1\ \text{ft}} \times \frac{1\ \text{m}}{39.37\ \text{in}} \times \frac{1\ \text{km}}{1000\ \text{m}} \times \frac{60\ \text{s}}{1\ \text{min}} \times \frac{60\ \text{min}}{1\ \text{h}} = 1.71\text{x}10^3\ \text{km/h}$$

1-23. (a) If 1 in = 2.54 cm, exactly, the no. in = $1\ \text{m} \times \frac{100\ \text{cm}}{1\ \text{m}} \times \frac{1\ \text{in}}{2.54\ \text{cm}} = 39.37007874\ \text{in}$

Therefore, 1 in = 2.54 cm and 1 m = 39.37 in cannot both be exact.

(b) As seen in part (a), to six significant figures, 1 m = 39.3701 m.

Temperature scales

1-24. high: $°\text{C} = \frac{5}{9}(118 - 32) = \frac{5}{9}(86) = 47.8°\text{C}$

low: $°\text{C} = \frac{5}{9}(17 - 32) = \frac{5}{9}(-15) = -8.3°\text{C}$

1-25. The upper limit of the thermometer, in °F, is:

$°\text{F} = \frac{9}{5}°\text{C} + 32 = (\frac{9}{5} \times 110) + 32 = 198 + 32 = 230$

This is just below the "soft ball" stage--234 to 240°F. The thermometer cannot be used.

1-26. $°\text{F} = \frac{9}{5}°\text{C} + 32 = \frac{9}{5}(-273.15) + 32.00 = -491.67 + 32.00 = -459.67°\text{F}$

Density

1-27. mass = 209.4 - 110.4 = 99.0 g; volume = 125 cm^3; density = $\frac{99.0\ \text{g}}{125\ \text{cm}^3} = 0.792\ \text{g/cm}^3$

1-28. mass = $298\ \text{lb} \times \frac{454\ \text{g}}{1\ \text{lb}} = 1.35\text{x}10^5\ \text{g}$; volume = $42.0\ \text{gal} \times \frac{3.785\ \text{L}}{1\ \text{gal}} \times \frac{1000\ \text{cm}^3}{1\ \text{L}} = 1.59\text{x}10^5\ \text{cm}^3$

$$\text{density} = \frac{1.35\text{x}10^5\ \text{g}}{1.59\text{x}10^5\ \text{cm}^3} = 0.849\ \text{g/cm}^3$$

1-29. mass of carbon tetrachloride = 283.2 - 121.3 = 161.9 g

$$\text{volume of carbon tetrachloride} = 161.9 \text{ g} \times \frac{1 \text{ cm}^3}{1.59 \text{ g}} = 102 \text{ cm}^3$$

volume capacity of vessel = volume of carbon tetrachloride = 102 cm^3

1-30. Determine the volume, in quarts, of 2.50 kg milk.

$$\text{no. qt} = 2.50 \text{ kg} \times \frac{1000 \text{ g}}{1 \text{ kg}} \times \frac{1 \text{ cm}^3}{1.03 \text{ g}} \times \frac{1 \text{ L}}{1000 \text{ cm}^3} \times \frac{1.06 \text{ qt}}{1 \text{ L}} = 2.57 \text{ qt}$$

Two qt and 1 pt = 2.5 qt. For the remaining 0.07 qt, no. cup = $0.07 \text{ qt} \times \frac{4 \text{ cups}}{1 \text{ qt}} = 0.28 = 0.3$ cup

The volume of milk required is 2 qt, 1 pt, 0.3 cup.

1-31. It is necessary to determine the mass, in grams of each object.

(a) $\text{no. g iron} = (162 \text{ cm} \times 1.1 \text{ cm} \times 0.70 \text{ cm}) \times \frac{7.86 \text{ g iron}}{1 \text{ cm3}} = 9.8\times10^2 \text{ g iron}$

(b) $\text{no. g aluminum} = (165.0 \text{ cm} \times 23.0 \text{ cm} \times 0.0980 \text{ cm}) \times \frac{2.70 \text{ g aluminum}}{1 \text{ cm3}} = 1.00\times10^3 \text{ g aluminum}$

(c) $\text{no. g water} = 1.00 \text{ L} \times \frac{1000 \text{ cm}^3}{1 \text{ L}} \times \frac{0.998 \text{ g water}}{1 \text{ cm3}} = 998 \text{ g water}$

The order of increasing mass is (a) < (c) < (b).

1-32. The addition of 100 pieces of shot to water causes a volume increase of 8.8 ml - 8.4 ml = 0.4 ml. The volume per shot is 4×10^{-3} ml. The mass per shot is:

$$m = V \cdot d = 4\times10^{-3} \text{ mL} \times \frac{8.92 \text{ g}}{1 \text{ mL}} = 4\times10^{-2} \text{ g}$$

Percent composition

1-33. $\%A = \frac{7 \text{ A}}{55 \text{ grades}} \times 100 = 13\% \text{ A}$ $\%B = \frac{14 \text{ B}}{55 \text{ grades}} \times 100 = 25\% \text{ B}$

$\%C = \frac{24 \text{ C}}{55 \text{ grades}} \times 100 = 44\% \text{ C}$ $\%D = \frac{7 \text{ D}}{55 \text{ grades}} \times 100 = 13\% \text{ D}$

$\%F = \frac{3 \text{ F}}{55 \text{ grades}} \times 100 = 5\% \text{ F}$

1-34. $\text{no. g ethanol} = 5.75 \text{ L soln} \times \frac{1000 \text{ cm}^3 \text{ soln}}{1 \text{ L soln}} \times \frac{0.981 \text{ g soln}}{1 \text{ cm3 soln}} \times \frac{9.5 \text{ g ethanol}}{100 \text{ g soln}} = 5.4\times10^2 \text{ g ethanol}$

1-35. $\text{no. L soln} = 1.65 \text{ kg sodium hydroxide} \times \frac{1000 \text{ g sodium hydroxide}}{1 \text{ kg sodium hydroxide}} \times \frac{100 \text{ g soln}}{12.0 \text{ g sodium hydroxide}} \times$

$$\frac{1 \text{ cm}^3 \text{ soln}}{1.131 \text{ g soln}} \times \frac{1 \text{ L soln}}{1000 \text{ cm3 soln}} = 12.2 \text{ L soln}$$

Algebraic equations

1-36. Substitute %N = 1.15 into the following equation.

$$d(\text{g/cm}^3) = \frac{1}{1.153 - 0.00182 \ (\%N) + 1.08\times10^{-6} \ (\%N)^2}$$

$$= \frac{1}{1.153 - 0.00209 + 1.43\times10^{-6}} = \frac{1}{1.151} = 0.8688 \text{ g/cm}^3$$

1-37. Solve the following equation for t:

$$d(\text{g/cm}^3) = 1.5794 - 1.836\times10^{-3}\ (t - 15)$$

$$t = \frac{1.5794 - d}{1.836\times10^{-3}} + 15$$

Substitute $d = 1.543\ \text{g/cm}^3$

$$t = \frac{1.5794 - 1.543}{1.836\times10^{-3}} + 15 = \frac{0.036}{1.836\times10^{-3}} + 15 = 20 + 15 = 35°\text{C}$$

1-38. Volume of cube = l^3; volume of sphere = $\frac{4}{3}\pi r^3$

$$l^3 = \frac{4}{3}\pi r^3 = 4.19\ r^3;\ \frac{l^3}{r^3} = 4.19;\ \frac{l}{r} = (4.19)^{1/3} = 1.61$$

1-39. $°\text{C} = \frac{5}{9}\ (°\text{F} - 32)$. Substitute °F = °C. $°\text{C} = \frac{5}{9}\ (°\text{C} - 32);\ °\text{C} - \frac{5}{9}\ °\text{C} = \frac{5}{9}\ (-32);\ \frac{4}{9}\ °\text{C} = \frac{5}{9}\ (-32);$

$°\text{C} = \frac{9}{4} \times \frac{5}{9} \times (-32) = -40°$ To check this result, determine the Fahrenheit temperature that is equal to -40°C. You will find it to be -40°F.

1-40. One method is to calculate the density of water at several temperatures in the temperature interval 0° to 20°C to see whether the density varies continuously, i.e., increasing or decreasing continuously.

At 0°C: $d = \frac{0.99984}{1} = 0.99984\ \text{g/cm}^3$

At 5°C: $d = \frac{0.99984 + 5 \times 1.6945\times10^{-2} - (5)^2 \times 7.987\times10^{-6}}{1 + (5 \times 1.6880\times10^{-2})} = \frac{1.0844}{1.0844} = 1.000\ \text{g/cm}^3$

At 10°C: $d = \frac{0.99984 + 10 \times 1.6945\times10^{-2} - (10)^2 \times 7.987\times10^{-6}}{1 + (10 \times 1.6880\times10^{-2})} = \frac{1.1685}{1.1688} = 0.9997\ \text{g/cm}^3$

At 20°C: $d = 0.9985\ \text{g/cm}^3$

The density decreases continuously from 5°C to 20°C, but increases from 0°C to 5°C. We conclude that the density reaches a maximum at about 5°C (actually, 4°C).

A second method, for those familiar with calculus, involves taking the derivative of d with respect to t, setting this derivative equal to zero (as the condition for a maximum or mimimum), and solving the resulting equation for t. The final quadratic equation that must be solved is $t^2 + 118.5\ t - 504.4 = 0$. Its solution by the quadratic formula is $t = 4.1°\text{C}$.

Self-test Questions

1. (*c*) The mass 14.7 g is expressed to the nearest 0.1 g; 14.72 g to the nearest 0.01 g; 14.721 g to the nearest 0.001 g (0.001 g = 1 mg); 14.7213 g to the nearest 0.0001 g (one-tenth mg).

2. (*a*) Proceed in one of two ways. Convert all the lengths to a common unit (say, meters) and choose the largest. A simpler method is to compare the lengths two at a time. Since one meter is slightly longer than a yard (recall Figure 1-2), 4.0 m is greater than 12.0 ft: (a) > (c). The length 12 ft = 144 in: (c) > (b) and (a) > (c) > (b). One thousandth of a kilometer is simply one meter; length (d) is much less than (a).

3. (*d*) Again there are two possibiliites. One is to express all temperatures on the same scale (i.e., either °F or °C), and then choose the highest. Alternatively, temperatures (c) and (d) are above the boiling point of water (100°C = 212°F) and (a) and (b) are below. This eliminates (a) and (b). Temperature (c) is 5°F above the steam point and temperature (d) is 5°C above. Since five degrees of Celsius temperature is equivalent to nine degrees of Fahrenheit, temperature (d) must be the highest.

4. (*b*) You are expected to know the density of water: $1.00\ \text{g/cm}^3$. The density of the alcohol-water mixture is $0.83\ \text{g/cm}^3$. The density of the wood is less than $1.00\ \text{g/cm}^3$ (since a $10.0\ \text{cm}^3$ sample weighs less than 10.0 g). The density of chloroform is greater than $1.00\ \text{g/cm}^3$ (since a $100.0\ \text{cm}^3$ sample weighs more than 100.0 g). Note that no complete density calculation was required.

5. (*c*) Assign a volume to each part and select the largest. A 380 g sample of water has a volume of 380 cm^3. The volume of the chloroform is V = m/d = 600 g/1.5 g cm^{-3} = 400 cm^3. The 0.50 L of milk has a volume of 500 mL = 500 cm^3. The volume of steel is given--100 cm^3.

6. (*b*) The number 16.07 has four significant digits; 0.0140 has three (the final zero is significant); 1.070 has four significant digits; 0.016 has two. The number 200 has one, two, or three significant digits --we cannot be certain because the final zeros are used simply to locate the decimal point. The case of 0.0140 is without doubt, however.

7. (*d*) Since the fertilizer is only 20% N, more than 1 lb of fertilizer is required to provide 1 lb N. This fact eliminates items (b) and (c). 20% of 20 lb fertilizer would provide 4 lb N. 20% of 5 lb fertilizer provides 1 lb N.

8. (*a*) Oxygen is an element and, therefore, a substance. Sucrose is a compound, also a substance. Salad dressing is a homogeneous mixture of oil, water, egg whites, etc. Acid rain is the homogeneous mixture or solution. It is rainwater with dissolved acidic substances, such as oxides of sulfur and nitrogen.

9. (*a*) An element is one of a class of about 100 (actually 106) substances that cannot be reduced to simpler substances, either by physical or chemical changes. A compound is comprised of two or more elements and can be decomposed into these elements by appropriate chemical changes.

(*b*) The components of a heterogeneous mixture separate into physically distinct regions having different properties. In a homogeneous mixture the components are mixed uniformly--composition and properties are constant throughout the mixture.

(*c*) Mass is a measure of the quantity of matter in a sample. Density is the ratio of mass to volume and thus indicates the quantity of matter packed into a unit volume of a sample.

10. $(19.541 + 1.05 - 3.6) \times 651 = 17.0 \times 651 = 1.11 \times 10^4$
The sum of terms in the parenthetical expression must be stated to the nearest 0.1. The product of the three-digit numbers (17.0 and 651) must be expressed to three significant figures (1.11); the power of ten (10^4) locates the decimal point.

11. $$\text{no. lb of ethyl alcohol} = 55.0\text{ gal} \times \frac{3.78\text{ L}}{1\text{ gal}} \times \frac{1000\text{ cm}^3}{1\text{ L}} \times \frac{0.789\text{ g}}{1\text{ cm}^3} \times \frac{1\text{ lb}}{454\text{ g}} = 361\text{ lb ethyl alcohol}$$

total mass = 75.0 lb drum + 361 lb ethyl alcohol = 436 lb

12. First determine the area of the room in square yards.

$$\text{no. yd}^2 = 6.52\text{ m} \times 4.18\text{ m} \times \frac{(39.37\text{ in})^2}{1\text{ m}^2} \times \frac{1\text{ yd}^2}{(36\text{ in})^2} = 32.6\text{ yd}^2$$

$$\text{cost} = 32.6\text{ yd}^2 \times \frac{\$18.50}{1\text{ yd}^2} = \$603$$

Chapter 2

Development of the Atomic Theory

Review Problems

2-1. The total mass of substances before the reaction is 8.50 + 2.20 = 10.70 g. After the reaction the total mass must also be 10.70 g. The only product is 6.69 g zinc sulfide. The mass of unreacted zinc is 10.70 - 6.69 = 4.01 g.

2-2. (a) Determine the ratio by mass of carbon to carbon dioxide in the three cases.

$$\frac{1.07\text{ g carbon}}{3.92\text{ g carbon dioxide}} = 0.273 \qquad \frac{1.96\text{ g carbon}}{7.18\text{ g carbon dioxide}} = 0.273$$

$$\frac{2.03\text{ g carbon}}{7.44\text{ g carbon dioxide}} = 0.273$$

The constancy of these ratios establishes that carbon dioxide has a constant composition.

(b) $\%C = \frac{\text{no. g carbon}}{\text{no. g carbon dioxide}} \times 100 = 0.273 \times 100 = 27.3\%$. Since there are only two elements present, $\%O = 100.0 - 27.3 = 72.7\%$

2-3. (a) SO_2: $\frac{(2 \times 16)\text{g O}}{32\text{ g S}} = 1.0$ $\qquad SO_3$: $\frac{(3 \times 16)\text{g O}}{32\text{ g S}} = 1.5$ $\qquad$ Ratio $= \frac{1.5}{1.0} = \frac{3}{2}$

(b) H_2O: $\frac{(2 \times 1)\text{g H}}{16\text{ g O}} = \frac{2}{16}$ $\qquad H_2O_2$: $\frac{(2 \times 1)\text{g H}}{(2 \times 16)\text{g O}} = \frac{2}{32}$ $\qquad$ Ratio $= \frac{2/32}{2/16} = \frac{1}{2}$

(c) PCl_3: $\frac{31\text{ g P}}{(3 \times 35.5)\text{g Cl}}$ $\qquad PCl_5$: $\frac{31\text{ g P}}{(5 \times 35.5)\text{g Cl}}$ $\qquad$ Ratio $= \frac{31/(5 \times 35.5)}{31/(3 \times 35.5)} = \frac{3}{5}$

2-4. (a) Since ${}^{1}_{1}H$ is a neutral atom, there is no <u>net</u> charge on the atom and no net charge on 1.50×10^{15} atoms.

(b) The Br^- ion has an excess of one electron over the number of protons. Its net charge is the electronic charge, -1.602×10^{-19} C. For 1.50×10^{15} Br^- ions,

$$\text{no. C} = \frac{-1.602\times10^{-19}\text{C}}{Br^-} \times 1.50\times10^{15}\ Br^- = -2.40\times10^{-4}\text{ C}$$

(c) The ${}^{22}_{10}Ne^{2+}$ ion carries a net positive charge of +2.

$$\text{no. C} = \frac{+2 \times 1.602\times10^{-19}\text{ C}}{Ne^{2+}} \times 1.50\times10^{15}\ Ne^{2+} = +4.81\times10^{-4}\text{ C}$$

2-5.

Name	Symbol	Number Protons	Number Electrons	Number Neutrons	Mass Number
Sodium	${}^{23}_{11}Na$	11	11	12	23
Silicon	${}^{28}_{14}Si$	14	14	14	28
Rubidium	${}^{85}_{37}Rb$	37	37	48	85
Potassium	${}^{40}K$	19	19	21	40
Arsenic	${}^{75}_{33}As$	33	33	42	75
Neon Ion	${}^{20}Ne^{2+}$	10	8	10	20
Bromine*	${}^{80}_{35}Br$	35	35	45	80
Lead*	${}^{208}_{82}Pb$	82	82	126	208

*The information given is not enough to characterize a specific nuclide; several possibilities exist.

2-6. (a) Since all of these species are neutral atoms, the numbers of electrons are the subscript numerals. The symbols must be arranged in order of increasing value of these subscripts.

$${}^{40}_{18}Ar < {}^{39}_{19}K < {}^{58}_{27}Co < {}^{59}_{29}Cu < {}^{120}_{48}Cd < {}^{112}_{50}Sn < {}^{122}_{52}Te$$

(b) The order here is in terms of increasing neutron number, A-Z. This is the difference between the superscript and subscript numerals.

$${}^{39}_{19}K < {}^{40}_{18}Ar < {}^{59}_{29}Cu < {}^{58}_{27}Co < {}^{112}_{50}Sn < {}^{122}_{52}Te < {}^{120}_{48}Cd$$

(c) The increasing order is by mass number, the superscripts

$${}^{39}_{19}K < {}^{40}_{18}Ar < {}^{58}_{27}Co < {}^{59}_{29}Cu < {}^{112}_{50}Sn < {}^{120}_{48}Cd < {}^{122}_{52}Te$$

2-7. (a) The number of protons = 55 and the total number of particles in the nucleus is 133. The number of neutrons = 133 - 55 = 78.

$$\% \text{ neutrons} = \frac{78}{133} \times 100 = 58.6\%$$

(b) Use as approximate masses of both the proton and the neutron 1.00 u. Assume the nuclidic mass is very nearly 133 u.

$$\%P, \text{ by mass} = \frac{55 \text{ u}}{133 \text{ u}} \times 100 = 41\%$$

2-8. Each of the listed nuclidic masses must be divided by the nuclidic mass of ${}^{12}_{6}C$ (12.00000 u).

(a) ${}^{9}_{4}Be/{}^{12}_{6}C$ = 9.01218 u/12.00000 u = 0.751015

(b) ${}^{31}_{15}P/{}^{12}_{6}C$ = 30.97376 u/12.00000 u = 2.581147

(c) ${}^{94}_{40}Zr/{}^{12}_{6}C$ = 93.9061 u/12.00000 u = 7.82551

2-9. The mass of ${}^{19}_{9}F$ is 1.5832 times the mass of ${}^{12}_{6}C$, i.e., 1.5832 x 12.00000 u = 18.998. The mass of ${}^{35}_{17}Cl$ is 1.8406 x 18.998 = 34.968. The mass of ${}^{81}_{35}Br$ is 2.3140 x 34.968 = 80.916.

2-10. Use the method of Example 2-6.

at. wt. U = (0.9927 x mass of ${}^{238}U$) + (0.0072 x mass of ${}^{235}U$) + (0.00006 x mass of ${}^{234}U$)

= (0.9927 x 238.05) + (0.0072 x 235.04) + (0.00006 x 234.04)

= (236.3 + 1.7 + 0.01) = 238.0 u

Exercises

Law of conservation of mass

2-1. In the rusting of the iron the two principal reactants are iron and oxygen gas, but only one of these (the iron) is weighed initially. In contrast, all of the products of the reaction (the rusted iron) are recovered and weighed. Therefore, the mass of the iron increases. In the burning of the match, most of the products of the reaction (carbon dioxide gas and water vapor) escape and are not weighed. All that is weighed is the unburned residue of the match; the mass of the match decreases. In both cases, however, the total mass of the products is equal to the total mass of the reactants. The law of conservation of mass does apply.

2-2. Total the mass of materials present before and after the reaction, and show that this remains unchanged.

before: $10.00 \text{ g} + (100.0 \text{ cm}^3 \times \frac{1.148 \text{ g}}{1 \text{ cm}^3}) = 10.00 \text{ g} + 114.8 \text{ g} = 124.8 \text{ g}$

after: $120.40 \text{ g} + (2.22 \text{ L} \times 1.9769 \text{ g/L}) = 120.40 \text{ g} + 4.39 \text{ g} = 124.79 \text{ g}$

Law of constant composition

2-3. If the law of definite composition is followed, the percentage composition of the compound must be constant. In the first experiment 1.76 g sodium → 4.47 g sodium chloride.

$$\% \text{ sodium} = \frac{1.76 \text{ g sodium}}{4.47 \text{ g sodium chloride}} \times 100 = 39.4\% \ (\% \text{ chlorine} = 60.6)$$

In the second experiment 1.00 g chlorine → 1.65 g sodium chloride.

$$\% \text{ chlorine} = \frac{1.00 \text{ g chlorine}}{1.65 \text{ g sodium chloride}} \times 100 = 60.6\% \ (\% \text{ sodium} = 39.4)$$

2-4. The approach is the same as in Exercise 2-3--establish the constancy of the percent composition of the compound. In the first experiment 2.10 g hydrogen → 18.77 g water.

$$\% \text{ hydrogen} = \frac{2.10 \text{ g hydrogen}}{18.77 \text{ g water}} \times 100 = 11.2\% \ (\% \text{ oxygen} = 88.8)$$

In the second experiment, a sample of water weighing 12.96 g (11.51 g + 1.45 g) yields 1.45 g hydrogen upon electrolysis.

$$\% \text{ hydrogen} = \frac{1.45 \text{ g hydrogen}}{12.96 \text{ g water}} \times 100 = 11.2\% \ (\% \text{ oxygen} = 88.8)$$

2-5. Dalton's formulas would have indicated that the proportion of oxygen to hydrogen (both by mass and numbers of atoms) is greater in hydrogen peroxide than in water. He would have written

water: OH hydrogen peroxide: O_2H

2-6. Example 2-1 establishes the mass ratio of magnesium to magnesium oxide--0.60. That is, in 1.00 g of the oxide there is 0.60 g magnesium (and 0.40 g oxygen). The mass ratio of magnesium to oxygen is 0.60/0.40 = 1.5. Dalton would have assumed that magnesium oxide consists of one magnesium atom for every oxygen atom. Thus we can write

$$\frac{\text{at. wt. Mg}}{\text{at. wt. O}} = \frac{\text{at. wt. Mg}}{16.0} = 1.5 \qquad \text{at. wt. Mg} = 24$$

Law of multiple proportions

2-7. Establish the ratio of g H/g C in each compound. Either Dalton's atomic weights or currently accepted values (as shown below) can be used for this purpose; the choice is immaterial.

$$CH_4: \frac{4 \text{ g H}}{12 \text{ g C}}; \qquad C_2H_4: \frac{4 \text{ g H}}{(2 \times 12) \text{ g C}}; \qquad \text{Ratio: } \frac{4/12}{4/24} = 2$$

2-8. Dalton probably would have assigned the formulas: NO, N_2O, and NO_2. Determine the ratio g nitrogen/g oxygen for each of the three compounds and compare. If these ratios can be expressed as a series of small whole numbers, the law of multiple proportions is verified. Modern atomic weights are used below.)

$$NO: \frac{14 \text{ g N}}{16 \text{ g O}} = 0.88; \qquad N_2O: \frac{(2 \times 14) \text{ g N}}{16 \text{ g O}} = 1.75; \qquad NO_2: \frac{14 \text{ g N}}{(2 \times 16) \text{ g O}} = 0.44$$

The values 0.88, 1.75, and 0.44 are in the ratio, 2:4:1.

2-9. Let us denote one of the oxides as compound A and the other as compound B.

Compound A

$$\frac{96.2 \text{ g mercury}}{3.8 \text{ g oxygen}} = \frac{25 \text{ g mercury}}{\text{g oxygen}}$$

Compound B

$$\frac{92.6 \text{ g mercury}}{7.4 \text{ g oxygen}} = \frac{13 \text{ g mercury}}{\text{g oxygen}}$$

$$\text{ratio} = \frac{\dfrac{25 \text{ g mercury}}{\text{g oxygen}}}{\dfrac{13 \text{ g mercury}}{\text{g oxygen}}} \simeq 2$$

One of the oxides (A) has twice as many mercury atoms per oxygen as does the other (B). Possible formulas are Hg_2O and HgO. Note that these formulas are consistent with an atomic weight of 16 for O, 200 for Hg, and the combining ratios given here (i.e., the atomic weight of mercury should be about 12.5 times that of oxygen: 12.5 x 16 = 200).

Fundamental particles

2-10. Perhaps the most convincing proof that electrons are fundamental particles of all matter lies in the observation that (a) the properties of cathode rays are independent of the cathode material from which they are produced and (b) the e/m ratio of all cathode rays and β particles are identical.

2-11. Cathode rays are small negatively charged particles that are emitted by a cathode material (iron, platinum, and so on). Canal rays are produced through the bombardment of gaseous atoms and molecules by cathode rays (electrons). Canal rays are positively charged and much more massive than cathode rays. All cathode rays are identical in mass and charge. Canal rays have varying masses and charges (and e/m ratios).

2-12. The properties of cathode rays prove to be independent of the cathode material, residual gas in the cathode ray tube, and so on. This would have to be the case for fundamental particles of matter. Since the masses, electric charges and identities of canal ray particles depend on the experimental conditions under which they are produced, they cannot be unique particles fundamental to all matter.

2-13. Methods used to characterize electrons depended on the effects of electric and magnetic fields on these particles. Since neutrons carry no electric charge, they are unaffected by electric and magnetic fields. Altogether different approaches were required for the detection and characterization of neutrons.

Fundamental charges and charge-to-mass ratios

2-14. First determine the charges on the 10 drops.

Drop 1	1.28×10^{-18} C	(8)	Drop 6	3.84×10^{-18} C	(24)
Drop 2	0.64×10^{-18}	(4)	Drop 7	3.84×10^{-18}	(24)
Drop 3	0.64×10^{-18}	(4)	Drop 8	2.56×10^{-18}	(16)
Drop 4	0.32×10^{-18}	(2)	Drop 9	2.56×10^{-18}	(16)
Drop 5	5.12×10^{-18}	(32)	Drop 10	1.28×10^{-18}	(8)

These data are consistent with the value of the fundamental electronic charge given in the text (1.602×10^{-19} C). Each value is divisible by the electronic charge an integral number of times (numbers given in parentheses above). Millikan could not have inferred the fundamental electronic charge from these data. The largest common denominator of all the values listed is 3.2×10^{-19} C--twice the electronic charge.

2-15. (a) $\text{no. electrons} = -3.8\times10^{-14}\ \text{C} \times \frac{1\ \text{electron}}{-1.602\times10^{-19}\ \text{C}} = 2.4\times10^{5}$ electrons

(b) $\text{no. electrons} = +1.7\times10^{-11}\ \text{C} \times \frac{1\ \text{electron}}{-1.602\times10^{-19}\ \text{C}} = -1.1\times10^{8}$ electrons or 1.1×10^{8} electrons deficient

2-16. (a) The mass of a hydrogen atom is essentially that of a proton. Thus,

$$\frac{\text{mass H}}{\text{mass e}^-} = \frac{\text{mass P}}{\text{mass e}^-} = \frac{1.0073\ \text{amu}}{0.00055\ \text{amu}} = 1830$$

A hydrogen atom has 1830 times the mass of an electron, or an electron has about 1/2000 of the mass of a hydrogen atom.

(b) The highest charge-to-mass for positive rays is that of ${}^{1}_{1}H^{+}$.

$$\frac{\text{charge}}{\text{mass}} = \frac{1.602\times10^{-19}\ \text{C}}{1.673\times10^{-24}\ \text{g}} = 9.576\times10^{4}\ \text{C/g}$$

This ratio is considerably smaller than that for an electron: 1.759×10^{8} C/g.

2-17. We are to compare *absolute* values of e/m ratios. This means that we should treat all ratios as if they were positive quantities (even those involving negative charges). The charges are: proton, +1; electron, -1; neutron, 0; α particle, +2; ${}^{40}_{18}Ar$, 0; and ${}^{37}_{17}Cl^-$, -1. The approximate masses, expressed in atomic mass units, are: proton, 1; electron, 0.00055; neutron, 1; α particle, 4; ${}^{40}_{18}Ar$, 40; and ${}^{37}_{17}Cl^-$, 37. The order of increasing value of 3/m is

	neutron =	${}^{40}_{18}Ar$ <	${}^{37}_{17}Cl^-$ <	α particle <	proton <	electron
e/m =	0/1 =	0/40 <	1/37 <	2/4 <	1/1 <	1/0.00055
e/m =	0 =	0 <	0.027 <	0.50 <	1.0 <	1830

Atomic models

2-18. (a) He +2 ⊖ ⊖ (b) O 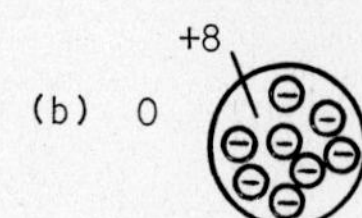 (c) N^+ +7 (d) F^- 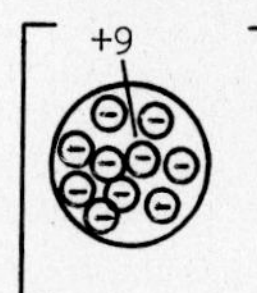

2-19. (a) 2p 2n $2e^-$ (b) 8p 8n $8e^-$ (c) 7p 7n $6e^-$ (d) 9p 10n $10e^-$

Atomic number, mass number, nuclides, and isotopes

2-20. X is the chemical symbol of an element. The numerical value of Z is the atomic number of the element (i.e., the number of protons in the nucleus). The numerical value of A is the mass number of a particular nuclide of the element (i.e., the sum of the number of protons and number of neutrons in the nucleus).

2-21. (a) ${}^{107}_{47}Ag$ has 47p, 47e, and 60n.

(b) The mass ratio ${}^{107}_{47}Ag/{}^{12}_{6}C$ is 106.90509/12.000000 = 8.9087575

(c) The mass of ${}^{16}_{8}O$ determined in Example 2-5 is 15.9948 u. The mass ratio ${}^{107}_{47}Ag/{}^{16}_{8}O$ is 106.90509/15.9948 = 6.68374

2-22. The symbols ${}^{35}_{17}Cl$ and ${}^{35}Cl$ convey the same information; the element Cl can have but one atomic number --17. The symbols ${}^{35}_{17}Cl$ and ${}_{17}Cl$ *do not* convey the same information. Although all atoms of Cl have an atomic number of 17, the mass number is not limited to 35.

2-23. Protium, deuterium, and tritium nuclei each contain but a single proton. They differ in numbers of neutrons, having 0, 1, and 2, respectively. Protium is the most abundant of the isotopes because the atomic weight of naturally occurring hydrogen is just slightly greater than one.

Atomic mass units, atomic masses

2-24. Use data from Table 2-1 to formulate a conversion factor between atomic mass units and grams.

$$\text{no. g} = 1.000 \text{ u} \times \frac{1.673\times10^{-24}\text{ g}}{1.0073\text{ u}} = 1.661\times10^{-24}\text{ g}$$

2-25. (a) $$\text{no. g} = 1.00\times10^{15}\text{ atoms Br-79} \times \frac{78.9183\text{ u}}{1\text{ atom Br-79}} \times \frac{1.673\times10^{-24}\text{ g}}{1.0073\text{ u}} = 1.31\times10^{-7}\text{ g}$$

(b) $$\text{no. g} = 1.00\times10^{15}\text{ atoms Br-81} \times \frac{80.9163\text{ u}}{1\text{ atom Br-81}} \times \frac{1.673\times10^{-24}\text{ g}}{1.0073\text{ u}} = 1.34\times10^{-7}\text{ g}$$

(c) $$\text{no. g} = (1.00\times10^{15}\text{ atoms Br} \times \frac{50.54\text{ Br-79}}{100\text{ atoms Br}} \times \frac{78.9183\text{ u}}{1\text{ atom Br-79}} \times \frac{1.673\times10^{-24}\text{ g}}{1.0073\text{ u}}) +$$

$$(1.00\times10^{15}\text{ atoms Cl} \times \frac{49.46\text{ Br-81}}{100\text{ atoms Br}} \times \frac{80\text{-}9163\text{ u}}{1\text{ atom Br-81}} \times \frac{1.673\times10^{-24}\text{ g}}{1.0073\text{ u}})$$

$$= 6.62\times10^{-8}\text{ g} + 6.65\times10^{-8}\text{ g} = 1.327\times10^{-7}\text{ g}$$

Atomic weights

2-26. Masses of individual nuclides, with the exception of C-12, are nonintegral (not whole numbers). On the other hand, they do not differ from whole numbers by a great deal. (See, for example, the nuclides in Exercise 33.) These facts suggest that a mass of 63.546 for copper must be an average of the masses of two or more isotopes. Thus, we would expect that *no* individual copper atoms have a mass of 63.546.

2-27. Although for many elements there does exist a nuclide with a mass nearly equal to the atomic weight (e.g., ^{12}C, ^{14}N, ^{16}O, ^{40}Ca), there are some instances where this is not the case. One case is cited in Example 2-7. Bromine has *two* naturally occurring isotopes, ^{79}Br and ^{81}Br, but the whole number closest to the atomic weight of bromine is 80. The nuclide ^{80}Br does not occur naturally (although it can be produced artificially). Thus, it is possible for the weighted coverage of the nuclidic masses to come close to being an integral number without there being a naturally occurring isotope with that particular integral number as its mass number. (By analogy, consider an average grade in a test taken by 100 students to be 72. It is possible that no single grade on the test was 72.)

2-28. (a) % natural abundance = 100.00 - 10.13 - 11.17 = 78.70%

(b) Since the mass number of the two isotopes listed are 25 and 26, respectively, and the average atomic weight of magnesium is 24.305, the mass number of the third isotope must be 24. (If it were 23, the weighted average atomic weight would be less than 24.)

(c) $\text{at. wt.} = 24.305 = 0.7870 \times \text{mass of } ^{24}Mg + (0.1013 \times 24.98584) + (0.1117 \times 25.98259)$

$$24.305 = (0.7870 \times \text{mass of } ^{24}Mg) + 2.531 + 2.902$$

$$\text{mass of } ^{24}Mg = \frac{24.305 - 2.531 - 2.902}{0.7870} = 23.98 \text{ u}$$

2-29. (a) Li-7 is the more abundant. This is so because the atomic weight of the naturally occurring element is much closer to 7 than to 6.

(b) Because the atomic weight of 6.941 is about nine-tenths of the way along the interval between the masses of the two isotopes (6.01513 and 7.01601 amu), we should expect about 9 parts Li-7 to 1 part Li-6.

(c) Let x = fraction of atoms that are Li-7 and $(1 - x)$ = fraction of Li-6. Express the experimentally determined atomic weight in terms of these fractions and the masses of the individual nuclides. Solve for x.

$$\text{At. wt.} = x(7.01601) + (1 - x)\cdot(6.01513) = 6.941$$

$$7.016x + 6.015 - 6.015x = 6.941$$

$$1.001x = 0.926; \quad x = 0.925; \quad 1 - x = 0.075$$

Abundances of the isotopes: 92.5% $^{7}_{3}Li$ and 7.5% $^{6}_{3}Li$

2-30. Let x = fraction of atoms that are N-15. Proceed as in Exercise 2-29(c). The atomic weight of naturally occurring nitrogen is 14.0067.

$$\text{At. wt.} = x(15.0001) + (1 - x)\cdot(14.0031) = 14.0067$$

$$15.0001x + 14.0031 - 14.0031x = 14.0067$$

$$0.997x = 0.0036; \quad x = 0.0036; \quad \%\ ^{15}_{7}N = 0.36$$

2-31. The total of the percent abundance of the two minor isotopes is 100.0 - 92.21 = 7.79%. Let the % ^{29}Si = x and the % ^{30}Si = $7.79 - x$. Now formulate the usual expression for the weighted average. (Note that the fractional part represented by each isotope is one-hundredth of its percentage.)

$$\text{At. wt.} = 28.086 = (0.9221 \times 27.97693) + 0.01x(28.97649) + (0.0779 - 0.01x)(29.97376)$$

$$28.086 = 25.80 + 0.2898x + 2.33 - 0.2997x$$

$$0.2997x - 0.2898x = 25.80 + 2.33 - 28.086$$

$$0.0099x = 0.04 \qquad x = 4\% \qquad 7.79 - x = 4\%$$

The percent abundances of the two minor isotopes are each about 4%. The precision of the calculation is poor because in two instances a small difference between two numbers of nearly equal magnitudes must be obtained. Significant figures are lost in this subtraction process. (For example, in the summation 25.80 + 2.33 - 28.086, the numbers involved have four, three, and five significant figures, respectively, but the summation carries only one significant figure.)

2-32. (a) Six different molecules of HCl are possible: $^{1}H^{35}Cl$, $^{2}H^{35}Cl$, $^{3}H^{35}Cl$, $^{1}H^{37}Cl$, $^{2}H^{37}Cl$, and $^{3}H^{37}Cl$.

(b) The mass numbers of the nuclei listed in (a) are 36, 37, 38, 38, 39, and 40; respectively.

(c) By far the most abundant of the hydrogen isotopes is H-1. The more abundant of the two chlorine isotopes is Cl-35. As a result the most abundant of the HCl molecules is $^{1}H^{35}Cl$, and the second most abundant, $^{1}H^{37}Cl$.

2-32. (d) $^{1}H^{35}Cl$ $^{2}H^{35}Cl$ $^{1}H^{37}Cl$ $^{2}H^{37}Cl$ $^{3}H^{37}Cl$

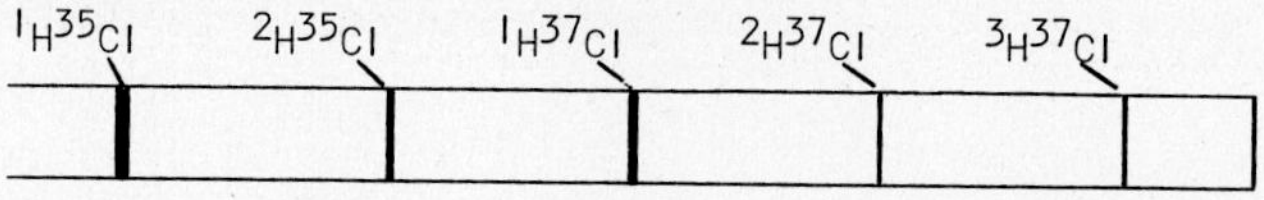

The line due to $^{3}H^{35}Cl$ is covered by the much stronger one for $^{1}H^{37}Cl$. The faintest of the five lines is that of $^{3}H^{37}Cl$.

2-33. In the usual manner, the atomic weight is a weighted average of the nuclidic masses (weighted according to their natural abundances).

At. wt. = (0.00146 × 195.9658) + (0.1002 × 197.9668) + (0.1684 × 198.9683) + (0.2313 × 199.9683)
+ (0.1322 × 200.9703) + (0.2980 × 201.9706) + (0.0685 × 203.9735)
= 0.29 + 19.84 + 33.51 + 46.25 + 26.57 + 60.19 + 14.0
= 200.6

Self-test Questions

1. (*d*) The existence of isotopes disapproves statement (*a*). The existence of radioactivity establishes that not all atoms are indivisible and indestructible. Oxygen is now known to have an atomic weight of 16, not 7.

2. (*c*) The ratio 16:1 (item *a*) has no significance. The ratio 85.5:16 = 5.34, but this is the ratio of g Rb to g O. Its inverse is 0.187 g O/g Rb, which is the same as the one given. Item (*b*) is incorrect. The ratio 0.374:1 is just twice the value given and is plausible according to the line of multiple properties. Item (*c*) is correct. Item (*d*) is incorrect because two of the other three responses have been shown to be incorrect.

3. (*b*) Cathode rays are fundamental particles of negative electrical charge--electrons. So too are β particles. Cathode rays are not electromagnetic radiation, and their properties are independent of their source.

4. (*a*) The other responses can be eliminated in terms of earlier discoveries. For example, Thomson established that electrons are fundamental particles of all matter, and Millikan, that all electrons have the same charge.

5. (*c*) $^{32}_{16}S$ has 16 electrons; $^{35}_{17}Cl^{-}$ has 18; $^{34}_{16}S^{+}$ has 15; and $^{35}_{16}S^{2-}$ has 18. An argon atom has 18 electrons but the ion, $^{40}_{18}Ar^{2+}$ has 16 electrons.

6. (*b*) 12.01115 amu is the atomic weight of naturally occurring carbon and is established by an appropriate averaging of the masses of the isotopes of carbon. These isotopes are C-12, C-13, and C-14 and their masses are given in (*a*), (*c*), and (*d*).

7. (*d*) Since the isotope $^{113}_{49}In$ has a mass of 112.9043 amu and the atomic weight of In is 114.82 amu, the second isotope must have a mass greater than 114.82 amu. Of the possibilities listed, only $^{115}_{49}In$ meets this requirement.

8. (*c*) The proposed change in atomic weight standard would <u>increase</u> all nuclidic masses by the factor 84.00000/83.9115. Item (*a*) is incorrect because it shows the nuclidic mass of C-12 as being <u>lowered</u>. Item (*b*) is incorrect for the same reason. Item (*c*) is correct. That is, 12.00000 × (84.00000/83.9115) = 12.1027. Item (*d*), although having a value larger than 12,00000, is still incorrect. It treats the change as if it were additive (i.e., 84.00000 - 83.9115) rather than multiplicative (84.00000/83.9115).

9. When the strip of magnesium is burned in open air, the original mass is that of the pure metal and the final mass is that of the magnesium oxide formed by its combustion. Since the oxygen gas consumed in the reaction (deived from the atmosphere) is not included in the original mass, an increase in mass of the strip is observed. In the case of the photoflash bulb, the oxygen gas that is to be consumed in the reaction is weighed along with the strip of metal initially. Thus, there is no change in mass.

10. Electrons were fairly easy to characterize in their free state becuase, having a negative electric charge, they are strongly affected by both electric and magnetic fields. Neutrons, because they carry no electric charge, are unaffected by electric and magnetic fields. They are more difficult to obtain in the free state. (They must be expelled from the nucleus.) And once obtained in the free state, they are more difficult to characterize than electrons. Thus, their discovery came after that of electrons and protons.

11. $$e/m = \frac{-1.602\times10^{-19}\ C}{36.966\ amu \times \frac{1.673\times10^{-24}\ g}{1.0073\ amu}} = -2.609\times10^{3}\ C/g$$

12. The atomic weight is expressed in terms of the masses of the two isotopes and their relative abundances. (Note that the abundance of $^{109}_{47}Ag$ must be 48.18%; there are only two isotopes to consider.)

$$107.87 = (0.5182 \times 106.9) + 0.4818x$$

$$x = \frac{107.87 - 55.40}{0.4818} = 108.9 \qquad \text{Mass of } ^{109}_{47}\text{Ag} = 108.9 \text{ u}$$

Chapter 3

Stoichiometry I: Elements and Compounds

Review Problems

3-1. (a) no. Cu atoms = $3.85 \text{ mol Cu} \times \frac{6.02\times10^{23} \text{ Cu atoms}}{1 \text{ mol Cu}} = 2.32\times10^{24}$ Cu atoms

(b) no. Ne atoms = $0.0163 \text{ mol Ne} \times \frac{6.02\times10^{23} \text{ Ne atoms}}{1 \text{ mol Ne}} = 9.81\times10^{21}$ Ne atoms

(c) no. Pu atoms = $3.4\times10^{-9} \text{ mol Pu} \times \frac{6.02\times10^{23} \text{ Pu atoms}}{1 \text{ mol Pu}} = 2.0\times10^{15}$ Pu atoms

3-2. (a) no. mol Al = $8.21\times10^{24} \text{ Al atoms} \times \frac{1 \text{ mol Al}}{6.02\times10^{23} \text{ Al atoms}} = 13.6$ mol Al

(b) no. g Cl_2 = $4.18 \text{ mol } Cl_2 \times \frac{70.9 \text{ g } Cl_2}{1 \text{ mol } Cl_2} = 296 \text{ g } Cl_2$

(c) no. kg Zn = $6.15\times10^{27} \text{ Zn atoms} \times \frac{1 \text{ mol Zn}}{6.02\times10^{23} \text{ Zn atoms}} \times \frac{65.4 \text{ g Zn}}{1 \text{ mol Zn}} \times \frac{1 \text{ kg Zn}}{1000 \text{ g Zn}} = 668$ kg Zn

(d) no. Fe atoms = $35.3 \text{ cm}^3 \text{ Fe} \times \frac{7.86 \text{ g Fe}}{1 \text{ cm}^3 \text{ Fe}} \times \frac{1 \text{ mol Fe}}{55.85 \text{ g Fe}} \times \frac{6.02\times10^{23} \text{ Fe atoms}}{1 \text{ mol Fe}} = 2.99\times10^{24}$ Fe atoms

(e) no. Li^+ ions = $1.51 \text{ kg } Li_2S \times \frac{1000 \text{ g } Li_2S}{1 \text{ kg } Li_2S} \times \frac{1 \text{ mol } Li_2S}{45.9 \text{ g } Li_2S} \times \frac{2 \text{ mol } Li^+ \text{ ions}}{1 \text{ mol } Li_2S} \times \frac{6.02\times10^{23} \text{ } Li^+ \text{ ions}}{1 \text{ mol } Li^+ \text{ ions}}$

$= 3.96 \times 10^{25}$ Li^+ ions

3-3. (a) mol wt = $(5 \times 12.0) + (11 \times 1.01) + 14.0 + (2 \times 16.0) + 32 = 149.2$

(b) no. mol H = $3.17 \text{ mol } C_5H_{11}NO_2S \times \frac{11 \text{ mol H}}{1 \text{ mol } C_5H_{11}NO_2S} = 34.9$ mol H

(c) no. C atoms = $1.53 \text{ mol } C_5H_{11}NO_2S \times \frac{5 \text{ mol C}}{1 \text{ mol } C_5H_{11}NO_2S} \times \frac{6.02\times10^{23} \text{ C atoms}}{1 \text{ mol C}} = 4.61\times10^{24}$ C atoms

(d) In methionine there are two moles of O per mol N. This leads to

$$\frac{2 \text{ mol O}}{1 \text{ mol N}} = \frac{2 \text{ mol O} \times \frac{16.0 \text{ g O}}{1 \text{ mol O}}}{1 \text{ mol N} \times \frac{14.0 \text{ g N}}{1 \text{ mol N}}} = 2.29 \text{ g O/g N}$$

3-4. (a) In one formula unit there are 7 C + 5 H + 3 N + 6 O = 21 atoms.

(b) The ratio of H to N atoms is 5 mol H/3 mol N. Since one mole of atoms is a fixed number, regardless of the element, the ratio of H to N atoms, by number, is also 5:3 = 1.67:1.

(c) Start with the ratio in moles and convert to a mass ratio.

$$\frac{6 \text{ mol O}}{7 \text{ mol O}} = \frac{6 \text{ mol O} \times \frac{16.0 \text{ g O}}{1 \text{ mol O}}}{7 \text{ mol C} \times \frac{12.0 \text{ g C}}{1 \text{ mol C}}} = 1.14 \text{ g O/g C}$$

(d) To determine the %N, by mass, start with the ratio mol N/mol $C_7H_5N_3O_6$.

$$\frac{3 \text{ mol N}}{1 \text{ mol } C_7H_5N_3O_6} = \frac{3 \text{ mol N} \times \frac{14.0 \text{ g N}}{1 \text{ mol N}}}{227.1 \text{ g } C_7H_5N_3O_6} = 0.185 \text{ g N/g } C_7H_5N_3O_6$$

The %N, by mass, is 100 x the mass ratio, i.e., 100 x 0.185 = 18.5%N

3-5. The molar mass of quinine is (20 x 12.01) + (24 x 1.008) + (2 x 14.01) + (2 x 16.00) = 324.42g/mol
The mass percents are:

$$\% \, C = \frac{20 \times 12.01 \text{ g C}}{324.42\text{g compd.}} \times 100 = 74.04\% \text{ C} \qquad \% \, H = \frac{24 \times 1.008 \text{ g H}}{324.42\text{g compd.}} \times 100 = 7.46\% \text{ H}$$

$$\% \, N = \frac{2 \times 14.01 \text{ g N}}{324.42\text{g compd.}} \times 100 = 8.64\% \text{ N} \qquad \% \, O = \frac{2 \times 16.00 \text{ g O}}{324.42\text{g compd.}} \times 100 = 9.86\% \text{ O}$$

3-6. In 100.0 g of the oxide there is present 71.06 g Co and 28.94 g O. On a molar basis these quantities are

$$\text{no. mol Co} = 71.06 \text{ g Co} \times \frac{1 \text{ mol Co}}{58.9 \text{ g Co}} = 1.21 \text{ mol Co} \qquad \text{no. mol O} = 28.94 \text{ g O} \times \frac{1 \text{ mol O}}{16.0 \text{ g O}} = 1.81 \text{ mol O}$$

The empirical formula is $Co_{1.21}O_{1.81} = Co_{\frac{1.21}{1.21}}O_{\frac{1.81}{1.21}} = CoO_{1.5} = Co_2O_3$

3-7. First, determine the % O. It is 100.00 - 59.96% C - 13.42% H = 26.62% O. In a 100.0 g sample,

$$\text{no. mol C} = 59.96 \text{ g C} \times \frac{1 \text{ mol C}}{12.0 \text{ g C}} = 4.99 \text{ mol C}$$

$$\text{no. mol H} = 13.42 \text{ g H} \times \frac{1 \text{ mol H}}{1.01 \text{ g H}} = 13.3 \text{ mol H}$$

$$\text{no. mol O} = 26.62 \text{ g O} \times \frac{1 \text{ mol O}}{16.0 \text{ g O}} = 1.66 \text{ mol O}$$

The empirical formula of isopropyl alcohol is

$$C_{4.99}H_{13.3}O_{1.66} = C_{\frac{4.99}{1.66}}H_{\frac{13.3}{1.66}}O_{\frac{1.66}{1.66}} = C_3H_8O$$

3-8. Again, determine % O. It is 100.00 - 57.83% C - 3.64% H = 38.53% O. In a 100.0 g sample,

$$\text{no. mol C} = 57.83 \text{ g C} \times \frac{1 \text{ mol C}}{12.01 \text{ g C}} = 4.815 \text{ mol C}$$

$$\text{no. mol H} = 3.64 \text{ g H} \times \frac{1 \text{ mol H}}{1.01 \text{ g H}} = 3.60 \text{ mol H}$$

$$\text{no. mol O} = 38.53 \text{ g O} \times \frac{1 \text{ mol O}}{16.00 \text{ g O}} = 2.408 \text{ mol O}$$

The empirical formula is $C_{4.815}H_{3.60}O_{2.408} = C_{\frac{4.815}{2.408}}H_{\frac{3.60}{2.408}}O_{\frac{2.408}{2.408}} = C_2H_{1.5}O = C_4H_3O_2$. The formula weight of terephthalic acid is (4 x 12.01) + (3 x 1.008) + (2 x 16.00) = 83.1. The experimentally determined molecular weight (166) is almost exactly twice the formula weight. The subscripts in the empirical formula should be doubled to obtain the molecular formula, which is $C_8H_6O_4$.

3-9. Determine the no. mol I^- present in the sample

$$\text{no. mo. } I^- = 1.186 \text{ g } I^- \times \frac{1 \text{ mol } I^-}{126.9 \text{ g I}} = 0.009346 \text{ mol } I^-$$

Since the compound has the formula XI, the no. mol X in the sample is also 0.009346. The mass of X in the sample of compound is 1.552 - 1.186 = 0.366 g X. The molar mass of X is 0.366 g X/0.009346 mol X = 39.2 g X/mol X. The atomic weight of X is numerically equal to the molar mass, i.e., 39.2. The element having this atomic weight (or very nearly so) is potassium (K).

3-10. This question requires using the principles embodied in Tables 3-2, 3-3, and 3-4.
(a) KI = potassium iodide
(b) $CaCl_2$ = calcium chloride
(c) KCN = potassium cyanide
(d) $Mg(NO_3)_2$ = magnesium nitrate
(e) ICl_3 = iodine trichloride
(f) ClO_2 = chlorine dioxide
(g) PCl_5 = phosphorus pentachloride

3-11. Each piece of information given provides a clue(s) to assist in supplying the missing information. For example, in part (a) tin in $SnCl_4$ is in the oxidation state +4. In SnF_2 tin is in the oxidation state +2. If Sn(IV) carries the name stann<u>ic</u> then we might expect that Sn(II) is stann<u>ous</u>.

(a) SnF_2, stannous fluoride
(b) PbO_2, lead(IV) oxide
(c) Co_2S_3, cobalt(III) sulfide
(d) KIO_4, potassium periodate
(e) AuCl, aurous chloride

In parts (b) and (c), since the stock system of nomenclature is called for no clues are required.

3-12. (a) 0. The oxidation state of an element in its free state is zero.

(b) -2. The oxidation state of K is +1. In order that the total of the oxidation numbers in K_2S be zero, that of S must be -2.

(c) +4. The oxidation state of O is -2; the sum of the oxidation numbers in NO_2 must be zero, that of N is +4.

(d) +5. The oxidation state of F is always -1. If the molecule BrF_5 is to have all its oxidation numbers total to zero, then the oxidation state of Br must be +5.

(e) +5. The oxidation state of H is +1, and of each of the three O atoms, -2. The oxidation state of N must be +5.

(f) +6. Each of the four O atoms is in the oxidation state -2, and each of the K atoms, +1. This requires Mn to be in the oxidation state +6.

(g) +3. The oxidation state of a monatomic ion must be equal to its charge.

(h) +6. The total of the oxidation numbers for 7 O atoms is -14. The total of all the oxidation numbers in the ion $Cr_2O_7^{2-}$ must be -2. The total for two Cr atoms must be +12; or for each Cr atom, +6.

3-13. This question requires that information from Tables 3-2 and 3-4 be used. (a) calcium oxide = CaO; (b) strontium fluoride = SrF_2; (c) magnesium hydroxide = $Mg(OH)_2$; (d) cesium carbonate = Cs_2CO_3; (e) mercury(II) nitrate = $Hg(NO_3)_2$; (f) iron(III) sulfide = Fe_2S_3; (g) magnesium perchlorate = $Mg(ClO_4)_2$; (h) potassium hydrogen sulfate = $KHSO_4$.

3-14. The binary acids are named according to the method described in the text, and the ternary oxoacids with the aid of Table 3-5.

(a) HBr = hydrobromic acid
(b) $HClO_2$ = chlorous acid
(c) HIO_3 = iodic acid
(d) HNO_2 = nitrous acid
(e) H_2SO_4 = sulfuric acid
(f) H_2S = hydrosulfuric acid

3-15. The formula weight of $ZnSO_4 \cdot 7H_2O$ is 65.38 + 32.06 + (4 x 16.00) + (7 x 18.02) = 287.6. (It is useful here to consider H_2O as a unit, which contributes (2 x 1.008) + 16.00 = 18.02 to the formula weight per mol H_2O.)

$$7\ H_2O = \frac{7 \times 18.02\ g\ H_2O}{287.6\ g\ ZnSO_4 \cdot 7H_2O} \times 100 = 43.86\%\ H_2O$$

Exercises

Terminology

3-1. (a) A formula unit is the smallest collection of atoms from which the formula of a compound can be established. A molecule is the smallest collection of atoms that can exist as a separate identifiable unit. In some cases the molecule and formula unit are identical; in some, the molecule consists of a number of formula units. For ionic compounds in the solid state there are no identifiable molecules.

(b) The empirical formula is the formula established from the formula unit of a compound. The molecular formula is the formula based on a molecule of compound. In some cases the molecular formula is the same as the empirical formula. In others, the molecular formula is a multiple of the empirical formula.

(c) A cation is a positive ion formed by the loss of an electron(s) by one or a group of atoms. An anion is a negative ion produced when one or a group of atoms gains an electron(s).

(d) A binary compound is made up to two elements; a ternary compound consists of three elements.

3-1. (e) An oxoacid has H and O atoms joined to a nonmetal atom such as P, S, or U, e.g., H_3PO_4, H_2SO_4, and $HClO_4$. The loss of one or more H atoms as H^+ ions leaves the remaining part of the molecule as an oxoanion, e.g., $H_2PO_4^-$, SO_4^{2-}, and ClO_4^-.

(f) The combination of hypo and ous refers to an oxoacid with the central nonmetal atom having the lowest oxidation state for the series of oxoacids. The per and ate refer to an oxoanion with the central nonmetal atom in its highest oxidation state for the series.

Avogadro's number and the mole

3-2. (a) $\text{no. S atoms} = 3.85 \text{ mol S} \times \frac{6.02\text{x}10^{23} \text{ S atoms}}{1 \text{ mol S}} = 2.32\text{x}10^{24} \text{ S atoms}$

(b) $\text{no. S atoms} = 0.0163 \text{ mol } S_8 \times \frac{8 \text{ mol S}}{1 \text{ mol } S_8} \times \frac{6.02\text{x}10^{23} \text{ S atoms}}{1 \text{ mol S}} = 7.85\text{x}10^{22} \text{ S atoms}$

(c) $\text{no. S atoms} = 3.4\text{x}10^{-9} \text{ mol } H_2S \times \frac{1 \text{ mol S}}{1 \text{ mol } H_2S} \times \frac{6.02\text{x}10^{23} \text{ S atoms}}{1 \text{ mol S}} = 2.0\text{x}10^{15} \text{ S atoms}$

(d) $\text{no. S atoms} = 0.162 \text{ mol } CS_2 \times \frac{2 \text{ mol S}}{1 \text{ mol } CS_2} \times \frac{6.02\text{x}10^{23} \text{ S atoms}}{1 \text{ mol S}} = 1.95\text{x}10^{23} \text{ S atoms}$

3-3. (a) $\text{no. mol } C_3H_7OH = 4.15\text{x}10^{24} \text{ } C_3H_7OH \text{ molecules} \times \frac{1 \text{ mol } C_3H_7OH}{6.02\text{x}10^{23} \text{ } C_3H_7OH \text{ molecules}} = 6.89 \text{ mol } C_3H_7OH$

(b) $\text{no. mol C} = 4.15\text{x}10^{24} \text{ } C_3H_7OH \text{ molecules} \times \frac{1 \text{ mol } C_3H_7OH}{6.02\text{x}10^{23} \text{ } C_3H_7OH \text{ molecules}} \times \frac{3 \text{ mol C}}{1 \text{ mol } C_3H_7OH} = 20.7 \text{ mol C}$

(c) $\text{no. mol H} = 4.15\text{x}10^{24} \text{ } C_3H_7OH \text{ molecules} \times \frac{1 \text{ mol } C_3H_7OH}{6.02\text{x}10^{23} \text{ } C_3H_7OH \text{ molecules}} \times \frac{8 \text{ mol C}}{1 \text{ mol } C_3H_7OH} = 55.1 \text{ mol H}$

(d) $\text{no. mol O} = 4.15\text{x}10^{24} \text{ } C_3H_7OH \text{ molecules} \times \frac{1 \text{ mol } C_3H_7OH}{6.02\text{x}10^{23} \text{ } C_3H_7OH \text{ molecules}} \times \frac{1 \text{ mol O}}{1 \text{ mol } C_3H_7OH} = 6.89 \text{ mol O}$

3-4. (a) There are three ions in a formula unit of Li_2S, two Li^+ and one S^{2-}. Thus, regardless of the sample size, one-third of all the ions in Li_2S are S^{2-}: 1/3 = 0.33.

(b) We need to determine the formula weight of Li_2S and then determine the fraction of this that is contributed by S^{2-}.

$\text{Fraction S, by mass} = \frac{\text{mass 1 mol S}}{\text{mass of 1 mol } Li_2S} = \frac{32.1 \text{ g S}}{[(2 \times 6.94) + 32.1] \text{ g } Cs_2S} = \frac{32.1}{46.0} = 0.698$

(c) $\text{no. } Li^+ \text{ ions} = 0.355 \text{ mol } Li_2S \times \frac{2 \text{ mol } Li^+}{1 \text{ mol } Li_2S} \times \frac{6.02\text{x}10^{23} \text{ } Li^+ \text{ ions}}{1 \text{ mol } Li^+} = 4.27\text{x}10^{23} \text{ } Li^+ \text{ ions}$

3-5. Determine the number of grams of tin, lead, and bismuth in the sample, and then add together these three masses.

$\text{no. g Sn} = 5.57\text{x}10^{25} \text{ atoms} \times \frac{2 \text{ Sn atoms}}{9 \text{ atoms}} \times \frac{1 \text{ mol Sn}}{6.02\text{x}10^{23} \text{ Sn atoms}} \times \frac{118.7 \text{ g Sn}}{1 \text{ mol Sn}} = 2.52\text{x}10^{3} \text{ g Sn}$

$\text{no. g Pb} = 5.75\text{x}10^{25} \text{ atoms} \times \frac{4 \text{ Pb atoms}}{9 \text{ atoms}} \times \frac{1 \text{ mol Pb}}{6.02\text{x}10^{23} \text{ Pb atoms}} \times \frac{207.2 \text{ g Pb}}{1 \text{ mol Pb}} = 8.80\text{x}10^{3} \text{ g Pb}$

$\text{no. g Bi} = 5.75\text{x}10^{25} \text{ atoms} \times \frac{3 \text{ Bi atoms}}{9 \text{ atoms}} \times \frac{1 \text{ mol Bi}}{6.02\text{x}10^{23} \text{ Bi atoms}} \times \frac{209.0 \text{ g Bi}}{1 \text{ mol Bi}} = 6.65\text{x}10^{3} \text{ g Bi}$

$\text{Total mass} = 2.52\text{x}10^{3} \text{ g Sn} + 8.80\text{x}10^{3} \text{ g Pb} + 6.65\text{x}10^{3} \text{ g Bi} = 1.797\text{x}10^{4} \text{ g}$

3-6. $\text{no. Ag atoms} = 65.2 \text{ g} \times \frac{92.5 \text{ g Ag}}{100 \text{ g}} \times \frac{1 \text{ mol Ag}}{108 \text{ g Ag}} \times \frac{6.02\text{x}10^{23} \text{ Ag atoms}}{1 \text{ mol Ag}} = 3.36\text{x}10^{23} \text{ Ag atoms}$

3-7. First determine the volume of the length of wire. Think of the wire as being a tall cylinder with volume = $\pi r^2 h$.

$V = \pi \times (\frac{0.03196}{2} \text{ in} \times \frac{2.54 \text{ cm}}{1 \text{ in}})^2 \times 1.00 \text{ m} \times \frac{100 \text{ cm}}{1.00 \text{ m}} = 0.518 \text{ cm}^3$

Next, determine the mass of this wire, the number of moles of copper, and then the number of Cu atoms.

$\text{no. Cu atoms} = 0.518 \text{ cm}^3 \text{ Cu} \times \frac{8.92 \text{ g Cu}}{1 \text{ cm3 Cu}} \times \frac{1 \text{ mol Cu}}{63.55 \text{ g Cu}} \times \frac{6.02\text{x}10^{23} \text{ Cu atoms}}{1 \text{ mol Cu}} = 4.38\text{x}10^{22} \text{ Cu atoms}$

3-8. As a first step, consider the mass of Pb in 500 ml of air.

$$\text{no. g Pb} = 500\ \text{mL} \times \frac{1\ \text{L}}{1000\ \text{mL}} \times \frac{1000\ \text{cm}^3}{1\ \text{L}} \times \frac{1\ \text{m}^3}{(100)^3\ \text{cm}^3} \times \frac{3.01\ \mu\text{g Pb}}{1\ \text{m}^3} \times \frac{1\ \text{g Pb}}{10^6\ \mu\text{g Pb}} = 1.50\times10^{-15}\ \text{g Pb}$$

Now convert from mass of Pb to number of Pb atoms.

$$\text{no. Pb atoms} = 1.50\times10^{-15}\ \text{g Pb} \times \frac{1\ \text{mol Pb}}{207\ \text{g Pb}} \times \frac{6.02\times10^{23}\ \text{Pb atoms}}{1\ \text{mol Pb}} = 4.36\times10^{12}\ \text{Pb atoms}$$

3-9. (a) $\text{no. mol } S_8 = 6.15\ \text{mm}^3 \times \frac{1\ \text{cm}^3}{(10\ \text{mm})^3} \times \frac{2.07\ \text{g}}{1\ \text{cm}^3} \times \frac{1\ \text{mol } S_8}{(8 \times 32.06)\text{g}} = 4.96\times10^{-5}\ \text{mol } S_8$

(b) $\text{no. S atoms} = 4.96\times10^{-5}\ \text{mol } S_8 \times \frac{8\ \text{mol S}}{1\ \text{mol } S_8} \times \frac{6.02\times10^{23}\ \text{S atoms}}{1\ \text{mol S}} = 2.39\times10^{20}\ \text{S atoms}$

3-10. (a) $\text{no. } CHCl_3 \text{ molecules} = 250\ \text{g water} \times \frac{1.00\ \text{g } CHCl_3}{1\times10^9\ \text{g water}} \times \frac{1\ \text{mol } CHCl_3}{119\ \text{g } CHCl_3} \times \frac{6.02\times10^{23}\ CHCl_3 \text{ molecules}}{1\ \text{mol } CHCl_3}$

$= 1.3\times10^{15}\ CHCl_3 \text{ molecules}$

(b) $\text{no. g } CHCl_3 = 250\ \text{g water} \times \frac{1.00\ \text{g } CHCl_3}{1\times10^9\ \text{g water}} = 2.5\times10^{-7}\ \text{g } CHCl_3$

This mass of $CHCl_3$ is much below the limit of detectability by an ordinary analytical balance.

Chemical formulas

3-11. (a) Incorrect: In CO the proportions of C to O, by mass, is 12:16. These same proportions will be found in $C_6H_{12}O_6$ (because there are six O atoms for every six C atoms). However, their percentages, by mass, cannot be the same in $C_6H_{12}O_6$ as in CO because of the introduction of the third element--hydrogen.

(b) Correct: In both $C_6H_{12}O_6$ and H_2O the ratio, by *number*, of H to O atoms is 2:1.

(c) Correct: The mass of 6 mol O is 96 g, compared to 72 g for 6 mol C and 12 g for 12 mol H. Since oxygen contributes most to the total mass of the compound, it is present in the highest percentage.

(d) Incorrect: Carbon and oxygen are present in the same proportions by number, 6:6 = 1:1. Because the masses of C and O differ, their mass ratio cannot be 1:1.

3-12. Because they are each multiples of the simplest formulas that could be written, the formulas (a) C_2H_6 and (d) N_2O_4 must be molecular formulas. The case of (b) Cl_2O and (c) CH_4O is uncertain. Each is an empirical formula, but whether it is also a molecular formula cannot be determined with the information in this exercise alone.

3-13. For the compound $Ge[S(CH_2)_4CH_3]_4$

(a) The total number of atoms in one formula unit is 1 Ge + 4 S + 20 C + 44 H = 69 atoms.

(b) The ratio of C to H by number of atoms can be based on the formula unit: 20 C/44 H = 0.45:1.00.

(c) The ratio of Ge to S can just be expressed by moles and then converted to a mass ratio.

$$\frac{1\ \text{mol Ge}}{4\ \text{mol S}} = \frac{1\ \text{mol Ge} \times \frac{72.6\ \text{g Ge}}{1\ \text{mol Ge}}}{4\ \text{mol S} \times \frac{32.1\ \text{g S}}{1\ \text{mol S}}} = 0.565\ \text{g Ge/g S}$$

(d) One mole of compound contains 4 mol S, which has the mass

$$\text{no. g S} = 4\ \text{mol S} \times \frac{32.1\ \text{g S}}{1\ \text{mol S}} = 128\ \text{g S}$$

(e) Here we need to know the molar mass of the compound.

molar mass = 72.6 g Ge + (4 x 32.1) g S + (20 x 12.0) g C + (44 x 1.01) g H = 485.4 g/mol

Then we can write

$$\text{no. g compd.} = 1.00\ \text{g Ge} \times \frac{1\ \text{mol Ge}}{72.6\ \text{g Ge}} \times \frac{1\ \text{mol compd.}}{1\ \text{mol Ge}} \times \frac{485\ \text{g compd.}}{1\ \text{mol compd.}} = 6.68\ \text{g compd.}$$

3-13. (f) no. C atoms = $2.500 \text{ g cpd} \times \frac{1 \text{ mol cpd}}{485.4 \text{ g cpd}} \times \frac{20 \text{ mol C}}{1 \text{ mol cpd}} \times \frac{6.022\times10^{23} \text{ C atoms}}{1 \text{ mol C}} = 6.203\times10^{23}$ C atoms

Percent composition of compounds

3-14. formula weight = (2 x 63.6) + (2 x 16.0) + (2 x 1.01) + 12.0 + (3 x 16.00) = 221.2

$$\% \text{ O} = \frac{5 \text{ mol O} \times \frac{16.0 \text{ g O}}{1 \text{ mol O}}}{221.2 \text{ g compound}} \times 100 = 36.17\% \text{ O}$$

3-15. Each of the compounds contains one sulfur atom per formula unit. The one that has the smallest formula weight must have the highest % S, that is, the smaller the denominator in the quantity,

$\% \text{ S} = \frac{32.1 \text{ g S}}{? \text{ g cpd}} \times 100$, the greater the % S. Of the compounds listed Li_2S has the smallest formula weight and the greatest percent S.

3-16. The formula weights of the compounds are: $CO(NH_2)_2$ = 60; NH_4NO_3 = 80; $HNC(NH_2)_2$ = 59. Urea and NH_4NO_3 have two N atoms per formula unit; quanidine has three. This, together with the fact that it has the smallest formula weight of the three, means that guanidine has the highest percent nitrogen. Alternatively, determine the percent N in each compound and compare the results.

3-17. no. lb $(NH_4)_2SO_4$ = $1.0 \text{ lb N} \times \frac{454 \text{ g N}}{1 \text{ lb N}} \times \frac{1 \text{ mol N}}{14.0 \text{ g N}} \times \frac{1 \text{ mol } (NH_4)_2SO_4}{2 \text{ mol N}} \times \frac{132 \text{ g } (NH_4)_2SO_4}{1 \text{ mol } (NH_4)_2SO_4}$

$$\frac{1 \text{ lb } (NH_4)_2SO_4}{454 \text{ g } (NH_4)_2SO_4} = 4.7 \text{ lb } (NH_4)_2SO_4$$

Note that in addition to the ususal cancellation of units the term "454" also cancels. A still simpler approach expresses the relative proportion of N in $(NH_4)_2SO_4$ in pound rather than grams.

no. lb $(NH_4)_2SO_4$ = $1.0 \text{ lb N} \times \frac{132 \text{ lb } (NH_4)_2SO_4}{(2 \times 14.0) \text{ lb N}} = 4.7 \text{ lb } (NH_4)_2SO_4$

Chemical formulas from percent composition

3-18. Based on 100.00g compound,

no. mol C = $93.71 \text{ g C} \times \frac{1 \text{ mol C}}{12.01 \text{ g C}} = 7.81$ mol C

no. mol H = $6.29 \text{ g H} \times \frac{1 \text{ mol H}}{1.01 \text{ g H}} = 6.23$ mol H

The empirical formula is

$$C_{7.81}H_{6.23} = C_{\frac{7.81}{6.23}}H_{\frac{6.23}{6.23}} = C_{1.25}H$$

Multiply both subscripts by 4, to obtain the empirical formula C_5H_4. The formula weight based on this empirical formula is (5 x 12.01) + (4 x 1.008) = 64.08. The molecular weight (128) is twice this value, so that the molecular formula is $C_{10}H_8$.

3-19. The first selenium oxide has 28.8% O and 100.0 - 28.8 = 71.2% Se. In 100.0 g of this oxide

no. mol Se = $71.2 \text{ g Se} \times \frac{1 \text{ mol Se}}{79.0 \text{ g Se}} = 0.901$ mol Se

no. mol O = $28.8 \text{ g O} \times \frac{1 \text{ mol O}}{16.0 \text{ g O}} = 1.80$ mol O

The empirical formula is $S_{0.90}O_{1.80} = SeO_2$; selenium dioxide or selenium(IV) oxide.

The second selenium oxide has 37.8% O and 100.0 - 37.8 = 62.2% Se. In 100.0 g of this oxide

no. mol Se = $62.2 \text{ g Se} \times \frac{1 \text{ mol O}}{79.0 \text{ g Se}} = 0.787$ mol Se

$$\text{no. mol O} = 37.8 \text{ g O} \times \frac{1 \text{ mol O}}{16.0 \text{ g O}} = 2.36 \text{ mol O}$$

The empirical formula is $Se_{0.787}O_{2.36} = Se_{\frac{0.787}{0.787}}O_{\frac{2.36}{0.787}} = SeO_3$; selenium trioxide or selenium(VI) oxide.

3-20. (a) In 100.00 g Warfarin

$$\text{no. mol C} = 74.01 \text{ g C} \times \frac{1 \text{ mol C}}{12.01 \text{ g C}} = 6.17 \text{ mol C}$$

$$\text{no. mol H} = 5.23 \text{ g H} \times \frac{1 \text{ mol H}}{1.01 \text{ g H}} = 5.18 \text{ mol H}$$

$$\text{no. mol O} = 20.76 \text{ g O} \times \frac{1 \text{ mol O}}{16.0 \text{ g O}} = 1.30 \text{ mol O}$$

Empirical formula $= C_{6.17}H_{5.18}O_{1.30} = C_{\frac{6.17}{1.30}}H_{\frac{5.18}{1.30}}O_{\frac{1.30}{1.30}} = C_{4.74}H_4O$

Searching for a multiplier: $2 \times 4.74 = 9.48$; $3 \times 4.74 = 14.22$; $4 \times 4.74 = 18.96 \cong 19.0$. Multiply all subscripts by 4 to obtain $C_{19}H_{16}O_4$.

(b) In 100.0 g citric acid

$$\text{no. mol C} = 37.51 \text{ g C} \times \frac{1 \text{ mol C}}{12.01 \text{ g C}} = 3.13 \text{ mol C}$$

$$\text{no. mol H} = 4.20 \text{ g H} \times \frac{1 \text{ mol H}}{1.01 \text{ g H}} = 4.16 \text{ mol H}$$

$$\text{no. mol O} = 58.29 \text{ g O} \times \frac{1 \text{ mol O}}{16.0 \text{ g O}} = 3.67 \text{ mol O}$$

Empirical formula $= C_{3.13}H_{4.16}O_{3.64} = C_{\frac{3.13}{3.13}}H_{\frac{4.16}{3.13}}O_{\frac{3.64}{3.13}} = CH_{1.33}O_{1.16}$

Multiply subscripts by 3 and this results in two of them being whole numbers: $C_3H_4O_{3.48}$. Now double all subscripts to obtain $C_6H_8O_7$.

3-21. The freon in question has the formula $CH_xCl_yF_z$. Where $x + y + z = 4$, i.e., there are four substituents attached to the C atom. The simplest approach, perhaps, is to consider several possible compounds and find the one with 31.43% F. For example,

CH_2ClF (mol. wt. = 68.5): $\% \text{ F} = \frac{19.0 \text{ g F}}{68.5 \text{ g cpd}} \times 100 = 27.7\% \text{ F}$

$CHCl_2F$ (mol. wt. = 103.0): $\% \text{ F} = \frac{19.0 \text{ g F}}{103.0 \text{ g cpd}} \times 100 = 18.4\% \text{ F}$

$CHClF_2$ (mol. wt. = 86.5): $\% \text{ F} = \frac{(2 \times 19.0) \text{ g F}}{86.5 \text{ g cpd}} \times 100 = 43.9\% \text{ F}$

CCl_2F_2 (mol. wt. = 120.9): $\% \text{ F} = \frac{(2 \times 19.0) \text{ g F}}{120.9 \text{ g cpd}} \times 100 = 31.4\% \text{ F}$

The component in question is CCl_2F_2.

Oxidation states

3-22. (a) -4. Since, by convention, the oxidation state of each of the H atoms is +1, that of C must be -4.

(b) +4. Each of the four F atoms is in the oxidation state -1. The oxidation state of S is +4.

(c) -1. Each Na atom is in the oxidation state +1, and each O atom, -1, to account for a sum of all oxidation numbers equal to zero.

(d) 0. The three H atoms are in the oxidation state of +1, for a total of +3. The two O atoms are in the oxidation state -2, for a total of -4. The sum of +3 -4 = -1, equal to the charge on the ion. The oxidation state of the two C atoms must total to zero and each C atom is assigned an oxidation state of 0.

(e) +6. The total for four O atoms is =8, requiring that the oxidation state of Fe by +6, so that the sum of all oxidation numbers = -2.

(f) +2.5. The total for six O atoms is -12. The total for four S atoms must be +10 (so that the sum of all oxidation numbers be -2). For each S atom this means an oxidation state of +2.5.

3-23. Assign oxidation states to S in each of the species.

SO_3^{2-} (ox. state of S = +4); $S_2O_3^{2-}$ (+2); $S_2O_8^{2-}$ (+7); HSO_4^- (+6); HS^- (-2); $S_4O_6^{2-}$ (+2.5). The order of increasing oxidation state of S is $HS^- < S_2O_3^{2-} < S_4O_6^{2-} < SO_3^{2-} < HSO_4^- < S_2O_8^{2-}$

(For a more complete analysis of these oxidation states see section 21-2.)

3-24. In each of the formulas the oxidation state of O is -2. That of N is the value specified. The formula is written so that the sum of all oxidation numbers is zero.

+1, N_2O; +2, NO; +3, N_2O_3; +4, NO_2; +5, N_2O_5.

Nomenclature

3-25. (a) MgS = magnesium sulfide
(b) ZnO = zinc oxide
(c) K_2CrO_4 = potassium chromate
(d) Cs_2SO_4 = cesium sulfate
(e) Cr_2O_3 = chromium(III) oxide
(f) $FeSO_4$ = iron(II) sulfate
(g) $Ca(HCO_3)_2$ = calcium hydrogen carbonate
(h) K_2HPO_4 = potassium monohydrogen phosphate
(i) NH_4I = ammonium iodide
(j) $Cu(OH)_2$ = copper(II) hydroxide
(k) HNO_2 = nitrous acid
(l) $HBrO_3$ = bromic acid
(m) $KClO_3$ = potassium chlorate
(n) KIO = potassium hypoiodite

3-26. (a) ICl = iodine (mono)chloride
(b) ClF_3 = chlorine trifluoride
(c) SF_4 = sulfur tetrafluoride
(d) BrF_5 = bromine pentafluoride

3-27. (a) aluminum sulfate = $Al_2(SO_4)_3$
(b) ammonium chromate = $(NH_4)_2CrO_4$
(c) silicon tetrafluoride = SiF_4
(d) zinc acetate = $Zn(C_2H_3O_2)_2$
(e) iron(II) oxide = FeO
(f) tricarbon disulfide = C_3S_2
(g) chromium(II) chloride = $CrCl_2$
(h) lithium sulfide = Li_2S
(i) chlorine dioxide = ClO_2
(j) calcium dihydrogen phosphate = $Ca(H_2PO_4)_2$
(k) tin(IV) oxide = SnO_2
(l) chlorous acid = $HClO_2$
(m) hydrobromic acid = HBr

Hydrates

3-28. The hydrate with the greatest % H_2O will be the one with the highest ratio of mass of water to that of the anhydrous compound.

(a) $\frac{5\ H_2O}{CuSO_4} = \frac{5 \times 18.0}{160} = \frac{90}{160}$

(b) $\frac{18\ H_2O}{Cr_2(SO_4)_3} = \frac{18 \times 18.0}{392} = \frac{324}{392}$

(c) $\frac{6\ H_2O}{MgCl_2} = \frac{6 \times 18.0}{95.2} = \frac{108}{95.2}$

(d) $\frac{2\ H_2O}{LiC_2H_3O_2} = \frac{2 \times 18.0}{66.0} = \frac{36.0}{66.0}$

The correct answer is (c). Only in $MgCl_2 \cdot 6H_2O$ does the mass of water exceed that of the anhydrous compound. $MgCl_2 \cdot 6H_2O$ has the highest % H_2O.

3-29. In 5.018 g of hydrate there is 2.449 g $MgSO_4$ and 5.018 - 2.449 = 2.569 g H_2O. On a molar basis,

$$\text{no. mol } MgSO_4 = 2.449 \text{ g } MgSO_4 \times \frac{1 \text{ mol } MgSO_4}{120 \text{ g } MgSO_4} = 0.0204 \text{ mol } MgSO_4$$

$$\text{no. mol } H_2O = 2.569 \text{ g } H_2O \times \frac{1 \text{ mol } H_2O}{18.0 \text{ g } H_2O} = 0.143 \text{ mol } H_2O$$

Formula: $(MgSO_4)_{0.0204}(H_2O)_{0.143} = (MgSO_4)_{\frac{0.0204}{0.0204}}(H_2O)_{\frac{0.143}{0.0204}} = (MgSO_4)(H_2O)_7 = MgSO_4 \cdot 7H_2O$

3-30. $$\text{no. g } Na_2SO_4 \cdot 10H_2O = 1.00 \text{ g } Na_2SO_4 \times \frac{1 \text{ mol } Na_2SO_4}{142 \text{ g } Na_2SO_4} \times \frac{1 \text{ mol } Na_2SO_4 \cdot 10H_2O}{1 \text{ mol } Na_2SO_4} \times \frac{322 \text{ g } Na_2SO_4 \cdot 10H_2O}{1 \text{ mol } Na_2SO_4 \cdot 10H_2O}$$

$$= 2.27 \text{ g } Na_2SO_4 \cdot 10H_2O$$

The increase in mass is 2.27 g - 1.00 g = 1.27 g.

3-31. (a) First determine the number of grams of C in the CO_2 and grams of H in the H_2O. These elements are derived from the hydrocarbon.

$$\text{no. g C} = 0.5008 \text{ g } CO_2 \times \frac{1 \text{ mol } CO_2}{44.01 \text{ g } CO_2} \times \frac{1 \text{ mol C}}{1 \text{ mol } CO_2} \times \frac{12.01 \text{ g C}}{1 \text{ mol C}} = 0.1367 \text{ g C}$$

$$\text{no. g H} = 0.1282 \text{ g } H_2O \times \frac{1 \text{ mol } H_2O}{18.02 \text{ g } H_2O} \times \frac{2 \text{ mol H}}{1 \text{ mol } H_2O} \times \frac{1.008 \text{ g H}}{1 \text{ mol H}} = 0.01434 \text{ g H}$$

The percent composition of the hydrocarbon is

$$\%\ C = \frac{0.1367 \text{ g C}}{0.1510 \text{ g cpd}} \times 100 = 90.53\% \qquad \%\ H = \frac{0.01434 \text{ g H}}{0.1510 \text{ g cpd}} \times 100 = 9.497 = 9.50\%$$

(b) In 100 grams of the compound there is 90.53 g C and 9.47 g H.

$$\text{no. mol C} = 90.53 \text{ g C} \times \frac{1 \text{ mol C}}{12.01 \text{ g C}} = 7.54 \text{ mol C}$$

$$\text{no. mol H} = 9.50 \text{ g H} \times \frac{1 \text{ mol H}}{1.01 \text{ g H}} = 9.41 \text{ mol H}$$

The empirical formula is $C_{7.54}H_{9.41} = C_{\frac{7.54}{7.54}}H_{\frac{9.41}{7.54}} = CH_{1.25} = C_4H_5$

(c) Since the formula weight based on the empirical formula is 53, just one half of the molecular weight, the molecular formula must be C_8H_{10}.

3-32. From the combustion data we determine the percent composition of P-cresol. The method is the same as that of the preceding exercise, except that three conversion factors are converted into a single one.

$$\text{no. g C} = 1.3077 \text{ g } CO_2 \times \frac{12.01 \text{ g C}}{44.01 \text{ g } CO_2} = 0.3569 \text{ g C}$$

$$\text{no. g H} = 0.3061 \text{ g } H_2O \times \frac{(2 \times 1.008) \text{ g H}}{18.02 \text{ g } H_2O} = 0.03425 \text{ g H}$$

$$\%\ C = \frac{0.3569 \text{ g C}}{0.4590 \text{ g cpd}} \times 100 = 77.76\ \%\ C \qquad \%\ H = \frac{0.03425 \text{ g H}}{0.4590 \text{ g cpd}} \times 100 = 7.462\%\ H = 7.46\%\ H$$

$\%\ O = 100.00 - \%\ C - \%\ H = 100.00 - 77.76\% - 7.46\% = 14.78\%\ O$

In 100.0 g compound,

$$\text{no. mol C} = 77.76 \text{ g C} \times \frac{1 \text{ mol C}}{12.01 \text{ g C}} = 6.47 \text{ mol C} \qquad \text{no. mol H} = 7.46 \text{ g H} \times \frac{1 \text{ mol H}}{1.008 \text{ g H}} = 7.40 \text{ mol H}$$

$$\text{no. mol O} = 14.78 \text{ g O} \times \frac{1 \text{ mol O}}{16.00 \text{ g O}} = 0.924 \text{ mol O}$$

Empirical formula = $C_{6.47}H_{7.40}O_{0.924} = C_{\frac{6.47}{0.924}}H_{\frac{7.40}{0.924}}O_{\frac{0.924}{0.924}} = C_7H_8O$

3-33. The percent composition of the compound is determined as in the preceding exercise:

$$\text{no. g C} = 0.305 \text{ g } CO_2 \times \frac{12.0 \text{ g C}}{44.0 \text{ g } CO_2} = 0.0832 \text{ g C} \qquad \text{no. g H} = 0.249 \text{ g } H_2O \times \frac{(2 \times 1.01)\text{g H}}{18.0 \text{ g } H_2O} = 0.0279 \text{ g H}$$

$$\% \text{ C} = \frac{0.0832 \text{ g C}}{0.208 \text{ g cpd}} \times 100 = 40.0\% \qquad \% \text{ H} = \frac{0.0279 \text{ g H}}{0.208 \text{ g cpd}} \times 100 = 13.4\%$$

% N can be obtained by difference: 100.0 - 40.0 - 13.4 = 46.6, or directly: $\% \text{ N} = \frac{0.163 \text{ g N}}{0.350 \text{ g cpd}} \times 100 = 46.6\%$.

Based on 100 g of dimethylhydrazine we may write:

$$\text{no. mol C} = 40.0 \text{ g C} \times \frac{1 \text{ mol C}}{12.0 \text{ g C}} = 3.33 \text{ mol C} \qquad \text{no. mol H} = 13.4 \text{ g H} \times \frac{1 \text{ mol H}}{1.01 \text{ g H}} = 13.3 \text{ mol H}$$

$$\text{no. mol N} = 46.6 \text{ g N} \times \frac{1 \text{ mol N}}{14.0 \text{ g N}} = 3.33 \text{ mol N}$$

The empirical formula is CH_4N.

3-34. The compound has the formula C_xH_y. Calculate the number of grams of CO_2 and grams of H_2O derived from a given mass of the compound, say 1.00 g. These quantities will be expressed in terms of x and y. The ratio of g CO_2 to H_2O must be 1.955:1. Find values of x and y consistent with this ratio.

$$\text{no. g } CO_2 = 1.00 \text{ g } C_xH_y \times \frac{(12x) \text{ g C}}{(12x + 1y) \text{ g } C_xH_y} \times \frac{44 \text{ g } CO_2}{12 \text{ g C}} = \frac{44x}{12x + y}$$

$$\text{no. g } H_2O = 1.00 \text{ g } C_xH_y \times \frac{(1y) \text{ g H}}{(12x + 1y) \text{ g } C_xH_y} \times \frac{18 \text{ g } H_2O}{2 \text{ g H}} = \frac{9y}{12x + y}$$

$$\frac{\text{no. g } CO_2}{\text{no. g } H_2O} = \frac{\frac{44x}{12x + y}}{\frac{9y}{12x + y}} = \frac{44x}{9y} = 1.955; \quad \frac{x}{y} = \frac{9}{44} \times 1.955 = 0.40 = \frac{4}{10} = \frac{2}{5}$$

The ratio of carbon atoms (x) to hydrogen atoms (y) in the hydrocarbon is 2:5. The empirical formula is C_2H_5.

Precipitation analysis

3-35. $\text{no. g AI} = 0.3518 \text{ g KI} \times \frac{1 \text{ mol KI}}{166.0 \text{ g KI}} \times \frac{1 \text{ mol I}}{1 \text{ mol KI}} \times \frac{1 \text{ mol AgI}}{1 \text{ mol I}} \times \frac{234.8 \text{ g AgI}}{1 \text{ mol AgI}} = 0.4976 \text{ g AgI}$

3-36. Determine the number of grams of tin in 0.245 g SnO_2, grams of lead in 0.115 g $PbSO_4$, and grams of zinc in 0.246 g $Zn_2P_2O_7$. Once these masses are determined the percent composition, by mass, follows easily.

$$\text{no. g Sn} = 0.245 \text{ g } SnO_2 \times \frac{118.7 \text{ g Sn}}{150.7 \text{ g } SnO_2} = 0.193 \text{ g Sn}$$

$$\text{no. g Pb} = 0.115 \text{ g } PbSO_4 \times \frac{207.2 \text{ g Pb}}{303.3 \text{ g } PbSO_4} = 0.0786 \text{ g Pb}$$

$$\text{no. g Zn} = 0.246 \text{ g } Zn_2P_2O_7 \times \frac{(2 \times 65.4) \text{ g Zn}}{305 \text{ g } Zn_2P_7O_7} = 0.105 \text{ g Zn}$$

$$\% \text{ Sn} = \frac{0.193 \text{ g Sn}}{1.713 \text{ g brass}} \times 100 = 11.3; \quad \% \text{ Pb} = 4.59 \quad \% \text{ Zn} = 6.13 \quad \% \text{ Cu} = 78.0$$

3-37. Determine the no. mol SO_4^{2-} in the zinc sulfate sample.

$$\text{no. mol } SO_4^{2-} = 0.8223 \text{ g } BaSO_4 \times \frac{1 \text{ mol } BaSO_4}{233.4 \text{ g } BaSO_4} \times \frac{1 \text{ mol } SO_4^{2-}}{1 \text{ mol } BaSO_4} = 3.523 \times 10^{-3} \text{ mol } SO_4^{2-}$$

$$\text{no. mol } Zn^{2+} = \text{no. mol } SO_4^{2-} = \text{no. mol } ZnSO_4 = 3.523 \times 10^{-3} \text{ mol } ZnSO_4$$

$$\text{no. g } ZnSO_4 = 3.523 \times 10^{-3} \text{ mol } ZnSO_4 \times \frac{161.4 \text{ g } ZnSO_4}{1 \text{ mol } ZnSO_4} = 0.5686 \text{ g } ZnSO_4$$

no. g H_2O = 1.013 g sample - 0.5686g $ZnSO_4$ = 0.444 g H_2O

no. mol H_2O = 0.444 g $H_2O \times \frac{1 \text{ mol } H_2O}{18.0 \text{ g } H_2O} = 0.0247 = 24.7\times10^{-3}$ mol H_2O

The mol ratio, $H_2O:ZnSO_4$ = 7:1. The formula of the hydrate is $ZnSO_4 \cdot 7H_2O$.

Atomic weight determinations

3-38. The two compounds can be represented as XCl_y and XCl_z, where y and z are integral (whole) numbers.

XCl_y: $\% \text{ Cl} = \frac{y(35.5)}{137} \times 100 = 77.5$ $y = \frac{77.5 \times 137}{35.5 \times 100} = 3$

XCl_3 has a molecular weight of X + 3(35.5) = 137. The atomic weight of X = 137 - 3(35.5) = 137 - 106.5 = 30.5. X is phosphorus.

XCl_z: $\% \text{ Cl} = \frac{z(35.5)}{208} \times 100 = 85.1$ $z = \frac{85.1 \times 208}{35.5 \times 100} = 5$

XCl_5 has a molecular weight of X + 5(35.5) = 208. The atomic weight of X = 208 - 5(35.5) = 208 - 177.5 = 30.5. X is phosphorus.

3-39. Proceed as in Example 3-14 of the textbook.

no. mol SO_4 = 0.2193 g $BaSO_4 \times \frac{1 \text{ mol } BaSO_4}{233.4 \text{ g } BaSO_4} \times \frac{1 \text{ mol } SO_4}{1 \text{ mol } BaSO_4} = 9.396\times10^{-4}$ mol SO_4

no. g SO_4 = 9.396×10^{-4} mol $SO_4 \times \frac{96.06 \text{ g } SO_4}{1 \text{ mol } SO_4} = 9.026\times10^{-2}$ g SO_4

no. g M = 0.1131 g MSO_4 - 0.09026 g SO_4 = 0.0228 g M

no. mol M = no. mol SO_4 = 9.396×10^{-4}

molar mass = $\frac{0.0228 \text{ g M}}{9.396\times10^{-4} \text{ mol M}}$ = 24.3 g M/mol M

The atomic weight of M is numerically equal to the molar mass: 24.3.

3-40. Proceed as in Example 3-15 of the textbook.

no. mol SO_4 = 1.511 g $BaSO_4 \times \frac{1 \text{ mol } BaSO_4}{233.4 \text{ g } BaSO_4} \times \frac{1 \text{ mol } SO_4}{1 \text{ mol } BaSO_4} = 6.474\times10^{-3}$ mol SO_4

no. g SO_4 = 6.474×10^{-3} mol $SO_4 \times \frac{96.06 \text{ g } SO_4}{1 \text{ mol } SO_4}$ = 0.6219 g SO_4

no. g M = 0.738 g $M_2(SO_4)_3$ - 0.622 g SO_4 = 0.116 g M

no. mol M = 6.474×10^{-3} mol $SO_4 \times \frac{2 \text{ mol M}}{3 \text{ mol } SO_4} = 4.316\times10^{-3}$ mol M

molar mass = $\frac{0.116 \text{ g M}}{4.316\times10^{-3} \text{ mol M}}$ = 26.9 g M/mol M

The atomic weight of M is numerically equal to the molar mass: 26.9.

3-41. Set up an equation in the usual conversion-factor format, to show how many grams of MS can be obtained from the given mass of M_2O_3. Of course, the number of grams of MS is not unknown; it is 0.685. What is unknown is the atomic weight of the metal M; call this x and solve for x.

no. g MS = 0.622 g $M_2O_3 \times \frac{1 \text{ mol } M_2O_3}{[2x + (3 \times 16.0)]\text{g } M_2O_3} \times \frac{2 \text{ mol M}}{1 \text{ mol } M_2O_3} \times \frac{1 \text{ mol MS}}{1 \text{ mol M}} \times \frac{(x + 32.1)\text{g MS}}{1 \text{ mol MS}}$ = 0.685g MS

$$\frac{0.622 \times 2 \times (x + 32.1)}{2x + 48.0} = 0.685 \qquad 1.244x + 39.9 = 1.370x + 32.9$$

$$0.126x = 7.0 \qquad x = 56$$

Self-test Questions

1. (*c*) One mole of F_2 weighs 38.0 g and contains 6.02×10^{23} F_2 molecules. In 6.02×10^{23} F_2 molecules there are $2 \times 6.02 \times 10^{23} = 1.20 \times 10^{24}$ F atoms. The answer (d) is one of those "foolish" errors always to be avoided. Avogadro's number is a number of atoms, molecules, etc.; it is not a number of grams.

2. (*b*) If all the subscripts in a formula are divisible by the same integer, the formula must be a molecular formula. Divide the subscripts of N_2O_4 by two and obtain NO_2. The other formulas are not divisible by integers.

3. (*d*) This result can be obtained by eliminating the first three choices and then proving that the fourth is correct. The compound contains 17 atoms per formula unit, but $17 \times 6.02 \times 10^{23}$ atoms per mole. The compound contains equal numbers of C and H atoms; but since the C atom is so much more massive than H, the percentages by mass are not equal. Again, there are twice as many O atoms as N atoms; but since an O atom is heavier than an N atom, the % O by mass is more than twice the % N. Each N atom contributes 14 units to the mass of the compound; each H atom 1 unit; and every 7 H atoms, 7 units. Since the mass contribution of N is almost exactly twice that of H, the % N, by mass, will be twice % H.

4. (*a*) We could simply calculate the number of N atoms in each sample and choose the largest. It is simpler, however, to work in moles. For example, 17 g NH_3 = 1 mol NH_3 = 1 mol N, but 1 mol N_2 = 2 mol N; therefore (d) > (b). The mol. wt. of N_2O = 44. In 50.0 g N_2O there is a more than 1 mol N_2O, and, therefore, more than 2 mol N; (a) > (d). The mol. wt. of C_6H_5N = 91. A 150 cm^3 sample will weigh less than 150 g (because the density is less than 1.00 g/cm^3), contain less than 2 mol of C_6H_5N, and less than 2 mol N.

5. (*c*) The contribution of 3 F atoms to the formula weight of XF_3 is $3 \times 19.0 = 57.0$. This represents 65% of the mass of the compound. $(57/x) \times 100 = 65$. $x = 5700/65 = 88$. The compound XF_3 has a formula weight of 88. The at. wt. of X = 88 − 57 = 31. An alternative method is to assume each of the four possible values for the at. wt. of X, determine the formula weight, and then the % F. Choose the value of X that yields 65% F.

6. (*c*) For six O atoms the sum of the oxidation numbers is −12. For the ion to have a charge of −1 requires the sum of the positive oxidation numbers to be +11. For four H atoms the total is +4. The oxidation state of I must be +7.

7. (*c*) The chlorite anion is ClO_2^-. Calcium chlorite contains two ClO_2^- ions for every Ca^{2+} ion. The formula is $Ca(ClO_2)_2$.

8. $$\text{no. cm}^3\ CHBr_3 = 3.40 \times 10^{24} \text{ molecules } CHBr_3 \times \frac{1 \text{ mol } CHBr_3}{6.02 \times 10^{23} \text{ molecules } CHBr_3} \times \frac{253 \text{ g } CHBr_3}{1 \text{ mol } CHBr_3}$$

$$\times \frac{1 \text{ cm}^3\ CHBr_3}{2.89 \text{ g } CHBr_3} = 494 \text{ cm}^3$$

9. (a) CaI_2 = calcium iodide
   (b) $Fe_2(SO_4)_3$ = iron(III) sulfate
   (c) SO_3 = sulfur trioxide
   (d) BrF_5 = bromine pentafluoride
   (e) NH_4CN = ammonium cyanide
   (f) $Ca(ClO_2)_2$ = calcium chlorite
   (g) $LiHCO_3$ = lithium hydrogen carbonate

10. (a) $$\% \text{ Cu} = \frac{(2 \times 63.55)}{(2 \times 63.55) + 12.01 + (5 \times 16.00) + (2 \times 1.008)} \times 100 = \frac{127.1}{221.1} \times 100 = 57.48\% \text{ Cu}$$

    (b) $$\text{no. g CuO} = 1000 \text{ g } CuCO_3 \cdot Cu(OH)_2 \times \frac{1 \text{ mol } CuCO_3 \cdot Cu(OH)_2}{221 \text{ g } CuCO_3 \cdot Cu(OH)_2} \times \frac{2 \text{ mol Cu}}{1 \text{ mol } CuCO_3 \cdot Cu(OH)_2}$$

    $$\times \frac{1 \text{ mol CuO}}{1 \text{ mol Cu}} \times \frac{79.6 \text{ g CuO}}{1 \text{ mol CuO}} = 720 \text{ g CuO}$$

11. In 100.00g hexachlorophene,

$$\text{no. mol C} = 38.37 \text{ g C} \times \frac{1 \text{ mol C}}{12.01 \text{ g C}} = 3.195 \text{ mol C}$$

$$\text{no. mol H} = 1.49 \text{ g H} \times \frac{1 \text{ mol H}}{1.008 \text{ g H}} = 1.48 \text{ mol H}$$

$$\text{no. mol Cl} = 52.28 \text{ g Cl} \times \frac{1 \text{ mol Cl}}{35.453 \text{ g Cl}} = 1.475 \text{ mol Cl}$$

$$\text{no. mol O} = 7.86 \text{ g O} \times \frac{1 \text{ mol O}}{16.00 \text{ g O}} = 0.491 \text{ mol O}$$

The empirical formula is $C_{\frac{3.195}{0.491}}H_{\frac{1.48}{0.491}}Cl_{\frac{1.475}{0.491}}O_{\frac{0.491}{0.491}} = C_{6.50}H_3Cl_3O$. Multiply all subscripts by two, to obtain $C_{13}H_6Cl_6O_2$.

12. If the hydrate contains almost exactly 50% H_2O, by mass, then the contribution of H_2O to the formula weight must be almost exactly the same as that of Na_2SO_3 (sodium sulfite). The formula weight of Na_2SO_3 = 126. A contribution of 126 would also be made by 126/18 = 7 molecules of H_2O. The hydrate formula is $Na_2SO_3 \cdot 7H_2O$.

Chapter 4

Stoichiometry II. Chemical Reactions

Review Problems

4-1. (a) $2\ Mg(s) + O_2(g) \rightarrow 2\ MgO(s)$

(b) $S(s) + O_2(g) \rightarrow SO_2(g)$

(c) $CH_4(g) + 2\ O_2(g) \rightarrow CO_2(g) + 2\ H_2O(\ell)$

(d) $Ag_2SO_4(aq) + BaI_2(aq) \rightarrow BaSO_4(s) + 2\ AgI(s)$

4-2. (a) $Na_2SO_4(s) + 2\ C(s) \rightarrow Na_2S(s) + 2\ CO_2(g)$

(b) $4\ HCl(g) + O_2(g) \rightarrow 2\ H_2O(\ell) + 2\ Cl_2(g)$

(c) $PCl_3(\ell) + 3\ H_2O(\ell) \rightarrow H_3PO_3(aq) + 3\ HCl(aq)$

(d) $3\ PbO(s) + 2\ NH_3(g) \rightarrow 3\ Pb(s) + N_2(g) + 3\ H_2O(\ell)$

(e) $Mg_3N_2(s) + 6\ H_2O(\ell) \rightarrow 3\ Mg(OH)_2(s) + 2\ NH_3(g)$

4-3. (a) $Zn(s) + 2\ Ag^+(aq) \rightarrow Zn^{2+}(aq) + 2\ Ag(s)$

(b) $Mn^{2+}(aq) + H_2S(aq) \rightarrow MnS(s) + 2\ H^+(aq)$

(c) $2\ Al(s) + 6\ H^+(aq) \rightarrow 2\ Al^{3+}(aq) + 3\ H_2(g)$

4-4. (a) $C_5H_{12} + 8\ O_2(g) \rightarrow 5\ CO_2(g) + 6\ H_2O(\ell)$

(b) $2\ C_2H_6O_2 + 5\ O_2(g) \rightarrow 4\ CO_2(g) + 6\ H_2O(\ell)$

(c) $HI(aq) + NaOH(aq) \rightarrow NaI(aq) + H_2O$

(d) $Pb(NO_3)_2(aq) + 2\ KI(aq) \rightarrow PbI_2(s) + 2\ KNO_3(aq)$

4-5. $\text{no. mol } FeCl_3 = 3.15 \text{ mol } Cl_2 \times \frac{2 \text{ mol } FeCl_3}{3 \text{ mol } Cl_2} = 2.10 \text{ mol } FeCl_3$

4-6. $\text{no. g } Cl_2 \text{ consumed} = 0.382 \text{ mol } PCl_3 \times \frac{6 \text{ mol } Cl_2}{4 \text{ mol } PCl_3} \times \frac{70.9 \text{ g } Cl_2}{1 \text{ mol } Cl_2} = 40.6 \text{ g } Cl_2$

$\text{no. g } P_4 \text{ consumed} = 0.382 \text{ mol } PCl_3 \times \frac{1 \text{ mol } P_4}{4 \text{ mol } PCl_3} \times \frac{(4 \times 30.97) \text{g } P_4}{1 \text{ mol } P_4} = 11.8 \text{ g } P_4$

4-7. (a) $\text{no. mol } H_2 = 312 \text{ g } CaH_2 \times \frac{1 \text{ mol } CaH_2}{42.1 \text{ g } CaH_2} \times \frac{2 \text{ mol } H_2}{1 \text{ mol } CaH_2} = 14.8 \text{ mol } H_2$

(b) $\text{no. g } H_2O = 88.5 \text{ g } CaH_2 \times \frac{1 \text{ mol } CaH_2}{42.1 \text{ g } CaH_2} \times \frac{2 \text{ mol } H_2O}{1 \text{ mol } CaH_2} \times \frac{18.0 \text{ g } H_2O}{1 \text{ mol } H_2O} = 75.7 \text{ g } H_2O$

(c) $\text{no. g } CaH_2 = 3.12\times10^{25} \text{ } H_2 \text{ molecules} \times \frac{1 \text{ mol } H_2}{6.02\times10^{23} \text{ } H_2 \text{ molecules}} \times \frac{1 \text{ mol } CaH_2}{2 \text{ mol } H_2} \times \frac{42.1 \text{ g } CaH_2}{1 \text{ mol } CaH_2}$

$= 1.09\times10^3 \text{ g } CaH_2$

4-8. (a) $\frac{2.17 \text{ mol } C_2H_5OH}{5.12 \text{ L soln}} = 0.424 \text{ M } C_2H_5OH$

(b) $\frac{12.35 \text{ mmol } CH_3OH}{50.00 \text{ mL soln}} = 0.2470 \text{ M } CH_3OH$

(c) $\frac{14.3 \text{ g } (CH_3)_2CO \times \frac{1 \text{ mol } (CH_3)_2CO}{58.1 \text{ g } (CH_3)_2CO}}{0.125 \text{ L soln}} = 1.97 \text{ M } (CH_3)_2CO$

(d) $\frac{12.2 \text{ mL} \times \frac{1.26 \text{ g } C_3H_8O_3}{1 \text{ mL}} \times \frac{1 \text{ mol } C_3H_8O_3}{92.1 \text{ g } C_3H_8O_3}}{0.375 \text{ L soln}} = 0.445 \text{ M } C_3H_8O_3$

4-9. (a) no. mol KCl = 3.55×10^{3} L × $\frac{0.250 \text{ mol KCl}}{\text{L}}$ = 0.888 mol KCl

(b) no. g Na_2SO_4 = 0.625 L × $\frac{0.415 \text{ mol } Na_2SO_4}{\text{L}}$ × $\frac{142 \text{ g } Na_2SO_4}{1 \text{ mol } Na_2SO_4}$ = 36.8 g Na_2SO_4

(c) no. mg KOH = 1.00 mL × $\frac{0.105 \text{ mmol KOH}}{\text{mL}}$ × $\frac{56.1 \text{ mg KOH}}{\text{mmol KOH}}$ = 5.89 mg KOH

4-10. no. mol KOH in final soln = 0.5000 L × $\frac{0.312 \text{ mmol KOH}}{\text{L}}$ = 0.156 mol KOH

no. L of conc. soln = 0.156 mol KOH × $\frac{1 \text{ L soln}}{0.800 \text{ mol KOH}}$ = 0.195 L

4-11. no. mmol HNO_3 = 25.00 mL × $\frac{0.1132 \text{ mmol } HNO_3}{\text{mL}}$ = 2.830mmol HNO_3

no. mmol NaOH required for titration = 2.830mmol HNO_3 × $\frac{1 \text{ mmol NaOH}}{1 \text{ mmol } HNO_3}$ = 28.30mmol NaOH

no. mL NaOH soln = 2.830mmol NaOH × $\frac{1 \text{ mL}}{0.1035 \text{ mmol NaOH}}$ = 27.34 mL

4-12. Determine the amount of each reactant and compare them to determine the limiting reagent.

no. mol Cu = 0.500 mol Cu

no. mol HNO_3 = 0.125 L × $\frac{6.0 \text{ mol } HNO_3}{\text{L}}$ = 0.75 mol HNO_3

no. mol HNO_3 required to dissolve all the Cu = 0.500 mol Cu × $\frac{8 \text{ mol } HNO_3}{3 \text{ mol Cu}}$ = 1.33 mol HNO_3

There is not enough HNO_3 available to dissolve all the Cu.

4-13. (a) The theoretical yield is

no. g C_6H_{10} = 100.0 g $C_6H_{12}O$ × $\frac{1 \text{ mol } C_6H_{12}O}{100.2 \text{ g } C_6H_{12}O}$ × $\frac{1 \text{ mol } C_6H_{10}}{1 \text{ mol } C_6H_{12}O}$ × $\frac{82.1 \text{ g } C_6H_{10}}{1 \text{ mol } C_6H_{10}}$ = 81.9 g C_6H_{10}

(b) % yield = $\frac{69.0 \text{ g } C_6H_{10}}{81.9 \text{ g } C_6H_{10}}$ × 100 = 84.2%

(c) no. g $C_6H_{12}O$ = 100.0 g C_6H_{10} made × $\frac{100.0 \text{ g } C_6H_{10} \text{ theo}}{84.2 \text{ g } C_6H_{10} \text{ made}}$ × $\frac{100.2 \text{ g } C_6H_{12}O}{82.1 \text{ g } C_6H_{10}}$ = 145 g $C_6H_{12}O$

4-14. (a)

$\overset{+2}{NO} + \overset{0}{H_2} \longrightarrow \overset{-3\,+1}{NH_3} + \overset{+1}{H_2O}$

decrease of 5 in O.S. per N

increase of 1 in O.S. per H

$2\,NO + 5\,H_2 \longrightarrow 2\,NH_3 + 2\,H_2O$

decrease of 10 in O.S.

increase of 10 in O.S.

(b) $\overset{0}{Cu} + H^+ + \overset{+5}{NO_3^-} \longrightarrow \overset{+2}{Cu^{2+}} + H_2O + \overset{+2}{NO}$

increase of 2 in O.S. per Cu

decrease of 3 in O.S. per N

$3\ Cu + H^+ + 2\ NO_3^- \longrightarrow 3\ Cu^{2+} + H_2O + 2\ NO$

increase of 6 in O.S.

decrease of 6 in O.S.

$3\ Cu + 8\ H^+ + 2\ NO_3^- \rightarrow 3\ Cu^{2+} + 4\ H_2O + 2\ NO$

(c) $\overset{0}{Zn} + H^+ + \overset{+5}{NO_3^-} \longrightarrow \overset{+2}{Zn^{2+}} + H_2O + \overset{+1}{N_2O}$

increase of 2 in O.S. per Zn

decrease of 4 in O.S. per N

$4\ Zn + H^+ + 2\ NO_3^- \longrightarrow 4\ Zn^{2+} + H_2O + N_2O$

increase of 8 in O.S.

decrease of 8 in O.S.

$4\ Zn + 10\ H^+ + 2\ NO_3^- \rightarrow 4\ Zn^{2+} + 5\ H_2O + N_2O$

(d) $\overset{-2}{Fe_2S_3} + H_2O + \overset{0}{O_2} \longrightarrow \overset{-2}{Fe(OH)_3} + \overset{0}{S}$

increase of 2 in O.S. per S

decrease of 2 per O

$2\ Fe_2S_3 + H_2O + 3\ O_2 \longrightarrow 4\ Fe(OH)_3 + 6\ S$

increase of 12 in O.S.

decrease of 12 in O.S.

$2\ Fe_2S_3 + 6\ H_2O + 3\ O_2 \rightarrow 4\ Fe(OH)_3 + 6\ S$

(e) $\overset{-1}{H_2O_2} + \overset{+7}{MnO_4^-} + H^+ \longrightarrow \overset{+2}{Mn^{2+}} + H_2O + \overset{0}{O_2}$

increase of 1 in O.S. per O

decrease of 5 in O.S. per Mn

$5\ H_2O_2 + 2\ MnO_4^- + H^+ \longrightarrow 2\ Mn^{2+} + H_2O + 5\ O_2$

increase of 10 in O.S.

decrease of 10 in O.S.

$5\ H_2O_2 + 2\ MnO_4^- + 6\ H^+ \rightarrow 2\ Mn^{2+} + 8\ H_2O + 5\ O_2$

4-15. The oxidizing agents experience a decrease in oxidation state for one of their elements. As indicated in Review Problem 14, these would be (a) NO; (b) NO_3^-; (c) NO_3^-; (d) O_2; (e) MnO_4^-. The reducing agents experience an increase in oxidation state for one of their elements: (a) H_2; (b) Cu; (c) Zn; (d) Fe_2S_3; (e) H_2O_2.

Exercises

4-1. (a) $H_2O(g) + C(s) \xrightarrow{\Delta} CO(g) + H_2(g)$

(b) $2\ Al(s) + 3\ Cu^{2+}(aq) \rightarrow 2\ Al^{3+}(aq) + 3\ Cu(s)$

(c) $ZnS(s) + 2\ H^+(aq) \rightarrow Zn^{2+}(aq) + H_2S(g)$

(d) $2\ Cl_2(g) + 2\ H_2O(g) \rightarrow 4\ HCl(g) + O_2(g)$

4-2. (a) $Cl_2(g) + H_2O(\ell) \rightarrow HCl(aq) + HOCl(aq)$

(b) $6\ P_2H_4(\ell) \rightarrow 8\ PH_3(g) + P_4(s)$

(c) $3\ NO_2(g) + H_2O(\ell) \rightarrow 2\ HNO_3(aq) + NO(g)$

(d) $6\ S_2Cl_2 + 16\ NH_3 \rightarrow N_4S_4 + 12\ NH_4Cl + S_8$

(e) $SO_2Cl_2 + 8\ HI \rightarrow H_2S + 2\ H_2O + 2\ HCl + 4\ I_2$

4-3. (a) $2\ Fe^{3+}(aq) + 3\ H_2S(aq) \rightarrow Fe_2S_3(s) + 6\ H^+(aq)$

(b) $Al^{3+} + 3\ NH_3 + 3\ H_2O \rightarrow Al(OH)_3(s) + 3\ NH_4^+(aq)$

(c) $S_2O_3^{2-} + 2\ H^+ \rightarrow H_2O + S(s) + SO_2(g)$

(d) $MnO_2(s) + 4\ H^+ + 2\ Cl^- \rightarrow Mn^{2+} + 2\ H_2O + Cl_2(g)$

4-4. (a) $2\ C_6H_6 + 15\ O_2 \rightarrow 12\ CO_2 + 6\ H_2O$

(b) $2\ C_3H_7OH + 9\ O_2 \rightarrow 6\ CO_2 + 8\ H_2O$

(c) $2\ C_6H_5COOH + 15\ O_2 \rightarrow 14\ CO_2 + 6\ H_2O$

(d) $C_6H_5COSH + 9\ O_2 \rightarrow 7\ CO_2 + 3\ H_2O + SO_2$

Stoichiometry of chemical reactions

4-5. (a) $\text{no. mol } O_2 = 25.5 \text{ g } KClO_3 \times \frac{1 \text{ mol } KClO_3}{123 \text{ g } KClO_3} \times \frac{3 \text{ mol } O_2}{2 \text{ mol } KClO_3} = 0.311 \text{ mol } O_2$

(b) $\text{no. mol } O_2 \text{ molecules} = 0.311 \text{ mol } O_2 \times \frac{6.02\times10^{23}\ O_2 \text{ molecules}}{1 \text{ mol } O_2} = 1.87\times10^{23}\ O_2 \text{ molecules}$

(c) $\text{no. g KCl} = 25.5 \text{ g } KClO_3 \times \frac{1 \text{ mol } KClO_3}{123 \text{ g } KClO_3} \times \frac{2 \text{ mol KCl}}{2 \text{ mol } KClO_3} \times \frac{74.6 \text{ g KCl}}{1 \text{ mol KCl}} = 15.5 \text{ g KCl}$

4-6. $\text{no. g } Na_2CO_3 \text{ in mixture} = 128 \text{ g mixt.} \times \frac{61.3 \text{ g } Na_2CO_3}{100 \text{ g mixt.}} = 78.5 \text{ g } Na_2CO_3$

$$\text{no. g } CO_2 = 78.5 \text{ g } Na_2CO_3 \times \frac{1 \text{ mol } Na_2CO_3}{106 \text{ g } Na_2CO_3} \times \frac{1 \text{ mol } CO_2}{1 \text{ mol } Na_2CO_3} \times \frac{44.0 \text{ g } CO_2}{1 \text{ mol } CO_2} = 32.6 \text{ g } CO_2$$

4-7. The key to this problem is to determine how much *pure* Fe_2O_3 is required to produce 515 kg Fe; the quantity will be less than 878 kg. The percent Fe_2O_3 in the ore is determined in the usual way.

$$\text{no. kg } Fe_2O_3 = 515 \text{ kg Fe} \times \frac{1000 \text{ g Fe}}{1 \text{ kg Fe}} \times \frac{1 \text{ mol Fe}}{55.8 \text{ g Fe}} \times \frac{1 \text{ mol } Fe_2O_3}{2 \text{ mol Fe}} \times \frac{160 \text{ g } Fe_2O_3}{1 \text{ mol } Fe_2O_3} \times \frac{1 \text{ kg } Fe_2O_3}{1000 \text{ g } Fe_2O_3}$$

$$= 738 \text{ kg } Fe_2O_3$$

$$\%\ Fe_2O_3 = \frac{738 \text{ kg } Fe_2O_3}{878 \text{ kg ore}} \times 100 = 84.1\%$$

Rather than convert from kg Fe to g Fe and, later, from g Fe_2O_3 to kg Fe_2O_3, consider the concept of a *kilo*mole:

1kmol Fe = 55.8 kg Fe　　　　1 kmol Fe_2O_3 = 160 kg Fe_2O_3

$$\text{no. kg } Fe_2O_3 = 515 \text{ kg Fe} \times \frac{1 \text{ kmol Fe}}{55.8 \text{ kg Fe}} \times \frac{1 \text{ kmol } Fe_2O_3}{2 \text{ kmol Fe}} \times \frac{160 \text{ kg } Fe_2O_3}{1 \text{ kmol } Fe_2O_3} = 738 \text{ kg } Fe_2O_3$$

4-8. First write a balanced equation for the reaction, and then proceed as in Exercise 4-7. That is, determine the mass of *pure* Ag_2O required to produce 0.183 g O_2.

$$2\ Ag_2O(s) \rightarrow 4\ Ag(s) + O_2(g)$$

$$\text{no. g } Ag_2O = 0.183 \text{ g } O_2 \times \frac{1 \text{ mol } O_2}{32.0 \text{ g } O_2} \times \frac{2 \text{ mol } Ag_2O}{1 \text{ mol } O_2} \times \frac{232 \text{ g } Ag_2O}{1 \text{ mol } Ag_2O} = 2.65 \text{ g } Ag_2O$$

$$\%\ Ag_2O = \frac{2.65 \text{ g } Ag_2O}{2.95 \text{ g sample}} \times 100 = 89.8\%$$

4-9. $$\text{no. mol Al that dissolves} = (5.11 \text{ in.} \times 3.23 \text{ in.} \times 0.0381 \text{ in.}) \times \frac{(2.54)^3 \text{ cm}^3}{(1)^3 \text{ in}^3} \times \frac{270 \text{ g Al}}{1 \text{ cm}^3} \times \frac{1 \text{ mol Al}}{27.0 \text{ g Al}}$$

$$= 1.03 \text{ mol Al}$$

$$\text{no. g } H_2 \text{ produced} = 1.03 \text{ mol Al} \times \frac{3 \text{ mol } H_2}{2 \text{ mol Al}} \times \frac{2.02 \text{ g } H_2}{1 \text{ mol } H_2} = 3.12 \text{ g } H_2$$

4-10. First determine the no. g $KMnO_4$ required.

$$\text{no. g } KMnO_4 = 9.13 \text{ g KI} \times \frac{1 \text{ mol KI}}{166 \text{ g KI}} \times \frac{2 \text{ mol } KMnO_4}{10 \text{ mol KI}} \times \frac{158 \text{ g } KMnO_4}{1 \text{ mol } KMnO_4} = 1.74 \text{ g } KMnO_4$$

Now use the solution composition as a conversion factor from g $KMnO_4$ to mL solution.

$$\text{no. mL soln} = 1.74 \text{ g } KMnO_4 \times \frac{1 \text{ L soln}}{12.6 \text{ g } KMnO_4} \times \frac{1000 \text{ mL}}{1 \text{ L}} = 138 \text{ mL}$$

Molar concentration

4-11. (a) $$\text{no. mol } CO(NH_2)_2 = 132 \text{ g solid} \times \frac{98.3 \text{ g } CO(NH_2)_2}{100 \text{ g solid}} \times \frac{1 \text{ mol } CO(NH_2)_2}{60.1 \text{ g } CO(NH_2)_2} = 2.16 \text{ mol } CO(NH_2)_2$$

$$\text{molarity} = \frac{2.16 \text{ mol } CO(NH_2)_2}{0.500 \text{ L}} = 4.32 \text{ M } CO(NH_2)_2$$

(b) $$\text{no. mol } (C_2H_5)_2O = 13.0 \text{ mg} \times \frac{1.00 \text{ g}}{1000 \text{ mg}} \times \frac{1 \text{ mol } (C_2H_5)_2O}{74.1 \text{ g } (C_2H_5)_2O} = 1.75\times10^{-4} \text{ mol } (C_2H_5)_2O$$

$$\text{no. L} = 3.00 \text{ gal} \times \frac{3.78 \text{ L}}{1 \text{ gal}} = 11.3 \text{ L}$$

$$\text{molarity} = \frac{1.75\times10^{-4} \text{ mol } (C_2H_5)_2O}{11.3 \text{ L}} = 1.55\times10^{-5} \text{ M } (C_2H_5)_2O$$

(c) Convert from ppm Na to mol NaCl per million grams of solution.

$$\text{no. mol NaCl} = 1.52 \text{ g Na} \times \frac{1 \text{ mol Na}}{23.0 \text{ g Na}} \times \frac{1 \text{ mol NaCl}}{1 \text{ mol Na}} = 0.0661 \text{ mol NaCl}$$

$$\text{no. L soln} = 1.00\times10^6 \text{ g soln} \times \frac{1.00 \text{ cm}^3 \text{ soln}}{1.00 \text{ g soln}} \times \frac{1 \text{ L soln}}{1000 \text{ cm}^3 \text{ soln}} = 1.00\times10^3 \text{ L soln}$$

$$\text{molarity} = \frac{0.0661 \text{ mol NaCl}}{1.00\times10^3 \text{ L}} = 6.61\times10^{-5} \text{ M NaCl}$$

4-12. (a) $$\text{no. cm}^3\ CH_3OH = 4.80 \text{ L} \times \frac{0.318 \text{ mol } CH_3OH}{\text{L}} \times \frac{32.0 \text{ g } CH_3OH}{1 \text{ mol } CH_3OH} \times \frac{1 \text{ cm}^3\ CH_3OH}{0.792 \text{ g } CH_3OH} = 61.7 \text{ cm}^3\ CH_3OH$$

(b) Convert from gallons of solution to mol C_2H_5OH required.

$$\text{no. mol } C_2H_5OH = 55.0 \text{ gal} \times \frac{3.78 \text{ L}}{1 \text{ gal}} \times \frac{1.50 \text{ mol } C_2H_5OH}{\text{L}} = 312 \text{ mol } C_2H_5OH$$

Now convert from mol C_2H_5OH to gal C_2H_5OH.

$$\text{no. gal } C_2H_5OH = 312 \text{ mol } C_2H_5OH \times \frac{46.1 \text{ g } C_2H_5OH}{1 \text{ mol } C_2H_5OH} \times \frac{1.00 \text{ cm}^3 \text{ } C_2H_5OH}{0.789 \text{ g } C_2H_5OH} \times \frac{1 \text{ L } C_2H_5OH}{1000 \text{ cm}^3 \text{ } C_2H_5OH}$$

$$\times \frac{1 \text{ gal } C_2H_5OH}{3.78 \text{ L } C_2H_5OH} = 4.82 \text{ gal } C_2H_5OH$$

(c) $$\text{no. mg } Ca(NO_3)_2 = 50.0 \text{ L soln} \times \frac{1000 \text{ cm}^3 \text{ soln}}{1.00 \text{ L soln}} \times \frac{1.00 \text{ g soln}}{1.00 \text{ cm}^3 \text{ soln}} \times \frac{1.21 \text{ g Ca}}{1.00\times10^6 \text{ g soln}} \times \frac{1 \text{ mol Ca}}{40.1 \text{ g Ca}}$$

$$\times \frac{1 \text{ mol } Ca(NO_3)_2}{1 \text{ mol Ca}} \times \frac{164 \text{ g } Ca(NO_3)_2}{1 \text{ mol } Ca(NO_3)_2} \times \frac{1000 \text{ mg } Ca(NO_3)_2}{1.00 \text{ g } Ca(NO_3)_2} = 247 \text{ mg } Ca(NO_3)_2$$

4-13. Determine the total number of moles of sucrose and divide by the total solution volume, in liters.

$$\text{no. mol } C_{12}H_{22}O_{11} = 0.158 \text{ L} \times \frac{1.50 \text{ mol } C_{12}H_{22}O_{11}}{\text{L}} = 0.237 \text{ mol } C_{12}H_{22}O_{11}$$

$$\text{no. mol } C_{12}H_{22}O_{11} = 0.273 \text{ L} \times \frac{1.25 \text{ mol } C_{12}H_{22}O_{11}}{\text{L}} = 0.341 \text{ mol } C_{12}H_{22}O_{11}$$

$$\text{total mol } C_{12}H_{22}O_{11} = 0.237 + 0.341 = 0.578 \text{ mol } C_{12}H_{22}O_{11}$$

$$\text{final molarity} = \frac{0.578 \text{ mol } C_{12}H_{22}O_{11}}{(0.158 + 0.237) \text{ L}} = 1.34 \text{ M } C_{12}H_{22}O_{11}$$

4-14. The mass of pure HCl required in the solution is determined in the usual way. But this mass must then be converted to mass of concentrated acid, by using the *reciprocal* of the percent composition (100/36.0), and then to volume of concentrated acid, by using the *reciprocal* of the density (1.00/1.18).

$$\text{no. mL conc. acid} = 20.0 \text{ L soln} \times \frac{0.125 \text{ mol HCl}}{1 \text{ L soln}} \times \frac{36.5 \text{ g HCl}}{1 \text{ mol HCl}} \times \frac{100 \text{ g conc. acid}}{36.0 \text{ g HCl}}$$

$$\times \frac{1 \text{ mL conc. acid}}{1.18 \text{ g conc. acid}} = 215 \text{ mL conc. acid}$$

4-15. Determine the no. mol $MgSO_4$ in the original solution.

$$\text{no. mol } MgSO_4 = 0.0850 \text{ L} \times \frac{0.512 \text{ mol } MgSO_4}{\text{L}} = 0.0435 \text{ mol } MgSO_4$$

This much solute remains in 61.5 mL of solution, leading to a molar concentration of

$$\frac{0.0435 \text{ mol } MgSO_4}{0.0615 \text{ L}} = 0.707 \text{ M } MgSO_4$$

Chemical reactions in solutions

4-16. $$\text{no. g } NaHCO_3 = 0.155 \text{ L} \times \frac{0.245 \text{ mol } Cu(NO_3)_2}{\text{L}} \times \frac{2 \text{ mol } NaHCO_3}{1 \text{ mol } Ca(NO_3)_2} \times \frac{84.0 \text{ g } NaHCO_3}{1 \text{ mol } NaHCO_3} = 6.38 \text{ g } NaHCO_3$$

4-17. (a) $$\text{no. g } Ca(OH)_2 = 0.325 \text{ L} \times \frac{0.410 \text{ mol HCl}}{\text{L}} \times \frac{1 \text{ mol } Ca(OH)_2}{2 \text{ mol HCl}} \times \frac{74.1 \text{ g } Ca(OH)_2}{1 \text{ mol } Ca(OH)_2} = 4.94 \text{ g } Ca(OH)_2$$

(b) $$\text{no. kg } Ca(OH)_2 = 215 \text{ L soln} \times \frac{1000 \text{ cm}^3 \text{ soln}}{1 \text{ L soln}} \times \frac{1.15 \text{ g soln}}{1 \text{ cm}^3 \text{ soln}} \times \frac{30.12 \text{ g HCl}}{100 \text{ g soln}} \times \frac{1 \text{ mol HCl}}{36.5 \text{ g HCl}}$$

$$\times \frac{1 \text{ mol } Ca(OH)_2}{2 \text{ mol HCl}} \times \frac{74.1 \text{ g } Ca(OH)_2}{1 \text{ mol } Ca(OH)_2} \times \frac{1 \text{ kg } Ca(OH)_2}{1000 \text{ g } Ca(OH)_2} = 75.6 \text{ kg } Ca(OH)_2$$

4-18. (a) The number of moles of NH_3 present in the 5.00 mL sample is

$$\text{no. mol } NH_3 = 31.20 \text{ mL acid} \times \frac{1 \text{ L acid}}{1000 \text{ mL acid}} \times \frac{0.9918 \text{ mol HCl}}{1 \text{ L acid}} \times \frac{1 \text{ mol } NH_3}{1 \text{ mol HCl}} = 0.0309 \text{ mol } NH_3$$

$$\text{molarity of } NH_3 = \frac{0.0309 \text{ mol } NH_3}{0.00500 \text{ L}} = 6.18 \text{ M } NH_3$$

(b) $$\frac{\text{no. g } NH_3}{\text{L}} = \frac{6.18 \text{ mol } NH_3}{1 \text{ L}} \times \frac{17.0 \text{ g } NH_3}{1 \text{ mol } NH_3} = \frac{105 \text{ g } NH_3}{\text{L}}$$

The mass of one liter of solution is: $m = d \cdot V$.

$$\text{no. g} = 0.96 \text{ g/mL} \times 1000 \text{ mL} = 960 \text{ g} \qquad \% \, NH_3 = \frac{105 \text{ g } NH_3}{9.6\times10^2 \text{ g soln}} \times 100 = 11\%$$

4-19. (a) $$\text{no. mol HCl} = 20 \text{ L} \times \frac{0.25 \text{ mol HCl}}{\text{L}} = 5.0 \text{ mol HCl}$$

The volume of concentrated HCl(aq) to produce 5.0 mol HCl is calculated next.

$$\text{no. mL conc. acid} = 5.0 \text{ mol HCl} \times \frac{36.5 \text{ g HCl}}{1 \text{ mol HCl}} \times \frac{100 \text{ g conc. acid}}{38 \text{ g HCl}} \times \frac{1 \text{ mL conc. acid}}{1.19 \text{ g conc. acid}}$$

$$= 4.0\times10^2 \text{ mL conc. acid}$$

(b) $$\text{no. mol HCl} = 0.03010 \text{ L} \times \frac{0.2000 \text{ mol NaOH}}{\text{L}} \times \frac{1 \text{ mol HCl}}{1 \text{ mol NaOH}} = 6.020\times10^{-3} \text{ mol HCl}$$

$$\text{molarity} = \frac{6.020\times10^{-3} \text{ mol HCl}}{0.02500 \text{ L}} = 0.2408 \text{ M HCl}$$

(c) Measurements required to produce the diluted acid from concentrated HCl(aq) cannot be made with sufficient precision to yield four significant figures in the calculated molarity. That is, the volume of concentrated acid cannot easily be measured to within 0.1 mL and the 20 L final solution volume cannot easily be measured to the nearest 10 mL. Equally important is the fact that the percent composition of the acid is usually not known to the nearest 0.01%, nor can we be assured that its concentration does not change during storage. Titration, on the other hand, easily yields four significant figures for solution concentrations.

4-20. From the titration data first determine the mass of Fe in the ore sample.

$$\text{no. g Fe} = 0.02972 \text{ L} \times \frac{0.0410 \text{ mol } K_2Cr_2O_7}{\text{L}} \times \frac{6 \text{ mol Fe}}{1 \text{ mol } K_2Cr_2O_7} \times \frac{55.85 \text{ g Fe}}{1 \text{ mol Fe}} = 0.408 \text{ g Fe}$$

$$\% \text{ Fe} = \frac{0.408 \text{ g Fe}}{0.8313 \text{ g ore}} \times 100 = 49.1\% \text{ Fe}$$

4-21. The quantity to determine first is the number of moles of HCl that must react to change the concentration in the manner desired.

Initial solution

$$\text{no. mol HCl} = 0.2500 \text{ L} \times \frac{1.017 \text{ mol HCl}}{1 \text{ L}} = 0.2542 \text{ mol HCl}$$

Final solution

$$\text{no. mol HCl} = 0.2500 \text{ L} \times \frac{1.000 \text{ mol HCl}}{1 \text{ L}} = 0.2500 \text{ mol HCl}$$

The amount of HCl that reacts is 0.2542 - 0.2500 = 0.0042 mol HCl. The quantity of magnesium required for the reaction may now be calculated.

$$\text{no. mg Mg} = 0.0042 \text{ mol HCl} \times \frac{1 \text{ mol Mg}}{2 \text{ mol HCl}} \times \frac{24.3 \text{ g Mg}}{1 \text{ mol Mg}} \times \frac{1000 \text{ mg Mg}}{1 \text{ g Mg}} = 51 \text{ mg Mg}$$

Determining the limiting reagent

4-22. The reaction of interest is: $3\ CS_2 + 6\ NaOH \rightarrow 2\ Na_2CS_3 + Na_2CO_3 + 3\ H_2O$

(a) The CS_2 is the excess reactant because only 0.50 mol CS_2 is consumed along with 1.00 mol NaOH, whereas 1.00 mol of each is available originally.

$$\text{no. mol } CS_2 \text{ consumed} = 1.00 \text{ mol NaOH} \times \frac{3 \text{ mol } CS_2}{6 \text{ mol NaOH}} = 0.50 \text{ mol } CS_2$$

The amounts of the products must be related either to the 1.00 mol NaOH or to the 0.50 mol CS_2 consumed.

$$\text{no. mol } Na_2CS_3 = 0.50 \text{ mol } CS_2 \times \frac{2 \text{ mol } Na_2CS_3}{3 \text{ mol } CS_2} = 0.33 \text{ mol } Na_2CS_3$$

$$\text{no. mol } Na_2CO_3 = 0.50 \text{ mol } CS_2 \times \frac{1 \text{ mol } Na_2CO_3}{3 \text{ mol } CS_2} = 0.17 \text{ mol } Na_2CO_3$$

$$\text{no. mol } H_2O = 0.50 \text{ mol } CS_2 \times \frac{3 \text{ mol } H_2O}{3 \text{ mol } CS_2} = 0.50 \text{ mol } H_2O$$

(b) First determine the limiting reagent. For example, calculate the number of mol CS_2 in 1.00 cm^3 of the liquid.

$$\text{no. mol } CS_2 = 100.0 \text{ cm}^3\ CS_2 \times \frac{1.26 \text{ g } CS_2}{1 \text{ cm}^3\ CS_2} \times \frac{1 \text{ mol } CS_2}{76.1 \text{ g } CS_2} = 1.66 \text{ mol } CS_2$$

Now compare this to the available NaOH: $\text{no. mol } CS_2 = 3.50 \text{ mol NaOH} \times \frac{3 \text{ mol } CS_2}{6 \text{ mol NaOH}} = 1.75 \text{ mol } CS_2$

To react with all of the NaOH, 1.75 mol CS_2 is required. However, only 1.66 mol CS_2 is available. CS_2 is the limiting reagent.

$$\text{no. g } Na_2CS_3 = 1.66 \text{ mol } CS_2 \times \frac{2 \text{ mol } Na_2CS_3}{3 \text{ mol } CS_2} \times \frac{154 \text{ g } Na_2CS_3}{1 \text{ mol } Na_2CS_3} = 1.70 \times 10^2 \text{ g } Na_2CS_3$$

4-23. Determine which is the limiting reagent.

$$\text{no. mol Al} = 1.75 \text{ g Al} \times \frac{1 \text{ mol Al}}{27.0 \text{ g Al}} = 0.0648 \text{ mol Al}$$

$$\text{no. mol HCl} = 0.0750 \text{ L} \times \frac{2.50 \text{ mol HCl}}{\text{L}} = 0.188 \text{ mol HCl}$$

According to equation (4.16)

$$\text{no. mol HCl required} = 0.0648 \text{ mol Al} \times \frac{6 \text{ mol HCl}}{2 \text{ mol Al}} = 0.194 \text{ mol HCl}$$

Since not this much HCl is available, HCl is the limiting reagent.

$$\text{no. g } H_2 = 0.188 \text{ mol HCl} \times \frac{3 \text{ mol } H_2}{6 \text{ mol HCl}} \times \frac{2.02 \text{ g } H_2}{1 \text{ mol } H_2} = 0.190 \text{ g } H_2$$

4-24. First, write a balanced equation for the reaction: $2\ NH_4Cl + Ca(OH)_2 \rightarrow CaCl_2 + 2\ H_2O + 2\ NH_3(g)$
Next, determine the no. mol of each reactant.

$$\text{no. mol } NH_4Cl = 15.0 \text{ g } NH_4Cl \times \frac{1 \text{ mol } NH_4Cl}{53.5 \text{ g } NH_4Cl} = 0.280 \text{ mol } NH_4Cl$$

$$\text{no. mol } Ca(OH)_2 = 15.0 \text{ g } Ca(OH)_2 \times \frac{1 \text{ mol } Ca(OH)2}{74.1 \text{ g } Ca(OH)_2} = 0.202 \text{ mol } Ca(OH)_2$$

From the balanced equation we see that 2 mol NH_4Cl is consumed for every mol $Ca(OH)_2$ that reacts. However, the amount of NH_4Cl available is less than twice the amount of $Ca(OH)_2$. NH_4Cl is the limiting reagent.

$$\text{no. g } NH_3 = 0.280 \text{ mol } NH_4Cl \times \frac{2 \text{ mol } NH_3}{2 \text{ mol } NH_4Cl} \times \frac{17.0 \text{ g } NH_3}{1 \text{ mol } NH_3} = 4.76 \text{ g } NH_3$$

4-25. Again, first determine the limiting reagent.

$$\text{no. mol BaS} = 0.0851 \text{ L} \times \frac{0.150 \text{ mol BaS}}{\text{L}} = 0.0128 \text{ mol BaS}$$

$$\text{no. mol } ZnSO_4 = 0.0432 \text{ L} \times \frac{0.355 \text{ mol } ZnSO4}{\text{L}} = 0.0153 \text{ mol } ZnSO_4$$

Since the two reactants combine in a 1:1 mol ratio, BaS is the limiting reagent. Now, determine the mass of ZnS and of $BaSO4$ produced by the reaction, and add the two quantities together.

$$\text{no. g ZnS} = 0.0128 \text{ mol BaS} \times \frac{1 \text{ mol ZnS}}{1 \text{ mol BaS}} \times \frac{97.4 \text{ g ZnS}}{1 \text{ mol ZnS}} = 1.25 \text{ g ZnS}$$

$$\text{no. g } BaSO_4 = 0.0128 \text{ mol BaS} \times \frac{1 \text{ mol } BaSO4}{1 \text{ mol BaS}} \times \frac{233 \text{ g } BaSO4}{1 \text{ mol } BaSO_4} = 2.98 \text{ g } BaSO_4$$

$$\text{no. g mixed precipitate} = 1.25 \text{ g ZnS} + 2.98 \text{ g } BaSO_4 = 4.23 \text{ g ppt.}$$

4-26. (a) $2 H_2 + O_2 \rightarrow 2 H_2O$

(b) First is must be determined which reactant is consumed and which is in excess. For example, determine the number of grams of H_2 required to react completely with all of the available O_2.

$$\text{no. g } H_2 = 36.40 \text{ g } O_2 \times \frac{1 \text{ mol } O2}{32.00 \text{ g } O_2} \times \frac{2 \text{ mol } H2}{1 \text{ mol } O_2} \times \frac{2.016 \text{ g } H2}{1 \text{ mol } H_2} = 4.586 \text{ g } H_2$$

no. g H_2 available = 4.800. Therefore, excess H_2 remains and all of the O_2 is consumed.

$$\text{no. g } H_2O \text{ produced} = 36.40 \text{ g } O_2 \times \frac{1 \text{ mol } O2}{32.00 \text{ g } O_2} \times \frac{2 \text{ mol } H_2O}{1 \text{ mol } O_2} \times \frac{18.02 \text{ g } H_2O}{1 \text{ mol } H_2O} = 41.00 \text{ g } H_2O$$

no. g H_2 left = 4.800 g H_2 available - 4.586 g H_2 consumed = 0.214 g H_2

no. g O_2 left = 0

(c) Original mass = 36.40 g O_2 + 4.800 g H_2 = 41.20 g total

Final mass = 41.00 g H_2O + 0.214 g H_2 + 0.00g O_2 = 41.21 g total

Within the limits of experimental error, the law of conservation mass is verified.

4-27. Again it is necessary to express the amount of each reactant in number of moles. Percentage composition and density enter into these calculations as conversion factors.

$$\text{no. mol } K_2Cr_2O_7 = 60.7 \text{ g sample} \times \frac{96.2 \text{ g } K_2Cr_2O_7}{100 \text{ g sample}} \times \frac{1 \text{ mol } K_2Cr_2O_7}{294 \text{ g } K_2Cr_2O_7} = 0.199 \text{ mol } K_2Cr_2O_7$$

$$\text{no. mol HCl} = 318 \text{ ml acid} \times \frac{1.15 \text{ g acid}}{1 \text{ mL acid}} \times \frac{30.1 \text{ g HCl}}{100 \text{ g acid}} \times \frac{1 \text{ mol HCl}}{36.5 \text{ g HCl}} = 3.02 \text{ mol HCl}$$

Inspection of the balanced chemical equation reveals that 14 mol HCl are consumed for every mol $K_2Cr_2O_7$. To react with the 0.199 mol $K_2Cr_2O_7$, 2.79 mol HCl is required. More than this amount of HCl is available. It is the reactant in excess; $K_2Cr_2O_7$ is the limiting reagent.

$$\text{no. g } Cl_2 = 0.199 \text{ mol } K_2Cr_2O_7 \times \frac{3 \text{ mol } Cl_2}{1 \text{ mol } K_2Cr_2O_7} \times \frac{70.9 \text{ g } Cl2}{1 \text{ mol } Cl_2} = 42.3 \text{ g } Cl_2$$

Theoretical, actual, and percent yield

4-28. (a) The theoretical yield is obtained with a conversion factor from the balanced equation. Also required is a determination of the limiting reagent.

$$\text{no. mol } C_4H_9OH = 13.0 \text{ g } C_4H_9OH \times \frac{1 \text{ mol } C_4H_9OH}{74.1 \text{ g } C_4H_9OH} = 0.175 \text{ mol } C_4H_9OH$$

$$\text{no. mol NaBr} = 21.6 \text{ g NaBr} \times \frac{1 \text{ mol NaBr}}{103 \text{ g NaBr}} = 0.210 \text{ mol NaBr}$$

$$\text{no. mol } H_2SO_4 = 33.8 \text{ g } H_2SO_4 \times \frac{1 \text{ mol } H_2SO_4}{98.1 \text{ g } H_2SO_4} = 0.345 \text{ mol } H_2SO_4$$

Since the reactants combine in the ratio 1:1:1, C_4H_9OH is the limiting reagent.

$$\text{no. g } C_4H_9Br = 0.175 \text{ mol } C_4H_9OH \times \frac{1 \text{ mol } C_4H_9Br}{1 \text{ mol } C_4H_9OH} \times \frac{137 \text{ g } C_4H_9Br}{1 \text{ mol } C_4H_9Br} = 24.0 \text{ g } C_4H_9Br$$

theoretical yield = 24.0 g C_4H_9Br

(b) actual yield = 16.8 g C_4H_9Br

(c) $\% \text{ yield} = \frac{16.8 \text{ g } C_4H_9Br}{24.0 \text{ g } C_4H_9Br} \times 100 = 70.0\%$

4-29. First, we must determine which of the two reactants, nitrobenzene or triethylene glycol, is the limiting reagent. Next, we must determine the quantity of azobenzene that should theoretically be produced in the reaction. The ratio of the actual yield to this calculated yield (multiplied by 100) gives the percent yield.

$$\text{no. mol nitrobenzene} = 0.10 \text{ L} \times \frac{1000 \text{ mL}}{1 \text{ L}} \times \frac{1.20 \text{ g}}{1 \text{ mL}} \times \frac{1 \text{ mol } C_6H_5NO_2}{123 \text{ g } C_6H_5NO_2} = 0.98 \text{ mol } C_6H_5NO_2$$

$$\text{no. mol } C_6H_{14}O_4 = 0.30 \text{ L} \times \frac{1000 \text{ mL}}{1 \text{ L}} \times \frac{1.12 \text{ g}}{1 \text{ mL}} \times \frac{1 \text{ mol } C_6H_{14}O_4}{150 \text{ g } C_6H_{14}O_4} = 2.2 \text{ mol } C_6H_{14}O_4$$

The reaction requires twice the number of moles of $C_6H_{14}O_4$ as $C_6H_5NO_2$; more than this is available. $C_6H_5NO_2$ is the limiting reagent.

$$\text{no. g } (C_6H_5N)_2 = 0.98 \text{ mol } C_6H_5NO_2 \times \frac{1 \text{ mol } (C_6H_5N)_2}{2 \text{ mol } C_6H_5NO_2} \times \frac{182 \text{ g } (C_6H_5N)_2}{1 \text{ mol } (C_6H_5N)_2} = 89 \text{ g } (C_6H_5N)_2 \text{ (theoretical yield)}$$

$$\% \text{ yield} = \frac{\text{actual yield}}{\text{theoretical yield}} \times 100 = \frac{55 \text{ g}}{89 \text{ g}} \times 100 = 65\%$$

4-30. $\text{no. g acetic acid} = 50.0 \text{ g } C_2H_3OCl \times \frac{100 \text{ g theoretical}}{70 \text{ g actual}} \times \frac{1 \text{ mol } C_2H_3OCl}{78.5 \text{ g } C_2H_3OCl} \times \frac{3 \text{ mol } C_2H_4O_2}{3 \text{ mol } C_2H_3OCl} \times \frac{60.1 \text{ g } C_2H_4O_2}{1 \text{ mol } C_2H_4O_2}$

$$\times \frac{100 \text{ g commer. acid}}{97 \text{ g } C_2H_4O_2} = 56 \text{ g commer. acid}$$

It is necessary to balance the equation before proceeding. $3\ C_2H_4O_2 + PCl_3 \rightarrow 3\ C_2H_3OCl + H_3PO_3$

Simultaneous reactions

4-31. Determine the number of moles of HCl required to dissolve each portion of the sample. Add these two amounts together and convert the total number of moles to a mass in grams.

$$\text{no. mol HCl} = 413 \text{ g sample} \times \frac{38.2 \text{ g } MgCO_3}{100 \text{ g sample}} \times \frac{1 \text{ mol } MgCO_3}{84.3 \text{ g } MgCO_3} \times \frac{2 \text{ mol HCl}}{1 \text{ mol } MgCO_3} = 3.74 \text{ mol HCl}$$

$$\text{no. mol HCl} = 413 \text{ g sample} \times \frac{61.8 \text{ g } Mg(OH)_2}{100 \text{ g sample}} \times \frac{1 \text{ mol } Mg(OH)_2}{58.3 \text{ g } Mg(OH)_2} \times \frac{2 \text{ mol HCl}}{1 \text{ mol } Mg(OH)_2} = 8.76 \text{ mol HCl}$$

$$\text{no. g HCl} = (3.74 + 8.76) \text{ mol HCl} \times \frac{36.5 \text{ g HCl}}{1 \text{ mol HCl}} = 456 \text{ g HCl}$$

4-32. The procedure here is similar to that of Exercise 4-31, except that an equation must be written for each combustion reaction.

$C_3H_8 + 5\ O_2 \rightarrow 3\ CO_2 + 4\ H_2O$ $\qquad$ $2\ C_4H_{10} + 13\ O_2 \rightarrow 8\ CO_2 + 10\ H_2O$

$$\text{no. mol } CO_2 = 613 \text{ g mixture} \times \frac{68.2 \text{ g } C_3H_8}{100 \text{ g mixture}} \times \frac{1 \text{ mol } C_3H_8}{44.1 \text{ g } C_3H_8} \times \frac{3 \text{ mol } CO_2}{1 \text{ mol } C_3H_8} = 28.4 \text{ mol } CO_2$$

$$\text{no. mol } CO_2 = 613 \text{ g mixture} \times \frac{31.8 \text{ g } C_4H_{10}}{100 \text{ g mixture}} \times \frac{1 \text{ mol } C_4H_{10}}{58.1 \text{ g } C_4H_{10}} \times \frac{8 \text{ mol } CO_2}{2 \text{ mol } C_4H_{10}} = 13.4 \text{ mol } CO_2$$

$$\text{total no. mol } CO_2 = 28.4 + 13.4 = 41.8 \text{ mol } CO_2$$

4-33. Again, we begin by writing balanced equations for the reaction.

$2\ CH_3OH + 3\ O_2 \rightarrow 2\ CO_2 + 4\ H_2O$ $\qquad$ $C_2H_5OH + 3\ O_2 \rightarrow 2\ CO_2 + 3\ H_2O$

The simplest approach is to calculate the number of grams of CO_2 that would be obtained by burning 0.220 g of a pure alcohol. If one of the results corresponds to the observed mass of CO_2--0.352 g--then the liquid is a pure alcohol. If neither of the results yields 0.352 g CO_2, the liquid must be a mixture.

If pure CH_3OH

$$\text{no. g } CO_2 = 0.220 \text{ g } CH_3OH \times \frac{1 \text{ mol } CH_3OH}{32.0 \text{ g } CH_3OH} \times \frac{2 \text{ mol } CO_2}{2 \text{ mol } CH_3OH} \times \frac{44.0 \text{ g } CO_2}{1 \text{ mol } CO_2} = 0.302 \text{ g } CO_2$$

If pure C_2H_5OH

$$\text{no. g } CO_2 = 0.220 \text{ g } C_2H_5OH \times \frac{1 \text{ mol } C_2H_5OH}{46.1 \text{ g } C_2H_5OH} \times \frac{2 \text{ mol } CO_2}{1 \text{ mol } C_2H_5OH} \times \frac{44.0 \text{ g } CO_2}{1 \text{ mol } CO_2} = 0.420 \text{ g } CO_2$$

Since neither of the results is 0.352 g CO_2, the liquid must be a mixture. (Can you determine the percent composition of this mixture?)

Consecutive reactions

4-34. $\text{no. mol } Cl_2 = 18.2 \text{ mol } CCl_2F_2 \times \frac{1 \text{ mol } CCl_4}{1 \text{ mol } CCl_2F_2} \times \frac{4 \text{ mol } Cl_2}{1 \text{ mol } CCl_4} = 72.8 \text{ mol } Cl_2$

4-35. In addition to the equation given for the precipitation of $BaCO_3$, we need an equation to represent the combustion of C_6H_6.

$2\ C_6H_6 + 15\ O_2 \rightarrow 12\ CO_2 + 6\ H_2O$ $\qquad$ $CO_2 + Ba(OH)_2(aq) \rightarrow BaCO_3(s) + H_2O$

Factors from these equations are used together with volume and density data to achieve the following series of conversions:

$\text{mL } C_6H_6 \rightarrow \text{g } C_6H_6 \rightarrow \text{mol } C_6H_6 \rightarrow \text{mol } CO_2 \rightarrow \text{mol } BaCO_3 \rightarrow \text{g } BaCO_3$

$$\text{no. g } BaCO_3 = 25.00 \text{ mL } C_6H_6 \times \frac{0.879 \text{ g } C_6H_6}{1 \text{ mL } C_6H_6} \times \frac{1 \text{ mol } C_6H_6}{78.1 \text{ g } C_6H_6} \times \frac{12 \text{ mol } CO_2}{2 \text{ mol } C_6H_6} \times \frac{1 \text{ mol } BaCO_3}{1 \text{ mol } CO_2} \times \frac{197 \text{ g } BaCO_3}{1 \text{ mol } BaCO_3}$$

$$= 333 \text{ g } BaCO_3$$

4-36. Factors from three equations are required, starting with I_2 in the final step and working backwards to $AgNO_3$ in the first step.

$$\text{no. g } AgNO_3 = 1000 \text{ g } I_2 \times \frac{1 \text{ mol } I_2}{254 \text{ g } I_2} \times \frac{2 \text{ mol } FeI_2}{2 \text{ mol } I_2} \times \frac{2 \text{ mol } AgI}{1 \text{ mol } FeI_2} \times \frac{1 \text{ mol } AgNO_3}{1 \text{ mol } AgI} \times \frac{170 \text{ g } AgNO_3}{1 \text{ mol } AgNO_3}$$

$$= 1.34\times10^3 \text{ g } AgNO_3 = 1.34 \text{ kg } AgNO_3$$

4-37. First, all equations must be balanced.

$Fe + Br_2 \rightarrow FeBr_2$ $\qquad$ $3\ FeBr_2 + Br_2 \rightarrow Fe_3Br_8$

$Fe_3Br_8 + 4\ Na_2CO_3 \rightarrow 8\ NaBr + 4\ CO_2 + Fe_3O_4$

$$\text{no. kg Fe} = 5.00\times10^3 \text{ kg NaBr} \times \frac{1000 \text{ g NaBr}}{1 \text{ kg NaBr}} \times \frac{1 \text{ mol NaBr}}{103 \text{ g NaBr}} \times \frac{1 \text{ mol } Fe_3Br_8}{8 \text{ mol NaBr}} \times \frac{3 \text{ mol } FeBr_2}{1 \text{ mol } Fe_3Br_8}$$

$$\times \frac{1 \text{ mol Fe}}{1 \text{ mol } FeBr_2} \times \frac{55.85 \text{ g Fe}}{1 \text{ mol Fe}} \times \frac{1 \text{ kg Fe}}{1000 \text{ g Fe}} = 1.02\times10^3 \text{ kg Fe}$$

4-38. (a)

$$\overset{+2}{S_2O_3^{2-}} + H_2O + \overset{0}{Cl_2(g)} \longrightarrow SO_4^{2-} + Cl^- + H^+$$

increase in O.S. of 4 per S atom

decrease in O.S. of 1 per Cl

$$S_2O_3^{2-} + H_2O + 4\ Cl_2(g) \rightarrow 2\ SO_4^{2-} + 8\ Cl^- + H^+$$

$$S_2O_3^{2-} + 5\ H_2O + 4\ Cl_2(g) \rightarrow 2\ SO_4^{2-} + 8\ Cl^- + 10\ H^+$$

(b)

$$(\overset{-3}{N}H_4)_2\overset{+6}{Cr_2}O_7(s) \longrightarrow Cr_2O_3(s) + N_2(g) + H_2O(g)$$

increase of 3 in O.S. per N

decrease of 3 per Cr

$$(NH_4)_2Cr_2O_7(s) \rightarrow Cr_2O_3(s) + N_2(g) + 4\ H_2O(g)$$

(c)

$$\overset{0}{P_4(s)} + H^+ + \overset{+5}{NO_3^-} + H_2O \longrightarrow \overset{+5}{H_2PO_4^-} + \overset{+2}{NO(g)}$$

increase of 5 per P atom

decrease of 3 per N atom

$$3\ P_4(s) + H^+ + 20\ NO_3^- + H_2O \rightarrow 12\ H_2PO_4^- + 20\ NO(g)$$

$$3\ P_4(s) + 8\ H^+ + 20\ NO_3^- + 8\ H_2O \rightarrow 12\ H_2PO_4^- + 20\ NO$$

(d)

$$\overset{+7}{MnO_4^-} + \overset{+3}{NO_2^-} + H^+ \longrightarrow \overset{+2}{Mn^{2+}} + \overset{+5}{NO_3^-} + H_2O$$

increase in O.S. of 2 per N

decrease in O.S. of 5 per Mn

$$2\ MnO_4^- + 5\ NO_2^- + H^+ \rightarrow 2\ Mn^{2+} + 5\ NO_3^- + H_2O$$

$$2\ MnO_4^- + 5\ NO_2^- + 6\ H^+ \rightarrow 2\ Mn^{2+} + 5\ NO_3^- + 3\ H_2O$$

(e)

$$\overset{0}{S_8(s)} + OH^- \longrightarrow \overset{-2}{S^{2-}} + \overset{+2}{S_2O_3^{2-}} + H_2O$$

decrease in O.S. of 2 per S atom

increase in O.S. of 2 per S atom

$$S_8 + OH^- \rightarrow 4\ S^{2-} + 2\ S_2O_3^{2-} + H_2O$$

$$S_8 + 12\ OH^- \rightarrow 4\ S^{2-} + 2\ S_2O_3^{2-} + 6\ H_2O$$

(f) $\overset{0}{P_4}(s) + OH^- + H_2O \longrightarrow \overset{+1}{H_2PO_2^-} + \overset{-3}{PH_3}(g)$

increase in O.S. of 1 per P atom

decrease in O.S. of 3 per P atom

$P_4(s) + OH^- + H_2O \rightarrow 3\ H_2PO_2^- + PH_3(g)$

$P\ (s) + 3\ OH^- + 3\ H_2O \rightarrow 3\ H_2PO_2^- + PH_3(g)$

(g) $\overset{+3\ -2}{As_2S_3} + H^+ + \overset{+5}{NO_3^-} + H_2O \longrightarrow \overset{+5}{H_3AsO_4} + \overset{0}{S}(s) + \overset{+2}{NO}(g)$

increase in O.S. of 2 per As atom

increase in O.S. of 2 per S atom

decrease in O.S. of 3 per N atom

Per formula unit of As_2S_3, the total increase in O.S. is 10. The ratio of NO_3^- to As_2S_3 must be 10:3.

$3\ As_2S_3 + H^+ + 10\ NO_3^- + H_2O \rightarrow 6\ H_3AsO_4 + 9\ S(s) + 10\ NO(g)$

$3\ As_2S_3 + 10\ H^+ + 10\ NO_3^- + 4\ H_2O \rightarrow 6\ H_3AsO_4 + 9\ S(s) + 10\ NO(g)$

(h) $\overset{-2}{C_2H_5OH}(aq) + \overset{+7}{MnO_4^-}(aq) \longrightarrow \overset{0}{C_2H_3O_2^-}(aq) + \overset{+4}{MnO_2}(s) + H_2O + OH^-(aq)$

increase in O.S. of 2 per C atom

decrease in O.S. of 3 per Mn atom

$3\ C_2H_5OH(aq) + 4\ MnO_4^-(aq) \rightarrow 3\ C_2H_3O_2^-(aq) + 4\ MnO_2(s) + H_2O + OH^-(aq)$

$3\ C_2H_5OH(aq) + 4\ MnO_4^-(aq) \rightarrow 3\ C_2H_3O_2^-(aq) + 4\ MnO_2(s) + 4\ H_2O + OH^-(aq)$

4-39. (a) $\overset{-3}{NH_3}(g) + \overset{0}{O_2}(g) \longrightarrow \overset{+2}{NO}(g) + \overset{-2}{H_2O}(g)$

increase in O.S. of 5 per N atom

decrease in O.S. of 2 per O atom

$4\ NH_3(g) + 5\ O_2(g) \rightarrow 4\ NO(g) + 6\ H_2O(g)$

(b) $\overset{+2}{Fe^{2+}} + \overset{+7}{MnO_4^-} + H^+ \longrightarrow \overset{+3}{Fe^{3+}} + \overset{+2}{Mn^{2+}} + H_2O$

increase in O.S. of 1 per Fe atom

decrease in O.S. of 5 per Mn atom

$5\ Fe^{2+} + MnO_4^- + H^+ \rightarrow 5\ Fe^{3+} + Mn^{2+} + H_2O$

$5\ Fe^{2+} + MnO_4^- + 8\ H^+ \rightarrow 5\ Fe^{3+} + Mn^{2+} + 4\ H_2O$

(c) $\overset{-2}{H_2S} + \overset{+4}{SO_2} \longrightarrow \overset{0}{S} + H_2O$

increase in O.S. of 2 per S

decrease in O.S. of 4 per S

$2\ H_2S + SO_2 \rightarrow 3\ S + 2\ H_2O$

4-40. $\text{no. L acid} = 2.50\times10^3 \text{ kg NaCl} \times \frac{1000 \text{ g NaCl}}{1 \text{ kg NaCl}} \times \frac{1 \text{ mol NaCl}}{58.5 \text{ g NaCl}} \times \frac{1 \text{ mol } H_2SO_4}{2 \text{ mol NaCl}} \times \frac{98.1 \text{ g } H_2SO_4}{1 \text{ mol } H_2SO_4}$

$$\times \frac{100 \text{ g acid}}{80 \text{ g } H_2SO_4} \times \frac{1 \text{ cm}^3 \text{ acid}}{1.73 \text{ g acid}} \times \frac{1 \text{ L acid}}{1000 \text{ cm}^3 \text{ acid}} = 1.5\times10^3 \text{ L acid}$$

4-41. Let us make all comparisons on a mole basis, and let us again begin by calculating the number of mol Cl_2 in 1.00 kg Cl_2.

$$\text{no. mol } Cl_2 = 1.00 \text{ kg } Cl_2 \times \frac{1000 \text{ g Cl}}{1 \text{ kg } Cl_2} \times \frac{1 \text{ mol } Cl_2}{70.9 \text{ g } Cl_2} = 14.1 \text{ mol } Cl_2$$

14.1 mol Cl_2 yields 14.1 mol NaOCl, which in turn yields 14.1 mol NH_2Cl, and finally, 14.1 mol N_2H_4. Whatever excess NH_3 remains in the reaction mixture from the second reaction becomes part of the NH_3 required in the third reaction. Determine the amount of NH_3 required to maintain a 30:1 mol ratio in the third reaction.

$$\text{no. mol } NH_3 \text{ required} = 14.1 \text{ mol } NH_2Cl \times \frac{30 \text{ mol } NH_3}{1 \text{ mol } NH_2Cl} = 423 \text{ mol } NH_3$$

The amount of NH_3 consumed in the third reaction is

$$\text{no. mol } NH_3 = 14.1 \text{ mol } NH_2Cl \times \frac{1 \text{ mol } NH_3}{1 \text{ mol } NH_2Cl} = 14.1 \text{ mol } NH_3$$

The quantity of recoverable NH_3 is

$$\text{no. kg } NH_3 = (423 - 14) \text{ mol } NH_3 \times \frac{17.0 \text{ g } NH_3}{1 \text{ mol } NH_3} \times \frac{1 \text{ kg } NH_3}{1000 \text{ g } NH_3} = 6.95 \text{ kg } NH_3$$

4-42. (a) $2\ CH_2CHCH_3 + 2\ NH_3 + 3\ O_2 \rightarrow 2\ CH_2CHCN + 6\ H_2O$

(b) First calculate the theoretical yield

$$\text{no. lb } CH_2CHCN = 1.00 \text{ lb } CH_2CHCH_3 \times \frac{454 \text{ g } CH_2CHCH_3}{1 \text{ bl } CH_2CHCH_3} \times \frac{1 \text{ mol } CH_2CHCH_3}{42.1 \text{ g } CH_2CHCH_3} \times \frac{2 \text{ mol } CH_2CHCN}{2 \text{ mol } CH_2CHCH_3}$$

$$\times \frac{53.1 \text{ g } CH_2CHCN}{1 \text{ mol } CH_2CHCN} \times \frac{1 \text{ lb } CH_2CHCN}{454 \text{ g } CH_2CHCN} = 1.26 \text{ lb } CH_2CHCN$$

$$\%\text{ yield: } \frac{0.73 \text{ lb (actual)}}{1.26 \text{ lb (theoretical)}} \times 100 = 58\%$$

(c) $\text{no. lb } NH_3 = 2000 \text{ lb } CH_2CHCN \times \frac{454 \text{ g } CH_2CHCN}{1 \text{ lb } CH_2CHCN} \times \frac{1 \text{ mol } CH_2CHCN}{53.1 \text{ g } CH_2CHCN} \times \frac{2 \text{ mol } NH_3}{2 \text{ mol } CH_2CHCN} \times \frac{17.0 \text{ g } NH_3}{1 \text{ mol } NH_3}$

$$\times \frac{1 \text{ lb } NH_3}{454 \text{ g } NH_3} \times \frac{100 \text{ lb } NH_3 \text{ (actual)}}{58 \text{ lb } NH_3 \text{ (theoretical)}} = 1.1\times10^3 \text{ lb } NH_3$$

Self-test Questions

1. (*c*) The relationships from the balanced equation are that 2 mol $H_2S \Leftrightarrow 1$ mol $SO_2 \Leftrightarrow 3$ mol S $\Leftrightarrow 2$ mol H_2O. Since 2 mol $H_2S \Leftrightarrow 2$ mol H_2O is the same as 1 mol $H_2S \Leftrightarrow 1$ mol H_2O, the correct answer is (c).

2. (*d*) Because of the relationship 1 mol $CaCN_2 \Leftrightarrow 3$ mol H_2O, the limiting reagent must be H_2O. Now consider the relationship between NH_3 and H_2O: 2 mol $NH_3 \Leftrightarrow 3$ mol H_2O. This yields the conversion factor, 2 mol NH_3/3 mol H_2O. From 1 mol H_2O one obtains 0.67 mol NH_3.

3. (*a*) The combining ratio of O_2 to NH_3 is 5 mol O_2/4 mol NH_3. In a mixture of 1.0 mol each of NH_3 and O_2, there is not enough O_2 available to react with all of the NH_3 (1.25 mol O_2 would be required). O_2 is entirely consumed in the reaction; NH_3 is in excess. Statement (b) is incorrect because 4.0 mol NH_3 and 5 mol O_2 would be required to produce 4 mol NO. Statement (c) is incorrect because to produce 1.50 mol H_2O would require that 1.00 mol NH_3 be consumed, but NH_3 is not the limiting reagent.

4. (*a*) To increase the concentration of KCl from 0.40 M to 0.50 M requires either that solute be added or water evaporated. Statement (b), involving the addition of water, must be incorrect. In 100 mL of 0.40 M KCl there is 0.04 mol KCl, and in 100 mL of 0.50 M KCl, 0.05 mol KCl. The increase in concentration would require 0.01 mol KCl = 0.75 g KCl. Statement (a)--0.75 g KCl--is correct, and (c)--0.10 mol KCl--incorrect. Evaporating 10 mL water would produce an increase in concentration to (100/90) $\times$ 0.40 = 0.44 M.

5. *(b)* To complete the titration of 10.00 mL 0.0500 M NaOH requires 0.0100 L x 0.0500 mol NaOH/L x 1 mol H_2SO_4/2 mol NaOH = 2.5×10^{-4} mol H_2SO_4. The number of mol H_2SO_4 in the four solutions given are (a) 5.0×10^{-4}; (b) 2.5×10^{-4}; (c) 1.0×10^{-3}; (d) 1.0×10^{-3}.

6. *(d)* The oxidation state of S in SO_4^{2-} is +6. Oxygen in the free state is in the oxidation state -2. Nitrogen in N_2H_4 is in the oxidation state -2. In $S_2O_3^{2-}$ the oxidation state of S is +2.

7. *(a)* The reducing agent is the substance containing an element which undergoes an increase in oxidation state during the course of the reaction. The oxidation of Fe^{2+} to Fe^{3+} is accompanied by an increase in oxidation state. Fe^{2+} is the reducing agent. O_2 is the oxidizing agent. The oxidation state of H is unchanged between H^+ and H_2O.

8. *(c)* From 2.0 mol CCl_4 the theoretical yield of CCl_2F_2 is 2.0 mol. [Statements (a) and (b) are incorrect.] Furthermore, the theoretical yield is independent of how much HF is present in excess. [Statement (d) is in error.] The percent yield of the reaction is (1.70/2.00) x 100 = 85%.

9. (a) $Hg(NO_3)_2(s) \rightarrow Hg(\ell) + 2\ NO_2(g) + O_2(g)$

(b) $Na_2CO_3(aq) + 2\ HCl(aq) \rightarrow 2\ NaCl(aq) + H_2O + CO_2(g)$

(c) The percent composition data must be used to establish the formula of malonic acid. Then an equation can be written.

In 100.0 g malonic acid:

$$\text{no. mol C} = 34.62\text{ g C} \times \frac{1\text{ mol C}}{12.0\text{ g C}} = 2.88\text{ mol C}$$

$$\text{no. mol H} = 3.88\text{ g H} \times \frac{1\text{ mol H}}{1.01\text{ g H}} = 3.84\text{ mol H}$$

$$\text{no. mol O} = 61.50\text{ g O} \times \frac{1\text{ mol O}}{16.0\text{ g O}} = 3.84\text{ mol O}$$

$$\text{empirical formula: } C_{2.88}H_{3.84}O_{3.84} = C_{\frac{2.88}{2.88}}H_{\frac{3.84}{2.88}}O_{\frac{3.84}{2.88}} = CH_{1.33}O_{1.33} = C_3H_4O_4$$

combustion reaction: $C_3H_4O_4 + 2\ O_2 \rightarrow 3\ CO_2 + 2\ H_2O$

10. Consider the following two-step procedure. Determine the no. mol $Ba(OH)_2$ required for the titration, and then the volume of solution containing this much $Ba(OH)_2$.

$$\text{no. mol } Ba(OH)_2 = 0.01000\text{ L} \times \frac{0.0526\text{ mol } HNO_3}{\text{L}} \times \frac{1\text{ mol } Ba(NO)_2}{2\text{ mol } HNO_3} = 2.63\times10^{-4}\text{ mol } Ba(OH)_2$$

$$\text{no. mL soln} = 2.63\times10^{-4}\text{ mol } Ba(OH)_2 \times \frac{1\text{ L soln}}{0.0102\text{ mol } Ba(OH)_2} \times \frac{1000\text{ mL soln}}{1\text{ L soln}} = 25.8\text{ mL soln}$$

11. The product of the first two terms in the following setup represents the no. mol NaOH that is to be produced in the reaction.

$$\text{no. g Na} = 0.125\text{ L} \times \frac{0.250\text{ mol NaOH}}{\text{L}} \times \frac{2\text{ mol Na}}{2\text{ mol NaOH}} \times \frac{23.0\text{ g Na}}{1\text{ mol Na}} = 0.719\text{ g Na}$$

12. A less-than-100% yield of desired product in synthesis reactions is almost always the case. This is because of side reactions yielding products other than the desired one (by-products) and because of the loss of material in various steps of the process. Almost by definition, a chemical reaction used in analyzing a compound must have a 100% yield. If an unknown quantity of product is lost in the reaction, errors enter into the analytical results. For this reason only certain carefully selected reactions can be used in analytical chemistry.

Chapter 5

Gases

Review Problems

5-1. Use appropriate factors from expression (5.3) to perform the following conversion.

(a) $\text{no. atm} = 737 \text{ mmHg} \times \frac{1 \text{ atm}}{760 \text{ mmHg}} = 0.970 \text{ atm}$

(b) $\text{no. atm} = 68.3 \text{ cm Hg} \times \frac{10 \text{ mmHg}}{1 \text{ cm Hg}} \times \frac{1 \text{ atm}}{760 \text{ mmHg}} = 0.899 \text{ atm}$

(c) $\text{no. atm} = 1215 \text{ torr} \times \frac{1 \text{ atm}}{760.0 \text{ torr}} = 1.599 \text{ atm}$

(d) $\text{no. atm} = 28 \text{ psi} \times \frac{1 \text{ atm}}{14.7 \text{ psi}} = 1.9 \text{ atm}$

5-2. Again, factors from expression (5.3) are required. Also required in part (c) is expression (5.2) and the method of Example 5-1.

(a) $\text{no. mmHg} = 1.35 \text{ atm} \times \frac{760 \text{ mmHg}}{1 \text{ atm}} = 1.03\times10^3 \text{ mmHg} = 103 \text{ cm}$

(b) $\text{no. mmHg} = 618 \text{ torr} \times \frac{1 \text{ mmHg}}{1 \text{ torr}} = 618 \text{ mmHg}$

(c) $\not{g} \times \underline{h}_{Hg} \times \underline{d}_{Hg} = \not{g} \times \underline{h}_{H_2O} \times \underline{d}_{H_2O}$

$$\underline{h}_{Hg} = \frac{\underline{h}_{H_2O} \times \underline{d}_{H_2O}}{\underline{d}_{Hg}} = \frac{138 \text{ ft} \times \frac{12 \text{ in}}{1 \text{ ft}} \times \frac{2.54 \text{ cm}}{1 \text{ in}} \times \frac{10 \text{ mm}}{1 \text{ cm}} \times 1.00 \text{ g/cm}^3}{13.6 \text{ g/cm}^3}$$

$\underline{h}_{Hg} = 3.09\times10^3 \text{ mm} = 3.09 \text{ m}$

5-3. The initial gas volume is 31.7 L at 753 mmHg.

(a) If the pressure is lowered to 487 mmHg, the volume should increase.

$$V = 31.7 \text{ L} \times \frac{753 \text{ mmHg}}{487 \text{ mmHg}} = 49.0 \text{ L}$$

(b) If the pressure is increased to 3.15 atm, the volume should decrease. Both pressures (initial and final) must be expressed in the same unit.

$$V = 31.7 \text{ L} \times \frac{753 \text{ mmHg}}{3.15 \text{ atm} \times \frac{760 \text{ mmHg}}{1 \text{ atm}}} = 9.97 \text{ L}$$

5-4. In order to reduce the gas volume from 10.5 L to 832 cm^3, the pressure must be increased from its initial value of one standard atmosphere (STP).

$$P = 1.00 \text{ atm} \times \frac{10.5 \text{ L}}{0.832 \text{ L}} = 12.6 \text{ atm}$$

5-5. All temperatures must be expressed in kelvins and the initial volume (138 cm^3) multipled by a ratio of kelvin temperatures that causes the final volume to be larger or smaller than the initial volume, depending on whether the temperature is raised or lowered.

(a) $V = 138 \text{ cm}^3 \times \frac{(85 + 273)\text{K}}{(30 + 273)\text{K}} = 163 \text{ cm}^3$

(b) $V = 138 \text{ cm}^3 \times \frac{(0 + 273)\text{K}}{(30 + 273)\text{K}} = 124 \text{ cm}^3$

5-6. Here the final (kelvin) temperature is the initial (kelvin) temperature multipled by the volume ratio 118 cm^3/87.5 cm^3. The volume ratio must have a value greater than one, since an increase in temperature is necessary to produce an increase in volume.

$$T = (23 + 273)\text{K} \times \frac{118 \text{ cm}^3}{87.5 \text{ cm}^3} = 399\text{K } (126°\text{C})$$

5-7. (a) A decrease in pressure while the temperature and amount of gas are held constant produces an increase in volume.

$$V_f = V_i \times \frac{3\ \text{atm}}{1\ \text{atm}} = 3\ V_i$$

(b) Lowering the temperature of a fixed amount of gas at a constant pressure causes a decrease in volume.

$$V_f = V_i \times \frac{100\text{K}}{400\text{K}} = 0.250\ V_i$$

(c) The effect of raising the temperature is to cause the volume to increase by a factor of 300K/200K, but the increase in pressure from 2 to 3 atm causes a decrease by the factor, 2 atm/3 atm. The final volume is equal to the initial volume.

$$V_f = V_i \times \frac{300\text{K}}{200\text{K}} \times \frac{2\ \text{atm}}{3\ \text{atm}} = V_i$$

5-8. Convert 62.3 g CO to mol CO, and then apply the ideal gas equation.

$$\text{no. mol CO} = 62.3\ \text{g CO} \times \frac{1\ \text{mol CO}}{28.0\ \text{g CO}} = 2.22\ \text{mol CO}$$

$$PV = nRT;\quad V = \frac{nRT}{P} = \frac{2.22\ \text{mol} \times 0.0821\ \text{L atm mol}^{-1}\text{K}^{-1} \times (33 + 273)\text{K}}{728\ \text{mmHg} \times (1\ \text{atm}/760\ \text{mmHg})} = 58.2\ \text{L}$$

5-9. Again, convert 45.7 g SO_2 to mol SO_2, and apply the ideal gas equation.

$$\text{no. mol } SO_2 = 45.7\ \text{g } SO_2 \times \frac{1\ \text{mol } SO_2}{64.1\ \text{g } SO_2} = 0.713\ \text{mol } SO_2$$

$$P = \frac{nRT}{V} = \frac{0.713\ \text{mol} \times 0.0821\ \text{L atm mol}^{-1}\text{K}^{-1} \times (22 + 273)\text{K}}{23.5\ \text{L}} = 0.735\ \text{atm}$$

5-10. In this problem the modification of the ideal gas equation given in expression (5.10) is useful. Solve the expression for the molar mass M and substitute the appropriate data.

$$PV = nRT;\quad PV = \frac{m}{M}RT;\quad M = \frac{mRT}{PV}$$

$$M = \frac{0.341\ \text{g} \times 0.0821\ \text{L atm mol}^{-1}\text{K}^{-1} \times (98.7 + 273.15)\text{K}}{743\ \text{mmHg} \times (1\ \text{atm}/760\ \text{mmHg}) \times 0.355\ \text{L}} = 30.0\ \text{g/mol}$$

The molecular weight is numerically equal to the molar mass: 30.0.

5-11. Use the same form of the ideal gas equation as in Review Problem 10, but rearrange it to represent density (m/V) as a function of T and P.

$$\frac{m}{V} = \frac{MP}{RT} = \frac{44.0\ \text{g/mol} \times 787\ \text{mmHg} \times (1\ \text{atm}/760\ \text{mmHg})}{0.0821\ \text{L atm mol}^{-1}\text{K}^{-1} \times (28.7 + 273.15)\text{K}} = 1.84\ \text{g/L}$$

5-12. First determine the no. mol CO_2 that can be removed per kg NaOH. Then express this quantity as a volume of gas at the given T and P.

$$\text{no. mol } CO_2 = 1.00\ \text{kg NaOH} \times \frac{1000\ \text{g NaOH}}{1.00\ \text{kg NaOH}} \times \frac{1\ \text{mol NaOH}}{40.0\ \text{g NaOH}} \times \frac{1\ \text{mol } CO_2}{2\ \text{mol NaOH}} = 12.5\ \text{mol } CO_2$$

$$V = \frac{nRT}{P} = \frac{12.5\ \text{mol} \times 0.0821\ \text{L atm mol}^{-1}\text{K}^{-1} \times (25.8 + 273.15)\text{K}}{749\ \text{mmHg} \times (1\ \text{atm}/760\ \text{mmHg})} = 311\ \text{L}$$

5-13. Since both gases are measured at STP, the usual mole-ratio conversion factor can be written as a volume ratio. The first step, however, is to balance the equation for the combustion reaction.

$$C_3H_8(g) + 5\ O_2(g) \rightarrow 3\ CO_2(g) + 4\ H_2O(\ell)$$

$$V = 32.2\ \text{L } C_3H_8(g) \times \frac{5\ \text{L } O_2(g)}{1\ \text{L } C_3H_8(g)} = 151\ \text{L } O_2(g)$$

5-14. Determine the number of moles of each gas in the mixture and use this total number in the ideal gas equation, PV = nRT.

$$\text{no. mol Ne} = 16.5\ \text{g Ne} \times \frac{1\ \text{mol Ne}}{20.18\ \text{g Ne}} = 0.818\ \text{mol Ne}$$

$$\text{no. mol Ar} = 34.5\ \text{g Ar} \times \frac{1\ \text{mol Ar}}{39.95\ \text{g Ar}} = 0.864\ \text{mol Ar}$$

total no. moles gas = 0.818 mol Ne + 0.864 mol Ar = 1.682 mol

$$V = \frac{nRT}{P} = \frac{1.682 \text{ mol} \times 0.0821 \text{ L atm mol}^{-1}\text{K}^{-1} \times (37.8 + 273.15)\text{K}}{11.2 \text{ atm}} = 3.83 \text{ L}$$

5-15. Determine the partial pressure of $O_2(g)$ in the "wet" gas.

$$P_{tot.} = P_{bar.} = 752 \text{ mmHg} = P_{O_2} + P_{H_2O}$$

$$P_{O_2} = 752 \text{ mmHg} - P_{H_2O} = 752 \text{ mmHg} - 19.8 \text{ mmHg} = 732 \text{ mmHg}$$

$$n = \frac{PV}{RT} = \frac{732 \text{ mmHg} \times (1 \text{ atm}/760 \text{ mmHg}) \times 0.0767 \text{ L}}{0.0821 \text{ L atm mol}^{-1}\text{K}^{-1} \times (22 + 273)\text{K}} = 0.00305 \text{ mol } O_2$$

5-16. The ratio of effusion times is equal to the square root of a ratio of molecular weight. Since SO_2 has a lower molecular weight (64.1) than Cl_2 (70.9), the effusion time for SO_2 should be shorter than the effusion time for Cl_2 (32.5s). In the set up below we use the ratio of molecular weights that is less than one.

$$\frac{\text{effusion time for } SO_2}{\text{effusion time for } Cl_2} = \frac{x}{32.5\text{s}} = \sqrt{\frac{\text{M.W. } SO_2}{\text{M.W. } Cl_2}} = \sqrt{\frac{64.1}{70.9}} = 0.904 = 0.951$$

x = effusion time for SO_2 = 0.951 × 32.5s = 30.9s

Exercises

Pressure and its measurement

5-1. Necessary conversion factors can be found in expression (5.3).

(a) $\text{no. atm} = 1387 \text{ mmHg} \times \frac{1 \text{ atm}}{760.0 \text{ mmHg}} = 1.825 \text{ atm}$

(b) $\text{no. atm} = 7.14 \text{ kg/cm}^2 \times \frac{1 \text{ atm}}{1.0333 \text{ kg/cm}^2} = 6.91 \text{ atm}$

(c) $\text{no. atm} = 314 \text{ kPa} \times \frac{1 \text{ atm}}{101.325 \text{ kPa}} = 3.10 \text{ atm}$

(d) $\text{no. atm} = 992 \text{ mb} \times \frac{1 \text{ atm}}{1013.25 \text{ mb}} = 0.979 \text{ atm}$

(e) $\text{no. atm} = 2.53\times10^5 \text{ N/m}^2 \times \frac{1 \text{ atm}}{101{,}325 \text{ N/m}^2} = 2.50 \text{ atm}$

5-2. These calculations require the use of equation (5.2) or a variation of it.

(a) $\cancel{g} \times h_{gly} \times d_{gly} = \cancel{g} \times h_{CCl_4} \times d_{CCl_4}$

$$h_{gly} = \frac{h_{CCl_4} \times d_{CCl_4}}{d_{gly}} = \frac{2.85 \text{ m} \times 1.59 \text{ g/cm}^3}{1.26 \text{ g/cm}^3} = 3.60 \text{ m}$$

(b) Perhaps the simplest approach is to convert the given pressure to atm and then to the height of a mercury column. The height of the benzene column can be compared to the height of the mercury column.

$$\text{no. mmHg} = 3.14\times10^4 \text{ N/m}^2 \times \frac{1 \text{ atm}}{101.325 \text{ N/m}^2} \times \frac{760 \text{ mmHg}}{1 \text{ atm}} = 236 \text{ mmHg}$$

$$h_{benz} = \frac{h_{Hg} \times d_{Hg}}{d_{benz}} = \frac{236 \text{ mm} \times 13.6 \text{ g/cm}^3}{0.879 \text{ g/cm}^3} = 3.65\times10^3 \text{ mm} = 3.65 \text{ m}$$

(c) A pressure of 12.5 lb/in^2 can first be converted to a pressure in mmHg.

$$\text{no. mmHg} = 12.5\ lb/in^2 \times \frac{1\ atm}{14.7\ lb/in^2} \times \frac{760\ mmHg}{1\ atm} = 646\ mmHg$$

Now proceed as in part (b).

$$d_{unk} = \frac{h_{Hg} \times d_{Hg}}{h_{unk}} = \frac{646\ mm \times 13.6\ g/cm^3}{15.0\ ft \times \frac{12\ in}{1\ ft} \times \frac{2.54\ cm}{1\ in} \times \frac{10\ mm}{1\ cm}} = 1.92\ g/cm^2$$

5-3. The pressure of the gas is greater than barometric pressure by the amount 283 - 38 = 245 mmHg. The pressure of the gas is

$$P_{gas} = P_{bar} + \Delta P = 745\ mmHg + 245\ mmHg = 990\ mmHg$$

$$\text{no. atm} = 990\ mmHg \times \frac{1\ atm}{760\ mmHg} = 1.30\ atm$$

5-4. The pressure inside the gas container is less than barometric pressure by 3.8 cm of water pressure. The equivalent pressure in mmHg is

$$h_{Hg} = \frac{h_{H_2O} \times d_{H_2O}}{d_{Hg}} = \frac{3.8\ cm \times \frac{10\ mm}{1\ cm} \times 1.00\ g/cm^3}{13.6\ g/cm^3} = 2.8\ mmHg$$

$$P_{gas} = P_{bar} + \Delta P = 753.5\ mmHg - 2.8\ mmHg = 750.7\ mmHg$$

The simple gas laws

5-5. To produce an expansion of the gas required a reduction of its pressure.

$$P_f = 746\ mmHg \times \frac{467\ cm^3}{569\ cm^3} = 619\ mmHg$$

5-6. Consider the reverse process. Compress an initial 2425 + 15.5 = 2441 L of gas, initially at 712 mmHg, into a volume of 15.5 L.

$$P = 712\ mmHg \times \frac{1\ atm}{760\ mmHg} \times \frac{2441\ L}{15.5\ L} = 148\ atm$$

5-7. If the volume and amount of gas are held constant, kelvin temperature and pressure are directly proportional (recall Example 5-8). The temperature must be lowered to produce a decrease in pressure to 1.00 atm = 760 mmHg.

$$T = (23.2 + 273.15)K \times \frac{760\ mmHg}{814\ mmHg} = 277K = 4°C$$

5-8. The pressure of a gas at constant volume is directly proportional to the amount of gas and to the kelvin temperature. Both the amount of gas and the temperature are increased, requiring each ratio to be greater than one.

$$P = 762\ mmHg \times \frac{12.3\ g}{10.0\ g} \times \frac{(51 + 273)K}{(35 + 273)K} = 986\ mmHg$$

5-9. (a) A decrease in temperature and an increase in pressure both cause a decrease in volume.

$$V_{STP} = 385\ cm^3 \times \frac{273K}{(22 + 273)K} \times \frac{748\ mmHg}{760\ mmHg} = 351\ cm^3$$

(b) Use the molar volume of a gas at STP as a converstion factor.

$$\text{no. mol } O_2 = 0.351\ L\ O_2 \times \frac{1\ mol\ O_2(g)}{22.4\ L} = 0.0157\ mol\ O_2(g)$$

5-10. A two-step approach works well here. First calculate the final pressure that would be attained by raising the temperature with the amount of gas held constant.

$$P = P_i \times \frac{(195 + 273)K}{(22 + 273)K} = 1.59\ P_i$$

Now determine what the final amount of gas must be, starting with 12.5 g, if the pressure is to be reduced back to P_i (from 1.59 P_i). The pressure and amount of gas are directly proportional at constant T.

$$\text{no. g gas} = 12.5\ g \times \frac{P_i}{1.59\ P_i} = 7.86\ g\ gas$$

The quantity of gas to be released = 12.5 g - 7.9 g = 4.6 g

Ideal gas equation

5-11. Solve the ideal gas equation for T and substitute the appropriate data.

$$T = \frac{PV}{nR} = \frac{1.85\ atm \times 47.3\ L}{1.62\ mol \times 0.0821\ L\ atom\ mol^{-1}K^{-1}} = 658K = 385°C$$

5-12. First determine the no. mol Kr.

$$n = \frac{PV}{RT} = \frac{8.61\ atm \times 18.5\ L}{0.0821\ L\ atm\ mol^{-1}K^{-1} \times (24.8 + 273.15)K} = 6.51\ mol$$

$$\text{no. g Kr} = 6.51\ mol\ Kr \times \frac{83.8\ g\ Kr}{1\ mol\ Kr} = 546\ g\ Kr$$

5-13. One approach is to determine how the 4.18 L volume changes with a change in temperature and pressure. An alternative is to use the ideal gas equation twice, first to determine the number of moles of gas and then, the final gas volume.

$$n = \frac{PV}{RT} = \frac{732\ mmHg \times (1\ atm/760\ mmHg) \times 4.18\ L}{0.0821\ Latm\ mol\text{-}1K\text{-}1 \times (29.7 + 273.15)K} = 0.162\ mol$$

$$V = \frac{nRT}{P} = \frac{0.162\ mol \times 0.0821\ L\ atm\ mol^{-1}K^{-1} \times (24.8 + 273.15)K}{756\ mmHg \times (1\ atm/760\ mmHg)} = 3.98\ L$$

5-14. Determine the mass of N_2 present in 25.0 L at 12°C and 1.65 atm. The modified form of the ideal gas equation (5.10) can be used.

$$m = \frac{MPV}{RT} = \frac{28.0\ g/mol \times 1.65\ atm \times 25.0\ L}{0.0821\ L\ atm\ mol^{-1}K^{-1} \times (12 + 273)K} = 49.4\ g\ N_2$$

The quantity of $N_2(g)$ present initially is 128 g. The quantity of $N_2(g)$ to be released is 128 g - 49 g = 79 g N_2.

5-15. To calculate the final gas pressure we need to know the amount of gas. The amount of gas is determined for 1.00 ft^3 at STP. This volume must also be converted from ft^3 to L.

$$n = \frac{PV}{RT} = \frac{1\ atm \times 1.00\ ft^3 \times \frac{(12)^3\ in^3}{1.00\ ft^3} \times \frac{(2.54)^3\ cm^3}{(1)^3\ in^3} \times \frac{1\ L}{1000\ cm^3}}{0.0821\ L\ atm\ mol^{-1}K^{-1} \times 273K} = 1.26\ mol$$

$$P = \frac{nRT}{V} = \frac{1.26\ mol \times 0.0821\ L\ atm\ mol^{-1}K^{-1} \times (-20 + 273)K}{75.0\ L} = 0.349\ atm$$

5-16. (a) Use the conversion factor from (5.3) to convert from atm to kPa and also use the fact that 1 dm^3 = 1 L.

$$\frac{0.082057\ \not{L}\ \not{atm}}{mol\ K} \times \frac{101.325\ kPa}{1\ \not{atm}} \times \frac{1\ dm^3}{1\ \not{L}} = 8.314\ kPa\ dm^3\ mol^{-1}K^{-1}$$

(b) $$R = \frac{8.314\ kPa \times \frac{1000\ Pa}{1\ kPa} \times \frac{1\ N\ m^{-2}}{1\ Pa} \times 1\ dm^3 \times \frac{(0.1)^3\ m^3}{1\ dm^3}}{mol\ K}$$

$$R = \frac{8.314 \text{ N m}^{-2} \times 1000 \times 1\times10^{-3} \text{ m}^3}{\text{mol K}} = \frac{8.314 \text{ N m}}{\text{mol K}}$$

$$= 8.314 \text{ J mol}^{-1}\text{K}^{-1}$$

(c) Use the ideal gas equation with R expressed as in (a).

$$P = \frac{nRT}{V} = \frac{(1205 \text{ g} \times \frac{1 \text{ mol CO}}{28.0 \text{ g}}) \times 8.314 \text{ kPa cm}^3 \text{ mol}^{-1}\text{K}^{-1} \times 291\text{K}}{1.56 \text{ m}^3 \times \frac{(10)^3 \text{ dm}^3}{1 \text{ m}^3}}$$

$$= \frac{1205 \times 8.314 \times 291}{44.0 \times 1.56 \times 1\times10^3} \text{ kPa} = 66.7 \text{ kPa}$$

Molecular weight determination

5-17. Use the alternate form of the ideal gas equation presented in (5-10).

$$M = \frac{mRT}{PV} = \frac{0.185 \text{ g} \times 0.0821 \text{ L atm mol}^{-1}\text{K}^{-1} \times (26 + 273)\text{K}}{743 \text{ mmHg} \times (1 \text{ atm}/760 \text{ mmHg}) \times 0.110 \text{ L}} = 42.2 \text{ g/mol}$$

The molecular weight is numerically equal to the molar mass: 42.2. The molecular formula cannot have more than three C atoms (which contribute 36 to the molecular weight). Actually, the most likely formula is C_3H_6. If the molecule has only two C atoms, it would also have to contain 18 H atoms, i.e., C_2H_{18}. (In the study of chemical bonding in Chapters 9 and 10 we will see that it is impossible to bond 18 H atoms to only 2 C atoms.)

5-18. First we can calculate the molecular weight of the gas by the method of the previous exercise.

$$M = \frac{mRT}{PV} = \frac{2.650 \text{ g} \times 0.0821 \text{ L atm mol}^{-1}\text{K}^{-1} \times (24.3 + 273.15)\text{K}}{742.3 \text{ mmHg} \times (1 \text{ atm}/760 \text{ mmHg}) \times 0.428 \text{ L}} = 155 \text{ g/mol}$$

Next determine the empirical formula (based on 100.0 g cpd).

$$\text{no. mol C} = 15.5 \text{ g C} \times \frac{1 \text{ mol C}}{12.01 \text{ g C}} = 1.29 \text{ mol C}$$

$$\text{no. mol Cl} = 23.0 \text{ g Cl} \times \frac{1 \text{ mol Cl}}{35.453 \text{ g Cl}} = 0.649 \text{ mol Cl}$$

$$\text{no. mol F} = 61.5 \text{ g F} \times \frac{1 \text{ mol F}}{19.00 \text{ g F}} = 3.24 \text{ mol F}$$

$$\text{Empirical formula} = C_{1.29}Cl_{0.649}F_{3.24} = C_{\frac{1.29}{0.649}}Cl_{\frac{0.649}{0.649}}F_{\frac{3.24}{0.649}} = C_2ClF_5$$

The formula weight = (2 x 12.01) + 35.453 + (5 x 19.00) = 154.47

The empirical formula and molecular formulas are the same: C_2ClF_5

5-19. The mass of Freon-113 to fill the vessel is 264.2931 - 56.1035 = 208.1896 g. The volume of Freon-113, and thus the volume of the vessel is

$$V = 208.1896 \text{ g} \times \frac{1 \text{ cm}^3}{1.576 \text{ g}} = 132.1 \text{ cm}^3 = 0.1321 \text{ L}$$

Now use equation (5.10). The mass of gas = 56.2445 - 56.1035 = 0.1410 g

$$M = \frac{mRT}{PV} = \frac{0.1410 \text{ g} \times 0.08206 \text{ L atm mol}^{-1}\text{K}^{-1} \times (20.02 + 273.15)\text{K}}{749.3 \text{ mmHg} \times (1 \text{ atm}/760 \text{ mmHg}) \times 0.1321 \text{ L}} = 26.05 \text{ g/mol}$$

The molecular weight of acetylene is numerically equal to its molar mass: 26.05.

Gas densities

5-20. A useful equation for gas densities is equation (5.10), rearranged to the form

$$\frac{m}{V} = d = \frac{MP}{RT}$$

Solve this equation for pressure, P.

$$P = \frac{d \times R \times T}{M} = \frac{1.50 \text{ g } N_2/L \times 0.0821 \text{ L atm mol}^{-1}K^{-1} \times (25.0 + 273.15)K}{28.0 \text{ g } N_2/\text{mol}}$$

5-21. Solve equation (5.10) for the molar mass, M.

$$M = \frac{mRT}{PV} = \frac{2.64 \text{ g} \times 0.0821 \text{ L atm mol}^{-1}K^{-1} \times (310 + 273)K}{(775/760) \text{ atm} \times 1.00 \text{ L}} = 124 \text{ g/mol}$$

Since the atomic weight of P = 31.0, a molecular weight of 124 corresponds to a molecular formula of P_4.

Cannizzaro's method

5-22.

Substance	Mol. wt. (relative to H = 1)	X (%, by mass)	Relative mass of X per molecule
Hydrogen	2.0	--	--
Nitryl fluoride	65.01	49.4	32.1
Nitrosyl fluoride	49.01	32.7	16.0
Thionyl fluoride	86.07	18.6	16.0
Sulfuryl fluoride	102.07	31.4	32.0

The atomic weight of X is the smallest of the relative masses per molecule: 16. The element is oxygen.

Gases in chemical reactions

5-23. We can first determine the no. mol $SO_2(g)$. Then we can calculate the volume occupied by this gas under the given conditions.

$$\text{no. mol } SO_2 = 2.0\times10^6 \text{ lb coal} \times \frac{454 \text{ g coal}}{1 \text{ lb coal}} \times \frac{2.32 \text{ g S}}{100 \text{ g coal}} \times \frac{1 \text{ mol S}}{32.1 \text{ g S}} \times \frac{1 \text{ mol } SO2}{1 \text{ mol S}} = 6.6\times10^5 \text{ mol } SO_2$$

$$V = \frac{nRT}{P} = \frac{6.6\times10^5 \text{ mol} \times 0.0821 \text{ L atom mol}^{-1}K^{-1} \times 298K}{(749/760) \text{ atm}} = 1.6\times10^7 \text{ L}$$

5-24. Use the ideal gas equation to calculate no. mol O_2.

$$n = \frac{PV}{RT} = \frac{(741/760) \text{ atm} \times 0.0902 \text{ L}}{0.0821 \text{ L atm mol}^{-1}K^{-1} \times 296K} = 3.62\times10^{-3} \text{ mol}$$

Next, calculate the mass of $KClO_3$ that must be decomposed to have produced this much O_2.

$$\text{no. g } KClO_3 = 3.62\times10^{-3} \text{ mol } O_2 \times \frac{2 \text{ mol } KClO_3}{3 \text{ mol } O_2} \times \frac{123 \text{ g } KClO_3}{1 \text{ mol } KClO_3} = 0.297 \text{ g } KClO_3$$

$$\text{Finally, } \% \; KClO_3 = \frac{0.297 \text{ g } KClO3}{2.71 \text{ g sample}} \times 100 = 11.0\% \; KClO_3$$

5-25. (a) Because the gases are at identical temperatures and pressures, we can write a conversion factor based on volume.

$$\text{no. L } NH_3(g) = 371 \text{ L } H_2(g) \times \frac{2 \text{ L } NH_3(g)}{3 \text{ L } H_2(g)} = 247 \text{ L } NH_3(g)$$

(b) The gas temperatures and pressures are not the same for the two gases. We must first determine the no. mol H_2 in 371 L at 525°C and 515 atm.

$$n = \frac{PV}{RT} = \frac{515 \text{ atm} \times 371 \text{ L}}{0.0821 \text{ L atm mol}^{-1}\text{K}^{-1} \times 798\text{K}} = 2.92\times10^{3} \text{ mol } H_2$$

$$\text{no. mol } NH_3 = 2.92\times10^{3} \text{ mol } H_2 \times \frac{2 \text{ mol } NH_3}{3 \text{ mol } H_2} = 1.95\times10^{3} \text{ mol } NH_3$$

Now calculate the volume of this amount of NH_3 at STP.

$$\text{no. L } NH_3 = 1.95\times10^{3} \text{ mol } NH_3 \times \frac{22.4 \text{ L } NH_3}{1 \text{ mol } NH_3} = 4.37\times10^{4} \text{ L } NH_3$$

5-26. First calculate the number of moles of each gas.

H_2S: $$n = \frac{PV}{RT} = \frac{(735/760) \text{ atm} \times 1.50 \text{ L}}{0.0821 \text{ L atm mol}^{-1}\text{K}^{-1} \times (273 + 23)\text{K}} = 5.97\times10^{-2} \text{ mol } H_2S$$

O_2: $$n = \frac{PV}{RT} = \frac{(750/760) \text{ atm} \times 4.45 \text{ L}}{0.0821 \text{ L atm mol}^{-1}\text{K}^{-1} \times (273 + 26)\text{K}} = 1.79\times10^{-1} \text{ mol } O_2$$

(a) It is necessary to determine the limiting reactant. Consider the number of moles of O_2 required to react with all the H_2S. (O_2 is the reactant in excess.)

$$\text{no. mol } O_2 = 5.97\times10^{-2} \text{ mol } H_2S \times \frac{3 \text{ mol } O_2}{2 \text{ mol } H_2S} = 8.96\times10^{-2} \text{ mol } O_2$$

$$\text{no. mol } SO_2 = 5.97\times10^{-2} \text{ mol } H_2S \times \frac{2 \text{ mol } SO_2}{2 \text{ mol } H_2O} = 5.97\times10^{-2} \text{ mol } SO_2$$

(b) no. mol $SO_2(g) = 5.97\times10^{-2}$ no. mol $H_2O(g)$ = no. mol $SO_2(g) = 5.97\times10^{-2}$

Excess no. mol $O_2(g) = 17.9\times10^{-2} - 8.96\times10^{-2} = 8.9\times10^{-2}$

Total no. mol gas $= 5.97\times10^{-2} + 5.97\times10^{-2} + 8.9\times10^{-2} = 2.08\times10^{-1}$

$$V = \frac{nRT}{P} = \frac{0.208 \text{ mol} \times 0.0821 \text{ L atom mol}^{-1}\text{K}^{-1} \times 393\text{K}}{(748/760) \text{ atm}} = 6.82 \text{ L}$$

Mixtures of gases

5-27. We need to determine three molar quantities--the no. mol N_2, the total no. mol gas in the mixture, and their difference.

$$\text{no. mol } N_2 = \frac{PV}{RT} = \frac{32.5 \text{ atm} \times 55.0 \text{ L}}{0.0821 \text{ L atm mol}^{-1}\text{K}^{-1} \times 296\text{K}} = 73.6 \text{ mol } N_2$$

$$\text{no. mol gas} = \frac{PV}{RT} = \frac{65.0 \text{ atm} \times 55.0 \text{ L}}{0.0821 \text{ L atm mol}^{-1}\text{K}^{-1} \times 296\text{K}} = 147 \text{ mol gas}$$

$$\text{no. g Ne} = (147 - 74) \text{ mol Ne} \times \frac{20.2 \text{ g Ne}}{1 \text{ mol Ne}} = 1.5\times10^{3} \text{ g Ne}$$

5-28. Again we must determine the amount of each gas and add these amounts together to obtain the total no. mol gas. In addition, the final gas volume is the sum of the volumes of the two containers.

$$n_{H_2} = \frac{PV}{RT} = \frac{(777/760) \text{ atm} \times 1.85 \text{ L}}{0.0821 \text{ L atm mol}^{-1}\text{K}^{-1} \times 298\text{K}} = 0.0773 \text{ mol } H_2$$

$$n_{He} = \frac{PV}{RT} = \frac{(742/760) \text{ atm} \times 2.52 \text{ L}}{0.0821 \text{ L atm mol}^{-1}\text{K}^{-1} \times 298\text{K}} = 0.101 \text{ mol He}$$

$$P_{tot.} = \frac{n_{tot.}RT}{V_{tot.}} = \frac{(0.101 + 0.077) \text{ mol} \times 0.0821 \text{ L atm mol}^{-1}\text{K}^{-1} \times 298\text{K}}{(1.85 + 2.52) \text{ L}} = 0.997 \text{ atm} = 758 \text{ mmHg}$$

5-29. In order for the gas volume to double at STP, the number of moles of gas must be doubled. This doubling of gas volume is produced by adding 10.0 g H_2 = 5.0 mol H_2. Thus there must have been 5.0 mol gas present initially. Of this gas, 4.0 g H_2 = 2.0 mol was hydrogen. The remainder must have been He: 5.0 mol total - 2.0 mol H_2 = 3.0 mol He.

$$\text{no. g He} = 3.0 \text{ mol He} \times \frac{4.0 \text{ g He}}{1 \text{ mol He}} = 12 \text{ g He}$$

5-30. (a) Proceed in the same fashion as in Example 5-18 of the textbook.

$$\text{apparent mol. wt.} = (0.742 \text{ mol } N_2 \times \frac{28.0 \text{ g } N_2}{1 \text{ mol } N_2}) + (0.152 \text{ mol } O_2 \times \frac{32.0 \text{ g } O_2}{1 \text{ mol } O_2})$$

$$+ (0.038 \text{ mol } CO_2 \times \frac{44.0 \text{ g } CO_2}{1 \text{ mol } CO_2}) + (0.059 \text{ mol } H_2O \times \frac{18.0 \text{ g } H_2O}{1 \text{ mol } H_2O}) + (0.009 \text{ mol Ar} \times \frac{40.0 \text{ g Ar}}{1 \text{ mol Ar}})$$

$= 28.7$ g/mol

(b) From equation (5.11) we see that at a given T and P gas density is directly proportional to mol. wt. Since the apparent mol. wt. of expired air (28.7) is less than that of normal air (29.0), its density is also less.

(c) Here we use equation (5.16). Assume a total pressure of 1.00 atm for expired air and for normal air.

$$\text{expired air:} \quad P_{CO_2} = \frac{V_{CO_2}}{V_{tot.}} \times P_{tot.} = 0.038 \times 1.00 \text{ atm} = 0.038 \text{ atm}$$

$$\text{normal air:} \quad P_{CO_2} = \frac{V_{CO_2}}{V_{tot.}} \times P_{tot.} = 0.0003 \times 1.00 \text{ atm} = 0.0003 \text{ atm}$$

ratio of partial pressures of CO_2, expired air/ordinary air = 0.038 atm/0.0003 atm ≈ 130

Collection of gases over water

5-31. To determine the no. mol $H_2(g)$ use information from the balanced equation.

$$\text{no. mol } H_2 = 1.93 \text{ g Al} \times \frac{1 \text{ mol Al}}{27.0 \text{ g Al}} \times \frac{3 \text{ mol } H_2}{2 \text{ mol Al}} = 0.107 \text{ mol } H_2$$

The partial pressure of this $H_2(g)$ is $P_{gas} = P_{bar} - P_{H_2O} = 738 - 25.2 = 713$ mmHg

Finally, use the ideal gas equation.

$$V = \frac{nRT}{P} = \frac{0.107 \text{ mol} \times 0.0821 \text{ L atm mol}^{-1}\text{K}^{-1} \times 299\text{K}}{(713/760) \text{ atm}} = 2.80 \text{ L}$$

5-32. Originally, $P_{Ar} = 755$ mmHg; but after saturation of the gas with water vapor, $P_{Ar} + P_{H_2O} = P_{bar} = 755$ mmHg. Thus, in the final gas mixture, $P_{Ar} = 755 - 25.2 = 730$ mmHg. In effect, the pressure of the original 243 cm^3 Ar has been allowed to decrease from 755 mmHg to 730 mmHg. The volume of the Ar (and hence of the gaseous mixture) must be

$$V = 243 \text{ cm}^3 \times \text{ratio of pressure} = 243 \text{ cm}^3 \times \frac{755 \text{ mmHg}}{730 \text{ mmHg}} = 251 \text{ cm}^3$$

5-33. Barometric pressure is given by $P_{O_2} + P_{H_2O} = (P_{O_2} + 23.8)$ mmHg. To determine P_{O_2} use data in the ideal gas equation written in the form of equation (5.10).

$$P_{O_2} = \frac{mRT}{MV} = \frac{1.58 \text{ g} \times 0.0821 \text{ L atm mol}^{-1}\text{K}^{-1} \times 298\text{K}}{32.0 \text{ g mol}^{-1} \times 1.28 \text{ L}} = 0.944 \text{ atm} = 717 \text{ mmHg}$$

$P_{bar} = (717 + 23.8)$ mmHg = 741 mmHg

Kinetic molecular theory

5-34. The expressions referred to in this exercise are:

$$PV = \frac{2}{3} \cdot n' \cdot \overline{\varepsilon_k} \qquad (5.19) \qquad\qquad \overline{\varepsilon_k} = \frac{3}{2} \frac{R}{N} \cdot T = \frac{3}{2} \cdot kT \qquad (5.21)$$

Boyle's Law

Equation (5.21) establishes the proportionality between the absolute temperature and the average kinetic energy of molecules ($\overline{\varepsilon_k}$). When temperature is held constant, $\overline{\varepsilon_k}$ = constant. If the amount of gas is also held constant, n' = constant. This results in the following form of equation (5.19):

$$PV = \frac{2}{3} n' \overline{\varepsilon_k} = \text{constant}$$

This is the same expression as Boyle's law, equation (5.4).

Charles' Law

Equation (5.19) can be rearranged to the form

$$V = \frac{2}{3} \cdot \frac{n'}{P} \cdot \overline{\varepsilon_k}$$

and this combined with equation (5.21) to yield

$$V = \text{constant} \cdot \frac{n'}{P} \cdot T$$

If the amount of gas and the gas pressure are held constant, then volume is seen to be proportional to absolute temperature; this is the same as equation (5.6)--Charles' law: V = constant·T.

5-35. (a) at 273 K: $u_{rms} = \sqrt{\frac{3 \times R \times T}{M}} = \sqrt{\frac{3 \times R \times 273}{M}} = 1.84 \times 10^3$ m/s

at T: $u_{rms} = 3.68 \times 10^3 \text{ m/s} = 2 \times 1.84 \times 10^3 \text{ m/s} = 2\sqrt{\frac{3 \times R \times 273}{M}}$

$$= 4 \times \sqrt{\frac{3 \times R \times 273}{M}} = \sqrt{\frac{4 \times 3 \times R \times 273}{M}} = \sqrt{\frac{3 \times R \times (4 \times 273)}{M}}$$

The temperature T = (4 x 273) = 1.09×10^3 K

(b) Two approaches are possible here. One is a direct substitution into equation (5.22), as in Example 5-20 of the textbook. The other follows the lines of the derivation of equation (5.24).

$$\frac{(u_{rms})_{N_2}}{(u_{rms})_{H_2}} = \sqrt{\frac{3\ RT/M_{N_2}}{3\ RT/M_{H_2}}} = \sqrt{\frac{M_{N_2}}{M_{H_2}}} = \sqrt{\frac{2.016}{28.01}} = 0.2683$$

$$(u_{rms})_{N_2} = 0.2683 \times (u_{rms})_{H_2} = 0.2683 \times 1.84 \times 10^3 \text{ m/s} = 494 \text{ m/s}$$

5-36. Here a direct application of equation (5.22) is required, rather than just a comparison to another gas. Use the method of Example 5-20 in the textbook.

$$(u_{rms})_{Cl_2} = \sqrt{\frac{3\ RT}{M}} = \sqrt{\frac{3 \times 8.314 \text{ kg m}^2 \text{ s}^{-2} \text{ mol}^{-1} \text{K}^{-1} \times 298\text{K}}{0.0709 \text{ kg mol}^{-1}}} = 324 \text{ m/s}$$

Effusion of gases

5-37. Use equation (5.25) to set up the desired ratios of rates of diffusion, as follows:

$$\frac{\text{rate } H_2}{\text{rate } O_2} = \sqrt{\frac{\text{mol. wt. } O_2}{\text{mol. wt. } H_2}} = \sqrt{\frac{32.0}{2.02}} = 3.98 \qquad \frac{\text{rate } H_2}{\text{rate } D_2} = \sqrt{\frac{\text{mol. wt. } D_2}{\text{mol. wt. } H_2}} = \sqrt{\frac{4.0}{2.0}} = 1.4$$

$$\frac{\text{rate } ^{235}UF_6}{\text{rate } ^{238}UF_6} = \sqrt{\frac{\text{mol. wt. } ^{238}UF_6}{\text{mol. wt. } ^{235}UF_6}} = \sqrt{\frac{352.0}{349.0}} = 1.004$$

5-38. Because the molar mass of HCl is greater than that of NH_3, we should expect less of it to effuse.

$$\frac{\text{amt. HCl}}{\text{amt. } NH_3} = \frac{\text{amt. HCl}}{0.00251 \text{ mol}} = \sqrt{\frac{M_{NH_3}}{M_{HCl}}} = \sqrt{\frac{17.0}{36.5}} = 0.682; \text{ amt. HCl} = 0.682 \times 0.00251 = 0.00171 \text{ mol HCl}$$

5-39. Following Example 5-22 of the textbook,

$$\frac{\text{effusion time for unknown}}{\text{effusion time for } N_2(g)} = \frac{55 \text{ s}}{38 \text{ s}} = \sqrt{\frac{M_{unknown}}{M_{N_2}}} = \sqrt{\frac{M_{unk.}}{28.0}}$$

$$\frac{M_{unk.}}{28.0} = \left(\frac{55}{38}\right)^2 = 2.1 \qquad M_{unk.} = 28.0 \times 2.1 = 59$$

Nonideal gases

5-40. (a) $P = \frac{nRT}{V} = \frac{1.00 \text{ mol} \times 0.0821 \text{ L atom mol}^{-1}\text{K}^{-1} \times (273 + 30) \text{ K}}{0.855 \text{ L}} = 29.1 \text{ atm}$

(b) $(P + a/V^2)(V - b) = RT$ (for one mole of gas)

$$P = \frac{RT}{V-b} - \frac{a}{V^2} = \frac{0.0821 \text{ L atm mol}^{-1}\text{K}^{-1} \times 303 \text{ K}}{(0.855 - 0.043) \text{ L mol}^{-1}} - \frac{3.61 \text{ L}^2 \text{ atm mol}^{-2}}{(0.855)^2 \text{ L}^2 \text{ mol}^{-2}}$$

$P = 30.6 \text{ atm} - 4.9 \text{ atm} = 25.7 \text{ atm}$

(c) The two values differ by about 3.4 atm. The pressure calculated by the van der Waals equation is lower than that calculated by the ideal gas equation, because the van der Waals equation accounts for forces of attraction between molecules (intermolecular forces). These forces cause gas molecules to strike a container wall with less force than if the gas were ideal. The real gas pressure is expected to be less than the calculated ideal gas pressure.

5-41. (a) Rearrange equation (5.26) to an expression involving powers of V.

$$(P + \frac{n^2a}{V^2})(V - nb) = nRT$$

$$(PV^2 + n^2a)(V - nb) = nRTV^2$$

$$PV^3 - nbPV^2 + n^2aV - n^3ab = nRTV^2$$

$$PV^3 - (nbP + nRT)V^2 + n^2aV - n^3ab = 0$$

$$V^3 - (\frac{nbP + nRT}{P})V^2 + \frac{n^2a}{P}V - \frac{n^3ab}{P} = 0$$

$$V^3 - n(\frac{RT + bP}{P})V^2 + (\frac{n^2a}{P})V - \frac{n^3ab}{P} = 0$$

(b) $n = 132 \text{ g } CO_2 \times \frac{1 \text{ mol } CO_2}{44.0 \text{ g } CO_2} = 3 \text{ mol } CO_2$

$P = 10$ atm $T = 280$ K

$$V^3 - 3(\frac{(0.0821 \times 280) + (0.043 \times 10)}{10})V^2 + (\frac{3^2 \times 3.61}{10})V - \frac{3^3 \times 3.61 \times 0.043}{10} = 0$$

$$V^3 - 7.03V^2 + 3.25 V - 0.419 = 0$$

Solve by successive approximations. As a first attempt try the volume corresponding to ideal gas behavior, i.e., try $V = \frac{nRT}{P} = 6.90$

$(6.90)^3 - 7.03(6.90)^2 + 3.25(6.90) - 0.419 = 15.8 > 0$

Try V = 6.0

$(6.0)^3 - 7.03(6.0)^2 + 3.25(6.0) - 0.419 = -18.0 < 0$

Try V = 6.5

$(6.5)^3 - 7.03(6.5)^2 + 3.25(6.5) - 0.419 = -1.69 < 0$

Try V = 6.6

$(6.6)^3 - 7.03(6.6)^2 + 3.25(6.6) - 0.419 = 2.30 > 0$

Try V = 6.55

$(6.55)^3 - 7.03(6.55)^2 + 3.25(6.55) - 0.419 = 0.28 > 0$

Try V = 6.54

$(6.54)^3 - 7.03(6.54)^2 + 3.25(6.55) - 0.419 = -0.12 < 0$

The volume of $CO_2(g)$ is very nearly 6.54 L.

A way of approaching the final result more quickly in this particular case is to note that for most of the approximations only the first three terms of the equation were significant. This is, drop the term "-0.419" and write

$$V^3 - 7.03\ V^2 + 3.25\ V = 0$$

Which simplifies to

$$V^2 - 7.03\ V + 3.25$$

According to the quadratic formula

$$V = \frac{+7.03 \pm \sqrt{(7.03)^2 - 4 \times 3.25}}{2} = \frac{+7.03 \pm 6.03}{2} = \frac{13.06}{2} = 6.53\ L$$

Still another general approach is a graphical one. Determine the value of the function $V^3 - 7.03\ V^2 + 3.25\ V - 0.419$ for several values of V between 6.0 and 7.0. Plot the values of this function against V (as indicated below), and select the value of V for which the function is zero.

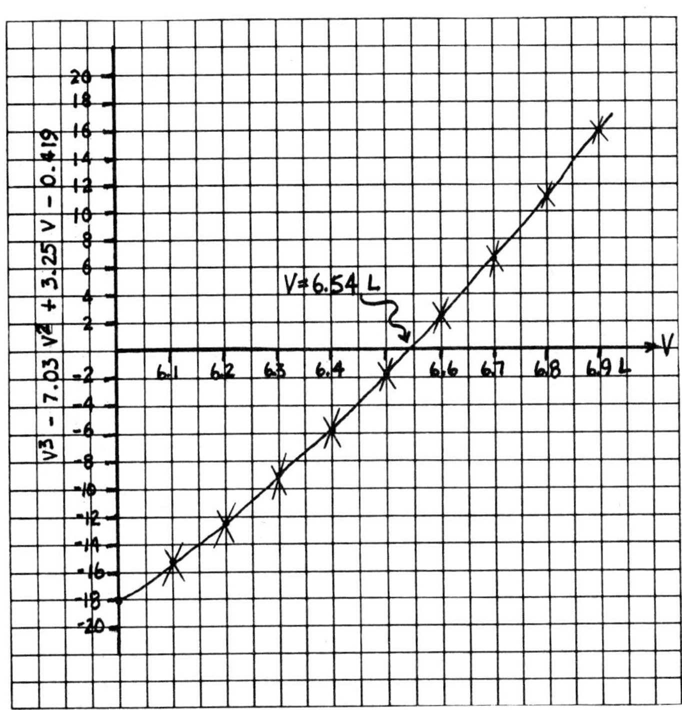

5-42. (a) Substitute the data given into the virial equation to obtain

$$P = \frac{RT}{V}\left\{1 + \frac{B}{V} + \frac{C}{V^2}\right\}$$

$$= \frac{82.1\ cm^3\ atm\ mol^{-1}K^{-1} \times 273K}{500\ cm^3/mol}\left\{1 - \frac{21.89\ cm^3/mol}{500\ cm^3/mol} + \frac{1230\ cm^6/mol^2}{(500)^2 cm^6/mol^2}\right\}$$

$$= 44.8\ atm \left\{1 - 0.0438 + 0.00492\right\} = 44.8\ atm \times 0.961 = 43.1\ atm$$

(b) According to Figure 5-17, in the pressure range of about 50 atm, we should expect the compressibility factor for one mole of gas, PV/RT to be less than one. The value of PV/RT found in part (a) is 0.961. The virial equation produces a result that is consistent with expectations.

Self-test Questions

1. (*b*) The pressure corresponding to 10.0 g H_2 at STP is standard pressure--1.00 atm. Each of the other pressures is less than 1.00 atm. That is, in (a) 75.0 cm Hg is less than standard (76.0 cm Hg). In (c), a column of air 10 mi high does not extend to the full height of the atmosphere; it would exert less than 1 atm pressure. In (d), since 60.0 cm Hg exerts less than 1 atm pressure, so too must 60.00 cm CCl_4--a liquid much less dense than Hg.

2. (*a*) A temperature of 100°C = 373K. Thus, the gas temperature is *lowered* from 373 K to 200 K. The gas volume must *decrease*. Only answer (a) represents a decrease in volume.

3. (*c*) Since $P_{O_2} = P_{bar} - P_{H_2O} = 751 - 21 = 730$ mmHg, items (a) and (b) must be incorrect. A pressure of 730 mmHg is less than atmospheric (760 mmHg). Item (d) must be wrong; and, by a process of elimination, item (c) must be correct.

4. (*a*) The gases cannot have equal volumes or equal effusion rates because their amounts differ, as do their molecular weights. When different gases are compared at the same temperature, their average molecular kinetic energies are equal. Their average molecular speeds in this case would be equal only if their molecular weights were equal (and for H_2 and He they are not).

5. (*b*) 4.48 L $NH_3(g)$ at 0.500 atm could be compressed to 2.24 L at 1.00 atm. 2.24 L $NH_3(aq)$ at STP corresponds to 0.10 mol NH_3, which weighs 1.70 g. Items (a) and (d) are incorrect for they speak of 0.20 and 0.40 mol NH_3. Item (c) is incorrect because this number of molecules represents one mole.

6. (*c*) For every mol Al that results, 1.5 mol = 33.6 L $H_2(g)$ at STP is produced. Item (a) is incorrect. Equating volume ratios to make ratios works only for gases, not for a liquid solution; item (b) is incorrect. For every mol of HCl consumed, 0.50 mol H_2 = 11.2 L $H_2(g)$ at STP is produced. Item (c) is correct. Item (d) would be correct only at STP, not at just and T and P.

7. (*c*) Because of its lower molecular weight we should expect $H_2(g)$ to effuse more rapidly than $SO_2(g)$. In the gas that *remains* the proportion of H_2 decreases with time. This means that the partial pressure of H_2 will be less than that of SO_2 (they were equal originally).

8. (*d*) Use the molar volume at STP--22.4 L/mol--to make comparisons. In 2.24 L $O_2(g)$ at STP, there is 0.10 mol O_2. If the pressure is to be 2.00 atm, the amount of gas must be 0.20 mol. The initial amount of oxygen is 1.60/32.0 = 0.05 mol; 0.15 mol, gas must be added. Adding another 1.60 g O_2 represents only an additional 0.05 mol; item (a) is incorrect. Item (b) is incorrect because it speaks of *releasing* $O_2(g)$. Item (c) refers to 2.00 g He; but this is 0.50 mol He. By the process of elimination, item (d) is correct. But also, 0.60/4.00 = 0.15 mol He, and this is the amount of gas that we indicated must be added.

9. This problem can be solved in several ways. One method involves determining the total number of moles of gas.

$$\text{no. mol } H_2 = 2.24 \text{ L } H_2(\text{STP}) \times \frac{1 \text{ mol } H_2}{22.4 \text{ L } H_2(\text{STP})} = 0.100 \text{mol } H_2$$

total no. mol gas = 0.100mol H_2 + 0.10 mol He = 0.20 mol

Now use the ideal gas equation, with n = 0.20 mol; P = 1 atm; and T = 373 K.

$$V = \frac{nRT}{P} = \frac{0.20 \text{ mol} \times 0.0821 \text{ L atm mol}^{-1}\text{K}^{-1} \times 373 \text{ K}}{1 \text{ atm}} = 6.1 \text{ L}$$

10. The pressure exerted by a liquid column depends only on the height of the column and the density of the liquid. P = ghd. In the derivation of this equation, (5.2 in the textbook), the cross-sectional area of the liquid column, A, does appear. Moreover, the cross-sectional area of a barometer tube does depend on the diameter. However, because the term A appears both in the numerator and denominator, it cancels out.

11. Because the gases are not measured at the same T and P, comparisons must be on a mole basis.

$$\text{no. mol CO} = 30.0 \text{ L (STP)} \times \frac{1 \text{ mol CO}}{22.4 \text{ L (STP)}} = 1.34 \text{ mol CO}$$

$$\text{no. mol } H_2 = 1.34 \text{ mol CO} \times \frac{7 \text{ mol } H_2}{3 \text{ mol CO}} = 3.13 \text{ mol } H_2$$

$$V = \frac{nRT}{P} = \frac{3.13 \text{ mol} \times 0.0821 \text{ L atm mol}^{-1}\text{K}^{-1} \times 295 \text{ K}}{(745/760) \text{ atm}} = 77.3 \text{ L } H_2(g)$$

12. From the density data we can calculate the molar mass of gas.

$$M = \frac{mRT}{PV} = (\frac{m}{V})\frac{RT}{P} = d \times \frac{RT}{P} = \frac{2.35 \text{ g/L} \times 0.0821 \text{ L atm mol}^{-1}\text{K}^{-1} \times 298 \text{ K}}{(752/760) \text{ atm}} = 58.1 \text{ g/mol}$$

From the percent composition data we can determine the empirical formula. In a 100.0 g sample,

$$\text{no. mol C} = 82.7 \text{ g C} \times \frac{1 \text{ mol C}}{12.0 \text{ g C}} = 6.89 \text{ mol C}$$

$$\text{no. mol H} = 17.3 \text{ g H} \times \frac{1 \text{ mol H}}{1.01 \text{ g H}} = 17.1 \text{ mol H}$$

$$\text{Empirical formula} = C_{6.89}H_{17.1} = C_{\frac{6.89}{6.89}}H_{\frac{17.1}{6.89}} = CH_{2.5} = C_2H_5$$

The formula weight = (2 x 12.01) + (5 x 1.108) = 29.06. The molecular weight = 58.1. The molecular formula is C_4H_{10}.

Chapter 6

Thermochemistry

Review Problems

6-1. (a) Heat is lost when water is cooled

$$q = 325\ g\ H_2O \times \frac{1\ cal}{g\ H_2O\ °C} \times (-2.92°C) = -949\ cal$$

(b) Heat is gained as the temperature of $CHCl_3$ is raised.

$$q = 43.5\ kg\ CHCl_3 \times \frac{1000\ g\ CHCl_3}{1\ kg\ CHCl_3} \times \frac{0.232\ cal}{g\ CHCl_3\ °C} \times (22.3 - 16.8)°C = +5.6\times10^4\ cal = +56\ kcal$$

(c) Since the $C_2H_4Cl_2$ is being cooled, heat must be lost.

$$q = 12.6\ L\ C_2H_4Cl_2 = \frac{1000\ cm^3\ C_2H_4Cl_2}{1\ L\ C_2H_4Cl_2} \times \frac{1.253\ g\ C_2H_4Cl_2}{1\ cm^3\ C_2H_4Cl_2} \times \frac{0.310\ cal}{g\ C_2H_4Cl_2°C} \times \frac{4.184\ J}{1\ cal} \times \frac{1\ kJ}{1000\ J}$$

$$\times\ (24.7 - 48.3)°C = -483\ kJ$$

6-2. All parts of this problem can be solved with the expression q = mass x specific heat x temperature change ($\Delta T = T_f - T_i$). In each case the unknown is T_f.

(a) $q = +107\ cal = 7.15\ g\ H_2O \times \frac{1.00\ cal}{g\ H_2O\cdot C} \times \Delta T$

$\Delta T = T_f - T_i = 15.0°C \qquad T_f = T_i + 15.0 = 18.2 + 15.0 = 33.2°C$

(b) $q = -167\ kcal = 3.56\ kg\ S \times \frac{1000\ g\ S}{1\ kg\ S} \times \frac{0.173\ cal}{g\ S\ °C} \times \frac{1\ kcal}{1000\ cal} \times \Delta T$

$\Delta T = \frac{-167 \times 1000}{3.56 \times 1000 \times 0.173} = -271°C \qquad T_f = T_i + \Delta T = 43.2 - 271 = -228°C$

(c) $q = -4.2\times10^3\ kJ = 205\ L \times \frac{1000\ cm^3}{L} \times \frac{0.866\ g\ C_7H_8}{1\ cm^3} \times \frac{0.40\ cal}{g\ C_7H_8\ °C} \times \frac{4.184\ J}{1\ cal} \times \frac{1\ kJ}{1000\ J} \times \Delta T$

$\Delta T = \frac{-4.2\times10^3 \times 1000}{205 \times 1000 \times 0.866 \times 0.40 \times 4.184} = -14°C \qquad T_f = T_i + \Delta T = 32.2 - 14°C$

6-3. The set up required for each of the calculations is that of Example 6-2.

$q\ water = 50.0\ g\ water \times \frac{1.00\ cal}{g\ water\ °C} \times (T_f - 22.0)°C$

and $q\ metal = 150\ g\ metal \times sp.\ ht.\ metal \times (T_f - 100)°C = -q\ water$

(a) $q\ water = 50.0 \times 1.00 \times (39.0 - 22.0) = 850\ cal$

$sp.\ ht.\ Zn = \frac{-850\ cal}{150\ g\ Zn \times (39.0 - 100)°C} = 0.0929\ cal\ (g\ Zn)^{-1}\ °C^{-1}$

(b) $q\ water = 50.0 \times 1.00 \times (55.3 - 22.0) = 1.66\times10^3\ cal$

$sp.\ ht.\ Mg = \frac{-1.66\times10^3\ cal}{150\ g\ Mg \times (55.3 - 100)°C} = 0.248\ cal\ (g\ Mg)^{-1}\ °C^{-1}$

(c) $q\ water = 50.0 \times 1.00 \times (41.8 - 22.0) = 990\ cal$

$sp.\ ht.\ Fe = \frac{-990\ cal}{150\ g\ Fe \times (41.8 - 100)°C} = 0.113\ cal\ (g\ Fe)^{-1}\ °C^{-1}$

6-4. (a) $\frac{5.36\ kJ}{0.107\ g\ C_2H_2} \times \frac{26.0\ g\ C_2H_2}{1\ mol\ C_2H_2} = 1.30\times10^3\ kJ/mol\ C_2H_2$

(b) $\frac{2.601\ kcal}{1.030\ g\ CO(NH_2)_2} \times \frac{60.06\ g\ CO(NH_2)_2}{1\ mol\ CO(NH_2)_2} \times \frac{4.184\ kJ}{1\ kcal} = 634.6\ kJ/mol\ CO(NH_2)_2$

(c) $\frac{6.22\ kcal}{1.05\ cm^3\ (CH_3)_2CO} \times \frac{1.00\ cm^3\ (CH_3)_2CO}{0.791\ g\ (CH_3)_2CO} \times \frac{58.1\ g\ (CH_3)_2CO}{1\ mol\ (CH_3)_2CO} \times \frac{4.184\ kJ}{1\ kcal} = -182\times10^3\ kJ/mol\ (CH_3)_2CO$

6-5. (a) The heat of combustion is

$$0.5060\ g\ C_6H_{12}O \times \frac{1\ mol\ C_6H_{12}O}{100.2\ g\ C_6H_{12}O} \times \frac{-890.7\ kcal}{1\ mol\ C_6H_{12}O} \times \frac{4.184\ kJ}{1\ kcal} \times \frac{1000\ J}{1\ kJ} = -1.882\times10^4\ J$$

This quantity of heat is absorbed between the water and the calorimeter assembly.

$$1.882\times10^4\ J = 1000\ g\ H_2O \times \frac{4.184\ J}{g\ H_2O\cdot C} \times \Delta T + (827\ J/°C \times \Delta T)$$

$$1.882\times10^4\ J = (4184 + 827)\frac{J}{°C}\ \Delta T = 5011\ \frac{J}{°C}\ \Delta T$$

$$\Delta T = \frac{1.882\times10^4\ J}{5011\ J/°C} = 3.756°C \qquad T_f = T_i + \Delta T = 24.98 + 3.76 = 28.74°C$$

(b) The heat of combustion is

$$0.853\ g\ C_{10}H_{14}O \times \frac{1\ mol\ C_{10}H_{14}O}{150.2\ g\ C_{10}H_{14}O} \times \frac{-5.65\times10^3\ kJ}{1\ mol\ C_{10}H_{14}O} \times \frac{1000\ J}{1\ kJ} = -3.21\times10^4\ J$$

The negative of this quantity is substituted into the final equation of part (a).

$$\Delta T = \frac{3.21\times10^4\ J}{5011\ J/°C} = 6.41°C \qquad T_f = T_i + \Delta T = 24.98 + 6.41 = 31.39°C$$

(c) The heat of combustion is

$$1.25\ cm^3 \times \frac{0.901\ g\ C_4H_8O_2}{1\ cm^3} \times \frac{1\ mol\ C_4H_8O_2}{88.1\ g\ C_4H_8O_2} \times \frac{-2246\ kJ}{mol\ C_4H_8O_2} \times \frac{1000\ J}{1\ kJ} = -2.87\times10^4\ J$$

As in part (b),

$$\Delta T = \frac{2.87\times10^4\ J}{5011\ J/°C} = 5.73°C \qquad T_f = T_i + \Delta T = 24.98 + 5.73 = 30.71°C$$

6-6. The set up here is similar to that in Review Problem 5. First determine the heat of solution of the solute and then its effect on the temperature of the water.

(a) $q = 1.00\ g\ LiCl \times \frac{1\ mol\ LiCl}{42.4\ LiCl} \times \frac{-35.0\ kJ}{1\ mol\ LiCl} \times \frac{1000\ J}{1\ kJ} = -825\ J$

The 100.0 g H_2O gains 825 J of heat.

$$825\ J = 100.0\ g\ H_2O \times \frac{4.184\ J}{g\ H_2O\ °C} \times \Delta T$$

$$\Delta T = \frac{825}{100.0 \times 4.184} = 1.97°C \qquad T_f = T_i + \Delta T = 24.8 + 2.0 = 26.8°C$$

(b) $q = 1.00\ g\ MgSO_4 \times \frac{1\ mol\ MgSO_4}{120\ g\ MgSO_4} \times \frac{-84.9\ kJ}{1\ mol\ MgSO_4} \times \frac{1000\ J}{1\ kJ} = -708\ J$

$$\Delta T = \frac{708}{100.0 \times 4.184} = 1.69°C \qquad T_f = T_i + \Delta T = 24.8 + 1.7 = 26.5°C$$

(c) $q = 1.00 \text{ g } Pb(NO_3)_2 \times \frac{1 \text{ mol } Pb(NO3)2}{331 \text{ g } Pb(NO_3)_2} \times \frac{+31.8 \text{ kJ}}{1 \text{ mol } Pb(NO_3)_2} \times \frac{1000 \text{ J}}{1 \text{ kJ}} = +96.1 \text{ J}$

$\Delta T = \frac{-96.1 \text{ J}}{100.0 \times 4.184} = -0.23°C \qquad T_f = T_i + \Delta T = 24.8 - 0.2 = 24.6°C$

6-7. Reverse equation (a) and add it to (b).

-(a):	$CO(g) \rightarrow C(\text{graphite}) + \frac{1}{2}O_2(g)$	$\Delta\overline{H}° = 110.54$ kJ/mol
(b):	$C(\text{graphite}) + O_2(g) \rightarrow CO_2(g)$	$\Delta\overline{H}° = -393.51$ kJ/mol
	$CO(g) + \frac{1}{2}O\ (g) \rightarrow CO_2(g)$	$\Delta\overline{H}° = -282.97$ kJ/mol

6-8. Note that to achieve the net equation, reaction (a) must be doubled and reaction (c) must be reversed.

2(a):	$2\ H_2(g) + \cancel{O_2(g)} \rightarrow 2\ \cancel{H_2O(\ell)}$	$\Delta\overline{H}° = 2 \times (-285.85) = -571.7$ kJ/mol
(b):	$C_3H_4(g) + \cancel{4\ O_2(g)} \rightarrow \cancel{3\ CO_2(g)} + \cancel{2\ H_2O(\ell)}$	$\Delta\overline{H}° = -1941$ kJ/mol
-(c):	$\cancel{3\ CO_2(g)} + \cancel{4\ H_2O(\ell)} \rightarrow C_3H_8(g) + \cancel{5\ O_2(g)}$	$\Delta\overline{H}° = +2220$ kJ/mol
	$C_3H_4(g) + 2\ H_2(g) \rightarrow C_3H_8(g)$	$\Delta\overline{H}° = -293$ kJ/mol

6-9. (a) $\Delta\overline{H}°_{rxn} = \Delta\overline{H}°_f[C_2H_6(g)] + \Delta\overline{H}°_f[CH_4(g)] - \Delta\overline{H}°_f[C_3H_8(g)] - \Delta\overline{H}°_f[H_2(g)]$

$= -84.68 \text{ kJ/mol} - 74.85 \text{ kJ/mol} - (-103.85 \text{ kJ/mol}) - 0$

$= -55.68 \text{ kJ/mol}$

(b) $\Delta\overline{H}°_{rxn} = 2\ \Delta\overline{H}°_f[SO_2(g)] + 2\ \Delta\overline{H}°_f[H_2O(\ell)] - 2\ \Delta\overline{H}°_f[H_2S(g)] - 3\ \Delta\overline{H}°_f[O_2(g)]$

$= 2 \times (-296.90 \text{ kJ/mol}) + 2\times(-285.85 \text{ kJ/mol}) - 2\times(-20.17 \text{ kJ/mol}) - 3 \times 0$

$= -593.80 \text{ kJ/mol} - 571.70 \text{ kJ/mol} + 40.34 \text{ kJ/mol} = -1125.16 \text{ kJ/mol}$

(c) $\Delta\overline{H}°_{rxn} = 2\ \Delta\overline{H}°_f[Fe_3O_4(s)] + \Delta\overline{H}°_f[H_2O(\ell)] - 3\ \Delta\overline{H}°_f[Fe_2O_3(s)] - \Delta\overline{H}°_f[H_2(g)]$

$= 2 \times (-1117.13 \text{ kJ/mol}) + (-285.85 \text{ kJ/mol}) - 3 \times (-822.16 \text{ kJ/mol}) - 0$

$= -2234.26 \text{ kJ/mol} - 285.85 \text{ kJ/mol} + 2466.48 \text{ kJ/mol} = -53.63 \text{ kJ/mol}$

6-10. (a) $(CH_3)_2O(g) + 3\ O_2(g) \rightarrow 2\ CO_2(g) + 3\ H_2O(\ell);\ \Delta\overline{H}° = -1.454\times10^3 \text{ kJ/mol}$

$\Delta\overline{H}° = -1.454\times10^3 \text{ kJ/mol} = 2\ \Delta\overline{H}°_f[CO_2(g)] + 3\ \Delta\overline{H}°_f[H_2O(\ell)] - \Delta\overline{H}°_f[(CH_3)_2O(g)]$

$= 2 \times (-393.51)\text{kJ/mol} + 3 \times (-285.85)\text{kJ/mol} - \Delta H°_f[(CH_3)_2O(g)]$

$H°_f[(CH_3)_2O(g)] = -787.02 \text{ kJ/mol} - 857.55 \text{ kJ/mol} + 1.454\times10^3 \text{ kJ/mol}$

$= -191 \text{ kJ/mol } (CH_3)_2O(g)$

(b) $C_7H_6O(\ell) + 8\ O_2(g) \rightarrow 7\ CO_2(g) + 3\ H_2O(\ell) \qquad \Delta\overline{H}° = -3.520\times10^3 \text{ kJ/mol}$

$\Delta\overline{H}° = -3.520\times10^3 \text{ kJ/mol} = 7 \times (-393.51) + 3 \times (-285.85) - \Delta\overline{H}°_f[C_7H_6O(\ell)]$

$\Delta\overline{H}°_f[C_7H_6O(\ell)] = -2754.57 \text{ kJ/mol} - 857.55 \text{ kJ/mol} + 3.520\times10^3 \text{ kJ/mol}$

$= -93 \text{ kJ/mol } C_7H_6O(\ell)$

(c) $C_6H_{12}O_6(s) + 6\ O_2(g) \rightarrow 6\ CO_2(g) + 6\ H_2O(\ell) \qquad \Delta\overline{H}° = -2.816\times10^3 \text{ kJ/mol}$

$\Delta\overline{H}° = -2.816\times10^3 \text{ kJ/mol} = 6 \times (-393.51) + 6 \times (-285.85) - \Delta\overline{H}°_f[C_6H_{12}O_6(s)] - 6\ \Delta\overline{H}°_f[O_2(g)]$

$\Delta\overline{H}°_f[C_6H_{12}O_6(s)] = -2361.06 \text{ kJ/mol} - 1715.10 \text{ kJ/mol} + 2.816\times10^3 \text{ kJ/mol}$

$= -1260 \text{ kJ/mol } C_6H_{12}O_6(g)$

Exercises

Specific heat

6-1. (a) no. J = 583 g × $\frac{0.4 \text{ J}}{\text{g °C}}$ × (20 − 535)°C = -1.2×10^5 J = -1×10^5 J

(b) The heat given off by the burner is absorbed by a mass of water, *m*.

no. J = 1×10^5 J = $m \times \frac{4.184 \text{ J}}{\text{g } H_2O \text{ °C}}$ × (100 − 20)°C

$$m = \frac{1\times10^5}{4.184 \times (100 - 20)} \text{ g } H_2O = 3\times10^2 \text{ g } H_2O$$

6-2. q water = 125 cm^3 H_2O × $\frac{1.00 \text{ g } H_2O}{cm^3 \; H_2O}$ × $\frac{1.00 \text{ cal}}{\text{g } H_2O \text{ °C}}$ × (40.3 − 24.8)°C = 1.94×10^3 cal

$q_{steel} = -q_{water} = -1.94\times10^3$ cal − 1.94×10^3 cal = $m \times \frac{0.12 \text{ cal}}{\text{g °C}}$ × (40.3 − 152)°C

$$m = \frac{-1.94\times10^3}{0.12 \times (40.3 - 152)} = 1.4\times10^2 \text{ g}$$

This method of mass determination is inaccurate due to the loss of heat to the air, etc. See answer to Exercise 3.

6-3. From the data given we are able to calculate both the heat lost by Mg and gained by water. These quantities can then be compared.

q_{Mg} = 70.0 g Mg × $\frac{1.04 \text{ J}}{\text{g Mg °C}}$ × (47.2 − 99.8)°C = −3830 J

q_{water} = 50.0 g water × $\frac{4.184 \text{ J}}{\text{g water °C}}$ × (47.2 − 30.0)°C = 3600 J

According to the law of conservation of energy we should expect $q_{Mg} + q_{water} = 0$. In this case however, −3830 J + 3600 J = −230 J. There appears to be a net loss of 230 J. The "problem" here is not with the failure of the law of conservation but failure to account for all of the heat lost by the Mg. For example, some might go toward vaporizing a small quantity of water (see Chapter 11) and some may simply be given off from the magnesium-water mixture to the surroundings.

6-4. Since $q_{Cu} = -q_{glyc.}$, we may write

74.8 g Cu × $\frac{0.393 \text{ J}}{\text{g Cu °C}}$ × (31.1 − 143.2)°C = −[165 cm^3 × $\frac{1.26 \text{ g glyc.}}{cm^3}$ × sp. ht. × (31.1 − 24.8)°C]

$$\text{sp. ht.} = \frac{74.8 \times 0.393 \times (31.1 - 143.2) \text{ J}}{-165 \times 1.26 \text{ g glyc.} \times (31.1 - 24.8)\text{°C}} = \frac{2.5 \text{ J}}{\text{g glyc. °C}}$$

$$\text{molar heat capacity} = \frac{2.5 \text{ J}}{\text{g glyc. °C}} \times \frac{92.1 \text{ g glyc.}}{1 \text{ mol glyc.}} = \frac{2.3\times10^2 \text{ J}}{\text{mol glyc. °C}}$$

6-5. Again, the basic expression required is $q_{brass} + q_{water} = 0$, or $q_{brass} = -q_{water}$. In this case, however, the unknown is the final temperature, *t*.

$(6.80 \text{ mm} \times \frac{1.00 \text{ cm}}{10.00 \text{ mm}})^3 \times \frac{8.40 \text{ g brass}}{cm^3} \times \frac{0.385 \text{ J}}{\text{g brass °C}} \times (t - 92.1)\text{°C}$

= −[17.5 g water × $\frac{4.184 \text{ J}}{\text{g water °C}}$ × (t − 25.2)°C]

1.02 t − 93.7 = 73.2 t + 1845 74.2 t = 1939 t = 26.1 °C

6-6. (a) no. kJ = $\frac{-350 \text{ kJ}}{\text{mol}} \times \frac{1 \text{ mol}}{56.1 \text{ g CaO}}$ = −6.24 kJ/g CaO

(b) no. kJ = 2000 lb $Ca(OH)_2$ × $\frac{454 \text{ g } Ca(OH)_2}{1 \text{ lb } Ca(OH)_2}$ × $\frac{1 \text{ mol } Ca(OH)_2}{74.1 \text{ g } Ca(OH)_2}$ × $\frac{-350 \text{ kJ}}{1 \text{ mol } Ca(OH)_2}$ = -4.29×10^6 kJ

6-7. (a) Consider this two-step approach. First determine the quantity of heat released in the combustion reaction.

no. kJ = 2.55×10^3 g CH_4 × $\frac{1 \text{ mol } CH_4}{16.0 \text{ g } CH_4}$ × $\frac{-890 \text{ kJ}}{1 \text{ mol } CH_4}$ = -1.42×10^5 kJ

Now calculate the mass of water that could be heated.

$$\text{no. J} = m_{H_2O} \times \frac{4.184 \text{ J}}{\text{g } H_2O \text{ °C}} \times (60.3 - 21.8)\text{°C} = 1.42\times10^5 \text{ kJ} \times \frac{1000 \text{ J}}{1 \text{ kJ}}$$

$$m_{H_2O} = \frac{1.42\times10^8}{4.184 \times (60.3 - 21.8)} \text{ g } H_2O = 8.82\times10^5 \text{ g } H_2O$$

Finally, assuming a density of H_2O = 1.00 g/cm^3

$$\text{no. L } H_2O = 8.82\times10^5 \text{ g } H_2O \times \frac{1 \text{ cm}^3 \ H_2O}{1.00 \text{ g } H_2O} \times \frac{1 \text{ L } H_2O}{1000 \text{ cm}^3} = 882 \text{ L } H_2O$$

(b) $$\text{no. mol } CH_4 = -1.00\times10^6 \text{ kJ} \times \frac{1 \text{ mol } CH_4}{-890 \text{ kJ}} = 1.12\times10^3 \text{ mol } CH_4$$

Now use the ideal gas equation to solve for V.

$$V = \frac{nRT}{P} = \frac{1.12\times10^3 \text{ mol} \times 0.0821 \text{ L atm mol}^{-1}\text{K}^{-1} \times 295.0 \text{ K}}{(748/760) \text{ atm}} = 2.76\times10^4 \text{ L}$$

6-8. Use the ideal gas equation to calculate the total number of moles of gas.

$$n = \frac{PV}{RT} = \frac{(753/760) \text{ atm} \times 312 \text{ L}}{0.0821 \text{ L atm mol}^{-1}\text{K}^{-1} \times 299 \text{ K}} = 12.6 \text{ mol gas}$$

Now determine the quantity of heat liberated by the combustion of each gas and add these quantities together.

$$\text{no. kJ} = 12.6 \text{ mol gas} \times \frac{83.0 \text{ mol } CH_4}{100 \text{ mol gas}} \times \frac{-890 \text{ kJ}}{1 \text{ mol } CH_4} = -9.31\times10^3 \text{ kJ}$$

$$\text{no. kJ} = 12.6 \text{ mol gas} \times \frac{11.2 \text{ mol } C_2H_6}{100 \text{ mol gas}} \times \frac{-1559 \text{ kJ}}{1 \text{ mol } C_2H_6} = -2.20\times10^3 \text{ kJ}$$

$$\text{no. kJ} = 12.6 \text{ mol gas} \times \frac{5.8 \text{ mol } C_3H_8}{100 \text{ mol gas}} \times \frac{-2219 \text{ kJ}}{1 \text{ mol } C_3H_8} = -1.62\times10^3 \text{ kJ}$$

$$\text{total heat} = -9.31\times10^3 - 2.20\times10^3 - 1.62\times10^3 = -13.13\times10^3 \text{ kJ} = -1.313\times10^4 \text{ kJ}$$

6-9. Both reactants are consumed. The heat of the reaction raises the temperature of the products only and not excess reactants.

$$\text{heat of reaction} = 1.00 \text{ mol } Fe_2O_3 \times \frac{-850 \text{ kJ}}{1 \text{ mol } Fe_2O_3} = -850 \text{ kJ}$$

$$\text{mass of products} = (1.00 \text{ mol } Al_2O_3 \times \frac{102 \text{ g } Al_2O_3}{1 \text{ mol } Al_2O_3}) + (2.00 \text{ mol Fe} \times \frac{55.85 \text{ g Fe}}{1 \text{ mol Fe}}) = 214 \text{ g}$$

heat absorbed by products = -heat of reaction = -(-850) kJ = 850 kJ

$$q = 850 \text{ kJ} \times \frac{1000 \text{ J}}{1 \text{ kJ}} = 214 \text{ g} \times \frac{0.2 \text{ cal}}{\text{g °C}} \times \frac{4.184 \text{ J}}{1 \text{ cal}} \times (t - 25)\text{°C}$$

$$\frac{8.50\times10^5}{214 \times 0.2 \times 4.184} = 4750 = t - 25 \qquad t \approx 4800\text{°C}$$

Because of heat loss to the surroundings, the temperature probably does not come close to reaching 4800°C, but it would certainly seem to be high enough to melt iron (1530°C).

6-10. (a) $q_{KOH} + q_{water} = 0$ and $q_{KOH} = -q_{water}$

$$q_{water} = 45.0 \text{ g } H_2O \times \frac{4.184 \text{ J}}{\text{g } H_2O \text{ °C}} \times (24.9 - 24.1)\text{°C} = 1.5\times10^2 \text{ J}$$

$$q_{KOH} = \frac{-1.5\times10^2 \text{J}}{0.150 \text{ g KOH}} \times \frac{56.1 \text{ g KOH}}{1 \text{ mol KOH}} = -5.6\times10^4 \text{ J/mol KOH} = -56 \text{ kJ/mol KOH} = -6\times10^1 \text{ kJ/mol KOH}$$

(b) The precision could be increased to three significant figures by using a large enough sample of KOH to produce a temperature increase in the water of more than 10°C.

6-11. The quantity of heat to be lost by the water is

$$q = 1400 \text{ ml} \times \frac{1.00 \text{ g } H_2O}{1.00 \text{ ml}} \times \frac{4.184 \text{ J}}{\text{g } H_2O \text{ °C}} \times (10 - 25)\text{°C} = -8.8\times10^4 \text{ J}$$

The quantity of NH_4NO_3 necessary to gain 8.8×10^4 J is

$$\text{no. kg } NH_4NO_3 = 8.8\times10^4 \text{ J} \times \frac{1 \text{ kJ}}{1000 \text{ J}} \times \frac{1 \text{ mol } NH_4NO_3}{26 \text{ kJ}} \times \frac{80 \text{ g } NH_4NO_3}{1 \text{ mol } NH_4NO_3} \times \frac{1 \text{ kg } NH_4NO_3}{1000 \text{ g } NH_4NO_3} = 0.27 \text{ kg } NH_4NO_3$$

6-12. $\text{no. mol NaOH} = 500 \text{ ml} \times \frac{1 \text{ L}}{1000 \text{ ml}} \times \frac{7.0 \text{ mol NaOH}}{1 \text{ L}} = 3.5 \text{ mol NaOH}$

$$\text{no. cal} = 3.5 \text{ mol NaOH} \times \frac{-10 \text{ kcal}}{1 \text{ mol NaOH}} \times \frac{1000 \text{ cal}}{1 \text{ kcal}} = -35{,}000 \text{ cal}$$

Assume that the NaOH solution weighs 500 g and has a specific heat capacity of about 1.00 cal/g °C.

$$35{,}000 \text{ cal} = 500 \text{ g} \times \frac{1.00 \text{ cal}}{\text{g °C}} \times \Delta t \qquad \Delta t = \frac{35{,}000}{500} = 70\text{°C}$$

Since the original temperature was 21°C, the final temperature should be about 90°C.

Bomb calorimetry

6-13. The heat absorbed by the water in the calorimeter is

$$\text{no. J} = 1155 \text{ g } H_2O \times \frac{4.184 \text{ J}}{\text{g } H_2O \text{ °C}} \times 3.68\text{°C} = 1.78\times10^4 \text{ J}$$

The heat absorbed by the reast of the calorimeter assembly is

$$2.09\times10^4 \text{ J} - 1.78\times10^4 \text{ J} = 0.31\times10^4 \text{ J} = 3.1\times10^3 \text{ J}$$

The heat capacity of the bomb is $\frac{3.1\times10^3 \text{ J}}{3.68\text{°C}} = 8.4\times10^2 \text{ J/°C}$

6-14. In the basic equation for bomb calorimetry, the unknown is Δt. $q_{rxn} = -(q_{water} + q_{calorim})$

$$0.242 \text{ g } C_{10}H_8 \times \frac{1 \text{ mol } C_{10}H_8}{128 \text{ g } C_{10}H_8} \times \frac{-5.15\times10^6 \text{ J}}{1 \text{ mol } C_{10}H_8} = -\left\{1025 \text{ g } H_2O \times \frac{4.184 \text{ J}}{\text{g } H_2O \text{ °C}} \times \Delta t\right\} - \left\{\frac{802 \text{ J}}{\text{°C}} \times \Delta t\right\}$$

$$-9.74\times10^3 = -4.29\times10^3 \, \Delta t - 802 \, \Delta t \qquad \Delta t = \frac{9.74\times10^3}{(4.29\times10^3) + (0.802\times10^3)} = 1.91\text{°C}$$

6-15. (a) $q_{rxn} = (-q_{water} + q_{calorim}) = -\left\{[980.0 \text{ g water} \times \frac{4.184 \text{ J}}{\text{g water °C}} \times (31.41 - 24.92)\text{°C}]\right.$

$$\left. + [\frac{812 \text{ J}}{\text{°C}} \times (31.41 - 24.92)\text{°C}]\right\} = -\left\{2.66\times10^4 \text{ J} + 5.27\times10^3 \text{ J}\right\} = -3.19\times10^4 \text{ J}$$

$$\text{Heat of combustion} = \frac{-3.19\times10^4 \text{ J}}{2.051 \text{ g } C_6H_{12}O_6} \times \frac{180 \text{ g } C_6H_{12}O_6}{1 \text{ mol } C_6H_{12}O_6} \times \frac{1 \text{ kJ}}{1000 \text{ J}} = -2.80\times10^3 \text{ kJ/mol } C_6H_{12}O_6$$

(b) $C_6H_{12}O_6(s) + 6\,O_2(g) \rightarrow 6\,CO_2(g) + 6\,H_2O(\ell) \qquad \Delta\bar{H} = -2.80\times10^3 \text{ kJ/mol}$

6-16. From the data for the first experiment we can calculate the heat capacity of the bomb.

$$q_{rxn} = -(q_{water} + q_{calorim})$$

$$1.148 \text{ g } C_7H_6O_2 \times \frac{-26.42 \text{ kJ}}{\text{g } C_7H_6O_2} = -\left\{1181 \text{ g water} \times \frac{4.184 \text{ J}}{\text{g water °C}} \times \frac{1 \text{ kJ}}{1000 \text{ J}} \times (30.25 - 24.96)\text{°C}\right\}$$

$$-\left\{\text{ht. cap. calorim} \times (30.25 - 24.96)\text{°C}\right\}$$

$$-30.33 \text{ kJ} = -26.1 \text{ kJ} - (\text{ht. cap. calorim} \times 5.29\text{°C})$$

$$\text{ht. cap. bomb} = \frac{-(30.33 - 26.1)\text{kJ}}{-5.29\text{°C}} \times \frac{1000 \text{ J}}{1 \text{ kJ}} = 7.9\times10^2 \text{ J/°C}$$

Next, determine the heat of combustion per g coal.

$$q_{rxn} = -(q_{water} + q_{calorim})$$

$$q_{rxn} = -\left\{1162 \text{ g } H_2O \times \frac{4.184 \text{ J}}{\text{g } H_2O \text{ °C}} \times (29.81 - 24.98)\text{°C}\right\} - \left\{\frac{790 \text{ J}}{\text{°C}} \times (29.81 - 24.98)\text{°C}\right\}$$

$$= -2.35\times10^4 \text{ J} - 3.8\times10^3 \text{ J} = -2.73\times10^4 \text{ J} = -27.3 \text{ kJ}$$

$$\text{Heat of combustion of the coal} = \frac{-27.3 \text{ kJ}}{0.895 \text{ g coal}} = -30.5 \text{ kJ/g coal}$$

Finally, the required quantity of coal is calculated through the setup

$$\text{no. m ton coal} = -2.15\times10^9 \text{ kJ} \times \frac{1 \text{ g coal}}{-30.5 \text{ kJ}} \times \frac{1 \text{ kg coal}}{1000 \text{ g coal}} \times \frac{1 \text{ m ton coal}}{1000 \text{ kg coal}} = 70.5 \text{ m ton coal}$$

6-17. If there is no water in the calorimeter, all the heat liberated in the combustion reaction is absorbed by the bomb: $q_{rxn} = -q_{calorim}$. In Example 6-4, $q_{rxn} = -26.42$ kJ

$$-26.42 \text{ kJ} = -(\text{ht. cap.} -q_{calorim} \times \Delta t)$$

$$-26.42 \text{ kJ} \times \frac{1000 \text{ J}}{1 \text{ kJ}} = -7.9\times10^2 \text{ J/°C} \times \Delta t$$

$$\Delta t = \frac{26.42 \times 1000}{7.9\times10^2} \text{°C} = 33\text{°C}$$

The temperature will increase to about 33°C above room temperature.

Functions of state

6-18. Our interest in enthalpy (H) always centers on enthalpy *change* (ΔH). Because enthalpy is a function of state, the difference in enthalpy between two states has a unique value, regardless of the actual enthalpy values in the two states. Thus, it is not necessary to know absolute values of enthalpy at all.

6-19. Because diamond and graphite are different solids, all properties of state, such as enthalpy, must be different for the two. The final state of the combustion--$CO_2(g)$ and $H_2O(\ell)$--is the same in the two cases, but the initial states are different. As a result, the enthalpy changes (ΔH) for the two combustion processes must be different. (As a matter of fact, this difference proves to be slight; the heat of combustion of diamond is about 2 kJ/mol greater than that of graphite.)

Hess's law

6-20. Since $N_2H_4(\ell)$ is a reactant, we should expect the first equation to appear as written, that is,

$$N_2H_4(\ell) + O_2(g) \rightarrow N_2(g) + 2\ H_2O(\ell) \qquad \Delta\overline{H} = -622.33 \text{ kJ/mol}$$

The third equation must be reversed and doubled to obtain 2 mol $H_2O_2(\ell)$ on the left.

$$2\ H_2O_2(\ell) \rightarrow 2\ H_2(g) + 2\ O_2(g) \qquad \Delta\bar{H} = -2 \times (-187.78) = 375.56\ kJ/mol$$

To provide for the proper cancellation of terms [i.e., $O_2(g)$ and $H_2(g)$] the second equation must be doubled.

$$2\ H_2(g) + O_2(g) \rightarrow 2\ H_2O(\ell) \qquad \Delta\bar{H} = 2 \times (-285.85) = -571.70\ kJ/mol$$

The sum of these three equations and their $\Delta\bar{H}$ values yields

$$N_2H_4(\ell) + 2\ H_2O_2(\ell) \rightarrow N_2(g) + 2\ H_2O(\ell) \qquad \Delta\bar{H} = -622.33 + 375.56 - 571.70$$

$$= -818.47\ kJ/mol$$

6-21. The five equations listed must be combined in the following way if their sum is to yield the desired net equation.

$$CS_2(\ell) + 3\ O_2(g) \rightarrow CO_2(g) + 2\ SO_2(g) \qquad \Delta\bar{H} = -1077\ kJ/mol$$

$$2\ S(s) + Cl_2(g) \rightarrow S_2Cl_2(\ell) \qquad \Delta\bar{H} = -60.2\ kJ/mol$$

$$C(s) + 2\ Cl_2(g) \rightarrow CCl_4(\ell) \qquad \Delta\bar{H} = -135.4\ kJ/mol$$

$$2\ SO_2(g) \rightarrow 2\ S(s) + 2\ O_2(g) \qquad \Delta\bar{H} = -2 \times (-296.9) = +593.8\ kJ/mol$$

$$CO_2(g) \rightarrow C(s) + O_2(g) \qquad \Delta\bar{H} = -(-393.5) = +393.5\ kJ/mol$$

$$CS_2(\ell) + 3\ Cl_2(g) \rightarrow CCl_4(\ell) + S_2Cl_2(\ell) \qquad \Delta\bar{H} = -1077 - 6.02 - 135.4 + 593.8 + 393.5$$

$$= -285\ kJ/mol$$

6-22. First, let us write equations for the three combustion reactions.

$$C_4H_6(g) + 11/2\ O_2(g) \rightarrow 4\ CO_2(g) + 3\ H_2O(\ell) \qquad \Delta\bar{H} = -2543.5\ kJ/mol$$

$$C_4H_{10}(g) + 13/2\ O_2(g) \rightarrow 4\ CO_2(g) + 5\ H_2O(\ell) \qquad \Delta\bar{H} = -2878.6\ kJ/mol$$

$$H_2(g) + 1/2\ O_2(g) \rightarrow H_2O(\ell) \qquad \Delta\bar{H} = -285.85\ kJ/mol$$

Now we must combine these to yield as a net equation:

$$C_4H_6(g) + 2\ H_2(g) \rightarrow C_4H_{10}(g) \qquad \Delta\bar{H} = ?$$

This is accomplished as follows:

$$C_4H_6(g) + 11/2\ O_2(g) \rightarrow 4\ CO_2(g) + 3\ H_2O(\ell) \qquad \Delta\bar{H} = -2543.5\ kJ/mol$$

$$2\ H_2(g) + O_2(g) \rightarrow 2\ H_2O(\ell) \qquad \Delta\bar{H} = 2 \times (-285.85)\ kJ/mol$$

$$4\ CO_2(g) + 5\ H_2O(\ell) \rightarrow 13/2\ O_2(g) + C_4H_{10}(g) \qquad \Delta\bar{H} = -(-2878.6)\ kJ/mol$$

$$C_4H_6(g) + 2\ H_2(g) \rightarrow C_4H_{10}(g) \qquad \Delta\bar{H} = -2543.5 - (2 \times 285.85) + 2878.6$$

$$= -236.6\ kJ/mol$$

6-23. Combine the data listed in the following way:

$$CO(g) + 1/2\ O_2(g) \rightarrow CO_2(g) \qquad \Delta\bar{H} = -282.97\ kJ/mol$$

$$2\ H_2(g) + O_2(g) \rightarrow 2\ H_2O(\ell) \qquad \Delta\bar{H} = 2 \times (-285.85)\ kJ/mol$$

$$3\ C\ (graphite) + 6\ H_2(g) \rightarrow 3\ CH_4(g) \qquad \Delta\bar{H} = 3 \times (-74.85)\ kJ/mol$$

$$3\ CO(g) \rightarrow 3\ C\ (graphite) + 3/2\ O_2(g) \qquad \Delta\bar{H} = -3 \times (-110.54)\ kJ/mol$$

$$4\ CO(g) + 8\ H_2(g) \rightarrow 3\ CH_4(g) + CO_2(g) + 2\ H_2O(\ell)$$

$$\Delta\bar{H} = -282.97 - (2 \times 285.85) - (3 \times 74.85) + (3 \times 110.54)\ kJ/mol = -747.60\ kJ/mol$$

6-24. Here the required combination of equations is:

$CO(g) \rightarrow \cancel{C\,(graphite)} + \cancel{1/2\ O_2(g)}$	$\Delta\overline{H} = -(-110.54)$ kJ/mol
$2\ H_2(g) + \cancel{O_2(g)} \rightarrow \cancel{2\ H_2O(\ell)}$	$\Delta\overline{H} = 2 \times (-285.85)$ kJ/mol
$\cancel{CO_2(g)} + \cancel{2\ H_2O(\ell)} \rightarrow \cancel{3/2\ O_2(g)} + CH_3OH(\ell)$	$\Delta\overline{H} = -(-726.6)$ kJ/mol
$\cancel{C\,(graphite)} + \cancel{O_2(g)} \rightarrow \cancel{CO_2(g)}$	$\Delta\overline{H} = -393.51$ kJ/mol

$CO(g) + 2\ H_2(g) \rightarrow CH_3OH(\ell)$

$\Delta\overline{H} = \{110.54 - (2 \times 285.85) + 726.6 - 393.51\}$ kJ/mol $= -128.1$ kJ/mol

6-25. Equations (6.18), (6.19), and (6.22) must be combined in the following manner to yield the desired net equation:

2 x (6.18): $2\ C(s) + 2\ H_2O(g) \rightarrow \cancel{2\ CO(g)} + \cancel{2\ H_2(g)}$

(6.22): $\cancel{CO(g)} + \cancel{3\ H_2(g)} \rightarrow CH_4(g) + \cancel{H_2O(g)}$

(6.19): $\cancel{CO(g)} + \cancel{H_2O(g)} \rightarrow CO_2(g) + \cancel{H_2(g)}$

$2\ C(s) + 2\ H_2O(g) \rightarrow CH_4(g) + CO_2(g)$

Enthalpies (heats) of formation

6-26. For the reaction $2\ Cl_2(g) + 2\ H_2O(\ell) \rightarrow 4\ HCl(g) + O_2(g)$

$\Delta\overline{H}^\circ_{rxn} = 4\ \Delta\overline{H}^\circ_f[HCl(g)] - 2\ \Delta\overline{H}^\circ_f[H_2O(\ell)] = 4 \times (-92.30\ \text{kJ/mol}) - 2 \times (-285.85\ \text{kJ/mol}) = +202.5\ \text{kJ/mol}$

6-27. First write a balanced equation for the reaction. Then use heat of formation data in equation (6.17) to calculate $\Delta\overline{H}^\circ_{rxn}$.

$C_2H_5OH(\ell) + 3\ O_2(g) \rightarrow 2\ CO_2(g) + 3\ H_2O(\ell)$

$\Delta\overline{H}^\circ_{rxn} = 2\ \Delta\overline{H}^\circ_f[CO_2(g)] + 3\ \Delta\overline{H}^\circ_f[H_2O(\ell)] - \Delta\overline{H}^\circ_f[C_2H_5OH(\ell)]$

$= 2 \times (-393.51) + 3 \times (-285.85) - (-277.65) = -1366.92$ kJ/mol

6-28. Equation (6.17) is written with data given in the exercise and in Table 6-1. The heat of formation of $CCl_4(g)$ appears as an unknown in this equation.

$CH_4(g) + 4\ Cl_2(g) \rightarrow CCl_4(g) + 4\ HCl(g)$ $\Delta\overline{H}^\circ = -402$ kJ/mol

$\Delta\overline{H}^\circ_{rxn} = -402\ \text{kJ/mol} = \Delta\overline{H}^\circ_f[CCl_4(g)] + 4\ \Delta\overline{H}^\circ_f[HCl(g)] - \Delta\overline{H}^\circ_f[CH_4(g)]$

$-402\ \text{kJ/mol} = \Delta\overline{H}^\circ_f[CCl_4(g)] + 4 \times (-92.30\ \text{kJ/mol}) - (-74.85\ \text{kJ/mol})$

$\Delta\overline{H}^\circ_f[CCl_4(g)] = -402\ \text{kJ/mol} + 369.2\ \text{kJ/mol} - 74.85\ \text{kJ/mol} = -108\ \text{kJ/mol}$

6-29. $2\ C(s) + 2\ H_2O(g) \rightarrow CH_4(g) + CO_2(g)$

$\Delta\overline{H}^\circ_{rxn} = \Delta\overline{H}^\circ_f[CH_4(g)] + \Delta\overline{H}^\circ_f[CO_2(g)] - 2\ \Delta\overline{H}^\circ_f[H_2O(g)] = -74.85\ \text{kJ/mol} - 393.51\ \text{kJ/mol} - 2 \times (-241.84\ \text{kJ/mol})$

$= +15.32$ kJ/mol

6-30. Express the desired reaction as the sum of the following two:

$C_2H_4(g) + 3\ O_2(g) \rightarrow 2\ CO_2(g) + \cancel{2\ H_2O(\ell)}$ $\Delta\overline{H}^\circ = -1410.8$ kJ/mol

$\cancel{2\ H_2O(\ell)} \rightarrow 2\ H_2O(g)$ $\Delta\overline{H}^\circ = 2\ \Delta\overline{H}^\circ_f[H_2O(g)] - 2\ \Delta\overline{H}^\circ_f[H_2O(\ell)]$

$= 2 \times (-241.84) - 2 \times (-285.85)$

$= 88.02$ kJ/mol

$C_2H_4(g) + 3\ O_2(g) \rightarrow 2\ CO_2(g) + 2\ H_2O(g)$ $\Delta\overline{H}^\circ = -1322.8$ kJ/mol

Consider that if the reaction first occurred to produce $H_2O(\ell)$, some of the liberated heat would have to be reabsorbed to vaporize the water. This would make $\Delta\overline{H}^\circ$ smaller (less negative) when $H_2O(g)$ is produced than when $H_2O(\ell)$ is formed, as the above calculation shows.

6-31. Equation (6.17) can be used twice, first to establish $\Delta\bar{H}_f^\circ[C_5H_{12}(\ell)]$ and then to calculate $\Delta\bar{H}^\circ_{rxn}$ for the given reaction.

$C_5H_{12} + 8\ O_2(g) \rightarrow 5\ CO_2(g) + 6\ H_2O(\ell)$ $\quad\quad$ $\Delta\bar{H}^\circ = -3534$ kJ/mol

$\Delta\bar{H}^\circ_{rxn} = -3534\ \text{kJ/mol} = 5\ \Delta\bar{H}_f^\circ[CO_2(g)] + 6\ \Delta\bar{H}_f^\circ[H_2O(\ell)] - \Delta\bar{H}_f^\circ[C_5H_{12}(\ell)]$

$-3534\ \text{kJ/mol} = 5 \times (-393.51\ \text{kJ/mol}) + 6 \times (-285.85\ \text{kJ/mol}) - \Delta\bar{H}_f^\circ[C_5H_{12}(\ell)]$

$\Delta\bar{H}_f^\circ[C_5H_{12}(\ell)] = -1967.55 - 1715.10 + 3534 = -149\ \text{kJ/mol}$

$5\ CO(g) + 11\ H_2(g) \rightarrow C_5H_{12}(\ell) + 5\ H_2O(\ell)$

$\Delta\bar{H}^\circ_{rxn} = \Delta\bar{H}_f^\circ[C_5H_{12}(\ell)] + 5\ \Delta\bar{H}_f^\circ[H_2O(\ell)] - 5\ \Delta\bar{H}_f^\circ[CO(g)]$

$= -149\ \text{kJ/mol} + 5 \times (-285.85\ \text{kJ/mol}) - 5 \times (-110.54\ \text{kJ/mol}) = -1026\ \text{kJ/mol}$

6-32. (a) First, determine $\Delta\bar{H}^\circ$ for the reaction $CaCO_3(s) \rightarrow CaO(s) + CO_2(g)$

$\Delta\bar{H}^\circ_{rxn} = \Delta H_f^\circ[CaO(s)] + \Delta\bar{H}_f^\circ[CO_2(g)] - \Delta\bar{H}_f^\circ[CaCO_3(s)]$

$= -635.5\ \text{kJ/mol} - 393.51\ \text{kJ/mol} - (-1027.1\ \text{kJ/mol}) = +178.1\ \text{kJ/mol}$

$$\text{no. kJ} = 755\ \text{kg}\ CaCO_3 \times \frac{1000\ \text{g}\ CaCO_3}{1\ \text{kg}\ CaCO_3} \times \frac{1\ \text{mol}\ CaCO_3}{100\ \text{g}\ CaCO_3} \times \frac{178.1\ \text{kJ}}{1\ \text{mol}\ CaCO_3} = 1.34\times10^6\ \text{kJ}$$

(b) Now, determine $\Delta\bar{H}^\circ$ for the reaction $CH_4(g) + 2\ O_2(g) \rightarrow CO_2(g) + 2\ H_2O(\ell)$

$\Delta\bar{H}^\circ_{rxn} = \Delta\bar{H}_f^\circ[CO_2(g)] + 2\ \Delta\bar{H}_f^\circ[H_2O(\ell) - \Delta\bar{H}_f^\circ[CH_4(g)]$

$= -393.51\ \text{kJ/mol} + 2 \times (-285.85\ \text{kJ/mol}) - (-74.85\ \text{kJ/mol}) = -890.36\ \text{kJ/mol}$

$$\text{no. mol CH (g)} = 1.34\times10^6\ \text{kJ} \times \frac{1\ \text{mol}\ CH_4}{890.36\ \text{kJ}} = 1.51\times10^3$$

$$V = \frac{nRT}{P} = \frac{1.51\times10^3 \times 0.0821\ \text{L atom mol}^{-1}\text{K}^{-1} \times 291\ \text{K}}{(737/760)\ \text{atm}} = 3.72\times10^4\ \text{L}$$

6-33. (a) The enthalpy of formation of a substance depends on the state in which the substance is found (i.e., enthalpy is a function of state). Pure $H_2SO_4(\ell)$ and H_2SO_4 in aqueous solutions of differing molar concentrations all represent different conditions or states for a substance. As a result there are different values of ΔH_f.

(b) All of the heats of formation are negative, more negative the more dilute the solution. Whether dilute $H_2SO_4(aq)$ solution is prepared by mixing pure $H_2SO_4(\ell)$ with water or by adding water to a more concentrated solution, $\Delta H < 0$. Thus the dilution processes are exothermic. Heat of dilution is absorbed by the water (surroundings) in which dilution occurs; the solution temperature rises.

(c) Express the dilution process as follows:

$H_2SO_4(\ell) + aq \rightarrow H_2SO_4(0.02\ M)$

$\Delta\bar{H} = \Delta\bar{H}_f[H_2SO_4(0.02\ M)] - \Delta\bar{H}_f[H_2SO_4(\ell)] = -897\ \text{kJ/mol} - (-814\ \text{kJ/mol}) = -83\ \text{kJ/mol}$

The amount of H_2SO_4 involved in the dilution is that contained in 250 ml 0.02 M H_2SO_4

$$\text{no. mol}\ H_2SO_4 = 0.250\ \text{L} \times \frac{0.02\ \text{mol}\ H_2SO_4}{\text{L}} = 0.005\ \text{mol}\ H_2SO_4$$

$$\text{no. J} = 0.005\ \text{mol}\ H_2SO_4 \times \frac{-83\ \text{kJ}}{\text{mol}\ H_2SO_4} \times \frac{1000\ \text{J}}{1\ \text{kJ}} = -415\ \text{J} = -4\times10^2\ \text{J}$$

The heat gained by the solution is +415 J, and

$\text{no. g}\ H_2O \times \text{sp. ht.}\ H_2O \times \Delta t = 415\ \text{J}$ $\quad\quad$ $250\ \text{g}\ H_2O \times \frac{4.184\ \text{J}}{\text{g}\ H_2O\ ^\circ C} \times \Delta t = 415\ \text{J}$

$\Delta t = 0.4°C$

The solution temperature increases by 0.4°C.

6-34. Determine the heat of combustion of the pure gases CH_4 and C_2H_6. Then calculate the mol percent of each gas in the mixture. Volume and mole percents are identical (recall equation 5.16). The molar heat of combustion of $CH_4(g)$ calculated in Exercise 6-32 is -890.36 kJ/mol.

$C_2H_6(g) + 7/2\ O_2(g) \rightarrow 2\ CO_2(g) + 3\ H_2O(\ell)$

$\Delta\overline{H}° = 2\ \Delta\overline{H}_f°[CO_2(g)] + 3\ \Delta\overline{H}_f°[H_2O(\ell)] - \Delta\overline{H}_f°[C_2H_6(g)]$

$= 2 \times (-393.51\ kJ/mol) + 3 \times (-285.85\ kJ/mol) - (-84.68\ kJ/mol)$

$= -787.02\ kJ/mol - 857.55\ kJ/mol + 84.68\ kJ/mol$

$= -1559.89\ kJ/mol$

In 1.00 L of gas at STP there is a total of 1.00/22.4 = 0.0446 mol gas. If we let the no. mol $CH_4(g) = x$ and no. mol $C_2H_6(g) = y$, $x + y = 0.0446$ and

$$(x\ mol\ CH_4 \times \frac{-890.36\ kJ}{mol\ CH_4}) + (y\ mol\ C_2H_6 \times \frac{-1559.89\ kJ}{mol\ C_2H_6}) = -43.6\ kJ \qquad \text{or}$$

$-890.36x - 1559.89y = -43.6$

Substitute $y = 0.0446 - x$ to obtain

$-890.36\ x - 1559.89(0.0446 - x) = -43.6$

$-890.36\ x - 69.6 + 1559.89\ x = -43.6$

$669.53\ x = 69.6 - 43.6 = 26.0$

$x = 0.0388 \qquad y = 0.0446 - 0.0388 = 0.0058$

Mole percents (volume percents)

$\%\ CH_4$: $\frac{0.0388\ mol}{0.0446\ mol} \times 100 = 87.0\%$ $\qquad$ $\%\ C_2H_6 = 100.0 - 87.0 \times 100 = 13.0\%$

6-35. Estimate $\Delta\overline{H}_f°[C_7H_{16}(\ell)]$ from data given in Appendix D and the rule stated in the exercise. That is,

	$C_3H_8(g)$	$C_4H_{10}(g)$	$C_5H_{12}(g)$	$C_6H_{14}(g)$	$C_7H_{16}(g)$
$\Delta\overline{H}_f°$, kJ/mol	-103.85	-125	-146	-167	-188

For the reaction $C_6H_{16}(g) + 11\ O_2(g) \rightarrow 7\ CO_2(g) + 8\ H_2O(\ell)$

$\Delta\overline{H}° = 7\ \Delta\overline{H}_f°[CO_2(g)] + 8\ \Delta\overline{H}_f°[H_2O(\ell)] - \Delta\overline{H}_f°[C_7H_{16}(g)]$

$= 7 \times (-393.51\ kJ/mol) + 8 \times (-285.85\ kJ/mol) - (-188\ kJ/mol)$

$= -2754.57\ kJ/mol - 2286.80\ kJ/mol + 188\ kJ/mol$

$= -4853\ kJ/mol$

Self-test Questions

1. (*c*) Item (a) is the definition of a calorie, not a kilocalorie. To raise the temperature of 1.00 L water by 1°C requires 1 kcal; by 10°C, 10 kcal--item (b) is incorrect. When the temperature of 100 cm^3 water (100 g water) is lowered by 10°C, 1 kcal of heat is released. Item (d) is incorrect because 1 kcal = 1.0×10^3 cal, not 1.0×10^6 cal.

2. (*a*) We can immediately eliminate (c) and (d) by reasoning that the highest temperature must be associated either with the highest or lowest value of the specific heat. The heat evolved by the metals as they cool is q = mass x sp. ht. x ΔT. For comparable quantities of heat (q), and equal masses of metal are being compared. The metal with the largest specific heat can give off this heat for the smallest value of ΔT and therefore have the highest final temperature. Alternatively, final temperatures can be calculated and compared, but this requires considerable algebraic manipulation.

3. (*d*) Again, the final temperature can be calculated algebraically. A much simpler method is to note that if equal volumesof 30° and 50° water are mixed the final temperature is 40°C. Since there is a larger volume of the hotter water, the final temperature must be greater than 40°C.

4. (*a*) If $\overline{\Delta H}_{soln}$ in NaOH is -41.6 kJ/mol, the NaOH gives off heat when it dissolves. The water must absorb this heat and its temperature will increase. Although the exact amount by which the temperature increases depends on the quantities of NaOH and water, since the process is exothermic some temperature increase is expected.

5. (*b*) The value given is the heat of combustion per g CS_2. Convert this to heat of combustion per mole of CS_2 = 3.24 x 76.1 = -247 kcal/mol. None of the answers have this value, but if we convert to kJ/mol we obtain -1.03×10^3 kJ/mol CS_2. Since the reaction whose $\overline{\Delta H}$ value we are seeking involves burning 1 mol CS_2, the correct answer is (b).

6. (*b*) Only the enthalpies of formation of elements can be 0, not those of compounds; item (a) is incorrect. The combustion of C (graphite) produces $CO_2(g)$. This combustion reaction is the same as the formation reaction for CO_2. Since the molar enthalpy of formation of $O_2(g)$ is zero, item (c) would make the molar enthalpies of formation of CO(g) and $CO_2(g)$ equal. Combustion of CO(g) does not correspond to making $CO_2(g)$ from its elements.

7. (*c*) The reaction given is the ***reverse*** of the reaction for the formation of NH_3 (ΔH = 46 kJ) and it is doubled (ΔH = +92 kJ).

8. (*d*) For the given reaction, $\overline{\Delta H}°$ = -3534 kJ/mol

$$= 5\ \overline{\Delta H}^\circ_f[CO_2(g)] + 6\ \overline{\Delta H}^\circ_f[H_2O(\ell)] - \overline{\Delta H}^\circ_f[C_5H_{12}(\ell)]$$

$$\overline{\Delta H}^\circ_f[C_5H_{12}(\ell)] = 5\ \overline{\Delta H}^\circ_f[CO_2(g)] + 6\ \overline{\Delta H}^\circ_f[H_2O(\ell)] + 3534$$

$$\overline{\Delta H}^\circ_f[C_5H_{12}(\ell)] = 5 \times (-394) + 6 \times (-286) + 3534$$

9. (a) Specific heat is the quantity of heat required to change the temperature of one gram of substance by one degree C. Molar heat capacity refers to the quantity of heat required to change the temperature of one mole of substance to the same extent.

(b) An endothermic reaction absorbs heat from its surroundings, and an exothermic reaction gives off heat to the surroundings.

(c) The enthalpy of formation of $C_4H_{10}(g)$ refers to the enthalpy change in the reaction where C_4H_{10} is formed from the most stable form of its elements at 1 atm pressure.

$$4\ C\ (graphite) + 5\ H_2(g) \rightarrow C_4H_{10}(g)$$

The heat of combustion is ΔH for the combustion reaction:

$$2\ C_4H_{10}(g) + 13\ O_2(g) \rightarrow 8\ CO_2(g) + 10\ H_2O(\ell)$$

10. Use the expression $q_{iron} = -q_{water}$, solve for Δt and then the initial temperature of the iron.

$$1500\ g\ Fe \times \frac{0.59\ J}{g\ Fe\ °C} \times \Delta t = -\ \{755\ g\ water \times \frac{4.184\ J}{g\ water\ °C} \times (38.6 - 21.3)°C\}$$

$$\Delta t = \frac{-755 \times 4.184 \times 17.3}{1500 \times 0.59}\ °C = -62°C$$

$\Delta t = t_f - t_i = 38.6°C - t_i = -62°C$ $\qquad$ $t_i = 101°C$

11. $C_6H_5OH(s) + 7\ O_2(g) \rightarrow 6\ CO_2(g) + 3\ H_2O(\ell)$ $\qquad$ $\overline{\Delta H} = -3.063\times10^3$ kJ/mol

To determine the value of $\overline{\Delta H}$

$$\overline{\Delta H} = \frac{-32.55\ kJ}{g\ C_6H_5OH} \times \frac{94.11\ g\ C_6H_5OH}{1\ mol\ C_6H_5OH} = -3063\ kJ/mol\ C_6H_5OH$$

12. From the reaction $CO(g) + Cl_2(g) \rightarrow COCl_2(g)$ $\qquad$ $\overline{\Delta H}$ = -108 kJ/mol, we can write

$$\overline{\Delta H} = \overline{\Delta H}^\circ_f[COCl_2(g)] - \overline{\Delta H}^\circ_f[CO(g)] = -108\ kJ/mol$$

But we need a value of $\overline{\Delta H}^\circ_f[CO(g)]$, which we obtain from

C (graphite) + $O_2(g) \rightarrow CO_2(g)$	$\overline{\Delta H}°$ = -393.51 kJ/mol
$CO_2(g) \rightarrow CO(g) + 1/2\ O_2(g)$	$\overline{\Delta H}°$ = - (-282.97) = +282.97 kJ/mol
C (graphite) + $1/2\ O_2(g) \rightarrow CO(g)$	$\overline{\Delta H}^\circ_f$ = -393.51 + 282.97 = -110.54 kJ/mol

$$\overline{\Delta H}^\circ_f[COCl_2(g)] = -108\ kJ/mol + (-110.54\ kJ/mol) = -219\ kJ/mol$$

Chapter 7

Electrons in Atoms

Review Problems

7-1. (a) $\text{no. nm} = 3015\ \text{Å} \times \frac{1\times10^{-10}\ \text{m}}{1\ \text{Å}} \times \frac{1\ \text{nm}}{1\times10^{-9}\ \text{m}} = 301.5\ \text{nm}$

(b) $\text{no. cm} = 1.56\ \mu\text{m} \times \frac{1\times10^{-6}\ \text{m}}{1\ \mu\text{m}} \times \frac{100\ \text{cm}}{1\ \text{m}} = 1.56\times10^{-4}\ \text{cm}$

(c) $\text{no. nm} = 3.92\ \text{cm} \times \frac{1\ \text{m}}{100\ \text{cm}} \times \frac{1\ \text{nm}}{1\times10^{-9}\ \text{m}} = 3.92\times10^{7}\ \text{nm}$

(d) $\text{no. m} = 376\ \text{nm} \times \frac{1\times10^{-9}\ \text{m}}{1\ \text{nm}} = 3.76\times10^{-7}\ \text{m}$

(e) $\text{no. }\mu\text{m} = 1812\ \text{nm} \times \frac{1\times10^{-9}\ \text{m}}{1\ \text{nm}} \times \frac{1\ \mu\text{m}}{1\times10^{-6}\ \text{m}} = 1.812\ \mu\text{m}$

(f) $\text{no. Å} = 2.18\ \mu\text{m} \times \frac{1\times10^{-6}\ \text{m}}{1\ \mu\text{m}} \times \frac{1\ \text{Å}}{1\times10^{-10}\ \text{m}} = 2.18\times10^{4}\ \text{Å}$

7-2. (a) $\lambda = c/\nu = \frac{3.00\times10^{8}\ \text{m s}^{-1}}{6.2\times10^{13}\ \text{s}^{-1}} = 4.8\times10^{-6}\ \text{m}$

(b) $\lambda = c/\nu = \frac{3.00\times10^{8}\ \text{m s}^{-1}}{4.5\times10^{16}\ \text{s}^{-1}} = 6.7\times10^{-9}\ \text{m}$

(c) $\lambda = c/\nu = \frac{3.00\times10^{8}\ \text{m s}^{-1}}{5.50\times10^{5}\ \text{s}^{-1}} = 5.45\times10^{2}\ \text{m}$

7-3. (a) $\nu = c/\lambda = \frac{3.00\times10^{8}\ \text{m s}^{-1}}{1.3\times10^{-4}\ \text{cm} \times \frac{1\ \text{m}}{100\ \text{cm}}} = 2.3\times10^{14}\ \text{s}^{-1}$

(b) $\nu = c/\lambda = \frac{3.00\times10^{8}\ \text{m s}^{-1}}{8.87\ \mu\text{m} \times \frac{1\times10^{-6}\ \text{m}}{1\ \mu\text{m}}} = 3.38\times10^{13}\ \text{s}^{-1}$

(c) $\nu = c/\lambda = \frac{3.00\times10^{8}\ \text{m s}^{-1}}{371\ \text{Å} \times \frac{1\times10^{-10}\ \text{m}}{1\ \text{Å}}} = 8.09\times10^{15}\ \text{s}^{-1}$

(d) $\nu = c/\lambda = \frac{3.00\times10^{8}\ \text{m s}^{-1}}{335\ \text{cm} \times \frac{1\ \text{m}}{100\ \text{cm}}} = 8.96\times10^{7}\ \text{s}^{-1}$

7-4. (a) $\nu = 3.28881\times10^{15}\ \text{s}^{-1}\ (\frac{1}{2^2} - \frac{1}{5^2}) = 3.28881\times10^{15}\ \text{s}^{-1}\ (0.25000 - 0.04000)$

$= 0.21000 \times 3.28881\times10^{15}\ \text{s}^{-1} = 6.9050\times10^{14}\ \text{s}^{-1}$

(b) $\nu = 3.28881\times10^{15}\ \text{s}^{-1}\ (\frac{1}{2^2} - \frac{1}{7^2}) = 3.28881\times10^{15}\ \text{s}^{-1}\ (0.25000 - 0.02041)$

$= 0.22959 \times 3.28881\times10^{15}\ \text{s}^{-1} = 7.5491\times10^{14}\ \text{s}^{-1}$

$\lambda = c/\nu = \frac{2.9979\times10^{8}\ \text{m s}^{-1}}{7.5491\times10^{14}\ \text{s}^{-1}} = 3.9712\times10^{-7}\ \text{m} = 397.12\ \text{nm}$

(c) First, calculate ν corresponding to the line at 380 nm.

$\nu = c/\lambda = \frac{3.00\times10^{8}\ \text{m s}^{-1}}{380\ \text{nm} \times \frac{1\times10^{-9}\ \text{m}}{1\ \text{nm}}} = 7.89\times10^{14}\ \text{s}^{-1}$

Now solve the Balmer equation for $1/n^2$, substituting $\nu = 7.89\times10^{14}\ \text{s}^{-1}$.

$\frac{7.89\times10^{14}\ \text{s}^{-1}}{3.28881\times10^{15}\ \text{s}^{-1}} = 0.240 = (\frac{1}{2^2} - \frac{1}{n^2}) = (0.2500 - 1/n^2)$

$1/n^2 = 0.2500 - 0.240 = 0.010$

$n^2 = 1/0.010 = 100; \quad n = \sqrt{n^2} = \sqrt{100} = 10$

7-5. (a) $E = h\nu = 6.626\times10^{-34}$ Js $\times$ 3.10×10^{15} s^{-1} = 2.05×10^{-18} J/photon

(b) $E = h\nu = 6.626\times10^{-34}\ \dfrac{\text{Js}}{\text{photon}} \times 4.26\times10^{14}\ s^{-1} \times \dfrac{6.02\times10^{23}\ \text{photon}}{1\ \text{mol}} \times \dfrac{1\ \text{kJ}}{1000\ \text{J}} = 1.70\times10^{2}\ \text{kJ/mol}$

(c) $\nu = E/h = \dfrac{3.54\times10^{-20}\ \text{J/photon}}{6.626\times10^{-34}\ \text{Js/photon}} = 5.34\times10^{13}\ s^{-1}$

(d) First solve for ν.

$$\nu = E/h = \frac{185\ \text{kJ mol}^{-1} \times \dfrac{1000\ \text{J}}{1\ \text{kJ}} \times \dfrac{1\ \text{mol}}{6.02\times10^{23}\ \text{photon}}}{6.626\times10^{-34}\ \text{Js/photon}} = 4.64\times10^{14}\ s^{-1}$$

$$\lambda = c/\nu = \frac{3.00\times10^{8}\ \text{m s}^{-1}}{4.64\times10^{14}\ s^{-1}} = 6.47\times10^{-7}\ \text{m} = 647\ \text{nm}$$

7-6. (a) The only allowable value of m_ℓ is zero, since $\ell = 0$.

(b) Since $m_\ell = -1$, ℓ cannot be zero; ℓ can be 1 or 2. (Because n = 3, ℓ cannot be greater than 2.)

(c) Since $\ell = 1$, n must be an integer greater than 1. That is, n = 2, 3, ...

(d) If $\ell = 2$, n must be an integer larger than 2. That is, n = 3, 4, ... The value of m_ℓ can range from $-\ell$ to $+\ell$, i.e., m = -2, -1, 0, +1, or +2.

7-7. (a) 4s: n = 4 $\ell = 0$ $m_\ell = 0$

(b) 3p: n = 3 $\ell = 1$ m_ℓ = -1, 0, or +1

(c) 5f: n = 5 $\ell = 3$ m_ℓ = -3, -2, -1, 0, +1, +2, or +3

(d) 3d: n = 3 $\ell = 2$ m_ℓ = -2, -1, 0, +1, or +2

7-8. (a) This set is allowable; no rules are violated.

(b) This set is not allowable; the value of ℓ (3 in this case) cannot exceed the value of n (2).

(c) This set is not allowable; m_ℓ can only range from $-\ell$ to $+\ell$ and $\ell = 0$ here.

(d) This set is allowable.

(e) This set is not allowable; $\ell \neq n$.

(f) This set is allowable.

7-9. The order of filling of the orbitals is that of increasing orbital energy: 3p, 3d, 4p, 5s, 6s, 6p, 5f.

7-10. (b) The element with this electron configuration has (2 + 2 + 6 + 2 + 3) 15 electrons. The element with Z = 15 is phosphorus.

(c) There are 40 electrons to be assigned; the Kr configuration accounts for 36. Of the remaining 4 electrons, two are in the 5s orbital. This means that the remaining part of the configuration must be $4d^2$.

(d) Before the 5p subshell receives any electrons, the 4d subshell must fill, i.e., $4d^{10}$. This means that the electron configuration is $[Kr]4d^{10}5s^25p^4$. The element is Z = 52 (tellurium, Te).

(e) As in part (d) we reason that before the 4p subshell receives any electrons the 4s and 3d subshells must fill. The electron configuration becomes

As (Z = 33) $[Ar]3d^{10}4s^24p^3$

(f) 83 electrons are to be accounted for; 54 of these are in the Xe configuration. The maximum capacities of the 4f and 5d subshells are 14 and 10, respectively. This still leaves 5 electrons to be distributed between the 6s and 6p subshells; Bi (Z = 83) $[Xe]4f^{14}5d^{10}6s^26p^3$.

7-11. (a) This electron configuration corresponds to the element with Z = 5, boron.

(b) Here the element is in the first transition series (filling of the 3d sublevel) and has an atomic number of Z = 18 + 3 + 2 = 23. The element is vanadium.

(c) This is the element with atomic number Z = 14--silicon.

7-12. (a) Al: $1s^22s^22p^63s^23p^1$ or $[Ne]3s^23p^1$

(b) Rb: $1s^22s^22p^63s^23p^63d^{10}4s^24p^65s^1$ or $[Kr]5s^1$

(c) Cd: $1s^22s^22p^63s^23p^63d^{10}4s^24p^64d^{10}5s^2$ or $[Kr]4d^{10}5s^2$

(d) Sb: $1s^22s^22p^63s^23p^63d^{10}4s^24p^24d^{10}5s^25p^3$ or $[Kr]4d^{10}5s^25p^3$

(e) Pb: $1s^22s^22p^63s^23p^63d^{10}4s^24p^64d^{10}4f^{14}5s^25p^65d^{10}6s^26p^2$ or $[Xe]4f^{14}5d^{10}6s^26p^2$

(f) Xe: $1s^22s^22p^63s^23p^63d^{10}4s^24p^64d^{10}5s^25p^6$ or $[Kr]4d^{10}5s^25p^6$

Exercises

Electromagnetic radiation

7-1. (a) Correct: The wavelength of the radiation in question is 285.2 nm. This is a shorter wavelength than 315 nm. From the relationship, $c = \nu \cdot \lambda$, we see that the shorter the wavelength, the greater the frequency.

(b) Incorrect: The visible region of the spectrum ranges from about 380 to 760 nm (see Figure 7-3).

(c) Incorrect: All forms of electromagnetic radiation have the same speed in vacuum.

(d) Correct: The radiation in question, 285.2 nm, is in the ultraviolet range of the spectrum. X-rays have shorter wavelengths than ultraviolet light.

7-2. The number of periods per second is simply the number of vibrations per second, that is, the frequency of the radiation.

$\nu = 9.192631770\times10^9\ s^{-1}$

$$\lambda = \frac{c}{\nu} = \frac{2.997925\times10^8\ m\ s^{-1}}{9.192631770\times10^9\ s^{-1}} = 3.261226\times10^{-2}\ m = 3.261226\ cm$$

7-3. $$\text{no. min} = 93\times10^6\ mi \times \frac{5280\ ft}{1\ mi} \times \frac{12\ in}{1\ ft} \times \frac{1\ m}{39.37\ in} \times \frac{1\ s}{3.00\times10^8\ min} \times \frac{1\ min}{60\ s} = 8.3\ min$$

Atomic spectra

7-4. The longest wavelength component corresponds to a value of n = 3. The other three lines correspond to values of n = 4, n = 5, and n = 6. Use the form of the Balmer equation given in equation (7.2).

longest wavelength line: $n = 3$

$$\lambda(\text{in } \text{Å}) = 3645.6\left(\frac{n^2}{n^2 - 4}\right) = 3645.6\left(\frac{9}{9 - 4}\right) = 6562.1\ \text{Å}\ (=656.21\ nm)$$

$n = 4$

$$\lambda = 3645.6\left(\frac{16}{16 - 4}\right) = 4860.8\ \text{Å}\ (=486.08\ nm)$$

$n = 5$

$$\lambda = 3645.6\left(\frac{25}{25 - 4}\right) = 4340.0\ \text{Å}\ (=434.00\ nm)$$

$n = 6$

$$\lambda = 3645.6\left(\frac{36}{36 - 4}\right) = 4101.3\ \text{Å}\ (=410.13\ nm)$$

7-5. The longest wavelength for a line in the Balmer series is obtained for $n = 3$. In Exercise 4 this was calculated to be 656.21 nm. The spectral line at 1880 nm is of still longer wavelength. This line cannot be in the Balmer series.

7-6. By referring to Exercise 4, we see that the wavelengths corresponding to the spectral lines get shorter as n becomes larger. The shortest wavelength will be for $n = \infty$.

$$\lambda(\text{in Å}) = 3645.6\left(\frac{n^2}{n^2 - 4}\right) = 3645.6\left(\frac{\infty^2}{\infty^2 - 4}\right) = 3645.6 \text{ Å } (=364.56 \text{ nm})$$

7-7. (a) The maximum wavelength corresponds to n = 2.

$$\nu = 3.2881\times10^{15}\ s^{-1}\left(\frac{1}{1^2} - \frac{1}{2^2}\right) = 0.75000 \times 3.2881\times10^{15}\ s^{-1}$$

$$= 2.4661\times10^{15}\ s^{-1}$$

$$\lambda = c/\nu = \frac{2.9979\times10^{8}\ m\ s^{-1}}{2.4661\times10^{15}\ s^{-1}} = 1.2156\times10^{-7}\ m = 121.56\ nm$$

The minimum wavelength corresponds to n = ∞.

$$\nu = 3.2881\times10^{15}\ s^{-1}\left(\frac{1}{1^2} - \frac{1}{\infty^2}\right) = 3.2881\times10^{15}\ s^{-1}$$

$$\lambda = c/\nu = \frac{2.9979\times10^{8}\ m\ s^{-1}}{3.2881\times10^{15}\ s^{-1}} = 9.1174\times10^{-8}\ m = 91.174\ nm$$

(b) The Lyman series is in the ultraviolet region of the spectrum.

(c) The spectral line at 95.0 nm corresponds to a frequency of

$$\tilde{\nu} = c/\lambda = \frac{3.00\times10^{8}\ m\ s^{-1}}{95.0\ nm \times \frac{1\times10^{-9}\ m}{1\ nm}} = 3.16\times10^{15}\ s^{-1}$$

Now solve for n in the equation.

$$\tilde{\nu} = 3.16\times10^{15}\ s^{-1} = 3.2881\times10^{15}\ s^{-1}\left(\frac{1}{1^2} - \frac{1}{n^2}\right)$$

$$\left(1.000 - \frac{1}{n^2}\right) = \frac{3.16\times10^{15}\ s^{-1}}{3.2881\times10^{15}\ s^{-1}} = 0.961$$

$$\frac{1}{n^2} = 1.000 - 0.961 = 0.039 \qquad n^2 = 25.6 \qquad n = 5$$

(d) One approach is to proceed as in part (c) and determine the value of n corresponding to a hypothetical line at 108.5 nm. If the calculated value of n is an integer, the line exists; but if the required value of n is nonintegral there would be no such line.

$$\tilde{\nu} = \frac{3.00\times10^{8}\ m\ s^{-1}}{108.5\ nm \times \frac{1\times10^{-9}\ m}{1\ nm}} = 2.76\times10^{15}\ s^{-1}$$

$$\left(1.000 - \frac{1}{n^2}\right) = \frac{2.76\times10^{15}\ s^{-1}}{3.2881\times10^{15}\ s^{-1}} = 0.839 \qquad \frac{1}{n^2} = 1.000 - 0.839 = 0.161;\ n^2 = 6.2;\ n = 2.5$$

Since the value of n is nonintegral, there is no line at 108.5 nm.

Quantum theory

7-8. (a) First determine the frequency corresponding to 285 nm.

$\nu = c/\lambda = 3.00\times10^{8}\ ms^{-1}/2.85\times10^{-7}\ m = 1.05\times10^{15}\ s^{-1}$

energy per photon = $E = h\nu = 6.626\times10^{-34}\ J\ s^{-1} \times 1.05\times10^{15}\ s^{-1} = 6.96\times10^{-19}$ J/photon

(b) energy per mol photons = $\frac{6.96\times10^{-19} J}{photon} \times \frac{6.02\times10^{23}\ photon}{1\ mol} = 4.19\times10^{5}$ J/mol = 419 kJ/mol

7-9. Convert from the energy content of 125 kcal/mol to a value of J/photon. Following this, apply the Planck equation, $E = h\nu$, solving for ν. Finally, determine the corresponding wavelength with the expression $c = \nu\lambda$.

$$\text{no. J/photon} = \frac{125\ kcal}{1\ mol} \times \frac{4.184\ kJ}{1\ kcal} \times \frac{1000\ J}{1\ kJ} \times \frac{1\ mol}{6.02\times10^{23}\ photon} = 8.69\times10^{-19}\ \text{J/photon}$$

$$\nu = \frac{E}{h} = \frac{8.69\times10^{-19}\ J}{6.62\times10^{-34}\ Js} = 1.31\times10^{15}\ s^{-1} \qquad \lambda = c/\nu = \frac{3.00\times10^{8}\ m/s}{1.31\times10^{15}\ s^{-1}} = 2.29\times10^{-7}\ m = 229\ nm$$

Light of this wavelength is in the ultraviolet region of the electromagnetic spectrum.

7-10. Convert the wavelength limits λ = 390 nm and λ = 770 nm to frequency limits through the expression $c = \nu\lambda$. Then calculate the energies with Planck's equation, $E = h\nu$.

Short wavelength limit

$$\nu = c/\lambda = \frac{3.00\times10^{8}\ m/s}{390\ nm \times \frac{1\times10^{-9}\ m}{1\ nm}} = 7.69\times10^{14}\ s^{-1}$$

$$E = h\nu = 6.626\times10^{-34}\ Js \times 7.69\times10^{14}\ s^{-1}$$
$$= 5.10\times10^{-19}\ \text{J/photon} \times 6.02\times10^{23}\ \text{photon/mol}$$
$$= 307\ \text{kJ/mol}$$

Long wavelength limit

$$\nu = \frac{3.00\times10^{8}\ m/s}{770\ nm \times \frac{1\times10^{-9}\ m}{1\ nm}} = 3.90\times10^{14}\ s^{-1}$$

$$E = h\nu = 6.626\times10^{-34}\ Js \times 3.90\times10^{14}\ s^{-1}$$
$$= 2.58\times10^{-19}\ \text{J/photon} \times 6.02\times10^{23}\ \text{photon/mol}$$
$$= 155\ \text{kJ/mol}$$

The photoelectric effect

7-11. (a) The energy per photon corresponding to a frequency of $1.3\times10^{15}\ s^{-1}$ is

$E = h\nu = 6.626\times10^{-34}\ Js \times 1.3\times10^{15}\ s^{-1} = 8.6\times10^{-19}$ J

(b) Refer to Exercise 7-10 for the energy limits corresponding to visible light: 2.58×10^{-19} to 5.10×10^{-19} J/photon. The energy calculated in part (a) is greater than 5.10×10^{-19}. The photoelectric effect will not be displayed by platinum with visible light, and certainly not with infrared (which is still less energetic than visible light). Somewhere within the ultraviolet range, platinum will display the photoelectric effect (corresponding to $E = 8.6\times10^{-19}$ J; $\nu = 1.3\times10^{15}\ s^{-1}$; and λ = 231 nm).

7-12. In the photoelectric effect a single photon transfers all its energy to a single electron. Any energy in excess of that required to just secure the release of an electron appears as excess kinetic energy of the ejected electron. A single photon cannot transfer its energy to two different electrons, nor can two different photons cooperatively transfer their energy to a single electron.

The Bohr atom

7-13. (a) Radius of a Bohr orbit, $r = a_o n^2$ (where a_o = 0.53 Å)

For the orbit, n = 6, 0.53 Å × $(6)^2$ = 19 Å

(b) Energy corresponding to n = 6 (use equation 7.5):

$$E = \frac{-B}{n^2} = \frac{-2.179\times10^{-18}\ J}{(6)^2} = -6.053\times10^{-20}\ J$$

7-14. (a) The distance between the Bohr orbits n = 1 and n = 4 is $r_4 - r_1$, where $r_4 = a_o(4)^2$ and $r_1 = a_o(1)^2$.

$$r_4 - r_1 = a_o[(4)^2 - (1)^2] = 15\ a_o = 15\times0.53\ \text{Å} = 8.0\ \text{Å}$$

(b) Similarly, the difference in energy ΔE is $E_4 - E_1$, where $E_4 = -B/(4)^2$ and $E_1 = -B/(1)^2$.

$$\Delta E = \frac{-B}{16} - (-B) = \frac{15B}{16} = \frac{15 \times 2.179\times10^{-18}\ \text{J}}{16} = 2.043\times10^{-18}\ \text{J}$$

7-15. (a) Calculate the energies corresponding to an electron in the Bohr orbit n = 6 and n = 4. Then determine the difference in energy ΔE, which is also equal to $h\nu$.

$$E_4 = \frac{-B}{(4)^2} = \frac{-B}{16} \qquad E_6 = \frac{-B}{(6)^2} = \frac{-B}{36}$$

$$\Delta E = E_4 - E_6 = \frac{-B}{16} - (\frac{-B}{36}) = \frac{-B}{16} + \frac{B}{36} = -0.06250\,B + 0.02778\,B$$

$\Delta E = -0.03472\,B$ (The negative sign signifies that energy is emitted.)

$$\nu = \frac{\Delta E}{h} = \frac{0.03472\times 2.179\times10^{18}\ \text{J}}{6.626\times10^{-34}\ \text{Js}} = 1.142\times10^{14}\ \text{s}^{-1}$$

(b) $\lambda = c/\nu = \dfrac{2.998\times10^{8}\ \text{m/s}}{1.142\times10^{14}\ \text{s}^{-1}} = 2.625\times10^{-6}\ \text{m} = 2.625\ \mu\text{m} = 2.625\times10^{-4}\ \text{cm}$

(c) Radiation of this wavelength is in the infrared.

7-16. Energy must be absorbed if an electron is to be moved from one orbit to another of higher quantum number. This would be the case for (a), (b), and (c), but not for (d), where energy is released ($n = \infty \rightarrow n = 1$). For (a), (b), and (c) determine the energy difference between quantum levels in terms of B, and then compare these differences.

(a) $\Delta E = \dfrac{-B}{(2)^2} - \dfrac{-B}{(1)^2} = B - \dfrac{B}{4} = 3\ B/4$

(b) $\Delta E = \dfrac{-B}{(4)^2} - \dfrac{-B}{(2)^2} = \dfrac{B}{4} - \dfrac{B}{16} = 3\ B/16$

(c) $\Delta E = \dfrac{-B}{(6)^2} \quad \dfrac{-B}{(3)^2} = \dfrac{B}{9} - \dfrac{B}{36} = 3\ B/36$

The greatest energy requirement is for case (a)--an electronic transition from n = 1 to n = 2 (recall also, Figure 7-13).

7-17. The simplest approach is first to convert from 2170 nm to the corresponding ν

$$\nu = c/\lambda = 3.00\times10^{8}\ \text{m s}^{-1}/2.170\times10^{-6}\ \text{m} = 1.38\times10^{14}\ \text{s}^{-1}$$

Then use an expression like (7.7) between two energy levels. The upper level is n = 7 and the lower level is the unknown n.

$$\nu = 1.38\times10^{14}\ \text{s}^{-1} = 3.289\times10^{15}\ \text{s}^{-1}\left(\frac{1}{n^2} - \frac{1}{7^2}\right) = 3.289\times10^{15}\ \text{s}^{-1}\left(\frac{1}{n^2} - 0.0204\right)$$

$$\left(\frac{1}{n^2} - 0.0204\right) = \frac{1.38\times10^{14}\ \text{s}^{-1}}{3.289\times10^{15}\ \text{s}^{-1}} = 0.0420 \qquad \frac{1}{n^2} = 0.0420 + 0.0204 = 0.0624 \qquad n^2 = 16.0 \qquad n = 4$$

7-18. (a) For He^{+}: $E_1 = \dfrac{-(2)^2 \times 2.179\times10^{-18}\ \text{J}}{(1)^2} = -8.716\times10^{-18}\ \text{J}$

(b) For Li^{2+}: $E_3 = \dfrac{-(3)^2 \times 2.179\times10^{-18}\ \text{J}}{(3)^2} = -2.179\times10^{-18}\ \text{J}$

7-19. Lines in the Balmer series correspond to transitions of an electron from a higher level to the level, n = 2. The higher the level from which the electron drops, the shorter the wavelength of the emitted light. Converging lines in the violet region of the Balmer series correspond to transitions from the converging series of energy levels in the energy-level diagram. The more closely spaced these upper energy levels, the more closely spaced the spectral lines.

The uncertainty principle

7-20. The radius of a Bohr orbit is exact ($r = a_o \cdot n^2$); this means a precise statement of the location of the electron. The dynamic properties of an electron, such as velocity, are assumed to have their precise classical values. Thus the Bohr model permits a precise statement of both the position and momentum of an electron. This is in violation of the Heisenberg uncertainty principle.

7-21. Einstein believed strongly in the law of cause and effect. He felt that the need to use probability and chance ("the playing of dice") in a description of atomic structure resulted because a suitable theory had not been developed to permit accurate predictions. He believed that such a theory could be developed, whereas Heisenberg and Bohr argued that it was an inherent law of nature that uncertainty must always exist in the behavior of subatomic particles.

7-22. The quantity $h/2\pi$ has a numerical value of $6.626\times10^{-34}/2\pi = 1\times10^{-34}$. Expressed in kg, the mass of an automobile is about 1×10^3. If the position of an automobile were known to the nearest 0.001 m, and its velocity to the nearest 0.001 m/sec. we would surely say that its behavior was precisely known. Yet, under these conditions, $\Delta x \cdot \Delta p = \Delta x \cdot m \cdot \Delta v = 1\times10^{-3} \times 1\times10^{3} \times 1\times10^{-3} = 1\times10^{-3}$, which is far in excess of 1×10^{-34}. The velocity and position of an automobile could be defined with astonishing precision and still the Heisenberg uncertainty limit would not be reached.

7-23. Start with the statement: $\Delta x \cdot \Delta p > h/2$, where $\Delta p = m\Delta v$. Thus, $\Delta x > h/2\pi \cdot m\Delta v$. Δx is the uncertainty in position of the proton. Convert the unit J to kg $m^2\ s^{-2}$. The uncertainty in velocity of the proton is 1% of 0.1 c = 0.001 c = 3.00×10^5 m/s.

$$\Delta x > \frac{6.626\times10^{-34}\ \text{kg m}^2\ \text{s}^{-2}\ \text{s}}{2\pi \times 1.67\times10^{-24}\ \text{g} \times \frac{1\ \text{kg}}{1000\ \text{g}} \times 3.00\times10^5\ \text{m/s}} > 2.10\times10^{-13}\ \text{m}$$

Wave-particle duality

7-24. The deBroglie wavelength is $\lambda = h/p = h/mg$; h is Planck's constant. If the electron and proton are to have the same λ, they must have the same momentum ($p = mv$). Since this electron mass is so much smaller than the proton mass, the electron would have to possess a greater velocity to have the same momentum as a proton.

7-25. The deBroglie equation, $\lambda = h/mv$, must be solved for v. The wavelength, λ, must be expressed in m (1 nm = 1×10^{-9} m). The joule = 1 kg $m^2\ s^{-2}$.

$$v = \frac{h}{m\lambda} = \frac{6.626\times10^{-34}\ \text{kg m}^2\ \text{s}^{-1}}{9.110\times10^{-28}\ \text{g} \times \frac{1\ \text{kg}}{1000\ \text{g}} \times 1\times10^{-9}\ \text{m}} = 7\times10^5\ \text{m/s}$$

Wave mechanics

7-26.

	Bohr Theory	*Wave Mechanics*
(a)	The electron may be found only in one of a fixed set of closed circular orbits.	The electron may be found anywhere outside the nucleus, but there are certain regions in which the highest probabilities exist; these are called orbitals.
(b)	The electron orbit is a planar figure--a circle.	Electron orbitals correspond to three-dimensional regions--spherical shells, dumbbell-shaped regions, and so on.
(c)	The position and velocity of the electron can be described with a high degree of certainty.	Certainty is replaced by probability in describing the position and velocity of the electron.

The principal similarity is that the radii of the Bohr orbits do correspond to the distances from the nucleus at which there is a high probability of finding an electron.

7-27. For any *single* point the probability of finding an electron in a 1s orbital is greatest at the nucleus of the atom. At greater distances from the nucleus, however, there are many points equivalent to one another. Although the probability at any one of these points is considerably less than at the nucleus, the total probability is based on the sum of the probabilities at equivalent points. Thus, the region of greatest probability becomes a spherical shell at some distance from the nucleus, in fact, having a radius equal to the first Bohr orbit.

Quantum numbers and electron orbitals

7-28. Whether the spin quantum number, m_s, is +1/2 or -1/2 does not depend on the values of the other quantum numbers. The magnetic quantum number, m_l, cannot exceed the value of the orbital quantum number, l. Therefore, if $m_l = 2$, $l \neq 1$. Neither can $l = 0$. The only allowable value of $l = 2$. The correct answer is (d).

7-29.

principal shell	N															
n =	4	4	4	4	4	4	4	4	4	4	4	4	4	4	4	4
l =	0	1	1	1	2	2	2	2	2	3	3	3	3	3	3	3
m_l =	0	-1	0	+1	-2	-1	0	+1	+2	-3	-2	-1	0	+1	+2	+3
orbital designation:	4*s*	4*p*	4*p*	4*p*	4*d*	4*d*	4*d*	4*d*	4*d*	4*f*	4*f*	4*f*	4*f*	4*f*	4*f*	4*f*
number of orbitals in subshell:	1		3				5						7			
total number of orbitals:								16								

7-30. Apply the statements describing the relationships among quantum numbers given in Section 7-8 of the textbook in each of the following cases:

(a) All the required relationships are met by the combination, $n = 2$, $l = 1$, $m_l = 0$. This is an allowable set.

(b) This set is not allowable; l cannot be equal to n. Thus, if $n = 2$, $l = 1$ or $l = 0$.

(c) This is an allowable set. All the rules are obeyed.

(d) This is also an allowable set.

(e) This set is not allowable. If $l = 0$, m_l must also be equal to zero.

(f) This set is not allowable. The value of $l(3)$ is not permitted to exceed that of $n(2)$.

7-31. (a) $l = 1$ specifies a *p* orbital; $n = 2$, the second principal shell. The orbital is a 2*p* orbital.

(b) $l = 2$ specifies a *d* orbital; $n = 4$, the fourth principal shell. The orbital is a 4*d* orbital.

(c) $l = 0$ specifies a *s* orbital; $n = 5$, the fifth principal shell. The orbital is a 5*s* orbital.

7-32. The l quantum number designations establish the orbital type (s, p, d, f) and the n quantum numbers, the principal electronic shell in which the orbital is found. 3*s*: $n = 3$, $l = 0$; 4*p*: $n = 4$, $l = 1$; 5*f*: $n = 5$, $l = 3$; 6*d*: $n = 6$, $l = 2$.

7-33. For each orbital type establish the value of l. The number of possibilities for m_l and, therefore, the number of possible orbitals of the given type is $2l + 1$.

(a) 2*s*: $l = 0$, $2l + 1 = 1$; there is one 2*s* orbital.

(b) 3*f*: *f* orbitals require that $l = 3$, but $l \neq 3$ if $n = 3$. There can be no 3*f* orbitals.

(c) 4*p*: $l = 1$, $2l + 1 = 3$; there are three 4*p* orbitals.

(d) 5*d*: $l = 2$, $2l + 1 = 5$; there are five 5*d* orbitals.

7-34. For an electron with $n = 4$ and $m_l = -2$,

(a) Correct: The principal shell number is determined by n.

(b) Correct: The electron *may* be associated with a 4*d* orbital. It will if $l = 2$. However, if $l = 3$ and $m_l = -2$, this will correspond to a 4*f* orbital.

(c) Incorrect: The m_l quantum numbers associated with *p* orbitals are -1, 0, and +1.

(d) Incorrect: The spin quantum number does not depend on the values of n and m_l. That is, m_s may either be +1/2 or -1/2.

Electron configurations

7-35. Probably the most direct approach is to convert each set of quantum numbers to an orbital designation, and then arrange the orbitals in accordance with expression (7.13) and Figure 7-24.

(a) *5s*; (b) *3p*; (c) *3d*; (d) *3d*; (e) *3s*

order of increasing energy: *3s* < *3p* < *3d* = *3d* < *3s*
(e) < (b) < (c) = (d) < (a)

7-36. (a) In the configuration $2s^3$ all electrons would have $n = 2$, $l = 0$, $m_l = 0$. Since there are only two possible values of the m_s quantum number, two of the three electrons would have the same value of m_s; and, therefore, all four quantum numbers alike. This would isolate the Pauli exclustion principle. The correct configuration is B: $1s^22s^22p^1$.

(b) A 2d electron would have $n = 2$ and $l = 2$; but if $n = 2$, l can only be 0 or 1. There can be no 2d electrons. The correct electron configuration is Na: $1s^22s^22p^63s^1$.

(c) In K the 4s subshell receives an electron rather than the 3d. The correct configuration is K: $[Ar]4s^1$.

(d) After the 4s subshell begins to fill the next subshell to do so is the 3d. The correct configuration is Ti: $[Ar]3d^24s^2$.

(e) Following Kr it is the 4d subshell, not the 5d, that fills is a series of transition elements. The configuration of Xe is $[Kr]4d^{10}5s^25p^6$.

(f) The 4f subshell has a capacity of 14 electrons, not 10. There are no 6p electrons in Hg. The correct electron configuration is Hg: $[Xe]4f^{14}5d^{10}6s^2$.

7-37. (a) In this configuration both electrons in the 3s orbital have the same spin. This isolates the Pauli exclusion principle.

(b) The unpaired electrons in the 3p subshell should all have the same spin.

(c) This configuration is correct.

(d) This configuration, by placing an electron pair in one 3p orbital while leaving one 3p orbital empty, violates Hund's rule of maximum multiplicity.

7-38. (a) The electron configuration of Si is $1s^22s^22p^63s^23p^2$. The <u>two</u> electrons in the 3p subshells are unpaired.

(b) The electron configuration of S is $1s^22s^22p^63s^23p^4$. Sulfur has <u>no</u> 3d electrons.

(c) The electron configuration of As is $1s^22s^22p^63s^23p^63d^{10}4s^24p^3$. As has <u>three</u> 4p electrons.

(d) The filling of the 3s subshell occurs with Na (Z = 11) and Mg (Z = 12). All elements with atomic number greater than 12 (including Sr) have <u>two</u> 3s electrons.

(e) Au follows the series of elements (lanthanoids) in which the 4f subshell is filled. Therefore, Au has <u>14</u> 4f electrons.

7-39. (a) The normal configuration for four electrons would be $1s^22s^2$. One of the 2s electrons has been promoted to an empty 2p orbital in the configuration shown.

(b) The element in question has Z = 16 (sulfur); its normal electron configuration is $[Ne]3s^23p^4$. The configuration shown involves the promotion of two 3p electrons to 3d orbitals.

(c) Although the 4s orbital may have only a single electron for some atoms with a partially filled 3d subshell, once the 3d subshell is closed (with 10 electrons) the 4s must also be closed (with 2 electrons) before the filling of the 4p begins. The configuration shown involves the promotion of a 4s electron to a 4p orbital.

7-40. (a) We must assign 55 electrons into the usual orbitals but increase the capacity of each orbital to three electrons, that is, $1s^3$, $2s^3$, and so on.

$1s^32s^32p^93s^33p^93d^{15}4s^34p^95s^1$

(b) The effect of permitting $l = n$ is that of adding a *p* orbital to the first shell, *d* to the second, *f* to the third, and so on. Under these conditions the order of filling might be:

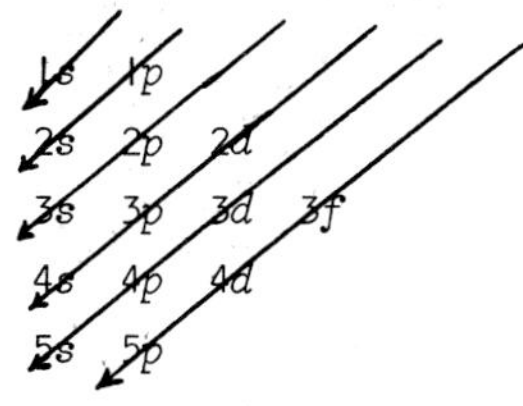

We must assign 55 electrons according to the order suggested above and observe all the normal rules of electron configurations while doing so.

$1s^2 1p^6 2s^2 2p^6 2d^{10} 3s^2 3p^6 3d^{10} 3f^1 4s^2 4p^6 5s^2$

Self-test Questions

1. (*b*) Convert all wavelengths to a common unit, e.g., meters, and choose the shortest.

   (a) 735 nm x (1×10^{-9} m/1 nm) = 7.35×10^{-7} m

   (b) 6.3×10^{-5} cm x (1×10^{-2} m/1 cm) = 6.3×10^{-7} m

   (c) 1.05 μm x (1×10^{-6} m/1 μm) = 1.05×10^{-6} m

   (d) 3.5×10^{-6} m

2. (*a*) The shorter the wavelength the higher the frequency of electromagnetic radiation. Radiation of 200 nm wavelength must have a higher frequency than 400 nm radiation. Item (b) is incorrect because 200 nm radiation is in the ultraviolet region; (c) is incorrect because the speed of light in vacuum is independent of wavelength; (d) is incorrect because 100 nm radiation has a greater frequency, and therefore, a higher energy content than 200 nm radiation.

3. (*d*) An electron cannot have its l quantum number equal to its n quantum number. If $n = 2$, l can only be equal to 1 or 0, not 2. The set of quantum numbers $n = 2$, $l = 2$, $m_l = 0$ is not allowed.

4. (*b*) If $n = 3$, the quantum number m_l may be 0, but it is not required to be. Since the maximum value of l is $n - 1$, in this case the maximum value of l is 2. The quantum number m_s does not depend on the value of n, l, or m_l. Therefore, may be either +1/2 or -1/2. With $n = 3$, the possible values of m_l are -3, -2, -1, 0, +1, +2, or +3--seven values in all.

5. (*b*) For a 5d orbital, $n = 5$ and $l = 2$. The maximum value of m_l is 2, not 5. The quantum numbers +1/2 and -1/2 describe m_s, not m_l. The only possible value listed for m_l is 3.

6. (*d*) The electron configuration of Cl is $1s^2 2s^2 2p^6 3s^2 3p^5$. The number of 2p electrons is *six*. (Do not confuse the 2p electrons with the 3p electrons, of which there are five.)

7. (*c*) Because scandium has an odd atomic number (Z = 21) there must be at least one unpaired electron. The electron configuration of Ca (Z = 20) has all electrons paired, $1s^2 2s^2 2p^6 3s^2 3p^6 4s^2$. The additional electron in building up Sc from Ca goes into a 3d orbital, unpaired.

8. (*a*) Only (a) and (d) correspond to light emission. Items (b) and (c) represent energy absorption. Transactions of electrons to the orbit n = 3 represent emission (and longer wavelengths) than transitions to n = 1.

9. Use the Planck equation E = hν to determine the energy per photon. Then convert to a mole basis.

$$E = h\nu = 6.626 \times 10^{-34} \text{ Js} \times 4.00 \times 10^{14} \text{ s}^{-1} = 2.65 \times 10^{19} \text{ J/photon}$$

$$\text{kJ/mol} = \frac{2.65 \times 10^{-19} \text{ J}}{\text{photon}} \times \frac{6.02 \times 10^{23} \text{ photons}}{1 \text{ mol}} \times \frac{1 \text{ kJ}}{1000 \text{ J}} = 1.60 \times 10^{2} \text{ kJ/mol}$$

10. The line at the shorter wavelength, 589.0 nm, has a higher frequency and higher energy content than the longer wavelength line. To obtain the *difference* in energy between the two, determine the frequencies for these lines and then use the Planck equation.

589.0 nm line

$$\nu = c/\lambda = \frac{3.00 \times 10^{8} \text{ m/s}}{589.0 \times 10^{-9} \text{ m}}$$

589.6 nm line

$$\nu = \frac{3.00 \times 10^{8} \text{ m/s}}{589.6 \times 10^{-9} \text{ m}}$$

$$\Delta E = (h\nu)_1 - (h\nu)_2 = 6.626\times10^{-34}\ \text{Js} \times 3.00\times10^{8}\ \text{m/s} \times \left(\frac{1}{589.0\times10^{-9}} - \frac{1}{589.6\times10^{-9}} \right) \text{m}^{-1}$$

$$\Delta E = 6.626\times10^{-34} \times 3.00\times10^{8} \times (1.698\times10^{6} - 1.696\times10^{6})\text{J}$$

$$\Delta E = 4\times10^{-22}\ \text{J}$$

11. Find the frequency corresponding to 434 nm; substitute this into the equation given; and solve for n.

$$\nu = c/\lambda = 3.00\times10^{8}\ \text{m s}^{-1}/434\times10^{-9}\ \text{m} = 6.91\times10^{14}\ \text{s}^{-1}$$

$$6.91\times10^{14}\ \text{s}^{-1} = 3.2881\times10^{15}\ \text{s}^{-1}\left(\frac{1}{2^2} - \frac{1}{n^2}\right)$$

$$\left(\frac{1}{4} - \frac{1}{n^2}\right) = 6.91\times10^{14}/3.2881\times10^{15} = 0.210$$

$$0.250 - \frac{1}{n^2} = 0.210 \qquad \frac{1}{n^2} = 0.250 - 0.210 = 0.040 \qquad n^2 = \frac{1}{0.040} \qquad n^2 = 25 \qquad n = 5$$

12. (a) Se: $1s^2 2s^2 2p^6 3s^2 3p^6 3d^{10} 4s^2 4p^4$

(b) I: [Kr] 4d [↑↓|↑↓|↑↓|↑↓|↑↓] 5s [↑↓] 5p [↑↓|↑↓|↓]

Chapter 8

Atomic Properties and the Periodic Table

Review Problems

8-1. (a) The group IIIA elements are B, Al, Ga, In and Tl. The one that is also in the fifth period (the period beginning with Rb) is indium, In.

(b) Elements similar to sulfur would also be in group VIA of the periodic table, e.g., Se and Te. Most other elements are unlike sulfur, e.g., all the metallic elements.

(c) The most active metals are those in groups IA and IIA. The metal(s) in question would be in one of these groups and also in the sixth period, i.e., Cs or Ba.

(d) The halogen element in the fifth period is the element that is both in the fifth period and in groups VIIA--iodine.

(e) Element 18 is argon, a noble gas. The other noble gases with $Z > 50$ are xenon ($Z = 54$) and radon ($Z = 86$).

8-2. (a) This electron configuration has seven electrons in the shell of highest principal quantum number. The element is in group VIIA (chlorine).

(b) This electron configuration corresponds to group IVA (the element is Ge).

(c) This electron configuration, with a closed s subshell in the electronic shell of highest principal quantum number, corresponds to group IIA (Mg).

(d) This element comes at the end of the first transition series, where the 3d subshell is filled. The element is Cu and it is in group IB.

(e) This configuration is that of a transition element (partial filling of a d subshell). Since the 5d subshell is the one being filled, this element is in the third transition series (following the lanthanoid elements). The atomic number of the element is 74; the element is tungsten, and its group number is VIB.

8-3. (a) In: $[Kr]4d^{10}5s^25p^1$ (group IIIA)

(b) Y: $[Kr]4d^15s^2$ (note similarity to Sc)

(c) Sb: $[Kr]4d^{10}5s^25p^3$ (group VA)

(d) Au: $[Xe]4f^{14}5d^{10}6s^1$ (follows the lanthanoid series and is added to the end of the third transition series)

8-4. In each case start with the electron configuration of the neutral atom and then add or remove the requisite number of electrons.

(a) Rb^+: $1s^22s^22p^63s^23p^63d^{10}4s^24p^6$ *or* $[Ar]3d^{10}4s^24p^6$ *or* $[Kr]$

(b) Br^-: $1s^22s^22p^63s^23p^63d^{10}4s^24p^6$ *or* $[Ar]3d^{10}4s^24p^6$ *or* $[Kr]$

(c) O^{2-}: $1s^22s^22p^6$ *or* $[Ne]$

(d) Ba^{2+}: $1s^22s^22p^63s^23p^63d^{10}4s^24p^64d^{10}5s^25p^6$ *or* $[Kr]4d^{10}5s^25p^6$ or $[Xe]$

(e) Zn^{2+}: $1s^22s^22p^63s^23p^63d^{10}$ *or* $[Ar]3d^{10}$

(f) Ag^+: $1s^22s^22p^63s^23p^63d^{10}4s^24p^64d^{10}$ *or* $[Kr]4d^{10}$

(g) Bi^{3+}: $1s^22s^22p^63s^23p^63d^{10}4s^24p^64d^{10}4f^{14}5s^25p^65d^{10}6s^2$ *or* $[Xe]4f^{14}5d^{10}6s^2$

8-5. (a) 1: K is in group IA. Its atoms have one outershell *s* electron.

(b) 5: I is in group VIIA and the fifth period. Its outershell electron configuration must be $5s^25p^5$.

(c) 10: Zn comes at the end of the first transition series, which features the filling of the 3d subshell.

(d) 6: S is in the third period. The 2p subshell is filled in the second period. Every atom with atomic number 10 (Ne) or larger has its 2p subshell filled.

(e) 14: Pb follows the lanthanoid and third transition series of elements. Its atoms have a filled 4f subshell.

(f) 8: Ni is in the first transition series, in which the 3d level is being filled. We know this much of its configuration $1s^2 2s^2 2p^6 3s^2 3p^6 3d^{(?)} 4s^2$. The atomic number of Ni is 28. The other subshells account for 20 electrons, leaving 8 for the 3d.

8-6. (a) As is larger than Br. It is further left in the fourth period.

(b) Sr is larger than Mg. It is further down group IIA.

(c) Cs is larger than Ca, both because it has more electronic shells (six compared to four) and because it is further to the left in the periodic table.

(d) Xe is larger than Ne. Both are noble gases but Xe has five electronic shells, whereas Ne has only two.

(e) C is larger than O. Both are in the second period, but C is situated further to the left.

(f) Hg is larger than Cl, both because it has more electronic shells and because it is located further to the left in the periodic table.

8-7. Here are the species that are being compared: Al, Ar, As, Cs^+, F, I^-, and N. We are not attempting to arrange them in order but only to find the smallest and the largest. Consider I^-. I^- is isoelectric with Xe and larger than Xe because I^- carries a net negative charge. Since Xe is larger than Ar then I^- must be larger than Ar. Cs^+ is also isoelectric with Xe, but it is smaller than Xe. I^- is the largest species. Two of the atoms, N and F, are in the second period. For all the others the period numbers are larger than two. F is found further to the right than N. F is the smallest of the group.

8-8. Use the relationship summarized in Figure 8-10 to note that I_1 for Sr must be greater than I_1 for Cs. (Sr is further to the right and higher in its group of the periodic table.) I_1 for F must be greater than I_1 for S (again because it is farther to the right and higher in its group). And so on. The order obtained is Cs < Sr < As < S < F.

8-9. From Table 8-5 the sum of I_1 and I_2 for Mg is 737.7 + 1451 = 2189 kJ/mol.

$$\text{no. J} = 1.86\times10^{-6} \text{ mol } Mg^{2+} \times \frac{2189 \text{ kJ}}{\text{mol } Mg^{2+}} \times \frac{1000 \text{ J}}{1 \text{ kJ}} = 4.07 \text{ J}$$

8-10. Again (as in Review Problem 8) use relationships summarized in Figure 8-10.

(a) Only two of the five elements are active metals--Mg and Ba. Since Ba is the larger atom (farther down the group), it is more metallic than Mg and the most metallic of the five elements.

(b) Only one of the five elements is a clear-cut nonmetal--S. Arsenic is a metalloid and Bi has some metallic character.

(c) From what has been said in parts (a) and (b), the lowest electronegativities are those of Ba and Mg, and the highest, those of S and As. The intermediate electronegativeity is that of Bi.

8-11. First the group into metals: Sc, Fe, Rb, Ca; nonmetals: Br, O, F; and metalloid: Te. Now arrange the metals in decreasing metallic order: Rb > Ca > Sc > Fe; the nonmetals in increasing nonmetallic order: Br < O < F. The metalloid falls between the two groupings, and the overall order of decreasing metallic character is Rb > Ca > Se > Fe > Te > Br > O > F.

8-12. K^+ is isoelectronic with Ar; it must be diamagnetic. Cr^{3+} has 21 electrons; this is an odd number of electrons and Cr^{3+} must be paramagnetic. Zn^{2+} has the electron configuration $1s^2 2s^2 2p^6 3s^2 3p^6 3d^{10}$; all electrons are paired and the ion is diamagnetic. Cd has the electron configuration $[Kr]4d^{10}5s^2$; all electrons are paired and the atom is diamagnetic. Co^{3+} contains 24 electrons and we cannot be certain whether the ion is diamagnetic or paramagnetic from this information alone. However, if we write out the electron configurations for Co, $[Ar]3d^7 4s^2$, and Co^{3+}, $[Ar]3d^6$, we see that there are 6 electrons in the five 3d orbitals. Two of the six are paired and four are unpaired; Co^{3+} is paramagnetic. The electron configuration of Sn^{2+} is $[Kr]3d^{10}4s^2$. All electrons are paired and the ion is diamagnetic. The atomic number of Br is 35, and from this part alone we can conclude that the atom is paramagnetic. (The electron configuration, $[Ar]3d^{10}4s^2 4p^5$, shows one unpaired electron in a 4p orbital.)

Exercises

The periodic law

8-1. Use Figure 8-1 to estimate the atomic volume of Fr (Z = 87)--say 85 cm^3/mol.

$$d = \frac{\text{no. g}}{cm^3} = \frac{\text{at. wt.}}{\text{at. vol.}} = \frac{223\ g/mol}{85\ cm^3/mol} = 2.6\ g/cm^3$$

8-2. Determine the atomic volumes of elements similar to Z = 114 (Group IVA) from Figure 8-1: Ge, 15 cm^3/mol; Sn, 18 cm^3/mol; Pb, 19 cm^3/mol. From this trend we would expect the atomic volume of A = 114 to be about 20 cm^3/mol. The expected density is

$$d = \frac{\text{at. wt.}}{\text{at. vol.}} = \frac{298\ g/mol}{20\ cm^3/\ mol} = 15\ g/cm^3$$

8-3. Assign atomic numbers to each of the listed elements and plot melting points against atomic numbers. The melting points are seen to rise to a maximum with the Group IVA element, drop to comparatively low values in Groups VIA and VIIA, and reach a minimum for Group 0 (noble gases). Because no data are given for transition elements, generalizations encompassing the entire range of atomic numbers cannot be made.

The periodic table

8-4. To product additional elements within the body of the periodic table would require adding protons in *fractions*, but this would be contrary to all experience--atomic numbers must be integers. Since all the integral atomic numbers from 1 to 106 have been assigned to known elements, there appears to be no possibility of discovering new elements with atomic numbers less than 106.

8-5. The expected atomic number of the noble gas following radon is 86 + 32 = 118, and of the alkali metal following francium, 87 + 32 = 119. Both of these elements would have atomic weights of about 310.

8-6. These are the inversions found in the periodic table: Ar-K, Co-Ni, Te-I. The periodic table is based on electron configurations, and these in turn are related to *atomic numbers*. Therefore, the arrangement of elements in the periodic table must be by increasing atomic number, even if in a few cases this results in an inverse order by atomic weight.

8-7. (a) The predicted formula from Mendeleev's table is Al_2O_3, and this is the same formula predicted from Table 3-2.

(b) The formula predicted from Mendeleev's table is SO_3. Although SO_3 is a well-known oxide of sulfur, another oxide of sulfur that we have encountered (see text, page 64) is SO_2, and this is not predicted by Mendeleev's table.

(c) The predicted formula is $SiCl_4$, another well-known compound.

(d) If we consider that H can be replaced by Cl in the formula RH_3, Mendeleev's table predicts PCl_3. [Another compound of phosphorus and chlorine is PCl_5 (see Review Problem 3-10g).]

(e) From Mendeleev's table the oxide of iron is predicted to be FeO_4, but from the tables in Chapter 3 we see that the expected formulas are FeO and Fe_2O_3.

Periodic table and electron configurations

8-8. The length of a period is determined by the number of elements that intervene before a similar electron configuration is attained. Thus, following Li ($1s^22s^1$) the next element with one electron is an s orbital of the outer shell is Na ($1s^22s^22p^63s^1$). The length of the second period is 8. K ($1s^22s^22p^63s^23p^64s^1$) is also eight elements removed from its earlier family member, Na; the third period is eight members long. But before the $5s$ orbital acquires an electron, the filling of $4s$, $3d$, and $4p$ must occur. This required 18 electrons; the fourth period is 18 members long, and so is the fifth. In the sixth period the orbitals that must be filled are $6s$, $4f$, $5d$, and $6p$, making the period 32 members long.

8-9. To accomodate all the known elements within the body of the periodic table would require a width of 32 elements, corresponding to the filling of $6s$, $4f$, $5d$, and $6p$ orbitals in the sixth period and $7s$, $5f$, $6d$, and $7p$, in the seventh.

H																															He
Li	Be																									B	C	N	O	F	Ne
Na	Mg																									Al	Si	P	S	Cl	Ar
K	Ca	Sc															Ti	V	Cr	Mn	Fe	Co	Ni	Cu	Zn	Ga	Ge	As	Se	Br	Kr
Rb	Sr	Y															Zr	Nb	Mo	Tc	Ru	Rh	Pd	Ag	Cd	In	Sn	Sb	Te	I	Xe
Cs	Ba	La	Ce	Pr	Nd	Pm	Sm	Eu	Gd	Tb	Dy	Ho	Er	Tm	Yb	Lu	Hf	Ta	W	Re	Os	Ir	Pt	Au	Hg	Tl	Pb	Ri	Po	At	Rn
Fr	Ra	Ac	Th	Pa	U	Np	Pu	Am	Cm	Bk	Cf	Es	Fm	Md	No	Lr	Ku	Ha													

8-10. (a) The elements in question have atomic numbers 89, 104, and 105.

(b) One of these elements, Ac, comes before the inner transition series in which $5f$ orbitals are filled. Its electron configuration is expected to resemble that of La, which is in the period preceding it. The other two elements, Z = 104 and Z = 105, follow the actinide series. Their $5f$ orbitals are filled. The differentiating electrons in these atoms are expected to go into $6d$ orbitals, and the elements are expected to resemble Hf and Ta.

8-11. (a) 5: There are five outer-shell electrons in a Sb atom (Group VA): $5s^2 5p^3$.

(b) 6: Platinum is in the sixth period of elements; its atoms contain electrons in six principal electronic shells.

(c) 5: These are the five elements of Group VIA--O, S, Se, Te, Po.

(d) 2: Tellurium is in periodic Group VIA and in the fifth period. Its outer-shell electron configuration is $5s^2 5p^4$. The two $5s$ electrons are paired; one of the three $5p$ orbitals is doubly occupied and two are singly occupied by unpaired electrons.

(e) 24: The sixth period (from Ba to Xe) includes an inner transition series of 14 elements, in which the $4f$ subshell fills, and a regular transition series of 10 elements, in which the $5d$ subshell fills.

8-12. (a) Starting with the electron configuration of Xe (Z = 54), for Pb (Z = 82) we may write Pb: $[Xe]4f^{14}5d^{10}6s^2 6p^2$

(b) By applying the rules for electron configurations from Chapter 7, we might write for element, Z = 114

Z = 114: $[Xe]4f^{14}5d^{10}6s^2 6p^6 7s^2 5f^{14}6d^{10}7p^2$ <u>or</u>

Z = 114: $[Rn]5f^{14}6d^{10}7s^2 7p^2$

Note the similarity of this electron configuration to that for Pb.

8-13. Modify the electron configuration by adding or removing electrons to produce an ion of the appropriate charge. In most cases, the affected orbitals are s and p, but in one instance (d) loss of an electron from a d orbital is involved as well.

(a) Sr^{2+}: $[Ar]3d^{10}4s^2 4p^6$

(b) Y^{3+}: same as (a)

(c) Se^{2-}: same as (a)

(d) Cu^{2+}: $[Ar]3d^9$

(e) Ni^{2+}: $[Ar]3d^8$

(f) Ga^{3+}: $[Ar]3d^{10}$

(g) Ti^{2+}: $[Ar]3d^2$

Atomic sizes

8-14. The basic reason why the sizes of atoms do not increase uniformly with atomic number stems from the electronic shell structure of atoms. If electrons are added to the same outermost shell while protons are added to the nucleus, the increased nuclear charge results in a greater attraction for outer-shell electrons and a decrease in size (moving from left to right across a period in the periodic table). If electrons enter an inner electronic shell (transition elements) there is little effect on atomic size in moving from one atom to the next. If an electron enters a new electronic shell (as in moving from a noble gas atom to an alkali metal atom), there is a large increase in size.

8-15. (a) The smallest atoms are located at the top of the periodic groups. The smallest atom in Group IIIA is boron (B).

(b) All of the atoms except Po are in the fifth period; Po is in the sixth. Within a period, atoms decrease in size from left to right. Of the elements listed, Te is farthest to the right. Po is in the same group as Te, but farther toward the bottom. Te should be the smallest of the atoms listed.

8-16. We should expect the hydrogen ion, H^+, to be the smallest of atomic species, since it consists of a lone proton. Certainly H^+ should be smaller than He, which has one shell with two electrons. The hydride ion, H^-, should be larger than the He atom. Although H^- and He both have two electrons ($1s^2$), in H^- there is only one proton to offset the negative charge; in He there are two protons. Repulsion between the electrons is more significant in H^- than in He, resulting in a larger atomic size.

8-17. The basic principle in the following arrangement of atomic sizes is that if the number of protons exceeds the number of electrons (positive ion) the atomic size is smaller than in the corresponding neutral species (in this case, the noble gas, Kr). If the number of electrons exceeds the number of protons (negative ion), the atomic size is larger. Our comparisons then must be based on the noble gas Kr.

$Y^{3+} < Sr^{2+} < Rb^+ < Kr < Br^- < Se^{2-}$

8-18. Consider Li^+ and Y. Li^+ is in the second period and Y is in the fifth. We expect a large decrease in size when an Li atom becomes an Li^+ ion. With respect to their atomic sizes $Li^+ < Y$. Now consider Se and Br. The neutral atoms should be similar in size, but there is a large increase in size when Br is converted to Br^- (i.e., comparing the covalent and ionic radii of Br). Therefore, with respect to their atomic sizes, $Se < Br^-$. Based on the relationship of atomic size to position in a period of elements, we should expect Se < Y. The expected order of increasing size is $Li^+ < Se < Y < Br^-$. A further check on this prediction is possible by referring to Figures 8-5 and 8-7.

8-19. Begin by comparing certain species two or three at a time. Then combine these small clusters of comparisons into a single grand comparison. For example, the Br^- ion is larger than the Br atom, which in turn is larger than the Cl atom. This leads to the grouping $Cl < Br < Br^-$. Comparing B and Li^+ we would conclude that Li^+, being isoelectronic with He and having just one shell of electrons ($1s^2$), is smaller than B (which has two electronic shells ($1s^22s^22p^1$): $Li^+ < B$. Whether the boron atom is larger or smaller than the chlorine atom is a point of uncertainty. This leads to $Li^+ < B ? Cl < Br < Br^-$. Certainly we expect P to be slightly larger than Cl, but we cannot immediately say whether it is larger or smaller than Br. This leads to the order: $Li^+ < B ? Cl < P ? Br < Br^-$. Rb begins the fifth period and we expect it to be larger than Kr; but whether it is also larger than Br^- is less certain. However, we can say that Y is smaller than Rb and that both are larger than Br: Br < Y < Rb. We can put all this information together into the order: $Li^+ < B ? Cl < P ? Br < Y < Rb ? Br^-$. Now let us assess these sizes in terms of data from the chapter. From Figure 8-5 we see that B is somewhat smaller than Cl and that P and Br are very nearly equal in size. The atomic radius of Rb is about 215 pm and that of Y, about 160 pm. From Figure 8-7, the radius of Br^- is 195 pm. The order of increasing size then is $Li^+ < B < Cl < P < Br < Y < Br^- < Rb$.

8-20. If we divide the molar volume by Avogadro's number we obtain the volume associated with a single atom, but this cannot be the actual volume of the atom. If the atoms are taken to be spherical in shape there must exist empty space among the atoms in the bulk solid. The calculated volume per atom does not take into account this empty space. (The calculation treats the atoms as if they were cubes rather than spheres.) The true volume of an atom is less than this calculated volume.

Ionization energies, electron affinities

8-21. I_2 measures the energy required to move an electron away from a species that acquires a +2 charge; I_1, from a species that acquires a +1 charge. For any given element $I_2 > I_1$.

8-22. The first electron lost by a Na atom is a 3s electron. The same is true for Mg. However, because of its larger size, this 3s electron is extracted more easily from Na than from Mg. That is, I_1 (Na) < I_1 (Mg). The second electron lost by a Mg atom is also a 3s electron, whereas that lost by Na must now come from the extremely stable Ne core ($2s^22p^6$). As a result, I_2 (Na) > I_2 (Mg).

8-23. In each case the ionization energy is that required to remove an electron from an especially stable electron configuration, that of the noble gas Ne, $1s^22s^22p^6$. With Ne the electron is moved away from a unipositive ion, Ne^+; the required energy is I_1 (Ne). With sodium, removal of the electron converts Na^+ to Na^{2+}. To move an electron away from a dipositive ion requires more energy than from a unipositive ion. Thus, I_2 (Na) > I_1 (Ne).

8-24. The total energy requirement to remove all five outershell electrons from P atoms is (from Table 8-5): 1012 + 1903 + 2912 + 4957 + 6274 = 17,058 kJ/mol P. To determine the energy requirement for a <u>single</u> P atom in the unit eV, we must use equation (8.4) in a conversion factor.

$$\text{no. eV/P atom} = \frac{17{,}058 \text{ kJ}}{\text{mol P}} \times \frac{1 \text{ eV/P atom}}{96.49 \text{ kJ/mol P}} = 176.8 \text{ eV/P atom}$$

8-25. Here we form a conversion factor from the first ionization energy of Cs (see Table 8-4).

$$\text{no. } Cs^+ \text{ ions} = 1.000 \text{ J} \times \frac{1 \text{ kJ}}{1000 \text{ J}} \times \frac{1 \text{ mol Cs}}{375.7 \text{ kJ}} \times \frac{6.022\times10^{23} \text{ } Cs^+ \text{ ions}}{1 \text{ mol Cs}} = 1.603\times10^{18} \text{ } Cs^+ \text{ ions}$$

8-26. Convert 1.00 mg Cl to mol Cl and use the electron affinity from equation (8.5) as a conversion factor.

$$\text{no. J} = 1.00 \text{ mg Cl} \times \frac{1 \text{ g Cl}}{1000 \text{ mg Cl}} \times \frac{1 \text{ mol Cl}}{35.453 \text{ g Cl}} \times \frac{-348.8 \text{ kJ}}{1 \text{ mol Cl}} \times \frac{1000 \text{ J}}{1 \text{ kJ}} = -9.84 \text{ J}$$

8-27. From Figure 7-13, the ionization energy of H is $E_\infty - E_1 = 0 - \frac{-B}{(1)^2} = +B = +2.179\times10^{-18}$ J

$$\text{no. eV/atom} = \frac{2.179\times10^{+18} \text{ J}}{\text{atom}} \times \frac{1 \text{ kJ}}{1000 \text{ J}} \times \frac{6.022\times10^{23} \text{ H atoms}}{1 \text{ mol H}} \times \frac{1 \text{ eV/H atom}}{96.49 \text{ kJ/mol H}} = 13.60 \text{ eV/H atom}$$

8-28. Neither I_2 for Ba nor I_3 for Sc should be unusually large because in each case this is the final ionization leading to a noble gas electron configuration. I_2 for Ba should be smaller than I_3 for Sc because Ba atom is larger than Sc and the positive ion from which the electron is being extracted is not so highly charged: I_2 (Ba) < I_3 (Sc). Both I_2 for Na and I_3 for Mg are expected to be considerably larger than the ionization energies just described for Ba and Sc^{3+}. This is because the electron is being removed from a noble gas electron core. Because of the higher positive charge produced on Mg by the removal of its third electron (compared to the second for Na), we expect I_2 (Na) < I_3 (Mg). For the four metals, I_2 (Ba) < I_3 (Sc) < I_2 (Na) < I_3 (Mg). I_1 for F is expected to be large compared to I_1 for metals but we are comparing it here with I_2 and I_3 values. Certainly to remove an electron from a noble gas electron core requires more energy than to singly ionize an F atom. Whether more or less energy is required to ionize one electron from F than a second or third from a metal, we cannot easily say. I_2 (Ba) < I_3 (Sc) ? I_1 (F) < I_2 (Na) < I_3 (Mg).

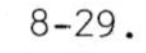
8-29.

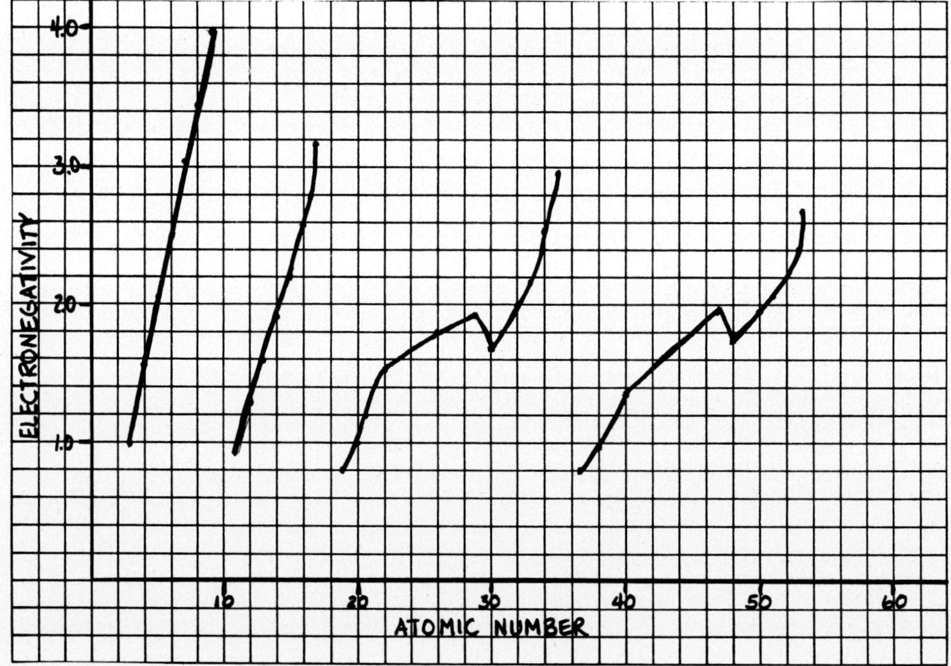

In a general way this plot does conform to the periodic law, and it is not surprising that it should. Electronegativity is related to ionization energy, which in turn is related to atomic size. We have previously seen atomic size to be a periodic property of the elements. The graph shown here, for example, illustrates that electronegativities of the most active metals have low values, and of the active nonmetals, high values.

8-30. We need to compare Table 8-6 (selected electronegativities) with Figure 8-8 (ionization energies as a function of atomic number). The most active metals, those with electronegativities of about 1 or less, all have low ionization energies--I_I < 5 eV. I_I is a good criterion for these elements. The highest ionization energies are those of the noble gases, but these are not the most nonmetallic elements. Ionization energy as a measure of nonmetallic character does not work for them. The most nonmetallic elements according to electronegativity data--the smaller members of Groups VIA and VIIA--do have high ionization energies (I_I > 12 eV). There is a reasonably good correlation between ionization energy and nonmetallic character for them. The greatest shortcoming of ionization energy as a criterion of metallic/nonmetallic character is with those elements that lie in the "middle". For example, Zn, which is a fairly active metal has $I_I \approx 9$ eV (see Figure 8-8), whereas Si, which is a metalloid, has I_I = 8.2 eV (see Table 8-5).

Magnetic properties

8-31. Three of the species listed have the electron configuration of a noble gas and thus all electrons are paired. The three are F^- ($1s^22s^22p^6$); Ca^{2+} ($1s^22s^22p^63s^23p^6$); and S^{2-} ($1s^22s^22p^63s^23p^6$). By a process of elimination we are led to conclude that Fe^{2+} has unpaired electrons. But we can also arrive at this conclusion through the electron configuration. Fe^{2+} $1s^22s^22p^63s^23p^63d^6$. If the 3d electrons enter orbitals singly before pairing up, we conclude that there are *four* unpaired electrons (six electrons distributed among five 3d orbitals).

8-32. V^{3+}: $1s^22s^22p^63s^23p^63d^2$. The two 3d electrons are unpaired.

Cu^{2+}: $1s^22s^22p^63s^23p^63d^9$. Of the nine 3d electrons there are four pairs and a single unpaired electron in the remaining 3d orbital.

Cr^{3+}: $1s^22s^22p^63s^23p^63d^3$. The three 3d electrons are all unpaired.

8-33. All atoms with an odd atomic number must be paramagnetic, for there is no way to pair up all the electrons in an odd-numbered set. However, it does not follow that all atoms with an even atomic number are diamagnetic. The electrons would all be paired if they entered orbitals in pairs, but Hund's rule states that the tendency is for orbitals to be singly occupied where possible. As an illustration consider oxygen. It has an even atomic number (8) and two unpaired electrons.

	1s	2s	2p		
O	↑↓	↑↓	↑↓	↑	↑

8-34. With properties such as atomic size, ionization energy, and electronegativity, trends are regular. For example, metals have low electronegativities and nonmetals, higher electronegativities. With unpaired electrons, the most active metals (IA) have one, and so do the most active nonmetals (VIIA). The Group IIA metals have none, and the Group VA elements, three. About the only useful generalization is that the maximum number of unpaired electrons occurs among transition elements.

Predictions based on periodic relationships

8-35. Dobereiner's method works only when the numbers of intervening elements between the first pair and the second pair of elements in the triad are the same. For example, in (a) this separation is 8 elements: Li (Z = 3), Na (Z = 11), and K (Z = 19); but in (c) the first pair--C (Z = 6) and Si (Z = 14)--are separated by 8 elements and the second pair--Si (Z = 14) and Ge (Z = 32)--by 18 elements.

(a) $\underset{\text{(est.)}}{\text{at. wt. Na}} = \dfrac{\text{at. wt. Li + at. wt. K}}{2} = \dfrac{6.94 + 39.10}{2} = 23.02$ actual: 22.99

(b) $\underset{\text{(est.)}}{\text{at. wt. Br}} = \dfrac{\text{at. wt. Cl + at. wt. I}}{2} = \dfrac{35.45 + 126.90}{2} = 81.18$ actual: 79.90

(c) $\underset{\text{(est.)}}{\text{at. wt. Si}} = \dfrac{\text{at. wt. C + at. wt. Ge}}{2} = \dfrac{12.01 + 72.59}{2} = 42.30$ actual: 28.09

(d) $\underset{\text{(est.)}}{\text{at. wt. Sb}} = \dfrac{\text{at. wt. As + at. wt. Bi}}{2} = \dfrac{74.92 + 208.98}{2} = 141.95$ actual: 121.75

(e) $\underset{\text{(est.)}}{\text{at. wt. Ga}} = \dfrac{\text{at. wt. B + at. wt. Tl}}{2} = \dfrac{10.81 + 204.37}{2} = 107.59$ actual: 69.72

8-36. Assuming a regular relationship within each group of substances, we might estimate boiling points of about -60°C for SnH_4 and about -100°C for H_2O. The true boiling points are -52°C for SnH_4 and 100°C for H_2O. The estimation is quite good for SnH_4 but greatly in error for H_2O.

8-37. (a) Since Figure 8-1 is based on atomic numbers we must use the atomic number of Ga in this estimation. Its atomic number is 31 and its atomic volume is approximately 12 cm^3/mol. But let us use Mendeleev's estimate of the atomic weight of Ga--68.

$$\text{estimated density} = \frac{68 \text{ g/mol}}{12 \text{ cm}^3\text{/mol}} = 5.7 \text{ g/cm}^3$$

(b) Because Ga is expected to resemble Al, we would predict the formula R_2O_3 (see Table 8-1), i.e., Ga_2O_3.

$$\% \text{ Ga} = \frac{(2 \times 68) \text{ g Ga}}{[(2 \times 68) + (3 \times 16)] \text{g Ga}_2\text{O}_3} \times 100 = 74\% \text{ Ga and } 26\% \text{ O}$$

Self-test Questions

1. (*b*) The element in question is in Group VA of the periodic table (it is Sb). As such, it resembles Bi, which is also in Group VA. The element does not bear a particular resemblance to Te [eliminating (c) as an answer], nor is it a transition element [eliminating (d)].

2. (*c*) Arsenic is in Group VA. An atom with five 4p electrons would be in Group VIIA. Atoms with the 4d subshell filled are found in the fifth period, but As is in the fourth period. All atoms following Ar have six 3p electrons; this includes As. No atom may have three electrons in an s subshell.

3. (*d*) The species listed are isoelectronic. All have the electron configuration of Ar. The positive ions are smaller than Ar, and the negative ion, Cl^- is larger.

4. (*b*) We can eliminate the two metals, Cs and Li, which have low ionization energies, and compare the two nonmetals. The more nonmetallic of the two, and therefore the more difficult to ionize, is Cl.

5. (*a*) The most nonmetallic of the four elements is the one found farthest to the right and farthest up in the periodic table. This element will have the highest electronegativity; it is Br.

6. (*d*) Of the four elements S is clearly the most nonmetallic and should have the greatest affinity for an electron.

7. (*c*) Use the idea that the most metallic elements are found far to the left and toward the bottom of the periodic table. Group IA metals are more metallic than IIA.

8. (*b*) The electron configuration of Br is $[Ar]3d^{10}4s^24p^5$. Of the five 4p electrons, there are two pairs and one, unpaired.

9. All the questions can be answered by reference to a periodic table.

   (a) There are 34 protons in ${}^{79}_{34}Se$.

   (b) The number of neutrons is 79 - 34 = 45.

   (c) The third principal shell is filled with 18 electrons in Se. That is, the electron configuration is $[Ar]3d^{10}4s^24p^4$.

   (d) There are *two* electrons in the 2s orbital of all atoms starting with Be.

   (e) From the electron configuration written in (c), we see that there are *four* 4p electrons.

   (f) The number of electrons in the shell of highest principal quantum number corresponds to the group numeral for a representative element. Se is in Group VIA. It has *six* outer-shell electrons.

10. The element with the highest electronegativity must be a nonmetal. There are two nonmetals in the listing: Br and As. Bromine is the more nonmetallic of the two (being in Group VIIA).

11. (a) C: The element at the top of a group has the smallest atoms.

   (b) Rb: The alkali metal that opens a period of elements has the largest atomic radius in the period.

   (c) At: The members at the bottom of a group are more metallic (less nonmetallic) than those higher in the group; they have lower electronegativities.

12. (a) The theoretical basis of the periodic table is to bring together in vertical columns elements with similar electron configurations. Electronic shells have differing capacities for electrons, and the subshells fill in a particular order. Because of this, the number of elements that must intervene before an element similar to a previous one is encountered is not constant, being 2 in one case, 8 in two cases, and 18 and 32 in others.

(b) The modern periodic table arranges elements according to increasing atomic number. In a few cases the order by atomic weight is not the same as by atomic number, for example Ar (Z = 18) has a higher atomic weight than K (Z = 19).

Chapter 9

Chemical Bonding I: Basic Concepts

Review Problems

9-1. Locate each atom in the periodic table, and use the table as a guide in determining numbers of valence shell electrons. For ions, make the appropriate changes in valence-shell configurations before writing Lewis symbols.

(a) H· (b) :Kr: (c) ·Ge· (d) $[Mg]^{2+}$ (e) $[:Br:]^{-}$ (f) ·Ga· (g) $[Sc]^{3+}$

(h) Cs· (i) $[:S:]^{2-}$

9-2. (a) $[Na]^{+}[:F:]^{-}$ (b) $[Mg]^{2+}[:O:]^{2-}$ (c) $^{-}[:I:]\ [Sr]^{2+}[:I:]^{-}$

9-3. (a) :Br-Br: (b) :I-Cl: (c) :F-O(-F:) (d) :I-N(-I:)-I: (e) H-Te:(-H)

9-4. Use rule (c) of page 225 of the text. That is, first provide each terminal atom with an octet and then shift electron pairs into the bonds to the central atom, if it is necessary to do so.

(a) :S-C-S: This structure uses all 16 valence electrons and completes the octet for each S atom, but the C atom is two pair of electrons short of an octet. When lone-pair electrons are moved from the S atoms to bonds with the C atom, the S atoms retain their octets and the C atom acquires one.

:S=C=S:

(b) :O-O-O: This structure uses all 18 valence electrons but leaves the central atom with 6 rather than 8 valence electrons. The situation is corrected by forming one oxygen-to-oxygen double bond.

:O-O=O:

(c) H-C(-H)-O: This trial structure is modified through a carbon-to-oxygen double bond. H-C(-H)=O:

9-5. (a) One of the H atoms is shown to form two bonds and the N atom has only six valence electrons. These problems are corrected by bonding both H atoms directly to the N atom.

H-N(-H)-O-H

(b) The structure shown has 20 valence electrons, but there should only be 19 (6 + 6 + 7). ClO_2 is an odd-electron species; it has one unpaired electron.

:O-Cl-O·

(c) The C atom has an incomplete octet and two unpaired electrons. Both matters are corrected by pairing up the electrons and showing a carbon-to-electron triple bond.

$[:C\equiv N:]^{-}$

(d) The structure shown suggests a calcium-to-oxygen single covalent bond, but the combination of an active metal and nonmetal should produce an ionic bond.

$[Ca]^{2+}[:O:]^{2-}$

9-6. (a) In the structure for I_2 each I atom is assigned seven valence electrons, the same number found in the isolated I atom. There are no formal charges in

$:\ddot{I}-\ddot{I}:$

(b) The O atom that is singly bonded to the S atom is assigned 7 valence electrons in the structure (three pairs of lone-pair electrons and one half of a bond pair). The number of valence electrons in the isolated O atom is 6. This O atom carries a formal charge of -1. The sulfur atom has only 5 electrons associated with its valence shell in the structure; its formal charge is +1. The second O atom has no formal charge

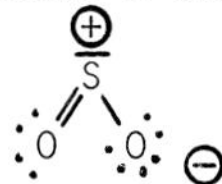

(c) Each of the O atoms that is singly bonded to the C atom carries a formal charge of -1. Neither the central C atom nor the doubly bonded O atom carries a formal charge. Note that the sum of the formal charges is equal to the charge on the ion.

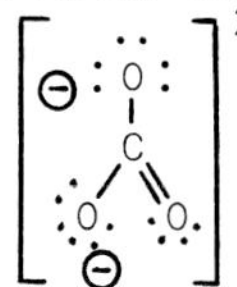

(d) There are no formal charges in the structure for $COCl_2$.

(e) Here, the terminal O atom carries a formal charge of -1.

$[H-O-O]^-$

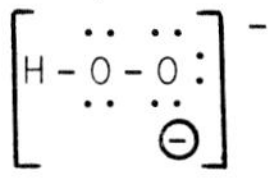

(f) The singly bonded O atom has a formal charge of -1 and the N atom, of +1.

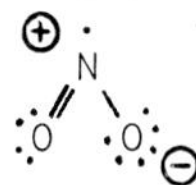

9-7. Combine the structures of the polyatomic ions from Section 9-5 with the appropriate simple cations and anions.

(a) 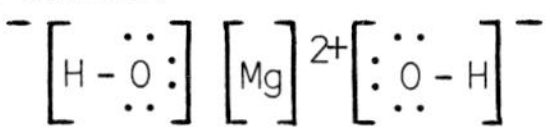 $[H-O]^-\,[Mg]^{2+}\,[O-H]^-$ (b) 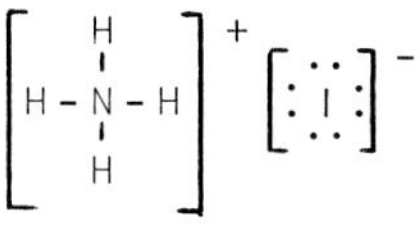 $[H_4N]^+\,[I]^-$ (c) 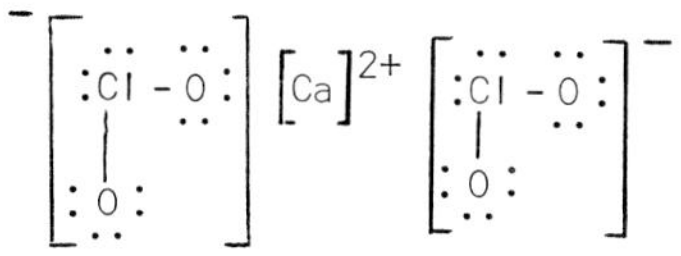 $[Cl-O(-O)]^-\,[Ca]^{2+}\,[Cl-O(-O)]^-$

9-8. Determine which species contain an odd number of electrons, and therefore at least one unpaired electron. These will be the paramagnetic species.

(a) OH^- (6 + 1 + 1 = 8 valence electrons): diamagnetic

(b) OH (6 + 1 = 7 valence electrons): paramagnetic

(c) NO_3 [7 + (6 x 3) = 25 valence electrons]: paramagnetic

(d) SO_3 [6 + (6 x 3) = 24 valence electrons]: diamagnetic

(e) SO_3^{2-} [6 + (6 x 3) + 2 = 26 valence electrons]: diamagnetic

(f) HO_2 [1 + (6 x 2) = 13 valence electrons]: paramagnetic

9-9. (a) To bond five F atoms to a central Br atom requires five pairs of electrons. The number of electrons about the central Br atom must be 10.

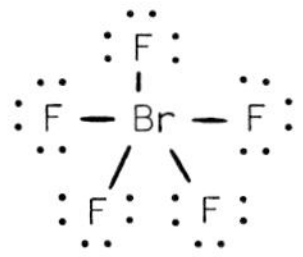

BrF_5

(b) PF_3 (like PCl_3 in structure 9.24) can be represented without an octet expression.

```
 ..  ..  ..
:F - P - F:
 ..  |   ..
    :F:
     ..
```

(c) The ICl_3 structure requires that 7 + (3 x 7) = 28 electrons be shown. If each Cl atom has an octet, this leaves 28 - (3 x 8) = 4 electrons that must exist as lone-pair electrons on the central I atom. Add to these, two electrons for each of the I-Cl bonds and this leads to 10 around the I atom.

```
 ..  ..  ..
:Cl - I - Cl:
 ..   |   ..
    :Cl:
     ..
```

(d) Reasoning as in part (c), the total number of valence electrons is 6 + (4 x 7) = 34. Of these 4 x 8 = 32 are associated with octets around the F atoms. This leaves 34 - 2 = 2 lone-pair electrons on the S atom. These two, together with 8 electrons found in the S-F bonds, lead to 10 valence shell electrons for S.

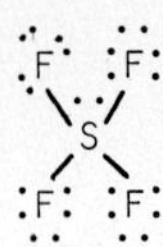

9-10. Either the method of Example 9-10 or 9-12 can be used. Using that of 9-10, i.e., by first writing a Lewis structure, we obtain

(a) :C ≡ O: Linear. (A structure is not necessary since this is a diatomic molecule.)

(b)

```
      ..
     :Cl:
 ..   |   ..
:Cl - Si - Cl:
 ..   |   ..
     :Cl:
      ..
```

Tetrahedral. (The four electron pairs about the Si atoms are distributed tetrahedrally; all are bond pairs.)

(c)

```
      ..
     :Cl:
 ..   |   ..
:Cl - Sb - Cl:
 ..   /\  ..
     /  \..
 :Cl:   Cl:
  ..    ..
```

Trigonal bipyramidal. (This structure involves the distribution of five pairs; all bond pairs.)

(d)

```
 ..
:Se - H
 |
 H
```

Angular. (Four electron pairs are distributed in a tetrahedral fashion. Two are bond pairs and two are lone pairs, AX_2E_2.)

(e)

```
 ..  ..  ..
:Cl - I - Cl:
 ..   |   ..
    :Cl:
     ..
```

T-shaped. [The five central pairs are distributed in a trigonal bipyramidal fashion. Three are bond pairs and two are lone pairs, AX_2E_2 (see also Table 9-2).]

(f) $[AlF_6]^{3-}$ Try the method of Example 9-12 here.

no. valence electron pairs = $\frac{3 + (6 \times 7) + 3}{2} = \frac{48}{2} = 24$

no. bond pairs = 7 - 1 = 6 pairs

no. central pairs = 24 - (3 x 6 terminal atoms) = 6 pairs

no. lone pairs = 6 - 6 = 0

The structure is AX_6--octahedral

(g)

```
   ..
  :O:
   |
   S
  // \
:O    .O:
```

Trigonal planar. [We count the double bond as if only a single electron pair were involved. The distribution of the three central pairs, all bond pairs, is trigonal planar.]

9-11.

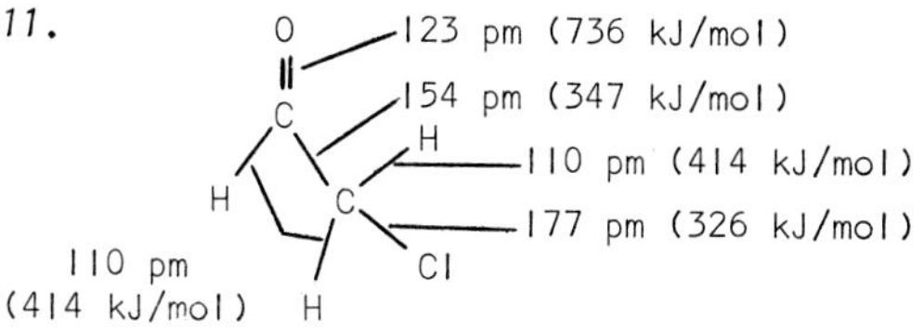

O —123 pm (736 kJ/mol)
C —154 pm (347 kJ/mol)
H —110 pm (414 kJ/mol)
C —177 pm (326 kJ/mol)
H Cl
110 pm (414 kJ/mol) H

(b) Per mol of compound:

1 C=O: 736 kJ/mol
1 C-C: 347
1 C-Cl: 326
3 C-H: (3 x 414) = 1242 kJ/mol

$\Delta H = 736 + 347 + 326 + 1242 = 2651$ kJ/mol

Per molecule: $2651 \frac{\text{kJ}}{\text{mol}} \times \frac{1 \text{ mol}}{6.02 \times 10^{23} \text{ molecules}} \times \frac{1000 \text{ J}}{1 \text{ kJ}} = 4.40 \times 10^{-18}$ J/molecule

9-12. Assess the energies of the bonds to be broken and formed and determine whether the net change (ΔH) is positive or negative.

(a) Break 1 C-H bond = +414 kJ/mol
Form 1 H-I bond = -297 kJ/mol
net ΔH = +117 kJ/mol > 0, endothermic

(b) Break 1 H-H bond = +435 kJ/mol
1 I-I bond = +151 kJ/mol
Form 2 H-I bond = -2 x 297 kJ/mol
net ΔH = -8 kJ/mol, exothermic

(c) Break 1 C-H bond = +414 kJ/mol
1 Cl-Cl bond = +243 kJ/mol
Form 1 Cl-Cl bond = -326 kJ/mol
1 H-Cl bond = -431 kJ/mol
net ΔH = -100 kJ/mol < 0, exothermic

9-13. (a) No dipole moment. A homonuclear diatomic molecule.

(b) Dipole moment. A diatomic molecule with an electronegativity difference between N and O.

(c) No dipole moment. BF_3 is a trigonal planar molecule. The symmetrical distribution of the three F atoms around the B atom causes the bond dipole moments to cancel.

(d) Dipole moment. This is an electronegativity difference between H and Br in the diatomic molecule HBr.

(e) Dipole moment. The situation with $HCBr_3$ is very similar to that of $HCCl_3$ pictured in Figure 9-15.

(f) No dipole moment. Here the situation (for $SiCl_4$) resembles that for CCl_4 in Figure 9-15.

(g) Dipole moment. Unlike the case of CO_2 discussed in the text, substitution of an S for an O atom would result in a slight alternation of electrons to the remaining O atom. That is

:S̈ = C ⇸= Ö:

9-14. In each case an element is bonded to As, and the geometrical shapes of the molecules (AsX_3) are all alike--trigonal pyramidal. The magnitude of the dipole moment is determined primarily by electronegativity differences, i.e., $EN_F - EN_{As}$, $EN_{Cl} - EN_{As}$, etc. The smallest dipole moment is for AsH_3, the largest, AsF_3. The order of increasing dipole moment is

$AsH_3 < AsI_3 < AsBr_3 < AsCl_3 < AsF_3$

9-15. As in Review Problem 14, the order is by increasing electronegativity difference.

C-H < Br-H < F-H < Na-Cl < K-F

Exercises

Lewis theory

9-1. The Lewis structure of an ionic compound is written to emphasize that electrons are ***transferred*** from metal to nonmetal atoms. This is done by indicating net ionic charges on the Lewis symbols. The principal requirement of the Lewis structure of a covalent bond is that the ***sharing*** of an electron pair(s) be represented. For most covalent compounds and many ionic compounds the electron configurations of the bonded atoms become those of a noble gas, but there are important exceptions to this idea.

9-2. There are numerous exceptions to the statement that all atoms in a Lewis structure have an octet in their valence shells. This can never be the case for H, which cannot accommodate more than two electrons. Nor is it usually the case for certain atoms that form electron deficient molecules--Be in $BeCl_2$, B in BF_3. Another important set of exceptions is for nonmetals in the third and higher periods that can accommodate 10 or 12 electrons, such as P in PCl_5, S in SF_6, and I in ICl_3.

Ionic bonding

9-3. (a) Li_2O $[Li]^+$ $[:\ddot{O}\times]^{2-}$ $[Li]^+$ (b) NaBr $[Na]^+$ $[:\ddot{Br}:]^-$

(c) SrF_2 $[:\ddot{F}:]^-$ $[Sr]^{2+}$ $[:\ddot{F}:]^-$ (d) $ScCl_3$ $[:\ddot{Cl}\times]^-$ $[Sc]^{3+}$ $[\times\ddot{Cl}:]^-$ $[:\ddot{Cl}:]^-$

9-4. The nonmetals are representative elements whose atoms lack one (Group VIIA), two (Group VIA) or, occasionally, three (N in Group VA) electrons from having the electron configurations of noble gases. When the nonmetals gain these small numbers of electrons, they form anions with noble gas electron configurations. Among metals, which tend to lose electrons to form cations, noble gas electron configurations are achieved only for those metals with one, two, or possibly three electrons beyond a noble gas core. But there are many metals (e.g., transition elements) for which the underlying core is not that of a noble gas (see Table 9-1).

9-5. Solid NaCl consists of an array of positive (Na^+) and negative (Cl^-) ions. There are no identifiable small groups of ions that exist separate from others--there are no entities that could be called molecules. In *gaseous* NaCl discrete ion pairs (such as the one pictured in Figure 9-2) can exist. These might be referred to as molecules.

Lewis structures

9-6. Several of these terms are described in the section, "Some New Terms". What is intended here is a description of terms in relation to Lewis structures.

(a) Valence electrons are the outer-shell electrons of an atom, represented by dots in a Lewis structure.

(b) An octet refers to eight electrons in the outer shell of an atom (represented as eight dots surrounding a chemical symbol in a Lewis structure).

(c) Unshared electron pairs or nonbonding electron pairs or lone pairs are identified as belonging exclusively to one atom in a Lewis structure.

(d) Multiple bonds result from the sharing of more than one pair of electrons between two atoms (shown in Lewis structures as double (=) and triple (≡) bonds).

(e) A coordinate covalent bond is one in which all the electrons being shared are identified as coming from a single atom in the bonded pair of atoms.

(f) Resonance describes a condition in which more than a single plausible Lewis structure can be written for a species.

(g) An odd electron species is one in which the total number of valence electrons is an odd number, which means that not all electrons in the Lewis structure can be paired.

(h) An expanded octet results whenever more than eight valence electrons (e.g., 10 or 12) are represented for one of the atoms (the central atom) in a Lewis structure.

9-7. (a) RbCl, rubidium chloride, f. wt. = 120.9, (ionic): [Rb]$^+$ [$\times$Cl:]$^-$

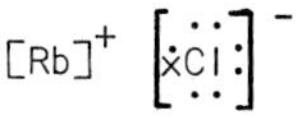

(b) H_2Se, hydrogen selenide, f. wt. = 80.98, (covalent): H-Se: with H

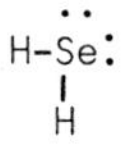

(c) BCl_3, boron trichloride, f. wt. = 117.2, (covalent): :Cl - B - Cl: with :Cl:

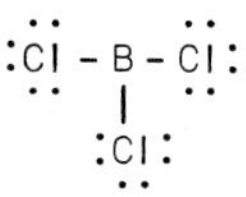

(d) Cs_2S, cesium sulfide, f. wt. = 297.9, (ionic): [Cs]$^+$ [:S$\times$]$^{2-}$ [Cs]$^+$

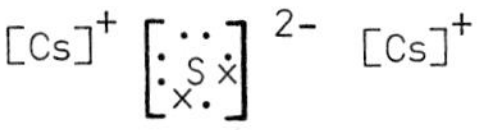

(e) SrO, strontium oxide, f. wt. = 103.6, (ionic): [Sr]$^{2+}$ [$\times$O:]$^{2-}$

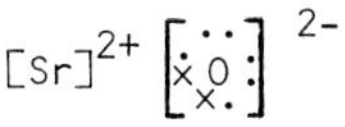

(f) OF_2, oxygen fluoride, f. wt. = 54.00, (covalent): : O - F : with : F :

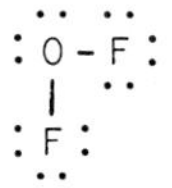

9-8. (a) H_2NOH. N and O are central atoms and H, terminal atoms. 14 valence electrons must appear in the structure. All bonds are single bonds.

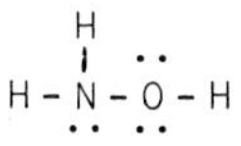

H - N - O - H (with H on N)

(b) N_2F_2. Based on electronegativities, we expect the ***less*** electronegative N atoms to be the central atoms. The number of valence electrons to be assigned is (2 x 5) + (2 x 7) = 24. The structure must include some multiple bonding.

Initial Attempt : F - N - N - F :

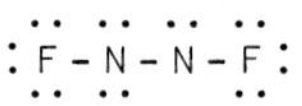

Change To : F - N = N - F :

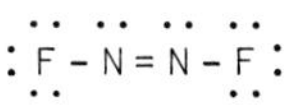

(c) HONO. Atoms are bonded together in the order shown. The number of valence electrons to be assigned is 1 + 6 + 5 + 6 = 18. Again, some multiple bonding is required.

Initial Attempt H - O - N - O :

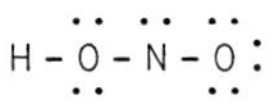

Change To H - O - N = O :

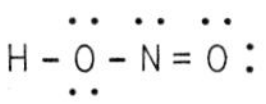

(d) H_2NNO_2. The two N atoms are central atoms and the two H and two O atoms are terminal atoms. The number of valence electrons to be assigned is (2 x 1) + 5 + 5 + (2 x 6) = 24. Multiple bonding is once more required.

Initial Attempt :O: | H - N - N - O : | H

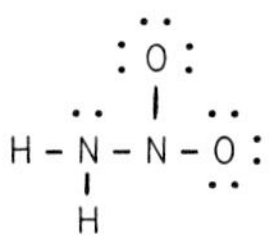

Change To :O: | H - N - N = O : | H

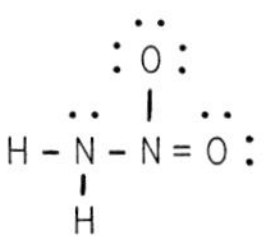

9-9. (a) The structure has the correct number of valence electrons, 6 + 4 + 5 + 1 = 16; but the C atom has an incomplete octet. Two possibilities for correcting the structure are

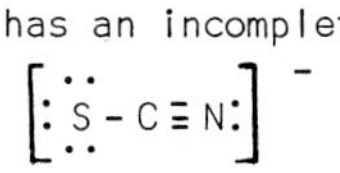

[: S - C ≡ N:]$^-$ or [: S ≡ C = N :]$^-$

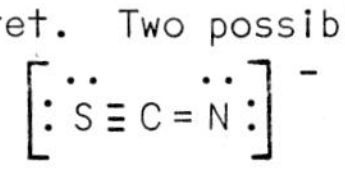

(b) The structure shown suggests ionic bonding, but Cl and O are two active nonmetals. A compound formed from them should be covalent, not ionic.

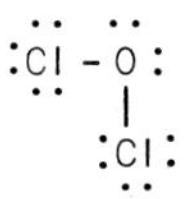

:Cl - O : | :Cl:

(c) The structure shows 16 valence electrons, but there should be 17 (6 + 5 + 6). The structure should have an unpaired electron. To accommodate this electron (say on the central N atom) also requires that one of the double bonds be changed to a single bond (to avoid 9 electrons around the N atom).

```
 ..  .  ..
:O = N - O:
         ..
```

(d) This structure has the correct total number of valence electrons, 26; but the central N atom is surrounded by 10 electrons and it can only accommodate 8.

```
 ..  ..  ..
:Cl - N - Cl:
 ..   |   ..
     :Cl:
      ..
```

9-10. In the species H_3, one H atom would have to be bonded to two others simultaneously. This would place four electrons in the valence shell (n = 1) of that H atom, an impossibility. The species HHe would require three electrons in the outer shell of the He atom, H·He, but this shell (n = 1) can accommodate only two. In the species He_2 there would be four electrons in the outer shell of one of the atoms (He:He); this, too, is an impossibility. The structure H_3O would require 9 electrons in the outer most shell of the O atom, still another impossibility.

Formal charge

9-11. (a) In the structure shown, 5 electrons are assigned to Cl; this produces a formal charge of +2 on Cl. Each of the three O atoms is assigned 7 valence electrons in the structure and has, therefore, a formal charge of -1. (Note that the sum of the formal charges +2-1-1-1 = -1, which is the net charge on the ion.)

(b) Each of the three atoms have 6 valence electrons assigned to it in the structure and have 6 valence electrons in its free, isolated state. There are no formal charges in this structure.

(c) The F atoms do not carry a formal charge; but the central B atom, by having 4 electrons assigned to it in this structure, carries a formal charge of -1. This is also the net charge on the BF_4^- ion.

(d) Each O atom is assigned 6 valence electrons; the P atom, 5; and each H atom, 1. There are no formal charges in this structure.

(e) The central N atom is assigned 4 valence electrons, and so it carries a formal charge of +1. Each of the terminal N atoms is assigned 6 valence electrons and carries a formal charge of -1.

(f) In this alternative to structure (e), the central N atom has a formal charge of +1; the N atom on the left, -2; and the N atom on the right, 0.

(g) In the ion NO_2^+ the O atoms have no formal charges and the central N atom carries a formal charge of +1, which is equal to the charge on the ion.

9-12. Write Lewis structures for each of the possibilities listed. Choose as the most plausible the one that best meets the requirements stated in the text.

```
(a)      ..  ..            ..  ..
     H - N - O - H     H - O - N - H
         |   ..            |   ..
         H                 H
```

The structure on the left has no formal charges. In the structure on the right the O atom has a formal charge of +1 and the N atom, -1. The structure on the left is the more plausible.

```
(b)   ..      ..      ..      ..
     :S = C - S:     :C = S = S:
              ..
```

The structure on the left has no formal charges. For the structure on the right, formal charges are -2 for C, +2 for the central S atom, and 0 for the S atom on the right. The structure on the left is the more plausible.

```
(c)   ..  ..  ..      ..  ..  ..
     :N = O - Cl:    :O = N - Cl:
              ..              ..
```

The structure on the left has formal charges on N(-1) and O(+1). The structure on the right has no formal charges. The structure on the right is the more plausible.

(d) $\left[:\ddot{S}=C=\ddot{N}:\right]^-$ $\left[:\ddot{C}=N=\ddot{S}:\right]^-$ $\left[:\ddot{C}=S=\ddot{N}:\right]^-$

Whatever structure is written for this ion, at least one atom must carry a formal charge of -1 (to account for the charge on the ion). The structure on the left has a formal charge of -1 on the N atom (the most electronegative) and no other formal charges. This structure seems quite plausible. The structure in the center has a formal charge of +1 on N and -2 on C. The (large) negative formal charge is not the most electronegative atom. The structure on the right seems least plausible of all. The sulfur carries a formal charge of +2, the carbon, -2, and the nitrogen, -1. The best structure is the one of the left. (Note: Several other structures could be written, but none with a better distribution of formal charges than the one on the left.)

9-13. Use the expression for formal charge given in the footnote on page 227. Let FC = formal charge; N = number of valence electrons in the isolated atom; B = number of bonding electrons joining the atom in question to others to which it is bonded; LP = number of lone pair electrons on the atom in question.

$$FC = N - \frac{1}{2}B - LP$$

In the <u>absence</u> of coordinate covalent bonds, each atom contributes equally to all bonds that it forms. This means that each atom contributes $\frac{1}{2}$ to the number of bonding electrons, i.e., $\frac{1}{2}B$. The number of lone-pair electrons is the number of valence electrons minus the number of bonding electrons for the atom. In the <u>absence</u> of coordinate covalent bonding, $LP = N - \frac{1}{2}B$ and $FC = N - \frac{1}{2}B - (N - \frac{1}{2}B) = 0$. If coordinate covalent bonding is involved then one atom contributes less than B/2 to the bonding electrons and another atom, more than B/2. For both atoms $LP \neq N - B/2$ and $FC \neq 0$.

Polyatomic ions

9-14. Consider the Lewis symbol of a sulfur atom to be not the normal $:\dot{\ddot{S}}\cdot$, but $:\dot{\ddot{S}}:$; that of S^{2-} is $\left[:\ddot{\underset{..}{S}}:\right]^{2-}$. The number of valence electrons for an S atom is 6, and for an S^{2-} ion, 8. Thus, in the Lewis structure of S_2^{2-} 14 electrons must be assigned; in S_3^{2-}, 20; S_4^{2-}, 26; S_5^{2-}, 32.

(a) $S_2^{2-} = \left[:\ddot{\underset{..}{S}}:\ddot{\underset{..}{S}}:\right]^{2-}$ (b) $S_3^{2-} = \left[:\ddot{\underset{..}{S}}:\ddot{\underset{..}{S}}:\ddot{\underset{..}{S}}:\right]^{2-}$

(c) $S_4^{2-} = \left[:\ddot{\underset{..}{S}}:\ddot{\underset{..}{S}}:\ddot{\underset{..}{S}}:\ddot{\underset{..}{S}}:\right]^{2-}$ (d) $S_5^{2-} = \left[:\ddot{\underset{..}{S}}:\ddot{\underset{..}{S}}:\ddot{\underset{..}{S}}:\ddot{\underset{..}{S}}:\ddot{\underset{..}{S}}:\right]^{2-}$

9-15. (a) $\left[:\ddot{\underset{..}{Cl}}-\ddot{\underset{..}{O}}:\right]^-$ (b) $\left[:\ddot{O}=N-\ddot{\underset{..}{O}}:\right]^-$ or $\left[:\ddot{\underset{..}{O}}-N=\ddot{O}:\right]^-$

(c) $\left[:\ddot{O}-\ddot{Br}-\ddot{O}:\ \text{with}\ :\ddot{O}:\ \text{bonded below Br}\right]^-$

9-16. (a) $[K]^+\left[:\ddot{O}-\ddot{I}-\ddot{O}:\ \text{with}\ :\ddot{O}:\ \text{bonded below I}\right]^-$ (b) $^-\left[:\ddot{\underset{..}{O}}-\ddot{\underset{..}{Cl}}:\right]$ $[Ca]^{2+}$ $\left[:\ddot{\underset{..}{O}}-\ddot{\underset{..}{Cl}}:\right]^-$

(c) $\left[H-N-H\ \text{with H above and H below N}\right]^+\left[:\ddot{O}-Cl-\ddot{O}:\ \text{with}\ :\ddot{O}:\ \text{above and}\ :\ddot{O}:\ \text{below Cl}\right]^-$

Resonance

9-17. In the Lewis structure of SO_3 there are 24 valence-shell electrons to be depicted. The three O atoms are bonded to a central S atom.

$:\ddot{\underset{..}{O}}-S-\ddot{\underset{..}{O}}:$ with $:\ddot{O}:$ bonded above S

The difficulty with the structure just written is that the S atom lacks an octet. This can be remedied by shifting an electron pair into one of the S-O bonds. There are three ways of doing so.

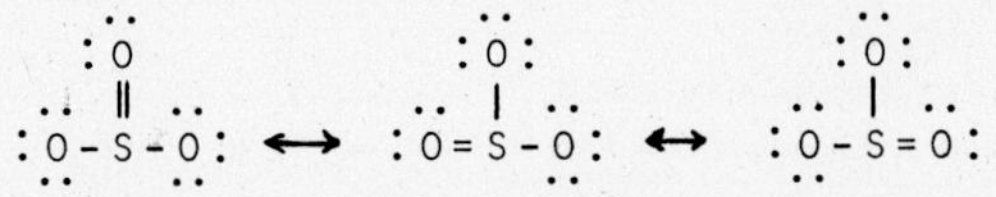

:O=S(–O:)–O: ↔ :O=S(–O:)–O: ↔ :O–S(–O:)=O:

9-18. [:O=N–O:]⁻ or [:O–N=O:]⁻

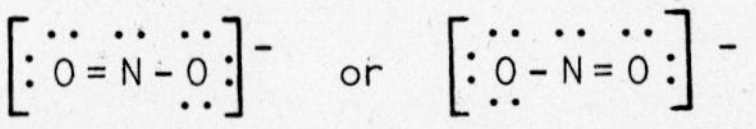

9-19. Two equally plausible Lewis structures for ozone are illustrated below. The true structure is a resonance hybrid with equal contributions from the following.

:O–O=O: ↔ :O=O–O:

9-20. Structures (a) and (b) are equivalent. Each has the N atom with a formal charge of +1 and one O atom with a formal charge of -1. In structure (c) the N atom again has a formal charge of +1, as does one O atom. The other two O atoms have a formal charge of -1. Structure (c) is much less plausible than (a) and (b).

Odd-electron species

9-21. The total number of valence electrons is 11 (five from N and six from O). These will appear as five pairs and one unpaired electron. A plausible structure is

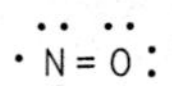

·N=O:

This structure has no formal charges. Another structure, which seems less plausible because it involves formal charges and places a positive formal charge on the more electronegative atom (O), is

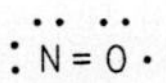

:N=O·

9-22. The dimer, N_2O_4, is diamagnetic. All electrons are paired.

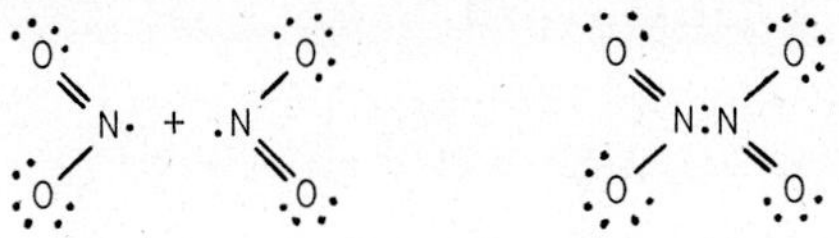

$O_2N\cdot + \cdot NO_2$ → $O_2N:NO_2$

9-23. (a) 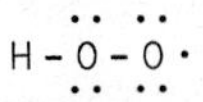 H–O–O·

(b) 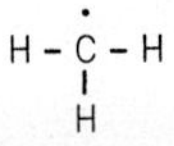 H–Ċ–H (with H below C)

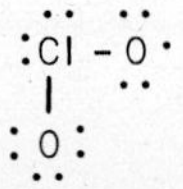

(c) 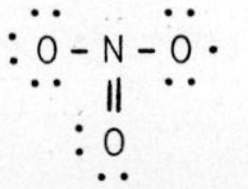 :Cl–O· (with :O: below Cl)

(c) :O–N–O· (with :O double bonded below N)

9-24. A valence-shell octet is based on the filling of *s* and *p* orbitals. The use of an expanded octet depends on the availability of *d* orbitals in the central atom. Third-period elements have such orbitals available (3*d*) but second-period elements do not (no 2*d*). Since As and Se do have *d* orbitals available they should resemble P and S more nearly than they do N and O.

9-25. (a) The structure below does not use an expanded octet and has the minimum formal charges possible, -1 on each of the singly bonded O atoms.

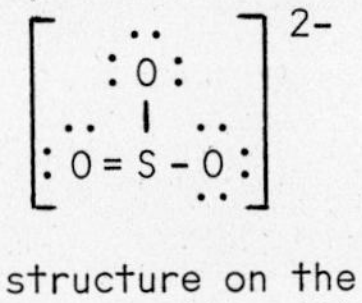

[:O=S(–O:)–O:]²⁻

(b) The structure on the left does not use an expanded octet, but it has more formal charges than do the ones on the right, which use expanded octets.

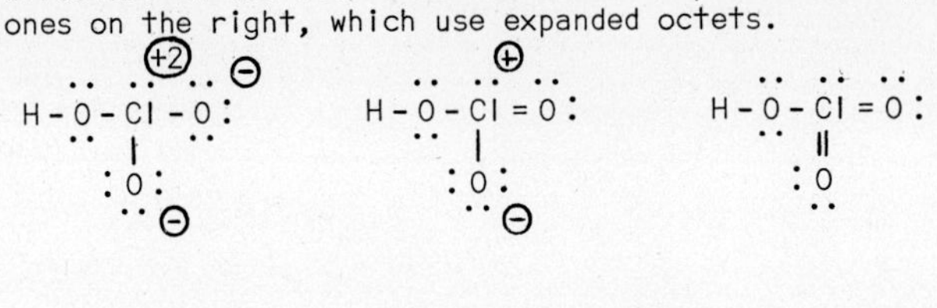

H–O–Cl(+2)(–O:⊖)–O:⊖ H–O–Cl⊕(–O:⊖)=O: H–O–Cl(=O)=O:

(c) An expanded octet is not possible with second period elements, but neither is it necessary to achieve the following structure having no formal charges.

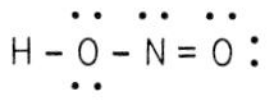

```
     ..  ..  ..
  H - O - N = O :
     ..
```

(d) Placement of 10 electrons around the P atom produces an acceptable structure.

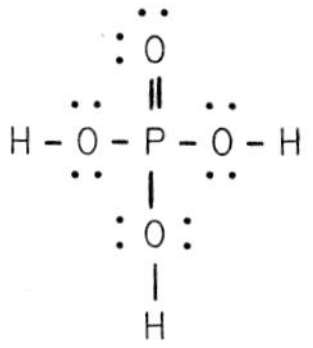

```
          ..
         : O
     ..   ||   ..
  H - O - P - O - H
     ..   |    ..
         : O :
           |
           H
```

(e)

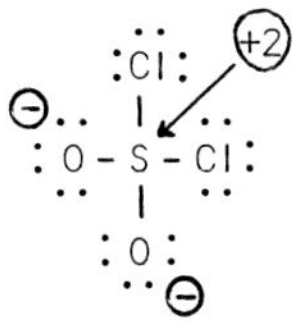

```
       ..
      :Cl:  (+2)
  (-)..  |  ..
   : O - S - Cl :      better structure:    : O = S - Cl :
     ..  |   ..                                   ||
       : O :                                     : O
        ..(-)
```

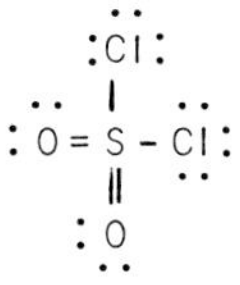

(f)

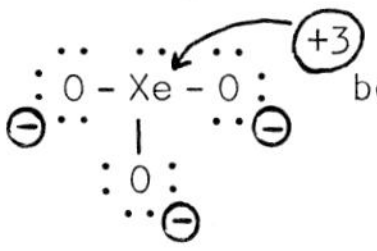

```
   ..    ..   (+3)
  : O - Xe - O :       better structure:    : O - Xe = O :
 (-)..   |   ..(-)                                ||
       : O :                                     : O
        ..(-)
```

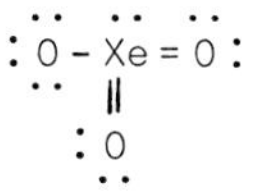

9-26. Write three Lewis structures with S as the central atom. In one of the structures use a sulfur-to-nitrogen single bond; in another, a double bond; and in the third, a triple bond. Comment on the plausibility of each structure.

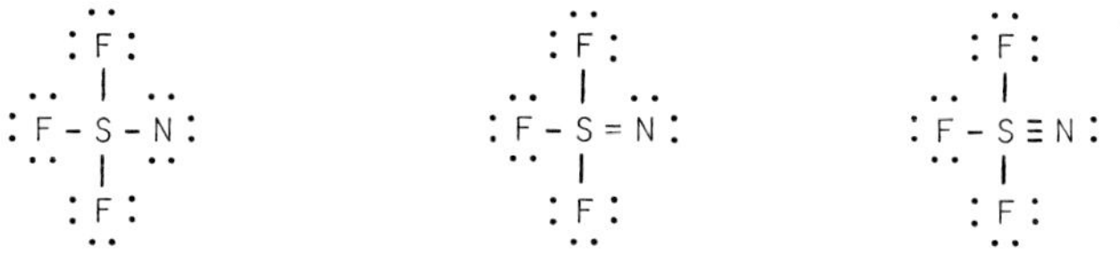

```
      ..               ..               ..
     : F :            : F :            : F :
  ..   |  ..       ..   |  ..       ..   |
 : F - S - N :    : F - S = N :    : F - S ≡ N :
  ..   |  ..       ..   |           ..   |
     : F :            : F :            : F :
      ..               ..               ..
```

In the single-bond structure, each atom has an outer shell octet but the S atom has a formal charge of +2 and N, -2. In the double-bond structure the formal charges are reduced to +1 and -1, though the S atom must now have an expanded octet--10 valence electrons. In the triple-bond structure, the S atom has 12 valence electrons, but there are no formal charges. An expanded octet for S is permitted in a Lewis structure, and if formal charges are to be minimized, it seems most probable that the nitrogen-to-sulfur bond has multiple bond character.

9-27. The seven Lewis structures referred to here are

```
      I              II             III            IV             V              VI             VII
     ..             ..             ..             ..             ..             ..             ..
    : O            : O :          : O :          : O :          : O            : O            : O
 ..  ||  ..     ..   |  ..     ..   |  ..     ..   |  ..     ..  ||  ..     ..  ||  ..     ..  ||  ..
: O - S - O :  : O = S - O :  : O - S = O :  : O = S = O :  : O = S - O :  : O - S = O :  : O = S = O :
 ..      ..             ..     ..                                   ..     ..
```

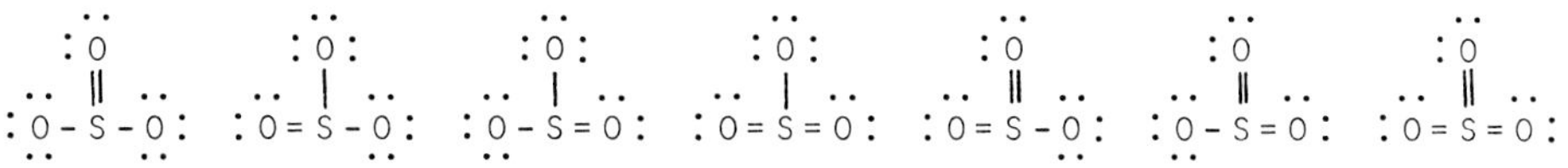

In structures I, II, and III, the S atom has a formal charge of +2; two of the O atoms have a formal charge of -1; and the third O atom, 0. In structures IV, V and VI, the formal charge of S is reduced to +1 and only one O atom carries a formal charge (-1). In structure VII there are no formal charges. Based on formal charge considerations alone, structure VII is the most plausible. Measurement of bond energies and distances could be used to establish whether the sulfur-to-oxygen bonds are double (confirming structure VII) or less-than-double (suggesting some contributions to the resonance hybrid from structures I-VI).

Molecular shapes

9-28. (a) 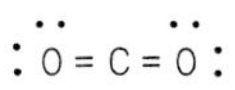 $:\ddot{O}=C=\ddot{O}:$ With the equivalent of two electron pairs for distribution about the central C atom, we predict a linear shape.

(b) 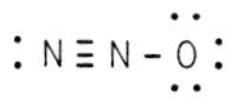 $:N\equiv N-\ddot{\underset{..}{O}}:$ or $:\ddot{N}=N=\ddot{O}:$

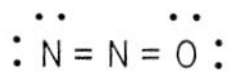

In either case there are the equivalent of two electron pairs distributed about the central N atom. The shape is linear.

(c) $:\ddot{N}=\ddot{S}-\ddot{F}:$ or $N\equiv\ddot{S}-\ddot{F}:$

This situation is just like that described in (b). The molecule is linear.

(c) $:\ddot{Cl}-N=\ddot{O}:$ with $:\ddot{O}:$ bonded to N — Here we see the equivalent of three electron pairs (all bond pairs) distributed about the central N atom. The molecule is trigonal planar.

9-29. The VSEPR notation for this molecule would be AXE_3. Its Lewis structure would be $:\ddot{A}-X$. Because the molecule is diatomic, it would have to be linear. An example would be a hydrogen halide, such as $:\ddot{F}-H$.

9-30. Tetrahedral: The total number of valence electrons is 8 (3 from B, one each from 4 F atoms, and one additional electron which corresponds to the -1 charge). The distribution of four electron pairs is tetrahedral, and each pair is a bond pair. Therefore, the ion has a tetrahedral shape.

9-31. See the solution to Exercise 9-27 for the seven Lewis structures referred to here. Electrons in multiple bonds are treated as if a single electron pair were involved. For each of the seven structures, the distribution corresponds to three electron pairs. Moreover, none of the structures involves lone-pair electrons. All have the VSEPR notation AX_3, and all lead to a prediction of trigonal planar geometry.

9-32. For this exercise let us use the method of Example 9-12.

(a) OSF_2. The number of valence electrons = 6 + 6 + (2 x 7) = 26.

total no. electron pairs = $\frac{26}{2}$ = 13

no. bond pairs = 4 - 1 = 3

no. central pairs = 13 - (3 x 3) = 4

no. lone pairs = 4 - 3 = 1

VSEPR designation: AX_3E. Molecular shape: trigonal pyramid.

(b) O_2SF_2. The number of valence electrons = (2 x 6) + 6 + (2 x 7) = 32

total no. electron pairs = $\frac{32}{2}$ = 16

no. bond pairs = 5 - 1 = 4

no. central pairs = 16 - (3 x 4) = 4

no. lone pairs = 0

VSEPR designation: AX_4. Molecular shape: tetrahedral.

(c) XeF_4. Number of valence electrons = 8 + (4 x 7) = 36

total no. electron pairs = 36/2 = 18

no. bond pairs = 5 - 1 = 4

no. central pairs = 18 - (3 x 4) = 6

no. lone pairs = 6 - 4 = 2

VSEPR notation: AX_4E_2. Molecular shapes: square planar.

(d) ClO_4^-. Number of valence electrons = 7 + (4 x 6) + 1 = 32

total no. electron pairs = 32/2 = 16

no. bond pairs = 5 - 1 = 4

no. central pairs = 16 - (3 x 4) = 4

no. lone pairs = 4 - 4 = 0

VSEPR notation: AX_4. Molecular shape: tetrahedral

(e) I_3^-. Number of valence electrons = (3 x 7) + 1 = 22

total no. electron pairs = 22/2 = 11

no. bond pairs = 3 - 1 = 2

no. central pairs = 11 - (3 x 2) = 5

no. lone pairs = 5 - 2 = 3

VSEPR notation: AX_2E_3. Molecular shape: linear

Bond distances

9-33. The nitrogen-to-oxygen bonds with a 121 pm bond distance have more multiple bond character than does the 140 pm bond. In the Lewis structure below one nitrogen-to-oxygen bond is shown as a single bond. The other two nitrogen-to-oxygen bonds, because of resonance, are essentially 1-1/2 bonds.

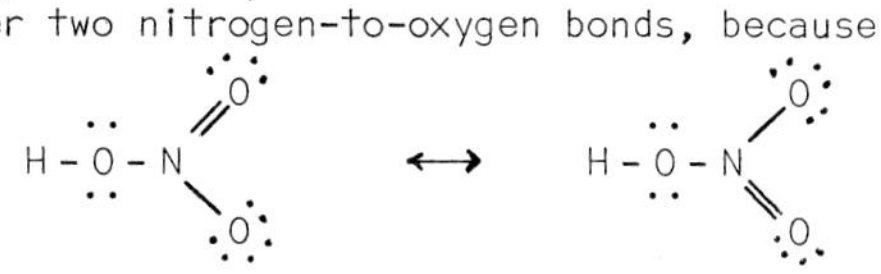

H - O - N(=O)(-O) ⟷ H - O - N(-O)(=O)

9-34. Use bond distances between like atoms to establish covalent radii. Then add covalent radii for different atoms to estimate the length of the bond between them.

(a) covalent radius of H = 1/2(H - H) = 1/2(74) = 37 pm

covalent radius of Cl = 1/2(Cl - Cl) = 1/2(199) = 100 pm

H - Cl bond distance = 137 pm

(b) C - N bond distance = 1/2(C - C) + 1/2(N - N) = 1/2(154) + 1/2(145) = 149 pm

(c) C - Cl bond distance = 1/2(C - C) + 1/2(Cl - Cl) = 1/2(154) + 1/2(199) = 177 pm

(d) C - F bond distance = 1/2(C - C) + 1/2(F - F) = 1/2(154) + 1/2(128) = 141 pm

(e) N - I bond distance = 1/2(N - N) + 1/2(I - I) = 1/2(145) + 1/2(266) = 205 pm

9-35. First draw a plausible Lewis structure.

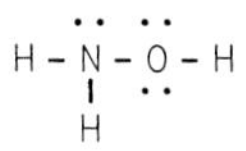

H - N(-H) - O - H

Now, predict the geometrical distribution of electron pairs expected about the N atom and about the O atom. In each case there are four pairs of electrons. For N this corresponds to AX_3E and for O, AX_2E_2. The predicted bond angles are the tetrahedral bond angle, 109.5°. The O-H bond distance is listed in Table 9-3. The N-H distance can be determined by the method of Exercise 9-34: 1/2(145 + 74) = 110 pm. Data are not available for estimating the N-O bond distance.

Bond energies

9-36. In C_2H_6 the bonds are

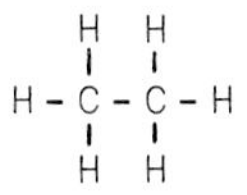

H₃C-CH₃ (H - C(H)(H) - C(H)(H) - H)

1 C - C = 347 kJ/mol
6 C - H = (6 x 414)kJ/mol

Total: 2.83×10^3 kJ/mol

In C_2H_4 the bonds are

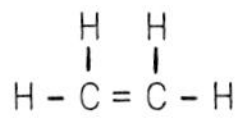

H - C(H) = C(H) - H

Compared to C_2H_6, there is one C = C in place of C - C. This corresponds to (611 - 347)kJ/mol = 264 kJ/mol of additional energy. However, the molecule has only 4 C - H bonds instead of 6, resulting in a reduction in bond energy of 2 x 414 = 828 kJ/mol. Overall the bond energy in C_2H_4 (2.27×10^3 kJ/mol) is less than in C_2H_6 (2.83×10^3 kJ/mol) by 5.6×10^2 kJ/mol.

9-37. (a)

```
    H  H                          H  H
    |  |                          |  |
H - C - C - H  +  Cl - Cl  →  H - C - C - Cl  +  H - Cl
    |  |                          |  |
    H  H                          H  H
```

Bonds broken: 1 C - H = +414 kJ/mol

1 Cl - Cl = +243 kJ/mol

Bonds formed: 1 C - Cl = -326 kJ/mol

1 H - Cl = -431 kJ/mol

ΔH_{rxn} = 414 + 243 - 326 - 431 = -100 kJ/mol

(b)

```
    H  H                   H  H
    |  |                   |  |
H - C = C - H  +  H - H → H - C - C - H
                           |  |
                           H  H
```

Bonds broken: 1 C = C = +611 kJ/mol

1 H - H = +435 kJ/mol

Bonds formed: 1 C - C = -347 kJ/mol

2 C - H = 2 x (-414) kJ/mol

ΔH_{rxn} = 611 + 435 - 347 - 2 x 414 = -129 kJ/mol

9-38.

```
                               H
                               |
N ≡ N(g) + 3 H - H(g)  →  2 N - H(g)
                               |
                               H
```

Bonds broken: 1 N ≡ N = +946 kJ

3 H - H = +3 x 435 = $+1.03\times10^{3}$ kJ

Bonds formed: 6 N - H = -6 x 389 = -2.33×10^{3} kJ

ΔH_{rxn} = $(0.95 + 1.30 - 2.33)\times10^{2}$ kJ = -0.080×10^{3} kJ

per mol NH_3: ΔH_{rxn} = $-8.0\times10 \div 2 = -4.0\times10^{1}$ kJ tabulated value of $\Delta\overline{H}^\circ_f[NH_3(g)]$ = -46.19 kJ/mol

9-39. Combine the two equations.

$2\ C(s) + 3\ H_2(g) \rightarrow C_2H_6(g)$ ΔH = -84.61 kJ/mol

$C_2H_6(g) \rightarrow 2\ C(g) + 3\ H_2(g)$ ΔH = 1.53 MJ/mol

$2\ C(s) \rightarrow 2\ C(g)$ $\Delta H = (-0.085 + 1.53)\times10^{3} = +1.45\times10^{3}$ kJ/mol

$C(s) \rightarrow C(g)$ $\Delta H = \dfrac{1.45\times10^{3}\ \text{kJ/mol}}{2}$ = 725 kJ/mol

(Value from Example 9-16: +717 kJ/mol)

To obtain the value 1.53×10^{3} kJ/mol for the second equation in the summation, note that in that reaction

Bonds broken: 1 C - C = +347 kJ/mol

6 C - H = +(6 x 414) kJ/mol

Bonds formed: 3 H - H = -(3 x 435) kJ/mol

$\Delta H = (0.347 + 2.48 - 1.30)\times10^{3} = 1.53\times10^{3}$ kJ/mol

9-40. Base your calculation on the formation reaction, for which $\Delta\overline{H}^\circ_f[NO(g)]$ is listed in Appendix D.

$1/2\ N_2(g) + 1/2\ O_2(g) \rightarrow NO(g)$ $\Delta\overline{H}^\circ_f[NO(g)]$ = +90.37 kJ/mol

$1/2\ N{\equiv}N(g) + 1/2\ O_2(g) \rightarrow NO(g)$

Bonds broken: 1/2 N ≡ N = 1/2 x 946 = +473 kJ

$1/2\ O_2$ = 1/2 x 497 = +248 kJ

Bond formed: 1 NO = $-x$

$\Delta H_{rxn} = \Delta\overline{H}^\circ_f[NO(g)]$ = +90.37 kJ = 473 kJ + 248 kJ - x

x = bond energy of NO = 473 + 248 - 90 = 631 kJ/mol

Note that we did not have to speculate on the nature of the bonds in O_2 and NO (double, triple?), but we did have to know that the bond in N_2 is a triple bond to select the proper value from Table 9-3.

Polar molecules

9-41. Proceed as in the illustration for HCl in Section 9-10. Obtain the H - Br bond distance from Table 9-3--151 pm.

magnitude of charge x bond distance = 0.79 D = $0.79 \times 3.34\times10^{-30}$ C m

magnitude of charge x 1.51×10^{-10} m = $0.79 \times 3.34\times10^{-30}$ C m

magnitude of charge = $0.79 \times 3.34\times10^{-30}/1.51\times10^{-10} = 1.7\times10^{-20}$ C

fraction of electronic charge = 1.7×10^{-20} C/1.60×10^{-19} C = 0.11

percent ionic character of HBr ≈ 11%

9-42. The shape of each of these molecules has been described previously in the chapter. This information must be combined with knowledge of electronegativities.

(a) SO_2, V-shaped, polar

(b) NH_3, trigonal pyramidal, polar

(c) H_2S, angular, polar

(d) C_2H_4, $H_2C=CH_2$, symmetrical planar, nonpolar

(e) SF_6, octahedral, nonpolar. (This is a symmetrical molecule and the bond dipole moments cancel.)

(f) CH_2Cl_2, tetrahedral, polar. [Because of differences in the bond dipole moments for C - H and C - Cl, there is a nonsymmetrical pull on electrons (recall Figure 9-15).]

9-43. The H - O bonds in the H - O - O - H have bond dipole moments, but if the molecule were linear there would be no resultant dipole moment. H_2O_2 possesses a dipole moment of 2.13 debye; the molecule cannot be linear. (Also, recall Example 9-11.)

Partial ionic character of covalent bonds

9-44. Use the method of Example 9-18.

For HF: H - F (est.) = 1/2(H - H + F - F) = 1/2(435 + 155) = 295 kJ/mol

H - F (actual, from Table 9-3) = 565 kJ

Ionic resonance energy (IRE) = 565 - 295 = 270 kJ/mol

For HBr: H - Br (est.) = 1/2(H - H + Br - Br) = 1/2(435 + 192) = 314 kJ/mol

H - Br (actual, from Table 9-3) = 364 kJ/mol

IRE = 364 - 314 = 50 kJ/mol

The value derived in the text for HCl was IRE = 92 kJ/mol. We expect the HF bond to have the greatest percent ionic character and HBr the least when we compare HF, HCl and HBr. This is the same order obtained for ionic resonance energies, that is IRE(HF) > IRE(HCl) > IRE(HBr).

9-45. (a) $(\Delta EN)^2 = \frac{IRE}{96} = \frac{270}{96} = 2.81$ $\Delta EN = \sqrt{2.81} = 1.68$

$E_F - E_H = \Delta EN = 1.68$ (calculated)

$E_F - E_H = 3.98 - 2.20 = 1.78$ (from Table 8-6)

(b) $(\Delta EN)^2 = \frac{IRE}{96} = \frac{50}{96} = 0.52$ $\Delta EN = \sqrt{0.52} = 0.72$

$E_{Br} - E_H = \Delta EN = 0.72$ (calculated)

$E_{Br} - E_H = 2.96 - 2.20 = 0.76$ (from Table 8-6)

9-46. Use electronegativity differences (from Table 8-6) and equation (9.34) to calculate IRE for the N-O bond.

$\Delta EN = E_O - E_N = 3.44 - 3.04 = 0.40$ $IRE = 96(\Delta EN)^2 = 96 \times (0.40)^2 = 15$ kJ/mol

Since the measured N-O bond energy is 201 kJ/mol and the IRE is 15 kJ/mol, the average of N-N and O-O bond energies must be 201 - 15 = 186 kJ/mol.

$$E_{N-O} = \frac{E_{N-N} + E_{O-O}}{2} = \frac{163 \text{ kJ/mol} + E_{O-O}}{2} = 186 \text{ kJ/mol}$$

$E_{O-O} = (2 \times 186 - 163)$ kJ/mol = 209 kJ/mol

Polymers

9-47. (a) A monomer is a simple molecule that is able to join with others of a like kind to form giant molecules called polymers. A polymer, then, is a molecule comprised of a small repeating unit (the monomer).

(b) Elastomers are polymers that have the ability to regain their shapes after deformation (they are elastic). Fibers are polymers that have great strength along their long axis (they can be pulled without breaking).

(c) Natural rubber is a polymer hydrocarbon (an elastomer) based on the monomer shown in structure (9.40). A snythetic rubber is any of a number of elastomers having properties similar to natural rubber but based on different monomers.

(d) An organic polymer is based on a carbon-atom chain. Inorganic polymers have their chains based on atoms other than carbon, e.g., silicon or sulfur.

(e) A thermoplastic polymer becomes soft on heating and regains its shape and strength on cooling. A thermosetting plaster develops permanent crosslinks the first time it is heated. Upon reheating it does not regain its plastic condition. Instead, crosslinks break and the polymer degrades or decomposes.

9-48. (a) The repeating unit in Teflon is C_2F_4. Four units are shown below.

```
  F   F   F   F   F   F   F   F
  |   |   |   |   |   |   |   |
- C - C - C - C - C - C - C - C -
  |   |   |   |   |   |   |   |
  F   F   F   F   F   F   F   F
```

(b) Based on the monomer unit C_2F_4, the

$$\% \text{ F} = \frac{(4 \times 19.0) \text{ g F}}{[(2 \times 12.0) + (4 \times 19.0)] \text{ g } C_2F_4} \times 100 = 76.0\% \text{ F}$$

This calculation assumes that the polymer contains only C and F, and in the same proportions as found in the monomer C_2F_4. As can be seen from this four-unit structure in (a), some additional group(s) must be located at the ends of the chain. Suppose that in the four-unit chain these were the group X, having a formula weight of 50. That is, X-C-C ··· C-X. The formula weight of the four-unit chain would be (2 x 50) + (8 x 12) + (16 x 19) = 500. The % F in this chain would be

$$\% \text{ F} = \frac{(16 \times 19) \text{ g F}}{500 \text{ g}} \times 100 = 61\% \text{ F}$$

The % F does appear to depend on the chain length. However, as the chains get longer and longer, the proportions of C and F become more nearly the same as in the monomer C_2F_4 (the number of X groups remains fixed at two per chain). In actual practice, then, we would find the % F to be essentially 76.0% F, as calculated in (a).

9-49. (a) In polyformaldehyde the double bond between C and O opens up producing a chain with alternating C and O atoms. Ten units of the chain are illustrated below:

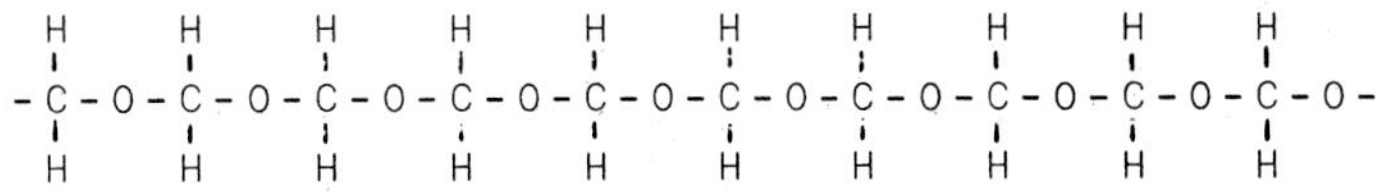

```
 H   H   H   H   H   H   H   H   H   H
 |   |   |   |   |   |   |   |   |   |
-C-O-C-O-C-O-C-O-C-O-C-O-C-O-C-O-C-O-C-O-
 |   |   |   |   |   |   |   |   |   |
 H   H   H   H   H   H   H   H   H   H
```

(b) Assume that the % C in the polymer is the same as in formaldehyde, CH_2O. Rephrase the question to what volume of $CO_2(g)$, measured at 25°C and 751 mmHg would be produced by the combustion of 1.05 g CH_2O.

$CH_2O + O_2(g) \rightarrow CO_2(g) + H_2O(\ell)$

$$\text{no. mol } CO_2(g) = 1.05 \text{ g } CH_2O \times \frac{1 \text{ mol } CH_2O}{30.0 \text{ g } CH_2O} \times \frac{1 \text{ mol } CO_2}{1 \text{ mol } CH_2O} = 0.0350 \text{ mol } CO_2$$

$$V = \frac{nRT}{P} = \frac{0.0350 \text{ mol} \times 0.0821 \text{ L atom mol}^{-1}\text{K}^{-1} \times 298 \text{ K}}{(751/760) \text{ atm}} = 0.867 \text{ L} = 867 \text{ cm}^3 \ CO_2(g)$$

9-50. Determine the number of monomers (C_3H_6) in the sample of propylene gas. Since each polymer contains 875 monomers, the number of polymer molecules is just 1/875 of the number of monomers.

$$n = \frac{PV}{RT} = \frac{(748/760) \text{ atm} \times 0.315 \text{ L}}{0.0821 \text{ L atom mol K} \times 293 \text{ K}} = 0.0129 \text{ mol } C_3H_6$$

$$\text{no. polymer molecules} = 0.0129 \text{ mol } C_3H_6 \times \frac{6.02 \times 10^{23} \ C_3H_6 \text{ molecules}}{1 \text{ mol } C_3H_6} \times \frac{1 \text{ polymer molecule}}{875 \ C_3H_6 \text{ molecules}}$$

$$= 8.88 \times 10^{18} \text{ polymer molecules}$$

Self-test Questions

1. (*b*) Eliminate (d) as being an ionic compound (or a covalent compound with three Al - Cl single bonds). CO_2 is discussed in several places in the text as having double bonds :O=C=O: The NO_3^- ion is given as an example of resonance with one N=O and two N-O bonds. The only possibility is CN^-, for which we can write [:C≡N:]⁻.

2. (*d*) The Lewis structure is $\left[\begin{array}{c} H \\ | \\ H-N-H \\ | \\ H \end{array}\right]^+$. Because there are four electron pairs (all bond pairs) distributed around the central N atom, the structure is tetrahedral (not square planar). The N-H bonds are covalent (only the bond between NH_4^+ and some anion is ionic). One of the N-H bonds can be considered coordinate covalent, but not all four.

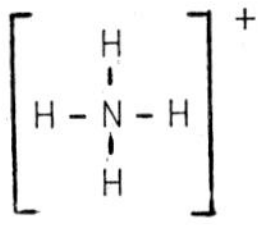

3. (*c*) Only the N atom carries a formal charge (+1), equal to the charge on the ion. The O atoms are assigned 6 electrons in the structure given, and this is the same as their number of valence electrons. The O atoms carry no formal charge.

4. (*a*) Sketch Lewis structures for each molecule to see how the valence shell electron pairs are distributed. Recall that multiple bonds are treated as if they were single bonds. The following are all linear: :O=C=O: , H-C≡N: , and H-C≡C-H. SO_2 is not

```
   ..
   S
 // \
:O   O:
```

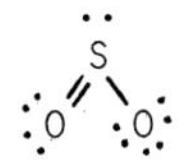

5. (*a*) The nonpolar molecule is the one whose symmetrical geometric shape causes bond dipole moments to cancel. BCl_3 is trigonal planar. It has no resultant dipole moment.

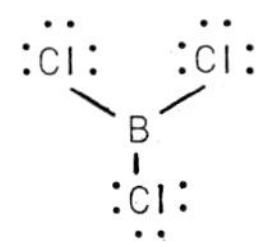

```
:Cl:   :Cl:
    \  /
     B
     |
    :Cl:
```

CH_2Cl_2 is tetrahedral and nonsymmetrical

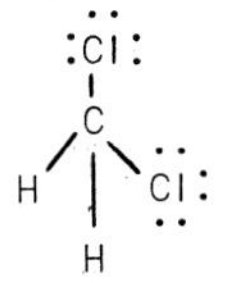

```
   :Cl:
     |
     C
   / | \
  H  |  Cl:
     H
```

6. (*c*) The structure of cyanate ion shows C with an incomplete octet. That of the carbide ion has ten electrons around one of the C atoms. The structure of NO has a negative formal charge on N and a positive formal charge on O. A more likely structure would be to move a pair of electrons from N to O and to place the unpaired electron on N. Of the structures shown, only that of hypochlorite ion seems correct in all details.

7. (a) Based on 100.0 g of compound,

$$\text{no. mol C} = 24.3\text{ g C} \times \frac{1\text{ mol C}}{12.0\text{ g C}} = 2.02\text{ mol C}$$

$$\text{no. mol Cl} = 71.6\text{ g Cl} \times \frac{1\text{ mol Cl}}{35.5\text{ g Cl}} = 2.02\text{ mol Cl}$$

$$\text{no. mol H} = 4.1\text{ g H} \times \frac{1\text{ mol H}}{1.01\text{ g H}} = 4.06\text{ mol H}$$

$$\text{empirical formula} = C_{2.02}H_{4.06}Cl_{2.02} = CH_2Cl$$

(b) Lewis structure:

```
     H
     |   ..
 H - C - Cl:
     .   ..
```

The C atom does not have a complete octet and there is one unpaired electron in the structure.

(c) A more plausible Lewis structure is based on the molecular formula $C_2H_4Cl_2$

```
       H   H
  ..   |   |   ..
 :Cl - C - C - Cl:
  ..   |   |   ..
       H   H
```

8. First draw plausible Lewis structures. Then try to deduce their geometric shapes.

I

```
 H - C = C = C - H
     |       |
     H       H
```

II

```
              H
              |
 H - C ≡ C -  C - H
              |
              H
```

In structure I the three C atoms lie along the same straight line, but the H atoms cannot lie on the same line. The distribution of three electron pairs about each end carbon atom is trigonal planar. The molecule is not linear. In structure II the three C atoms again lie along the same straight line. The H atom at the left of the structure is situated on the same line, but the H atoms at the right are not. [There is a tetrahedral distribution of the four electron pairs (bond pairs) about the C atom at the right.] Structure II is not linear.

9. The shortest nitrogen-to-nitrogen bond distance is expected for the structure with the most multiple bond character in the nitrogen-to-nitrogen bond.

(a)
```
 H - N - N - H
     |   |
     H   H
```

(b) :N ≡ N:

(c)
```
 :O:            :O:
    \\          /
      N  -  N
    /          \\
 :O:            :O:
```

(d)
```
              ..        ..       ..
 :N ≡ N - O:  <-->  :N = N = O:
              ..
```

There is but a single triple-bond structure for N_2. For N_2O the true structure is a hybrid of a triple bond structure and structures with less multiple bond character. The shortest nitrogen-to-nitrogen bond distance is expected to be that in N_2.

10. (a)
```
     ..
     S
   //  \
 :O:    :O:
```

VSEPR theory predicts the distribution of three electron pairs (with one pair being a lone pair). For the molecule AX_2E the geometry is V-shaped.

(b)
```
     ..
    :O
     ||
     S
    /  \
 :O:    :O:
```

Although other plausible structures can be written as well, all lead to a prediction of trigonal planar geometry.

(c)
```
 [     ..     ]2-        [     ..     ]2-        [     ..     ]2-
 [    :O:     ]          [    :O:     ]          [    :O      ]
 [ ..  |  ..  ]          [ ..  ||  .. ]          [ ..  ||  .. ]
 [:O - S - O: ]    or    [:O - S - O: ]    or    [:O - S = O: ]
 [ ..  |  ..  ]          [ ..  |   .. ]          [ ..  |      ]
 [    :O:     ]          [    :O:     ]          [    :O:     ]
 [     ..     ]          [     ..     ]          [     ..     ]
```

Whichever structure is used, all lead to the prediction of a tetrahedral distribution of four electron pairs (all bond pairs). The SO_4^{2-} ion has a tetrahedral shape.

11. $2\ NO(g) + 5\ H-H(g) \rightarrow 2\ \underset{H}{\overset{H}{N}}-H(g) + 2\ H-\overset{H}{O}(g)$

Bonds broken: $2\ NO = 2 \times 628 = 1.26\times10^3$ kJ/mol

$5\ H-H = 5 \times 435 = 2.18\times10^3$ kJ/mol

Bonds formed: $6\ N-H = -6 \times 389 = -2.33\times10^3$ kJ/mol

$4\ O-H = -4 \times 464 = -1.86\times10^3$ kJ/mol

$\Delta H = (1.26 + 2.18 - 2.33 - 1.86)\times10^3 = -0.75\times10^3$ kJ/mol - 0.75×10^3

12. (a) The three atoms of a triatomic molecule must always lie in the same plane, but sometimes they are situated along a straight line, as in CO_2. Some triatomic molecules, then, are better described as linear than as planar.

(b) An electronegativity difference between bonded atoms always results in a bond dipole moment, but whether there is a resultant dipole moment in the molecule depends on the shape of the molecule. Sometimes, because of the symmetry of a molecule, bond dipole moments cancel out and the molecule as a whole is nonpolar, as is the case with CCl_4, for example.

(c) Although Lewis structures without formal charges are generally preferred, at times the only plausible structures that can be written, or the ones that correspond best to experimental evidence, carry formal charges, for example,

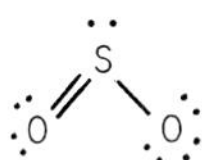

Chapter 10

Chemical Bonding II: Additional Aspects

Review Problems

10-1. (a) HCl: H $1s$ [↑]   Cl[Ne] $3s$ [↑↓] $3p$ [↑↓|↑↓|↑]

The $1s$ orbital of H overlaps the half-filled $3p$ orbital of Cl to produce a linear molecule.

(b) ICl: I [Kr]$4d^{10}$ $5s$ [↑↓] $5p$ [↑↓|↑↓|↑]   Cl [Ne] $3s$ [↑↓] $3p$ [↑↓|↑↓|↑]

The half-filled $3p$ orbital of Cl overlaps with the half-filled $5p$ orbital of I to produce a linear molecule.

(c) H_2Se: H $1s$ [↑]   Se [Ar]$3d^{10}$ $4s$ [↑↓] $4p$ [↑↓|↑|↑]

This structure is like that of H_2S pictured in Figure 10-2, except that the bonding orbitals of Se are $4p$, rather than $3p$, as in S.

(d) NI_3: N $1s$ [↑↓] $2s$ [↑↓] $2p$ [↑|↑|↑]   I [Kr]$4d^{10}$ $5s$ [↑↓] $5p$ [↑↓|↑↓|↑]

This structure is similar to that of NH_3 in Figure 10-3, with I atoms substituting for H atoms.

10-2. (a) The central C atom employs four $2sp^3$ hybrid orbitals; these have a tetrahedral symmetry. Each of the two H atoms employs a $1s$ orbital, and each of the two Cl atoms, a half-filled $3p$ orbital.

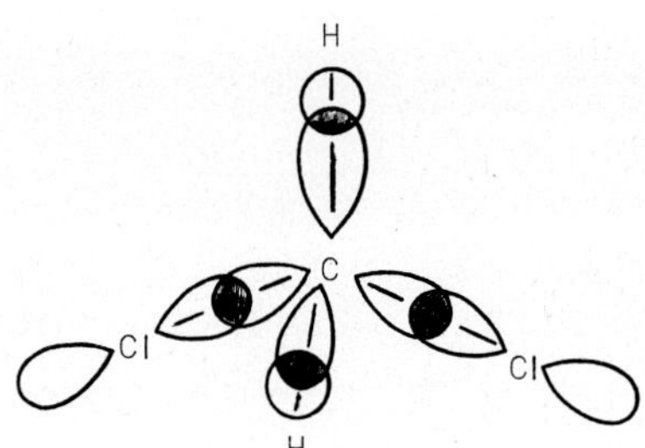

(b) The normal Be atom has the electron configuration $1s$ [↑↓] $2s$ [↑↓] $2p$ [| |]

Based on this ground-state electron configuration, we should expect no bond formation at all. To produce the half-filled orbitals required to form two covalent bonds requires an excited electron configuration.

Be $1s$ [↑↓] $2sp$ [↑|↑] $2p$ [|]

The overlap of the $2sp$ orbitals with the half-filled $3p$ orbitals of the two Cl atoms produces the linear molecule, $BeCl_2$.

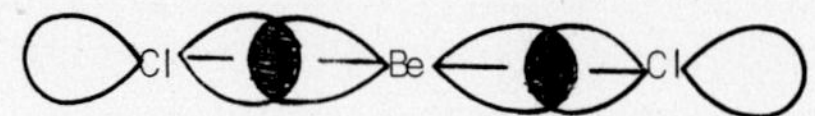

(c) The normal electron configuration of boron suggests an ability to form one covalent bond, not three.

	$1s$	$2s$	$2p$		
	↑↓	↑↓	↑		

Again, the hybridization of orbitals is required. The geometry is trigonal planar with F-B-F bond angles of 120°.

	$1s$	$2sp^2$			$2p$
B	↑↓	↑	↑	↑	

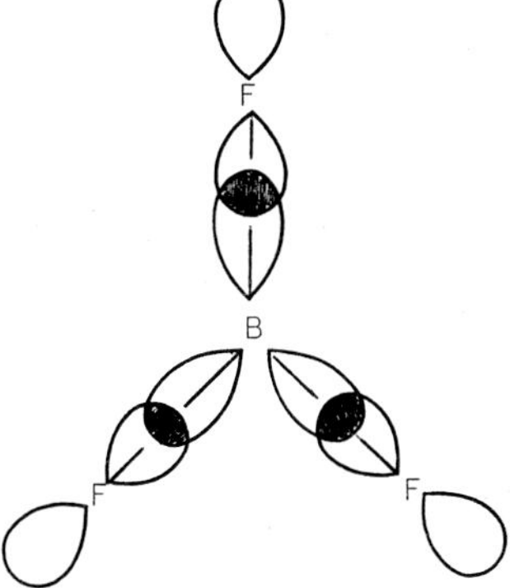

10-3. (a) SF_6 With VSEPR theory we predict an octahedral distribution of six pairs of electrons. The hybridization scheme which produces octahedral geometry is sp^3d^2.

(b) CS_2 The Lewis structure reveals two double bonds $:\ddot{S}=C=\ddot{S}:$ The predicted geometry is linear. (Recall that each double bond is treated as if it contained only a single electron pair for the purpose of predicting the geometrical shape of the molecule.) The hybridization scheme leading to a linear geometry is sp.

(c) SiF_4 The Lewis structure is

```
      ..
     :F:
 ..   |   ..
:F - Si - F:
 ..   |   ..
     :F:
      ..
```

The four bond pairs are arranged about the central Si atom in a tetrahedral fashion. The hybridization scheme to account for this is sp^3.

(d) NO_3^- The Lewis structure is

```
[ ..       ..  ]-
[:O - N - O:   ]
[ ..  ||   ..  ]
[    :O:       ]
```

By the valence-shell electron-pair repulsion method we treat this ion as if only three electron pairs were distributed about the N atom. This suggests trigonal planar geometry and an sp^2 hybridization scheme.

(e) AsF_5 This molecule has 10 valence electrons surrounding the As atom, as five bond pairs in a trigonal bipyramidal arrangement. The corresponding hybridization scheme is sp^3d.

10-4. The Lewis structure of NH_2OH is $H-\ddot{N}-\ddot{O}-H$ (with H bonded to N). Both from VSEPR theory and the bond angles given, the distribution of electron pairs about the N atom (three bond pairs and one lone pair) is tetrahedral, as is the distribution about the O atom (two bond pairs and two lone pairs). Both the N and O atoms employ sp^3 hybrid orbitals and the H atoms, $1s$.

10-5. (a)

```
      π
     /
H - C≡N:
   /\
  π  σ
```

(b)

```
      π
     / \
:N≡C - C≡N:
    \ σ /
     \ /
      π
```

(c)

```
    H   H   H   Cl
    |   |   |   |
H - C - C = C - C - Cl
    |   |       |
    H   π       Cl
```

(all other bonds are σ)

(d)

```
    ..  ..  π
H - O - N = O:
    ..  σ
```

10-6. First write a Lewis structure. Determine the molecular shape from VSEPR theory. They propose a hybridization scheme consistent with this shape.

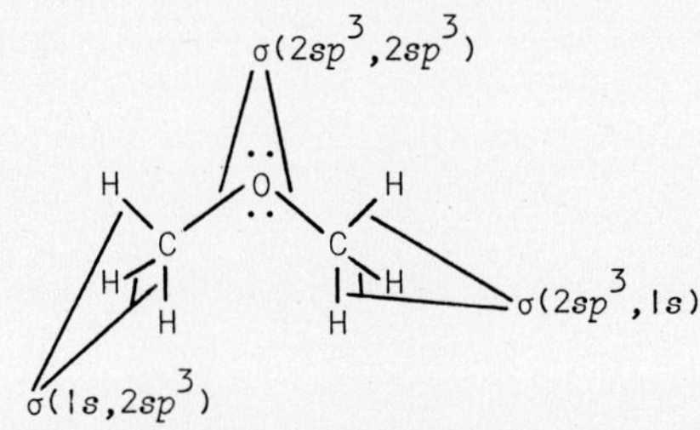

The two C and the O atoms each have four pair of valence electrons. The distribution of these electron pairs is tetrahedral, requiring sp^3 hybridization.

10-7. (a) H - C ≡ N: The distribution of the two σ bonds about the C atom must be *linear*, using *sp* hybrid orbitals.

(b) :N ≡ C - C ≡ N: Each C atom forms two σ and two π bonds. This requires *sp* hybridization at each C atom and a linear distribution of electron pairs at each C atom. The molecule is *linear*.

(c)
```
    ..
   :F:
 ..  |
:F - C - C ≡ N:
 ..  |
   :F:
    ..
```
The hybridization scheme at the C atom on the left is sp^3 (tetrahedral). At the other C atom, it is *sp* (linear). The molecule is neither linear nor planar; it has a three-dimensional shape.

(d)
```
H
 \         ..
  C = C = O:
 /
H
```
The C atom at the left must use sp^2 hybridization to form three σ and one π bond. The C atom at the right uses *sp* hybridization to form two σ and two π bonds. The C atom at the left establishes a planar shape, and the C = C = O linear end of the molecule is in the same plane.

10-8. One approach is to write a molecular orbital diagram for each species, in the manner of Example 10-8 and to determine if there are any unpaired electrons. Perhaps simpler still is to refer to Figure 10-18; add or subtract electrons for ions; and determine if there are any unpaired electrons.

(a) F_2: All electrons are paired--diamagnetic.

(b) N_2^+: The neutral molecule has all electrons paired; but when one electron is lost, this leaves one electron unpaired--paramagnetic.

(c) O_2^-: The neutral O_2 molecule has two unpaired electrons, when an electron is gained to form O_2^-, this still leaves one unpaired electron--paramagnetic.

10-9. (a) H_2^-:

σ^b_{1s}	σ^*_{1s}
↑↓	↑

(b) N_2^+: KK

σ^b_{2s}	σ^*_{2s}	π^b_{2p}	σ^b_{2p}	π^*_{2p}	σ^*_{2p}
↑↓	↑↓	↑↓ ↑↓	↑		

(c) F_2^-: KK

σ^b_{2s}	σ^*_{2s}	σ^b_{2p}	π^b_{2p}	π^*_{2p}	σ^*_{2p}
↑↓	↑↓	↑↓	↑↓ ↑↓	↑↓ ↑↓	↑

(d) Ne_2^+ is isoelectronic with F_2^- and has the molecular orbital diagram shown in (c).

10-10. Each Na atom contributes one 3*s* atomic orbital to form a set of molecular orbitals. There are as many molecular orbitals (energy level) in the 3*s* conduction band as there are atoms of Na in the crystal.

$$\text{no. energy levels} = \text{no. Na atoms} = 25.3 \text{ mg Ma} \times \frac{1 \text{ g Na}}{1000 \text{ mg Na}} \times \frac{1 \text{ mol Na}}{23.0 \text{ g Na}} \times \frac{6.02\times10^{23} \text{ Na atoms}}{1 \text{ mol Na}}$$

$$= 6.62\times10^{20} \text{ Na atoms} = 6.62\times10^{20} \text{ energy levels.}$$

Exercises

Valence bond method

10-1. The best description of the bond angle in H_2Se is, "less than in H_2S, but not less than 90°". In terms of atomic orbital overlap, the orbitals of the Se atom involved in the bonding are two $4p$ orbitals. These are at right angles to each other, suggesting a bond angle of 90°. For the analogous molecule, H_2O, mutual repulsion of the H atoms causes the bond angle to enlarge (to 104.5°). In H_2S the effect is greatly reduced and the bond angle is about 92°. In H_2Se this angle should be still closer to the predicted 90°.

10-2. The electron configuration of nitrogen is $1s$ [↑↓] $2s$ [↑↓] $2p$ [↑ | ↑ | ↑] and the Lewis structure is :N≡N: The lone pair electrons in this structure are the $2s$ electrons of N. The bonds involve the overlap of the three $2p$ orbitals of one N atom with those of the other N atom. Electron pairs are shared in the regions of overlap. The only difficulty in describing this structure is in depicting how the overlap occurs. Two of the p orbitals overlap in the end-to-end fashion pictured in Figures 10-2 and 10-3. The other p orbitals must overlap in a "sidewise" fashion (referred to as π bonds and described in Section 10-3).

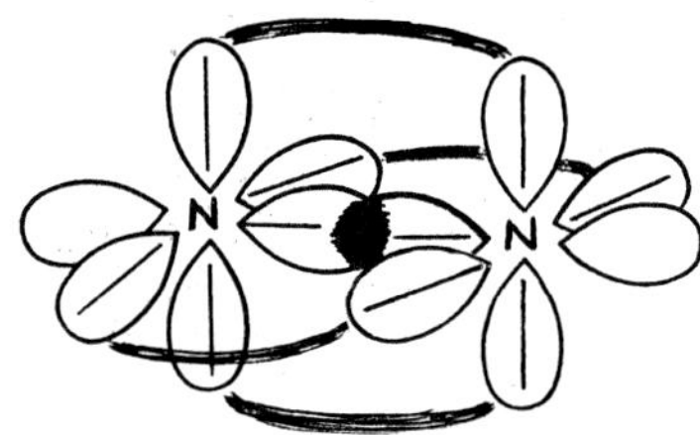

10-3. The principal advantage of the valence bond method over the use of Lewis structures in describing chemical bonding is that a prediction of the geometric structure of the molecule is made possible. Also, when the idea of hybridized atomic orbitals is included, the valence bond method leads to the prediction of multiple bonds in a more direct fashion than the arbitrary shifting of electron dots in an attempt to satisfy the octet rule.

10-4. From the Lewis structure we note four pairs of electrons around the central N atom, all bond pairs. The geometric shape is tetrahedral. The bonding scheme must involve sp^3 hybrid orbitals of the central N atom.

$$\left[\begin{array}{c} H \\ | \\ H-N-H \\ | \\ H \end{array}\right]^+$$

N $1s$ [↑↓] $2sp^3$ [↑↓ | ↑ | ↑ | ↑]

The overlap of $1s$ orbitals of H atoms with half-filled $2sp^3$ orbitals of the central N atom accounts for three of the four N-H bonds. The fourth N-H bond results from the overlap of an *empty* $1s$ orbital of H (that is, of H^+) with a *filled* $2sp^3$ orbital of N; this fourth bond is coordinate covalent.

10-5. There are several ways to proceed but one of the simplest is to write the best Lewis structure. This will identify the central atom, indicate the number of valence electron pairs, and differentiate between σ and π bonds. With this information it is not difficult to propose a bonding scheme for the central atom.

(a) CO_2: :Ö=C=Ö: The central C atom forms two σ and two π bonds. According to VSEPR theory, the molecule is linear. The hybridization scheme is sp.

(b) $ClNO_2$: :Ö=N(⊕)-C̈l: with :Ö:(⊖) bonded below N. Although structures can be written that have no formal charges, they do not involve a single central atom. This one puts the least electronegative atom as the central atom. The distribution of electrons about the N atom is trigonal planar. The hybridization of the N atom is sp^2.

(c) N_2O: :N̈=N=Ö: or :N≡N-Ö: Whichever structure is used, the central N atom forms two σ and two π bonds. This requires sp hybridization at the central N atom.

10-6. According to VSEPR theory we should predict a trigonal planar distribution of electron pairs about the central C atom, leading to 120° angles for the C - C = O and the O = C - O bonds. The hybridization scheme for the central C atom is sp^2. The C atom of the H_3C- group must use sp^3 hybrid orbitals (because it is able to form four single bonds). The situation with the O atom of the -OH group is not clear, since the C - O - H bond angle is not given. For example, structures could be written in which this O uses $2p$ orbitals or $2sp^2$ or $2sp^3$. In the structure shown below, $2sp^3$ orbitals are used. The orbital used by the O atom of C = O is also not certain, but for the reason stated in Example 10-2 of the text, a p orbital is used below.

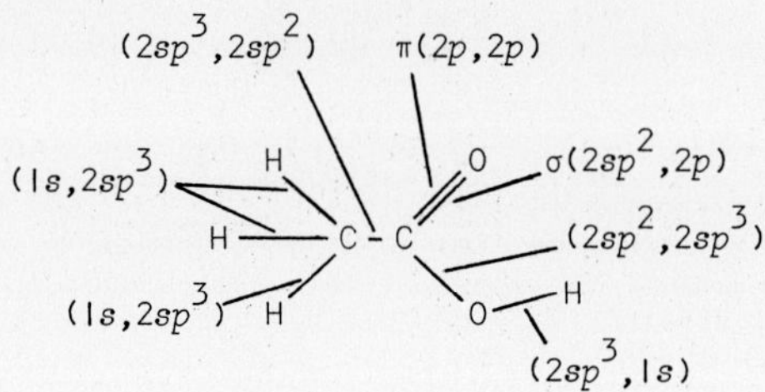

10-7. Begin with a Lewis structure. Use VSEPR theory to show this structure in three dimensions. Identify the orbitals consistent with this structure.

H, H, N - C - C, H, H, O, O - H

The C - C(=O)(-O) portion of the molecule is planar, with the central C atom using sp^2 hybrid orbitals. The C atom on the left uses sp^3 hybrid orbitals in a tetrahedral arrangement; so does the N atom. The O atom in the -OH group also uses sp^3 hybrid orbitals. The N atom and all of the H atoms are out of the plane of the C - C(=O)(-O) grouping of atoms.

10-8. The data provided agree reasonably well (compare with Table 9-3) with this Lewis structure.

$\pi(2p,2p)$ $\sigma(2sp^2,2p)$
H - C ≡ C - C(=O) - H $\sigma(2sp^2,1s)$
$\sigma(1s,2sp)$ $\sigma(2sp,2sp)$ $\sigma(2sp,2sp^2)$ $\pi(2p,2p)$

The C atom on the left and the one in the center use sp hybridization. The C atom on the right uses sp^2 hybridization. Overall, the molecule has a planar shape.

10-9. (a) Cl, C, Cl, Cl, Cl — Each of the bonds involves overlap of $2sp^3$ orbitals of C with $3p$ orbitals of Cl atoms.

(b) $\pi(2p,2p)$ $\sigma(2sp^2,3p)$ O = N - Cl $\sigma(2sp^2,2sp^2)$ — In this planar molecule the central N atom uses $2sp^2$ orbitals.

(c) 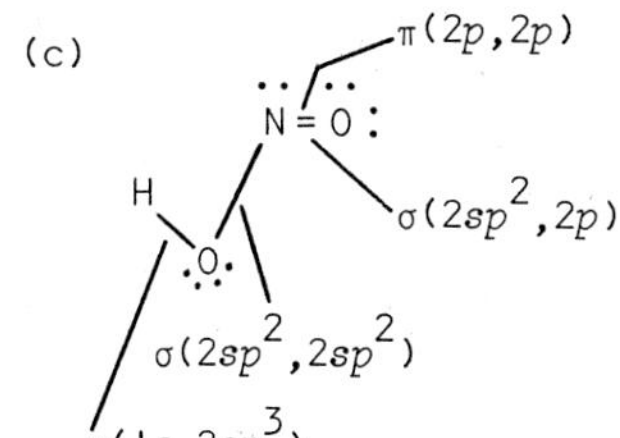

The central N atom uses sp^2 hybrid orbitals. The ONO group is planar and the H atom is out of the plane.

(d) $\pi(2p,2p)$; $\text{H}_3\text{C}-\text{C}\equiv\text{C}-\text{H}$; $\sigma(1s,2sp^3)$; $\pi(2p,2p)$; $\sigma(2sp^3,2sp)$; $\sigma(2sp,1s)$; $\sigma(2sp,2sp)$

Based on this Lewis structure the two C atoms on the right use sp hybridization. The C - C ≡ C - H grouping is linear. The C atom on the left uses sp^3 hybridization.

10-10. (a) XeF_4: Use VSEPR theory to determine the shape of this molecule. Then propose a bonding scheme.

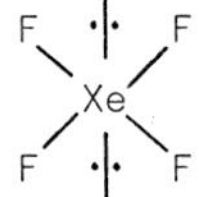

Each bond involves sp^3d^2 hybridization of Xe and a $2p$ orbital of F.

(b) C_3O_2:

$\pi(2p,2p)$; :O=C=C=C=O:; $\sigma(2p,2sp)$; $\sigma(2sp,2sp)$; $\sigma(2sp,2p)$

(c) XeF_2: Use VSEPR theory to determine the number of central electron pairs (and the shape) of this molecule.

The five electron pairs correspond to sp^3d hybridization of the Xe atom. The F atoms use $2p$ orbitals.

(d) I_3^-: Proceed as in part (c).

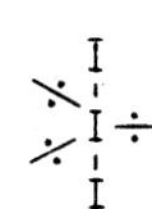

The central I atom uses sp^3d hybrid orbitals and the terminal I atoms, $5p$ orbitals.

(e) 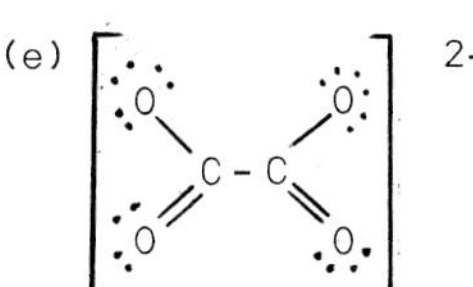

Each C atom uses sp^2 hybrid orbitals to form bonds with two O atoms and the other C atom. The O atoms use $2p$ orbitals to form σ bonds to the C atoms; and the two π bonds each involve $2p,2p$ orbital overlap.

10-11. In the structure ClF_3 there are 10 valence electrons around the central Cl atom (seven from Cl and one each from the F atoms). The VSEPR designation is AX_3E_2. A hybridization scheme consistent with this VSEPR designation is

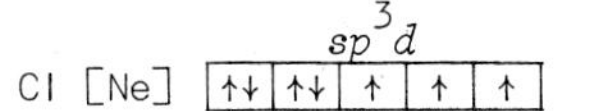

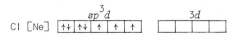

Three of the sp^3d orbitals are involved in the bonding scheme, and two contain lone-pair electrons, as indicated above. (The molecule is T-shaped.)

10-12. Use the VSEPR theory to determine the number of central pairs of electrons (and the geometric shape) of each molecule. For PF_5 the number of pairs is five--all are bond pairs. The molecule is trigonal bipyramidal and the hybridization is sp^3d. For BrF_5 there are six pairs--five bond and one lone pair --by an octahedral distribution. The hybridization is sp^3d^2 and the shape is square pyramidal.

10-13. As a first approximation the strength of a σ bond may be taken as the C - C single bond energy: 347 kJ/mol. The difference between the C = C and C - C bond energies should represent the strength of a π bond: 611 - 347 = 264 kJ/mol. If this assumption is correct, the C ≡ C bond energy should be stronger than the C = C bond by about 264 kJ/mol. Estimate of C ≡ C bond energy: 611 + 264 = 875 kJ/mol. The measured value of the C ≡ C bond energy (Table 9-3) is 837 kJ/mol. The agreement between these two values is only fair, but the line of reasoning is essentially correct.

Molecular orbital method

10-14. Of the diatomic species represented in Figure 10-18, those with more bonding than antibonding electrons are stable molecules, that is, Li_2, B_2, C_2, N_2, O_2 and F_2. Among the stable molecules, those with all electrons paired are diamagnetic and those with unpaired electrons are paramagnetic. Diamagnetic: Li_2, C_2, N_2, F_2; paramagnetic: B_2, O_2.

10-15. Start with the electron configuration of N_2 and determine how that configuration is altered in the formation of N_2^- and N_2^{2-}. The test of whether a species is stable is whether the number of bonding electrons exceeds the number of antibonding electrons.

N_2: KK 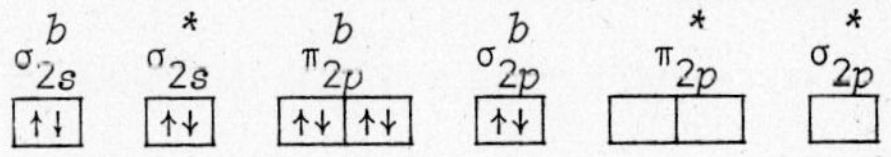 σ^b_{2s} [↑↓] σ^*_{2s} [↑↓] π^b_{2p} [↑↓|↑↓] σ^b_{2p} [↑↓] π^*_{2p} [|] σ^*_{2p} [ ] Bond order = 1/2(8 - 2) = 3

For N_2^- an electron enters a π^*_{2p} molecular orbital. This changes the bond order to 1/2(8 - 3) = 2 1/2; the ion is stable. For N_2^{2-} a second electron enters a π^*_{2p} orbital. This reduces the bond order to 2, but the ion is stable.

10-16. Refer to Figure 10-18. The bond order for O_2 is 1/2(8 - 4) = 2. If an electron is lost to produce O_2^+, we should expect this to be an electron from a π^*_{2p} molecular orbital. The bond order for O_2^+ is 1/2(8 - 3) = 2 1/2. Because we expect bond strength to increase with bond order, the bond in O_2^+ should be stronger than in O_2.

10-17. Consider what would be the electron configuration of B_2 if the σ^b_{2p} molecular orbital filled before π^b_{2p}.

B_2: KK σ^b_{2s} [↑↓] σ^*_{2s} [↑↓] σ^b_{2p} [↑↓] π^b_{2p} [|] π^*_{2p} [|] σ^*_{2p} [ ]

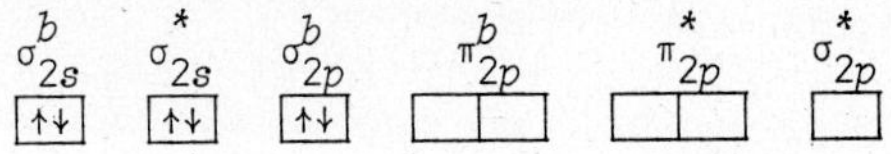

In this configuration all electrons are paired and the molecule should be diamagnetic. The fact that B_2 is paramagnetic provides experimental evidence for an order of filling in which π^b_{2p} precedes σ^b_{2p}. (Two electrons enter the two π^b_{2p} orbitals singly, producing a molecule with two unpaired electrons.)

10-18. To represent the molecule C_2 by a Lewis structure we would have to write C≣C, involving a quadruple bond between carbon atoms. That is, structures such as :C ≡ C would leave one of the C atoms with an incomplete octet. In Lewis structures electrons either are involved in forming bonds or are excluded totally from the process (lone pair electrons). Molecular orbital theory predicts the existence of antibonding electrons. These electrons actively detract from bond formation. In the molecular orbital diagram for C_2, eight valence electrons are assigned, but two of these are antibonding. The bond order in C_2 is 1/2(6 - 2) = 2.

10-19. The molecular orbitals for the first two electronic shells (through nitrogen) are diagrammed below in order of increasing energy:

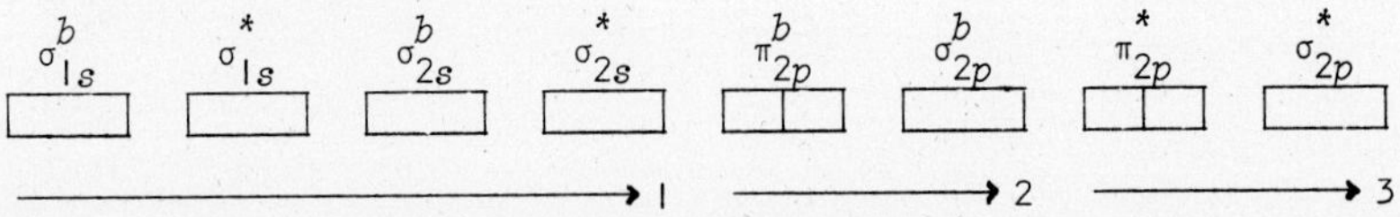

σ^b_{1s} [ ] σ^*_{1s} [ ] σ^b_{2s} [ ] σ^*_{2s} [ ] π^b_{2p} [|] σ^b_{2p} [ ] π^*_{2p} [|] σ^*_{2p} [ ]

——————————→ 1 ——————→ 2 ——————→ 3

In the filling of orbitals through the σ^*_{2s} orbital (to the point indicated as "1"), the bond order varies between 0 and 1 (specifically: 0 → 1/2 → 1 → 1/2 → 0 → 1/2 → 1 → 1/2 → 0). In the filling of orbitals through σ^b_{2p} (to point "2") the bond order increases progressively since all orbitals are bonding orbitals: 1/2 → 1 → 3/2 → 2 → 5/2 → 3. But the remaining orbitals are antibonding orbitals, and as additional electrons are added (to point "3"), the bond order again decreases: 3 → 5/2 → 2 → 3/2 → 1 → 1/2 → 0. Thus, in no instance is a bond order higher than three encountered. (Reversal of the σ^b_{2p} and π^b_{2p} levels in O_2, F_2 and Ne_2 does not alter this result in any way.)

10-20. In all cases the molecular-orbital diagram is assumed to be the same as in Figure 10-18. The electrons to be assigned to these orbitals are the valence shell electrons of the two atoms in the molecule. In this assignment we overlook the fact that the two atoms do not contribute equal numbers of valence electrons to the molecular orbitals:

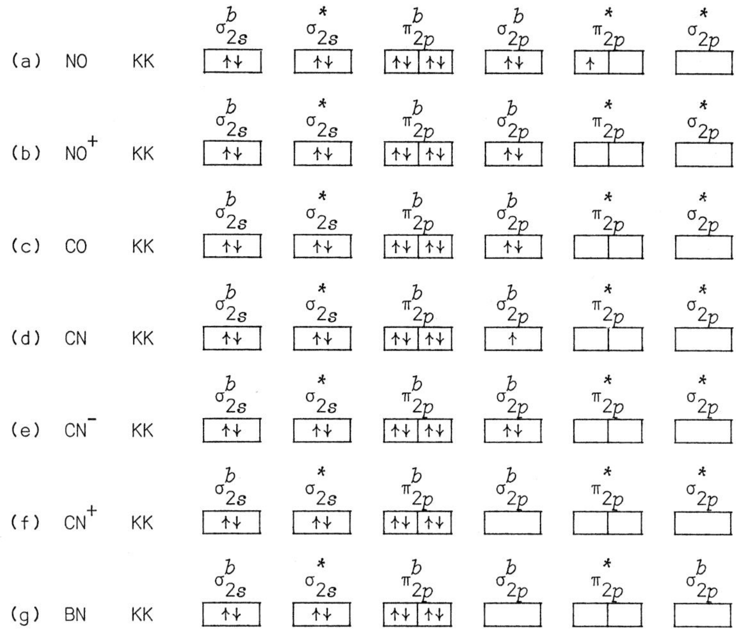

10-21. Isoelectronic species, each containing 8 valence electrons: CN^+ and BN. Isoelectronic species, each containing 10 valence electrons: CO, CN^- and NO^+.

10-22. Refer to Figure 10-17. If the nodal plane is perpendicular to the line joining the centers of the bonded atoms, there is no way for a high electron charge density to exist between the bonded atoms. A molecular orbital having this type of nodal plane must be antibonding. With the π^b_{2p} orbitals there exists a nodal plane between the two lobes of the orbital. High electron charge density is still possible between the bonded atoms, above and below the nodal plane.

Delocalized molecular orbitals

10-23. In writing Lewis structures or structures based on the valence bond method, all electrons are localized to a given bond (or lone pair). If a single structure is inadequate to represent a species, several alternate structures are written, each with a distribution of electrons different from the others. The true structure is a hybrid or "average" of the different plausible structures. Molecular orbital theory allows for the combination of atomic orbitals from several different atoms and the distribution of electrons among the resulting delocalized molecular orbitals. In a sense, the electrons entering these orbitals are "averaged" among several atoms. A plausible structure is written after this averaging has occurred, with the result that only a single structure is needed.

10-24. (a) [Lewis resonance structures of SO_3: O=S(–O)(–O) ⟷ O–S(=O)(–O) ⟷ O–S(–O)(=O)]

(b) The hybridization scheme for the central S atom that accounts for the σ bond framework of SO_3 is sp^2. Consider that the O atoms also employ sp^2 hybridization. All of the sp^2 orbitals are either used in σ bond formation or filled with lone pair electrons. This accounts for 18 of the 24 valence electrons of the S and O atoms. (Recall Figure 10-22a depicting the NO_3^- ion.) Each of the four atoms also has a p orbital to be accounted for. Assume that these four p orbitals are combined into four delocalized molecular orbitals--two bonding and two antibonding. Six electrons must be assigned to these molecular orbitals. Four go into bonding orbitals and two into antibonding orbitals. The result is a π bond with a bond order of 1/2(4 - 2) = 1. This bond is distributed among three sulfur-to-oxygen bonds, making each of these a 1/3 bond. The combination of a σ and 1/3 of a π bond leads to a sulfur-to-oxygen bond order of 1 1/3, the same as predicted from the Lewis structures.

10-25. We expect to find delocalized molecular orbitals in species containing multiple bonds and exhibiting resonance (that is, species for which more than a single Lewis structure must be written).

(a) H-C(H)=C(H)-H Delocalized molecular orbitals are not necessary to describe bonding in C_2H_4.

(b) $[CO_3]^{2-}$ Additional structures must be written in which the double bond is shifted to other O atoms. This is a situation involving resonance. The carbon-to-oxygen π bonding can be described through delocalized molecular orbitals.

(c) NO_2 ⟷ NO_2 The nitrogen-to-oxygen π bond can be described through a molecular orbital spread over the three atoms.

(d) $H_2C=O$ Delocalized molecular orbitals are not needed to describe this structure since multiple bonding is limited to the carbon-to-oxygen bond.

10-26. The line of reasoning here is similar to that of Exercise 10-24. We need to propose a bonding scheme consistent with the structures.

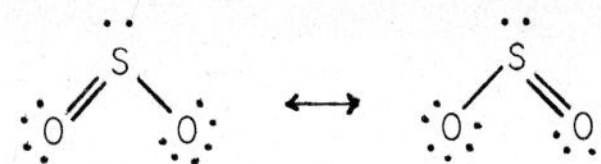

We should employ the hybridization scheme $sp^2 + p$ for each atom. The sp^2 hybrid orbitals, some with bond pairs and some with lone pair electrons, accommodate a total of 14 electrons. There are 18 valence electrons to be accounted for altogether. The three p orbitals (one from each atom) are combined into three delocalized molecular orbitals of the π type. One of these three orbitals is a bonding orbital and one is antibonding. The third orbital is a *nonbonding* orbital--it neither adds to nor detracts from bond formation. The order of increasing energy of these orbitals is bonding < nonbonding < antibonding. Of the four electrons to be assigned to these delocalized molecular orbitals, two enter the bonding orbital and two, the nonbonding. The bond order is 1/2(2) = 1 for a π bond distributed over three atoms. Each sulfur-to-oxygen bond consists of one σ and one half π bond for a bond order of 1.5, just as predicted from the Lewis structure.

Metallic bonding

10-27. (a) Although there exists a pattern for the atomic numbers within a group of the periodic table, there is no relationship between atomic number and metallic character of an element. That is there are no relationships of the sort that would say that metals are elements of low atomic number, or that the metallic character of the elements increases continuously with atomic number and so on.

(b) The situation concerning atomic weight and metallic character is the same as that discussed in part (a) for atomic number.

(c) Here there is a relationship. Metal atoms have small numbers (1, 2, cometimes 3) of valence electrons. Within the representative groups (IA and IIA, for example) the smaller the number of valence electrons the more metallic the element.

(d) Here there is also a relationship. In general, the greater the number of empty valence shell orbitals, the more metallic the element.

(e) Metals occur in every period of the periodic table (except the first period). Thus, there is no relationship between metallic properties and total number of electronic shells in an atom.

10-28. Here are the ground-state electron configurations of the metals in question:

Zn $[Ar]3d^{10}$ 4s [↓↑] 4p [ ][ ][ ]

The greatest possibility for the formation of hybrid orbitals and the filling of these orbitals with available electrons exists for Fe, where *3d*, *4s* and *4p* orbitals are all available. In Zn the *4s* and *4p* orbitals are available for hybridization and two electrons for sharing with other atoms. In Na, presumably the *3s* and *3p* orbitals are available for hybridization, but there is only one electron per atom. We should expect both hardness and melting point to increase in the order: Na < Zn < Fe.

Semiconductors

10-29. (a) Germanium is an intrinsic semiconductor. Its semiconductor properties derive solely from the narrow energy gap between the valence and conduction bonds.

(b) Germanium doped with Al is analogous to the Si-B combination shown in Figure 10-25b; it is a p-type semiconductor.

(c) Here the similarity is to Figure 10-25a (i.e., As is in the same group as P). The semiconductor is n-type.

(d) Silicon, like Ge, is an intrensic semiconductor.

10-30. (a) In an intrinsic semiconductor, postive holes and electrons are equally important in the conduction of electricity. The n-type semiconductor is produced from an intrinsic semiconductor by the appropriate doping. Although this doping greatly increases the number of electron carriers, the positive holes associated with the original intrinsic semiconductor are still present.

(b) For an intrinsic semiconductor the jumping of electrons across the narrow energy gap between valence and conduction bonds is the entire basis of its electrical conductivity. This passage of electrons is greatly facilitated by raising the temperature. With intrinsic semiconductors electrical conductivity is due to the dopant added rather than electrons having to jump an energy gap.

10-31. (a) Take the minimum energy content associated with visible light (minimum frequency). From Figure 7-3 we see this to be $\nu = 4\times10^{14}s^{-1}$. $E = h\nu = 6.626\times10^{-34}$ Js $\times$ $4\times10^{14}s^{-1} = 2.65\times10^{-19}$ J. Convert this to kJ/mol.

$$\frac{2.65\times10^{-19}\ \text{J}}{\text{photon}} \times \frac{6.02\times10^{23}\ \text{photons}}{\text{mol}} \times \frac{1\ \text{kJ}}{1000\ \text{J}} = 160\ \text{kJ/mol}$$

Now convert this to eV.

$$\text{no. eV} = \frac{160\ \text{kJ}}{\text{mol}} \times \frac{1\ \text{eV}}{96.49\ \text{kJ}} = 1.66\ \text{eV}.$$ The energy gap must be less than about 1.7 eV.

(b) Because there are so many energy levels in the conduction band, and because these levels are so closely spaced, any photon of light whose energy content exceeds the energy gap would have the proper energy to correspond to an allowable transition to an energy level in the conduction band. Thus a broad range of wavelengths of light can be used in a solar energy cell.

10-32. (a) Convert from the incident solar power to the power that falls on the 40 cm^2 cell, and include a factor for the efficiency of the cell.

$$\frac{1.00\ \text{kW}}{\text{m}^2} \times \frac{1\ \text{m}^2}{(100)^2\text{cm}^2} \times \frac{1000\ \text{W}}{\text{kW}} \times 40\ \text{cm}^2 \times \frac{15\ \text{W delivered power}}{100\ \text{W incident power}} = 0.60\ \text{W}$$

(b) The conversions required are

$0.60\ \text{W} = 0.60\ \text{Js}^{-1}$

one J = one volt coulomb

$0.60\ \text{W} = 0.60\ \text{VCs}^{-1}$

one Cs^{-1} is one amp (A) of current

$0.60\ \text{W} = 0.60\ \text{VA}$

If the 0.60 VA product is to be associated with a voltage of 0.45 V, 0.60 VA = 0.45 V x current

$$\text{current} = \frac{0.60\ \cancel{\text{V}}\text{A}}{0.45\ \cancel{\text{V}}} = 1.3\ \text{A}$$

Self-Test Questions

1. (*d*) Draw Lewis structures for the four molecules listed and determine which one requires a trigonal planar distribution of valence shell electron pairs. This distribution corresponds to sp^2 hybridization.

```
 ..                            .. .. ..          H     ..
H-N-H         :C≡O:          :Cl-S-Cl:            \C=O:
  |                            .. .. ..          H/
  H
```

2. (*b*) π bonds are found in multiple covalent bonds and arise from the sidewise overlap of *p* orbitals. In PCl_5 bonding orbitals of the P atom are sp^3d hybrid orbitals that overlap in an end-to-end fashion with $3p$ orbitals of the Cl atoms. In N_2 each N atom uses three $2p$ orbitals for bond formation. One pair of orbitals overlpas in an end-to-end (σ) fashion and the other two, sidewise (π). The orbital overlap in OF_2 involves $2p$ orbitals of F with $2p$ (or $2sp^3$) orbitals of O, but the overlap is end-to-end (σ). The species He_2 is not a stable molecule.

3. (*b*) Hydrogen cannot form multiple bonds. That is, H atoms can be bonded to other atoms only by σ bonds, not π bonds. Thus, (a) is incorrect and (b) is correct. Some carbon-to-carbon bonds are σ bonds (single covalent bonds) and some are combinations of one σ and one π (double bond) or one σ and two π (triple bond). Responses (c) and (d) are both incorrect.

4. (*d*) For the bond order to be 1, the number of bonding electrons must exceed the number of antibonding electrons by two. In H_2^+ there is only one electron. In H_2^- there are three electrons--two bonding and one antibonding. The bond order is 1/2. He_2 does not exist as a stable molecule; its bond order is zero. Li_2 meets the stated requirement. Of the six electrons in this species, four are in the closed KK shell. The remaining two electrons go into a bonding molecular orbital of the second shell, σ^b_{2s}.

5. (*c*) Delocalized molecular orbitals are expected for species which cannot be represented through a single Lewis structure. That is, they are expected for species that exhibit resonance, of the species listed here only CO_3^{2-} involves resonance, and therefore, delocalized molecular orbitals. This can be established either by writing Lewis structures for CO_3^{2-} or for H_2, HS^- and CH_4, which do not require resonance.

6. (*c*) Of the substances listed only Li is a metal. Ge and Si are both semiconductors, and $Br_2(\ell)$, a nonmetal, is a nonconductor.

7. Draw Lewis structures of the molecules. Use VSEPR theory to predict the distribution of electron pairs about the C atom. Then propose a hybridization scheme for the central atom that will account for this geometric shape.

(a) H-C≡N linear *sp* hybridization

(b) tetrahedral sp^3 hybridization

```
       H
 ..    |   ..
:Cl - C - Cl:
 ..    |   ..
      :Cl:
       ..
```

(c) tetrahedral sp^3 hybridization

```
    H
    |  ..
H - C - O - H
    |  ..
    H
```

(d) trigonal planar sp^2 hybridization

```
        ..
       :O
   ..   ||  ..
H - N - C - O - H
    |       ..
    H
```

8. 6σ and 2π bonds

```
     H     π
σ—-- |  .. / \ ..
  H - C - N = C = O:
     |   \ | /
σ —-/    σ
     H
```

9. (a) Lewis theory does not tell anything about the shape of the H_2O molecule. For example, both of the following structures are acceptable:

```
                  H
  ..              ..
H:O:H    and     :O:H
  ..              ..
```

(b) The O atom would use $2p$ orbitals in forming bonds with H atoms. The predicted bond angle is 90°.

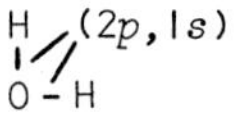

H (2p, 1s) | O – H

(c) The hybridization scheme for the O atom in H_2O is sp^3. This suggests a bent molecule with a bond angle of 109.5°.

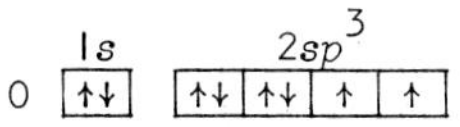

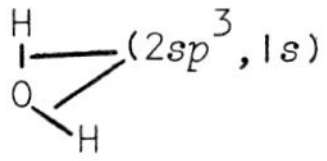

O $1s$ [↑↓] $2sp^3$ [↑↓|↑↓|↑|↑]

H — O — H $(2sp^3, 1s)$

(d) VSEPR predicts a tetrahedral distribution of four electron pairs--two bond pairs and two lone pairs--AX_2E_2. The molecule should be bent and have a bond angle of 109.5°.

10. Whether the bond energy of a diatomic molecule increases or decreases when an electron is lost depends on whether the loss occurs from a bonding or an antibonding orbital. In the case, $O_2 \rightarrow O_2^+ + e^-$, the electron is lost from an antibonding orbital, π^*_{2p}, and the bond energy increases. But in $N_2 \rightarrow N_2^+ + e^-$, the electron is lost from a bonding orbital. The bond energy decreases.

11. The C_6H_6 molecule has a unique structure but it cannot be represented either by Lewis theory (Kekule structures) or by the valence bond method. At least two structures must be drawn in either case. However, by using sp^2 hybrid orbitals of the six C atoms to construct the σ-bond framework, six p orbitals can be combined into six delocalized molecular orbitals, three bonding and three antibonding. Placement of the six $2p$ electrons into the three bonding orbitals produces three bonds with are distributed among the six carbon atoms. In this way a single structure can be drawn for C_6H_6.

12. A hybridization scheme for BrF_5 must be able to accommodate 12 valence electrons about the central Br atom--7 from Br and one each from 5 F atoms. The orbital set sp^3d can accommodate only 10 electrons. The required hybridization scheme is sp^3d^2, with five of the six hybrid orbitals receiving a bonding pair and the sixth, a lone pair. That is, the structure of BrF_5 corresponds to the VSEPR designation AX_5E and is depicted in Table 9-2.

Chapter 11

Liquids, Solids, and Intermolecular Forces

Review Problems

11-1. (a) $q = 1.25 \text{ kg } C_2H_5OH \times \frac{1000 \text{ g } C_2H_5OH}{1 \text{ kg } C_2H_5OH} \times \frac{-204.3 \text{ cal}}{\text{g } C_2H_5OH} = -2.55\times10^5 \text{ cal}$

(The negative sign signifies that heat is evolved in the condensation.)

(b) $\text{no. g } (CH_3)_2CO = 12.5 \text{ kJ} \times \frac{1000 \text{ J}}{1 \text{ kJ}} \times \frac{1 \text{ cal}}{4.184 \text{ J}} \times \frac{1.00 \text{ g } (CH_3)_2CO}{124.5 \text{ cal}} = 24.0 \text{ g } (CH_3)_2CO$

(c) $\Delta\overline{H}_{vap} = \frac{59.0 \text{ cal}}{\text{g } CHCl_3} \times \frac{4.184 \text{ J}}{1 \text{ cal}} \times \frac{1 \text{ kJ}}{1000 \text{ J}} \times \frac{119 \text{ g } CHCl_3}{1 \text{ mol } CHCl_3} = 29.4 \text{ kJ/mol}$

11-2. (a) The intersection of the line T = 50°C with the vapor pressure curve for C_6H_6 is at about 300 mmHg.

(b) Here the intersection is that of the line P = 760 mmHg with the vapor pressure curve of $C_4H_{10}O$; it comes at about 35°C.

11-3. Use the ideal gas equation to calculate the pressure of $Br_2(g)$ in the gas sample described. This will be the vapor pressure of $Br_2(\ell)$ at 25°C, since the vapor was in equilibrium with its liquid.

$$P = \frac{nRT}{V} = \frac{mRT}{\mathcal{M}V} = \frac{0.486 \text{ g} \times 0.0821 \text{ L atm mol}^{-1}\text{K}^{-1} \times 298 \text{ K}}{160 \text{ g mol}^{-1} \times 0.250 \text{ L}} = 0.297 \text{ atm} = 226 \text{ mmHg}$$

11-4. (a) The boiling point is at P = 760 mmHg; log P = log 760 = 2.88. We must find the intersection of the line log P = 2.88 with that representing the vapor pressure of aniline. This intersection (Figure 11-8) comes at about $(1/T)\times10^3 = 2.19$. Thus $(1/T) = 2.19\times10^{-3}$ and $T = 1/2.19\times10^{-3} = 457 \text{ K} = 184°\text{C}$.

(b) Here we must first convert the temperature 75°C to $(1/T)\times10^3$. T = 75°C = 348 K. $1/T = 2.87\times10^{-3}$, and $(1/T)\times10^3 = 2.87$. The intersection of the line $(1/T)\times10^3 = 2.87$ with the vapor pressure cume for toluene comes at about log P = 2.41. The vapor pressure we are seeking is antilog 2.41. $P = 2.6\times10^2$ mmHg.

11-5. (a) $\text{no. g Mg} = 985 \text{ cal} \times \frac{1 \text{ kcal}}{1000 \text{ cal}} \times \frac{1 \text{ mol Mg}}{2.140 \text{ kcal}} \times \frac{24.3 \text{ g Mg}}{1 \text{ mol Mg}} = 11.2 \text{ g Mg}$

(b) First determine the volume of the lead bar.

$$V = \left(8 \text{ in.} \times \frac{2.54 \text{ cm}}{1 \text{ in}}\right) \times \left(2 \text{ in.} \times \frac{2.54 \text{ cm}}{1 \text{ in}}\right) \times \left(1 \text{ in.} \times \frac{2.54 \text{ cm}}{1 \text{ in}}\right) = 262 \text{ cm}^3$$

Now determine the quantity of heat involved.

$$q = 262 \text{ cm}^3 \times \frac{11 \text{ g Pb}}{1 \text{ cm}^3} \times \frac{1 \text{ mol Pb}}{207 \text{ g Pb}} \times \frac{1.141 \text{ kcal}}{1 \text{ mol Pb}} = 16 \text{ kcal}$$

(c) $q = 2.12 \text{ kg Cu} \times \frac{1000 \text{ g Cu}}{1 \text{ kg Cu}} \times \frac{1 \text{ mol Cu}}{63.55 \text{ g Cu}} \times \frac{-3.120 \text{ kcal}}{1 \text{ mol Cu}} \times \frac{4.184 \text{ kJ}}{1 \text{ kcal}} = -435 \text{ kJ}$

(The negative sign signifies that heat is evolved.)

11-6. In each case determine the pressure that would be exerted if the water were completely vaporized. If this calculated pressure is *less* than the equilibrium vapor pressure, then the sample exists completely as vapor at this pressure. If the calculated pressure is *greater* than the vapor pressure, the sample exists as a liquid-vapor mixture at the equilibrium vapor pressure (See Table 11-1).

(a) $P = \frac{nRT}{V} = \frac{(0.180/18.0)\text{mol} \times 0.0821 \text{ L atm mol}^{-1}\text{K}^{-1} \times 303\text{K}}{2.50 \text{ L}} = 0.0995 \text{ atm} = 75.6 \text{ mmHg} > 31.8 \text{ mmHg}$

The sample exists as a liquid-vapor mixture at 31.8 mmHg pressure.

(b) $P = \frac{0.0100 \text{ mol} \times 0.0821 \text{ L atm mol}^{-1}\text{K}^{-1} \times 323 \text{ K}}{2.50 \text{ L}} = 0.106 \text{ atm} = 80.6 \text{ mmHg} < 92.5 \text{ mmHg}$

The sample exists entirely as a vapor, at 80.6 mmHg pressure.

(c) $P = \frac{0.0100 \text{ mol} \times 0.0821 \text{ L atm mol}^{-1}\text{K}^{-1} \times 343 \text{ K}}{2.50 \text{ L}} = 0.113 \text{ atm} = 85.6 \text{ mmHg} < 233.7 \text{ mmHg}$

The sample exists entirely as a vapor, at 85.6 mmHg pressure.

11-7. (a) The upper of the two regions labeled (?) is that of liquid P; the lower region, phosphorus vapor.

(b) The triple point pressure for red P (43 atm) is much higher than normal atmospheric pressure. This means that as a sample of solid red P is heated it will sublime away at 1 atm pressure at a temperature much below its triple point temperature (863 K).

(c) The sample begins as a solid red P and converts to liquid P when the pressure is reduced to a point on the fusion curve. At a still lower pressure (on the order of 40-50 atm) the liquid vaporizes. At point B the sample is completely in the vapor state.

11-8. (a) $C_{10}H_{22}$: The intermolecular forces in C_7H_{16} and $C_{10}H_{22}$ are of the same kind--London forces--but $C_{10}H_{22}$ has the higher molecular weight and the higher boiling point.

(b) $H_3C-O-CH_3$: The molecular weight of this substance (dimethyl ether) is 46. The molecular weight of C_3H_8 (44) is almost the same. But the dimethyl ether is a polar molecule and exhibits stronger intermolecular forces than C_3H_8.

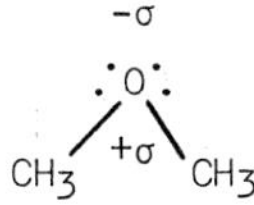

(c) CH_3CH_2OH: The molecules differ only in the O and S atoms. Both molecules are polar, but because the electronegativity difference between C and O is greater than between C and S, the dipole moment of CH_3CH_2OH is greater than that of CH_3CH_2SH. More important still, hydrogen bonding occurs in CH_3CH_2OH.

11-9. The basic principle in estimating lattice energies is that electrostatic attractions increase as the charges on ions *increase* and as ionic radii *decrease*. In establishing the order indicated below, we note that Mg^{2+}, Ca^{2+}, and O^{2-} each carries two units of net electrical charge, and that Cs^+ and I^- are the largest ions present.

$CsI < MgBr_2 < CaO$

11-10. (a) To obtain the lattice energy of LiF(s), add together the following five equations to obtain a net equation for the formation of LiF(s) from its elements in their standard states (i.e., $\Delta\overline{H}^\circ_f$[LiF(s)]). All the data in this summation are given in the problem except the value of U, the lattice energy. This must be solved for.

$Li(s) \rightarrow Li(g)$	$\Delta\overline{H}_1 = +160.7$ kJ/mol
$Li(g) \rightarrow Li^+(g) + e^-$	$\Delta\overline{H}_2 = +520.5$ kJ/mol
$\frac{1}{2}F(g) \rightarrow F(g)$	$\Delta\overline{H}_3 = +157.8/2$ kJ/mol
$F(g) + e^- \rightarrow F^-(g)$	$\Delta\overline{H}_4 = -328$ kJ/mol
$Li^+(g) + F^-(g) \rightarrow LiF(s)$	$U = ?$
$Li(s) + \frac{1}{2}F_2(g) \rightarrow LiF(s)$	$\Delta\overline{H}^\circ_f = -616.9$ kJ/mol

$\Delta\overline{H}^\circ_f = \Delta\overline{H}_1 + \Delta\overline{H}_2 + \Delta\overline{H}_3 + \Delta\overline{H}_4 + U = -616.9$ kJ/mol

$U = (-616.9 - 160.7 - 520.5 - 157.8/2 + 328)$ kJ/mol

$= -1049$ kJ/mol

(b) Write a similar set of five equations and obtain their sum. For the net reaction, $\Delta H = \Delta H^\circ_f[\text{NaF(s)}]$.

$\text{Na(s)} \rightarrow \text{Na(g)}$	$\Delta\bar{H} = +107.8$ kJ/mol
$\text{Na(g)} \rightarrow \text{Na}^+\text{(g)} + e^-$	$\Delta\bar{H} = +495.4$ kJ/mol
$\frac{1}{2}\text{F}_2\text{(g)} \rightarrow \text{F(g)}$	$\Delta\bar{H} = +157.8/2$ kJ/mol
$\text{F(g)} + e^- \rightarrow \text{F}^-\text{(g)}$	$\Delta\bar{H} = -328$ kJ/mol
$\text{Na}^+\text{(g)} + \text{F}^-\text{(g)} \rightarrow \text{NaF(s)}$	$U = -927.7$ kJ/mol
$\text{Na(s)} + \frac{1}{2}\text{F}_2\text{(g)} \rightarrow \text{NaF(s)}$	$\Delta\bar{H}^\circ_f = ?$

$\Delta\bar{H}^\circ_f = (107.8 + 495.4 + 157.8/2 - 328 - 927.7) = -574$ kJ/mol

11-11. (a) The diagonal (d) of the cell face pictured is four times the atomic radius (r), and the cell length (l) is $l = d/\sqrt{2}$. $l = 4 \times 128/\sqrt{2} = 362$ pm $= 3.62\times10^{-8}$ cm. 3.62×10^{-8} cm $= 3.62\times10^{-10}$ m $= 362\times10^{-12}$ m $= 362$ pm.

(b) volume of unit cell $= l^3 = (3.62\times10^{-8}\text{ cm})^3 = 4.74\times10^{-23}\text{ cm}^3$

(c) corner atoms = 1/8 x 8 = 1
face-centered atoms = 1/2 x 6 = 3 } atoms per unit cell = 4

(d) The mass contained in a unit cell is the mass of four Cu atoms.

$$\text{no. g Cu} = 4\text{ Cu atoms} \times \frac{1\text{ mol Cu}}{6.02\times10^{23}\text{ Cu atoms}} \times \frac{63.55\text{ g Cu}}{\text{mol Cu}} = 4.22\times10^{-22}\text{ g Cu}$$

(e) $$\text{density of Cu} = \frac{4.22\times10^{-22}\text{ g}}{4.74\times10^{-23}\text{ cm}^3} = 8.90\text{ g/cm}^3$$

11-12. CsCl has a body-centered-cubic unit cell. Consider that a Cs^+ ion is at the very center of the cell and belongs entirely to this cell. Each of the eight corners of the cell is occupied by a Cl^- ion, and each of these Cl^- ions is shared among eight unit cells. The number of Cl^- ions belonging to the cell in question is 8 x 1/8 = 1. The unit cell contains the equivalent of one Cs^+ and one Cl^- ion. This is consistent with the formula CsCl.

Exercises

Surface tension and related properties

11-1. (a) Surface tension is a force acting at the surface of a liquid that causes the liquid to assume a droplike shape (to minimize the surface area), to rise in a capillary tube, and to exhibit a number of other distinctive properties.

(b) An adhesive force is an intermolecular force of attraction between unlike molecules, such as the attraction between water molecules and the silicate structure of glass (responsible for the spreading of a film of water over glass).

(c) Capillary action refers to the ability of certain liquids to rise in capillary tubes to levels higher than the surrounding liquid. The liquid column is supported in the tube by the surface tension of the liquid.

(d) A wetting agent is a substance that reduces the surface tension of water so that it flows more freely over a surface.

(e) The meniscus is the interface between a liquid and the air above it. It is especially noticeable when the liquid is maintained in a tube of small diameter, and results from the same forces (surface tension) responsible for capillary action.

11-2. If a boot or tent is treated with a silicone oil, water, instead of forming a film on the surface that can be drawn into the material by capillary action, forms droplets that can be shaken off.

11-3. When the wick of the candle is ignited, some of the hydrocarbons (wax) at the tip of the candle melt; the liquid hydrocarbons are drawn into the wick through capillary action; the liquid vaporizes and the gaseous hydrocarbons burn. Heat given off in the combustion causes more wax to melt, to be drawn up the wick, to burn, and so on.

Vaporization

11-4. First determine the number of moles of vapor produced.

$$n = \frac{PV}{RT} = \frac{1 \text{ atm} \times 0.94 \text{ L}}{0.0821 \text{ L atom mol}^{-1}\text{K}^{-1} \times (273.2 + 80.1)\text{K}} = 0.032 \text{ mol } C_6H_6(g)$$

$$\Delta\overline{H}_{vap} = \frac{1.00 \text{ kJ}}{0.032 \text{ mol } C_6H_6(g)} = 31 \text{ kJ/mol } C_6H_6$$

11-5. The heat required to vaporize 1.00 L water at 100°C is

$$q_{water} = 4.55 \text{ L} \times \frac{1000 \text{ cm}^3}{1 \text{ L}} \times \frac{0.958 \text{ g}}{1 \text{ cm}^3} \times \frac{1 \text{ mol } H_2O}{18.0 \text{ g}} \times \frac{40.6 \text{ kJ}}{1 \text{ mol } H_2O} = 9.83\times10^3 \text{ kJ}$$

$$q_{water} + q_{combus.} = 0 \qquad q_{combus.} = -q_{water} = -9.83\times10^3 \text{ kJ}$$

$$\text{no. mol } CH_4 = -9.83\times10^3 \text{ kJ} \times \frac{1 \text{ mol } CH_4}{-890 \text{ kJ}} = 11.0 \text{ mol } CH_4$$

$$V = \frac{nRT}{P} = \frac{11.0 \text{ mol} \times 0.0821 \text{ L atm mol}^{-1}\text{K}^{-1} \times 298 \text{ K}}{(748/760)\text{atm}} = 273 \text{ L } CH_4(g)$$

11-6. (a) When steam condenses on the walls of the inner container, heat is evolved--the heat of condensation of the steam (-40.6 kJ/mol). This heat is transferred to the contents of the inner container where cooking occurs.

(b) The maximum temperature that can be reached by the contents of the inner container is 100°C--the temperature of the condensing steam.

11-7. When vaporization of a liquid occurs from a container that can exchange heat with the surroundings, the heat required to vaporize the liquid is drawn from the surroundings; the temperature remains essentially constant. When vaporization occurs from a thermally insulated container, the heat of vaporization is drawn from the liquid itself. This causes the average kinetic energy of the liquid molecules to decrease and the temperature of the liquid to drop. Very little heat can enter the liquid from the surroundings to offset the heat loss to the vapor.

Vapor pressure and boiling point

11-8. (a) From Table 11-1 estimate the temperature at which the vapor pressure of water is 600 mmHg. This is a temperature of about 93.5°C.

(b) Determine the vapor pressure of water at 89°C from Table 11-1. Note that in the interval 90°C to 91°C the vapor pressure increases by 20.2 mmHg, and that from 91°C to 92°C, the vapor pressure increases by 21 mmHg. Assume that in the interval from 89°C to 90°C the difference in vapor pressure is 19 mmHg. This leads to an approximate pressure of 525.8 - 19 = 507 mmHg.

11-9. The mass of aniline that vaporizes is 6.220 - 6.108 = 0.112 g. The number of moles of aniline in the 25.0 L vapor volume at 30°C is (0.112/93.1) = 1.20×10^{-3}. Use these data to calculate the partial pressure of aniline in the gaseous mixture. This value is the vapor pressure of liquid aniline at 30.0°C.

$$P = \frac{nRT}{V} = \frac{1.20\times10^{-3} \text{ mol} \times 0.0821 \text{ L atm mol}^{-1}\text{K}^{-1} \times 303 \text{ K}}{25.0 \text{ L}} = 1.19\times10^{-3} \text{ atm} = 0.904 \text{ mmHg}$$

11-10. In the gaseous mixture, $P_{tot.}$ = 742 mmHg = $P_{N_2} + P_{CCl_4}$. P_{CCl_4} is simply the vapor pressure of CCl_4 at 45°C--261 mmHg. P_{N_2} = 742 mmHg - 261 mmHg = 481 mmHg. Since the amount of N_2 and its temperature remain constant, we restate the question as a Boyle's law problem: What is the final volume of N_2 at 494 mmHg, if initially there was present 10.0 liters at 755 mmHg?

$$V_f = V_i \times \frac{P_i}{P_f} = 7.53 \text{ L} \times \frac{742 \text{ mmHg}}{481 \text{ mmHg}} = 11.6 \text{ L}$$

The Clausius-Clapeyron equation

11-11. The relevant data for phosphorus are

t, °C	T, K,	1/T, K^{-1}	P, mmHg	log P(mmHg)
76.6	349.8	2.86×10^{-3}	1	0.00
128.0	401.2	2.49×10^{-3}	10	1.00
166.7	439.9	2.27×10^{-3}	40	1.60
197.3	470.5	2.13×10^{-3}	100	2.00
251.0	524.2	1.91×10^{-3}	400	2.60

Plot log P vs. 1/T. Estimate the value of 1/T corresponding to log P = log 760 = 2.88

$1/T = 1.81\times10^{-3}$ $\quad T \simeq 553\ K = 280°C$

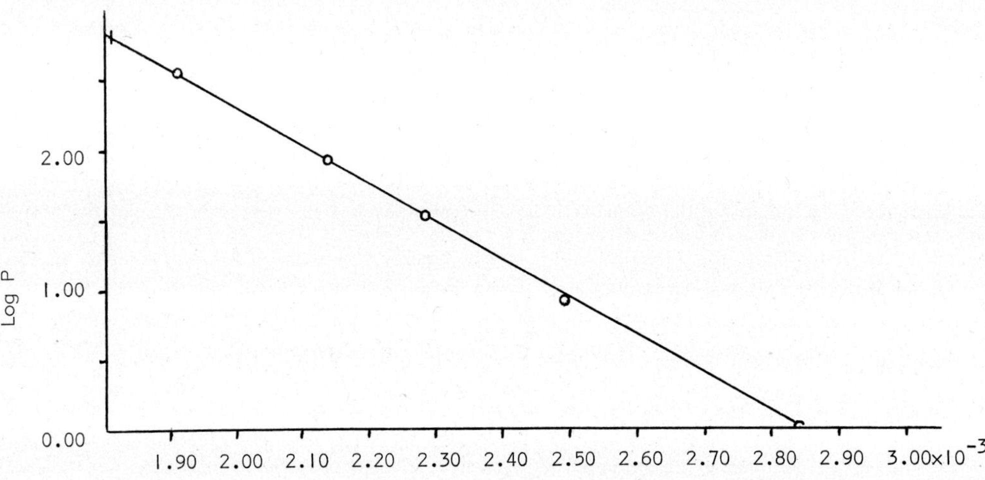

11-12. Use equation (11.3)

(a) $$\log \frac{100.0\ mmHg}{10.0\ mmHg} = \frac{\Delta\overline{H}_{vap}}{2.303 \times 8.314\ J\ mol^{-1}K^{-1}}\left(\frac{376.9 - 329.2}{329.2 \times 376.9}\right)K^{-1} = 1.00$$

$$\Delta\overline{H}_{vap} = \frac{1.00 \times 2.303 \times 8.314\ J\ mol^{-1} \times 329.2 \times 376.9}{(376.9 - 329.2)} = 4.98\times10^{4}\ J/mol$$

(b) Let T = the boiling point of cyclohexanol, expressed in kelvins. Use equation (11.3) again.

$$\log \frac{760\ mmHg}{100\ mmHg} = \frac{4.98\times10^{4}}{2.303 \times 8.314}\left(\frac{T - 376.9}{376.9T}\right) = 0.8808$$

$$4.98\times10^{4}\ T - 1.88\times10^{7} = 0.8808 \times 2.303 \times 8.314 \times 376.9\ T$$

$$4.98\times10^{4}\ T - 1.88\times10^{7} = 6.356\times10^{3}\ T \qquad T = 433\ K = 160°C$$

11-13. Let P_1 = 175 mmHg, P_2 = 760 mmHg, and T_2 = 125.8 + 273.2 = 399 K. Solve for T_1 in equation (11.3).

$$\log \frac{760 \text{ mmHg}}{175 \text{ mmHg}} = \frac{3.39\times10^4 \text{ J/mol}}{2.303 \times 8.314 \text{ J mol}^{-1}\text{K}^{-1}} \left(\frac{399 - T}{399\ T}\right) K^{-1} = 0.638$$

$$1.35\times10^7 - 3.39\times10^4\ T = 0.638 \times 2.303 \times 8.314 \times 399\ T$$

$$1.35\times10^7 - 3.39\times10^4\ T = 4.87\times10^3\ T \qquad T = 348 \text{ K} = 75°C$$

Critical point

11-14. Any substance whose critical temperature is above about 293 K can exist as a liquid at this temperature provided that a sufficient pressure is maintained. Of the substances in Table 11-2 these are CO_2, HCl, NH_3, SO_2, and H_2O.

11-15. SO_2 can be maintained as a liquid at 0°C under moderate pressures (77.7 atm is more than sufficient). Methane cannot be maintained as a liquid at 0°C and 100 atm, because its critical temperature (191.1 K) is well below 0°C.

Fusion

11-16. $$\text{no. kJ} = (30.5 \text{ cm})^3 \times \frac{0.92 \text{ g}}{1 \text{ cm}^3} \times \frac{1 \text{ mol}}{18.0 \text{ g}} \times \frac{6.02 \text{ kJ}}{1 \text{ mol}} = 8.7\times10^3 \text{ kJ}$$

11-17. $$q_{lead} = \left\{0.803 \text{ kg Pb} \times \frac{1000 \text{ g Pb}}{1 \text{ kg Pb}} \times \frac{0.134 \text{ J}}{\text{g Pb °C}} \times (327.4 - 25.0)°C \times \frac{1 \text{ kJ}}{1000 \text{ J}}\right\}$$

$$+ \left\{803 \text{ g Pb} \times \frac{1 \text{ mol Pb}}{207 \text{ g Pb}} \times \frac{4.774 \text{ kJ}}{1 \text{ mol Pb}}\right\} = 32.5 \text{ kJ} + 18.5 \text{ kJ} = 51.0 \text{ kJ}$$

States of matter and phase diagrams

11-18. (a) Calculate the pressure associated with the sample if it were all present as vapor.

$$P = \frac{mRT}{MV} = \frac{2.50 \text{ g} \times 0.0821 \text{ L atm mol}^{-1}\text{K}^{-1} \times 393 \text{ K}}{18.0 \text{ g/mol} \times 5.00 \text{ L}} = 0.896 \text{ atm}$$

The vapor pressure of water at 120°C is considerably greater than 1 atm; the vapor is unsaturated; and the calculated pressure, 0.896 atm, is the true pressure.

(b) If the vapor is cooled, at constant pressure, to the point at which the vapor pressure of water = 0.896 atm = 681 mmHg, the vapor will condense. This occurs at a temperature of about 97°C (see Table 11-1).

11-19. The $H_2(g)$ and $O_2(g)$ are in exactly stoichiometric amounts, that is, there is twice the number of moles of H_2 as of O_2, just as required by the balanced equation: $2\ H_2 + O_2 \rightarrow 2\ H_2O$. The net result of the reaction is the production of 0.100 mol H_2O, with no excess H_2 or O_2. Determine the pressure exerted by 0.100 mol H_2O vapor in 20.0 liters at 27°C.

$$P = \frac{nRT}{V} = \frac{0.100 \text{ mol} \times 0.0821 \text{ L atm mol}^{-1}\text{K}^{-1} \times 300 \text{ K}}{20.0 \text{ L}} = 0.123 \text{ atm} = 93.5 \text{ mmHg}$$

This calculated pressure is considerably in excess of the equilibrium vapor pressure of water at 27°C. Some of the water vapor condenses and the pressure drops to the vapor pressure of water at 27°C--26.7 mmHg.

11-20. Consider the following quantities of heat: q_{ice} is the heat required to raise the temperature of the block of ice to 0°C, to melt the ice, and to raise the temperature of the melted water to a final value of T; q_{water} is the heat associated with the change in temperature of the originial water from 32.0°C to T.

First, determine the mass of the ice.

$$\text{mass} = (8.0 \text{ cm} \times 2.5 \text{ cm} \times 2.7 \text{ cm}) \times \frac{0.917 \text{ g}}{1 \text{ cm}^3} = 50 \text{ g}$$

$$q_{ice} = \left\{ 50 \text{ g} \times \frac{2.01 \text{ J}}{\text{g } ^\circ\text{C}} \times [0 - (-25)]^\circ\text{C} \times \frac{1 \text{ kJ}}{1000 \text{ J}} \right\} + \left\{ 50 \text{ g} \times \frac{1 \text{ mol}}{18.0 \text{ g}} \times \frac{6.02 \text{ kJ}}{1 \text{ mol}} \right\}$$

$$+ \left\{ 50 \text{ g} \times \frac{4.18 \text{ J}}{\text{g } ^\circ\text{C}} \times (T - 0)^\circ\text{C} \times \frac{1 \text{ kJ}}{1000 \text{ J}} \right\} = 2.5 + 17 + 0.21 \text{ T}$$

$$q_{water} = 400.0 \text{ cm}^3 \times \frac{0.998 \text{ g}}{1 \text{ cm}^3} \times \frac{4.18 \text{ J}}{\text{g } ^\circ\text{C}} \times (T - 32.0)^\circ\text{C} \times \frac{1 \text{ kJ}}{1000 \text{ J}} = 1.67 \text{ T} - 53.4$$

$q_{ice} + q_{water} = 0$ $\qquad$ $2.5 + 17 + 0.21 \text{ T} + 1.67 \text{ T} - 53.4 = 0$

T = 18°C. The water is present as liquid only.

11-21. Dry ice maintains a constant temperature of -78.5°C, the temperature at which its sublimation pressure is 1 atm. The constant temperature maintained by ordinary ice is its melting point, 0°C. At a temperature of 0°C frozen foods would begin to thaw along with ordinary ice. At -78.5°C the frozen foods remain frozen as dry ice sublimes.

11-22. Both the melting point of ice and the boiling point of water are dependent on atmospheric pressure; this is especially true of the boiling point. Furthermore, at its normal freezing point water must be kept saturated with air (with which it is maintained in contact). The triple point is a unique fixed temperature point. The phases in equilibrium--ice, liquid water, and water vapor--are maintained out of contact with air and under their own unique equilibrium pressure (4.58 mmHg).

11-23. (a) No: Even if sufficiently low temperatures existed, the sublimation pressure of $CO_2(s)$ would be so great that it would sublime completely.

(b) No: The critical temperature of methane (191.1 K = -82°C) is much below ambient temperatures.

(c) Yes: SO_2 normally occurs in the atmosphere in trace amounts.

(d) No: The required temperature to melt iodine under normal atmospheric pressure (114°C) is very much higher than would be found anywhere on the earth's surface.

(e) No: The required temperature to maintain O_2 as a liquid is much too low and the pressure, much too high, for these conditions to be found on earth.

11-24. (a) The path we must trace on the phase diagram (Figure 11-12) is slightly below the broken line, PR. At a temperature just slightly greater than 0°C the ice will melt. (The melting point of ice under 760 mmHg is 0°C.) The sample remains in the liquid state up to a temperature of about 93.5°C (at which the vapor pressure of water is 600 mmHg). At this temperature complete vaporization of the water occurs. Following this, the temperature is free to rise again, with the volume increasing in accordance with Charles' law.

(b) The triple point for I_2 is at 91 mmHg and 114°C. At a slightly higher temperature (114.5°C) and lower pressure (90 mmHg), iodine exists as a vapor. A small increase in pressure is enough to condense the vapor to $I_2(\ell)$. The sample remains as a liquid for a time as the pressure is increased, but by the time the pressure has been increased to 100 atm, conversion to the solid is likely. (The fusion curve has a positive slope.)

(c) Since 35°C is above the critical temperature of CO_2, the sample must be completely gaseous at the start. At a temperature approaching -50°C, the gas condenses to pure liquid CO_2. The liquid freezes to $CO_2(s)$ at about -57°C. Further cooling produces no additional phase changes.

11-25. When the pressure reaches about 4.5 mmHg, the vapor condenses to ice. At a pressure somewhat above 1 atm, the ice melts to liquid water. The sample remains as a single liquid phase up to 100 atm pressure.

11-26. (a) Since 600 K is below the critical temperature of H_2O and 220 atm, above the critical pressure, water exists as liquid only under these conditions.

(b) A process in which a phase transition curve is crossed from low to high temperature is endothermic. The reverse situation applies here; the transition is exothermic.

(c) The data given are for point D in Figure 11-12. Since the fusion curve for ice III → liquid water has a positive slope, ice III is more dense than liquid water; its density is greater than 1.00 g/cm^3.

Van der Waals forces

11-27. $N_2 < F_2 < Ar < O_3 < Cl_2$

mol wt. 28 < 38 < 40 < 48 < 71

The intermolecular forces in these substances are of the London type. The strength of these forces increases with molecular weight. O_3 is out of place in the original listing.

11-28. The line of reasoning here is the same as illustrated in Figure 11-15. The more compact, symmetrical molecule is the less easily polarized of the two being compared. There are several different octane molecules based on a five-carbon chain, but each has a more compact structure than the straight-chain molecule.

```
       H     H
       |     |
     H-C-H H-C-H
   H   |   H   |   H              H   H   H   H   H   H   H   H
   |   |   |   |   |              |   |   |   |   |   |   |   |
H -C - C - C - C - C - H      H - C - C - C - C - C - C - C - C - H
   |   |   |   |   |              |   |   |   |   |   |   |   |
   H   |   H   H   H              H   H   H   H   H   H   H   H
     H-C-H
       |
       H
```

isooctane b. pt. 99.2°C   normal octane b. pt. 125.7°C

11-29. Plot a graph of boiling point versus chain length of the alkane hydrocarbons; extrapolate to the nine-carbon alkane, nonane, C_9H_{20}. The boiling point is slightly greater than 150°C.

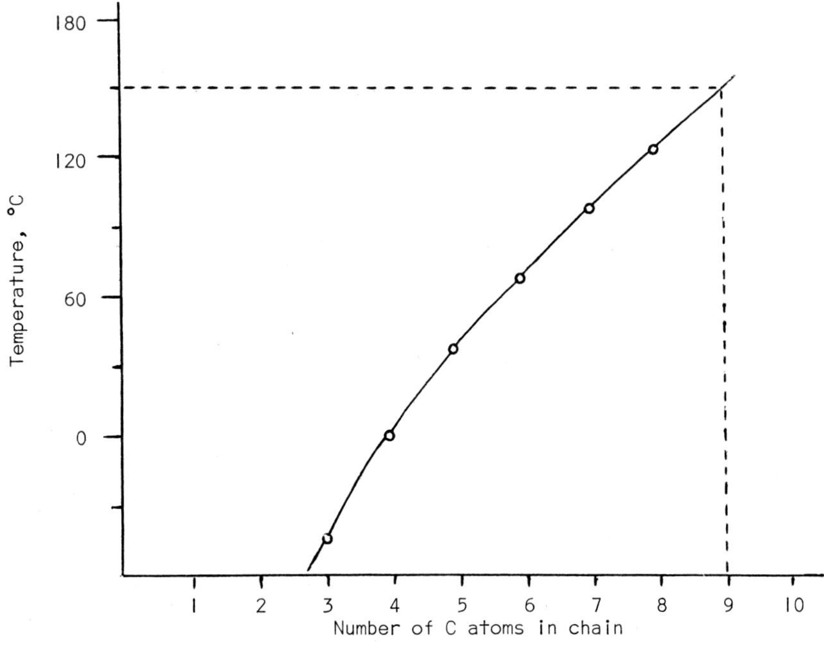

11-30. The substitution of heavier atoms, such as Cl and Br, for H produces an increase in molecular weight and hence in boiling point. This is to be expected. The substitution of -OH for H does not increase the molecular weight as much as does the substitution of either Cl or Br. The reason that C_6H_5OH has the highest boiling point of the group must involve a phenomenon other than the effect of molecular weight on dispersion forces. The new phenomenon encountered in C_6H_5OH is hydrogen bonding between the H atom of the -OH group of one molecule and the O atom of the -OH group of a neighboring molecule.

Hydrogen bonding

11-31. The primary condition required for hydrogen bond formation is that a hydrogen atom covalently bonded to one small electronegative atom (usually N, O, or F) be in close proximity to another small electronegative atom (usually N, O, or F) in the same or a neighboring molecule. This is a stronger intermolecular force than the London forces present in all covalent substances.

11-32. C_3H_8 and N_2 are both nonpolar. The intermolecular forces in these substances are of the London type. We have encountered N_2 and C_3H_8 before and know them to be gases at room temperature. CO, although weakly polar, is a gas at room temperature. It has a low molecular weight (28). This leaves CH_3OH as the liquid. Although its molecular weight is less than that of C_3H_8, we should recognize CH_3OH as a polar molecule and one that also presents a potential for hydrogen bonding. The stronger intermolecular forces in CH_3OH account for its existence as a liquid at room temperature.

11-33. *HCl* London forces and dipole-dipole interactions. The Cl atom is too large for hydrogen bonding to be a significant factor. However, the molecule is polar and dipole-dipole interactions should add to the usual London forces in establishing the intermolecular forces in HCl.

Br_2 London forces. Br_2 is a homonuclear, nonpolar molecule.

ICl London forces. Unlike the case of Br_2, the intermolecular forces are not limited to the London type, since the ICl molecule has a small dipole moment. Nevertheless, the dipole-dipole interactions do not contribute significantly to the properties of ICl. For example, its boiling point is roughly intermediate to those of Cl_2 and I_2, which is what would be expected even if London forces were the only intermolecular forces.

HF Hydrogen bonds. The requirements of atomic size and electronegativity that are necessary for hydrogen bonding are met ideally by the fluorine atom.

CH_4 London forces. This is a nonpolar molecule. Hydrogen bonding is not a factor because, although C is a small nonmetal atom, its electronegativity is not large (being about the same size as that of H).

11-34. The situation with NH_3 would be the same as with H_2O--four molecules can be hydrogen bonded to any given molecule.

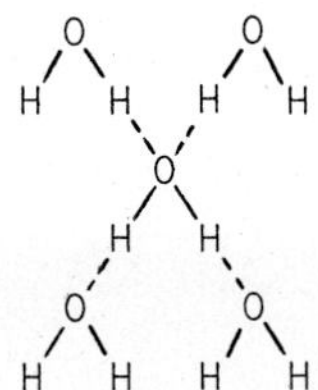

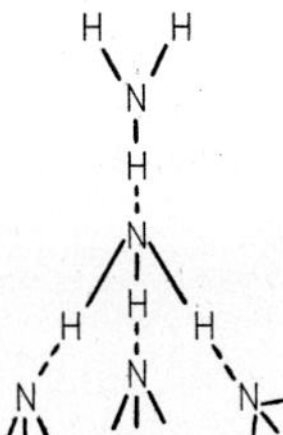

11-35. Were it not for the presence of hydrogen bonds in water, we would expect the data for group VIA hydrides to be in a regular progression (see Figure 11-17). (a) The boiling point of water would be at about 200 K. (b) The freezing point would be perhaps as low as 75 K. (c) The temperature of maximum density of water would be at its freezing point, as with most liquids. (d) The density of solid water (ice) would be greater than that of liquid water, again as with most substances.

11-36. Intramolecular hydrogen bonding requires that two nonmetal atoms having lone pair electrons and within the same molecule be bridged by an H atom. This will occur only if the intermolecular distances are just right.

(a) C_2H_6: no hydrogen bonding

(b) H_3CCH_2OH: Intermolecular hydrogen bonding. There is only one highly electronegative atom per molecule (the O atom).

(c) H_3CCOOH: Intermolecular hydrogen bonding. The distance between the two O atoms is too short for an H atom to form a bridge between them. (See also, Figure 11-20.)

(d) $C_6H_4(COOH)_2$: Intramolecular hydrogen bonding, as indicated below.

```
          O
          ‖
       —C - O
             \H
       —C = O'
        \
         O - H
```

Network covalent solids

11-37. (a) Localized electrons have their charge densities concentrated between two atoms. Delocalized electrons have their charge densities distributed among three or more atoms; the charge density of delocalized electrons is more "spread out".

(b) The delocalized electrons in graphite are the $2p$ electrons of the carbon atoms in hexagonal rings, one electron per carbon atom.

11-38. (a) Simply replace one half of the C atoms by Si atoms in the structures shown in Figure 11-21 in such a way that all the bonds become Si - C bonds.

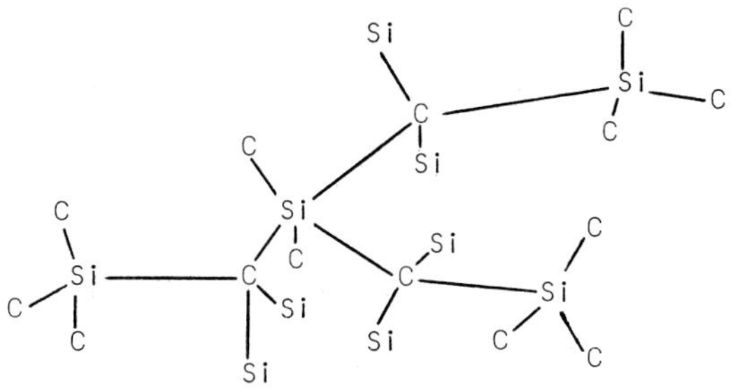

(b) Each B and N atom forms three bonds within a plane by using sp^2 hybrid orbitals. This produces layers of atoms arranged in hexagonal rings. Bonding between layers occurs through delocalized electrons derived from the p orbitals of nitrogen atoms.

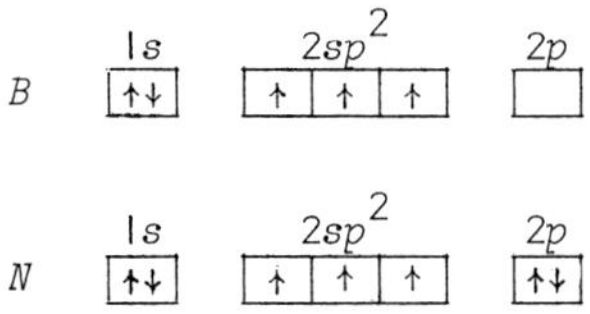

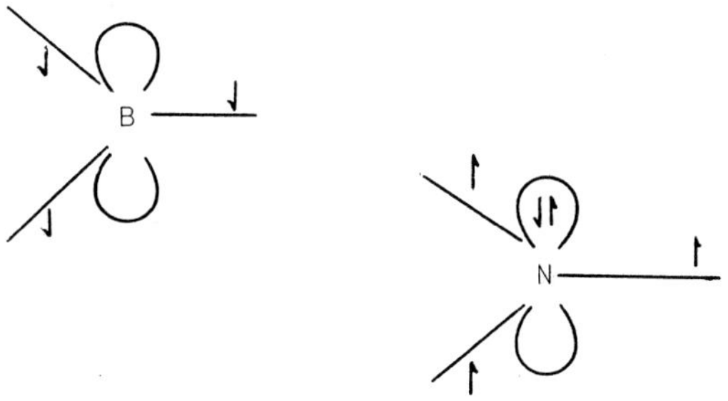

	$1s$	$2sp^2$	$2p$
B	↑↓	↑ ↑ ↑	
N	↑↓	↑ ↑ ↑	↑↓

(c) In order for π-bonds to form, as in graphite or BN, there must be an overlap of p orbitals that are oriented parallel to one another. This occurs with $2p$ orbitals; but with $3p$ orbitals, as would be required in the case of silicon, the orbitals are too far apart to overlap effectively.

11-39. Although the carbon atom within layers in graphite are joined by σ-bonds and have short bond distances, bonding between layers occurs only through π-bonds and these bond distances are greater. In diamond all the bonds are equivalent and are of the σ-type. Because the atoms in diamond are packed more closely, we should expect diamond to have the greater density.

11-40. The property of diamond that makes it useful in glass cutters is its extreme hardness. Graphite could not be used because bonding between layers of carbon atoms is weak. Any attempt to scratch glass with graphite would simply result in flaking of the graphite.

Ionic properties and bonding

11-41. Melting points parallel the strengths of interionic attractions, which in turn depend on the factors of ionic charge and size. Of the four compounds, each involves Na^+ and a halide ion. The halide ions are alike in charge but differ in ionic radius. We would predict that melting points *increase* in the order in which halide ion radii *decrease*.

$$NaI < NaBr < NaCl < NaF$$

(observed m. pt. 651 < 755 < 801 < 988°C)

11-42. (a) BaF_2: The ion, Ba^{2+}, is larger than Mg^{2+}. Interionic forces should be weaker in BaF_2 than in MgF_2, suggesting that BaF_2 dissolves to a greater extent in water than does MgF_2.

(b) $MgCl_2$: The ions are of similar types, but since Cl^- is larger than F^- we should expect weaker interionic forces and a higher water solubility for $MgCl_2$ than for MgF_2.

11-43. Coulomb's law is expressed as follows (see Appendix B): $F = Q_1Q_2/\varepsilon r^2$.

NaCl — relative force $= \dfrac{(+1)(-1)}{\varepsilon(276)^2}$

MgO — relative force $= \dfrac{(+2)(-2)}{\varepsilon(205)^2}$

$$\frac{\text{relative force (MgO)}}{\text{relative force (NaCl)}} = \frac{-4/(205)^2}{-1/(276)^2} = 7.25$$

The statement in Figure 11-23 is that intermolecular forces in MgO are about seven times as strong as in NaCl.

11-44. In addition to the data listed in the exercise, we need to use the ionization energy of Cs (from Table 8-4, 375.7 kJ/mol) the dissociation energy of $Cl_2(g)$ and the electron affinity of Cl(g). (These latter two quantities can be found in Section 11-12.)

$Cs(s) \rightarrow Cs(g)$ $\qquad \Delta\overline{H}_1 = 77.6$ kJ/mol

$Cs(g) \rightarrow Cs^+(g) + e^-$ $\qquad \Delta\overline{H}_2 = 375.7$ kJ/mol

$\frac{1}{2}\,Cl_2(g) \rightarrow Cl(g)$ $\qquad \Delta\overline{H}_3 = +121$ kJ/mol

$Cl(g) + e^- \rightarrow Cl^-(g)$ $\qquad \Delta\overline{H}_4 = -348$ kJ/mol

$Cs^+(g) + Cl^-(g) \rightarrow CsCl(s)$ $\qquad U = ?$

$Cs(s) + \frac{1}{2}\,Cl_2(g) \rightarrow CsCl(s)$ $\quad \Delta\overline{H}^\circ_f = -442.8$ kJ/mol

$\Delta\overline{H}^\circ_f = \Delta\overline{H}_1 + \Delta\overline{H}_2 + \Delta\overline{H}_3 + \Delta\overline{H}_4 + U = -442.8$ kJ/mol

$U = (-442.8 - 77.6 - 375.7 - 121 + 348)$ kJ/mol $= -669$ kJ/mol

11-45. The heat of sublimation of Na is obtained from Section 11-12; the ionization energy of Na, from Table 8-4; the heat of sublimation of $I_2(s)$, from Appendix D; the heat of dissociation of $I_2(g)$, from Table 9-3; and the heat of ionization (electron affinity) of I(g), from page 209.

$Na(s) \rightarrow Na(g)$; $\Delta\overline{H}_1 = 108$ kJ/mol

$Na(g) \rightarrow Na^+(g) + e^-$; $\Delta\overline{H}_2 = 496$ kJ/mol

$\frac{1}{2}\,I_2(s) \rightarrow \frac{1}{2}\,I_2(g)$; $\Delta\overline{H}_3 = \frac{1}{2} \times 62.26 = 31.13$ kJ/mol

$\frac{1}{2}\,I_2(g) \rightarrow I(g)$; $\Delta\overline{H}_4 = \frac{1}{2} \times 151 = 75.5$ kJ/mol

$I(g) + e^- \rightarrow I^-(g)$; $\Delta\overline{H}_5 = -295.4$ kJ/mol

$Na^+(g) + I^-(g) \rightarrow NaI(s)$; $\Delta\overline{H}_6 = U = ?$

---

net: $Na(s) + \frac{1}{2}\,I_2(s) \rightarrow NaI(s)$; $\Delta\overline{H} = \Delta\overline{H}_1 + \Delta\overline{H}_2 + \Delta\overline{H}_3 + \Delta\overline{H}_4 + \Delta\overline{H}_5 + U = \Delta\overline{H}^\circ_f[NaI(s)]$

$108 + 496 + 31.1 + 75.5 - 295.4 + U = -288$

lattice energy, $U = -703$ kJ/mol NaI

Crystal structures

11-46. (a) The closest-packing of spheres refers to an arrangement in which the voids or empty spaces among spheres are reduced to a minimum.

(b) Tetrahedral holes, pictured in Figure 11-28, may be visualized in this way: Bring three spheres into mutual contact. A "triangular" shaped opening exists among them.

Cover this region with a sphere nestled among the original three; close off the region from below in the same way. The volume bounded by these five spheres is a tetrahedral hole. The hole extends through three layers of spheres.

(c) In an octahedral hole the empty region that results from three spheres in contact is *partially* covered by another set of three spheres that is rotated 60° with respect to the first set, that is,

The resulting hole is closed off from above and below by a single sphere. The void among the eight spheres is an octahedral hole. It extends through four layers of spheres (see Figure 11-28).

11-47. When spheres are laid down in layers, two different kinds of voids are found among the spheres in a layer. These are the tetrahedral and octahedral holes referred to in the preceding exercise. Depending on whether one set of voids or the other is covered in the next layer, one of two different closest-packed structures is obtained.

11-48. Along the diagonal of a face of the *fcc* unit cell three atoms are in contact. The length of the face diagonal, $\ell\sqrt{2}$, is equal to four times the radius of an Al atom, $4r$. The atomic radius is 143.1 pm.

$$\ell\sqrt{2} = 4r = 4 \times 143.1 \text{ pm}$$

$$\ell = \frac{4 \times 143.1}{\sqrt{2}} = 404.7 \text{ pm} = 4.047\times10^{-10} \text{ m} = 4.047\times10^{-8} \text{ cm}$$

$$\ell^3 = (4.047\times10^{-8})^3\text{cm}^3 = 66.28\times10^{-24} \text{ cm}^3 = 6.628\times10^{-23} \text{ cm}^3$$

The mass of the unit cell is

$$m = 6.628\times10^{-23} \text{ cm}^3 \times \frac{2.6984 \text{ g Al}}{\text{cm}^3} = 1.788\times10^{-22} \text{ g Al}$$

The unit cell contains 4 atoms (see Table 11-4).

$$\text{mass Al atom} = \frac{1.788\times10^{-22} \text{ g Al}}{\text{unit cell}} \times \frac{1 \text{ unit cell}}{4 \text{ Al atoms}} = 4.470\times10^{-23} \text{ g}$$

The number of Al atoms in one mol (Avogadro's number) is then

$$\text{no. Al atoms} = \frac{26.98 \text{ g Al}}{1 \text{ mol Al}} \times \frac{1 \text{ atm Al}}{4.470\times10^{-23} \text{ g Al}} = 6.036\times10^{23}$$

11-49. In a *bcc* structure the relationship between the length of the cell ℓ and the atomic radius r is

$$\ell\sqrt{3} = 4r \quad \text{and} \quad \ell = \frac{4r}{\sqrt{3}}$$

The volume of a cube with this length is

$$V_{cube} = \ell^3 = \left(\frac{4r}{\sqrt{3}}\right)^3$$

The unit cell contains the equivalent of two atoms, whose combined volume is

$$V_{spheres} = 2 \times \frac{4}{3}\pi r^3$$

The *free volume* (voids) is the difference between these two quantities.

$$V_{free} = V_{cube} - V_{spheres} = \left(\frac{4r}{r_3}\right)^3 - \frac{8\pi}{3}r^3 = \left\{\frac{(4)^3}{(\sqrt{3})^3} - \frac{8\pi}{3}\right\} r^3$$

The % free volume is

$$\frac{V_{free}}{V_{cube}} \times 100 = \frac{\left\{\frac{(4)^3}{(\sqrt{3})^3} - \frac{8\pi}{3}\right\}\cancel{r^3}}{\frac{(4)^3}{(\sqrt{3})^3}\cancel{r^3}} \times 100 = \frac{(64/3\sqrt{3}) - (8\pi/3)}{64/3\sqrt{3}} \times 100 = 31.98\%$$

For a *fcc* structure,

$$\ell\sqrt{2} = 4r \quad \text{and} \quad \ell = \frac{4r}{\sqrt{2}}$$

$$V_{cube} = \ell^3 = \frac{(4)^3}{(\sqrt{2})^3} r^3$$

$$V_{spheres} = 4 \times \frac{4\pi}{3} r^3$$

$$\% \text{ Free Volume} = \frac{\left(\frac{64}{2\sqrt{2}} - \frac{4 \times 4\pi}{3}\right)\cancel{r^3}}{\frac{64}{2\sqrt{2}}\cancel{r^3}} \times 100 = 25.95\%$$

11-50. We have the following information about the unit cell of Mg:

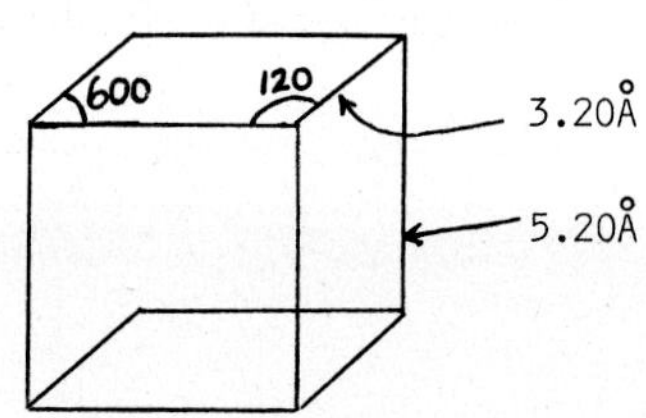

The volume of this parallelepiped is the area of the base multiplied by the height (5.20Å). The area of the base is the area of the two triangles.

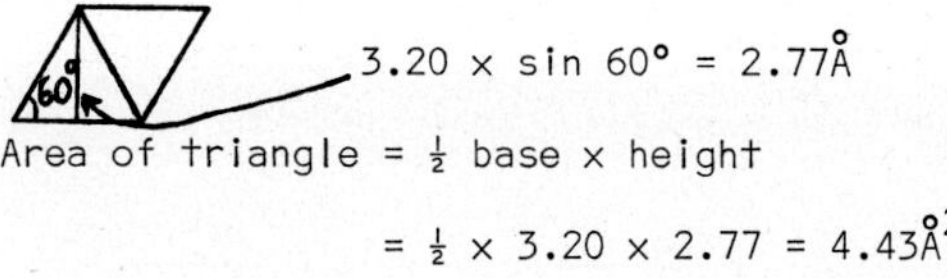

Area of triangle = $\frac{1}{2}$ base x height

$= \frac{1}{2} \times 3.20 \times 2.77 = 4.43\text{Å}^2$

For two triangles = $8.86\text{Å}^2 = 8.86\times10^{-16}\ \text{cm}^2$

The volume of the cell is

$$V = 8.86\times10^{-16}\ \text{cm}^2 \times 5.20\times10^{-8}\ \text{cm} = 4.61\times10^{-23}\ \text{cm}^3$$

The unit cell contains the equivalent of two Mg atoms (Table 11-4).

$$\text{mass of unit cell} = 2\ \text{Mg atom} \times \frac{24.305\ \text{g Mg}}{6.02\times10^{23}\ \text{Mg atom}} = 8.07\times10^{-23}\ \text{g}$$

$$\text{Density} = \frac{\text{mass}}{\text{volume}} = \frac{8.07\times10^{-23}\ \text{g}}{4.61\times10^{-23}\text{cm}^3} = 1.75\ \text{g/cm}^3$$

Ionic crystal structure

11-51. *CaF_2:* Each corner Ca^{2+} ion is shared among eight unit cells: 8 x 1/8 = 1. Each face-centered Ca^{2+} ion is shared between two unit cells: 6 x 1/2 = 3. The total number of Ca^{2+} ions in the unit cell is four. The eight F^- ions are pictured as belonging entirely to the unit cell. The ratio of Ca^{2+} to F^- ions is 4:8 = 1:2, consistent with the formula CaF_2.

TiO_2: Eight Ti^{4+} ions are situated at corners of the unit cell: 8 x 1/8 = 1; and one is located at the very center of the cell. Four of the O^{2-} ions are shared between two unit cells: 4 x 1/2 = 2. The remaining two are interior to the cell. Thus, the effective number of Ti^{4+} ions is two and O^{2-} ions, four. This ratio is consistent with the formula TiO_2.

11-52. (a) Volume of unit cell = length^3 = $(5.52\times10^{-8})^3\ \text{cm}^3 = 1.68\times10^{-22}\ \text{cm}^3$

(b) Each unit cell contains the equivalent of four formula units of NaCl.

$$\text{no. g NaCl} = 4\ \text{f. u. NaCl} \times \frac{1\ \text{mol NaCl}}{6.02\times10^{23}\ \text{f.u. NaCl}} \times \frac{58.5\ \text{g NaCl}}{1\ \text{mol NaCl}} = 3.89\times10^{-22}\ \text{g NaCl}$$

(c) $d = \frac{m}{V} = \frac{3.89\times10^{-22}\ g}{1.68\times10^{-22}\ cm^3} = 2.32\ g\ cm^{-3}$

Self-test Questions

1. (*d*) None of the substances listed is polar. The only type of intermolecular forces to consider are London or dispersion forces. The strengths of these forces increase with increasing molecular weight. Since Br_2 has the highest molecular weight of the substances listed, we should expect it to have the highest boiling point.

2. (*b*) None of the substances listed is a metal, so we should not expect any of them to be an especially good conductor. NaCl(s) is ionic but the ions are fixed in place in a crystalline lattice. Because the ions are not free to move throughout the crystal, NaCl(s) is a nonconductor. $Br_2(\ell)$ is a typical nonmetal and nonconductor of electricity. CO_2(s) is a molecular covalent solid with all electrons localized in covalent bonds. Si(s) is a metalloid (and semiconductor). It is the best conductor of the four substances listed.

3. (*c*) HF, CH_3OH, and N_2H_4 all meet the requirement for hydrogen bonding--a hydrogen atom bonded to a small electronegative atom, such as N, O, or F. In CH_4, although the C atom is small it is not highly electronegative. (C and H have about the same electronegativity.)

4. (*a*) We have not been given enough information in the chapter to say how we should expect ΔH_{vap} to vary with temperature. With few exceptions, we know that density should *decrease* with temperature. A point emphasized in this chapter, however, is that vapor pressure always increases with temperature (recall Figure 11-6).

5. (*c*) Graphite and diamond are both pure carbon. [Answer (b) is incorrect.] Diamond is harder than graphite, but graphite is an electrical conductor (because of its delocalized electrons) whereas diamond is not. Diamond has equal bond distances in all directions, but in graphite there is one C - C bond distance for atoms in the hexagonal planes and another C - C bond distance between planes.

6. Since wetting refers to the ability of water to flow over a surface, anything that enhances this ability is a wetting agent and makes water "wetter". There is a basis to the television commerical claim.

7. (a) The stronger the forces between molecules in a liquid, the lower the tendency for the molecules to pass into the vapor state--the lower the vapor pressure.

   (b) The vapor pressure of a liquid *does not* depend on the volume of liquid in the liquid-vapor equilibrium.

   (c) The vapor pressure of a liquid *does not* depend on the volume of vapor in the liquid-vapor equilibrium.

   (d) As long as liquid-vapor equilibrium is established, the vapor becomes saturated and its pressure is fixed, regardless of the size of the container.

   (e) Vapor pressure *does* depend on the temperature of the liquid, always increasing as temperature increases.

8. (*c*) Consider the heat loss when (a) 10 g steam is condensed at 100°C, (b) 10 g of liquid water is cooled from 100°C to 20°C and (c) 110 g water is cooled from 20°C to 0°C. This quantity of heat is greater than that required to melt 100 g ice. The final condition is one of liquid water only.

$$q_a + q_b + q_c = \left\{10\ g \times \frac{1\ mol}{18\ g} \times \frac{-40.7\ kJ}{1\ mol}\right\} + \left\{10\ g \times \frac{4.18\ J}{g\ ^\circ C} \times (-80^\circ C) \times \frac{1\ kJ}{1000\ J}\right\}$$
$$+ \left\{110\ g \times \frac{4.18\ J}{g\ ^\circ C} \times (-20^\circ C) \times \frac{1\ kJ}{1000\ J}\right\} = -35.2\ kJ$$

$$q_{ice} = 100\ g \times \frac{1\ mol}{18\ g} \times \frac{6.02\ kJ}{1\ mol} = +33.4\ kJ$$

9. If the rate of evacuation is fast enough, the heat of vaporization required to sustain vaporization of the water cannot be drawn from the surroundings and must be taken from the liquid itself. This causes the liquid temperature to drop. If the temperature drops to 0°C as a result, ice may begin to freeze from the water.

10. The student calculation is based on removing a sample of vapor from contact with the liquid with which it is in equilibrium at 20°C. The pressure of this vapor (gas) would then vary with temperature (with amount of gas and volume held constant) in the manner predicted from the gas laws. But to determine the vapor pressure at 50°C requires that the vapor remain *in contact with the liquid* as the temperature is raised from 20°C to 50°C. More liquid vaporizes and the vapor pressure increases to a much greater extent than indicated in the student's calculation. The variation of vapor pressure with temperature must be calculated with equation (11.3).

11. The *fcc* structure simply refers to the location of structural particles in a crystalline lattice but not to what these particles are or to the type of forces that exist among them. In Ar these particles are Ar atoms and in carbon dioxide, CO_2 molecules. In both these substances the only intermolecular forces are of the instantaneous dipole-induced dipole type. In sodium chloride the structural particles are Na^+ and Cl^- ions and the forces among them are interionic; copper atoms in copper metal are joined by metallic bonds. Thus, both sodium chloride and copper have much higher melting points than argon and carbon dioxide.

12. *Ionic:* The structural particules are ions. The intermolecular forces are interionic attractions. In general, ionic compounds have moderate to high melting points and are good electrical conductors in the molten state.

Network covalent: The structural particles are atoms. The bonds are covalent. All atoms are bonded together into a giant crystal. If bond strengths are equal in all directions, as in diamond and silica, the solid substance is hard, has a very high melting point, and is a nonconductor.

Molecular: The structural particles are discrete molecules. Intermolecular forces are generally weak, involving instantaneous, induced or permanent dipoles, and, in some cases, hydrogen bonds. Melting points are low, and the substances are nonconductors, both as solids and liquids.

Metallic: The structural particles are metal ions. These are bound together by a "sea of electrons". The mobility of these electrons accounts for the electrical and thermal conductivity of metals and for their malleability, ductility, and luster. An alternate description based on band theory is also consistent with the properties of metals.

Chapter 12

Solutions

Review Problems

12-1. The total solution mass is 144 g KI + 100 g H_2O = 244 g.

$$\%\text{KI, by mass} = \frac{144\ \text{g KI}}{244\ \text{g total}} \times 100 = 59.0\%\ \text{KI}$$

12-2. (a) $$\%\ CH_3OH\ \text{(vol/vol)} = \frac{11.3\ \text{mL}\ CH_3OH}{75.0\ \text{mL soln.}} \times 100 = 15.1\%\ CH_3OH$$

(b) $$\%\ CH_3OH\ \text{(mass/vol)} = \frac{11.3\ \text{mL}\ CH_3OH \times \frac{0.793\ \text{g}\ CH_3OH}{1.00\ \text{mL}\ CH_3OH}}{75.0\ \text{mL soln.}} \times 100 = 11.9\%\ CH_3OH$$

(c) $$\%\ CH_3OH\ \text{(mass/mass)} = \frac{11.3\ \text{mL}\ CH_3OH \times \frac{0.793\ \text{g}\ CH_3OH}{1.00\ \text{mL}\ CH_3OH}}{75.0\ \text{mL soln.} \times \frac{0.980\ \text{g soln.}}{1.00\ \text{mL soln.}}} \times 100 = 12.2\%\ CH_3OH$$

12-3. $$\text{no. g solute} = 0.2500\ \text{L soln.} \times \frac{0.0250\ \text{mol}\ AgNO_3}{\text{L soln.}} \times \frac{170\ \text{g}\ AgNO_3}{1\ \text{mol}\ AgNO_3} \times \frac{100.0\ \text{g solute}}{99.35\ \text{g}\ AgNO_3} = 1.07\ \text{g solute}$$

12-4. Consider 1.000 L of the solution. Its mass is

$$\text{no. g soln.} = 1.000\ \text{L} \times \frac{1000\ \text{cm}^3}{1\ \text{L}} \times \frac{1.005\ \text{g}}{\text{cm}^3} = 1005\ \text{g}$$

The mass of NaCl in this solution is

$$\text{no. g NaCl} = 1005\ \text{g soln.} \times \frac{0.90\ \text{g NaCl}}{100\ \text{g soln.}} = 9.0\ \text{g NaCl}$$

The molarity of the solution is

$$\frac{9.0\ \text{g NaCl} \times \frac{1\ \text{mol NaCl}}{58.5\ \text{g NaCl}}}{1.00\ \text{L soln.}} = 0.15\ \text{M NaCl}$$

12-5. $$\text{no. mol}\ C_6H_4Cl_2 = 2.15\ \text{g}\ C_6H_4Cl_2 \times \frac{1\ \text{mol}\ C_6H_4Cl_2}{147\ \text{g}\ C_6H_4Cl_2} = 0.0146\ \text{mol}\ C_6H_4Cl_2$$

$$\text{no. kg solv.} = 15.0\ \text{cm}^3 \times \frac{0.879\ \text{g}}{1\ \text{cm}^3} \times \frac{1\ \text{kg}}{1000\ \text{g}} = 0.0220\ \text{kg solv.}$$

$$\text{molality} = \frac{0.0146\ \text{mol}\ C_6H_4Cl_2}{0.0220\ \text{kg}\ C_6H_6} = 0.664\ \text{m}\ C_6H_4Cl_2$$

12-6. The total no. mol solution components = 2.13 + 1.79 + 3.11 = 7.03 mol

(a) $$X_{C_7H_{16}} = \frac{2.13\ \text{mol}\ C_7H_{18}}{7.03\ \text{mol total}} = 0.303;\ X_{C_8H_{18}} = \frac{1.79\ \text{mol}\ C_8H_{18}}{7.03\ \text{mol total}} = 0.255;\ X_{C_9H_{20}} = \frac{3.11\ \text{mol}\ C_9H_{20}}{7.03\ \text{mol total}} = 0.442$$

(b) Mole percents = 30.3% C_7H_{16}, 15.5% C_8H_{18}, 44.2% C_9H_{20}.

12-7. The data in Figure 12-7 are presented as g solute/100 g soln. The given data must first be converted to this unit.

$$\text{no. g } NH_4Cl = 0.80 \text{ mol } NH_4Cl \times \frac{53.5 \text{ g } NH_4Cl}{1 \text{ mol } NH_4Cl} = 43 \text{ g } NH_4Cl$$

$$\text{no. g soln.} = 43 \text{ g } NH_4Cl + 150 \text{ g } H_2O = 193 \text{ g soln.}$$

$$\frac{43 \text{ g } NH_4Cl}{193 \text{ g soln.}} = \frac{0.22 \text{ g } NH_4Cl}{1.00 \text{ g soln.}} \times \frac{100}{100} = \frac{22 \text{ g } NH_4Cl}{100 \text{ g soln.}}$$

At 25°C saturated $NH_4Cl(aq)$ is about 28 g NH_4Cl/100 g soln. The given solution is unsaturated.

12-8. (a) Convert the information about the quantity of $O_2(g)$ that dissolves to a molar basis.

$$n = \frac{PV}{RT} = \frac{1.00 \text{ atm} \times 0.0309 \text{ L}}{0.0821 \text{ L atm mol}^{-1}\text{K}^{-1} \times 298\text{K}} = 1.26\times10^{-3} \text{ mol } O_2$$

The molarity is $\dfrac{1.26\times10^{-3} \text{ mol } O_2}{1.00 \text{ L}} = 1.26\times10^{-3} \text{ M } O_2(aq)$

(b) For the expression $C = k \cdot P_{Gas}$ $\quad k = \dfrac{1.26\times10^{-3} \text{ M}}{1.00 \text{ atm}}$

For air (a mixture of gases) with 20.95% O_2, by volume, we can use equation (5.16) to write

$$\frac{n_{O_2}}{n_{tot.}} = \frac{P_{O_2}}{P_{tot.}} = \frac{V_{O_2}}{V_{tot.}} \qquad P_{O_2} = \frac{V_{O_2}}{V_{tot.}} \times P_{tot.} = \frac{20.95 \text{ L}}{100.0 \text{ L}} \times 1 \text{ atm} = 0.2095 \text{ atm}$$

$$C = k \cdot P_{O_2} = \frac{1.26\times10^{-3} \text{ M}}{\text{atm}} \times 0.2095 \text{ atm} = 2.64\times10^{-4} \text{ M } O_2(aq)$$

12-9. Convert the masses of solution components to numbers of moles and then determine the mole fraction composition of the solution.

$$\text{no. mol } C_6H_6 = 60.0 \text{ g } C_6H_6 \times \frac{1 \text{ mol } C_6H_6}{78.1 \text{ g } C_6H_6} = 0.768 \text{ mol } C_6H_6$$

$$\text{no. mol } C_7H_8 = 75.0 \text{ g } C_7H_8 \times \frac{1 \text{ mol } C_7H_8}{92.1 \text{ g } C_7H_8} = 0.814 \text{ mol } C_7H_8$$

$$X_{C_6H_6} = \frac{0.768 \text{ mol } C_6H_6}{(0.768 + 0.814) \text{ mol total}} = 0.485 \qquad X_{C_7H_8} = \frac{0.814 \text{ mol } C_7H_8}{(0.768 + 0.814) \text{ mol total}} = 0.515$$

Now use Raoult's law:

$$P_{C_6H_6} = 0.485 \times 95.1 \text{ mmHg} = 46.1 \text{ mmHg} \qquad P_{C_7H_8} = 0.515 \times 28.4 \text{ mmHg} = 14.6 \text{ mmHg}$$

$$P_{tot.} = 46.1 + 14.6 = 60.7 \text{ mmHg}$$

12-10. Use equation (5.16), that is, $\dfrac{n_A}{n_{tot}} = \dfrac{P_A}{P_{tot}} = \dfrac{V_A}{V_{tot}}$

In the *vapor*,

$$X_{C_6H_6} = \frac{46.1 \text{ mmHg}}{60.7 \text{ mmHg}} = 0.759 \qquad X_{C_7H_8} = \frac{14.6 \text{ mmHg}}{60.7 \text{ mmHg}} = 0.241$$

12-11. From the measured freezing point depression determine the molality of the solution.

$$\Delta T_f = K_f \cdot m \qquad m = \frac{\Delta T_f}{K_f} = \frac{(5.51-4.90)\,°C}{4.90\ \frac{\text{kg solv. } °C}{\text{mol solute}}} = 0.12\ \frac{\text{mol solute}}{\text{kg solv.}}$$

Now determine the number of moles of solute in the 0.12 m solution.

$$\text{no. mol solute} = 75.22\ \text{g solv.} \times \frac{1\ \text{kg solv.}}{1000\ \text{kg solv.}} \times \frac{0.12\ \text{mol solute}}{\text{kg solv.}} = 9.0\times10^{-3}\ \text{mol solute}$$

The molar mass is

$$M = \frac{1.10\ \text{g solute}}{9.0\times10^{-3}\ \text{mol solute}} = 1.2\times10^{2}\ \text{g solute/mol solute} \qquad \text{Molecular weight} = 1.2\times10^{2}$$

12-12. The expected freezing point depression of a 0.01 m aqueous solution of a nonelectrolyte is $\Delta T_f = K_f \cdot m = 1.86 \times 0.01 = 0.0186°C$. The measured freezing point depression is 0.072. This corresponds to a van't Hoff factor of

$$i = \frac{\text{measured freezing point depression}}{\text{freezing point depression for nonelectrolyte}} = \frac{0.072}{0.0186} = 3.9$$

The expected boiling point elevation for an 0.01 m aqueous solution is

$\Delta T_b = K_b \cdot m = 0.512 \times 0.01 = 0.005°C$. For the solution in question $\Delta T_b = i \times 0.005°C$.

$\Delta T_b = 3.9 \times 0.005 = 0.02$. The expected normal boiling point is $100.00 + 0.02 = 100.02°C$.

12-13. In the expression relating osmotic pressure and solution molality below, osmotic pressure is expressed in atm.

$$M = \frac{\pi}{RT} = \frac{(0.79/760)\ \text{atm}}{0.0821\ \text{L atm mol}^{-1}\text{K}^{-1} \times 298\ \text{K}} = 4.2\times10^{-5}\ \text{M}$$

$$\text{no. mol PVC} = 0.250\ \text{L} \times \frac{4.2\times10^{-5}\ \text{mol PVC}}{1\ \text{L}} = 1.0\times10^{-5}\ \text{mol}$$

$$\text{molar mass} = M = \frac{0.61\ \text{g}}{1.0\times10^{-5}\ \text{mols}} = 6.1\times10^{4}\ \text{g/mol}.$$ The molecular weight of the PVC is 6.1×10^{4}.

12-14. A solution that is 0.110 M KCl is 0.110 M K^+ and 0.110 M Cl^-. A solution that is 0.125 M $MgCl_2$ is 0.125 M Mg^{2+} and (2x0.125) = 0.250 M Cl^-. In the mixed solution

molarity of K^+ = 0.110 M

molarity of Mg^{2+} = 0.125 M

molarity of Cl^- = 0.110 M + 0.250 M = 0.360 M

12-15. The solution with the highest freezing point (that is, with the lowest freezing point depression) is the *nonelectrolyte*, C_2H_5OH. Next comes the weak electrolyte, $HC_2H_3O_2$. The remaining three solutions are all strong electrolytes. NaCl produces two mol ions per mole of component; $MgBr_2$, three, and $Al_2(SO_4)_3$, five. The order of *decreasing* freezing point is

$$C_2H_5OH > HC_2H_3O_2 > NaCl > MgBr_2 > Al_2(SO_4)_3$$

Exercises

Homogeneous and heterogeneous mixtures

12-1. (a) C_2H_5OH is the solute and H_2O is the solvent.

(b) Since equal masses of CH_3OH and H_2O are present it may be difficult to decide which is the solvent and which is the solute. However, since more molecules are present in 50 g H_2O than in 50 g CH_3OH, it is probably appropriate still to refer to H_2O as the solvent.

(c) The CCl_4 is the solute. The C_6H_6-C_7H_8 *mixture* can be thought of as the solvent.

(d) An aqueous solution is indicated. Water is the solvent and Na_2SO_4 is the solute.

12-2. If two substances are similar to one another, then it is likely that intermolecular forces between the different molecules will also be similar and the substances will form a solution. Thus, "like dissolves like". Oils are either hydrocarbons or hydrocarbon derivatives. Oil molecules (nonpolar) and water molecules (polar) do not exert sufficiently strong attractions for one another to keep these different molecules in a homogeneous mixture or solution. Thus, "oil and water don't mix".

12-3. The decrease in volume indicates that ethanol and water molecules exist in close proximity when mixed. This suggests strong intermolecular forces--case 2 of Section 12-1.

12-4. The anions in metal nitrates carry a charge of -1 (NO_3^-) and in metal sulfides, a charge of -2 (S^{2-}). We should expect the lattice energies of metal sulfides to be of greater magnitude than those of metal nitrates. In turn, this means that as a group metal sulfides are less soluble than metal nitrates. Among metal sulfides the most soluble should be those in which the cation is relatively large and has a charge of +1--NH_4^+ and the alkali metal ions, for example.

Percent concentration

12-5. $$\text{no. g } HC_2H_3O_2 = 0.750 \text{ L} \times \frac{1000 \text{ cm}^3}{1.00 \text{ L}} \times \frac{1.01 \text{ g soln.}}{1.00 \text{ cm}^3} \times \frac{6.10 \text{ g } HC_2H_3O_2}{100 \text{ g soln.}} = 46.2\ HC_2H_3O_2$$

12-6. Since the density of ethanol is less than that of water, whatever quantities of ethanol and water are mixed the proportion of ethanol to water by mass will always be less than its proportion by volume. As a consequence, the mass percent ethanol is always less than the volume percent in aqueous solutions. If the other component with which ethanol is mixed has a *smaller* density than ethanol, the mass percent ethanol exceeds its volume percent.

12-7. Percent by mass is based on a ratio of masses. Since mass is independent of temperature, so is percent by mass. Percent by volume is based on a ratio of volumes. Volume is temperature dependent, and so too is volume percent.

Molar concentration

12-8. In 1.000 L solution,

$$\text{no. mol } H_2SO_4 = 1000 \text{ cm}^3 \text{ soln.} \times \frac{1.831 \text{ g soln.}}{1 \text{ cm}^3 \text{ soln.}} \times \frac{94.0 \text{ g } H_2SO_4}{100 \text{ g soln.}} \times \frac{1 \text{ mol } H_2SO_4}{98.1 \text{ g } H_2SO_4} = 17.5 \text{ mol } H_2SO_4$$

Molarity = 17.5 mol H_2SO_4/1.00 L solution = 17.5 M H_2SO_4

12-9. $$\text{no. g } C_2H_5OH \text{ in final soln.} = 1.125 \text{ L} \times \frac{0.175 \text{ mol } C_2H_5OH}{1.00 \text{ L}} \times \frac{46.1 \text{ g } C_2H_5OH}{1 \text{ mol } C_2H_5OH} = 9.08 \text{ g } C_2H_5OH$$

From Example 12-1 we see that the original ethanol solution is 8.03%, by mass and has a density of 0.982 g/cm^3.

$$\text{no. mL original soln.} = 9.08 \text{ g } C_2H_5OH \times \frac{100.0 \text{ g soln.}}{8.03 \text{ g } C_2H_5OH} \times \frac{1.00 \text{ mL soln.}}{0.982 \text{ g soln.}} = 115 \text{ mL soln.}$$

12-10. Choose a particular solution sample size on which to base these calculations, say 100.0 g. In both cases the numerator represents the number of moles of solute and the denominator, the volume of solution in liters.

$$\textit{at 15°C:}\quad \text{molar conc.} = \frac{100.0\text{ g soln.} \times \dfrac{10.00\text{ g } C_2H_5OH}{100.0\text{ g soln.}} \times \dfrac{1\text{ mol } C_2H_5OH}{46.07\text{ g } C_2H_5OH}}{100.0\text{ g soln.} \times \dfrac{1.000\text{ mL soln.}}{0.9831\text{ g soln.}} \times \dfrac{1.000\text{ L soln.}}{1000\text{ mL soln.}}} = 2.134\text{ M } C_2H_5OH$$

$$\textit{at 25°C:}\quad \text{molar conc.} = \frac{100.0\text{ g soln.} \times \dfrac{10.00\text{ g } C_2H_5OH}{100.0\text{ g soln.}} \times \dfrac{1\text{ mol } C_2H_5OH}{46.07\text{ g } C_2H_5OH}}{100.0\text{ g soln.} \times \dfrac{1.000\text{ mL soln.}}{0.9804\text{ g soln.}} \times \dfrac{1.000\text{ L soln.}}{1000\text{ mL soln.}}} = 2.128\text{ M } C_2H_5OH$$

Molal concentration

12-11. In the setup below the molality concentration is written as a conversion factor between kg CCl_4 and mol I_2, i.e. 1 kg $CCl_4 \Leftrightarrow$ 0.158 mol I_2 or 1000 g $CCl_4 \Leftrightarrow$ 0.158 mol I_2

$$\text{no. g } I_2 = 125.0\text{ mL } CCl_4 \times \frac{1.595\text{ g } CCl_4}{1.000\text{ mL } CCl_4} \times \frac{0.158\text{ mol } I_2}{1000\text{ g } CCl_4} \times \frac{254\text{ g } I_2}{1.00\text{ mol } I_2} = 8.00\text{ g } I_2$$

12-12. Let us base this calculation on 1.000 liter of solution. We must determine the masses of solution and solvent and the number of moles of solute. From these data both the molarity and molality can be determined easily.

$$\text{no. g soln.} = 1.000\text{ L soln.} \times \frac{1000\text{ mL soln.}}{\text{L soln.}} \times \frac{1.101\text{ g soln.}}{1\text{ mL soln.}} = 1101\text{ g soln.}$$

$$\text{no. g HF} = 1101\text{ g soln.} \times \frac{30.0\text{ g HF}}{100\text{ g soln.}} = 330\text{ g HF}$$

$$\text{no. mol HF} = 330\text{ g HF} \times \frac{1\text{ mol HF}}{20.0\text{ g HF}} = 16.5\text{ mol HF}$$

$$\text{no. kg } H_2O = (1101 - 330)\text{g } H_2O \times \frac{1\text{ kg } H_2O}{1000\text{ g } H_2O} = 0.771\text{ kg } H_2O$$

$$\frac{16.5\text{ mol HF}}{1.00\text{ L soln.}} = 16.5\text{ M HF};\qquad \frac{16.5\text{ mol HF}}{0.771\text{ kg } H_2O} = 21.4\text{ m HF}$$

12-13. Consider 1.00 L of the solution. This solution weighs 1000 × 1.09 = 1090 g and contains 109.2 g KOH. The mass of water in the solution is 1090 - 109.2 = 980 g H_2O. Use these data to determine the molal concentration of the solution.

$$\text{molal conc.} = \frac{109.2\text{ g KOH} \times \dfrac{1\text{ mol KOH}}{56.1\text{ g KOH}}}{0.98\text{ kg } H_2O} = 2.0\text{ m KOH}$$

To convert this solution to 0.250 m KOH requires that water be added. Set up the following equation based on 100.0 cm^3 (109 g) of the original solution and let x = no. kg H_2O to be added.

$$0.250\text{ m} = \frac{10.92\text{ g KOH} \times \dfrac{1\text{ mol KOH}}{56.1\text{ g KOH}}}{(0.098 + x)\text{kg } H_2O}\qquad 0.250 \times (0.098 + x) = 0.195$$

$$x = \frac{0.195}{0.250} - 0.098\qquad x = 0.682\text{ kg } H_2O = 682\text{ g } H_2O$$

Mole fraction, mole percent

12-14. To calculate mole fractions we must be able to calculate the numbers of moles of each component. If this calculation can be performed exactly, then the calculated mole fraction is exact. If an assumption is required to calculate the number of moles of either component, then the mole fraction calculation becomes only approximate.

(a) 100.0 g of solution contains 12.2 g C_2H_5OH (and 87.8 g H_2O).

$$\text{no. mol } C_2H_5OH = 12.2 \text{ g } C_2H_5OH \times \frac{1 \text{ mol } C_2H_5OH}{46.1 \text{ g } C_2H_5OH} = 0.265 \text{ mol } C_2H_5OH$$

$$\text{no. mol } H_2O = 87.8 \text{ g } H_2O \times \frac{1 \text{ mol } H_2O}{18.0 \text{ g } H_2O} = 4.88 \text{ mol } H_2O$$

$$X_{C_2H_5OH} = \frac{0.265 \text{ mol } C_2H_5OH}{(0.265 + 4.88) \text{ mol total}} = \frac{0.265}{5.14} = 0.0516 \text{ (exact)}$$

(b) Per kg H_2O there is present 0.255 mol $CO(NH_2)_2$.

$$X_{CO(NH_2)_2} = \frac{0.255 \text{ mol } CO(NH_2)_2}{[0.255 + (1000/18.0)] \text{ mol total}} = \frac{0.255}{55.8} = 4.57\times10^{-3} \text{ (exact)}$$

(c) In 1.00 L of solution there is 0.050 mol $C_6H_{12}O_6$.
The mass of $C_6H_{12}O_6$ = 0.050 mol × 180 g/mol = 9.0 g $C_6H_{12}O_6$. To determine the mass of water we must know the mass of solution. In turn this requires knowing the density of the solution, but this is not given. We have to *estimate* the density of the solution; a density of 1.00 g/cm^3 is a fair estimate,

Mass of solution = 1000 g

Mass of water = 1000 g - 9.0 g = 991 g H_2O

$$X_{C_6H_{12}O_6} = \frac{0.050 \text{ mol } C_6H_{12}O_6}{[0.050 + (991/18.0)] \text{ mol total}} = \frac{0.050}{55.1} = 9\times10^{-4} \text{ (approximate)}$$

12-15. In Example 12-1 we learned that 100.0 mL of the ethanol solution weighed 98.2 g and consisted of 90.3 g H_2O (5.02 mol) and 7.89 g C_2H_5OH (0.171 mol). Solve the following equation for x, the number of moles of C_2H_5OH to be added.

$$X_{C_2H_5OH} = \frac{(0.171 + x)\text{mol } C_2H_5OH}{[(0.171 + x) + 5.02]\text{mol}} = 0.0500 \qquad 0.171 + x = 0.00855 + 0.0500x + 0.251$$

$$0.950x = 0.089 \qquad x = 0.094 \text{ mol } C_2H_5OH$$

$$\text{no. g } C_2H_5OH = 0.094 \text{ mol } C_2H_5OH \times \frac{46.1 \text{ g } C_2H_5OH}{1 \text{ mol } C_2H_5OH} = 4.3 \text{ g } C_2H_5OH$$

12-16. A 8.15 mol percent solution corresponds to $X_{C_3H_8O_3} = 0.0815$. Let V = no. mL glycerol required.

Determine the number of moles of glycerol in terms of V

$$\text{no. mol } C_3H_8O_3 = V \times \frac{1.26 \text{ g}}{1.00 \text{ mL}} \times \frac{1 \text{ mol } C_3H_8O_3}{92.1 \text{ g } C_3H_8O_3} = 0.0137V$$

Next, write an expression for the mole fraction.

$$X_{C_3H_8O_3} = \frac{0.0137V}{0.0137V + \left(1000 \text{ g } H_2O \times \frac{1 \text{ mol } H_2O}{18.0 \text{ g } H_2O}\right)} = \frac{0.0137V}{0.0137V + 55.6} = 0.0815$$

$$0.0137V - 0.00112V = 4.53; \qquad V = 3.60\times10^2 \text{ mL } C_3H_8O_3$$

Solubility equilibrium

12-17. The data in Figure 12-7 are expressed in g $KClO_4$/100 g soln., that is, percent $KClO_4$, by mass. We must first describe a 1.00 m $KClO_4$ soln. in % $KClO_4$, by mass.

For every 1000 g H_2O, the solution contains 1.00 mol $KClO_4$ = 139 g $KClO_4$.

$$\%\ KClO_4 = \frac{139\text{ g }KClO_4}{(1000 + 139)\text{g soln.}} \times 100 = 12.2\%\ KClO_4$$

A saturated solution is 12.2% $KClO_4$ at about 80°C.

12-18. (a) The solution in question has 26.0 g $KClO_4$ in (500.0 + 26.0) = 526.0 g solution. Its % $KClO_4$, by mass, is

$$\%\ KClO_4 = \frac{26.0\text{ g }KClO_4}{526.0\text{ g soln.}} \times 100 = 4.94\%\ KClO_4\ (4.94\text{ g }KClO_4/100\text{ g soln.})$$

This solution is supersaturated.

(b) The mass of $KClO_4$ per 100 g saturated solution at 20°C is about 2 g $KClO_4$. The mass of $KClO_4$ that should crystallize per 100 g of the given solution is about 3 g $KClO_4$ (that is, 4.94 - 2). For the 526 g supersaturated solution,

$$\text{no. g }KClO_4\text{ crystallizing} = \frac{3\text{ g }KClO_4}{100\text{ g soln.}} \times 526\text{ g soln.} \doteq 16\text{ g }KClO_4$$

12-19. (a) Determine the % by mass of each solute in the solution.

$$\%\ NH_4Cl = \frac{(50.0 \times 0.950)\text{g }NH_4Cl}{150.0\text{ g soln.}} \times 100 = 31.7\%\ NH_4Cl\ (31.7\ NH_4Cl/100\text{ g soln.})$$

$$\%\ (NH_4)_2SO_4 = \frac{(50.0 \times 0.050)\text{g }(NH_4)_2SO_4}{150.0\text{ g soln.}} \times 100 = 1.7\%\ (NH_4)_2SO_4\ [1.7\text{ g }(NH_4)_2SO_4/100\text{ g soln.}]$$

Reference to Figure 12-7 indicates that the solid sample will dissolve completely.

(b) At 0°C the solubility of NH_4Cl is about 23 g NH_4Cl/100 g soln. Let x = no. g NH_4Cl that crystallize. The original number of grams of NH_4Cl was 0.95 x 50 = 48 g. Neglect the presence of the $(NH_4)_2SO_4$.

$$\frac{(48 - x)\text{g }NH_4Cl}{[100 + (48 - x)]\text{g soln.}} = \frac{23\text{ g }NH_4Cl}{100\text{ g soln.}} \qquad 4800 - 100x = 2300 + 1100 - 23x$$

$$1400 = 77x \qquad x = 18\text{ g }NH_4Cl$$

(c) At 0°C the solubility of $(NH_4)_2SO_4$ is about 42 g $(NH_4)_2SO_4$/100 g soln. The original solution at 90°C contained 2.5 g $(NH_4)_2SO_4$ in 150 g soln. No $(NH_4)_2SO_4$ will crystallize when the solution is cooled to 0°C.

Solubility of gases

12-20. Based on data from Example 12-3, we conclude that at 740 mmHg the solubility of H_2S in water is

$$\text{molal conc.} = \frac{0.195\text{ m}}{\text{atm}} \times (740/760)\text{atm} = 0.190\text{ m }H_2S$$

Determine the % H_2S, by mass, in a saturated aqueous solution, i.e., in 0.190 m H_2S.

$$\frac{0.190\text{ mol }H_2S \times \dfrac{34.1\text{ g }H_2S}{1\text{ mol }H_2S}}{[(0.190 \times 34.1) + 1000]\text{g soln.}} \times 100 = 0.644\%\ H_2S$$

Since the natural water contains only 0.5% H_2S, by mass, it will dissolve more H_2S.

12-21. First, establish the value of k in the expression C = k x P.

$$\frac{0.02 \text{ g } CH_4}{\text{kg } H_2O} = k \times 1 \text{ atm}; \qquad k = \frac{0.02 \text{ g } CH_4}{\text{kg } H_2O \text{ atm}}$$

Under the stated conditions: $C = \frac{0.02 \text{ g } CH_4}{\text{kg } H_2O \text{ atm}} \times 20 \text{ atm} = \frac{0.4 \text{ g } CH_4}{\text{kg } H_2O}$

$$\text{no. g natural gas dissolving} = \frac{0.4 \text{ g } CH_4}{\text{kg } H_2O} \times 1.00\times10^3 \text{ kg } H_2O = 4\times10^2 \text{ g } CH_4$$

12-22. According to Henry's law (equation 12.4) the concentration of a dissolved gas is directly proportional to the pressure of the gas in contact with the solution. If this concentration is expressed as molality (mole gas/kg solvent), and if the quantity of solvent is fixed, then the number of moles of dissolved gas is directly proportional to the pressure of the gas. But the mass of a substance is itself directly proportional to the number of moles of a substance. Therefore, the mass of dissolved gas should be directly proportional to the gas pressure, as stated in this exercise.

12-23. Again, the equation in question is $C = k \times P_{gas}$, with the concentration expressed as the number of moles of dissolved gas in a fixed amount of solvent.

$$C = \frac{n}{\text{amt. solv.}} = k \times P_{gas}$$

Before dissolving, that is, while still in the gaseous state, the number of moles of gas could be related to the gas pressure and volume through the ideal gas equation.

$$n = \frac{P_{gas} \times V_{gas}}{RT} \qquad C = \frac{P_{gas} \times V_{gas}}{R \times T \times \text{amt. solvent}} = k \times P_{gas}$$

$$V_{gas} = k \times R \times T \times \text{amt. solvent} = \text{constant}$$

At a constant temperature and for a fixed amount of solvent, the volume of dissolved gas is a constant, independent of pressure. Recall, however, that as the gas pressure changes the number of moles in a fixed volume of gas changes proportionately. Thus, although the *volume* of gas that dissolves remains constant as the gas pressure changes, the *number of moles* of gas does change in the expected manner--the higher the gas pressure, the more moles of gas dissolve. However, this second statement is not expected to hold under conditions where the ideal fails. Neither is it expected to hold if the gas reacts with the solvent.

Raoult's law and liquid-vapor equilibrium

12-24. The solution has the composition $X_{benz.} = 0.300$ and $X_{tol.} = 0.700$. The partial pressure of the toluene is

$$P_{tol.} = X_{tol.} \cdot P^\circ_{tol.} = 0.700 \times 533 \text{ mmHg} = 373 \text{ mmHg}.$$

Since the solution boils at this temperature, $P_{tot.} = 760$ mmHg and

$$P_{tot.} = P_{tol.} + P_{benz.} = 373 \text{ mmHg} + P_{benz.} = 760 \text{ mmHg}$$

$$P_{benz.} = 760 \text{ mmHg} - 373 \text{ mmHg} = 387 \text{ mmHg}.$$

Now use Raoult's law for the benzene.

$$P_{benz.} = 387 \text{ mmHg} = X_{benz.} \cdot P^\circ_{benz.} = 0.300 \times P^\circ_{benz.}$$

$$P^\circ_{benz.} = \frac{387 \text{ mmHg}}{0.300} = 1.29\times10^3 \text{ mmHg}$$

12-25. First, calculate the mole fraction of H_2O in the urea solution.

$$\text{no. mol } CO(NH_2)_2 = 21.8 \text{ g } CO(NH_2)_2 \times \frac{1 \text{ mol } CO(NH_2)_2}{60.1 \text{ g } CO(NH_2)_2} = 0.363 \text{ mol } CO(NH_2)_2$$

$$\text{no. mol } H_2O = 525 \text{ g } H_2O \times \frac{1 \text{ mol } H_2O}{18.0 \text{ g } H_2O} = 29.2 \text{ mol } H_2O$$

$$X_{H_2O} = \frac{29.2 \text{ mol } H_2O}{(29.2 + 0.4) \text{ mol total}} = 0.986$$

Now use Raoult's law.

$$P_{H_2O} = X_{H_2O} \cdot P^\circ_{H_2O} = 0.986 \times 23.8 \text{ mmHg} = 23.5 \text{ mmHg}$$

12-26. If the vapor phase contains 62.0 mole % C_6H_6, then in the vapor

$$\frac{n_{C_6H_6}}{n_{tot.}} = \frac{P_{C_6H_6}}{P_{tot.}} = 0.620$$

In the above expression, $P_{C_6H_6} = P^\circ_{C_6H_6} \cdot X_{C_6H_6} = 95.1X_{C_6H_6}$

Also, $P_{C_7H_8} = 28.4X_{C_7H_8} = 28.4(1 - X_{C_6H_6})$. The mol fractions are those of the *liquid* solution.

$$P_{tot.} = P_{C_6H_6} + P_{C_7H_8} = 95.1X_{C_6H_6} + 28.4(1 - X_{C_6H_6})$$

Now substitute back into the first equation, and solve it for $X_{C_6H_6}$.

$$\frac{95.1X_{C_6H_6}}{28.4(1 - X_{C_6H_6}) + 95.1X_{C_6H_6}} = \frac{95.1X_{C_6H_6}}{28.4 + 66.7X_{C_6H_6}} = 0.620$$

$$95.1X_{C_6H_6} = 17.6 + 41.4X_{C_6H_6}; \qquad X_{C_6H_6} = 0.328$$

Freezing point depression and boiling point elevation

12-27. Two distinct aspects are involved in this problem. First the method of Review Problem 11 is used to determine the molecular weight of the compound. The empirical formula of the compound is established by the method introduced in Chapter 3. A simple comparison of these two results leads to the molecular formula.

Molecular weight determination

$$\Delta T_f = K_f \cdot m; \qquad m = \frac{(5.51 - 1.35)°C}{\dfrac{4.90 \text{ kg solv. } °C}{\text{mol solute}}} = 0.849 \frac{\text{mol solute}}{\text{kg solv.}}$$

$$\text{no. mol solute} = 50.0 \text{ mL } C_6H_6 \times \frac{0.879 \text{ g } C_6H_6}{1 \text{ mL } C_6H_6} \times \frac{1 \text{ kg } C_6H_6}{1000 \text{ g } C_6H_6} \times \frac{0.849 \text{ mol solute}}{1 \text{ kg } C_6H_6} = 0.0373 \text{ mol solute}$$

$$\text{molar mass} = \frac{6.45 \text{ g}}{0.0373 \text{ mol}} = 173 \text{ g/mol} \qquad \text{Molecular weight} = 173.$$

Empirical formula

In 100 grams of the unknown compound,

no. mol C = 42.4 g C × $\frac{1 \text{ mol C}}{12.0 \text{ g C}}$ = 3.53 mol C   no. mol H = 2.4 g H × $\frac{1 \text{ mol H}}{1.0 \text{ g H}}$ = 2.4 mol H

no. mol N = 16.6 g N × $\frac{1 \text{ mol N}}{14.0 \text{ g N}}$ = 1.19 mol N   no. mol O = 37.8 g O × $\frac{1 \text{ mol O}}{16.0 \text{ g O}}$ = 2.36 mol O

Empirical formula = $C_{3.53}H_{2.4}N_{1.19}O_{2.36} = C_3H_2NO_2$

Formula weight = (3 × 12.0) + (2 × 1.0) + 14.0 + (2 × 16.0) = 84.0

Molecular weight = 173. [Within the limits of experimental error this value is twice the formula weight.]

Molecular formula = $C_6H_4N_2O_4$

12-28. (a) Data concerning the solution composition are converted to molality, and expression (12.7) is solved for K_f.

$$\text{molality of } C_6H_6 \text{ in } C_6H_{12} = \frac{1.00 \text{ g } C_6H_6 \times \frac{1 \text{ mol } C_6H_6}{78.1 \text{ g } C_6H_6}}{80.00 \text{ g } C_6H_6 \times \frac{1 \text{ kg } C_6H_{12}}{1000 \text{ g } C_6H_{12}}} = 0.160 \text{ m}$$

$$K_f = \frac{\Delta T_f}{m} = \frac{(6.5 - 3.3)°C}{\frac{0.160 \text{ mol solute}}{\text{kg solvent}}} = 20 \frac{°C \text{ kg solvent}}{\text{mol solute}}$$

(b) Cyclohexane is a better solvent for molecular weight determinations than is benzene. Although the two solvents have about the same freezing point when pure, the addition of a given amount of solute to cyclohexane lowers the freezing point about four times as much as in benzene. [The values of K_f for cyclohexane and benzene are 20 and 4.9, respectively.]

12-29. To ensure protection of the cooling system of an automobile to -10°C requires adding enough ethylene glycol to water to produce ΔT_f = 10°C. The molality corresponding to this freezing point depression is,

$$m = \frac{\Delta T_f}{K_f} = \frac{10°C}{\frac{1.86 \text{ kg solv. } °C}{\text{mol solute}}} = 5.4 \frac{\text{mol solute}}{\text{kg solvent}}$$

Let us base the remainder of the calculation on 1 liter (1000 mL) of water.

$$\text{no. kg water} = 1000 \text{ mL water} \times \frac{1.00 \text{ g water}}{1 \text{ mL water}} \times \frac{1 \text{ kg water}}{1000 \text{ g water}} = 1.00 \text{ kg water}$$

The number of moles of $C_2H_6O_2$ that must be dissolved in this water to yield a 5.4 m solution is,

$$\text{no. mol } C_2H_6O_2 = 1.00 \text{ kg water} \times \frac{5.4 \text{ mol } C_2H_6O_2}{1 \text{ kg water}} = 5.4 \text{ mol } C_2H_6O_2$$

The volume of $C_2H_6O_2$ follows readily.

$$\text{no. mL } C_2H_6O_2 = 5.4 \text{ mol } C_2H_6O_2 \times \frac{62.1 \text{ g } C_2H_6O_2}{1 \text{ mol } C_2H_6O_2} \times \frac{1 \text{ mL } C_2H_6O_2}{1.12 \text{ g } C_2H_6O_2} = 3.0 \times 10^2 \text{ mL } C_2H_6O_2$$

The required proportions, by volume, are: 3 parts $C_2H_6O_2$ to 10 parts H_2O.

12-30. (a) Freeze damage to citrus occurs if the juice of the fruit is frozen. The juice is a water solution containing sugars, citric acid and other solutes. These solutes lower the freezing point of the juice to a few degrees below the normal freezing point of water. Thus, it is not necessary to begin freeze protection measures at exactly 0°C = 32°F.

(b) Orange juice has a higher sugar content than lemon juice. This higher solute concentration means that the freezing point of orange juice is lower than that of lemon juice.

12-31. Assume that the pressure remains constant, and determine the molality of $C_{12}H_{22}O_{11}$ required to produce ΔT_b = 100.00 - 99.85 = 0.15°C.

$$\Delta T_b = K_b \cdot m \qquad m = \frac{\Delta T_b}{K_b} = \frac{0.15°C}{0.512\ (\text{kg solv.})\text{mol}^{-1}\ °C} = 0.29\ m.$$

This solution has 0.29 mol × 342 g/mol = 99 g $C_{12}H_{22}O_{11}$ in 1000 g H_2O.

$$\text{The \% } C_{12}H_{22}O_{11}\text{, by mass} = \frac{99\ g\ C_{12}H_{22}O_{11}}{1099\ \text{g soln.}} \times 100 = 9.0\%\ C_{12}H_{22}O_{11}\text{, by mass}$$

Osmotic pressure

12-32. In both cases osmosis occurs. Water is transported through cell membranes from the more dilute solution within the cells into the more concentrated salt solution. This loss of water is manifested by the wilting of the flowers and the shriveling of the cucumber.

12-33. The simplest verification of the statement is from the osmotic pressure given for the 20% sucrose solution--15 atm. (Note: In Example 5-1 we established that the height of a water column equivalent to one atm is 10.3 m.)

$$\text{no. m water} = 15\ \text{atm} \times \frac{10.3\ m}{1\ \text{atm}} = 1.5 \times 10^2\ m$$

12-34. Use the expression π = MRT

$$M = \frac{\pi}{RT} = \frac{7.7\ \text{atm}}{0.0821\ \text{L atm mol}^{-1}\ K^{-1} \times 298\ K} = 0.31\ \text{mol/L}$$

A glucose solution isotonic with blood contains 0.31 mol $C_6H_{12}O_6$/L.

12-35. Determine the molarity required to produce the indicated osmotic pressure.

$$M = \frac{\pi}{RT} = \frac{(6.15/760)\ \text{atm}}{0.0821\ \text{L atm mol}^{-1}\ K^{-1} \times 298\ K} = 3.31 \times 10^{-4}\ M$$

Now determine the mass of hemoglobin.

$$\text{no. g} = 0.100\ L \times \frac{3.31 \times 10^{-4}\ \text{mol hemogl}}{L} \times \frac{6.84 \times 10^4\ g}{1\ \text{mol hemogl}} = 2.26\ g$$

12-36. Convert the height of liquid column to an equivalent mercury column. This is the osmotic pressure.

$$h_{soln.} \times d_{soln.} = h_{Hg} \times d_{Hg}$$

$$h_{Hg} = h_{soln.} \times \frac{d_{soln.}}{d_{Hg}} = 5.1\ mm \times \frac{0.88\ g/cm^3}{13.6\ g/cm^3} = 0.33\ \text{mmHg}$$

Use the expression π = MRT to solve for M.

$$M = \frac{\pi}{RT} = \frac{(0.33/760)\ \text{atm}}{0.0821\ \text{L atm mol}^{-1}\ K^{-1} \times 298\ K} = 1.8 \times 10^{-5}\ \text{mol/L}$$

Now use the definition of molarity, with molar mass $\mathcal{M}$ as an unknown.

$$1.8 \times 10^{-5}\ M = \frac{\frac{0.50}{\mathcal{M}}\ \text{mol}}{0.100\ L} \qquad \mathcal{M} = \frac{0.50}{0.100 \times 1.8 \times 10^{-5}} = 2.8 \times 10^5$$

Strong, weak, and nonelectrolytes

12-37. Determine the total number of moles of Cl^- in the final solution from the two sources indicated below:

(a) $$\text{no. mol } Cl^- = 250 \text{ mL} \times \frac{1 \text{ L}}{1000 \text{ mL}} \times \frac{0.217 \text{ mol } MgCl_2}{1 \text{ L}} \times \frac{2 \text{ mol } Cl^-}{1 \text{ mol } MgCl_2} = 0.108 \text{ mol } Cl^-$$

(b) $$\text{no. mol } Cl^- = 1.85 \text{ g } MgCl_2 \times \frac{1 \text{ mol } MgCl_2}{95.2 \text{ g } MgCl_2} \times \frac{2 \text{ mol } Cl^-}{1 \text{ mol } MgCl_2} = 0.0389 \text{ mol } Cl^-$$

Total number of moles of Cl^- = (a) + (b) = 0.108 + 0.0389 = 0.147 mol Cl^-

$$\frac{\text{no. mol } Cl^-}{\text{L}} = \frac{0.147 \text{ mol } Cl^-}{0.250 \text{ L}} = 0.588 \text{ M } Cl^-$$

12-38. The order of increasing ability to conduct electric current is the same as the order of increasing concentration of ions.

Ion concentrations

1.0 M C_2H_5OH = 0.

0.01 M $HC_2H_3O_2$ < 0.01 M. This weak acid is only slightly ionized.

0.01 M NaCl = 0.02 M. This strong electrolyte dissociates to produce two moles of ions per mole of compound.

1.0 M $MgCl_2$ = 3.0 M. Three moles of ions are produced per mole of compound when this electrolyte dissociates.

Order of increasing ability to conduct electricity: 1.0 M C_2H_5OH < 0.01 M $HC_2H_3O_2$ < 0.01 M NaCl < 0.01 M $MgCl_2$

12-39. Both solutes, NH_3 and $HC_2H_3O_2$, undergo only slight dissociation in aqueous solution; they are weak electrolytes. When mixed in the same solution they react to produce the salt, $NH_4C_2H_3O_2$, which dissociates completely into the ions NH_4^+ and $C_2H_3O_2^-$. The ammonium acetate is a strong electrolyte.

12-40. Begin by predicting the freezing point depression of a nonelectrolyte in an 0.10 m aqueous solution.

$\Delta T_f = K_f \cdot m = 1.86 \times 0.10 = 0.186°C.$

The observed freezing point should be about -0.19°C. Now estimate the freezing points of the several solutions by comparison to this value.

(a) urea = -0.19°C: Urea is an undissociated covalent compound--a nonelectrolyte.

(b) NH_4NO_3 = -0.38°C: NH_4NO_3 is an ionic compound that dissociates into the ions NH_4^+ and NO_3^-.

(c) $CaCl_2$ = -0.57°C: $CaCl_2$ dissociates to produce one mole of Ca^{2+} and two moles of Cl^- ions per mole of compound.

(d) $MgSO_4$ = -0.38°C: The ions produced by $MgSO_4$ are Mg^{2+} and SO_4^{2-}.

(e) ethanol = -0.19°C: Ethanol is a nonelectrolyte.

(f) HCl = -0.38°C: Although it is a covalent compound, HCl dissociates in water solution to produce one H^+ and one Cl^- ion per molecule.

(g) $HC_2H_3O_2$ = -0.20°C: According to Example 12-10 in the text, acetic acid is about 4% dissociated in aqueous solution. This corresponds to a van't Hoff factor, i, of 1.04. The freezing point should be about 1.04 × (-0.19°C) = -0.20°C.

12-41. (a) $i = \dfrac{\text{f. pt. depression observed}}{\text{f. pt. depression for nonelectrolyte}} = \dfrac{0.0986°C}{(1.86 \times 0.0500)°C} = 1.06$

(b) Every HNO_2 molecule that dissociates produces one H^+ and one NO_2^- ion. Since the number of moles of H^+ and of NO_2^- produced per liter are each 6.91×10^{-3} M. This must also be the number of moles per liter of HNO_2 ionized. The molarity of HNO_2 unionized = $0.100 - 6.91 \times 10^{-3}$. The total molarity of solute particles is $\underset{(HNO_2)}{0.100 - 6.9 \times 10^{-3}} + \underset{(H^+)}{6.9 \times 10^{-3}} + \underset{(NO_2^-)}{6.9 \times 10^{-3}}$.

= 0.1069 M. ($\simeq$ 0.107 m)

$$i = \frac{\Delta T_f \text{ (ionized soln.)}}{\Delta T_f \text{ (nonelectrolyte)}} = \frac{K_f \cdot 0.107 \text{ m}}{K_f \cdot 0.100 \text{ m}} \simeq 1.07$$

Colloidal mixtures

12-42. True solutions contain solute particles having dimensions of molecular size, say less than 1 nm. In colloidal mixtures the particles have one or more dimensions that exceed 1 nm, sometimes by a great deal. The particles in a colloidal mixture scatter light that is passed through the mixture (Tyndall effect). True solutions do not display the Tyndall effect. For a given mass of solute, true solutions have much more pronounced colligative properties (osmotic pressure, vapor pressure lowering, freezing point depression, and boiling point elevation) than do colloidal mixtures.

12-43. (a) An aerosol is a colloidal dispersion of small particles in a gas. These may be particles of a solid, as in cigarette smoke, or they may be droplets of liquid, as in a fog or mist.

(b) An emulsion is a colloidal dispersion of one liquid in another. One liquid is a continuous phase called a dispersion medium and is analogous to the solvent in a solution. The other liquid is distributed as tiny droplets throughout the dispersion medium. This phase, which is analogous to the solute in a solution, is called the dispersed phase.

(c) A foam is a dispersion of microscopic bubbles of a gas or air in a liquid medium. Common examples are soap foams.

(d) A hydrophobic colloid is a dispersion of particles in water in which the suspended particles (disperse phase) have no particular affinity for water. Such colloids owe their stabilities to the fact that the colloidal particles adsorb ions on their surfaces and repel one another because they carry like charges.

(e) Electrophoresis is a process in which the charged particles in a colloidal mixture are made to migrate in an electric field. If the particles have different sizes, shapes and charges, it may be possible to separate different kinds of particles by differences in their mobilities in an electric field.

12-44. (a) The boundary regions between a negatively charged arsenic trisulfide sol and an electrolyte in the electrophoresis apparatus of Figure 12-23 are expected to move as follows: The region in the left arm will move down (descend) and the one in the right arm will move up (ascend).

(b) The effective ion in coagulating a negatively charged sol is the cation (positive ion). The higher the charge on the cation, the more effective it should be. The cations in the three electrolytes listed are K^+, Mg^{2+}, and Al^{3+}. The most effective electrolyte should be $AlCl_3$.

12-45. (a) Determine the volume and mass of one colloidal gold particle. Next, calculate the number of such particles in 1.00 mg gold. Finally, from the number of particles and the area per particle determine the total surface area.

$$\text{volume} = \frac{4}{3}\pi r^3 = \frac{4}{3}\pi\left(100 \text{ nm} \times \frac{10^{-7} \text{ cm}}{1 \text{ nm}}\right)^3 = 4.19 \times 10^{-15} \text{ cm}^3$$

$$\text{mass} = \text{volume} \times \text{density} = 4.19 \times 10^{-15} \text{ cm}^3 \times \frac{19.3 \text{ g}}{1 \text{ cm}^3} = 8.09 \times 10^{-14} \text{ g}$$

$$\text{no. particles} = 1.00 \text{ mg} \times \frac{1 \text{ g}}{1000 \text{ mg}} \times \frac{1 \text{ particle}}{8.09 \times 10^{-14} \text{ g}} = 1.24 \times 10^{10} \text{ particles}$$

$$\text{surface area per particle} = 4\pi r^2 = 4\pi \left(100 \text{ nm} \times \frac{10^{-7}\text{ cm}}{1\text{ nm}}\right)^2 = \frac{1.26\times10^{-9}\text{ cm}^2}{1\text{ particle}}$$

$$\text{total surface area} = 1.24 \times 10^{10}\text{ particles} \times \frac{1.26\times10^{-9}\text{ cm}^2}{1\text{ particle}} = 16\text{ cm}^2$$

(b) A single cube of gold weighing 1.00 mg has the following volume and length.

$$\text{volume} = 1.00\text{ mg} \times \frac{1\text{ g}}{1000\text{ mg}} \times \frac{1\text{ cm}^3}{19.3\text{ g}} = 5.18\times10^{-5}\text{ cm}^3$$

$$\text{volume} = (\text{length})^3;\quad \text{length} = (5.18 \times 10^{-5}\text{ cm}^3)^{1/3} = 3.73\times10^{-2}\text{ cm}$$

$$\text{surface area of cube} = 6 \times (\text{length})^2 = 6 \times (3.73\times10^{-2}\text{ cm})^2 = 8.35\times10^{-3}\text{ cm}^2$$

Self-Test Questions

1. *(b)* In 0.01 M CH_3OH there is 0.01 mole CH_3OH (0.32 g) per L solution. A solution that is 0.01% (mass/vol.) would have 0.01 g CH_3OH per 100 cm^3 or 0.1 g CH_3OH per L. Item (a) is incorrect. Because the solution is so dilute (practically pure water), its density should be about 1.00 g/cm^2. One liter of solution would have a mass of about 1.00 kg and practically all of this would be water. A concentration of 0.01 mol/L (0.01 M) would be about the same as 0.01 mol/kg solvent (0.01 m). Item (b) is correct. There is 0.01 mol solute for every 1000/18 = 55.5 mol water. The mole fraction of CH_3OH is much less than 0.01 (it is 0.01/55.5 = 1.8×10^{-4}). The molarity of H_2O in the solution is far greater than 0.99 M (actually it is 55.5 mol/L). Also, note that it is mole fraction concentrations that must always total 1.00, not molarities, molalities, etc.

2. *(c)* Apply the rule that "like dissolves like". The substances C_6H_6(l) and $C_{10}H_8$(s) are hydrocarbons; they are unlike the polar covalent H_2O. C(s) is a network covalent solid; the covalent bonds in this substance cannot be broken by interactions with H_2O. CH_3OH(l), on the other hand, is a polar covalent substance with certain similarities to water. (Think of CH_3OH as a molecule in which a $-CH_3$ group substitutes for an $-H$ atom in H—O—H.)

3. *(d)* The most likely solution to be ideal is the one whose components are most similar. Strong similarity of components if found only for the hydrocarbon mixture C_7H_{16} and C_8H_{18}.

4. *(b)* Solubility of gases in water generally *decrease* with increased temperature, but always increase with an increase in gas pressure. Then (c) cannot be correct since it combines (a), which is incorrect with (b), which is correct. If more water is available, more gas dissolves, but the concentration of the saturated solution is unchanged.

5. *(a)* Ethanol and sucrose are nonelectrolytes. Acetic acid is a weak electrolyte and its aqueous solutions are only fair conductors of electricity. NaCl is a strong electrolyte and NaCl(aq) is a good electrical conductor.

6. *(d)* Determine the number of moles of particles present in each solution. In 0.01 m $MgSO_4$ this would be 0.02 m; in 0.01 m NaCl, 0.02 m; in 0.01 m C_2H_5OH, 0.01 m (a nonelectrolyte); 0.008 m MgI_2, 0.024 m. Since $T_f = K_f \times m$, the solution with the greatest molality of dissolved particles has the greatest freezing point depression and the lowest freezing point.

7. *(c)* Since the mole fractions of A and B in the liquid phase are equal, the mole fractions of A and B in the vapor can be equal (each equal to 0.50) only if the vapor pressures of A and B are equal--this is unlikely. Item (a) is incorrect. Item (b) is also incorrect; the mole fractions of A and B in the vapor could not be equal and have any value other than 0.50. The mole fractions of A and B in the vapor are not likely to be equal (because their vapor pressures are not likely to be equal). Both components are volatile; both will be present in the vapor.

8. *(b)* A polymeric substance has a high molecular weight. Of the colligative properties--osmotic pressure, freezing point depression, and boiling point elevation--only osmotic pressure measurements are useful for substances of high molecular weight (and very dilute solution concentration). Measurement of vapor density [i.e., based on the expression: M = mRT/PV] is feasible only for gases, and substances with high molecular weights are solids or liquids under normal conditions.

9. (a) no. g solution = 44.0 g C_6H_6 + 1.00 g $C_{10}H_8$ = 45.0 g solution

$$\% \ C_{10}H_8 = \frac{1.00 \text{ g } C_{10}H_8}{45.0 \text{ g solution}} \times 100 = 2.22\% \ C_{10}H_8$$

(b)
$$\text{molality} = \frac{1.00 \text{ g } C_{10}H_8 \times \dfrac{1 \text{ mol } C_{10}H_8}{128 \text{ g } C_{10}H_8}}{44.0 \text{ g C H} \times \dfrac{1 \text{ kg } C_6H_6}{1000 \text{ g } C_6H_6}} = 0.178 \text{ m } C_{10}H_8$$

(c)
$$\Delta T_f = K_f \times m = 4.90°\text{C kg } C_6H_6 \ (\text{mol } C_{10}H_8)^{-1} \times \frac{0.178 \text{ mol } C_{10}H_8}{1.00 \text{ kg } C_6H_6} = 0.87°\text{C}$$

$$T_f = 5.51 - 0.87 = 4.64°\text{C}$$

10. Determine the molality of the solution from the boiling point data.

$$\Delta T_b = K_b \times m \qquad m = \frac{(100.0 - 99.07)°\text{C}}{0.512°\text{C kg solv. (mol solute)}^{-1}} \qquad m = 1.8 \text{ mol solute/kg solv.}$$

Since NaCl dissociates into two ions per formula unit, the molality expressed in terms of NaCl is 0.90 m NaCl.

$$\frac{0.90 \text{ mol NaCl} \times \dfrac{58.5 \text{ g NaCl}}{1 \text{ mol NaCl}}}{(0.90 \times 58.5)\text{g NaCl} + 1000 \text{ g } H_2O} \times 100 = 5.0\% \text{ NaCl, by mass}$$

11. Neither *pure* liquid is an electrical conductor, but when HCl is dissolved in water it ionizes completely into the ions H^+ and Cl^-; HCl(aq) is a strong electrolyte.

12. First, determine the no. mol Cl^- in the original solution.

$$\text{no. mol } Cl^- = 0.350 \text{ L} \times \frac{0.250 \text{ mol NaCl}}{1 \text{ L}} \times \frac{1 \text{ mol } Cl^-}{1 \text{ mol NaCl}} = 0.0875 \text{ mol } Cl^-$$

Now determine the no. mol Cl^- in V mL of 0.250M $MgCl_2$

$$\text{no. mol } Cl^- = V \text{ mL} \times \frac{1 \text{ L}}{1000 \text{ mL}} \times \frac{0.250 \text{ mol } MgCl_2}{1 \text{ L}} \times \frac{2 \text{ mol } Cl^-}{1 \text{ mol } MgCl_2} = 5.00\times10^{-4}V$$

The total no. mol Cl^- in the final soln. = $0.0875 + 5.00\times10^{-4}V$

The final solution volume is 350 mL + V mL = [350 + V)/1000]L

The molarity of Cl^- in the final solution is

$$\frac{0.0875 + 5.00\times10^{-4}V}{(350 + V)/1000} = 0.30$$

Solve for : $0.0876 + 5.00\times10^{-4}V = 0.105 + 3.0\times10^{-4}V$

$$2.0\times10^{-4}V = 0.105 - 0.0875 = 0.017$$

$$V = 85 \text{ mL}$$

Chapter 13

An Introduction to Descriptive Chemistry; The First 20 Elements

Review Problems

13-1. (a) Allotropy refers to the existence of an element in more than one physical form within the same state of matter. For example, $O_2(g)$ and $O_3(g)$ are allotropic forms of oxygen in the gaseous state; and red and white phosphorus are different forms of phosphorus in the solid state. Allotropic forms of an element have different physical properties and sometimes different chemical properties. These differences stem from a different molecular form of the element (as in O_2 and O_3) or a different crystalline structure (as in rhombic and monoclinic sulfur).

(b) An ore is a mineral substance from which an element (usually a metal) can be extracted in a reasonably economical fashion.

(c) A diagonal relationship refers to similarities that are found between certain elements in the second and third periods of the periodic table. The second period member of one group bears a resemblance to the third period element in the next group (e.g., Li in group IIA resembles Mg in group IIA).

(d) Nitrogen fixation refers to a method, either natural or artificial, of converting $N_2(g)$ to a nitrogen compound.

(e) The thermite reaction involves powdered aluminum metal and certain metal oxides (e.g. Fe_2O_3 or Cr_2O_3). Al(s) reduces the metal oxide to the free metal [e.g., Fe(O) or Cr(O)] in a highly exothermic reaction.

(f) An interhalogen compound is a compound involving different halogen elements, such as I and Cl in I-Cl.

(g) A noble gas is an element in group 0 of the periodic table (i.e., He, Ne, Ar, Kr, Xe, Rn). Atoms of the noble gas elements have the outer-shell electron configuration ns^2np^6 ($1s^2$ for He).

13-2. (a) sodium peroxide = Na_2O_2

(b) magnesium nitride = Mg_3H_2

(c) $Ca(OH)_2$ = calcium hydroxide

(d) sodium perchlorate = $NaClO_4$

(e) $LiAlH_4$ = lithium aluminum hydride

(f) O_2^- = superoxide ion

(g) hydrogen difluoride ion = HF_2^-

(h) $NH_4H_2PO_4$

(i) H_2O_2 = hydrogen peroxide

(j) $NaHSO_3$ = sodium hydrogen sulfite

(k) ozone = O_3

(l) N_2O_4 = dinitrogen tetroxide

(m) calcium hydrogen carbonate = $Ca(HCO_3)_2$

(n) bromine trifluoride = BrF_3

13-3. (a) coke = carbon, C

(b) limestone = $CaCO_3$ (or a mixed $CaCO_3/MgCO_3$)

(c) quicklime = CaO

(d) slaked lime = $Ca(OH)_2$

(e) silica = SiO_2

(f) gypsum = $CaSO_4 \cdot 2H_2O$

(g) water glass = Na_2SiO_3

(h) plaster of paris = $CaSO_4 \cdot \frac{1}{2}H_2O$

(i) synthesis gas = a mixture of CO and H_2

13-4. (a) $LiH(s) + H_2O \rightarrow LiOH(aq) + H_2(g)$

(b) $2\,Na_2O_2(s) + 2\,H_2O \rightarrow 4\,NaOH(aq) + O_2(g)$

(c) $Ca(s) + 2\,H_2O \rightarrow Ca(OH)_2 + H_2(g)$

(d) $CaO(s) + H_2O \rightarrow Ca(OH)_2(s)$

(e) $P_4O_{10}(s) + 6\,H_2O \rightarrow 4\,H_3PO_4(aq)$

(f) $C(s) + H_2O \xrightarrow{\Delta} CO(g) + H_2(g)$

13-5. (a) $Mg(s) + 2HCl(aq) \rightarrow MgCl_2(aq) + H_2(g)$

(b) $NH_3(g) + HNO_3(aq) \rightarrow NH_4NO_3(aq)$

(c) $CaCO_3(s) + 2\ HCl(aq) \rightarrow CaCl_2(aq) + H_2O + CO_2(g)$

(d) $2NaF(s) + H_2SO_4(conc\ aq) \xrightarrow{\Delta} Na_2SO_4(s) + 2HF(g)$

(e) $3Ag(s) + 4HNO_3(aq) \rightarrow 3AgNO_3(aq) + 2H_2O + NO(g)$

13-6. (a) $H_2O_2 + NO_2^- \rightarrow H_2O + NO_3^-$

(b) $H_2O_2 + SO_2(g) + 2OH^- \rightarrow SO_4^{2-} + 2H_2O$

(c) $MnO_4^- + H_2O_2 \longrightarrow Mn^{2+} + O_2(g)$

decrease of 5 in O.S./Mn

increase of 1 in O.S./O

$2MnO_4^- + 5H_2O_2 + 6H^+ \rightarrow 2Mn^{2+} + 5O_2(g) + 8H_2O$

(d) $H_2O_2 + Cl_2(g) + 2OH^-(aq) \rightarrow 2H_2O + 2Cl^- + O_2(g)$

13-7. (a) N in the oxidation state +4: NO_2, N_2O_4

(b) O in the oxidation state -1: H_2O_2, Na_2O_2

(c) Cl in the oxidation state +1: HOCl, NaOCl

(d) S in the oxidation state -2: H_2S, Na_2S

(e) P in the oxidation state -3: PH_3

(f) H in the oxidation state -1: LiH, NaH, CaH_2

8. (13.3): The oxidation state (O.S.) of H decreases from +1 in H_2O to 0 in H_2. H_2O is the oxidizing agent. The O.S. of C increases from -4 in CH_4 to +2 in CO. CH_4 is the reducing agent.

(13.26): The O.S. of H decreases from +1 in H_2O to 0 in H_2. H_2O is the oxidizing agent. The O.S. of Mg increases from 0 to +2. Mg is the reducing agent.

(13.56): The oxidation state of N decreases from +5 in NO_3^- to +2 in NO. NO_3^- is the oxidizing agent. The O.S. of Cu increases from 0 to +2. Cu is the reducing agent.

(13.61): The O.S. of N in NO_2 is +4. In HNO_3 it is +5, and in NO, +2. NO_2 is both oxidized and reduced (that is, some is oxidized and some is reduced). NO_2 is both the oxidizing and the reducing agent.

(13.64): The O.S. of iodine increases from 0 in I_2 to +5 in HIO_3. I_2 is the reducing agent. The O.S. of nitrogen decreases from +5 in HNO_3 to +2 in NO. HNO_3 is the oxidizing agent.

(13.65): The oxidation state of carbon increases from 0 to +2 in CO. C is the reducing agent. The oxidation state of phosphorus decreases from +5 in $Ca_3(PO_4)_2$ to 0 in P_4. $Ca_3(PO_4)_2$ is the oxidizing agent.

(13.72): The oxidation state of oxygen increases from -1 in H_2O_2 to zero in O_2. O_2 is the reducing agent. The O.S. of manganese decreases from +7 in MnO_4^- to +2 in Mn^{2+}. MnO_4^- is the oxidizing agent.

9. For each of the following the necessary expression is

$$\Delta\bar{H}^\circ_{rxn} = \Sigma\ \Delta\bar{H}^\circ_f(\text{products}) - \Sigma\Delta\bar{H}^\circ_f(\text{reactants})$$

It is not necessary to obtain exact numerical values of $\Delta\bar{H}^\circ_{rxn}$, however. The calculation needs to be carried only to the point of determining whether $\Delta\bar{H}^\circ_{rxn}$ has a positive or negative value.

(13.2): $C(s) + H_2O(g) \rightarrow CO(g) + H_2(g)$

$$\Delta\bar{H}^\circ_{rxn} = \Delta\bar{H}^\circ_f[CO(g)] - \Delta\bar{H}^\circ_f[H_2O(g)] = [-110.54 - (-241.84)]\ \text{kJ/mol}$$

<0 exothermic

(13.3): $CH_4(g) + H_2O(g) \rightarrow CO(g) + 3H_2(g)$

$$\Delta\bar{H}^\circ_{rxn} = \Delta\bar{H}^\circ_f[CO(g)] - \Delta\bar{H}^\circ_f[H_2O(g)] - \Delta\bar{H}^\circ_f[CH_4(g)]$$

$$\Delta\bar{H}^\circ_{rxn} = [-110.54 - (-241.84) - (-74.85)]\text{kJ/mol}$$

> 0 (endothermic)

(13.13): $Fe_2O_3(s) + 3H_2(g) \rightarrow 2Fe(s) + 3H_2O(g)$

$$\Delta\bar{H}^\circ_{rxn} = 3 \times \Delta\bar{H}^\circ_f[H_2O(g)] - \Delta\bar{H}^\circ_f[Fe_2O_3(s)]$$

$$= [3 \times (-241.84) - (-822.16)]\text{kJ/mol}$$

> 0 (endothermic)

(13.27): $CaCO_3(s) \rightarrow CaO(s) + CO_2(g)$

$$\Delta\bar{H}^\circ_{rxn} = \Delta\bar{H}^\circ_f[CaO(s)] + \Delta\bar{H}^\circ_f[CO_2(g)] - \Delta\bar{H}^\circ_f[CaCO_3(s)]$$

$$= [-635.3 - 393.51 - (-1207.1)]\text{kJ/mol}$$

> 0 (endothermic)

(13.30): $Ca(OH)_2(s) + CO_2(g) \rightarrow CaCO_3(s) + H_2O(g)$

$$\Delta\bar{H}^\circ_{rxn} = \Delta\bar{H}^\circ_f[CoCO_3(s)] + \Delta\bar{H}^\circ_f[H_2O(g)] - \Delta\bar{H}^\circ_f[Ca(OH)_2(s)] - \Delta\bar{H}^\circ_f[CO_2(g)]$$

$$= [-1207.1 - 241.84 - (-98.66) - (-393.51)]\text{kJ/mol}$$

< 0 (exothermic)

(13.43): $CO(g) + 2H_2(g) \rightarrow CH_3OH(l)$

$$\Delta\bar{H}^\circ_{rxn} = \Delta\bar{H}^\circ_f[CH_3OH(l)] - \Delta\bar{H}^\circ_f[CO(g)]$$

$$= [-238.66 - (-110.54)]\text{kJ/mol}$$

< 0 (exothermic)

10. (a) The abundances refer to the data in Table 13-1. Of the nonmetals oxygen is most abundant (it is also the most abundant element in the earth's crust). Of the metals, aluminum is most abundant (third most abundant in the earth's crust).

(b) Hydrogen exists as $H_2(g)$ and helium as He(g) at STP. Elements Z = 3 through Z = 6 are solids. Elements Z = 7 (nitrogen), Z = 8 (oxygen), Z = 9 (fluorine) and Z = 10 (neon) are all gases. Elements Z = 11 (sodium) through Z = 16 (sulfur) are all solids. Elements, Z = 17 (chlorine) and Z = 18 (argon) are gases. K(Z = 19) and Ca(Z = 20) are both solids. In all, eight of the first 20 elements are gases.

(c) The best example of a semiconductor among the first 20 elements is silicon. Boron also has semiconductor properties. The other elements among the first 20 are either metals or nonmetals.

(d) The bext oxidizing agent among the first 20 elements is $F_2(g)$. The next best oxidizing agent is oxygen in the form of ozone (O_3).

(e) Allotropy is exhibited by carbon (graphite and diamond) oxygen (O_2 and O_3), phosphorus (red, white, black), and sulfur (monoclinic, rhombic and several other forms).

(f) The highest melting point of the first 20 elements is that of carbon (diamond).

(g) The element with the lowest critical point temperature is the one that has the weakest intermolecular forces and is therefore most difficult to condense--helium.

(h) All of the first 20 elements for compounds with oxygen except the noble gases He, Ne, and Ar.

Exercises

Groups and period trends

13-1. (a) Most metallic: The element farthest to the left and farthest down the group, the group IA IA element K.

(b) Most soluble: The group IIA carbonates are insoluble ($CaCO_3$). Because of the diagonal relationship, Li_2CO_3 is similar to $MgCO_3$ in solubility--its solubility is limited. Na_2CO_3 is rather highly soluble, as are most sodium compounds.

(c) Best oxidizing agent: O_3. All three compounds have oxidizing properties but O_3 is one of the best oxidizing agents of all substances.

(d) Most volatile liquid: $H_2S(\ell)$. Because of strong hydrogen bonding, both H_2O and HF persist in the liquid state to a much higher temperature than expected (H_2O is a liquid at room temperature). Hydrogen bonding in H_2S is about that expected for a substance of molecular weight 34.

(e) Hardest substance: SiO_2. The network covalent bonding in SiO_2 makes it the hardest of the three substances.

(f) Best electrical conductor: $LiF(\ell)$. $CH_4(\ell)$ and $F_2O(\ell)$ are covalent substances and therefore nonconductors of electricity. BeF_2 has some ionic character but not enough to be a good electrical conductor. LiF has a much greater ionic character and is therefore the best conductor of the group.

(g) Highest melting point: LiF(s). All the solids consist of unipositive cations and uninegative anions. The highest lattice energy (and highest melting point) is that of the compound with the *smallest* ions.

13-2. (a) The plot of single covalent radius as a function of atomic number is simply the second segment of Figure 8-5 (i.e., the segment from Na to Cl). As described in Chapter 8, there is a steady decrease in single covalent radius (atomic radius) from left to right in the period.

(b) The plot of first ionization energies would be a plot of the first row of values in Table 8-5 (marked by the colored stripe) as a function of atomic number. These data can also be seen in the portion of Figure 8-8 ranging from Na to Ar. In general ionization energies increase with atomic number in the period. The apparent exception in the case of aluminum is explained on page 209 of the text (ionization of an unpaired $3p$ electron in Al compared to a paired $3s$ electron in Mg).

(c)

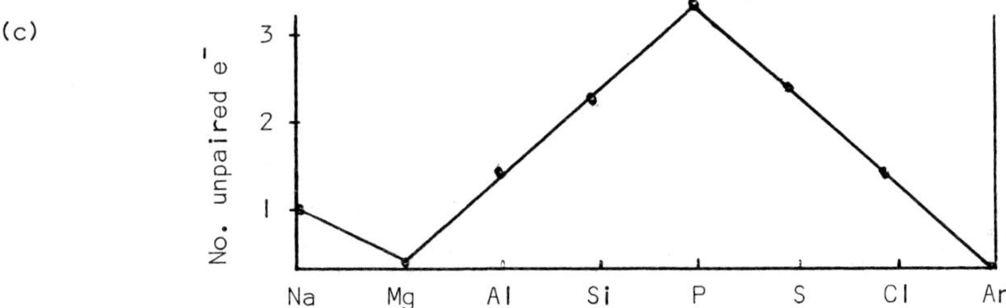

The "regularity" noted here is a reflection of the Aufban process for electron configurations. Since the periodic table is based on similarities in electron configurations the "regularity" appears here as well. But it is difficult to relate other properties of the elements simply to number of unpaired electrons (i.e., Na and Cl each have one unpaired electron but are not at all similar. Likewise there is no similarity between Mg and Ar, each with no unpaired electrons).

Hydrogen

13-3. (a) Hydrogen can be prepared by the reaction of an active metal with H_2O: e.g., $2Na(s) + 2H_2O \rightarrow 2Na^+ + 2OH^- + H_2(g)$

It can also be produced by the action of a reducing agent on H_2O, e.g., carbon (in the water gas reaction)

$$C(s) + H_2O(g) \rightarrow CO(g) + H_2(g)$$

And it can be prepared by the electrolysis of H_2O

$$2H_2O \xrightarrow{\text{electrolysis}} 2H_2(g) + O_2(g)$$

(b) The action of a metal (most, but not all, metals) on HI(aq) will produce $H_2(g)$. For example,

$$Zn(s) + 2HI(aq) \rightarrow ZnI_2(s) + H_2(g)$$

(c) Mg will liberate $H_2(g)$ from an acidic solution, e.g.,

$$Mg(s) + 2HCl(aq) \rightarrow MgCl_2(aq) + H_2(g)$$

(d) A water-gas reaction can be used to produce $H_2(g)$ from CO(g).

$$CO(g) + H_2O(g) \rightarrow CO_2(g) + H_2(g)$$

(e) A small number of metals are capable of liberating $H_2(g)$ from NaOH(aq). Aluminum metal is the one encountered in this chapter. (Recall reaction 13.34.)

$$2Al(s) + 2OH^-(aq) + 6H_2O \rightarrow 2Al(OH)_4^-(aq) + 3H_2(g)$$

13-4. Use the ideal gas equation to determine the no. mol $H_2(g)$.

$$n = \frac{PV}{RT} = \frac{(748/760)\text{ atm} \times 225\text{ L}}{0.0821\text{ L atm mol}^{-1}\text{ K}^{-1} \times 291\text{K}} = 9.27\text{ mol } H_2$$

Then determine the mass of CaH_2 based on the reaction

$$CaH_2 + 2H_2O \rightarrow Ca(OH)_2 + 2H_2(g)$$

$$\text{no. g } CaH_2 = 9.27\text{ mol } H_2 \times \frac{1\text{ mol } CaH_2}{2\text{ mol } H_2} \times \frac{42.1\text{ g } CaH_2}{1\text{ mol } CaH_2} = 195\text{ g } CaH_2$$

Helium, neon, argon

13-5. First determine the amount of Ar in the cylinder.

$$n = \frac{PV}{RT} = \frac{125\text{ atm} \times 55\text{ L}}{0.0821\text{ L atm mol}^{-1}\text{ K}^{-1} \times 298\text{K}} = 2.8 \times 10^2\text{ mol Ar}$$

The volume percent composition of air is the same as the mole percent composition (recall expression 5.16).

$$\text{no. mol air} = 2.8 \times 10^2\text{ mol Ar} \times \frac{100\text{ mol air}}{0.934\text{ mol Ar}} = 3.0 \times 10^4\text{ mol air}$$

$$\text{Vol air} = 3.0 \times 10^4\text{ mol air} \times \frac{22.4\text{ L (STP)}}{1\text{ mol air}} = 6.7 \times 10^5\text{ L air (STP)}$$

13-6. The natural gas has a ratio of He to gas of 0.3:100. In air the relative abundance is 5:1,000,000. The ratio of these two values is $0.003/5 \times 10^{-6} = 600$. Helium is 600 times more abundant in the natural gas than in air.

13-7. (a) Consider 22.4 L of the mixture at STP

$$\text{no. g He} = 22.4 \text{ L gas} \times \frac{79 \text{ L He}}{100 \text{ L air}} \times \frac{1 \text{ mol He}}{22.4 \text{ L He}} \times \frac{4.00 \text{ g He}}{1 \text{ mol He}} = 3.2 \text{ g He}$$

$$\text{no. g } O_2 = 22.4 \text{ L gas} \times \frac{21 \text{ L } O_2}{100 \text{ L air}} \times \frac{1 \text{ mol } O_2}{22.4 \text{ L air}} \times \frac{32.0 \text{ g } O_2}{1 \text{ mol } O_2} = 6.7 \text{ g } O_2$$

no. g gas = 9.9 g

$$\text{density} = \frac{9.9 \text{ g gas}}{22.4 \text{ L gas}} = 0.44 \text{ g/L}$$

(b) The volume percent is the same as the mole percent.

mol. wt. = (0.79 x at. wt. He) + (0.21 x mol. wt. O_2)

= (0.79 x 4.00) + (0.21 x 32.00) = 9.9

[Note that in part (a) the mass of 22.4 L (at STP) is 9.9 g. This is the molar mass of gas, and the molecular weight is numerically equal to the molar mass.]

Lithium, sodium, potassium

13-8. (a) The reaction of an active metal with H_2O yields the metal hydroxide and $H_2(g)$ as products.

$2\,K(s) + 2\,H_2O \rightarrow 2\,KOH(aq) + H_2(g)$

(b) The reaction of an active metal hydride with H_2O also produces the metal hydroxide and $H_2(g)$.

$KH(s) + H_2O \rightarrow KOH(aq) + H_2(g)$

(c) The reaction of a metal superoxide with water produces the metal hydroxide and oxygen gas.

$4\,KO_2(g) + 2\,H_2O \rightarrow 4\,KOH(aq) + 3\,O_2(g)$

13-9. The simplest test is probably a flame test. Li compounds produce a red flame and K compounds, violet (see Table 13-3). [Moreover, the violet K flame is easily obscured by traces of Na (yellow color) but the Li flame is not.] Another possibility is to prepare an aqueous solution of the chloride and treat it with a carbonate solution [e.g., $Na_2CO_3(aq)$]. If Li^+ is present, white Li_2CO_3 precipitates. K^+ does not precipitate as the carbonate.

13-10. $\Delta\bar{H}_f^\circ$ of LiF(s) is expressed as the sum of the following five steps. The source of the data for each step is indicated.

$Li(s) \rightarrow Li(g)$	$\Delta\bar{H} = 160.7$ kJ/mol	
$Li(g) \rightarrow Li^+(g) + e^-$	$\Delta\bar{H} = 520.3$ kJ/mol	(Table 8-4)
$\frac{1}{2}F_2(g) \rightarrow F(g)$	$\Delta\bar{H} = (\frac{1}{2} \times 155)$ kJ/mol	(Table 9-3
$F(g) + e^- \rightarrow F^-(g)$	$\Delta\bar{H} = -328.0$ kJ/mol	(Section 8-9)
$Li^+(g) + F^-(g) \rightarrow LiF(s)$	$\Delta\bar{H} = -1043$ kJ/mol	(Section 13-5)
$Li(s) + \frac{1}{2}F_2(g) \rightarrow LiF(s)$	$\Delta\bar{H}_f^\circ = (160.7 + 520.3 + 77.5 - 328.0 - 1043)$ kJ/mol	
	$= -612$ kJ/mol	

Beryllium, magnesium, calcium

13-11. (a) A group IIA metal carbonate decomposes to the oxide

$$MgCO_3(s) \xrightarrow{\Delta} MgO(s) + CO_2(g)$$

(b). Electrolysis of a molten group IIA chloride yields the metal and chlorine gas.

$$CaCl_2(\ell) \xrightarrow{\text{electrolysis}} Ca(\ell) + Cl_2(g)$$

(c) Be dissolves in NaOH(aq), liberating $H_2(g)$

$$Be + 2OH^-(aq) \rightarrow BeO_2^{2-} + H_2(g)$$

(d) A group IIA metal dissolves in HCl(aq) and liberates $H_2(g)$.

$$Ca + 2\,HCl(aq) \rightarrow CaCl_2(aq) + H_2(g)$$

13-12. (a) The simplest way to determine whether the compound can be pure MgO is to calculate the expected mass of MgO that would be obtained from 0.200 g Mg.

$$\text{no. g MgO} = 0.200 \text{ g Mg} \times \frac{1 \text{ mol Mg}}{24.3 \text{ g Mg}} \times \frac{1 \text{ mol MgO}}{1 \text{ mol Mg}} \times \frac{40.3 \text{ g MgO}}{1 \text{ mol MgO}} = 0.332 \text{ g MgO}$$

The amount of product actually obtained is only 0.305 g. This suggests that the product contains a substance with a higher percentage of Mg than MgO.

(b) A probable substance that is present is magnesium nitride--Mg_3N_2. (Note that this compound has 72.2% Mg compared to 60.3% in MgO.)

(c) Add water to the solid material. If $NH_3(g)$ is evolved (detectable by its odor), the presence of Mg_3N_2 can be inferred.

$$Mg_3N_2 + 6\,H_2O \rightarrow 3\,Mg(OH)_2 + 2\,NH_3(g).$$

Boron, aluminum

13-13. (a) The reaction of an acid metal and HI(aq) produces $H_2(g)$.

$$2\,Al(s) + 6\,HI(aq) \rightarrow 2\,AlI_3(aq) + 3\,H_2(g)$$

(b) $H_2SO_4(aq)$ is a moderately good oxidizing agent and is capable of oxidizing Al to Al^{3+}; the reduction product is $SO_2(g)$. [In very dilute $H_2SO_4(aq)$, the oxidizing agent would be H^+ and the reduction product, $H_2(g)$.]

$$2\,Al(s) + 12\,H^+(aq) + 3\,SO_4^{2-}(aq) \rightarrow 2\,Al^{3+}(aq) + 6\,H_2O + 3\,SO_2(g)$$

(c) Al is soluble in KOH(aq), producing $Al(OH)_4^-(aq)$ and $H_2(g)$.

$$2\,Al(s) + 2\,OH^-(aq) + 6\,H_2O \rightarrow 2\,Al(OH)_4^-(aq) + 3\,H_2(g)$$

(d) This is the thermite reaction

$$2\,Al(s) + Cr_2O_3(s) \rightarrow 2\,Cr(\ell) + Al_2O_3(s)$$

(e) Al is oxidized to the oxide by $O_2(g)$ at high temperatures.

$$2\,Al(s) + 3\,O_2(g) \rightarrow Al_2O_3(s)$$

(f) $2\,Al(s) + 3\,Cl_2(g) \rightarrow 2\,AlCl_3(s)$

13-14. (a) An electron-deficient molecule is one in which there is an insufficient number of electrons to provide each atom with an outer-shell octet of electrons, e.g., BF_3.

(b) A dimer is a molecule that is produced by two simpler molecules joining together. For example, the dimer Al_2Cl_6 is formed by the joining of two $AlCl_3$ monomers.

(c) An adduct is a molecule formed from two other molecules through a coordinate covalent bond. An electron deficient atom in one molecule shares a pair of electrons with an electron donating atom in another molecule, such as in $F_3B:NH_3$.

13-15. (a) $2\,Al + Fe_2O_3 \rightarrow 2\,Fe + Al_2O_3$

$$\Delta H° = \Delta\bar{H}_f°(Al_2O_3) - \Delta\bar{H}_f°(Fe_2O_3) = -1670 - (-824) = -846 \text{ kJ}$$

For one mole of iron, $\Delta\bar{H}° = -846/2 = -423$ kJ/mol Fe

(b) $4\,Al + 3\,MnO_2 \rightarrow 3\,Mn + 2\,Al_2O_3$

$$\Delta H° = 2\,\Delta\bar{H}_f°(Al_2O_3) - 3\,\Delta\bar{H}_f°(MnO_2) = 2 \times (-1670) - 3 \times (-519) = -1.78\times10^3 \text{ kJ}$$

For one mole of Mn, $\Delta\bar{H}° = -1.78\times10^3/3 = -593$ kJ/mol Mn

(c) $2\,Al + 3\,MgO \rightarrow 3\,Mg + Al_2O_3$

$$\Delta H° = \Delta\bar{H}_f°(Al_2O_3) - 3\,\Delta\bar{H}_f°(MgO) = -1670 - 3 \times (-602) = 1.4\times10^2 \text{ kJ}$$

For one mole of Mg, $\Delta\bar{H}° = 1.4\times10^2 = +47$ kJ/mol Mg. This reaction is endothermic; ΔH is positive.

Carbon

13-16. (a) $2\,C_8H_{18}(\ell) + 17\,O_2(g) \rightarrow 16\,CO(g) + 18\,H_2O(\ell)$

(b) $PbO(s) + CO(g) \rightarrow Pb(s) + CO_2(g)$

(c) $CO_2(g) + 2\,KOH(aq) \rightarrow K_2CO_3(aq) + H_2O$

(d) $MgCO_3(s) + 2\,HCl(aq) \rightarrow MgCl_2(aq) + H_2O + CO_2(g)$

13-17. $CH_4 + 4\,S(g) \rightarrow CS_2 + 2\,H_2S$

$CS_2 + 3\,Cl_2(g) \rightarrow CCl_4 + S_2Cl_2$

$CS_2 + 2\,S_2Cl_2 \rightarrow CCl_4 + 6\,S$

13-18. Form a factor between billions of metric tons of coal burned and ppm $CO_2(g)$ from the equivalence 0.7 ppm CO_2 ⇔ 5×10^9 m ton.

$$\text{total ppm } CO_2 \text{ increase} = 1.0 \times 10^4 \times 10^9 \text{ m ton} \times \frac{0.7 \text{ ppm } CO_2}{5\times10^9 \text{ m ton}} = 1.4\times10^3 \text{ ppm}$$

This approximate increase in ppm CO_2 should then be added to the current level of CO_2:

total ppm = 335 ppm + 1.4×10^3 ppm $\simeq$ 1.7×10^3 ppm

Silicon

13-19. The composition of mica is expressed in Table 13-5 as $KAl_2(AlSi_3O_{10})(OH)_2$

If we assign the usual oxidation states of +1 for K and H, +3 for Al, +4 for Si, and -2 for O, we see that the total of all the oxidation numbers is zero, as expected for a neutral formula unit.

$(+1) + 2\cdot(+3) + (+3) + 3\cdot(+4) + 10\cdot(-2) + 2\cdot(-2) + 2\cdot(+1)$

$= +1 \quad +6 \quad +3 \quad +12 \quad -20 \quad -4 \quad +2 \quad = 0$

13-20. (a) $3\,SiO_2 + 4\,Al \rightarrow 3\,Si + 2\,Al_2O_3$

(b) Potassium metasilicate is K_2SiO_3.

$SiO_2 + K_2CO_3 \rightarrow K_2SiO_3 + CO_2(g)$

(c) $SiCl_4 + LiAlH_4 \rightarrow SiH_4 + LiCl + AlCl_3$

(d) $Si_3H_8 + 5\,O_2 \rightarrow 3\,SiO_2 + 4\,H_2O$

13-21. Silicate anions exist in several different forms, the simplest being metasilicate, SiO_3^{2-}, and orthosilicate, SiO_4^{4-}. Carbonate anion exists only as CO_3^{2-}. Each of these anions combines with an acid (H^+) to form an insoluble acidic species, i.e., H_2CO_3, H_2SiO_3 and H_2SiO_4. These acidic species are unstable and decompose to water and the element oxide. In the case of H_2CO_3 the oxide is gaseous CO_2. In the case of the silicic acids the oxide is SiO_2, which is not a gaseous species. The SiO_2 may, under certain circumstances, remain suspended in a colloidal mixture. In other cases it may precipitate as gelatinous SiO_2 or it may produce a gel (solid suspended in liquid). The differences in the reactions of silicates and carbonates with acids are largely due to differences in the nature of the oxides, SiO_2 and CO_2.

Nitrogen

13-22. (a) $2\,NH_3(aq) + H_2SO_4(aq) \rightarrow (NH_4)_2SO_4(aq)$

(b) $Ag + 2H^+ + NO_3^- \rightarrow Ag^+ + H_2O + NO_2(g)$

or $[Ag + 2\,HNO_3(aq) \rightarrow AgNO_3(aq) + H_2O + NO_2(g)]$

(c) $3\,C + 4\,HNO_3 \rightarrow 3\,CO_2(g) + 4\,NO(g) + 2\,H_2O$

(d) The principal reactions from Table 13-6 referred to here are

$NO_2 + H_2 \rightarrow NO + O$

$O + O_2 \rightarrow O_3$

Oxides of nitrogen are also involved in some of the hydrocarbon reactions in Table 13-6 that also yield O_3.

13-23. The reactions are similar to those presented in the text.

$N_2(g) + O_2(g) \rightarrow 2\,NO(g)$

$2\,NO(g) + O_2(g) \rightarrow 2\,NO_2(g)$

$3\,NO_2(g) + H_2O \rightarrow 2\,HNO_3(aq) + NO(g)$

13-24. In the setup below, NO_x represents a mixture of oxides of nitrogen.

$$\text{no. t } NO_x = 75 \times 10^9 \text{ gal} \times \frac{15 \text{ mi}}{1 \text{ gal}} \times \frac{5 \text{ g } NO_x}{1 \text{ mi}} \times \frac{1 \text{ lb } NO_x}{454 \text{ g } NO_x} \times \frac{1 \text{ t } NO_x}{2000 \text{ lb } NO_x} = 6 \quad \times 10^6 \text{ t } NO_x$$

13-25. A balanced equation for the reduction of NO_3^- to NH_3 is written first.

$$\overset{0}{Zn}(s) + OH^- + NO_3^- + H_2O \qquad Zn(OH)_4^{2-} + NH_3(g)$$

increase of 2 in O.S. per Zn atom

decrease of 8 in O.S. per N atoms

$$4\,Zn(s) + 7\,OH^- + NO_3^- + 6\,H_2O \rightarrow 4\,Zn(OH)_4^{2-} + NH_3(g)$$

The no. mol HCl in 50.00 mL 0.1500 M HCl is

$$\text{no. mol HCl} = 0.05000\text{ L} \times \frac{0.1500\text{ mol HCl}}{1\text{ L}} = 7.500\times10^{-3}\text{ mol HCl}$$

The no. mol HCl in excess, that is, remaining after neutralizing the NH_3 is

$$\text{no. nol HCl in excess} = 0.03210\text{ L} \times \frac{0.1000\text{ mol NaOH}}{1\text{ L}} \times \frac{1\text{ mol HCl}}{1\text{ mol NaOH}} = 3.210\times10^{-3}\text{ mol HCl}$$

(Note: Hcl and NaOH react as follows: $NaOH + HCl \rightarrow NaCl + H_2O$)

$$\text{no. mol HCl that react with }NH_3 = 7.500\times10^{-3} - 3.210\times10^{-3} = 4.290\times10^{-3}\text{ mol HCl}$$

$$\text{no. mol }NH_3\text{ that react} = 4.290\times10^{-3}\text{ mol HCl} \times \frac{1\text{ mol }NH_3}{1\text{ mol HCl}} = 4.290\times10^{-3}\text{ mol }NH_3$$

(HCl and NH_3 react as follows: $NH_3 + HCl \rightarrow NH_4Cl$.)

$$\text{no. mol }NO_3^- = 4.290\times10^{-3}\text{ mol }NH_3 \times \frac{1\text{ mol }NO_3^-}{1\text{ mol }NH_3} = 4.290\times10^{-3}\text{ mol }NO_3^-$$

The molar concentration of NO_3^- in the original solution is

$$\frac{4.290\times10^{-3}\text{ mol }NO_3^-}{0.02500\text{ L}} = 0.1716\text{ M }NO_3^-$$

Phosphorus

13-26. The no. of mol of phosphorus to be prepared is determined first.

$$\text{no. mol }P_4 = 135\text{ kg P} \times \frac{1000\text{ g P}}{1\text{ kg P}} \times \frac{1\text{ mol }P_4}{(4 \times 31.0)\text{ g P}} = 1.09\times10^{3}\text{ mol }P_4$$

The required no. mol $Ca_3(PO_4)_2$ is (recall equation 13.65)

$$\text{no. mol }Ca_3(PO_4)_2 = 1.09\times10^{3}\text{ mol }P_4 \times \frac{2\text{ mol }Ca_3(PO_4)_2}{1\text{ mol }P_4} = 2.18\times10^{3}\text{ mol }Ca_3(PO_4)_2$$

The required mass of phosphate rock is

$$\text{no. kg rock} = 2.18\times10^{3}\text{ mol }Ca_3(PO_4)_2 \times \frac{310\text{ g }Ca_3(PO_4)_2}{1\text{ mol }Ca_3(PO_4)_2} \times \frac{100\text{ g rock}}{58.0\text{ g }Ca_3(PO_4)_2} \times \frac{1\text{ kg rock}}{1000\text{ g rock}}$$

$$= 1.17\times10^{3}\text{ kg}$$

13-27. Of the two common forms of phosphorus, white P melts at a low temperature (about 44°C) and red P at a much higher temperature (590°C and 43 atm). When liquid white P boils at 287°C it produces a vapor at a pressure of 1 atm. This vapor can easily come to equilibrium with *solid* red P at a *higher* temperature, the temperature at which the vapor pressure of solid red P is 1 atm.

13-28. (a) Lewis structures of H_2S and H_2O are quite similar. It is the existence of intermolecular hydrogen bonding among H_2O molecules that allows water to persist as a liquid at higher temperatures than is possible for hydrogen sulfide, which does not engage in hydrogen bonding.

(b) Although alternate structures are possible for ozone, each structure has all electrons paired, for example,

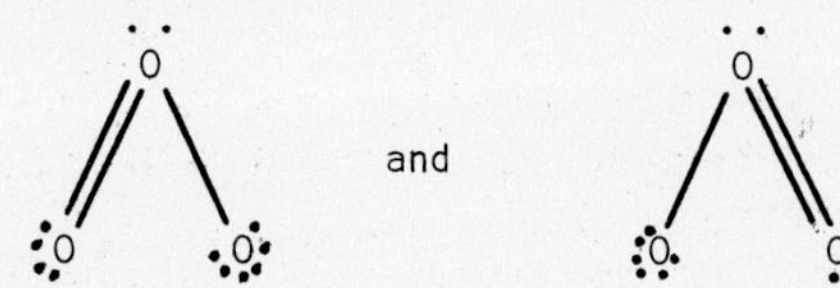

(c) The O-O bond in O_2, as indicated through molecular orbital theory, is essentially a double bond. In O_3, as suggested in part (b), the bond is intermediate between a single and a double bond. In H_2O_2 the bond is essentially a single bond.

H - O - O - H

We should expect the bond lengths to be in the order: $O_2 < O_3 < H_2O_2$.

13-29. The oxidation state of oxygen in H_2O_2 is -1. If this O.S. is to increase it can only be the value zero, in $O_2(g)$. [Oxygen exhibits positive oxidation states only in rare instances, e.g., OF_2.] The required decrease in O.S. must be to the oxidation state -2, as in H_2O. The products of the disproportionative of $H_2)_2$, then, are O_2 and H_2O.

$$2\,H_2O_2 \rightarrow 2\,H_2O + O_2(g)$$

13-30. (a) $O_3(g) + 2\,H^+ + 2\,I^- \rightarrow H_2O + I_2 + O_2(g)$

(b) $S + 3\,O_3(g) + H_2O \rightarrow H_2SO_4 + 3\,O_2(g)$

(c) $2\,[Fe(CN)_6]^{4-} + O_3(g) + H_2\,O \rightarrow 2\,[Fe(NH)_6]^{3-} + O_2(g) + 2OH^-$

13-31. Based on this Lewis structure: we consider that *three* electron pairs must be distributed around the central O atom. (Recall that in the valence-shell electron-pair repulsion theory the O-O double bond is treated as if only one pair of electrons were involved.) The distribution should be trigonal planar--three atoms in a plane with a 120° bond angle. The measured bond angle, as we have seen, is 116.5°.

13-32. The ozone layer must be thought of as a spherical shell with a radius of about 4000 mi and a thickness 0.3 cm. First, determine the surface area of the shell and then multiply by the thickness to obtain the approximate volume of $O_3(g)$. The remainder of the calculation follows easily.

$$A = 4\pi r^2 = 4\pi \times \left\{4000 \text{ mi} \times \frac{5280 \text{ ft}}{1 \text{ mi}} \times \frac{12 \text{ in.}}{1 \text{ ft}} \times \frac{1 \text{ m}}{39.37 \text{ in.}} \times \frac{100 \text{ cm}}{1 \text{ m}}\right\}^2 = 5.2\times10^{18} \text{ cm}^2$$

$$V = 5.2\times10^{18} \text{ cm}^2 \times 0.3 \text{ cm} = 2\times10^{18} \text{ cm}^3$$

$$\text{no. mol } O_3(g) = 2\times10^{18} \text{ cm}^3 \times \frac{1 \text{ L}}{1000 \text{ cm}^3} \times \frac{1 \text{ mol } O_3}{22.4 \text{ L}} = 9\times10^{13} \text{ mol } O_3$$

$$\text{no. molecules } O_3(g) = 9\times10^{13} \text{ mol } O_3 \times \frac{6.02\times10^{23} \text{ molecules } O_3}{1 \text{ mol } O_3} = 5\times10^{37} \text{ molecules } O_3$$

Sulfur

13-33. (a) In comparing the structures of H_2O_2 and S_2Cl_2, the S - S bond is analogous to the O - O bond. The Cl atoms bonded to the S atom are limited to forming one covalent bond, just as are the H atoms in H_2O_2.

The Lewis structures are:

H - O - O - H and :Cl - S - S - Cl:

The geometric structure of H_2O_2 is pictured in Figure 13-16, and that of S_2Cl_2 is similar.

(b) The structures of both SO_2 and SO_3 are based on the distribution of three electron pairs. They both have bond angles of about 120°, and, of course, the electronegativity difference between S and O is the same in each. However, because the structure of SO_3 is symmetrical whereas that of SO_2 is not, SO_3 is nonpolar and SO_2 is polar.

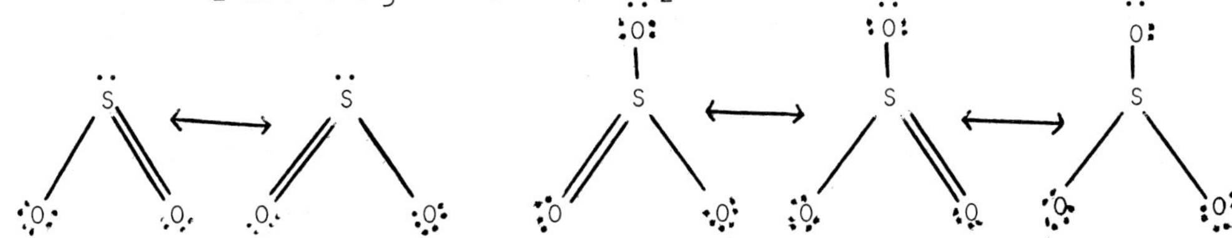

(c) We should expect a paramagnetic, diatomic molecule of sulfur to be similar to that of oxygen.

:O - O: and :S - S:

13-34. In sulfate compounds sulfur exists in its highest oxidation state. Sulfates cannot undergo air oxidation and they are quite stable compounds. Sulfites, on the other hand, have S in the oxidation state +4. Sulfites readily undergo air oxidation to sulfates.

13-35. $\text{no. mol } H_2S = 0.535 \text{ g PbS} \times \frac{1 \text{ mol PbS}}{239 \text{ g PbS}} \times \frac{1 \text{ mol } H_2S}{1 \text{ mol PbS}} = 2.24\times10^{-3} \text{ mol } H_2S$

$$\text{Volume of } H_2S = V = \frac{nRT}{P} = \frac{2.24\times10^{-3} \text{ mol} \times 0.0821 \text{ L atm mol}^{-1} \text{ K}^{-1} \times 298 \text{ K}}{(740/760)\text{atm}} = 5.63\times10^{-2} \text{ L } H_2S$$

$$\%\ H_2S, \text{ by volume} = \frac{5.63\times10^{-2}}{25.0} \times 100 = 0.225\%\ H_2S$$

13-36. (a) Superheated water is water that has been heated under pressure to temperatures above its normal boiling point.

(b) Sulfur melts at 119°C. The vapor pressure of water at 110°C is about 1.4 atm. We might estimate that at 119°C the vapor pressure of water is about 2 atm. This is the minimum pressure under which superheated water can be maintained to be hot enough to melt sulfur.

Fluorine, chlorine

13-37. Assuming that none of the bonds has a high degree of polarity, the indicated properties are estimated by averaging the values for the respective halogens.

BrCl:

$$\text{estimated melting point} = \frac{-101 - 7.2}{2} = -54\ °\text{C (actual: } -54°\text{C)}$$

$$\text{estimated boiling point} = \frac{-34.6 + 58.8}{2} = +12°\text{C (actual: } 5°\text{C)}$$

ICl:

estimated melting point $= \dfrac{-101 + 114}{2} = +6°C$ (actual: 14°C)

estimated boiling point $= \dfrac{+58.5 + 184.4}{2} = 122°C$ (actual: 97°C)

Alternatively, we might say that ICl (mol. wt. = 162) should have about the same physical properties as Br_2 (mol. wt. = 160); or, because of some polarity in the I-Cl bond, perhaps a somewhat higher melting and boiling point than Br_2. This does appear to be the case: m.p. of Br_2 = -7.2°C and b. pt. = +58.8°C.

13-38. Here, again, let us average properties. Also we used the actual melting and boiling points of F_2 listed in Example 13-4.

ClF:

estimated melting point $= \dfrac{-220 - 101}{2} = -160°C$ (actual: -156°C)

estimated boiling point $= \dfrac{-188 - 34.6}{2} = -111°C$ (actual: -90°C)

13-39. (a) $Ca(s) + Cl_2(g) \rightarrow CaCl_2(s)$

(b) $P_4(s) + 6Cl_2(g) \rightarrow 4PCl_3$

(c) $2F_2 + 2H_2O \rightarrow 4HF + O_2(g)$

(d) $2Al(s) + 3F_2(g) \rightarrow 2AlF_3(s)$

Self-Test Questions

1. *(b)* Group IA (e.g., Li, K) and the heavier group IIA (e.g. Ca) metals liberate $H_2(g)$ from cold water. Al does not.

2. *(a)* Of the compounds listed, $BeCl_2$ has the least ionic (or most covalent) character. We should expect it to be the poorest electrical conductor.

3. *(b)* To function as an oxidizing agent, a substance must contain an atom which can be reduced to a lower oxidation state. This condition exists for the other three but not for Cl^-. In Cl^- chlorine exists in its lowest possible oxidation state (-1).

4. *(d)* Only HCl has a significant polar character, but not enough to cause its boiling point to be much higher than expected for a low molecular weight substance (mol. wt. = 36.5). The highest boiling point corresponds to the substance of highest molecular weight, CCl_4.

5. *(c)* CaO reacts with H_2O to produce $Ca(OH)_2$ and *no* gaseous product. CaH_2 liberates $H_2(g)$. Na reacts with H_2O to produce $H_2(g)$ but Na_2O_2 produces $O_2(g)$. F_2 liberates $O_2(g)$ on reaction with water (see Table 13-8) and so does Na_2O_2. Li_3N liberates NH_3 when it reacts with H_2O, and LiH produces $H_2(g)$.

6. *(c)* Hydrogen does not occur in great abundance either in the earth's cove or crust. In H_2O is constitutes about 11% of the mass; but in the universe as a whole it accounts for about 75% of the mass.

7. *(b)* When heated, limestone ($CaCO_3$) yields calcium oxide (CaO). CaO reacts with water to produce $Ca(OH)_2$ --an inexpensive alkali. None of the other three yields an alkaline material.

8. *(d)* NH_3, $(NH_4)_2$, SO_4, and various phosphates are all fertilizers. Na_2CO_3 is not.

9. *(a)* lithium hydride = LiH | *(b)* $KHSO_4$ = potassium hydrogen sulfate

(c) ClO_2^- = chlorite ion | *(d)* SiO_2 = silicon dioxide

(e) potassium perchlorate = $KClO_4$ | *(f)* sodium nitrite = $NaNO_2$

(g) N_2O_5 = dinitrogen pentoxide | *(h)* BaO_2 = barium superoxide

10. (*a*) $MgCO_3(s) \xrightarrow{\Delta} MgO(s) + CO_2(g)$

(*b*) $MgCO_3(s) + 2\,HCl(aq) \rightarrow MgCl_2(aq) + H_2O + CO_2(g)$

(*c*) $P_4 + 5\,O_2(g) \rightarrow P_4O_{10}$

$P_4O_{10} + 6\,H_2O \rightarrow 4\,H_3PO_4$

(*d*) $H_2O_2 + Cl_2 \rightarrow 2\,Cl^0 + 2\,H^+ + O_2(g)$

(*e*) $4\,NH_3 + 5\,O_2 \rightarrow 4\,NO + 6\,H_2O$

(*f*) $3\,Cu + 8\,H^+ + 2\,NO_3^- \rightarrow 3\,Cu^{2+} + 4\,H_2O + 2\,NO(g)$

11. First convert from ft^3 (STP) to L (STP).

$$\text{no. L} = 3.58\times10^{11}\ ft^3 \times \frac{(12)^3\ in^3}{1\ ft^3} \times \frac{(2.54)^3\ cm^3}{1\ in^3} \times \frac{1\ L}{1000\ cm^3} = 1.01\times10^{13}\ L$$

$$\text{no. kg } O_2 = 1.01\times10^{13}\ L \times \frac{1\ mol\ O_2}{22.4\ L} \times \frac{32.0\ g\ O_2}{1\ mol\ O_2} \times \frac{1\ kg\ O_2}{1000\ g\ O_2} = 1.44\times10^{10}\ kg\ O_2$$

12. Air pollution control measures for automobiles must focus on a reduction of oxides of nitrogen, carbon monoxide, and unburned hydrocarbons. This can be brought about by more complete combustion and catalytic exhaust systems. Although some of the pollutants from power plants, such as oxides of nitrogen, are similar to those from automobiles, a most important pollutant from fossil-fuel power plants is sulfur dioxide. Air pollution control measures for these power plants must deal with a pollutant (SO_2) that is not a major source produced by automobiles.

Chapter 14

Chemical Kinetics

Review Problems

14-1. (a) The initial rate of reaction is the rate of change of concentration with time at a point early in the reaction. Here it is based on the first 53 s.

$$\text{rate} = \frac{-\Delta[A]}{\Delta t} = \frac{-(0.1168\ M - 0.1205\ M)}{53\,s} = \frac{+0.0037\ M}{53\,s} = 7.0\times10^{-5}\ M\ s^{-1}$$

$$= 7.0\times10^{-5}\ mol\ L^{-1}\ s^{-1}$$

(b) $7.0\times10^{-5}\ mol\ L^{-1}\ s^{-1} \times \frac{60\,s}{min} = 4.2\times10^{-3}\ mol\ L^{-1}\ min^{-1}$

14-2. (a) For the reaction $2A + B \rightarrow C + 3D$ the rate of reaction is minus one half the rate of disappearance of A.

$$\text{rate of reaction} = -\tfrac{1}{2} \times \text{rate disappearance A} = -\tfrac{1}{2} \times (-2.6\times10^{-4}\ mol\ L^{-1}\ s^{-1})$$

$$= 1.3\times10^{-4}\ mol\ L^{-1}\ s^{-1}$$

(b) The rate of formation of D is related to the rate of reaction as follows:

$$\text{rate of reaction} = \frac{1}{3}\ \text{rate formation of D}$$

$$\text{rate of formation of D} = 3 \times \text{rate of reaction} = 3\times1.3\times10^{-4}\ mol\ L^{-1}\ s^{-1}$$

$$= 3.9\times10^{-4}\ mol\ L^{-1}\ s^{-1}$$

14-3. (a) The volume of $MnO_4^-(aq)$ is directly proportional to $[H_2O_2]$. Since H_2O_2 disappears in the reaction, less H_2O_2 is present in each sample titrated. This means that less $MnO_4^-(aq)$ is required in each successive titration.

(b) For each titration the volume of $MnO_4^-(aq)$ is determined by a calculation of the following type (for $t = 600\,s$, $[H_2O_2] = 1.49\ M$).

$$\text{no. mL } MnO_4^-(aq) = 5.00\ mL \times \frac{1.49\ mmol\ H_2O_2}{mL} \times \frac{2\ mmol\ MnO_4^-}{5\ mmol\ H_2O_2} \times \frac{1.00\ mL\ MnO_4^-(aq)}{0.1000\ mol\ MnO_4^-}$$

$$= 29.8\ mL\ MnO_4^-(aq)$$

Other volumes of $MnO_4^-(aq)$ are

$$\text{no. mL } MnO_4^-(aq) = 5.00\ mL \times \frac{0.98\ mmol\ H_2O_2}{mL} \times \frac{2\ mmol\ MnO_4^-}{5\ mmol\ H_2O_2} \times \frac{1.00\ mL\ MnO_4^-(aq)}{0.1000\ mmol\ MnO_4^-}$$

$$= 19.6\ mL\ MnO_4^-(aq)$$

$$\text{no. mL } MnO_4^-(aq) = 5.00\ mL \times \frac{0.62\ mmol\ H_2O_2}{mL} \times \frac{2\ mmol\ MnO_4^-}{5\ mmol\ H_2O_2} \times \frac{1.00\ mL\ MnO_4^-(aq)}{0.1000\ mmol\ MnO_4^-}$$

$$= 12.4\ mL\ MnO_4^-(aq)$$

$$\text{no. mL } MnO_4^-(aq) = 5.00\ mL \times \frac{0.25\ mmol\ H_2O_2}{mL} \times \frac{2\ mmol\ MnO_4^-}{5\ mmol\ H_2O_2} \times \frac{1.00\ mL\ MnO_4^-(aq)}{0.1000\ mmol\ MnO_4^-}$$

$$= 5.0\ mL\ MnO_4^-(aq)$$

14-4. (a) Construct a tangent line to the curve of Figure 14-1 at t = 1000 s and determine its slope.

$\Delta[H_2O_2] \simeq -1.88$ M $\qquad \Delta t \simeq 2450$ s

$$\text{rate of reaction} = \frac{-(-1.88\ M)}{2450\ s} = 7.7\times10^{-4}\ mol\ L^{-1}\ s^{-1}$$

(b) $\Delta[H_2O_2] \simeq -1.28$ M $\qquad \Delta t \simeq 3400$ s

$$\text{rate of reaction} = \frac{-(-1.28\ M)}{3400\ s} = 3.7\times10^{-4}\ mol\ L^{-1}\ s^{-1}$$

14-5. Even though the stoichiometric coefficient of CH_3CHO in the decomposition equation is 1, since the reaction is second order $[CH_3CHO]$ must be raised to the second power.

$$\text{rate of reaction} = k[CH_3CHO]^2$$

14-6. (a) To determine the order with respect to A, take the ratio of the rate of expt. 4 to expt. 2.

$$\frac{\text{rate 4}}{\text{rate 2}} = \frac{5.13\times10^{-3}}{1.25\times10^{-3}} = \frac{[A]_4^m[B]_4^n}{[A]_2^m[B]_2^n} = \frac{(0.420)^m(0.230)^n}{(0.210)^m(0.230)^n} = 2^m = 4.10$$

$2^m \simeq 4 \qquad m = 2.$ The reaction is second order in A.

To determine the order with respect to B, take the ratio of the rate of expt. 2 to expt. 1.

$$\frac{\text{rate 2}}{\text{rate 1}} = \frac{1.25\times10^{-3}}{6.30\times10^{-4}} = \frac{(0.210)^m(0.230)^n}{(0.210)^m(0.115)^n} = 2^n = 1.98$$

$2^n \simeq 2 \qquad n = 1.$ The reaction is first order in B.

(b) The rate equation is rate of reaction = $k[A]^2[B]$. The overall reaction order is third order.

14-7. The data given correspond to an integral number of half lives. That is, after one half-life [A] = 1.00 M; after two half-lives, [A] = 0.50 M; and after three half-lives, [A] = 0.25 M. Three half-lives total 201 min.

(a) $3t_{\frac{1}{2}} = 201$ min. and $t_{\frac{1}{2}} = \frac{1}{3} \times 200$ min. $= 67.0$ min.

(b) $k = \frac{0.693}{t_{\frac{1}{2}}} = \frac{0.693}{67.0\ min.} = 0.0103\ min^{-1}$

14-8. Use equation (14.13); $\log \frac{[A]_t}{[A]_o} = \frac{-kt}{2.303}$

(a) $$\log \frac{[A]_t}{0.85} = \frac{-1.0\times10^{-3}\ s^{-1} \times \frac{60\ s}{1\ min.}}{2.303} = -0.26$$

$\frac{[A]_t}{0.85} = \text{antilog}(-0.26) = 0.55 \qquad [A]_t = 0.85 \times 0.55 = 0.47$ M

(b) $$\log \frac{0.25}{0.85} = \frac{-1.0\times10^{-3}\ s^{-1} \times t}{2.303} = -0.53$$

$$t = \frac{-0.53 \times 2.303}{-1.0 \times 10^{-3}}\ s = 1.2\times10^3\ s$$

(c) If 90% of the original reactant is consumed, 10% remains unreacted $[A]_t/[A]_o = 0.10$

$$\log 0.10 = \frac{-1.0\times10^{-3}\ s^{-1} \times t}{2.303} = -1.0$$

$$t = \frac{2.303}{1.0\times10^{-3}\ s^{-1}} = 2.3\times10^3\ s$$

14-9. For this problem use equation (14.29): $\log \frac{k_2}{k_1} = \frac{E_a (T_2 - T_1)}{2.303\ RT_1T_2}$

(a) Solve equation (14.29) for E_a and substitute the known data.

$$E_a = \frac{2.303\ RT_1T_2}{T_2-T_1} \log \frac{k_2}{k_1}$$

$$E_a = \frac{2.303 \times 8.314\ \text{K mol}^{-1}\ \text{K}^{-1} \times 556\ \text{K} \times 666\ \text{K}}{(666-556)\ \text{K}} \times \log \frac{3.8\text{x}10^{-2}}{1.2\text{x}10^{-4}}$$

$$E_a = \frac{(2.303 \times 8.314 \times 556 \times 666)\text{J mol}^{-1}}{110} \times \log (3.2\text{x}10^2)$$

$$= 1.61\text{x}10^5\ \text{J/mol} = 1.61\text{x}10^2\ \text{kJ/mol}$$

(b) Use $k_1 = 1.2\text{x}10^{-4}$ L mol^{-1} s^{-1}; $T_1 = 556$ K; $k_2 = 1.0\text{x}10^{-3}$ L mol^{-1} s^{-1}; T_2 = ?, and

$E_a = 1.61\text{x}10^5$ J/mol

$$\log \frac{k_2}{k_1} = \log \frac{1.0\text{x}10^{-3}}{1.2\text{x}10^{-4}} = 0.92 = \frac{1.61\text{x}10^5 (T_2 - 556)}{2.303 \times 8.314 \times 556\ T_2}$$

$$0.92 \times 2.303 \times 8.314 \times 556\ T_2 = 1.61\text{x}10^5\ T_2 - 8.95\text{x}10^7$$

$$1.51\text{x}10^5 T_2 = 8.95\text{x}10^7 \qquad T_2 = 5.93\text{x}10^2\ \text{K}$$

14-10. (a) Add the two steps to obtain the net equation

(slow) 2A → ~~I~~ + D

(fast) ~~I~~ + B → A + C

net: 2A + B → A + C + D

or: A + B → C + D

(b) For the slow step: rate = k $[A]^2$

This is the rate-determining step since the intermediate I reacts as fast as it forms. For the overall reaction, then,

rate of reaction = k $[A]^2$

14-11. (a) The overall reaction: $H_2 + 2ICl \rightarrow I_2 + 2HCl$ and the slow step: $H_2 + ICl \rightarrow HI + HCl$ are given.

The second step must be such that when added to the slow step the sum is the net equation. The following second step works.

$ICl + HI \rightarrow I_2 + HCl$

That is

(slow) $H_2 + ICl \rightarrow HI + HCl$

$ICl + HI \rightarrow I_2 + HCl$

net: $H_2 + 2ICl \rightarrow I_2 + 2HCl$

(b) The rate equation predicted from the first step is

rate of reaction = $k[H_2][ICl]$

Since the reaction is indeed found to be first order in both $[H_2]$ and $[ICl]$, the first step must be the slow step and the second reaction must occur rapidly (and thus not be significant in establishing the overall reaction rate).

14-12. (a) The data given in column II correspond to a zero-order reaction. This fact can be easily established in two ways: First, this is the only one of the three sets of data that shows the reaction going to completion. Also, the rate of decrease of [A] is constant; [A] decreases by 0.25 for every 25 s.

(b) The half-life of a first-order reaction is constant. Note the data in column III. The time required for [A] to decrease to 0.50, one half its original value, is 100 sec. But at 200 sec [A] = 0.33, not 0.25 as would be required in a first-order reaction. The first-order reaction must correspond to the data in column I.

(c) By a process of elimination it appears that the second-order reaction corresponds to the data in column III. An alternative approach is also possible. Show that 1/[A] plotted against time is a straight line. For example, at t = 0, 1/[A] = 1.00; t = 25, 1/[A] = 1.25; t = 50, 1/[A] = 1.50; t = 75, 1/[A] = 1.75; t = 100, 1/[A] = 2.00. The increase in 1/[A] over every 25-second interval is constant--0.25. The graph is linear.

14-13. Although graphical methods are best for determining the half-life of a reaction, since we are asked only to give an approximate value, a simple inspection of the data in column I is sufficient. The half-life is somewhat less than 75 s, say approximately 70 s. This appears to be the case because at 75 s [A], which was initially 1.00, has decreased to 0.47. The half-life would be the time at which [A] = 0.50. At t = 140 s, [A] should be 0.25. The data in column I indicate that at t = 150 s, [A] = 0.22. This fact is consistent with a half-life of 70 s. And we should expect a value of [A] = 0.125 at t = 210 s, which is consistent with the observed value of [A] = 0.14 at t = 200 s. Note, finally, that in the 75-second interval from t = 75 s to t = 150 s, [A] decreases from 0.47 to 0.22.

14-14. Use the first two data points in column III to estimate the initial rate of reaction.

$$\text{rate of reaction} = -\text{rate disappearance of A} = \frac{-\Delta[A]}{\Delta t} = \frac{-(0.80 - 1.00)\ M}{25\ s}$$

$$= 8\times10^{-3}\ \text{mol L}^{-1}\ \text{s}^{-1}$$

This estimate is probably not too good, however, because it is based on the first 20% of the reaction instead of the customary 2-3%. Another estimate is based on obtaining a value of k from the expression

$$t_{\frac{1}{2}} = \frac{1}{k[A]_0} \quad \text{or} \quad k = \frac{1}{t_{\frac{1}{2}}[A]_0}$$

For the half-life period in which [A] decreases from 1.00 M to 0.50 M.

$$k = \frac{1}{100\ s\ (1.00\ M)} = 1.00\times10^{-2}\ \text{L mol}^{-1}\ \text{s}^{-1}$$

Now use the expression

rate reaction = $k[A]^2$

with [A] = 1.00 M (the initial concentration of A)

$$\text{rate of reaction} = 1.00\times10^{-2}\ \text{L mol}^{-1}\ \text{s}^{-1} \times (1.00)^2\ \text{mol}^2/\text{L}^2 = 1.00\times10^{-2}\ \text{mol L}^{-1}\ \text{s}^{-1}$$

14-15. Although rates of reaction could be estimated in the same manner as in the preceding exercise, a different method is suggested here. The data, [A] versus time, are plotted below. The rates of reaction are simply the slopes of the tangent lines drawn at t = 75 s.

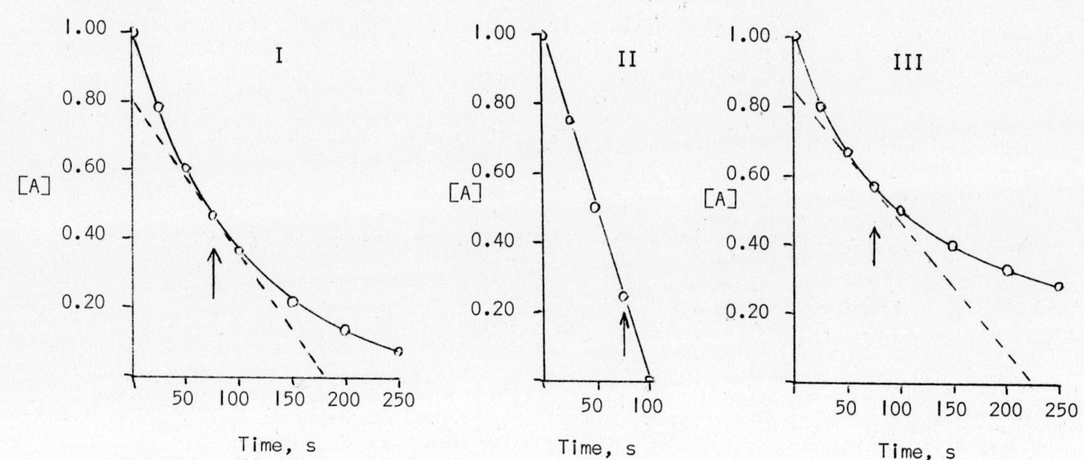

(a) *Zero-order reaction (II):* The graph is a straight line. The slope (and hence the rate of reaction) is a constant. Rate of reaction = $-\Delta[A]/\Delta t$ = 1.00 mol L^{-1}/100 s = 1.0×10^{-2} mol A L^{-1} s^{-1}.

(b) *First-order reaction (I):* The tangent drawn at t = 75 s intersects the y axis at 0.80 mol A/L and the x axis at 175 s. The slope of this tangent line is, Rate of reaction = $-\Delta[A]/\Delta t$ = 0.80 mol A L^{-1}/175 s = 4.6×10^{-3} mol A L^{-1} s^{-1}.

(c) *Second-order reaction (III):* The tangent line drawn at 75 s intersects the y axis at 0.84 mol A/L and the x axis at 225 s. Rate of reaction = $-\Delta[A]/\Delta t$ = 0.84 mol A L^{-1}/225 s = 3.7×10^{-3} mol A L^{-1} s^{-1}.

14-16. Because we are asked only for approximate values, an estimation of reaction rates can be based simply on the tabular data instead of a graph.

(a) The zero-order reaction is described by the data in column II. The reaction is concluded at 100 s. For this reason [A] = 0 at t = 110 s.

(b) The first-order reaction corresponds to the data in column I. The rate of the reaction at 110 s is nearly the same (through slightly less) than at 100 s. In the 25-second interval between t = 75 s and t = 100 s, $\Delta[A] = -0.10$, for an approximate rate of reaction of $-\Delta[A]/\Delta t = 0.10/25 = 0.0040$. In the 50-second interval between t = 100 s and t = 150 s, the corresponding rate of reaction is $-\Delta[A]/\Delta t = 0.15/50 = 0.0030$. Assume that the rate of reaction at t = 110 s is intermediate to these two values, that is, $-\Delta[A]/\Delta t$ = 0.0035 mol L^{-1} s^{-1}. In the 10-second interval between 100 and 110 s, [A] decreases by 10 x 0.0035 = 0.035. This suggests [A] at 110 s to be 0.37 - 0.035 = 0.33 M.

(c) Turn to the data in column III and follow the same procedure as in part (b). The approximate rate of reaction at 110 s is $-\Delta[A]/\Delta t$ = 0.0024 mol L^{-1} s^{-1}. In the 10-second interval between 100 and 110 s, [A] decreases by 10 x 0.0024 = 0.024. This suggests [A] at 110 s to be 0.50 - 0.02 = 0.48 M.

Exercises

Rates of reactions

14-1. (a) The rate of reaction during the first minute is

$$\text{rate of reaction} = \frac{-[A]}{t} \times \frac{-(0.1455 - 0.1503)\text{ M}}{1.00\text{ min}} = \frac{0.0048\text{ M}}{1.00\text{ min}}$$

$$= 4.8\times10^{-3}\text{ mol L}^{-1}\text{ min}^{-1}$$

During the second minute,

$$\text{rate of reaction} = \frac{-(0.1409 - 0.1455)}{(2.00 - 1.00)\text{ min}} = \frac{0.0046\text{ M}}{1.00\text{ min}} = 4.6\times10^{-3}\text{ mol L}^{-1}\text{ min}^{-1}$$

(b) The rates are not equal because the rate of reaction is proportional to some power of [A]. As the reactant A is consumed, the reaction must slow down. Thus the rate of reaction in the second minute is slightly less than in the first minute.

14-2. (a) rate of formation of C = 2 x rate of reaction

$$= 2 \times 2.50\times10^{-5} \text{ mol L}^{-1} \text{ s}^{-1} = 5.00\times10^{-5} \text{ mol L}^{-1} \text{ s}^{-1}$$

(b) If [A] decreases at a rate of 2.5×10^{-5} mol L^{-1} s^{-1}, then during a period of 1.00 min = 60 s

$\Delta[A] = 2.50\times10^{-5} \text{ mol L}^{-1} \text{ s}^{-1} \times 60 \text{ s} = 1.50\times10^{-3} \text{ mol L}^{-1}$

Since [A] = 0.5000 M at the start of the time interval, at the end of the interval,

[A] = 0.5000 - 0.0015 = 0.4985 M

(c) The required decrease in [A] = Δ[A] = 0.5000 - 0.4900 = 0.0100 M

$$\frac{\Delta[A]}{\Delta t} = 2.50\times10^{-5} \text{ mol L}^{-1} \text{ s}^{-1}$$

$$\Delta t = \frac{0.0100 \text{ mol L}^{-1}}{2.50\times10^{-5} \text{ mol L}^{-1} \text{ s}^{-1}} = 4.00\times10^{2} \text{ s}$$

14-3. The decomposition reaction is $H_2O_2 \rightarrow H_2O + \frac{1}{2}O_2(g)$.

rate of reaction = $\frac{1}{\frac{1}{2}}$ rate of formation of $O_2(g)$

rate of formation of $O_2(g)$ = $\frac{1}{2}$ rate of reaction

(a) rate of formation of $O_2(g)$ = $\frac{1}{2} \times 5.7\times10^{-4} \text{ mol L}^{-1} \text{ s}^{-1} = 2.8\times10^{-4} \text{ mol L}^{-1} \text{ s}^{-1}$

(b) rate of formation of $O_2(g)$ = $2.8\times10^{-4} \text{ mol L}^{-1} \text{ s}^{-1} \times \frac{60 \text{ s}}{1 \text{ min}} = 1.7\times10^{-2} \text{ mol L}^{-1} \text{ min}^{-1}$

(c) rate of formation of $O_2(g)$ = $1.00 \text{ L} \times 1.7\times10^{-2} \text{ mol L}^{-1} \text{ min}^{-1} \times \frac{22.4 \text{ L } O_2(g)}{1 \text{ mol}} \times \frac{1000 \text{ cm}^3 \text{ } O_2(g)}{1 \text{ L } O_2(g)}$

$$= 3.8\times10^{2} \text{ cm}^3 \text{ } O_2 \text{ (STP) min}^{-1}$$

14-4. (a) $[A]_0$ refers to the concentration of the reactant A, expressed in moles per liter, that exists at whatever time is taken to be the start of the reaction (t = 0).

(b) $[A]_t$ refers to the molar concentration of the reactant A at some time, t, after the start of of a chemical reaction.

(c) Δ[A] represents the difference in the molar concentration of the reactant A at two different times in a chemical reaction.

(d) The ratio of Δ[A] to Δt represents the rate of change of concentration of reactant A with time.

(e) The ratio of Δ[A] to Δt represents the rate of change of concentration with time. The negative of this ratio, that is, $-\Delta[A]/\Delta t$, is the rate of the reaction.

(f) $\Delta[B]/\Delta t$ is the rate of increase of concentration of product B with time. This quantity is also equal to the rate of the reaction.

(g) $t_{\frac{1}{2}}$ is the half-life of the reaction--the time required for one half of the quantity of A present initially to react.

14-5. (a) The initial rate of a reaction is the rate at which the reaction occurs just as it is beginning, i.e., immediately after the reactants are mixed.

(b) The instantaneous rate of reaction is the rate at which the reaction occurs at any specific instant of time during the course of the reaction. This rate is established by the tangent line to the graph of concentration vs. time at a particular time.

(c) A zero-order reaction is one that proceeds at a constant rate, regardless of the concentration of the reactant(s).

(d) The half-life of a reaction is the length of time required for one half of the reactant originally present to be consumed.

14-6. The rate of a reaction and the corresponding rate constant can best be compared through the rate equation: rate of reaction = $k[A]^n[B]^m$.... The rate constant k is the proportionality constant that relates the rate of reaction to powers of the reactant concentrations. If all the exponents in the rate equation are zero, that is, if the reaction is zero order, the rate of reaction and the rate constant are identical: rate of reaction = k.

14-7. If the reaction is zero order,

rate of reaction = $k[A]^\circ = k$

The units of k are the same as those of the rate of reaction: $\text{mol L}^{-1}\text{ s}^{-1}$

If the reaction is first order,

$$\text{rate of reaction (mol L}^{-1}\text{ s}^{-1}) = k[A] = k \times (\text{mol L}^{-1})$$

$$\text{The units of k are } \frac{\text{mol L}^{-1}\text{ s}^{-1}}{\text{mol L}^{-1}} = \text{s}^{-1}$$

If the reaction is second order,

$$\text{rate of reaction (mol L}^{-1}\text{ s}^{-1}):\ k[A]^2 = k \times (\text{mol L}^{-1})^2$$

$$\text{The units of k are } \frac{(\text{mol L}^{-1}\text{ s}^{-1})}{(\text{mol L}^{-1})^2} = \frac{\text{mol L}^{-1}\text{ s}^{-1}}{\text{mol}^2\text{ L}^{-2}} = \text{L mol}^{-1}\text{ s}^{-1}$$

14-8. (a) Assume that the rate given is maintained for 10 s

$$\frac{-\Delta[H_2O_2]}{\Delta t} = 1.7\times10^{-3}\text{ mol L}^{-1}\text{ s}^{-1}$$

$$\Delta[H_2O_2] = -1.7\times10^{-3}\text{ mol L}^{-1}\text{ s}^{-1} \times 10\text{ s} = -1.7\times10^{-2}\text{ mol L}^{-1}$$

$$[H_2O_2] = 1.55\text{ M} - 0.017\text{ M} = 1.53\text{ M}$$

(b) We are seeking the time Δt when $\Delta[H_2O_2] = 1.50\text{ M} - 1.55\text{ M} = -0.05\text{ M}$

$$\frac{-\Delta[H_2O_2]}{\Delta t} = \frac{-(0.05)\text{ mol L}^{-1}}{\Delta t} = 1.7\times10^{-3}\text{ mol L}^{-1}\text{ s}^{-1};\quad \Delta t = \frac{0.05}{1.7\times10^{-3}} = 3\times10^1\text{ s}$$

(c) First proved as in (a), but with t = 60 s.

$$\Delta[H_2O_2] = -60\text{ s} \times 1.7\times10^{-3}\text{ mol L}^{-1}\text{ s}^{-1} = 0.10\text{ mol L}^{-1}$$

$$\text{no. mol } H_2O_2 \text{ decomposed} = 0.10\text{ mol L}^{-1} \times 0.175\text{ L} = 0.018\text{ mol } H_2O_2$$

$$\text{no. cm}^3\ O_2(g) = 0.018\text{ mol } H_2O_2 \times \frac{\tfrac{1}{2}\text{ mol } O_2}{1\text{ mol } H_2O_2} \times \frac{22.4\text{ L } O_2}{1\text{ mol } O_2} = 0.20\text{ L } O_2\text{ (STP)}$$

14-9. Assume that the initial rate listed for Expt. 2 holds for 60 s.

$$\text{rate of reaction} = \frac{-\Delta[S_2O_8^{2-}]}{\Delta t} = 2.8\times10^{-5}\ \text{mol L}^{-1}\ \text{s}^{-1}$$

$$\Delta[S_2O_8^{2-}] = -2.8\times10^{-5}\ \text{mol L}^{-1}\ \text{s}^{-1} \times 60\ \text{s} = -1.7\times10^{-3}\ \text{mol L}^{-1}$$

$$\Delta[I^-] = 3\Delta[S_2O_8^{2-}] = 3 \times (-1.7\times10^{-3}\ \text{mol L}^{-1}) = -5.1\times10^{-3}\ \text{mol L}^{-1}$$

(a) After 60 s

$$[S_2O_8^{2-}] = 0.076\ \text{M} - 1.7\times10^{-3}\ \text{M} = 0.074\ \text{M}$$

(b) After 60 s

$$[I^-] = 0.060\ \text{M} - 5.1\times10^{-3}\ \text{M} = 0.055\ \text{M}$$

14-10. In the reaction $A(g) \rightarrow 2B(g) + C(g)$ three moles of gaseous products are formed for every mole of gaseous reactant consumed.

(a) The pressure of the gas is directly proportional to the number of moles of gas if the volume of gas and temperature are constant.

$$P_f = P_i \times \text{ratio of no. mol gas} = 1000\ \text{mmHg} \times \frac{3\ x\ \text{mol}}{1\ x\ \text{mol}} = 3000\ \text{mmHg}$$

(b) When P_A has decreased from 1000 mmHg to 800 mmHg, ΔP_A = 200 mmHg. The increase in pressure of the products is $\Delta P_{prod.} = 3 \times \Delta P_A$ = 3 x 200 mmHg = 600 mmHg.

$$P_{tot.} = P_A + P_{prod.} = 800\ \text{mmHg} + 600\ \text{mmHg} = 1400\ \text{mmHg}$$

14-11.

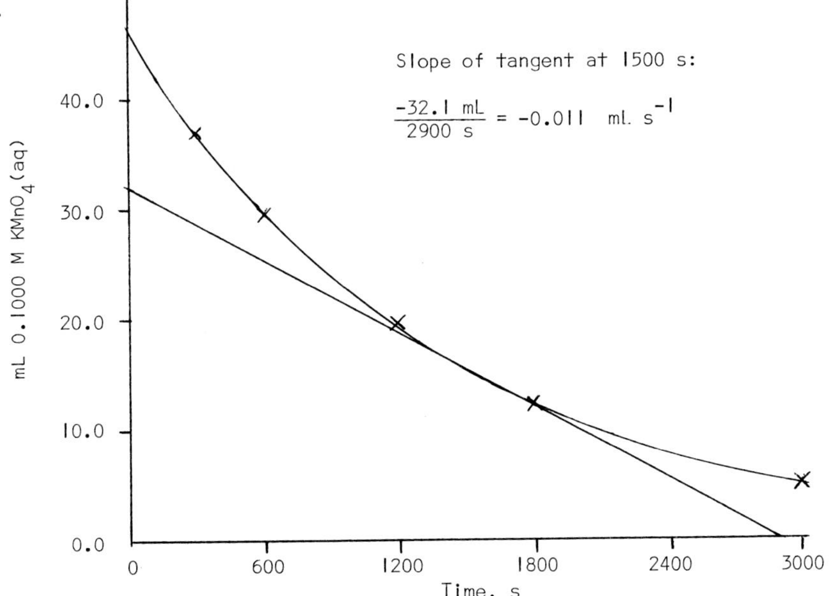

$$\text{rate of disappearance } H_2O_2 = -0.011\ \text{mL } MnO_4^-(aq)\ \text{s}^{-1} \times \frac{0.1000\ \text{mmol } MnO_4^-}{\text{mL } MnO_4^-} \times \frac{5\ \text{mmol } H_2O_2}{2\ \text{mmol } MnO_4^-} \times$$

$$\frac{1}{5.00\ \text{mL}} = -5.5\times10^{-4}\ \text{mmol mL}^{-1}\ \text{s}^{-1}$$

$$= -5.5\times10^{-4}\ \text{mol L}^{-1}\ \text{s}^{-1}$$

rate of reaction = -rate of disappearance of H_2O_2 = 5.5×10^{-4} mol L^{-1} s^{-1}
(The result obtained in Example 14-3 was 5.7×10^{-4} mol L^{-1} s^{-1}. This is good agreement.)

14-12.

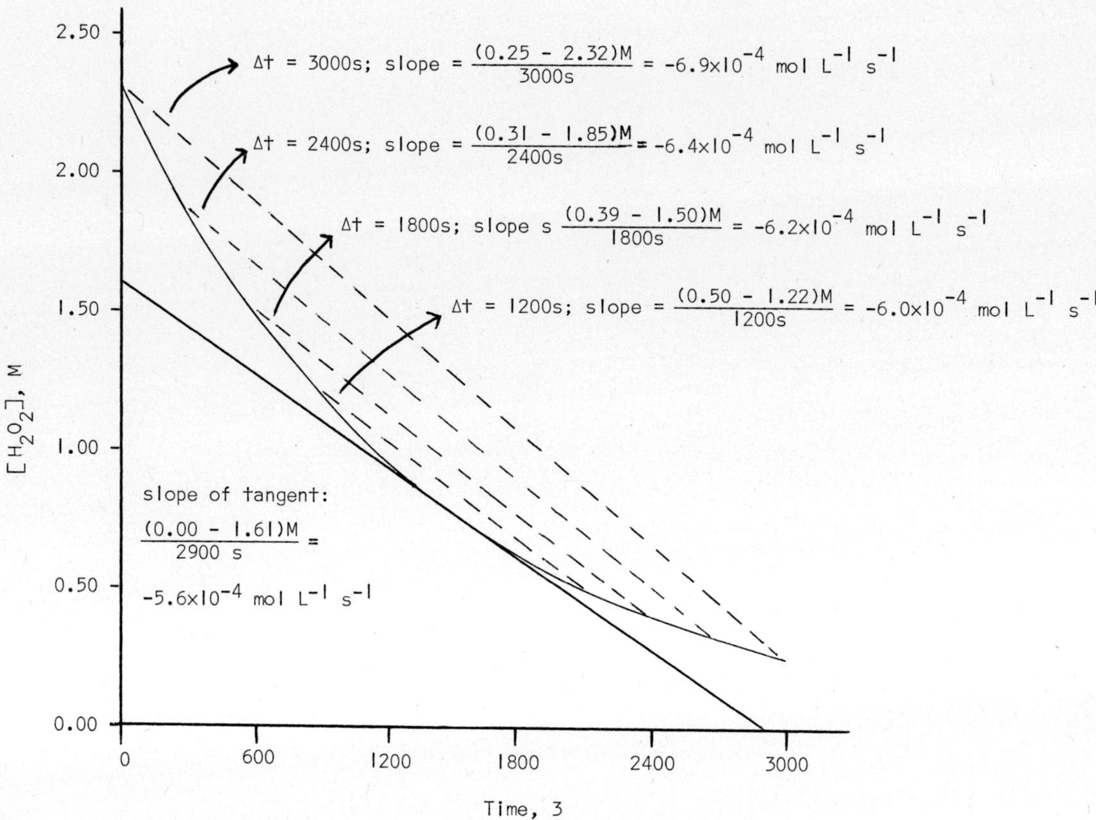

The data of Figure 14-1 are replotted above with the tangent line at 1500 s ($\Delta t \rightarrow 0$) and other lines for differing values of Δt. The slopes of the various lines are compared. The slope of each line differs, but it can be seen that the shorter the time interval Δt, the more nearly the slope of the line approaches the slope of the tangent line, -5.6×10^{-4} mol L^{-1} s^{-1}.

Method of initial rates

14-13. Compare the data for this fourth experiment with data listed for Expt. 3 in Table 14-3: $[I^-]$ = 0.015 M for Expt. 4 compared to $[I^-]$ = 0.030 M for Expt.3. With this change alone, we would expect the rate of Expt. 4 to be only one half that of Expt. 3. Reducing $[S_2O_8^{2-}]$ from 0.076 M in Expt. 3 to 0.019 M in Expt. 4 represents reducing $[S_2O_8^{2-}]$ in half, twice. Altogether, rate of reaction (Expt. 4) = $\frac{1}{2}\times\frac{1}{2}\times\frac{1}{2}$ rate (Expt. 3).

rate of reaction (Expt. 4) = $\frac{1}{8}\times 1.4\times10^{-5}$ mol L^{-1} s^{-1} = 1.8×10^{-6} mol L^{-1} s^{-1}

14-14. (a) rate = $k[HgCl_2]^m [C_2O_4^{2-}]^n$

$$\frac{\text{rate (2)}}{\text{rate (1)}} = \frac{k(0.105)^m (0.30)^n}{k(0.105)^m (0.15)^n} = \frac{\cancel{k(0.105)^m} (2)^n \cancel{(0.15)^n}}{\cancel{k(0.105)^m} \cancel{(0.15)^n}} = 2^n = \frac{7.1\times10^{-5}}{1.8\times10^{-5}} = 3.94 \simeq 4$$

n = 2; reaction is second order in $C_2O_4^{2-}$.

$$\frac{\text{rate (2)}}{\text{rate (3)}} = \frac{k(0.105)^m (0.30)^n}{k(0.052)^m (0.30)^n} = \frac{\cancel{k}(2)^m \cancel{(0.052)^m} \cancel{(0.30)^n}}{\cancel{k(0.052)^m} \cancel{(0.30)^n}} = 2^m = \frac{7.1\times10^{-5}}{3.5\times10^{-5}} = 2.02 \simeq 2$$

m = 1; reaction is first order in $HgCl_2$.

Overall reaction order = 2 + 1 = 3 (third order).

(b) Use representative data, such as for rate (1).

$$k = \frac{\text{rate (1)}}{[HgCl_2][C_2O_4^{2-}]^2} = \frac{1.8\times10^{-5}\ \text{mol L}^{-1}\ \text{min}^{-1}}{0.105\ \text{mol/L} \times (0.15\ \text{mol/L})^2} = 7.6\times10^{-3}\ \text{L}^2\ \text{mol}^{-2}\ \text{min}^{-1}$$

(c) Initial rate = $7.6\times10^{-3}\ \text{L}^2\ \text{mol}^{-2}\ \text{min}^{-1} \times 0.020\ \text{mol/L} \times (0.22\ \text{mol/L})^2 = 7.4\times10^{-6}\ \text{mol L}^{-1}\ \text{min}^{-1}$

14-15. (a) rate = $k(P_{H_2})^m(P_{NO})^n$. Use appropriate entries from the data listed.

$$\frac{\text{rate (2)}}{\text{rate (3)}} = \frac{k(400\ \text{mmHg})^m\ (300\ \text{mmHg})^n}{k(400\ \text{mmHg})^m\ (152\ \text{mmHg})^n} = \frac{\cancel{k(400\ \text{mmHg})^m}\ (2)^n\ \cancel{(150\ \text{mmHg})^n}}{\cancel{k(400\ \text{mmHg})^m}\ \cancel{(152\ \text{mmHg})^n}} = (2)^n = \frac{0.515\ \text{mmHg s}^{-1}}{0.125\ \text{mmHg s}^{-1}}$$

$= 4.12 \simeq 4;\ n = 2.$

The reaction is second order in NO.

$$\frac{\text{rate (1)}}{\text{rate (3)}} = \frac{k(289\ \text{mmHg})^m\ (400\ \text{mmHg})^n}{k(147\ \text{mmHg})^m\ (400\ \text{mmHg})^n} = \frac{\cancel{k}(2)^m\ \cancel{(144\ \text{mmHg})^m}\ \cancel{(400\ \text{mmHg})^n}}{\cancel{k(147\ \text{mmHg})^m}\ \cancel{(400\ \text{mmHg})^n}} = (2)^m = \frac{0.800\ \text{mmHg s}^{-1}}{0.395\ \text{mmHg s}^{-1}}$$

$= 2.02 \simeq 2;\ m = 1.$

The reaction is first order in H_2.

Overall reaction order = 2 + 1 = 3.

(b) rate = $k \times P_{H_2} \times (P_{NO})^2$

14-16. (a) rate = $k[OCl^-]^m\ [I^-]^n\ [OH^-]^o$

Order with respect to I^-: $\frac{\text{rate (2)}}{\text{rate (3)}} = \frac{k(0.0020)^m\ (0.0040)^n\ (1.00)^o}{k(0.0020)^m\ (0.0020)^n\ (1.00)^o} = \frac{5.0\times10^{-4}\ \text{mol L}^{-1}\ \text{s}^{-1}}{2.4\times10^{-4}\ \text{mol L}^{-1}\ \text{s}^{-1}}$

$2^n = 2;\ n = 1$; first order in I^-.

Order with respect to OH^-: $\frac{\text{rate (4)}}{\text{rate (5)}} = \frac{k(0.0020)^m\ (0.0020)^1\ (0.50)^o}{k(0.0020)^m\ (0.0020)^1\ (0.25)^o} = \frac{4.6\times10^{-4}\ \text{mol L}^{-1}\ \text{s}^{-1}}{9.4\times10^{-4}\ \text{mol L}^{-1}\ \text{s}^{-1}}$

$2^o = \frac{1}{2};\ o = -1$; order with respect to OH^- is -1.

Order with respect to OCl^-: $\frac{\text{rate (1)}}{\text{rate (3)}} = \frac{k(0.0040)^m\ (0.0020)^1\ (1.00)^{-1}}{k(0.0020)^m\ (0.0020)^1\ (1.00)^{-1}} = \frac{4.8\times10^{-4}\ \text{mol L}^{-1}\ \text{s}^{-1}}{2.4\times10^{-4}\ \text{mol L}^{-1}\ \text{s}^{-1}}$

$2^m = 2;\ m = 1$; first order in OCl^-.

(b) Overall reaction order: 1 - 1 + 1 = 1.

(c) rate = $k\frac{[OCl^-]\ [I^-]}{[OH^-]}$

Use the data for one of the reactions to determine k.

$$k = \frac{\text{rate}\ [OH^-]}{[OCl^-]\ [I^-]} = \frac{4.8\times10^{-4}\ \text{mol L}^{-1}\ \text{s}^{-1} \times 1.00\ \text{mol/L}}{0.0040\ \text{mol/L} \times 0.0020\ \text{mol/L}} = 60\ \text{s}^{-1}$$

14-17. (a) True. As more of the reactant is consumed, more of the products are formed, but at a rate that constantly decreases, since rate = k[A].

(b) False. The time for one half of substance A to react is a constant, regardless of the quantity of A remaining at the beginning of the half-life period.

(c) False. A plot of [A] versus time yields a straight line for a zero order reaction. For a first-order reaction it is log [A] that varies linearly with time.

(d) True. The rates are related as follows

rate of reaction = $-\frac{1}{2}$ rate of disappearance A = rate of formation of A

A disappears twice as fast as C is formed.

14-18. Use equation (14.13) with $[A]_t = 0.20\ [A]_o$. That is, if 80% of a reactant has decomposed then only 20% remains undecomposed, $[A]_o = 0.50$ M and $[A]_t = 0.20 \times 0.50$ M = 0.10 M.

$$\log \frac{[A]_t}{[A]_o} = \log \frac{0.10}{0.50} = -0.70 = \frac{-kt}{2.303} = \frac{-6.5\times10^{-4}\ s^{-1}\ t}{2.303}$$

$$t = \frac{2.303 \times 0.70}{6.5\times10^{-4}}\ s = 2.5\times10^{3}\ s$$

14-19. Use equation (14.13) to determine a value of k. Then determine $t_{\frac{1}{2}}$ from k. Note that $[A]_t = 0.01\ [A]_o$ (i.e., if 99% of A has dissociated only 1% remains).

$$k = \frac{-2.303 \log \frac{[A]_t}{[A]_o}}{t} = \frac{-2.303 \times \log \frac{0.01\ [A]_o}{[A]_o}}{185\ min} = \frac{-2.303 \times (-2.0)}{185\ min} = 0.025\ min^{-1}$$

$$t_{\frac{1}{2}} = 0.693/k = 0.693/0.0249\ min^{-1} = 28\ min.$$

14-20. (a) First determine $t_{\frac{1}{2}}$ from the value of k: $t_{\frac{1}{2}} = 0.693/k$.

$$t_{\frac{1}{2}} = 0.693/6.2\times10^{-4}\ s^{-1} = 1.1\times10^{3}\ s$$

Now notice that an integral number of half-life periods are involved in reducing 80.0 g N_2O_5 to 2.5 g.

$$80_g \xrightarrow{1} 40g \xrightarrow{2} 20_g \xrightarrow{3} 10g \xrightarrow{4} 5g \xrightarrow{5} 2.5g$$

The time required = $5 \times t_{\frac{1}{2}} = 5 \times 1.1 \times 10^{3} = 5.5\times10^{3}$ s

(b) The quantity of N_2O_5 decomposed is 80.0 - 2.5 = 77.5 g N_2O_5. Convert from g $N_2O_5 \longrightarrow$ mol $N_2O_5 \longrightarrow$ mol $O_2 \longrightarrow$ L $O_2(g)$.

$$\text{no. L } O_2(g) = 77.5\ g\ N_2O_5 \times \frac{1\ mol\ N_2O_5}{108\ g\ N_2O_5} \times \frac{0.5\ mol\ O_2}{1\ mol\ N_2O_5} \times \frac{22.4\ L\ O_2(g)}{1\ mol\ O_2} = 8.04\ L\ O_2(g)$$

14-21. (a) In Example 14-11 the half life is given as 80 min.

$$k = 0.693/t_{\frac{1}{2}} = 0.693/80\ min. = 8.7\times10^{-3}\ min^{-1}$$

Now use equation (14.16) with $(P_A)_t = 700$ mmHg and $(P_A)_o = 800$ mmHg.

$$\log \frac{700}{800} = \frac{-(8.7\times10^{-3}\ min^{-1})\ t}{2.303} = -0.058$$

$$t = \frac{2.303 \times 0.058}{8.7\times10^{-3}} = 15 \text{ min.}$$

(b) Denote the initial pressure of pure DTBP by P_i. Since 3 mol gas is produced for every mol DTBP consumed, when the partial pressure of DTBP drops by $P_i - P_{DTBP}$, the pressure of the product gases increases by $3(P_i - P_{DTBP})$. This means that at any time the total pressure can be expressed as $P_{tot} = P_{DTBP} + 3(P_i - P_{DTBP})$.

To determine the time at which P_{tot} = 2000 mmHg, first determine the partial pressure of DTBP. 2000 mmHg = P_{DTBP} + 3(800 - P_{DTBP}) mmHg; 2 P_{DTBP} = 2400 - 2000 = 400 mmHg; P_{DTBP} = 200 mmHg

For P_{DTBP} to drop from 800 mmHg to 200 mmHg, two half-life periods must elapse, $2\ t_{\frac{1}{2}}$ = 2 x 80 min = 160 min.

When the total gas pressure reaches 2100 mmHg: 2100 mmHg = P_{DTBP} + 3(800 - P_{DTBP})mmHg

2 P_{DTBP} = 2400 - 2100 = 300 mmHg; P_{DTBP} = 150 mmHg

$$\log\frac{150 \text{ mmHg}}{800 \text{ mmHg}} = \frac{-(0.693/80)\text{min}^{-1} \times t}{2.303} = -0.727 \qquad t = 193 \text{ min}$$

14-22. (a) Plot log $P_{(CH_3)_2O}$ vs. time to obtain a straight line

time, s	$P_{(CH_3)_2O}$, mmHg	log $P_{(CH_3)_2O}$
0	312	2.494
390	264	2.422
777	224	2.350
1195	187	2.272
3155	78.5	1.895

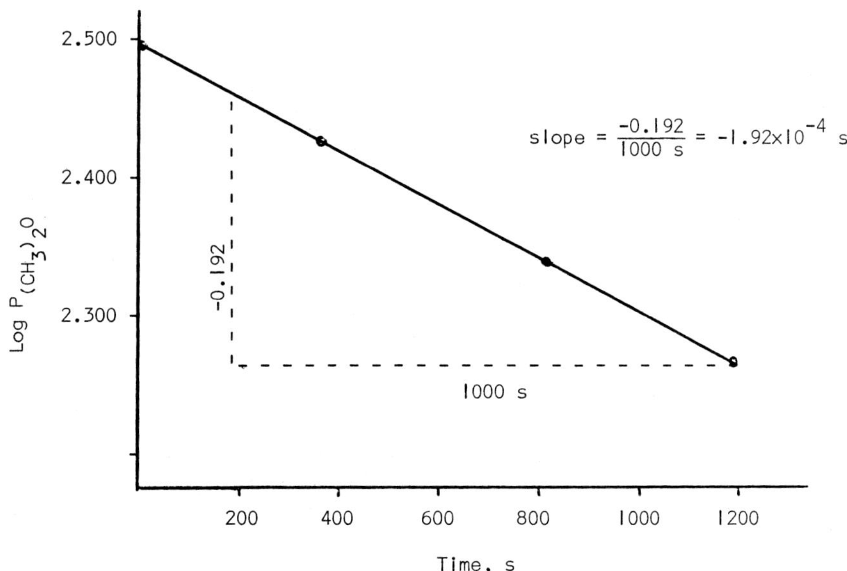

(b) For the straight line obtained in part (a), slope = -k/2.303

k = -2.303 x slope = -2.303 x (-1.92×10^{-4}) = $4.42\times10^{-4}\ s^{-1}$

(c) From the table we see that at 390 s, $P_{(CH_3)_2O}$ = 264 mmHg. $\Delta P_{(CH_3)_2O}$ = 312 - 264 = 48 mmHg.

$P_{products}$ = 3 x $\Delta P_{(CH_3)_2O}$ = 3 x 48 = 1.4×10^2 mmHg. P_{total} = $P_{(CH_3)_2O}$ + $P_{products}$ = 264 + 1.4×10^2 = 4.0×10^2

(d) The maximum gas pressure that could develop would be that corresponding to the complete dissociation of $(CH_3)_2O$, since three moles of products are produced for every mole of reactant. Maximum total pressure = 3 x initial $P_{(CH_3)_2O}$ = 3 x 312 = 936 mmHg.

(e) Here we must use equation (14.13) and solve for the partial pressure of $(CH_3)_2O$ at 1000 s. Use the value of k from part (b).

$$\log \frac{P}{P_o} = \frac{-kt}{2.303} \qquad \log \frac{P}{312} = \frac{-4.42\times10^{-4}\ s^{-1} \times 1000\ s}{2.303} = -0.192$$

$\log P = -0.192 + \log 312 = -0.192 + 2.494 = 2.302$ $\qquad P_{(CH_3)_2O} = 200$ mmHg

P_{total} = 200 mmHg + 3 x (312 - 200) mmHg = 536 mmHg

14-23. (a) Fraction of $C_6H_5N_2Cl$ decomposed = 44.3/58.3 = 0.76

Fraction of $C_6H_5N_2Cl$ remaining = 1.00 - 0.76 = 0.24

$[C_6H_5N_2Cl]_{remaining} = [C_6H_5N_2Cl]_{initially}$ x 0.24 = 0.071 x 0.24 = 0.017 M $C_6H_5N_2Cl$

(b) *Decomposition of $C_6H_5N_2Cl$*

I Time, min	*II* t, min	*III* $[C_6H_5N_2Cl]$ mol liter^{-1}	*IV* $\Delta[C_6H_5N_2Cl]$ mol liter^{-1}	*V* $\Delta[C_6H_5N_2Cl]/\Delta t$ mol liter^{-1} min^{-1}
0		0.071		
	3		-0.013	-4.3×10^{-3}
3		0.058		
	3		-0.011	-3.7×10^{-3}
6		0.047		
	3		-0.008	-2.7×10^{-3}
9		0.039		
	3		-0.007	-2.3×10^{-3}
12		0.032		
	3		-0.006	-2.0×10^{-3}
15		0.026		
	3		-0.005	-1.7×10^{-3}
18		0.021		
	3		-0.004	-1.3×10^{-3}
21		0.017		
	3		-0.003	-1.0×10^{-3}
24		0.014		
	3		-0.002	-0.7×10^{-3}
27		0.012		
	3		-0.002	-0.7×10^{-3}
30		0.010		

(c)

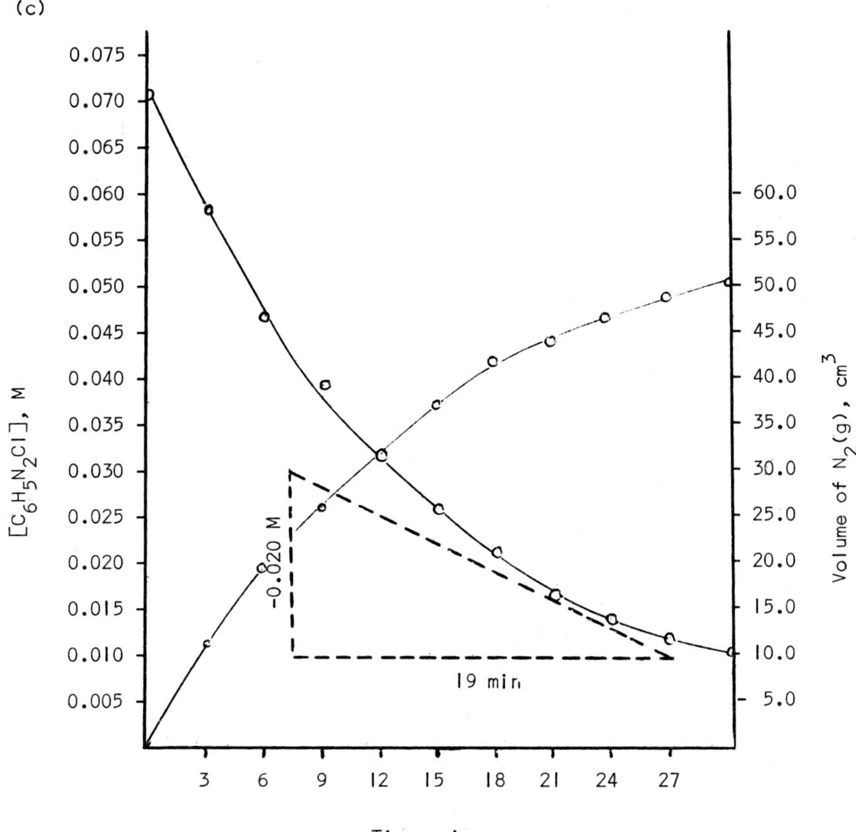

(d) Either draw a tangent line to the curve plotted in part (c) at t = 0, or use the first two data points from the table constructed in part (b).

$$\frac{\Delta[C_6H_5N_2Cl]}{\Delta t} = \frac{(0.058 - 0.071)M}{3 \text{ min}} = -4.3\times10^{-3} \text{ mol L}^{-1} \text{ min}^{-1}$$

initial reaction rate = $-\Delta[C_6H_5N_2Cl]/\Delta t = 4.3\times10^{-3}$ mol L^{-1} min^{-1}

(e) The rate of the reaction at t = 21 min is given by the negative of the slope of the tangent line (*t*).

$$\text{Slope} = \frac{\Delta[C_6H_5N_2Cl]}{\Delta t} = \frac{(0.010 - 0.030)\text{mol L}^{-1}}{(27.3 - 8.3) \text{ min}} = \frac{-0.020 \text{ mol L}^{-1}}{19.0 \text{ min}} = 1.1\times10^{-3} \text{ mol L}^{-1} \text{ min}^{-1}$$

Rate = 1.1×10^{-3} mol L^{-1} min^{-1}

(f) The rate law is: rate = $k[C_6H_5N_2Cl]$.

Based on initial data: $k = \dfrac{4.3\times10^{-3} \text{ mol L}^{-1} \text{ min}^{-1}}{0.071 \text{ mol L}^{-1}} = 6.1\times10^{-2} \text{ min}^{-1}$

Based on data at t = 21 min: $k = \dfrac{1.1\times10^{-3} \text{ mol L}^{-1} \text{ min}^{-1}}{0.017 \text{ mol L}^{-1}} = 6.5\times10^{-2} \text{ min}^{-1}$

(g) $t_{\frac{1}{2}} = \dfrac{0.693}{6.5\times10^{-2} \text{ min}^{-1}} = 11 \text{ min}$

From the graph of part (c), note that t = 0, $[C_6H_5N_2Cl]$ = 0.071. The concentration is half this value, that is, $[C_6H_5N_2Cl]$ = 0.0355, at about 10.5 min.

(h) Half of the substance is decomposed in 11 minutes. Half of what remains is decomposed in another 11 minutes. That is, after 22 minutes, three fourths of the sample is decomposed and one fourth remains.

(i)

t, min	$[C_6H_5N_2Cl]$	$\log[C_6H_5N_2Cl]$
0	0.071	- 1.15
3	0.058	- 1.24
6	0.047	- 1.33
9	0.039	- 1.41
12	0.032	- 1.49
15	0.026	- 1.59
18	0.021	- 1.68
21	0.017	- 1.77
24	0.014	- 1.85
27	0.012	- 1.92
30	0.010	- 2.00

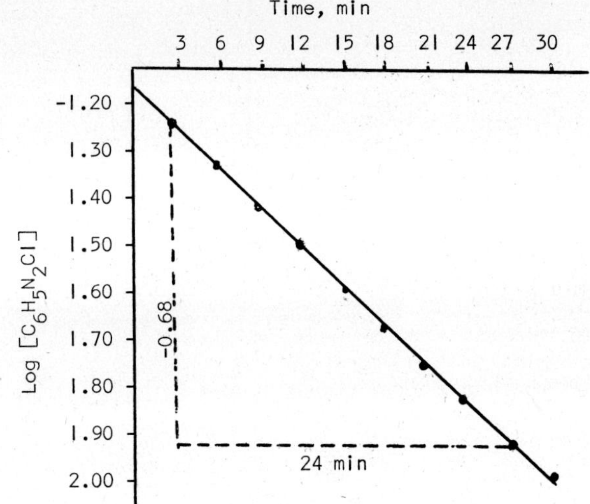

(j) Slope = -0.68/24 min = -0.028 min^{-1}

$$k = -2.303 \times \text{slope} = -2.303 \times (-0.028)min^{-1} = 6.4\times10^{-2}\ min^{-1}$$

14-24. (a) In the manner of Example 14-12(a) determine which of the three plots yields a straight line. Only the first-order plot does.

t, min	[A]	log [A]
0	0.80	-0.097
8	0.60	0.222
24	0.35	0.456
40	0.20	0.699

Time, min
8 16 24 32 40
-0.20
0.40
0.60
0.80
1.00
log [A]
-0.48
32 min

(b) Slope = $\frac{-0.48}{32\ \text{min}} = -1.5\times10^{-2}\ \text{min}^{-1}$

$k = -2.303 \times \text{slope} = -2.303 \times (-1.5\times10^{-2})\text{min}^{-1} = 3.5\times10^{-2}\ \text{min}^{-1}$

(c) First calculate [A] at t = 30 min by using equation (14.13).

$$\log\frac{[A]_t}{0.80} = \frac{-3.5\times10^{-2}\ \text{min}^{-1} \times 30\ \text{min}}{2.303} = -0.46 \qquad [A]_t/0.80 = 0.35;\quad [A]_t = 0.28\ \text{M}$$

Now use the rate equation to calculate the rate of disappearance of A when t = 30 min.

$$\text{rate} = k[A] = 3.5\times10^{-2}\ \text{min}^{-1} \times 0.28\ \text{M} = 9.8\times10^{-3}\ \text{mol L}^{-1}\ \text{min}^{-1}$$

The rate of formation of B is twice the rate of disappearance of A.

$$\text{rate} = \Delta[B]/\Delta t = 2.0\times10^{-2}\ \text{mol L}^{-1}\ \text{min}^{-1}$$

14-25. The reaction cannot be first order since $t_{\frac{1}{2}}$ is not a constant. If the reaction were zero order, we should expect $t_{\frac{1}{2}}$ to be smaller the smaller the value of $[A]_0$. That is, if a reaction proceeds at a constant rate (as does a zero order reaction), the smaller $[A]_0$ the more quickly it is consumed. This suggests that the reaction might be second order. The test for $t_{\frac{1}{2}}$ for a second order reaction is

$$t_{\frac{1}{2}} = \frac{1}{k[A]_0} \qquad \text{or} \qquad k = \frac{1}{t_{\frac{1}{2}}[A]_0}$$

Calculate three values of k:

$$k = \frac{1}{50\ \text{min} \times 1.00\ \text{M}} = 0.020\ \text{L mol}^{-1}\ \text{min}^{-1}$$

$$k = \frac{1}{25\ \text{min} \times 2.00\ \text{M}} = 0.020\ \text{L mol}^{-1}\ \text{min}^{-1}$$

$$k = \frac{1}{100\ \text{min} \times 0.50\ \text{M}} = 0.020\ \text{L mol}^{-1}\ \text{min}^{-1}$$

The reaction is second order, and

rate of reaction = $k[A]^2$ (where k = $0.020\ \text{L mol}^{-1}\ \text{min}^{-1}$).

14-26. (a) There are several ways to proceed: graph data, substitute into equations, etc. One simple method with which to begin is to calculate the rate based on each pair of data points. If the rate is *constant*, the reaction is zero order. If the rate is not constant then other means must be used to determine the reaction order.

$$\text{rate of reaction} = \frac{-(0.605 - 0.715)\text{M}}{22\ \text{s}} = 5.0\times10^{-3}\ \text{mol L}^{-1}\ \text{s}^{-1}$$

$$= \frac{-(0.345 - 0.605)\text{M}}{(74 - 22)\text{s}} = 5.0\times10^{-3}\ \text{mol L}^{-1}\ \text{s}^{-1}$$

$$= \frac{-(0.055 - 0.345)\text{M}}{(132 - 74)\text{s}} = 5.0\times10^{-3}\ \text{mol L}^{-1}\ \text{s}^{-1}$$

The reaction is zero order.

(b) Determine the time required for the reaction to go to completion. $t_{\frac{1}{2}}$ = one half time to completion (but only for a zero order reaction).

$$\frac{\Delta[A]}{\Delta t} = 5.0\times10^{-3}\ \text{mol L}^{-1}\ \text{s}^{-1}$$

$$\Delta t = \frac{\Delta[A]}{5.0\times10^{-3}\ \text{mol L}^{-1}\ \text{s}^{-1}} = \frac{0.715\ \text{mol L}^{-1}}{5.0\times10^{-3}\ \text{mol L}^{-1}\ \text{s}^{-1}} = 1.4\times10^{2}\ \text{s}$$

$$t_{\frac{1}{2}} = \tfrac{1}{2} \times 1.4\times10^{2}\ \text{s} = 7.0\times10^{1}\ \text{s}$$

Collision theory, activation energy

14-27. (a) In order for a collision between molecules to result in a chemical reaction, the molecules must be especially energetic and they must have the correct orientation. As a result, only a tiny fraction of all molecular collisions are effective in producing chemical reaction. The reaction rate is much smaller than the collision frequency.

(b) A critical factor in establishing the rate of a reaction is the fraction of all the molecules that are especially energetic, possessing energies in excess of the activation energy. This fraction, and hence the rate of the reaction itself, increases with temperature at a much faster rate than does the collision frequency.

(c) The effect of a catalyst is to lower the activation energy of a reaction by changing the mechanism of the reaction. If the activation energy is lowered, the fraction of the molecules that are energetic enough to react, and hence the rate of the reaction itself, increases profoundly, even though the temperature remains unchanged.

14-28. (a) Use expression (14.27): $\Delta H = E_a(\text{forward}) - E_a(\text{reverse})$. ΔH = 21 kJ/mol and E_a(forward = 84 kJ/mol.

$E_a(\text{reverse}) = E_a(\text{forward}) - \Delta H$ = 84 kJ/mol - 21 kJ/mol = 63 kJ/mol

(b)

E_a(forward) = 84 kJ/mol

A + B

C + D

ΔH = 21 kJ/mol

14-29. From the sketches below it can be seen that for the endothermic reaction, no matter what the value of H, the activation energy, E_a, must be greater. With exothermic reactions, however, there is no relationship between the numerical values of E_a and ΔH. For example, for the two cases shown, E_a is seen to be the same but ΔH is very much different.

Endothermic

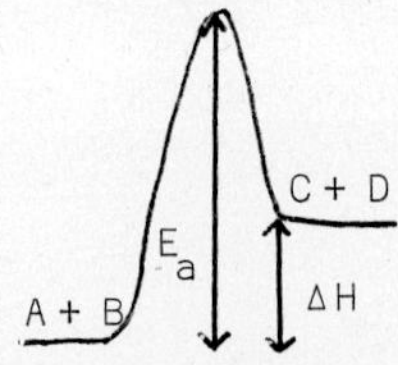

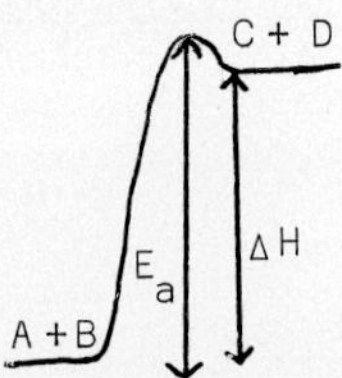

Exothermic

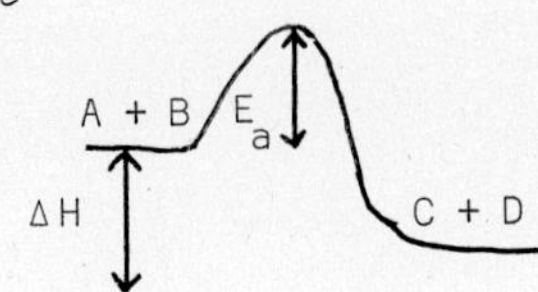

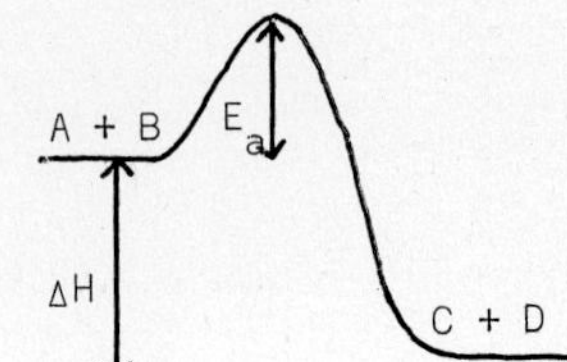

14-30. (a) The activation energy for the reaction of $H_2(g)$ and $O_2(g)$ is high. At room temperature the fraction of molecules having energies in excess of the activation energy is so small that a reaction cannot be sustained. The presence of a spark heats the mixture in one small portion. Enough molecules acquire energies in excess of the activation energy that a reaction is initiated.

(b) Once the reaction is initiated the exothermic heat of reaction is sufficient to raise large numbers of molecules to energies in excess of the activation energy and a rapid (explosive) reaction occurs throughout the mixture. Thus, the size of the initiating spark is immaterial.

14-31. Molecules acquire energies sufficient to undergo chemical reaction as a result of collisions with other molecules. Some of these other molecules may themselves be inert to reaction. Thus an inert gas can affect the rate of a reaction by moderating the collisions of the molecules that do react.

Effect of temperature on reaction rate

14-32. (a) Construct a graph of log k vs. 1/T.

t, °C	T, K	1/T, K^{-1}	k, L mol^{-1} s^{-1}	log K
3	276	3.62×10^{-3}	1.4×10^{-3}	-2.85
13	286	3.50×10^{-3}	2.9×10^{-3}	-2.54
24	297	3.37×10^{-3}	6.2×10^{-3}	-2.21
33	306	3.27×10^{-3}	1.2×10^{-2}	-1.92

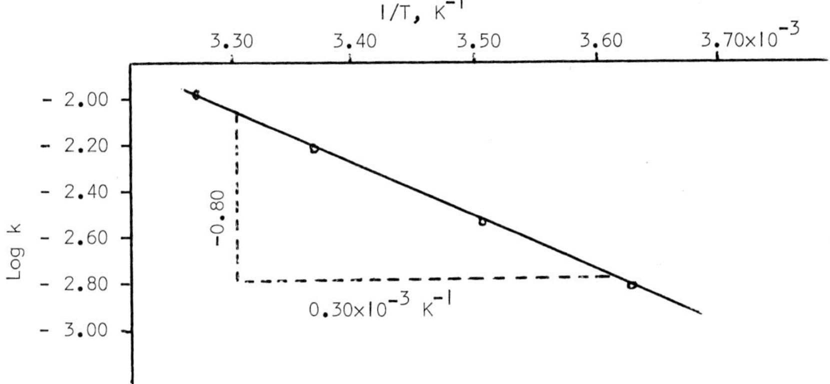

(b) Slope $= -E_a/2.303 \times R = -2.7\times10^3$ K

$E_a = 2.303 \times 8.314 \text{ J mol}^{-1} \text{ K}^{-1} \times 2.7 \times 10^3 \text{ K} = 5.2 \times 10^4 \text{ J/mol} = 52 \text{ kJ/mol}$

(c) Use the expression $\log \frac{k_2}{k_1} = \frac{E_a}{2.303\, R}\left(\frac{T_2 - T_1}{T_2T_1}\right)$ with $k_1 = 1.2\times10^{-2}$, $T_1 = 306$ K, $k_2 = ?$, and $T_2 = 313$ K.

From part (b) substitute $E_a = 5.2\times10^4$ J/mol

$$\log \frac{k_2}{1.2\times10^{-2}} = \frac{5.2\times10^4 \text{ J/mol}}{2.303 \times 8.314 \text{ J mol}^{-1} \text{ K}^{-1}} \left(\frac{313 - 306}{313 \times 306}\right) \text{K}^{-1}$$

$$\log k_2 - \log 1.2\times10^{-2} = 0.20$$

$$\log k_2 = 0.20 + \log 1.2\times10^{-2} = 0.20 - 1.92 = -1.72$$

$$k_2 = 1.9\times10^{-2} \text{ L mol}^{-1} \text{ s}^{-1}$$

(d) Proceed as in part (c) to determine k at 50°C. Then use the rate equation to calculate the initial rate.

$$\log \frac{k_2}{1.2\times10^{-2}} = \frac{5.2\times10^4 \text{ J/mol}}{2.303 \times 8.314 \text{ J mol}^{-1} \text{ K}^{-1}} \left(\frac{323 - 306}{323 \times 306}\right) \text{K}^{-1}$$

$\log k_2 - 0.47 - 1.92 = -1.45$ $\qquad k_2 = 3.5\times10^{-2} \text{ L mol}^{-1} \text{ s}^{-1}$

$$\text{initial rate} = k \times [S_2O_8{}^{2-}]_{init} \times [I^-]_{init} = 3.5\times10^{-2} \text{ L mol}^{-1} \text{ s}^{-1} \times 0.076 \text{ mol/L} \times 0.030 \text{ mol/L}$$

$$= 8.0\times10^{-5} \text{ mol L}^{-1} \text{ s}^{-1}$$

14-33. Use equation (14.29) to solve for one of the temperatures, say T_2 (the higher temperature).

$$\log \frac{1.0\times10^{-5}}{1.0\times10^{-10}} = \frac{111\times10^3 \text{ J/mol}}{2.303 \times 8.314 \text{ J mol}^{-1} \text{ K}^{-1}} \left(\frac{T_2 - 300}{300\ T_2}\right) \text{K}^{-1} = 5.00$$

$$1.11\times10^5\ T_2 - 3.33\times10^7 = 5.00 \times 2.303 \times 8.314 \times 300\ T_2$$

$$8.23\times10^4\ T_2 = 3.33\times10^7; \quad T_2 = 405 \text{ K}$$

14-34. If a reaction rate doubles for a temperature increase of 10°C, the value of k must also double. Assume two temperatures, such as 25°C and 35°C, and use equation (14.29)

(a) $$\log \frac{2\not{k}}{\not{k}} = \frac{E_a}{2.303 \times 8.314 \text{ J mol}^{-1} \text{ K}^{-1}} \times \left(\frac{308 - 298}{298 \times 308}\right) \text{K}^{-1} = 0.301$$

$E_a = 5.3\times10^4$ J/mol $= 53$ kJ/mol

(b) The rule of thumb would not apply well to the formation of HI from its elements. E_a for this reaction is shown in Figure 14-10 to be 171 kJ/mol.

14-35. Use data in Table 11-1 to estimate the temperature difference between the conditions of 1 atm and 2 atm vapor pressure of water. For P = 1 atm, T = 100°C; for P = 1489 mmHg (1.96 atm), T = 120°C. Using the rule of thumb that a reaction rate doubles for a 10°C temperature increase, for a 20°C increase the reaction rate should quadruple. The cooking rate in a pressure cooker is about four times as great as in ordinary boiling water.

14-36. First, determine the initial rate of decomposition for the 1.25 M solution. (Use the value of k listed for 0°C).

$$\text{initial rate} = 7.87\times10^{-7} \text{ s}^{-1} \times 1.25 \text{ M} = 9.84\times10^{-7} \text{ mol L}^{-1} \text{ s}^{-1}$$

Next, determine a value of k for which the initial rate of decomposition of an 0.15 M solution will be that just calculated.

$9.84\times10^{-7} \text{ mol L}^{-1} \text{ s}^{-1} = k \times 0.15 \text{ mol L}^{-1}$ $\qquad k = 6.6\times10^{-6} \text{ s}^{-1}$

Now, use equation (14.29) to find the appropriate temperature. Use t = 0°C as the second temperature and $E_a = 1.0\times10^5$ J/mol.

$$\log \frac{6.6\times10^{-6}}{7.87\times10^{-7}} = \frac{1.0\times10^5 \text{ J/mol}}{2.303 \times 8.314 \text{ J mol}^{-1} \text{ K}^{-1}} \times \left(\frac{T_2 - 273}{273\ T_2}\right) = 0.92$$

$1.0\times10^5\ T_2 - 2.73\times10^7 = 4.8\times10^3\ T_2$ $\qquad T_2 = 2.9\times10^2$ K

Catalysis

14-37. (a) A catalyst is a substance that speeds up a chemical reaction by entering into the reaction *in such a way that its composition remains unchanged.*

(b) The function of a catalyst is to *change the mechanism* of a chemical reaction to one requiring a lower activation energy.

14-38. An enzyme is a very specific catalyst, catalyzing one reaction and one alone. Also, the conditions of temperature and acid/base concentration under which the enzyme functions are very limited. A platinum catalyst, on the other hand, can catalyze a large number of different reactions and under various conditions of temperature and pressure.

14-39. The order of the reaction can be established by plotting the data in the form, [S] versus t, but it is also possible to do this simply by examining the data themselves. The concentration of S decreases by 0.10 mol/L for every 20 minutes. A reaction that proceeds at a *constant* rate is zero-order.

Reaction Mechanisms

14-40. An elementary process is a single step in the mechanism of an overall reaction. The molecularity of the elementary process indicates whether an activated molecule undergoes dissociation (unimolecular), whether reaction occurs as a result of a collision between two molecules (bimolecular), and so on. The order of an *elementary process* is related to its molecularity (i.e., unimolecular--first order; bimolecular--second order; etc.). However, the overall reaction generally proceeds by a mechanism involving more than one step and the order of the overall reaction is not related in any direct way to the molecularity of an individual elementary process.

14-41. The rate-determining step in a reaction mechanism is the slowest step and, as in any process that involves several steps, the slowest step acts as a bottleneck. The overall process can proceed no faster than the rate at which events occur in the bottleneck.

14-42. In order for a given molecule to dissociate in a unimolecular process, the molecule must first acquire energy in excess of the activation energy. This results from repeated collisions of the given molecule with other molecules. Thus, when dissociation does occur only a single molecule is involved, but molecular collisions must have preceded the dissociation.

14-43. Rate of formation of $N_2O_2 = k_1[NO]^2$ Rate of dissociation of $N_2O_2 = k_2[N_2O_2]$

Steady state condition: $k_1[NO]^2 = k_2[N_2O_2]$ $[N_2O_2] = \frac{k_1}{k_2}[NO]^2$

14-44. (a) The rate law, rate $= k[NO]^2\,[O_2]$, is consistent with the one-step mechanism,

$$2\ NO + O_2 \rightarrow 2\ NO_2.$$

(b) Despite the fact that the one-step mechanism of part (a) is consistent with the observed rate law, this mechanism is highly unlikely. The one-step mechanism is a thermolecular process, requiring the simultaneous collision of three molecules. Such collisions are much less probable than bimolecular collisions. Thus the mechanism of the reaction in question is more likely to involve a combination of uni- and bimolecular steps.

14-45. The steady-state condition in the mechanism has already been described in Exercise 43. The rate of formation of NO_2 is given by:

$$\text{Rate} = k_3[N_2O_2]\,[O_2] = \frac{k_1k_3}{k_2}\,[NO]^2[O_2] = k[NO]^2[O_2]$$

14-46. Consider the following two steps:

(fast) $$O_3 \underset{k_2}{\overset{k_1}{\rightleftharpoons}} O_2 + O \qquad \text{(a)}$$

(slow) $$O + O_3 \xrightarrow{k_3} 2\ O_2 \qquad \text{(b)}$$

Assume that the first step (a) reaches the steady-state condition.

rate formation O = rate disappearance O

Write expressions for these two rates based on the balanced equation (a) for the elementary process.

$$\text{rate formation O} = k_1[O_3] = k_2[O_2][O] = \text{rate disappearance O} \qquad \text{(c)}$$

Solve equation (c) for the steady-state concentration of the intermediate, that is, [O].

$$[O] = \frac{k_1[O_3]}{k_2[O_2]} \qquad \text{(d)}$$

Assume that the second step, (b), is the rate-determining step and write its rate equation.

$$\text{rate disappearance of } O_3 = k_3[O][O_3] \qquad \text{(e)}$$

Substitute the steady-state concentration of O from equation (d) into equation (e). Combine k_1, k_2, and k_3 into a single constant, k.

$$\text{rate disappearance } O_3 = \frac{k_1k_3[O_3][O_3]}{k_2[O_2]} = k\,\frac{[O_3]^2}{[O_2]}$$

Self-Test Questions

1. (*a*) Recall the relationship between the order of reaction and the function of [A] that yields a straight line. A plot of [A] vs. t is linear for a zero order reaction; log [A] vs $\cdot$ t for a first order reaction; and 1/[A] vs. t for a second order reaction.

2. (*b*) The half life of a reaction is not one half the time required for a reaction to go to completion [Item (a) is incorrect]. At the end of each 100 s interval, half of the reactant present at the beginning of the interval is consumed [Item (b) is correct]. If the same quantity were consumed in the second half-life period as in the first half-life period, the reaction would go to completion in two half-life periods. [Item (c) is incorrect for the same reason as item (a).] The reaction starts as soon as the reaction condition is established; there is no waiting time [item (d) is incorrect].

3. (*d*) The rate equation, rate = k[A]•[B], signifies a second order reaction. Item (a) is incorrect because the unit s^{-1} corresponds to k for a first order reaction. Item (b) is also incorrect because $t_{\frac{1}{2}}$ is a constant only for a first order reaction. Item (c) is incorrect because the value of k for a reaction depends on the particular reaction and the temperature, but not on the initial concentrations of reactants chosen. Item (d) id correct based on the stoichiometry of the reaction--two molecules of A are consumed for each molecule of C produced.

4. (*d*) Here we must recall the method of initial rates. For the second-order decomposition, rate = $k[A]^2$. The initial rate = $k[A]_0^2$. Its value does depend on the concentration chosen for $[A]_0$. [Item (a) is incorrect.] Now compare the initial rates at the following values: $[A]_0$ = 0.50 M, $[A]_0$ = 1.00 M, $[A]_0$ = 0.10 M, and $[A]_0$ = 0.25 M.

   Initial rates: (1) $k(0.50)^2 = 0.25\cdot k$; (2) $k(1.00)^2 = 1.00\cdot k$; (3) $k(0.10)^2 = 0.01\cdot k$; (4) $k(0.25)^2 = 0.0625\cdot k$

   Rate (1) is only one quarter of Rate (2). Item (b) is incorrect. Rate (1) is 25 times greater than Rate (3). Item (c) is incorrect. Rate (1) is four times Rate (4). Item (d) is correct.

5. (*b*) Although the average kinetic energy of gas molecules [item (d)] and collision frequency [item (a)] increase with temperature, they do not increase rapidly enough to account for the very sharp increase in reaction rate with temperature. Item (c) must be incorrect, because an increase in the activation energy for a reaction would cause the reaction to slow down. It is the fraction of the molecules that possess energies in excess of the activation energy that increases so rapidly with temperature.

6. (*c*) The activation energy of a reaction is always a positive quantity.[item (a) must be incorrect]. Because the reaction is endothermic ($\Delta H > 0$), the activation energy must equal or exceed ΔH. Items (b) and (d) are incorrect and (c) is correct.

7. (*d*) A catalyst can affect neither the average kinetic energies nor the collision frequencies of molecules. Items (a) and (b) are incorrect. A catalyst does *not* increase the activation energy of a reaction; this would slow down a reaction. A catalyst permits a reaction mechanism of *lower* activation energy; if E_a is lowered, the fraction of molecules with energies in excess of E_a increases. Item (d) is the correct response.

8. (*b*) The relationship between k and $t_{\frac{1}{2}}$ in item (a) is for a first-order reaction. If the reaction mechanism involves a *single step*, and if this step is bimolecular, the overall reaction order is second. The rate equation becomes k[A]·[B]. Item (b) is correct. The rate of appearance of C is twice the rate of disappearance of A [item (c) is incorrect]. The equation given in item (d) is for a first-order reaction, not second order.

9. The decrease in [A] with time is not constant (the reaction slows down with time); the reaction cannot be zero order. The concentration of A decreases to one half of its initial value in 500 s. The second half life is 1000 s. The reaction cannot be first order because the half life is not constant. By a process of elimination we conclude that the reaction is second order. (This conclusion could be verified by plotting 1/[A] vs. time.)

10. Since the reaction is first order, $t_{\frac{1}{2}}$ is constant. Based on the first and third points (where [A]^2 decreases from 0.88 M to 0.44 M) $t_{\frac{1}{2}} = 100$ s. Based on the second and fourth points (where [A] decreases from 0.62 M to 0.31 M) $t_{\frac{1}{2}}$ is also 100 s.

$$k = 0.693/t_{\frac{1}{2}} = 0.693/100\ \text{s} = 6.93\times10^{-3}\ \text{s}^{-1}$$

At 100 s, [A] = 0.44 M

The instantaneous rate of reaction at 100 s is

$$\text{rate of reaction} = k[A] = 6.93\times10^{-3}\ \text{s}^{-1} \times 0.44\ \text{M} = 3.0\times10^{-3}\ \text{mol L}^{-1}\ \text{s}^{-1}$$

11. (a) The reaction is first order, and the quantity of A decreases from 1.60 g to 0.40 g in 20.0 min. This decrease corresponds to two half-life periods: 1.60 g → 0.80 g → 0.40 g. The half-life of the reaction, $t_{\frac{1}{2}} = 10.0$ min.

(b) Use the value of $t_{\frac{1}{2}}$ to calculate k for the reaction.

$$k = 0.693/t_{\frac{1}{2}} = 0.693/10.0\ \text{min} = 0.0693\ \text{min}^{-1}$$

Now solve for the mass of A (call it A_t) at t = 33.2 min. (Note that we can deal with mass directly; we do not need to employ molar concentrations.)

$$\log\frac{A_t}{A_0} = \log\frac{A_t}{1.60} = \frac{-(0.0693\ \text{min}^{-1}) \times 33.2\ \text{min}}{2.303} \qquad \log A_t - \log 1.60 = -0.999$$

$$\log A_t = -0.999 + \log 1.60 = -0.999 + 0.204 = -0.795 \qquad A_t = 0.160\ \text{g}$$

12. Use the heat of formation data to establish $\Delta\bar{H}$ for the reaction.

$CH_3CHO(g) \rightarrow CH_4(g) + CO(g)$; $\Delta\bar{H} = ?$

$$\Delta\bar{H} = \Delta\bar{H}^\circ_f[CH_4(g)] + \Delta\bar{H}^\circ_f[CO(g)] - \Delta\bar{H}^\circ_f[CH_3CHO(g)]$$

$$= -74.9\ \text{kJ/mol} - 110.5\ \text{kJ/mol} - (-166)\ \text{kJ/mol}$$

$$= -19\ \text{kJ/mol}$$

continued on next page.

Uncatalyzed reaction:

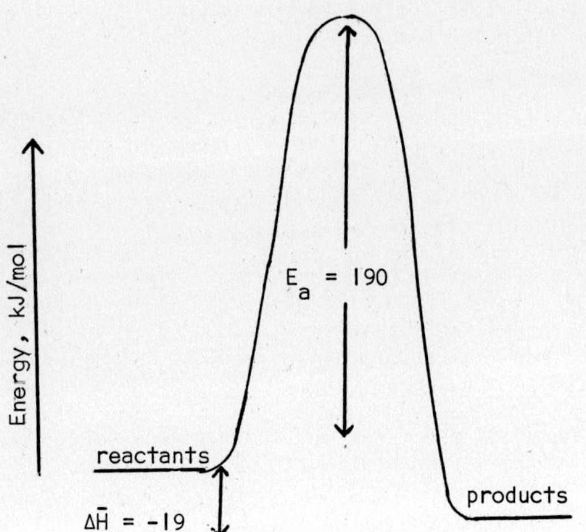

Catalyzed reaction:

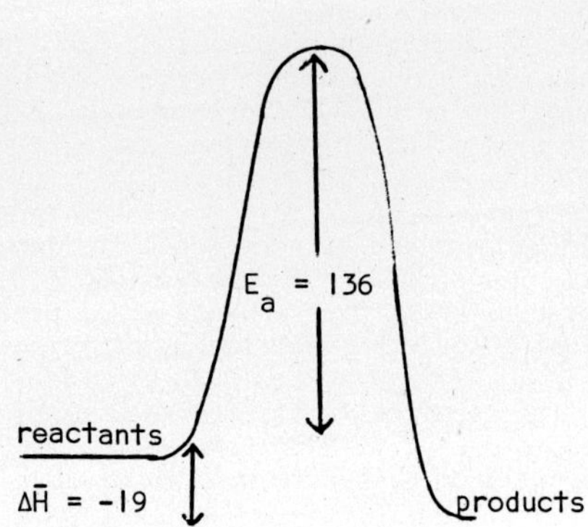

Chapter 15

Principles of Chemical Equilibrium

Review Problems

15-1. In the following expressions no concentration terms for solids appear.

(a) $K_c = \dfrac{[COCl_2(g)]}{[CO(g)][Cl_2(g)]}$ (b) $K_c = \dfrac{[NO_2(g)]^2}{[NO(g)]^2[O_2(g)]}$

(e) $K_c = [CO_2(g)]$ (d) $K_c = \dfrac{[CH_4(g)][H_2S(g)]^2}{[CS_2(g)][H_2(g)]^4}$

(e) $K_c = [CO_2(g)][H_2O(g)]$

15-2. (a) $K_c = \dfrac{[CO][H_2O]}{[CO_2][H_2]} = \dfrac{1}{23.2} = 4.31\times10^{-2}$

(b) $K_c = \dfrac{[SO_3]^2}{[SO_2]^2[O_2]} = (56)^2 = 3.1\times10^3$

(c) $K_c = \dfrac{[H_2][S_2]^{\frac{1}{2}}}{[H_2S]} = (2.3\times10^{-4})^{\frac{1}{2}} = 1.5\times10^{-2}$

(d) $K_c = \dfrac{[NO_2]}{[NO][O_2]^{\frac{1}{2}}} = \left(\dfrac{1}{1.8\times10^{-6}}\right)^{\frac{1}{2}} = 7.5\times10^2$

15-3. Reverse the first equation and divide the coefficients by two.

$N_2(g) + \frac{1}{2}O_2(g) \rightleftharpoons NO(g) \qquad K_c = \left(\dfrac{1}{2.4\times10^{30}}\right)^{\frac{1}{2}} = 6.5\times10^{-16}$

Add the second equation.

$NO(g) + \frac{1}{2}Br_2(g) \rightleftharpoons NOBr(g) \qquad K_c = 1.4$

Net: $\frac{1}{2}N_2(g) + \frac{1}{2}O_2(g) + \frac{1}{2}Br_2(g) \rightleftharpoons NOBr(g) \qquad K_c = 1.4 \times 6.5 \times 10^{-16} = 9.1\times10^{-16}$

15-4. $CO(g) + H_2O(g) \rightleftharpoons CO_2(g) + H_2(g)$; $K_c = 23.2$; $K_p = 23.2(0.0821 \times 600)^0 = 23.2$

$SO_2(g) + \frac{1}{2}O_2(g) \rightleftharpoons SO_3(g)$; $K_c = 56$; $K_p = 56(0.0821 \times 900)^{-\frac{1}{2}} = 6.5$

$2H_2S(g) \rightleftharpoons 2H_2(g) + S_2(g)$; $K_c = 2.3 \times 10^{-4}$; $K_p = 2.3 \times 10^{-4}(0.0821 \times 1405) = 2.7 \times 10^{-2}$

$2NO_2(g) \rightleftharpoons 2NO(g) + O_2(g)$; $K_c = 1.8 \times 10^{-6}$; $K_p = 1.8 \times 10^{-6}(0.0821 \times 457) = 6.8 \times 10^{-5}$

15-5. For the reaction stated,

$K_c = \dfrac{[C]^2}{[A][B]}$

Substitute the equilibrium concentrations and solve for K_c.

$$K_c = \frac{(0.36)^2}{(0.47)(0.55)} = 0.50$$

15-6. The information given can be conveniently set up as follows:

	2A +	B $\rightleftharpoons$	C
initial amounts:	1.18 mol	0.78 mol	-
changes:	-0.26 mol	-0.13 mol	+0.13 mol
equilibrium amounts:	0.92 mol	0.65 mol	0.13 mol
equilibrium concentrations	$\frac{0.92}{1.80}$ M	$\frac{0.65}{1.80}$ M	$\frac{0.13}{1.80}$ M

$$K_c = \frac{[C]}{[A]^2[B]} = \frac{(0.13/1.80)}{\left(\frac{0.92}{1.80}\right)^2\left(\frac{0.65}{1.80}\right)} = \frac{(0.13) \times (1.80)^2}{(0.92)^2 \times (0.65)} = 0.77$$

15-7. For the reaction, $CO(g) + H_2O(g) \rightleftharpoons CO_2(g) + H_2(g)$, $K_c = 1.00$ at 1100 K

$$K_c = \frac{[CO_2][H_2]}{[CO][H_2O]} = 1.00$$

From the expression for K_c it can be seen that $[CO_2][H_2] = [CO][H_2O] \times 1.00 = [CO][H_2O]$

Of the four statements about the equilibrium condition given in the text, only statement (b) is always correct. It is not necessarily true that any concentration terms must be equal (as suggested by statements a and c), nor that the numerator or denominator must themselves have a value of 1.00 (suggested by statement d).

15-8. Assume that a condition of equilibrium exists, so that

$$K_c = \frac{[SO_3]^2}{[SO_2]^2[O_2]} = 55.2$$

(a) If the numbers of moles of SO_2 and SO_3 are equal so too are $[SO_2]$ and $[SO_3]$ since cell gases are present in the same volume.

$$\frac{1}{[O_2]} = 55.2 \quad \text{and} \quad [O_2] = 1/55.2 = 0.0181 \text{ M}$$

The volume of the reaction vessel is 11.5 L

$$\text{no. mol } O_2 = 11.5 \text{ L} \times \frac{0.0181 \text{ mol } O_2}{1 \text{ L}} = 0.208 \text{ mol } O_2$$

(b) If no. mol SO_3 = 2 x no. mol SO_2 then $[SO_3] = 2 \times [SO_2]$

$$K_c = \frac{[SO_3]^2}{[SO_2]^3[O_2]} = \frac{\{2 \times [SO_2]\}^2}{[SO_2]^2[O_2]} = \frac{2^2}{[O_2]} = 55.2 \qquad [O_2] = \frac{4}{55.2} = 0.0725 \text{ M}$$

$$\text{no. mol } O_2 = 11.5 \text{ L} \times \frac{0.0725 \text{ mol } O_2}{1 \text{ L}} = 0.834 \text{ mol } O_2$$

15-9. (a) $K_c = \dfrac{[C]^2}{[A]^3[B]}$

(b) Substitute the data given into the K_c expression and compare the ratio of concentration terms to the numerical value of K_c.

$$K_c = \frac{\{1.55/1.15\}^2}{\{1.55/1.15\}^3\{1.55/1.15\}} = \frac{(1.15)^2}{(1.55)^2} = 0.550 \neq 8.8$$

The ratio of terms based on the concentrations of substances introduced into the flask is considerably smaller than K_c(8.8). The mixture is not at equilibrium.

(c) To establish equilibrium in the mixture described in (b) $[SO_3]$ must increase and $[SO_2]$ and $[O_2]$, decrease. A net reaction occurs to the right.

15-10. (a) The fact that I_2 is 5% dissociated into I atoms at 1200°C means that from an original 1.00 mol I_2 the equilibrium amounts become 0.95 mol I_2 and 0.10 mol I. Since the reaction volume is 1.00 L, the equilibrium concentrations are $[I_2] = 0.95$ and $[I] = 0.10$.

$$K_c = \frac{[I]^2}{[I_2]} = \frac{(0.10)^2}{0.95} = 1.1\times10^{-2}$$

(b) $K_p = K_c(RT)^{\Delta n} = 1.1\times10^{-2}(0.0821 \times 1473)^1 = 1.3$

15-11. Derive the equilibrium concentrations with the set up below.

	$H_2(g)$	+ $I_2(g)$	$2HI(g)$
Initial amounts	0.100 mol	0.100 mol	-
Changes:	$-x$ mol	$-x$ mol	$+2x$ mol
Equilibrium amounts:	$(0.100-x)$ mol	$(0.100-x)$ mol	$2x$ mol
Equilibrium concentrations:	$\left(\frac{0.100-x}{1.50}\right)$M	$\left(\frac{0.100-x}{1.50}\right)$M	$\left(\frac{x}{1.50}\right)$M

Solve for x by substituting into the equilibrium constant expression.

$$K_c = \frac{[HI]^2}{[H_2][I_2]} = \frac{(2x/1.50)^2}{\left(\frac{0.100-x}{1.50}\right)\left(\frac{0.100-x}{1.50}\right)} = \frac{4x^2}{(0.100-x)^2} = 50.2$$

Take the square root of both sides of the above equation:

$$\frac{2x}{0.100-x} = \sqrt{50.2} = 7.09 \qquad 2x = 0.709 - 7.09x$$

$9.09x = 0.709 \qquad x = 0.0780$

Equilibrium concentrations:

$$[HI] = \frac{2x}{1.50} = \frac{2 \times 0.0780}{1.50} = 0.104\text{ M} \qquad [H_2] = [I_2] = \frac{0.100 - x}{1.50} = \frac{0.100 - 0.078}{1.50} = 1.5\times10^{-2}\text{ M}$$

15-12. This problem is similar to Review Problem 11, except that the starting conditions involve the *product* and one reactant instead of two reactants.

	$H_2(g)$ +	$I_2(g) \rightleftharpoons$	$2HI(g)$
Initial amounts:	0.100 mol	-	0.100 mol
Changes:	$+x$ mol	$+x$ mol	$-2x$ mol
Equilibrium amounts:	$(0.100 + x)$ mol	x mol	$(0.100 - 2x)$ mol
Equilibrium concentrations:	$\left(\frac{0.100 + x}{1.50}\right)$ M	$\left(\frac{x}{1.50}\right)$M	$\left(\frac{0.100 - 2x}{1.50}\right)$M

Substitute into the equilibrium constant expression, K_c.

$$K_c = \frac{[HI]^2}{[H_2][I_2]} = \frac{(0.100 - 2x/1.50)^2}{(0.100 + x/1.50)(x/1.50)} = \frac{(0.100 - 2x)^2}{(0.100 + x)x} = 50.2$$

Solve for x:

As a simplifying assumption, try $x << 0.100$.

$$\frac{(0.100)^2}{(0.100)x} = 50.2 \qquad x = 0.100/50.2 = 2.0\times10^{-3}$$

The value of x (2.0×10^{-3}) is only 2% of 0.100 and the assumption is a reasonably good one.

no. mol I_2 present at equilibrium = x = 2.0×10^{-3} mol I_2

(A more exact answer based on solving a quadratic equation is $x = 1.8\times10^{-3}$.)

15-13. This problem requires use of the van't Hoff equation (15.26). The necessary value of E_a is obtained from Figure 15-6 and a known value of K_p can be selected from Table 15-4 (such as, $K_p = 9.1\times10^2$ at 800 K).

$$\log\frac{K_2}{K_1} = \frac{\Delta \bar{H}°(T_2 - T_1)}{2.303\ RT_2T_1} = \log\frac{9.1\times10^2}{K_1} = \frac{-1.8\times10^5(800 - 298)}{2.303 \times 8.314 \times 800 \times 298}$$

$$\log 9.1\times10^2 - \log K_1 = -20$$

$$\log K_1 = 2.96 + 20 = 23 \qquad K_1 \simeq 1\times10^{23}$$

15-14. The effect of pressure on the equilibrium condition is assessed by whether the number of moles of gaseous species represented by the balanced equation changes in the course of the reaction.

(a) One mole of gaseous reactant (and one mole of a solid reactant) yields two moles of gaseous product. Increased external pressure forms the reaction which leads to a decrease in number of moles of gases. The equilibrium condition shifts to the left.

(b) Equal numbers of moles of gas appear on each side of the equation. Pressure has no effect on this equilibrium.

(c) In this case a decrease in number of moles of gas occurs from left (5 mol gas) to right (4 mol gas). Increased pressure displaces equilibrium to the right.

15-15. (a) If additional $O_2(g)$ is introduced into the equilibrium mixture at constant volume, the equilibrium condition shifts in the direction in which some of the excess $O_2(g)$ is consumed--to the right. When equilibrium is re-established the equilibrium amount of HCl(g) will have been reduced.

(b) Removal of $Cl_2(g)$ has the same effect as described in (a), displacement of equilibrium to the right. The equilibrium amount of HCl(g) will be less than the amount initially present.

(c) An increase in volume of the reaction mixture has the same effect as reducing the external pressure. The mixture produces the maximum amount of gaseous substances possible. The equilibrium condition shifts to the left and the equilibrium amount of HCl increases.

(d) Although a catalyst speeds up the attainment of equilibrium, it has no effect of equilibrium amounts. The equilibrium condition is unaffected and the amount of HCl is unchanged.

15-16. In order that the percent dissociation increase with temperature, the equilibrium constant must increase with temperature. This, in turn, requires that $\Delta\bar{H}^\circ_{rxn} > 0$. That is, increased temperature favors the endothermic reaction. Of those listed, the reactions with endothermic heats of reaction are (b) and (d).

Exercises

15-1. Convert each verbal description to a chemical equation and then write the K_c expression.

(a) $3O_2(g) + 4NH_3(g) \rightleftharpoons 2N_2(g) + 6H_2O(g)$

$$K_c = \frac{[N_2]^2[H_2O]^2}{[O_2]^3[NH_3]^4}$$

(b) $7H_2(g) + 2NO_2(g) \rightleftharpoons 2NH_3(g) + 4H_2O(g)$

$$K_c = \frac{[NH_3]^2[H_2O]^4}{[H_2]^7[NO_2]^2}$$

(c) $(CH_3)_2CO(\ell) \rightleftharpoons (CH_3)_2CO(g)$ $\quad K_c = [(CH_3)_2CO(g)]$

(d) $3Cl_2(g) + CS_2(\ell) \rightleftharpoons CCl_4(\ell) + S_2Cl_2(\ell)$ $\quad K_c = 1/[Cl_2(g)]^3$

(e) $N_2(g) + Na_2CO_3(s) + 4C(s) \rightleftharpoons 2NaCN(s) + 3CO(g)$

$$K_c = \frac{[CO(g)]^3}{[N_2(g)]}$$

15-2. (a) $\frac{1}{2}N_2(g) + \frac{1}{2}O_2(g) \rightleftharpoons NO(g)$; $\quad K_c = \dfrac{[NO]}{[N_2]^{\frac{1}{2}}[O_2]^{\frac{1}{2}}}$

(b) $\frac{1}{2}H_2(g) + \frac{1}{2}Cl_2(g) \rightleftharpoons HCl(g)$; $\quad K_c = \dfrac{[HCl]}{[H_2]^{\frac{1}{2}}[Cl_2]^{\frac{1}{2}}}$

(c) $\frac{1}{2}N_2(g) + \frac{3}{2}H_2(g) \rightleftharpoons NH_3(g)$; $\quad K_c = \dfrac{[NH_3]}{[N_2]^{\frac{1}{2}}[H_2]^{3/2}}$

(d) $\frac{1}{2}Cl_2(g) + \frac{3}{2}F_2(g) \rightleftharpoons ClF_3(g)$; $\quad K_c = \dfrac{[ClF_3]}{[Cl_2]^{1/2}[F_2]^{3/2}}$

(e) $\frac{1}{2}N_2(g) + \frac{1}{2}O_2(g) + \frac{1}{2}Cl_2(g) \rightleftharpoons NOCl(g)$; $\quad K_c = \dfrac{[NOCl]}{[N_2]^{\frac{1}{2}}[O_2]^{\frac{1}{2}}[Cl_2]^{\frac{1}{2}}}$

15-3. The relationship between K_p and K_c is $K_p = K_c(RT)^{\Delta n}$. This leads to $K_c = \frac{K_p}{(RT)^{\Delta n}} = K_p(RT)^{-\Delta n}$

(a) $SO_2Cl_2(g) \rightleftharpoons SO_2(g) + Cl_2(g)$; $K_c = 2.9\times10^{-2}(0.0821 \times 303)^{-1} = 1.2\times10^{-3}$

(b) $2NO(g) + O_2(g) \rightleftharpoons 2NO_2(g)$; $K_c = 1.48\times10^{4}(0.0821 \times 457)^{1} = 5.55\times10^{5}$

(c) $Sb_2S_3(s) + 3H_2(g) \rightleftharpoons 2Sb(s) + 3H_2S(g)$; $K_c = K_p = 0.429$

15-4. The vaporization of water is represented as

$H_2O(\ell) \rightleftharpoons H_2O(g)$ $\quad K_p = P_{H_2O(g)} = 23.8 \text{ mmHg} = \frac{1 \text{ atm}}{760 \text{ mmHg}} = 0.0313 \text{ atm}.$

Since $K_p = K_c(RT)^{\Delta n}$, then $K_c = \frac{K_p}{(RT)^{\Delta n}} = K_p(RT)^{-\Delta n}$

For the vaporization of water, $\Delta n = 1$.

$K_c = 0.0313 \times (0.0821 \times 298)^{-1} = 1.28\times10^{-3}$

15-5. The following combination of equilibrium expressions yields the desired net equation. K_c values are combined in the appropriate manner.

$2N_2O(g) \rightleftharpoons 2N_2(g) + O_2(g)$ $\quad K_c = (1/3.4\times10^{-18})^2$

$4NO_2(g) \rightleftharpoons 2N_2O_4(g)$ $\quad K_c = (1/4.6\times10^{-3})^2$

$2N_2(g) + 4O_2(g) \rightleftharpoons 4NO_2(g)$ $\quad K_c = (4.1\times10^{-9})^4$

$2N_2O(g) + 3O_2(g) \rightleftharpoons 2N_2O_4(g)$ $\quad K_c = \frac{(4.1\times10^{-9})^4}{(3.4\times10^{-12})^2(4.6\times10^{-3})^2} = 1.2\times10^{6}$

15-6. This exercise, like the preceding one, requires that three equilibrium expressions be combined to yield a net equation and the corresponding value of K_p. An additional step, however, is to convert K_c to K_p values before then combining of equations. Expression (15.15) is used for this purpose.

2x(b): $2CO_2(g) + 2H_2(g) \rightleftharpoons 2CO(g) + 2H_2O(g)$ $\quad K_p = (1.4)^2 = 2.0$

2x(c): $2C(\text{graphite}) + O_2(g) \rightleftharpoons 2CO(g)$ $\quad K_p = (1\times10^{8})^2(0.0821 \times 1200)^{1} = 1\times10^{18}$

-2x(a): $4CO(g) \rightleftharpoons 2CO_2(g) + 2C(\text{graphite})$ $\quad K_p = \frac{1}{0.64}^2(0.0821 \times 1200)^{-2} = 2.5\times10^{-4}$

---

net: $2H_2(g) + O_2(g) \rightleftharpoons 2H_2O(g)$ $\quad K_p = 2.0 \times 1 \times 10^{18} \times 2.5 \times 10^{-4} = 5 \times 10^{14}$

Experimental determination of equilibrium constants

15-7. no. mol Cl_2, at equilibrium $= 0.25 \text{ g } Cl_2 \times \frac{1 \text{ mol } Cl_2}{70.9 \text{ g } Cl_2} = 3.5\times10^{-3} \text{ mol } Cl_2$

no. mol PCl_3, at equilibrium = no. mol Cl_2, at equilibrium $= 3.5\times10^{-3}$ mol PCl_3

$$\text{no. mol } PCl_5\text{, originally} = 1.00 \text{ g } PCl_5 \times \frac{1 \text{ mol } PCl_5}{208 \text{ g } PCl_5} = 4.81\times10^{-3} \text{ mol } PCl_5$$

$$\text{no. mol } PCl_5\text{, at equilibrium} = 4.81\times10^{-3} - 3.5\times10^{-3} = 1.3\times10^{-3} \text{ mol } PCl_5$$

$$K_c = \frac{[PCl_3][Cl_2]}{[PCl_5]} = \frac{(3.5\times10^{-3}/0.250)(3.5\times10^{-3}/0.250)}{(1.3\times10^{-3}/0.250)} = 3.8\times10^{-2}$$

15-8. First convert the quantities of initial reactants to moles.

$$\text{no. mol } H_2 = 1.00 \text{ g } H_2 \times \frac{1 \text{ mol } H_2}{2.016 \text{ g } H_2} = 0.496 \text{ mol } H_2$$

$$\text{no. mol } H_2S = 1.06 \text{ g } H_2S \times \frac{1 \text{ mol } H_2S}{34.08 \text{ g } H_2S} = 0.0311 \text{ mol } H_2S$$

Now tabulate the known information in the following manner:

	$2H_2(g)$	+ $S_2(g)$ $\rightleftharpoons$	$2H_2S(g)$
Initial amounts:	0.496 mol	-	0.0311 mol
Changes:	$+1.6\times10^{-5}$ mol	8.00×10^{-6} mol	-8.00×10^{-6} mol
Equilibrium amounts:	0.496 mol	8.00×10^{-6} mol	0.0311 mol
Equilibrium partial pressure (nRT/V):	$\frac{0.496 \text{ RT}}{V}$	$\frac{8.00\times10^{-6}\text{RT}}{V}$	$\frac{0.0311\text{RT}}{V}$

$$K_p = \frac{(P_{H_2S})^2}{(P_{H_2})^2(P_{S_2})} = \frac{(0.0311 \text{ RT/V})^2}{(0.496 \text{ RT/V})^2(8.00\times10^{-6} \text{ RT/V})} = \frac{V \times (0.0311)^2}{(RT) \times (0.496)^2 \times 8.00\times10^{-6}}$$

$$K_p = \frac{(0.500) \times (0.0311)^2}{(0.0821 \times 1670) \times (0.496)^2 \times 8.00\times10^{-6}} = 1.79$$

15-9. From the titration data we can establish the no. mol CH_3CO_2H remaining unreacted at equilibrium.

$$\text{no. mol } CH_3CO_2H = 0.02885 \text{ L} \times \frac{0.1000 \text{ mol Ba(OH)}_2}{\text{L}} \times \frac{2 \text{ mol } CH_3CO_2H}{1 \text{ mol Ba(OH)}_2} = 0.005770 \text{ mol } CH_3CO_2H$$

The sample titrated is one-hundredth of the equilibrium mixture. The total amount of CH_3CO_2H at equilibrium $\approx$ 0.577 mol CH_3CO_2H.

The amount of CH_3CO_2H consumed to reach equilibrium = 1.000 - 0.577 $\approx$ 0.423 mol. This is also the amount of C_2H_5OH consumed. The amount of C_2H_5OH present at equilibrium = 0.500 - 0.423 = 0.077 mol C_2H_5OH. The amounts of $CH_3CO_2C_2H_5$ and H_2O are both 0.423 mol. Assume that the volume of the reaction mixture is V(liters). Determine the equilibrium concentrations of the reactants and products, substitute into the K_c expression, and obtain a value of K_c.

$$K_c = \frac{[CH_3CO_2C_2H_5][H_2O]}{[C_2H_5OH][CH_3CO_2H]} = \frac{(0.423/V)(0.423/V)}{(0.077/V)(0.577/V)} = \frac{(0.423)^2}{0.077 \times 0.577} = 4.0$$

15-10. Five calculations of the following type are required to determine the concentrations of HI, H_2 and I_2 in each bulb at the time it is opened.

Bulb 1:

$$\text{no. mol } I_2 = 0.02096 \text{ L} \times \frac{0.0150 \text{ mol } Na_2S_2O_3}{\text{L}} \times \frac{1 \text{ mol } I_2}{2 \text{ mol } Na_2S_2O_3} = 1.57\times10^{-4} \text{ mol } I_2$$

$$\text{no. mol } H_2 = \text{no. mol } I_2 = 1.57\times10^{-4} \text{ mol } H_2$$

no. mol HI = original number moles HI - moles of HI consumed

$$= \left(0.300 \text{ g HI} \times \frac{1 \text{ mol HI}}{127.9 \text{ g HI}}\right) - \left(1.57\times10^{-4} \text{ mol } I_2 \times \frac{2 \text{ mol HI}}{1 \text{ mol } I_2}\right)$$

$$= 2.35\times10^{-3} \text{ mol HI} - 3.14\times10^{-4} \text{ mol HI} = 2.04\times10^{-3} \text{ mol HI}$$

Convert these amounts of reactants and products to concentrations and form the reaction quotient, Q.

$$Q = \frac{[H_2][I_2]}{[HI]^2} = \frac{(1.57\times10^{-4}/0.400)(1.57\times10^{-4}/0.400)}{(2.04\times10^{-3}/0.400)^2} = 5.92\times10^{-3}$$

Bulb 2:

$$\text{no. mol } I_2 = \text{no. mol } H_2 = 0.02790 \times 0.0150 \times \tfrac{1}{2} = 2.09\times10^{-4}$$

$$\text{no. mol HI} = (0.320/127.9) - (2 \times 2.09\times10^{-4}) = 2.08\times10^{-3}$$

$$Q = \frac{(2.09\times10^{-4}/0.400)^2}{(2.08\times10^{-3}/0.400)^2} = 1.01\times10^{-2}$$

Bulb 3:

$$\text{no. mol } I_2 = \text{no. mol } H_2 = 0.03231 \times 0.0150 \times \tfrac{1}{2} = 2.42\times10^{-4}$$

$$\text{no. mol HI} = (0.315/127.9) - (2 \times 2.42 \times 10^{-4}) = 1.98\times10^{-3}$$

$$Q = \frac{(2.42\times10^{-4}/0.400)^2}{(1.98\times10^{-3}/0.400)^2} = 1.49\times10^{-2}$$

Bulb 4:

$$\text{no. mol } I_2 = \text{no. mol } H_2 = 0.04150 \times 0.0150 \times \tfrac{1}{2} = 3.11\times10^{-4}$$

$$\text{no. mol HI} = (0.406/127.9) - (2 \times 3.11 \times 10^{-4}) = 2.55\times10^{-3}$$

$$Q = \frac{(3.11\times10^{-4}/0.400)^2}{(2.55\times10^{-3}/0.400)^2} = 1.49\times10^{-2}$$

Bulb 5:

$$\text{no. mol } I_2 = \text{no. mol } H_2 = 0.02868 \times 0.0150 \times \tfrac{1}{2} = 2.15\times10^{-4}$$

$$\text{no. mol HI} = (0.280/127.9) - (2 \times 2.15 \times 10^{-4}) = 1.76\times10^{-3}$$

$$Q = \frac{(2.15\times10^{-4}/0.400)^2}{(1.76\times10^{-3}/0.400)^2} = 1.49\times10^{-2}$$

The value of Q assumes a constant value of 1.49×10^{-2} in those bulbs opened after about 12 h or more. We conclude that equilibrium was attained in these cases and that $K_c = 1.49\times10^{-2}$.

Equilibrium relationships

15-11. (a) The term, V, representing the volume of the reaction mixture, appears the same number of times in the numerator and denominator. It cancels out of the expression for K_c.

$$K_c = \frac{[CO][H_2O]}{[CO_2][H_2]} = \frac{(0.224/\cancel{V})(0.224/\cancel{V})}{(0.276/\cancel{V})(0.276/\cancel{V})}$$

(b) $K_c = \frac{(0.224)(0.224)}{(0.276)(0.276)} = 6.59 \times 10^{-1}$

15-12. In the equilibrium mixture $[O_2] = 0.0148 \text{ mol}/0.755 \text{ L} = 0.0196 \text{ M}$

$$K_c = \frac{[NO_2]^2}{[NO]^2[O_2]} = \frac{[NO_2]^2}{[NO]^2 \times 0.0196} = 375$$

$$\frac{[NO_2]}{[NO]} = (0.0196 \times 375)^{\frac{1}{2}} = 2.71$$

and $\frac{[NO]}{[NO_2]} = \frac{1}{2.71} = 0.369$

15-13. Let the required volume of the reaction mixture be V.

$$K_c = \frac{[I]^2}{[I_2]} = \frac{(0.50/V)^2}{1.00/V} = \frac{0.25}{V^2 \times \frac{1.00}{V}} = 1.1 \times 10^{-2} \qquad V = \frac{0.25}{1.00 \times 1.1 \times 10^{-2}} = 23 \text{ L}$$

Direction and extent of chemical change

15-14. Determine the concentrations of the reactants.

$$[O_2] = \frac{3 \text{ mol}}{8.50 \text{ L}} = 0.35 \text{ M} \qquad [SO_2] = \frac{2 \text{ mol}}{8.50 \text{ L}} = 0.24 \text{ M} \qquad [SO_3] = \frac{6 \text{ mol}}{8.50 \text{ L}} = 0.71 \text{ M}$$

Now set up the reaction quotient.

$$Q = \frac{[SO_3]^2}{[SO_2]^2[O_2]} = \frac{(0.71)^2}{(0.24)^2 \times 0.35} = 25$$

Since $Q \neq K_c$, the mixture cannot be at equilibrium; and since $Q(25) < K_c(100)$, a net reaction occurs in the forward direction (to the right).

15-15.

The reaction is	$CO(g)$	+	$Cl_2(g)$	$\rightleftharpoons$	$COCl_2(g)$
Initial amounts, mol	1.00		-		1.00
Change; mol:	$+x$		$+x$		$-x$
Equilibrium amounts, mol:	$(1.00 + x)$		x		$(1.00 - x)$
Equilibrium concentrations, M:	$(1.00 + x)/1.75$		$x/1.75$		$(1.00 - x)/1.75$

$$K_c = \frac{[COCl_2]}{[CO][Cl_2]} = \frac{(1.00 - x)/1.75}{\left(\frac{1.00 + x}{1.75}\right)\left(\frac{x}{1.75}\right)} = \frac{1.75(1.00 - x)}{(1.00 + x)} = 1.2 \times 10^3$$

Assume $x << 1.00$, and $(1.00 - x) \approx (1.00 + x)$ 1.00

$$\frac{1.75}{x} = 1.2 \times 10^3 \qquad x = 1.5 \times 10^{-3} \text{ mol } Cl_2$$

15-16. In Example 15-6 it was established that the reaction must proceed to the left. The amounts of CO and H_2O increase (represented by $+x$) and those of CO_2 and H_2 decrease ($-x$).

The reaction is	$CO(g)$	+	$H_2O(g)$	$\rightleftharpoons$	$CO_2(g)$	+	$H_2(g)$
Initial amounts, mol:	1.00		1.00		2.00		2.00
Change, mol:	$+x$		$+x$		$-x$		$-x$
Equilibrium amounts, mol:	$1.00 + x$		$1.00 + x$		$2.00 - x$		$2.00 - x$
Equilibrium concentrations, M:	$(1.00 + x)/V$		$(1.00 + x)/V$		$(2.00 - x)/V$		$(2.00 - x)/V$

$$K_c = \frac{[CO_2][H_2]}{[CO][H_2O]} = \frac{(2.00 - x)(2.00 - x)}{(1.00 + x)(1.00 + x)} = 1.00$$

$$\frac{(2.00 - x)^2}{(1.00 + x)^2} = 1.00$$ Take the square root of both sides.

$$\frac{(2.00 - x)}{(1.00 + x)} = 1.00; \quad 2.00 - x = 1.00 + x; \quad 2x = 1.00; \quad x = 0.50 \text{ mol}$$

Equilibrium amounts: 1.50 mol CO, 1.50 mol H_2O, 1.50 mol CO_2, 1.50 mol H_2

15-17.

The reaction	$SbCl_5(g)$	$\rightleftharpoons$	$SbCl_3(g)$	+	$Cl_2(g)$
Initial amounts, mol:	--		3.00		1.00
Change, mol:	$+x$		$-x$		$-x$
Equilibrium amounts, mol:	x		$3.00 - x$		$1.00 - x$
Equilibrium concentrations, M	$x/5.00$		$(3.00 - x)/5.00$		$(1,00 - x)/5.00$

$$K_c = \frac{[SbCl_3][Cl_2]}{[SbCl_5]} = \frac{(3.00 - x)/5.00 \times (1.00 - x)5/00}{x/5.00} = 2.5\times10^{-2}$$

$$\frac{(3.00 - x)(1.00 - x)}{x} = 0.12$$

The simplifying assumption that $x << 1.00$ will not work here. (A very small value of x in the denominator would produce a large quotient, and 0.12 is not "large".)

$$3.00 - 4.00x + x^2 = 0.12x \qquad x^2 - 4.12x + 3.00 = 0$$

$$x = \frac{4.12 \pm \sqrt{(4.12)^2 - 4 \times 3.00}}{2} \qquad x = 0.94$$

Equilibrium mixture: 0.94 mol $SbCl_5$; 2.06 mol $SbCl_3$; 0.06 mol Cl_2

15-18. (a) Form the reaction quotient, Q, based on the given quantities and compare with $K_c = 2.8\times10^2$.

$$Q = \frac{[SO_3]^2}{[SO_2]^2[O_2]} = \frac{(0.6571/1.90)^2}{(0.390/1.90)^2(0.156/1.90)} = 34.6 < K_c$$

(b) A reaction must proceed in the forward direction (to the right) to establish equilibrium.

15-19. The reaction is $2SO_2(g) + O_2(g) \rightleftharpoons 2SO_3(g)$

	$2SO_2(g)$	$O_2(g)$	$2SO_3(g)$
Initial amounts, mol:	0.390	0.156	0.657
Change, mol:	$-2x$	$-x$	$+2x$
Equilibrium amounts, mol:	$0.390 - 2x$	$0.156 - x$	$0.657 + 2x$
Equilibrium concentrations, M:	$(0.390 - 2x)/1.90$	$(0.156 - x)/1.90$	$(0.657 + 2x)/1.90$

$$K_c = \frac{[SO_3]^2}{[SO_2]^2[O_2]} = \frac{(0.657 + 2x)^2(1.90)^2(1.90)}{(.190)^2(0.390 - 2x)^2(0.156 - x)} = 2.8\times10^2$$

$$1.90(0.432 + 2.63x + 4x^2) = (0.156 - x)(0.152 - 1.56x + 4x^2) = 2.8\times10^2$$

$$7.60x^2 + 5.00x + 0.821 = (0.0237 - 0.395x + 2.18x^2 - 4x^3)\ 2.8\times10^2$$

$$7.60x^2 + 5.00x + 0.821 = 6.64 - 111x + 610x^2 - 1120x^3$$

$$1120x^3 - 602x^2 + 116x - 5.82 = 0 \qquad x^3 - 0.538x^2 + 0.104x - 0.00520 = 0$$

Solve by a method of successive approximations. For example, first try $x = 0.100$.

$$(0.100)^3 - 0.538(0.100)^2 + 0.104(0.100) - 0.00520 = 0.00082 > 0$$

Try $x = 0.090$: $(0.090)^3 - 0.538(0.090)^2 + 0.104(0.090) - 0.00520 = 0.00053 > 0$

Try $x = 0.080$: $0.00019 > 0$

Try $x = 0.070$: $-0.0003 < 0$

Try $x = 0.075$: $-4.4\times10^{-6} < 0$

Try $x = 0.076$: $+3.5\times10^{-5} > 0$

Use $x = 0.075$ to obtain the equilibrium amounts: 0.240 mol SO_2; 0.081 mol O_2; 0.807 mol SO_3.

15-20. Convert the quantities of starting materials from grams to moles.

$$\text{no. mol CO} = 1.00\text{ g CO} \times \frac{1\text{ mol CO}}{28.01\text{ g CO}} = 0.0357\text{ mol CO}$$

$$\text{no. mol } H_2O = 1.00\text{ g } H_2O \times \frac{1\text{ mol } H_2O}{18.02\text{ g } H_2O} = 0.0555\text{ mol } H_2O$$

$$\text{no. mol } H_2 = 1.00\text{ g } H_2 \times \frac{1\text{ mol } H_2}{2.016\text{ g } H_2} = 0.496\text{ mol } H_2$$

Now establish the equilibrium concentrations of all species.

	$CO(g)$ +	$H_2O(g) \rightleftharpoons$	$CO_2(g)$ +	$H_2(g)$
Initial amounts:	0.0357 mol	0.0555 mol	-	0.496 mol
Changes:	$-x$ mol	$-x$ mol	$+x$ mol	$+x$ mol
Equilibrium amounts:	$(0.0357 - x)$ mol	$(0.0555 - x)$ mol	x mol	$(0.496 + x)$ mol
Equilibrium concentrations:	$\left(\frac{0.0357 - x}{1.41}\right)$M	$\left(\frac{0.0555 - x}{1.41}\right)$M	$\left(\frac{x}{1.41}\right)$ M	$\left(\frac{0.496 + x}{1.41}\right)$M

$$K_c = \frac{[CO_2][H_2]}{[CO][H_2O]} = \frac{x(0.496 + x)}{(0.0357 - x)(0.0555 - x)} = 23.2$$

$$x^2 + 0.496x = 23.2\ (1.98\times10^{-3} - 0.0912x + x^2)$$

$$x^2 + 0.496x = 0.0459 - 2.12x + 23.2x^2$$

$$22.2x^2 - 2.62x + 0.0459 = 0$$

$$x = \frac{2.62 \pm \sqrt{(2.62)^2 - 4 \times 0.0459 \times 22.2}}{2 \times 22.2} = \frac{2.62 \pm 1.67}{44.4} = 0.021$$

$$\text{no. g } CO_2 \text{ at equilibrium} = 0.021 \text{ mol } CO_2 \times \frac{44.0 \text{ g } CO_2}{1 \text{ mol } CO_2} = 0.92 \text{ g } CO_2$$

15-21. The reaction is

	$H_2(g)$	+ $I_2(g)$ $\rightleftharpoons$	$2HI(g)$
Initial amounts:	0.250 mol	0.250 mol	-
Changes:	$-x$ mol	$-x$ mol	$+2x$ mol
Equilibrium amounts:	$(0.250 - x)$ mol	$(0.250 - x)$ mol	$2x$ mol
Equilibrium concentrations:	$\left(\frac{0.250 - x}{4.10}\right)$ M	$\left(\frac{0.250 - x}{4.10}\right)$ M	$\left(\frac{2x}{4.10}\right)$ M

$$K_c = \frac{[HI]^2}{[H_2][I_2]} = \frac{4x^2}{(0.250 - x)^2} = 50.2$$

Take the square root of both sides.

$$\frac{2x}{0.250 - x} = \sqrt{50.2} = 7.09$$

$2x = 1.77 - 7.09x$ $\qquad$ $9.09x = 1.77$ $\qquad$ $x = 0.195$ mol

no. mol HI = $2x$ = 2 × 0.195 = 0.390 mol HI

no. mol H_2 = no. mol I_2 = 0.250 − x = 0.250 − 0.195 = 0.055

Total no. moles = 0.390 + 0.055 + 0.055 = 0.500

$$\text{mole \% HI} = \frac{0.390 \text{ mol HI}}{0.500 \text{ total mol}} \times 100 = 78.0 \text{ mole \% HI}$$

15-22. (a) The initial equilibrium condition is one in which the following amounts of reactants coexist in a 10.0-liter reaction vessel:

no. mol SO_3 = 0.68 $\qquad$ no. mol SO_2 = 0.32 $\qquad$ no. mol O_2 = 0.16

Immediately after the 1.00 mol SO_3 is added to the vessel there is present 1.68 mol SO_3, but according to Le Chatelier's principle some of the added SO_3 must decompose back into SO_2 and O_2. Call the amount of SO_3 that decomposes, $2x$. (This is done so that the additional amount of O_2 formed is x; *one* mole of O_2 is produced for every *two* moles of SO_3 that decompose.)

Equilibrium concentrations:

$[SO_3] = (1.68 - 2x)/10.0$ $\qquad [SO_2] = (0.32 + 2x)/10.0$ $\qquad [O_2] = (0.16 + x)/10.0$

$$K_c = \frac{[SO_3]^2}{[SO_2]^2[O_2]} = \frac{(1.68 - 2x)^2 \times (10.0)^2 \times 10.0}{(10.0)^2 \times (0.32 + 2x)^2 \times (0.16 + x)} = 280$$

$$\frac{2.82 - 6.72 + 4x^2}{(0.102 + 1.28x + 4x)^2(0.16 + x)} = \frac{280}{10.0} = 28.0$$

$$\frac{2.82 - 6.72x + 4x^2}{0.0163 + 0.205x + 0.64x^2 + 0.102x + 1.28x^2 + 4x^3} = 28.0$$

$$2.82 - 6.72x + 4x^2 = 0.456 + 5.74x + 17.9x^2 + 2.86x + 35.8x^2 + 112x^3$$

$$112x^3 + 49.7x^2 + 15.3x - 2.36 = 0 \qquad x^3 + 0.444x^2 + 0.137x - 0.0211 = 0$$

To solve this cubic equation, a method of successive approximations is required. The result obtained, to two significant figures, is $x = 0.11$.

Equilibrium amounts of reactants and products:

no. mol $SO_3 = 1.68 - 2x = 1.68 - (2 \times 0.11) = 1.46$ mol SO_3

no. mol $SO_2 = 0.32 + 2x = 0.32 + (2 \times 0.11) = 0.54$ mol SO_2

no. mol $O_2 = 0.16 + x = 0.16 + 0.11 = 0.27$ mol O_2

(b) The initial condition is one in which the following amounts of reactants coexist in a 10.0-liter reaction vessel:

no. mol $SO_3 = 0.68$ $\qquad$ no. mol $SO_2 = 0.32$ $\qquad$ no. mol $O_2 = 0.16$

When the reaction volume is reduced to 1.00 liter, the equilibrium condition must shift to the side of the equation representing the smaller number of moles of reactants--more SO_3 is formed at the expense of SO_2 and O_2. Call the additional number of moles of O_2 that react, x.

Equilibrium concentrations:

$[SO_3] = (0.68 + 2x)/1.00$ $\qquad [SO_2] = (0.32 - 2x)/1.00$ $\qquad [O_2] = (0.16 - x)/1.00$

$$K_c = \frac{[SO_3]^2}{[SO_2]^2[O_2]} = \frac{(0.68 + 2x)^2}{(0.32 - 2x)^2 \times (0.16 - x)} = 280$$

$$\frac{0.462 + 2.72x + 4x^2}{(0.102 - 1.28x + 4x^2)(0.16 - x)} = 280$$

$$\frac{0.462 + 2.72x + 4x^2}{0.0163 - 0.205x + 0.64x^2 - 0.102x + 1.28x^2 - 4x^3} = 280$$

$$0.462 + 2.72x + 4x^2 = 4.56 - 57.4x + 179x^2 - 28.6x + 358x^2 - 1120x^3$$

$$1120x^3 - 533x^2 + 88.7x - 4.10 = 0 \qquad x^3 - 0.476x^2 + 0.0792x - 0.00366 = 0$$

To solve this cubic equation, a method of successive approximations is again required. The result obtained, to two significant figures, is $x = 0.075$.

Equilibrium amounts of reactants and products:

no. mol SO_3 = $0.68 + 2x$ = $0.68 + (2 \times 0.075) = 0.83$ mol SO_2

no. mol SO_2 = $0.32 - 2x$ = $0.32 - (2 \times 0.075) = 0.17$ mol SO_2

no. mol O_2 = $0.16 - x = 0.16 - 0.075 = 0.085$ mol O_2

15-23.

The reaction is	$Ag^+(aq)$	+	$Fe^{2+}(aq) \rightleftharpoons$	$Fe^{3+}(aq)$	+	$Ag(s)$
Initial concentrations, M:	1.00		1.00	--		--
Changes, M:	$-x$		$-x$	$+x$		--
Equilibrium concentrations, M:	$1.00 - x$		$1.00 - x$	x		--

$$K_c = \frac{[Fe^{3+}]}{[Ag^+][Fe^{2+}]} = \frac{x}{(1.00 - x)^2} = 2.98$$

Note that the simplifying assumption that $x << 1.00$ does not work here. (This would lead to the result $x = 2.98$.)

$$x = 2.98(1.00 - 2x + x^2) \qquad 2.98x^2 - 6.9x + 2.98 = 0 \qquad x^2 - 2.34x + 1.00 = 0$$

$$x = \frac{2.34 \pm \sqrt{(2.34)^2 - 4.00}}{2} = 0.56$$

Equilibrium concentrations: $[Ag^+] = [Fe^{2+}] = 0.44$ M; $[Fe^{3+}] = 0.56$ M

15-24.

The reaction is	$2Cr^{3+}(aq)$	+	$Fe(s) \rightleftharpoons$	$2Cr^{2+}(aq)$	+	$Fe^{2+}(aq)$
Initial concentrations, M:	0.250			0.0500		0.00100
Changes, M:	$-2x$			$+2x$		$+x$
Equilibrium concentrations, M:	$0.250 - 2x$			$0.0500 + 2x$		$0.00100 + x$

$$K_c = \frac{[Cr^{2+}]^2[Fe^{2+}]}{[Cr^{3+}]^2} = \frac{(0.0500 + 2x)^2(0.00100 + x)}{(0.250 - 2x)^2} = 10.34$$

To determine whether a simplifying assumption is possible, let us first determine the value of Q based on initial concentrations and compare Q to K_c.

$$Q = \frac{(0.0500)^2(0.00100)}{(0.250)^2} = 4 \times 10^{-5} << K_c$$

In order that $Q = K_c$ the numerator in the above expression must become much larger and the denominator, smaller. This means that x is *not* an extremely small quantity. In fact, we might assume that $x >>$ 0.00100, so that $(0.00100 + x) \cong x$.

$$\frac{x(0.0500 + 2x)^2}{(0.250 - 2x)^2} = 10.34$$

$$x(0.00250 + 0.200x + 4x^2) = 10.34(0.0625 - x + 4x^2)$$

$$4x^3 + (0.200 - 41.36)x^2 + (0.00250 + 10.34)x - 0.646 = 0$$

$$4x^3 - 41.16x^2 + 10.34x - 0.646 = 0 \qquad x^3 - 10.29x^2 + 2.58x - 0.162 = 0$$

To solve this equation use a method of successive approximations. For example, try $x = 0.100$. (Note that x cannot exceed 0.125 or $[Cr^{3+}]$ would become negative.

$(0.100)^3 - 10.29(0.100)^2 + 2.58(0.100) - 0.162 = -0.0059 < 0$

Try $x = 0.11$: $(0.11)^3 - 10.29(0.11)^2 + 2.58(0.11) - 0.162 = -1.38\times10^{-3} < 0$

Try $x = 0.12$: $(0.12)^3 - 10.29(0.12)^2 + 2.58(0.12) - 0.162 = 1.15\times10^{-3} > 0$

The root we are seeking is $0.11 < x < 0.12$. Let us use $x = 0.11$

Equilibrium concentrations: $[Cr^{3+}] = 0.250 - 2 \times 0.11 = 0.03$ M

$[Cr^{2+}] = 0.0500 + 2 \times 0.11 = 0.27$ M

$[Fe^{2+}] = 0.11$ M

Partial pressure equilibrium constant, K_p

15-25. The reaction is: $SO_2Cl_2(g) \rightleftharpoons SO_2(g) + Cl_2(g)$

	$SO_2Cl_2(g)$	$SO_2(g)$	$Cl_2(g)$
Initial amounts:	--	1.00 mol	1.00 mol
Changes:	$+x$ mol	$-x$ mol	$-x$ mol
Equilibrium amounts:	x mol	$(1.00 - x)$RT	$(1.00 - x)$mol
Equalibrium partial pressure (nRT/V):	$\frac{xRT}{V}$	$\frac{(1.00 - x)RT}{V}$	$\frac{(1.00 - x)RT}{V}$

$$K_p = \frac{(P_{SO_2})(P_{Cl_2})}{P_{SO_2Cl_2}} = \frac{\{(1.00 - x)RT/V\}\{(1.00 - x)RT/V\}}{xRT/V} = 2.9\times10^{-2}$$

$$\frac{(1.00 - x)^2RT}{x\ V} = \frac{0.0821 \times 303 \times (1.00 - x)^2}{2.50x} = 2.9\times10^{-2}$$

$24.9(1.00 - 2x + x^2) = 0.072x$

$24.9 - 49.8x + 24.9x^2 = 0.072$

$24.9x^2 - 49.9x + 24.9 = 0$

$$x = \frac{49.9 \pm \sqrt{(49.9)^2 - 4 \times 24.9 \times 24.9}}{49.8} = \frac{49.9 \pm 3.2}{49.8} = 0.938$$

Partial Pressures:

$$P_{SO_2Cl_2} = \frac{0.938 \times 0.0821 \times 303}{2.50} = 9.33 \text{ atm}$$

$$P_{SO_2} = P_{Cl_2} = \frac{(1.00 - 0.938) \times 0.0821 \times 303}{2.50} = 0.6 \text{ atm}$$

(a) $P_{SO_2Cl_2} = 9.33$ atm

(b) $P_{tot.} = 9.33 + 0.6 + 0.6 = 10.5$ atm

15-26. Assume 1.00 mol of air originally. This consists of 0.79 mol N_2 and 0.21 mol O_2. (The presence of Ar and other gases is neglected.)

The reaction is	$N_2(g)$	+	$O_2(g)$	$\rightleftharpoons$	2NO(g)
Initial amounts, mol:	0.79		0.21		--
Changes, mol:	$-x$		$-x$		$+2x$
Equilibrium amounts:	$0.79 - x$		$0.21 - x$		$2x$
Equilibrium partial pressures:	$(0.79 - x)RT/V$		$(0.21 - x)RT/V$		$(2x)RT/V$

$$K_p = \frac{(P_{NO})^2}{(P_{N_2})(P_{O_2})} = \frac{\{(2x)RT/V\}^2}{\{(0.79 - x)RT/V\}\{(0.21 - x)RT/V\}} = \frac{4x^2}{(0.79 - x)(0.21 - x)}$$

To obtain a numerical value of K_p, it is first necessary to evaluate x.

The total amount of gas at equiplbrium = $(0.79 - x) + (0.21 - x) + 2x = 1.00$

mol % NO = $\frac{2x}{1.00} \times 100 = 1.8$ $\qquad x = 9.0\times10^{-3}$

$$K_p = \frac{4(9.0\times10^{-3})^2}{\{(0.79) - 9.0\times10^{-3}\}\{(0.21) - 9.0\times10^{-3}\}} = 2.1\times10^{-3}$$

15-27. The total gas pressure is the sum of the equilibrium partial pressures of NH_3 and H_2S. Let us find these quantities in the usual way.

The reaction is	$NH_4HS(s)$	$\rightleftharpoons$	$NH_3(g)$	+	$H_2S(g)$
Initial amounts, mol:	--		0.0100		--
Changes, mol:	--		$+x$		$+x$
Equilibrium amounts, mol:	--		$0.0100 + x$		x
Equilibrium partial pressures, atm:	--		$(0.0100 + x)RT/V$		xRT/V

$$K_p = (P_{NH_3})(P_{H_2S}) = \frac{x(0.0100 + x) \times (0.0821 \times 298)^2}{(1.60)^2} = 0.108$$

$$599x^2 + 5.99x - 0.276 = 0$$

$$x = \frac{-5.99 \pm \sqrt{(5.99)^2 + (4 \times 599 \times 0.276)}}{2 \times 599} = 0.0170$$

$$P_{NH_3} = \frac{0.0270 \times 0.0821 \times 298}{1.60} = 0.413 \text{ atm} \qquad P_{H_2S} = \frac{0.0170 \times 0.0821 \times 298}{1.60} = 0.260 \text{ atm}$$

$P_{tot} = 0.413 + 0.260 = 0.673$ atm.

15-28. (a) $K_p = (P_{CO_2})(P_{H_2O}) = 0.23$. Since the gases are produced in equal amounts, $P_{CO_2} = P_{H_2O}$.

$(P_{CO_2})^2 = 0.23$ $\quad\quad P_{H_2O} = P_{CO_2} = \sqrt{0.23} = 0.48$ atm

$P_{tot} = P_{H_2O} + P_{CO_2} = 0.48 + 0.48 = 0.96$ atm

(b) We need to compare Q with K_p.

$Q = (P_{CO_2})(P_{H_2O}) = (2.10)(715/760) = 1.98 > K_p$ (=0.23)

When equilibrium is established P_{CO_2} and P_{H_2O} will be smaller than their original values.

(c) Since H_2O and CO_2 enter into reaction in a 1:1 molar ratio, if x = no. mol CO_2 reacting, this is also no. mol H_2O reacting. For both gases we can express the decrease in pressure accompanying the reaction of x mol as $\Delta P = \frac{\Delta nRT}{V} = xRT/V$. We can just as well represent this change in pressure as $y = xRT/V$ and use only y in the expression below.

$K_p = (P_{CO_2})(P_{H_2O}) = (2.10 - y)(0.941 - y) = 0.23$

$1.98 - 3.04y + y^2 = 0.23$ $\quad\quad y^2 - 3.04y + 1.75 = 0$

$$y = \frac{3.04 \pm\sqrt{(3.04)^2 - 4 \times 1.75}}{2} = 0.77 \text{ atm.}$$

Equilibrium partial pressures: $P_{CO_2} = 2.10 - 0.77 = 1.33$ atm

$P_{H_2O} = 0.941 - 0.77 = 0.17$ atm

Dissociation reactions

15-29. Transferring the reaction mixture of Example 15-13 to a larger vessel (10.0 L compared to 0.372 L) would favor the forward reaction when equilibrium is restored. The percent dissociation would increase. This conclusion can be based on Le Chatelier's principle or on a comparison of Q and K_c. The equilibrium mixture described in Example 15-13 contains $2x$ mol $NO_2 = 2 \times 3.00 \times 10^{-3}$ $= 6.00\times10^{-3}$ mol NO_2 and $0.0240 - x = 0.0240 - 3.00\times10^{-3} = 0.0210$ mol N_2O_4

$$Q = \frac{[NO_2]^2}{[N_2O_4]} = \frac{(6.00\times10^{-3}/10.0)^2}{(0.0210/10.0)} = 1.71\times10^{-4} < K_p \;(= 4.61\times10^{-3})$$

Since $Q < K_p$, a reaction must proceed to the right. The percent dissociation increases.

15-30. The simplest approach here is to return to Example 15-13 and to restate the equilibrium concentrations as

$$[N_2O_4] = \frac{(0.0240 - x)\text{mol}}{10.0 \text{ L}} \quad \text{and} \quad [NO_2] = \frac{2x \text{ mol}}{10.0 \text{ L}}$$

$$K_c = \frac{[NO_2]^2}{[N_2O_4]} = \frac{(2x)^2}{(10.0)^2(0.0240 - x)/10} = \frac{4x^2}{10.0(0.0240 - x)} = 4.61\times10^{-3}$$

$4x^2 + 0.0461x - 0.00111 = 0$

$$x = \frac{-0.0461 \pm \sqrt{(0.0461)^2 + (4 \times 4 \times 0.00111)}}{8} = 0.012$$

$$\% \text{ dissoc.} = \frac{0.012 \text{ mol } N_2O_4 \text{ dissoc.}}{0.0240 \text{ mol } N_2O_4 \text{ initially}} \times 100 = 5.0\times10^{1}\ \%$$

15-31. For the reaction, $2HI(g) \rightleftharpoons H_2(g) + I_2(g)$; $K_c = 1/69 = 1.4\times10^{-2}$.

In this case, the percent dissociation does not depend on the numbers of moles of reactant and product. Consider, for example, a sample of 1.00 mole HI placed in a volume of V liter. Let x = no. mol H_2 = no. mol I_2 produced; $1.00 - 2x$ = no. mol HI at equilibrium.

$$K_c = \frac{[H_2][I_2]}{[HI]^2} = \frac{(x/V)(x/V)}{[(1.00 - 2x)/V]^2} = \frac{x^2}{1.00 - 4x + 4x^2} = 1.4\times10^{-2}$$

$$x^2 = 1.4\times10^{-2} - 5.6\times10^{-2}x + 5.6\times10^{-2}x^2 \qquad 0.94x^2 + 5.6\times10^{-2}x = 1.4\times10^{-2} = 0$$

$$x = \frac{-5.6\times10^{-2} \pm \sqrt{(5.6\times10^{-2})^2 + (4 \times 0.94 \times 1.4 \times 10^{-2})}}{2 \times 0.94}$$

$$x = \frac{-5.6\times10^{-2} \pm 2.36\times10^{-1}}{1.88} = 9.6\times10^{-2}$$

No. mol HI dissociated $= 2x = 2 \times 9.6 \times 10^{-2} = 19.2\times10^{-2} = 1.9\times10^{-1}$

$$\% \text{ dissociation of HI} = \frac{1.9\times10^{-1}}{1.00} \times 100 = 19\%$$

15-32. For simplicity, let us base this derivation on an original 1.00 mol PCl_5, of which the fraction dissociated is a.

The reaction is	$PCl_5(g)$ $\rightleftharpoons$	$PCl_3(g)$	+	$Cl_2(g)$
Initial amounts, mol:	1.00	--		--
Changes, mol:	$-a$	$+a$		$+a$
Equilibrium amounts, mol:	$1.00 - a$	a		a
Total moles of gas = $(1.00 - a) + a + a$	$= 1.00 + a$			
Equilibrium partial pressures in terms of total pressure P:	$\frac{(1.00 - a)}{(1.00 + a)}P$	$\frac{aP}{1.00 + a}$		$\frac{aP}{1.00 + a}$

$$K_p = \frac{(P_{PCl_3})(P_{Cl_2})}{(P_{Cl_5})} = \frac{\left(\frac{aP}{1.00 + a}\right)\left(\frac{aP}{1.00 + a}\right)}{\frac{(1.00 - a)}{(1.00 + a)}P} = \frac{a^2P^2}{(1.00 + a)^2 \; \frac{(1.00 - a)}{(1.00 + a)}P} = \frac{a^2P^2}{(1.00 + a)(1.00 - a)P}$$

$$= \frac{a^2P}{1.00 - a^2}$$

15-33. (a) $K_p = \frac{a^2P}{1-a^2} = \frac{a^2(1.00)}{1-a^2} = 1.78$ $a^2 = 1.78 \times 1.78^2$ $a^2 = 0.640$ $a = 0.800$

PCl_5 is 80.0% dissociated at 250°C.

(b) If dissociation is to be 10.0%, the value of = 0.100

$K_p = \frac{(0.100)^2P}{1-(0.100)^2} = 1.78$ $P = \frac{1.78 - 1.78\times10^{-2}}{1.00\times10^{-2}} = \frac{1.76}{1.00\times10^{-2}} = 1.76$ atm

Le Chatelier's principle

15-34. Removal of one of the reacting species in a chemical equilibrium causes a shift in the equilibrium condition that favors the production of an additional quantity of the species that has been removed. If the species removed is a product of the reaction, the equilibrium condition shifts in the direction of the forward reaction. If the species is continuously removed, the equilibrium condition is never reached as the forward reaction is constantly favored. The net effect is that the reaction goes to completion.

15-35. The following responses refer to the endothermic reaction $A(g) + B(g) \rightleftharpoons 2\,C(g)$.

(a) True: The number of moles of gaseous species is unchanged in the reaction. Increasing the volume of the reaction mixture has no effect on equilibrium amounts.

(b) False: The presence of a catalyst has no effect on the position of equilibrium in a reversible reaction.

(c) False: The forward reaction is endothermic; it is favored by an increase in temperature. A decrease in temperature (from 200 to 100°C) favors the reverse reaction and results in the production of *less* C(g) at equilibrium.

(d) True: The addition of an inert gas (He) has no effect on this equilibrium (and this is true whether the reaction volume is held constant or allowed to increase).

15-36. (a) Suppose we start with 1.00 mole of SO_2Cl_2 and allow it to come to equilibrium in a reaction vessel of volume, V. Let x = no. mol SO_2Cl_2 dissociated.

$$K_c = \frac{[SO_2][Cl_2]}{[SO_2Cl_2]} = \frac{(x/V)(x/V)}{(1.00-x)/V} = \frac{x^2}{V(1.00-x)} \qquad \frac{x^2}{1.00-x} = K_cV$$

Clearly the value of x will depend on the values of both K_c and V, and since percent dissociation of $SO_2Cl_2 = (x/1.00) \times 100$, the percent dissociation must depend on the volume, V.

This same conclusion is possible using Le Chatelier's principle. Increasing the volume, V, causes the equilibrium condition to be displaced to the right; the percent dissociation increases.

(b) By a similar method to that employed in part (a), we can establish the condition:

$$K_c = \frac{[S_2]}{[CS_2]} = \frac{x/V}{(1.00-x)/V} = \frac{x}{1.00-x}$$

Here the value of x depends only on the value of K_c, not on the volume, V.

One mole of gaseous reactant produces one mole of gaseous product. According to Le Chatelier's principlewe should expect no effect on equilibrium concentrations as a result of changing the reaction volume.

15-37. In reactions of the type, $A_2(g) \rightleftharpoons 2A(g)$, the bonds, A-A, are broken and no new bonds are formed.

Such reactions must be endothermic, and endothermic reactions are favored at high temperatures.

15-38. (a) Because the formation of NO(g) from its elements is an endothermic reaction, high temperatures favor an increase in the equilibrium amounts of NO(g).

(b) Not only is the equilibrium production of NO(g) favored by high temperatures, but the rate of attainment of equilibrium is greatly accelerated. This is because reaction rates increase so rapidly with temperature.

15-39. The application of pressure to ice causes it to melt because the liquid water occupies a smaller volume that a corresponding mass of ice. Stated in terms of Le Chatelier's principle, the application of a stress to an equilibrium mixture causes a displacement of the equilibrium in that direction in which the stress is relieved. The stress in this case is an increase in pressure and the relief comes through a reduction in volume. This behavior is not generally expected since in all but a few cases a liquid occupies a larger volume than the solid from which it is formed.

Effect of temperature on equilibrium constants

15-40. Use the van't Hoff equation (15.26) in both parts

(a) Let $K_2 = 0.113$ at $T_2 = 298$ K; $K_1 = ?$ at $T_1 = 273$ K and $\Delta\bar{H}° = +61.5\times10^3$ J/mol

$$\log\frac{0.113}{K_2} = \frac{61.5\times10^3(298-273)}{2.303\times8.314\times273\times298} = 0.99$$

$$\frac{0.113}{K_2} = \text{antilog } 0.99 = 9.8 \qquad K_2 = 0.113/9.8 = 1.2\times10^{-2}$$

(b) Let $K_2 = 1.00$ at $T_2 = ?$ $\quad K_1 = 0.113$ at $T_1 = 298$

$$\log\frac{1.00}{0.113} = \frac{61.5\times10^3\ (T_2 - 298)}{4.303\times8.314\times298\times T_2} = 0.947$$

$$61.5\times10^3\ T_2 - 1.83\times10^7 = 5.40\times10^3\ T_2$$

$$5.61\times10^4\ T_2 = 1.83\times10^7 \qquad T_2 = 326\text{ K}$$

15-41. Use the van't Hoff equation with $K_1 = 50.0$ @ $T_1 = 448 + 273 = 721$ K and $K_2 = 66.9$ @ T_2 $= 350 + 273 = 623$ K.

$$\log\frac{66.9}{50.0} = \frac{\Delta\bar{H}°}{2.303\times8.314\text{ J mol}^{-1}\text{ K}^{-1}}\ \frac{623-721}{623\times721}\text{ K}^{-1} = 0.126$$

$$\Delta\bar{H}° = \frac{0.126\times623\times721\times2.303\times8.314}{-98} = -1.1\times10^4\text{ J/mol} = -11\text{ kJ/mol}$$

(The value of $\Delta\bar{H}°$ represented in Figure 14-10 is -13 kJ/mol.)

15-42. Data from Appendix D are used to obtain $\Delta\bar{H}°_{rxn}$ for

$$2NO(g) + O_2(g) \rightleftharpoons 2NO_2(g)$$

$$\Delta\bar{H}°_{rxn} = 2\Delta\bar{H}°_f[NO_2(g)] - 2\Delta\bar{H}°_f[NO(g)]$$

$$= (2\times33.85) - (2\times90.37) = -113.0\text{ kJ/mol}$$

Data from Table 15-3 are used to supply a value of K_p (1.6×10^{12}) at 298 K.

$$\log \frac{K}{1.6\times10^{12}} = \frac{-113.0\times10^{3}(373-298)}{2.303\times8.314\times298\times373} = -4.0$$

$$\frac{K}{1.6\times10^{12}} = 1\times10^{-4} \qquad K = 2\times10^{8}$$

15-43. (a) Convert temperatures from °C to kelvins. Tabulate log K_p as a function of $1/T$ as indicated below.

t	T	1/T	K_p	log K_p
30°C	303 K	3.30×10^{-3}	1.66×10^{-5}	-4.780
50	323	3.10×10^{-3}	3.90×10^{-4}	-3.409
70	343	2.92×10^{-3}	6.27×10^{-3}	-2.203
100	373	2.68×10^{-3}	2.31×10^{-1}	-0.636

Plot log K_p vs $1/T$ and obtain the slope.

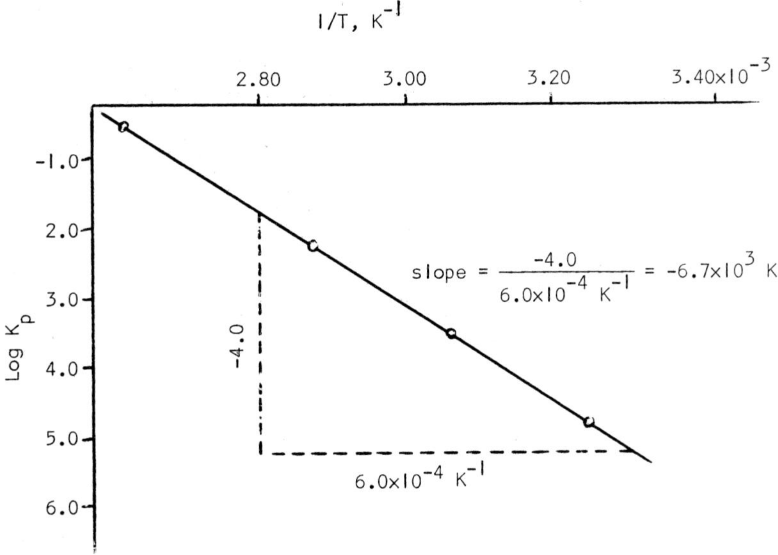

$\Delta\bar{H}° = -2.303 \times R \times \text{slope} = -2.303 \times 8.314\ \text{J mol}^{-1}\ \text{K}^{-1} \times (-6.7\times10^{3})\text{K} = 1.3\times10^{5}$ J/mol

$= 1.3\times10^{2}$

(b) If $P_{tot} = 2.00$ atm, then $P_{H_2O} = P_{CO_2} = 1.00$ atm.

$K_p = (P_{CO_2}) \times (P_{H_2O}) = (1.00) \times (1.00) = 1.00$

Calculate the temperature, T_2, at which $K_p = 1.00$, with $T_1 = 373$ K, $K_{p_1} = 0.231$, and $\Delta\bar{H}° = 1.3\times10^{5}$ J/mol.

$$\log \frac{1.00}{0.231} = \frac{1.3\times10^{5}\ \text{J/mol}}{2.303\times8.314\ \text{J mol}^{-1}\ \text{K}^{-1}}\left(\frac{T_2 - 373}{373\ T_2}\right)\text{K}^{-1}$$

$$\log 4.33 = 6.8\times10^{3}\left(\frac{T_2 - 373}{373\ T_2}\right) \qquad 0.636 = 6.8\times10^{3}\left(\frac{T_2 - 373}{373\ T_2}\right)$$

$$237\ T_2 = 6.8\times10^{3}\ T_2 - 2.5\times10^{6} \qquad 6.6\times10^{3}\ T_2 = 2.5\times10^{6} \qquad T_2 = 3.8\times10^{2}$$

15-44. (a) Since the variation of K_p with T takes the form

$$\log \frac{K_2}{K_1} = \frac{\Delta\bar{H}°(T_2 - T_1)}{2.303RT_1T_2}$$

if $\Delta\bar{H}° = 0$ then $\log (K_2/K_1) = 0$; $K_2/K_1 = 1$ and $K_2 = K_1$

That is we conclude that K_p does not vary with temperatures in this case.

(b) Although K_p is unchanged from 298 K to 400 K, if the reaction mixture is maintained in a constant volume, the total pressure increases as the temperature is raised. The effect of increasing the pressure on an equilibrium mixture is to shift the equilibrium condition to the side having the smaller number of moles of gas (the left side in the present case). As a result, we would expect the amount of D(g) present to *decrease*.

Another way to arrive at this same conclusion is to write the expression for K_p.

$K_p = (P_C)^2(P_D)^{\frac{1}{2}}$

If K_p maintains a constant value between 298 K and 400 K, so too must the product $(P_C)^2(P_D)^{\frac{1}{2}}$. But at the higher temperature the necessary pressures are maintained with a smaller amount of C and D in the gaseous state.

Self-test Questions

1. (*d*) Based on the availability of 1 mol I_2, if the reaction went to completion, 2 mol HI would form. Since the reaction is reversible, the equilibrium amount of HI is less than 2 mol.

2. (*c*) Write the equilibrium constant expression $K_c = \dfrac{[SO_3]^2}{[SO_2]^2[O_2]} = 100$

Because the number of moles of SO_2 and SO_3 have been made equal, $[SO_2] = [SO_3]$. The terms $[SO_3]^2$ and $[SO_2]^2$ cancel out in the K_c expression, leaving $1/[O_2] = 100$. This means that $[O_2] = 0.010$ M.

3. (*a*) An increase in reaction volume favors the side of the equation in which the greater number of moles of gas appears. This is the forward reaction. The amounts of $SO_2(g)$ and $Cl_2(g)$ increase.

4. (*c*) For the given condition, K = [C][D]/[A][B] = 10.0. The only one of the listed statements that can be derived from this expression is [A][B] = [C][D]/10.0 = 0.10 [C][D].

5. (*d*) A catalyst does not affect the position of equilibrium [item (a) is incorrect]. An increase of temperature favors the endothermic reaction--the reverse reaction. [Item (b) is incorrect because this would lead to a decrease in the amount of $H_2(g)$.] Transferring the equilibrium mixture to a larger container will have no effect on the amount of $H_2(g)$. The total number of moles of gas is the same on each side of the equation. The correct answer is (d).

6. (*b*) The value of Q for the initial conditions is

$$\frac{[CO_2(g)][H_2(g)]}{[CO(g)][H_2O(g)]} = \frac{(0.10)[H_2]}{(0.10)(0.10)} = 0 \text{ (because } [H_2(g)] = 0)$$

A net reaction must proceed to the right. The amount of $CO_2(g)$ would increase and the amounts of CO(g) and $H_2O(g)$ would decrease. Also, the amount of $H_2(g)$ produced would have to be less than one mole, since the reaction does not go to completion. The correct response is (b). Response (c) must be in error because there is no way for all amounts to exceed 1.00 mol. Response (d) is in error because the amount of $H_2(g)$ cannot exceed 1 mol.

7. (*b*) The given equation must be reversed and divided by 2 to obtain $NO(g) + \frac{1}{2}O_2(g) \rightleftharpoons NO_2(g)$.

$K_c = 1/\sqrt{1.8\times10^{-6}} = 7.5\times10^2$

8. (*a*) For this reaction $K_p = K_c(RT)^{\Delta n} = K_c(RT)^1$. K_p is larger than K_c (or K_c is smaller than K_p) at all temperatures.

9. A balanced equation, through its stoichiometric coefficients, provides the basis for writing an equilibrium constant expression. Also the balanced equation provides factors for relating the changes that occur among the amounts (or concentrations) of the various reactants and products as equilibrium is established. To determine the actual equilibrium concentrations, however, one must have a numerical value of the equilibrium constant. This value cannot be established in any way from the balanced equation alone. Experimental results are required.

10. (a) The rates of the forward and reverse reactions are equal when a reversible reaction reaches equilibrium. However, these rates must be established by the methods of chemical kinetics (Chapter 14) not simply by measuring equilibrium concentrations.

(b) The reaction quotient Q and the equilibrium constant expression K_c are set up in a very similar manner, but the concentrations substituted into the K_c expression must be those that exist at equilibrium. Any set of concentration terms can be substituted into the Q expression. (Depending on how the numerical values of Q and K_c compare one can make predictions about the direction of net chemical reaction.)

(c) The equilibrium constant expression K_c is based on molar concentrations of reactants and products and K_p on partial pressures of gases. The relationship between the two is $K_p = K_c(RT)^{\Delta n}$ (where Δn is the change in number of moles of gaseous species as the reaction proceeds as written, that is, from left to right.

11. The reaction is $S_2(g) \rightleftharpoons 2S(g)$

	$S_2(g)$	$2S(g)$
Initial amounts, mol:	0.0010	--
Changes, mol:	-0.5×10^{-11}	$+1.0\times10^{-11}$
Equilibrium amounts, mol:	$0.0010 - 5\times10^{-11}$	1.0×10^{-11}
Equilibrium concentrations, M:	0.0010/0.500	$1.0\times10^{-11}/0.500$

$$K_c = \frac{[S]^2}{[S_2]} = \frac{(1.0\times10^{-11}/0.500)^2}{0.0010/0.500} = \frac{1.0\times10^{-22}}{0.500 \times 0.0010} = 2.0\times10^{-19}$$

12. (a) For the conditions stated:

$$Q = \frac{[NOBr]^2}{[NO]^2[Br_2]} = \frac{(0.0100/1.00)^2}{(0.100/1.00)^2(0.100/1.00)} = 0.10 > K_c \; (= 1.32\times10^{-2})$$

A net reaction occurs to the left--in the reverse direction.

(b) The reaction is $2NO(g) + Br_2(g) \rightleftharpoons 2NOBr(g)$

	$2NO(g)$	$Br_2(g)$	$2NOBr(g)$
Initial amounts, mol:	0.100	0.100	0.0100
Changes, mol:	$+2x$	$+x$	$-2x$
Equilibrium amounts, mol:	$0.100 + 2x$	$0.100 + x$	$0.0100 - 2x$
Equilibrium concentrations, M:	$(0.100 + 2x)/1.00$	$(0.100 + x)/1.00$	$(0.0100 - 2x)/1.00$

$$K_c = \frac{[NOBr]^2}{[NO]^2[Br_2]} = \frac{(0.0100 - 2x)^2}{(0.100 + 2x)^2(0.100 + x)} = 1.32\times10^{-2}$$

By inspecting the numerator of the above expression we see that $x < 0.00500$. This means also that $x << 0.100$, and that $(0.100 + 2x) \simeq (0.100 + x) \simeq 0.100$.

$$\frac{(0.0100 - 2x)^2}{(0.100)^2(0.100)} = 1.32\times10^{-2}$$

$$(0.0100 - 2x)^2 = 1.32\times10^{-5}$$

$$0.0100 - 2x = \sqrt{1.32\times10^{-5}} = 3.63\times10^{-3}$$

$$x = \frac{0.0100 - 0.0036}{2}$$

$$x = 3.2\times10^{-3}$$

no. mol NO = $0.100 + (2 \times 3.2 \times 10^{-3}) = 0.106$

no. mol Br_2 = $0.100 + 3.2 \times 10^{-3} = 0.103$

no. mol NOBr = $0.0100 - (2 \times 3.2 \times 10^{-3}) = 0.0036$

total no. mol = $0.106 + 0.103 + 0.004 = 0.213$

total pressure = $nRT/V = 0.213 \times 0.0821 \times 1000/1.00 = 17.5$ atm

$$P_{NOBr} = \frac{0.0036 \text{ mol NOBr}}{0.213 \text{ mol total}} \times 17.5 \text{ atm} = 0.30 \text{ atm}$$

Chapter 16

Thermodynamics and Chemistry

Review Problems

16-1. In each case use the first law of thermodynamics, $\Delta E = q - w$, and the appropriate signs for q and w.

(a) $q = +58$ J $w = +58$ J

$\Delta E = q - w = (+58) - (+58) = 0$

(b) $q = +125$ J $w = +687$ J

$\Delta E = q - w = +125 - (+687) = -562$ J

(c) $q = -22$ J $w = -111$ J

$E = q - w = (-22) - (-111) - +89$ J

(d) $q = 0$ $w = +117$ J

$E = q - w = 0 - (+117) = -117$ J

16-2. In each case $\Delta\bar{H} = \Delta\bar{E} + 2.48\ \Delta\bar{n}_g$. If $\Delta\bar{n}_g$ is a positive quantity, $\Delta\bar{H} > \Delta\bar{E}$, that is, either more positive than $\Delta\bar{E}$ (if $\Delta\bar{E}$ is positive) or less negative than ΔE (if $\Delta\bar{E}$ is negative). If Δn_g is a negative quantity, $\Delta\bar{H} < \Delta\bar{E}$ (that is less positive or more negative than $\Delta\bar{E}$).

(a) $\Delta\bar{n}_g = 1$; $\Delta\bar{H} > \Delta\bar{E}$, i.e., $\Delta\bar{H} > +130$ kJ/mol

(b) $\Delta\bar{n}_g = -9$; $\Delta\bar{H} < \Delta\bar{E}$, i.e., $\Delta\bar{H} < -628$ kJ/mol

16-3. (a) A liquid vaporizes to produce a gas (a more disordered state). Entropy increases.

(b) Two moles of gas are consumed for every mole of $CuSO_4 \cdot 3H_2O$ that is converted to $CuSO_4 \cdot 5H_2O$. Entropy decreases.

(c) Here the number of moles of gaseous reactants and products are equal (two moles of each). Because there is no change in number of moles of gas, it can not be easily predicted whether entropy increases or decreases.

(d) First the equation must be balanced

$2H_2S(g) + 3O_2(g) \rightarrow 2H_2O(g) + 2SO_2(g)$

Five moles of gas are converted to four moles. We would expect entropy to decrease.

16-4. (a) ΔH must be positive (bonds are broken and not reformed); ΔS must also be positive (two moles of gaseous product formed from one mole of reactant). $\Delta H > 0$ and $\Delta S > 0$ corresponds to case 3 of Table 16-2.

(b) We are given that $\Delta\bar{H}° < 0$, and because the reaction is accompanied by a decrease in number of moles of gas (3 mol gas $\rightarrow$ 2 mol gas), we also expect $\Delta\bar{S}° < 0$. This corresponds to case 2 of Table 16-2.

(c) Here $\Delta\bar{H}° < 0$ but $\Delta\bar{S}° < 0$ (1 mol gas $\rightarrow$ 3 mol gas). This corresponds to case 1 of Table 16-2.

(d) $\Delta\bar{H}° > 0$ and $\Delta\bar{S}° < 0$ (4 mol gas $\rightarrow$ 1 mol liquid). This is case 4 of Table 16-2.

16-5. From the value of $\Delta\bar{G}_f°$ and $\Delta\bar{H}_f°$ determine $\Delta\bar{G}°_{rxn}$ and $\Delta\bar{H}°_{rxn}$ for $NH_3(g) + HCl(g) \rightarrow NH_4Cl(s)$.

$\Delta\bar{H}°_{rxn} = \Delta\bar{H}_f°[NH_4Cl(s)] - \Delta\bar{H}_f°[NH_3] - \Delta\bar{H}_f°[HCl(g)]$

$= -315.4$ kJ/mol $- (-46.2$ kJ/mol$) - (-92.3$ kJ/mol$) = -176.9$ kJ/mol

$$\Delta\bar{G}^\circ_{rxn} = \Delta\bar{G}^\circ_f[NH_4Cl(s)] - \Delta\bar{G}^\circ_f[NH_3] - \Delta\bar{G}^\circ_f[HCl(g)]$$

$$= -203.9 \text{ kJ/mol} - (-16.6 \text{ kJ/mol}) - (-95.3 \text{ kJ/mol}) = -92.0 \text{ kJ/mol}$$

Now use the expression $\Delta\bar{G}^\circ = \Delta\bar{H}^\circ - T\Delta\bar{S}^\circ$ and solve for $\Delta\bar{S}^\circ$.

$$\Delta\bar{S}^\circ = \frac{\Delta\bar{H}^\circ - \Delta\bar{G}^\circ}{T} = \frac{-176.9 \text{ kJ/mol} - (-92.0 \text{ kJ/mol})}{298 \text{ K}} = -0.285 \text{ kJ mol}^{-1} \text{ K}^{-1}$$

16-6. Use data from Appendix D in the form $\Delta\bar{G}^\circ_{rxn} = \Sigma\ \Delta\bar{G}^\circ_f(\text{products}) - \Sigma\ \Delta\bar{G}^\circ_f(\text{reactants})$.

(a) $N_2(g) + 3H_2(g) \rightarrow 2NH_3(g)$

$$\Delta\bar{G}^\circ = 2 \times \Delta\bar{G}^\circ_f[NH_3] - \Delta\bar{G}^\circ_f[N_2(g)] - 3 \times \Delta\bar{G}^\circ_f[H_2(g)]$$

$$= 2 \times (-16.65) = -33.30 \text{ kJ/mol}$$

(b) $C_2H_2(g) + 2H_2(g) \rightarrow C_2H_6(g)$

$$\Delta\bar{G}^\circ = (-32.89) - (209.20) - 2 \times 0 = -242.09 \text{ kJ/mol}$$

(c) $Fe_3O_4(s) + 4H_2(g) \rightarrow 3Fe(s) + 4H_2O(g)$

$$\Delta\bar{G}^\circ = 4 \times (-228.61) + (3 \times 0) - (-1014.20) - (4 \times 0) = +99.76 \text{ kJ/mol}$$

(d) $MgO(s) + 2HCl(g) \rightarrow MgCl_2(s) + H_2O(g)$

$$\Delta\bar{G}^\circ = (-592.33) + (-228.61) - (-569.57) - 2 \times (-95.27) = -60.8 \text{ kJ/mol}$$

16-7. (a) The process described, the conversion of liquid to gaseous water at 1 atm pressure, reaches equilibrium at the normal boiling point of water. The temperature at which the two line segments intersect is 100°C.

(c) Since the case described in (a) is one of equilibrium, $\Delta G = 0$.

16-8. (a) $$\Delta\bar{S}^\circ_{vap} = \frac{\Delta\bar{H}^\circ_{vap}}{T_{bp}} = \frac{3.86 \text{ kcal/mol} \times 1000 \text{ cal/kcal} \times 4.184 \text{ J/cal}}{(-85.05 + 273.15)\text{K}} = 85.9 \text{ J mol}^{-1} \text{ K}^{-1}$$

(b) $$\Delta\bar{S}^\circ_f = \frac{\Delta\bar{H}^\circ_f}{T_{mp}} = \frac{27.05 \text{ cal/g} \times 4.184 \text{ J/cal} \times 22.99\text{g/mol Na}}{(97.82 + 273.15)\text{K}} = 7.014 \text{ J mol}^{-1} \text{ K}^{-1}$$

(c) $$\Delta\bar{S}^\circ_{tr} = \frac{\Delta\bar{H}^\circ_{tr}}{T_{tr}} = \frac{96 \text{ cal/mol} \times 4.184 \text{ J/cal}}{(95.5 + 273.2)\text{K}} = 1.09 \text{ J mol}^{-1} \text{ K}^{-1}$$

16-9. $$\Delta\bar{G}^\circ = -2.303 \text{ RT log K} = -2.303 \times 8.314 \text{ J mol}^{-1} \text{ K}^{-1} \times 1000 \text{ K} \times \log(2.45\times10^{-7})$$

$$= -2.303 \times 8.314 \times 1000 \times (-6.611) = 1.27\times10^5 \text{ J/mol} = 127 \text{ kJ/mol}$$

16-10. (a) $$K = \frac{(a_{NO_2(g)})^2}{(a_{NO(g)})^2(a_{O_2})} = \frac{(P_{NO_2})^2}{(P_{NO})^2(P_{O_2})} = K_p$$

(b) $$K = \frac{(a_{MgO(s)})(a_{SO_2(g)})}{(a_{MgSO_3(s)})} = \frac{1 \times (P_{SO_2})}{1} = P_{SO_2} = K_p$$

(c) $K = \dfrac{(a_{H^+(aq)})(a_{C_2H_3O_2^-(aq)})}{(a_{HC_2H_3O_2(aq)})} = \dfrac{[H^+][C_2H_3O_2^-]}{[HC_2H_3O_2]} = K_c$

(d) $K = \dfrac{(a_{Na_2CO_3(s)})(a_{H_2O(g)})(a_{CO_2(g)})}{(a_{NaHCO_3(s)})^2} = \dfrac{1 \times (P_{H_2O})(P_{CO_2})}{(1)^2} = (P_{H_2O})(P_{CO_2}) = K_p$

(e) $K = \dfrac{(a_{Mn^{2+}(aq)})(a_{H_2O(l)})^2(a_{Cl_2(g)})}{(a_{MnO_2(s)})(a_{H^+(aq)})^4(a_{Cl^-(aq)})^2} = \dfrac{[Mn^{2+}] \times (1)^2 \times P_{Cl_2}}{(1) \times [H^+]^4[Cl^-]^2} = \dfrac{[Mn^{2+}] \times P_{Cl_2}}{[H^+]^4[Cl^-]^2} = K$

16-11. For the reaction $2SO_2(g) + O_2(g) \rightleftharpoons 2SO_3(g)$

(a) $K_p = K_c(RT)^{\Delta n} = 2.8\times10^2 \times (0.0821 \times 1000)^{-1} = 3.4$

(b) $\Delta\bar{G}° = -2.303\ RT \log K_p = -2.303 \times 8.314\ J\ mol^{-1}\ K^{-1} \times 1000\ K \times \log 3.4$

$\Delta\bar{G}° = -1.0\times10^4\ J\ mol^{-1} = -1.0\times10^1\ kJ/mol$

(c) Use $K_c = 2.8\times10^2$ and

$$Q = \frac{[SO_3]^2}{[SO_2]^2[O_2]} = \frac{(0.40/2.50)^2}{(0.72/2.50)^2(0.18/2.50)} = \frac{(0.40)^2 \times 2.50}{(0.72)^2 \times (0.18)} = 4.3$$

$Q = 4.3 < K_c$. A net reaction occurs to the right.

(d) Use $\Delta\bar{G} = \Delta\bar{G}° + 2.303\ RT \log Q$. Substitute partial pressures into Q.

$$Q = \frac{(P_{SO_3})^2}{(P_{SO_2})^2(P_{O_2})} = \frac{\left(\frac{0.72\ RT}{2.50}\right)^2}{\left(\frac{0.40\ RT}{2.50}\right)^2\left(\frac{0.18\ RT}{2.50}\right)} = \frac{(0.72)^2 \times 2.50}{(0.40)^2 \times (0.18) \times RT}$$

$$= \frac{(0.72)^2 \times 2.50}{(0.40)^2 \times (0.18) \times 0.0821 \times 1000} = 0.55$$

$\Delta\bar{G} = -1.0\times10^4\ J/mol + (2.303 \times 8.314 \times 1000 \times \log 0.55)\ J/mol$

$= -1.0\times10^4\ J/mol - 0.50\times10^4\ J/mol = -1.5\times10^4\ J/mol$

Since $\Delta\bar{G}$ for the stated conditions is negative, the forward reaction occurs. (If $\Delta\bar{G}$ had been positive we would have concluded that the reverse reaction occurs.)

16-12. The required data from Appendix D are values of $\Delta\bar{G}^\circ_f$.

$2NO(g) + Cl_2(g) \rightleftharpoons 2NOCl(g)$

$\Delta\bar{G}^\circ = 2 \times \Delta\bar{G}^\circ_f[NOCl(g)] - 2 \times \Delta\bar{G}^\circ_f[NO(g) - \Delta\bar{G}^\circ_f[Cl_2(g)]$

$= 2 \times (+66.36) - 2 \times (+86.69) - 0 = -40.66$ kJ/mol

$\Delta\bar{G}^\circ = -2.303\ RT \log K_p \qquad \log K_p = \dfrac{-\Delta\bar{G}^\circ}{2.303\ RT}$

$$\log K_p = \frac{-(-40.66\times10^3)\ J\ mol^{-1}}{2.303 \times 8.314\ J\ mol^{-1}\ K^{-1} \times 298.15\ K} = 7.122$$

$K_p = \text{antilog } 7.122 = 1.32\times10^7$

16-13. $CO(g) + H_2(g) \rightarrow CO_2(g) + H_2O(g)$

$\Delta\bar{S}^\circ = \bar{S}^\circ[CO_2(g)] + \bar{S}^\circ[H_2O(g)] - \bar{S}^\circ[CO(g)] - \bar{S}^\circ[H_2(g)]$

$= 213.64 + 188.74 - 197.90 - 130.58 = +73.90\ J\ mol^{-1}\ K^{-1}$

16-14. If $\Delta\bar{G}^\circ$ is negative or of a small positive magnitude (e.g., +10 kJ/mol) we would expect the forward reaction to occur to a significant extent. If $\bar{G}^\circ$ is a large positive quantity (e.g., +50 kJ/mol or more) we would expect the gorward reaction to be negligible.

(a) $3O_2(g) \rightarrow 2O_3(g)$

$\Delta\bar{G}^\circ = 2 \times (163.43) - 3 \times 0 = +326.86$ kJ/mol

The forward reaction occurs to a negligible extent.

(b) $N_2O_4(g) \rightarrow 2NO_2(g)$

$\Delta\bar{G}^\circ = 2 \times (51.84) - (98.28) = +5.4$ kJ/mol

The forward reaction is expected to occur to some extent.

(c) $Br_2(l) + Cl_2(g) \rightarrow 2BrCl(g)$

$\Delta\bar{E}^\circ = 2 \times (-0.88) - 0 - 0 = -1.8$ kJ/mol

The forward reaction is expected to occur to a significant extent.

Exercises

First law of thermodynamics

16-1. (a) As pictured in Figure 6-1 a gas does work when it expands, whether the gas is ideal or nonideal.

(b) The gas must absorb heat from its surroundings. $\Delta E = q - w$. If the internal energy of an ideal gas depends only on the temperature, and the temperature is held constant, $\Delta E = 0$. If $\Delta E = 0$ and $w > 0$ (part a) then $q > 0$.

(c) The temperature of the gas remains constant.

(d) $\Delta E = 0$, for the reason described in part (b).

16-2. (a) A gas must do work to expand against its surroundings, whether the expansion occurs isothermally (as in Exercise 16-1) or adiabatically.

(b) No heat is exchanged with the surroundings ($q = 0$). This means that $\Delta E = q - w = -w$. The internal energy of the gas decreases by an amount equal to the work that is done.

(c) Since the internal energy of an ideal gas depends only on temperature, if the internal energy decreases the temperature must fall.

16-3. (a) It is not possible for an ideal gas expanding at constant temperature to do any more work than the quantity of heat it absorbs. That is, since $\Delta E = q - w$ and $E = 0$, $q = w$.

(b) If a gas were being compressed at constant temperature it would have to give off heat to the surroundings. However if the temperature is allowed to increase during the compression heat might be absorbed. That is, it is possible simultaneously to compress a gas and to heat it.

16-4. The water at the top of the waterfall has more potential energy than that at the bottom. When the water falls, this potential energy is transformed into heat, which raises the water temperature.

Enthalpy and internal energy

16-5. This exercise is similar to Review Problem 2, except that we must first write chemical equations for the reactions in question to evaluate $\Delta\bar{n}_g$.

(a) $C_4H_9OH(l) + 6O_2(g) \rightarrow 4CO_2(g) + 5H_2O(l)$

$\Delta\bar{n}_g = -2$; $\Delta\bar{H} = \Delta\bar{E} + 2.48\Delta\bar{n}_g$; $\Delta\bar{H} < \Delta\bar{E}$.

(b) $C_6H_{12}O_6(s) + 6O_2(g) \rightarrow 6CO_2(g) + 6H_2O(l)$

$\Delta\bar{n}_g = 0$; $\Delta\bar{H} = \Delta\bar{E}$.

(c) $NH_4NO_3(s) \rightarrow N_2O(g) + 2H_2O(l)$

$\Delta\bar{n}_g = +1$ $\Delta\bar{H} = \Delta\bar{E} + 2.48\Delta\bar{n}_g$; $\Delta\bar{H} > \Delta\bar{E}$.

16-6. (a) $$\Delta\bar{E} = \frac{-33.41}{\text{g } C_3H_7OH} \times \frac{60.10 \text{ g } C_3H_7OH}{1 \text{ mol } C_3H_7OH} = -2008 \text{ kJ/mol } C_3H_7OH$$

(b) Use the expression $\Delta\bar{H}(\text{in kJ}) = \Delta\bar{E}(\text{in kJ}) + 2.48\ \Delta\bar{n}_g$

$$\bar{H} = -2008 + 2.48 \times (3 - 9/2) = -2008 - 3.72 = -2012 \text{ kJ/mol } C_3H_7OH$$

16-7. The first law of thermodynamics is $\Delta E = q - w$. This equation can be rearranged to $q = \Delta E + w$. This expression describes the heat of a reaction, regardless of how the reaction is carried out. Thus expression (c) is the most general. Expressions (a) and (d) are limited to describing the heat of a reaction at constant volume and (b) and (e), to a heat of reaction at constant pressure.

Entropy and disorder

16-8. (a) Decrease: Greater order exists in a solid than in a liquid.

(b) Increase: The gaseous state is more disordered than the solid state. Direct passage of molecules from the solid to the gaseous state (sublimation) should be accompanied by an increase in entropy.

(c) Increase: The burning of a rocket fuel produces large volumes of gases from solid or liquid fuels.

16-9. (a) 1 mol H_2O(g, 1 atm, 50°C). For a given amount of substance at a given temperature, gaseous matter has a greater entropy than liquid.

(b) 50.0 g Fe(s, 1 atm, 20°C). Entropy is an extensive property. There is more matter in 50.0 g Fe than in 0.80 mol Fe (44.7 g).

(c) 1 mol Br_2(l, 1 atm, 50°C). Both the fact that a liquid is being compared to a solid, and that entropy increases with temperature, suggest that Br_2(l) at 50°C will have a higher entropy than Br_2(s) at -10°C.

(d) 0.10 mol O_2(g, 0.10 atm, 25°C). The gas under this condition must occupy a larger volume (100 times larger) than at 10.0 atm. Expanding a gas produces a more disordered structure and thus, a higher entropy.

16-10. (a) If an exothermic process ($\Delta H < 0$) is accompanied by a *decrease* in entropy ($\Delta S < 0$), it might not be spontaneous. For example, the freezing of liquid water is an exothermic process accompanied by a decrease in entropy. The process is spontaneous below 0°C but nonspontaneous above 0°C.

(b) Increase in entropy alone may not be a sufficient factor to make a reaction occur spontaneously if a large positive ΔH is involved. Thus, the process $H_2O(l, 1\ atm) \rightarrow H_2O(g, 1\ atm)$, for which $\Delta S > 0$, occurs spontaneously only at elevated temperatures--100°C or above.

16-11. According to the first law of thermodynamics (the law of conservation of energy) energy can neither be created nor destroyed. Therefore the energy of the universe (world) must be a constant. [Recall expressions in Chapter 6 of the sort: $q_{reaction} + q_{water} + q_{bomb} = 0$.] According to the second law of thermodynamics, all natural (spontaneous) processes must be accompanied by an increase in total entropy, that is, the entropy of the universe must increase. The combination of all natural processes, then, should cause the entropy of the universe to attain a maximum value.

Free energy and spontaneous change

16-12. For the decomposition of nitrosyl bromide we expect $\Delta\bar{S}° > 0$, since 1.5 mol is produced from 1 mol gas. Apply the criterion $\Delta\bar{G}° = \Delta\bar{H}° - T\Delta\bar{S}°$, with $\Delta\bar{H}° > 0$ and $\Delta\bar{S}° > 0$. This corresponds to case (3) in Table 16-2. The higher the temperature, the larger the term $T\Delta\bar{S}°$. At low temperatures, $\Delta\bar{G}° > 0$, and at sufficiently high temperatures, $\Delta\bar{G}° < 0$. We should expect the decomposition to occur to a greater extent at high temperatures.

16-13. There is no heat effect on mixing ideal gases; $\Delta H = 0$. As discussed in connection with Figure 16-4 $\Delta S > 0$ for the mixing process. With this combination of ΔH and ΔS, $\Delta G < 0$. (Recall that $\Delta G = \Delta H - T\Delta S$.)

16-14. The results here would be exactly the same as described for the mixing of ideal gases in Exercise 13, and for the same reasons. In the formation of an ideal solution, $\Delta H = 0$ and $\Delta S > 0$. This means that $\Delta G = \Delta H - T\Delta S < 0$.

Standard free energy change

16-15. From the data given determine $\Delta\bar{S}°$.

$$\Delta\bar{S}° = 2 \times \bar{S}°[POCl_3(l)] - 2 \times \bar{S}°[PCl_3(l)] - \bar{S}°[O_2(g)]$$

$$= (2 \times 222) - (2 \times 217) - 205$$

$$= -195\ J\ mol^{-1}\ K^{-1} = 0.195\ kJ\ mol^{-1}\ K^{-1}$$

$$\Delta\bar{H}° = -555\ kJ/mol$$

$$\Delta\bar{G}° = \Delta\bar{H}° - T\Delta\bar{S}°$$

$$= -555 - 298(-0.195)$$

$$= -555 + 58 = -497\ kJ/mol$$

16-16. Here we substitute the data given for the reaction into the Gibbs equation.

$$\Delta \bar{G}° = \Delta \bar{H}° - T\Delta \bar{S}° \qquad T = \frac{\Delta \bar{H}° - \Delta \bar{G}°}{\Delta \bar{S}°}$$

$$T = \frac{-843.7 \text{ kJ/mol} - (-777.8) \text{ kJ/mol}}{-0.165 \text{ kJ mol}^{-1} \text{ K}^{-1}} \qquad T = 399 \text{ K}$$

16-17. The bond energy of the F_2 molecule should be $\Delta \bar{H}°$ for the reaction $F_2(g) \rightarrow 2F(g)$; $\Delta \bar{G}° = 123.85$ kJ/mol

$$\Delta S° = 2\bar{S}°[F(g)] - \bar{S}°[F_2(g)] = 2 \times (158.7) - 203.3 = 114.1 \text{ J mol}^{-1} \text{ K}^{-1}$$

$$\Delta \bar{G}° = \Delta \bar{H}° - T\Delta \bar{S}° \qquad \Delta \bar{H}° = \Delta \bar{G}° + T\Delta \bar{S}° = (123.85 \text{ kJ/mol}) + (298.2 \text{ K} \times 0.1141 \text{ kJ mol}^{-1} \text{ K}^{-1})$$

$$= 157.87 \text{ kJ/mol}$$

The value listed in Table 9-3 is 155 kJ/mol.

16-18. We need to combine two equations in the manner introduced in Chapter 6 (Hess's law). For the second of the two equations, free energies of formation from Appendix D are used to establish $\Delta \bar{G}°$.

$$C_8H_{18}(l) + 25/2\ O_2(g) \rightarrow 8CO_2(g) + 9H_2O(l);\ \Delta \bar{G}° = -5.28\times10^3 \text{ kJ/mol}$$

$$9H_2O(l) \rightarrow 9H_2O(g);\ \Delta \bar{G}° = 9 \times (-228.61) - 9 \times (-237.19) = +77.22 \text{ kJ/mol}$$

$$C_8H_{18}(l) + 25/2\ O_2(g) \rightarrow 8CO_2(g) + 9H_2O;\ \Delta \bar{G}° = -5.20\times10^3 \text{ kJ/mol}$$

Free energy and equilibrium

16-19. The condition referred to corresponds to liquid and gaseous water in equilibrium, with the vapor pressure of the water equal to 0.50 atm. This will occur at a temperature above room temperature, but below 100°C (where the vapor pressure is 1 atm). Vapor pressure increases continuously with temperature, and there is only one temperature at which the vapor pressure is 0.50 atm.

16-20. A temperature of 110°C is below the normal melting point of I_2. The process $I_2(s) \rightarrow I_2(l)$ is nonspontaneous at this temperature, meaning that $\Delta G > 0$. If $\Delta G > 0$ then solid I_2 must have a lower free energy than liquid I_2 at 110°C and 1 atm.

16-21. At -60°C and 1 atm carbon dioxide exists solely as a gas. The conversion of the gas to either a liquid or solid would be nonspontaneous, that is, having $\Delta G > 0$. This means that $CO_2(g)$ has a lower free energy at -60°C and 1 atm than do either $CO_2(l)$ or $CO_2(s)$.

Phase transitions

16-22. (a) All that is required here is to apply data from Appendix D to the process $H_2O(l) \rightleftharpoons H_2O(g)$

$$\Delta \bar{H}°_{vap} = \Delta \bar{H}°_f[H_2O(g)] - \Delta \bar{H}°_f[H_2O(l)] = -241.84 - (-285.85) = +44.01 \text{ kJ/mol}$$

$$\Delta \bar{S}°_{vap} = \bar{S}°[H_2O(g)] - \bar{S}°[H_2O(l)] = 188.74 - (69.96) = 118.78 \text{ J mol}^{-1} \text{ K}^{-1}$$

(b) The values listed in Example 16-7 for 100°C are $\Delta \bar{H}°_{vap} = 40.7$ kJ/mol and $\Delta \bar{S}°_{vap} = 109$ J mol^{-1} K^{-1}. Each of the values at 25°C [calculated in (a)] is somewhat larger than the corresponding value at 100°C. We can explain this in terms of strong intermolecular forces of attraction (hydrogen bonds) at the lower temperature. More energy is required for vaporization ($\Delta \bar{H}°_{vap}$ is larger) and more disorder is created in the process ($\Delta \bar{S}°_{vap}$ is larger.)

16-23. Of the three liquids, we should expect appreciable hydrogen bonding to occur in HF and in CH_3OH. This would produce higher-than-normal heats of vaporization and Trouton's rule would fail. In toluene, $C_6H_5CH_3$, the intermolecular forces are of the London type. We should expect this liquid to obey Trouton's rule.

16-24. (a) From Appendix D, determine $\Delta\bar{H}°$ for the process $Br_2(l) \rightleftharpoons Br_2(g)$.

$$\Delta\bar{H}°_{vap} = \Delta\bar{H}°_f[Br_2(g)] - \Delta\bar{H}°_f[Br_2(l)] = 30.71 \text{ kJ/mol} - 0 = 30.71 \text{ kJ/mol}$$

(b) Now use Trouton's rule to estimate T_{bp}.

$$\Delta\bar{S}°_{vap} = \frac{\Delta\bar{H}°_{vap}}{T_{bp}} = \frac{30.71\times10^3 \text{ J/mol}}{T_{bp}} = 88 \text{ J mol}^{-1}\text{ K}^{-1}$$

$$T_{bp} = (30.71\times10^3/88)\text{K} = 3.5\times10^2 \text{ K}$$

16-25. At 298 K for $Hg(l) \rightleftharpoons Hg(g)$ $\quad\Delta\bar{H}° = \Delta\bar{H}°_f[Hg(g)] = 60.84$ kJ/mol

Assuming Trouton's rule: $\Delta\bar{S}° = \dfrac{\Delta\bar{H}°}{T_{bp}} = \dfrac{60.84 \text{ kJ/mol}}{T_{bp}} = 0.088 \text{ kJ mol}^{-1}\text{ K}^{-1}$ $\quad T_{bp} = 60.84/0.088 = 6.9\times10^2$ K

Relationship of $\Delta\bar{G}$, $\Delta\bar{G}°$, Q and K

16-26. $\Delta\bar{G}°$ was given in equation (16-20): +8.58 kJ/mol.

The reaction quotient: $Q = \dfrac{a_{H_2O(g)}}{a_{H_2O(l)}} = \dfrac{P_{H_2O(g)}(\text{in atm})}{1} = \dfrac{10/760}{1} = 0.013$

Now use equation (16.23): $\Delta\bar{G} = \Delta\bar{G}° + 2.303\ RT \log Q$

$$\Delta\bar{G} = 8.58\times10^3 \text{ J/mol} + [2.303 \times 8.314 \text{ J mol}^{-1}\text{ K}^{-1} \times 298 \text{ K} \times \log(0.013)]$$

$$= 8.58\times10^3 + [2.303 \times 8.314 \times 298 \times (-1.89)] = 8.58\times10^3 - 10.8\times10^3 = -2.2\times10^3 \text{ J/mol}$$

The process in Figure 16-7 (c) is pictured to be spontaneous, and this value of $\Delta\bar{G}$ indicates that indeed it should be.

16-27. (a) We are seeking $\Delta\bar{G}°$ for the process $CCl_4(l, 1\text{ atm}) \rightarrow CCl_4(g, 1\text{ atm})$.

$$\Delta\bar{H}° = \Delta\bar{H}°_f[CCl_4(g)] - \Delta\bar{H}°_f[CCl_4(l)] = -106.7 - (-139.3) = +32.6 \text{ kJ/mol}$$

$$\Delta\bar{G}° = \Delta\bar{H}° - T\Delta\bar{S}° = +32.6 \text{ kJ/mol} - (298 \text{ K} \times 0.09498 \text{ kJ mol}^{-1}\text{ K}^{-1}) = +32.6 - 28.3 = +4.3 \text{ kJ/mol}$$

Since the free energy change for the stated process, $\Delta\bar{G}°$, is a positive quantity, the process does not occur spontaneously.

(b) Use the expression $\Delta\bar{G}° = -2.303 \cdot RT \cdot \log K$ to obtain a value of K. $\Delta\bar{G}°$ was calculated in part (a).

$$\log K = \frac{-4.3\times10^3 \text{ J/mol}}{2.303 \times 8.314 \text{ J mol}^{-1}\text{ K}^{-1} \times 298 \text{ K}} = -0.75 \qquad K = 0.18$$

For the process $CCl_4(l) \rightleftharpoons CCl_4(g)$

$$K = \frac{a_{CCl_4(g)}}{a_{CCl_4(l)}} = \frac{P_{CCl_4(g)}}{1} = 0.18$$

The equilibrium vapor pressure of CCl_4 = 0.18 atm = 1.4×10^2 mmHg

The thermodynamic equilibrium constant

16-28. The quantity K must be dimensionless so that $\Delta\bar{G}°$ has the units J/mol, and K will be dimensionless only if activities are used in its formulation. K_c can be used instead of K if reactants and products are in solution or appear as pure solid(s) and/or liquid(s). Substituting molar concentrations (without units) into K_c yields a numerical value that is the same as K. If reactants and products are gases or appear as pure solid(s) and/or liquid(s), K_p may be used. However an equilibrium constant K_c using molar concentrations of gases could not be substituted for K (unless $K_c = K_p$).

16-29. (a) $K = \dfrac{(a_{H_2})^4}{(a_{H_2O(g)})^4} = \dfrac{(P_{H_2})^4}{(P_{H_2O})^4}$

(b) Because the activities of Fe and Fe_3O_4 are unity, regardless of their amounts, the partial pressures of H_2 and H_2O are independent of the amounts of Fe and Fe_3O_4.

(c) The equilibrium pressures of $H_2(g)$ and $H_2O(g)$ are independent of the quantities of Fe(s) and $Fe_3O_4(s)$ as long as enough of these substances are present to satisfy the stoichiometric requirements of the reaction. That is, the oxygen atoms that must be removed from H_2O to produce H_2 must combine with Fe atoms to form Fe_3O_4. Enough Fe atoms must be present to permit this. Beyond this point the equilibrium condition is independent of how much excess Fe is present. (The same line of reasoning applies to the reverse reaction involving H_2 and Fe_3O_4.) Thus, the conversion of H_2O to H_2 cannot be carried out with total disregard of the amounts of Fe and Fe_3O_4 present.

$\Delta\bar{G}°$ and K

16-30. (a) For the reaction $H_2(g) + I_2(g) \rightleftharpoons 2HI(g)$ $K = K_c = K_p$, Thus we can write

$\Delta\bar{G}° = -2.303\ RT\ \log K_p = -2.303 \times 8.314 \times (445 + 273) \times \log 50.2$

$\Delta\bar{G}° = -2.34\times10^4\ J\ mol^{-1} = -23.4\ kJ/mol$

(b) For the reaction $N_2O(g) + \frac{1}{2}O_2(g) \rightleftharpoons 2NO(g)$

$K = K_p = K_c(RT)^{\frac{1}{2}} = 1.7\times10^{-13} \times (0.0821 \times 298)^{\frac{1}{2}} = 8.4\times10^{-13}$

$\Delta\bar{G}° = -2.303\ RT\ \log K_p = -2.303 \times 8.314 \times 298 \times \log(8.4\times10^{-13}) = +6.89\times10^4\ J\ mol^{-1} = +68.9\ kJ/mol$

(c) For the reaction $N_2O_4(g) \rightleftharpoons 2NO_2(g)$

$K = K_p = K_c(RT)^1 = 4.6\times10^{-3} \times (0.0821 \times 298)^1 = 1.1\times10^{-1}$

$\Delta\bar{G}° = -2.303\ RT\ \log K_p = -2.303 \times 8.314 \times 298 \times \log(1.1\times10^{-1}) = 5.5\times10^3\ J\ mol^{-1} = 5.5\ kJ/mol$

(d) For the reaction $2SO_2(g) + O_2(g) \rightleftharpoons 2SO_3(g)$

$K = K_p = 9.1\times10^2$

$\Delta\bar{G}° = -2.303\ RT\ \log K_p = -2.303 \times 8.314 \times (800 + 273) \times \log(9.1\times10^2)$

$= -6.1\times10^4\ J\ mol^{-1} = -61\ kJ/mol$

(e) For the reaction $2Fe^{3+} + Hg_2^{2+} \rightleftharpoons 2Fe^{2+} + 2Hg^{2+}$

$K = K_c = 9.14\times10^{-6}$

$\Delta\bar{G}° = -2.303\ RT\ \log K_c = -2.303 \times 8.314 \times 298 \times \log(9.14\times10^{-6}) = -2.88\times10^4\ J\ mol^{-1} = +28.8\ kJ/mol$

16-31. (a) From the large positive value of $\Delta\bar{G}°$ alone we can conclude that the forward reaction occurs hardly at all at 298 K. By converting $\Delta\bar{G}°$ to K (i.e., $\Delta\bar{G}° = -2.303$ RT log K), we obtain a value of $K = K_p = P_{CO_2} = 1.6\times10^{-23}$ atm. This is an even more explicit indication that the the forward reaction is negligible.

(b) Since $\Delta\bar{H}° > 0$ the decomposition reaction is endothermic. According to Le Chatelier's principle the endothermic reaction is favored by raising the temperature.

16-32. From the data given in this exercise we can calculate $\Delta\bar{G}°$.

$$\Delta\bar{G}° = -2.303 \times 8.314 \text{ J mol}^{-1}\text{ K}^{-1} \times 298 \text{ K} \times \log(5.64\times10^{35}) = -2.04\times10^{5} \text{ J/mol} = -204 \text{ kJ/mol}$$

Use data from Appendix D to proceed as follows:

$$CO(g) + Cl_2(g) \rightleftharpoons COCl_2(g)$$

$$\Delta\bar{G}° = \Delta\bar{G}_f°[COCl_2(g)] - \bar{G}_f°[CO(g)] = -204 \text{ kJ/mol}$$

$$\Delta\bar{G}_f°[COCl_2(g)] - (-137.28 \text{ kJ/mol}) = -204 \text{ kJ/mol}$$

$$\Delta\bar{G}_f°[COCl_2(g)] = -204 - 137 = -341 \text{ kJ/mol}$$

16-33. A value of K can be calculated with equation (16.25).

$$\log K = \frac{-\Delta\bar{G}°}{2.303\cdot RT} = \frac{-119.82\times10^{3} \text{ J/mol}}{2.303 \times 8.314 \text{ J mol}^{-1}\text{ K}^{-1} \times 298 \text{ K}} = -21.0$$

$$K = K_p = \frac{(P_{CO})^2}{(P_{CO_2})} = 1\times10^{-21} \qquad (P_{CO})^2 = 1\times10^{-21}(P_{CO_2}) \qquad P_{CO} = 3\times10^{-11}\sqrt{P_{CO_2}}$$

Thus, if $CO_2(g)$ is maintained at 298 K at a pressure of 1.00 atm in contact with C(s), the equilibrium partial pressure of CO would be 3×10^{-11} atm = 2×10^{-8} mmHg. Conversion of $CO_2(g)$ to CO(g) at room temperature is indeed a very limited reaction.

16-34. $Hg(l) \rightleftharpoons Hg(g)$

$$\Delta\bar{G}° = \Delta\bar{G}_f°[Hg(g)] - \Delta\bar{G}_f°[Hg(l)] = 31.76 - 0 = 31.76 \text{ kJ/mol}$$

$$\Delta\bar{G}° = -2.303 \text{ RT} \log K$$

$$\log K = \frac{-31.76\times10^{3} \text{ J mol}^{-1}}{2.303 \times 8.314 \text{ J mol}^{-1}\text{ K}^{-1} \times 298 \text{ K}} = -5.566$$

$K = 2.72\times10^{-6}$ But $K = K_p = P_{Hg}$

The vapor pressure of Hg at 298 K is 2.72×10^{-6} atm or 2.07×10^{-3} mmHg.

16-35. If K is very large, a reaction goes essentially to completion. If K is very small, a reaction proceeds in the forward direction only to a very limited extent. Let us say that for a range of K values from about 1×10^{-2} to 1×10^{2} we might generally expect significant equilibrium amounts of all reactants and products. Now calculate $\Delta\bar{G}°$ values corresponding to these values of K.

For $K = 1x10^{-2}$

$$\Delta\bar{G}° = -2.303\cdot RT\cdot\log K = -2.303 \times 8.314 \text{ J mol}^{-1}\text{ K}^{-1} \times 298 \text{ K} \log(1\times10^{-2})$$

$$\Delta\bar{G}° = 1\times10^{4} \text{ J/mol} = 10 \text{ kJ/mol}$$

For $K = 1x10^2$

$\Delta\bar{G}° = -2.303 \cdot RT \cdot \log K = -2.303 \times 8.314\ J\ mol^{-1}\ K^{-1} \times 298\ K \times \log(1x10^2)$

$\Delta\bar{G}° = -1x10^4\ J/mol = -10\ kJ/mol$

According to this calculation, we expect to find appreciable equilibrium amounts of all reactants and products only if $\Delta\bar{G}°$ is either a small positive or small negative value (say, ranging from about -10 to +10 kJ/mol).

The variation of $\Delta\bar{G}°$ with temperature

16-36. (a) $CO(g) + H_2O(g) \rightleftharpoons CO_2(g) + H_2(g)$

$\Delta\bar{H}° = \Delta\bar{H}_f°[CO_2(g)] - \Delta\bar{H}_f°[CO(g)] - \Delta\bar{H}_f°[H_2O(g)]$

$= -393.51 - (-100.54) - (-241.84) = -41.13\ kJ/mol$

$\Delta\bar{S}° = \bar{S}°[CO_2(g)] + \bar{S}°[H_2(g)] - \bar{S}°[CO(g)] - \bar{S}°[H_2O(g)]$

$= 213.64 + 130.58 - 197.90 - 188.74 = -42.42\ J\ mol^{-1}\ K^{-1}$

$\Delta\bar{G}° = \Delta\bar{G}_f°[CO_2(g)] - \Delta\bar{G}_f°[CO(g)] - \Delta\bar{G}_f°[H_2O(g)]$

$= -394.38 - (-137.28) - (-228.61) = -28.49\ kJ/mol$

Also, $\Delta\bar{G}° = \Delta\bar{H}° - T\Delta\bar{S}° = -41.13 - 298(-0.04242) = -41.13 + 12.64 = -28.49\ kJ/mol$

(b) at 1100 K: $\Delta\bar{G}° = -41.13\ kJ/mol - 1100 \times (-0.04242)\ kJ/mol$

$= -41.13\ kJ/mol + 46.66\ kJ/mol = +5.53\ kJ/mol$

$\log K = \frac{-\Delta\bar{G}°}{2.303 \cdot RT} = \frac{-5.53x10^3\ J/mol}{2.303 \times 8.314\ J\ mol^{-1}\ K^{-1} \times 1100\ K} = -0.26 \qquad K = K_p = 0.55$

(c) In Example 15-6 the value of K_c was given as 1.00. For this reaction $K_p = K_c(RT)^{\Delta n} = K_c(RT)^0$. $K_p = K_c = 1.00$. Agreement between this value and the one calculated in part (b) is rather good, given the long extrapolation of data from 298 K to 1100 K.

16-37. Use data from Appendix D to establish $\Delta\bar{H}°$ and $\Delta\bar{S}°$ at 298 K.

$\Delta\bar{H}° = 2\ \Delta\bar{H}_f°[SO_3(g)] - 2\ \Delta\bar{H}_f°[SO_2(g)] = 2 \times (-395.18\ kJ/mol) - 2 \times (-296.90\ kJ/mol) = -196.56\ kJ/mol$

$\Delta\bar{S}° = 2\ \bar{S}°[SO_3(g)] - 2\ \bar{S}°[SO_2(g)] - \bar{S}°[O_2(g)] = 2 \times (256.23) - 2 \times (248.53) - 205.02$

$= -189.62\ J\ mol^{-1}\ K^{-1}$

At the temperature in question, $\Delta\bar{G}° = \Delta\bar{H}° - T\Delta\bar{S}° = -2.303 \cdot RT \cdot \log K$

$\Delta\bar{G}° = (-196.56x10^3) - T(-189.62) = -2.303 \times 8.314 \cdot T \cdot \log 1.0x10^6$

$189.62\ T + 115\ T = 196.56x10^3 \qquad T = 644\ K$

The value calculated in Example 15-16 was 635 K. Agreement between the two methods is good.

16-38. (a) Determine $\Delta\bar{H}°$ for the reaction $C(s) + CO_2(g) \rightleftharpoons 2\ CO(g)$

$$\Delta\bar{H}° = 2\ \Delta\bar{H}_f°[CO(g)] - \Delta\bar{H}_f°[CO_2(g)] = 2 \times (-110.54) - (-393.51) = +172.43 \text{ kJ/mol}$$

Since the forward reaction is endothermic, the conversion of $CO_2(g)$ to CO(g) is favored at high temperatures.

(b) Determine $\Delta\bar{S}°$ for the reaction in part (a).

$$\Delta\bar{S}° = 2\ \bar{S}°[CO(g)] - \bar{S}°[CO_2(g)] - \bar{S}°[C(graphite)] = (2 \times 197.90) - 213.64 - 5.69$$

$$= 176.47 \text{ J mol}^{-1}\text{ K}^{-1}$$

We are looking for the equilibrium condition: $C(graphite) + CO_2(g, 1\text{ atm}) \rightleftharpoons 2\ CO(g, 1\text{ atm})$

$$K = K_p = \frac{(P_{CO})^2}{(P_{CO_2})} = \frac{(1)^2}{(1)} = 1.00 \qquad \Delta\bar{G}° = -2.303 \cdot RT \cdot \log K_p = -2.303\ RT \log 1.00 = 0$$

Now solve the following expression for T:

$$\Delta\bar{G}° = \Delta\bar{H}° - T\Delta\bar{S}° = 0 \qquad T\Delta\bar{S}° = \Delta\bar{H}°$$

$$T = \frac{\Delta\bar{H}°}{\Delta\bar{S}°} = \frac{172.43 \text{ kJ/mol}}{0.17647 \text{ kJ mol}^{-1}\text{ K}^{-1}} = 977 \text{ K}$$

16-39. From the data given on the % conversion of N_2 and O_2 to NO, determine the value of $K_p = K_c$. Base this determination on 1.00 mol each of N_2 and O_2 in a vessel of volume V.

	$N_2(g)$	+	$O_2(g)$	$\rightleftharpoons$	2 NO(g)
Initial amounts:	1.00 mol		1.00 mol		
Changes:	-0.01 mol		-0.01 mol		+0.02 mol
Equilibrium amounts:	0.99 mol		0.99 mol		0.02 mol
Equilibrium concentrations:	(0.99/V)M		(0.99/V)M		(0.02/V)M

$$K = K_p = K_c = \frac{[NO]^2}{[N_2][O_2]} = \frac{(0.02/V)^2}{(0.99/V)(0.99/V)} = \frac{(0.02)^2}{(0.99)^2} = 4\times10^{-4}$$

Now use this value of K to establish the corresponding $\Delta\bar{G}°$.

$$\Delta\bar{G}° = -2.303\ RT \log K = -2.303 \times 8.314 \times T \times \log(4\times10^{-4}) = 65\ T$$

For the reverse of the reaction in Example 16-12 we have, at 298 K

$\Delta\bar{H}° = +180.74$ kJ/mol and $\Delta\bar{S}° = +24.72$ J mol^{-1} K^{-1}

Assume that these values also hold at the temperature T.

$$\Delta\bar{G}° = \Delta\bar{H}° - T\Delta\bar{S}°$$

$$65\ T = 180.74\times10^3 - T(-24.72)$$

$$T = \frac{180.74\times10^3}{(65 + 25)} = 2.0\times10^3 \text{ K}$$

Heat engines

16-40. The thermodynamic limitation on the efficiency of the conversion of heat to work is

$$\% \text{ Efficiency} = \frac{T_h - T_l}{T_h} \times 100$$

Only in the case where $T_l \doteq 0$ would the ratio $(T_h - T_l/T_h) = 1.00$ and the efficiency be 100%. But it is not possible to operate a heat engine with its low temperature at 0 K (the absolute zero of temperature is unattainable). The efficiency of an engine could be increased by raising T_h to the point where $T_h \simeq T_h - T_l$ but there are also practical limits on how high a temperature one can use (e.g., containers would melt).

16-41. $$\text{efficiency} = \frac{T_h - T_l}{T_h} = \frac{(300 + 273) - (20 + 273)}{(300 + 273)} = 0.489$$

The maximum efficiency is 48.9%.

16-42. (a) $\text{efficiency} = \frac{T_h - T_l}{T_h} = \frac{T_h - 313}{T_h} = 0.36 \qquad T_h - 313 = 0.36\ T_h \qquad 0.64\ T_h = 313 \qquad T_h = 4.9 \times 10^2\ K$

(b) In order to offset other losses, e.g. between the turbine and the generator, the thermodynamic cycle would have to be more than 36% efficient if the overall conversion of heat to electrical work is to be 36% efficient. This higher efficiency for the thermodynamic cycle would require a higher steam temperature.

Self-Test Questions

1. (*b*) Use the expression $\Delta E = q - w$, with $\Delta E = +100$ J and $q = -100$ J. Under these conditions $w = q - \Delta E = -100 \text{ J} - 100 \text{ J} = -200$ J. The surroundings must do 200 J of work on the system.

2. (*d*) In this reaction 2 mol gaseous reactants yield 2 mol gaseous product--$\Delta n_g = 0$. Since $\Delta H = \Delta E + 2.48\ n_g$, if $\Delta n_g = 0$, $\Delta H = \Delta E$.

3. (*d*) It is not always the case that the entropy of a system increases in a spontaneous process (e.g., the freezing of water). Neither is it required that the entropy of the surroundings increase. Given these facts, a spontaneous process does not require that the entropy of both the system and surroundings increase. What is required is that the total entropy--the entropy of the universe--increase.

$$\Delta S_{univ.} = \Delta S_{sys.} + \Delta S_{surr.} > 0.$$

4. (*b*) The heat of a reaction is related to ΔH; the change in molecular disorder, to ΔS; and the rate of a reaction to the rate constant, k, and the rate law. The change in free energy, ΔG, is related to the direction of spontaneous change. If $\Delta G < 0$ the forward reaction is spontaneous, and if $\Delta G > 0$, the reverse reaction.

5. (*a*) Consider the expression $\Delta G = \Delta H - T\Delta S$. If $\Delta H < 0$ and $\Delta S < 0$, at low temperatures the magnitude of ΔH exceeds that of $T\Delta S$; $\Delta G < 0$ and the forward reaction is spontaneous. At high temperatures, $-T\Delta S$ is a positive quantity that exceeds ΔH; $\Delta G > 0$.

6. (*c*) If it is necessary to use an external agent (electricity) to produce a change, the change is nonspontaneous. For a nonspontaneous change, $\Delta G > 0$. The other values listed, ΔH and ΔS, might either be positive or negative.

7. (*b*) This is a dissociation reaction in which Br-Br bonds are broken ($\Delta H > 0$). The dissociation is accompanied by an increase in entropy ($\Delta S > 0$). The reaction is spontaneous at some temperatures ($\Delta G < 0$) but not at all temperatures.

8. (*d*) Since $\Delta\bar{G}° = -2.303 \cdot RT \cdot \log K$, if $\Delta\bar{G}° = 0$ then $\log K = 0$. $\log K = 0$ if $K = 1$. Based on the expression $\Delta\bar{G}° = \Delta\bar{H}° - T\Delta\bar{S}°$, if $\Delta\bar{G}° = 0$ then $\Delta\bar{H}° = T\Delta\bar{S}°$. This does not require that either $\Delta\bar{H}°$ or $\Delta\bar{S}°$ be equal to zero.

9. (a) The criterion for spontaneous change is $\Delta G = \Delta H - T\Delta S < 0$. If $\Delta S > 0$ and $\Delta H < 0$, then clearly $\Delta G < 0$; the change is spontaneous. However, even though $\Delta S > 0$, if $\Delta H > 0$, at low temperatures $\Delta G > 0$; the change is nonspontaneous. Thus, we must know both ΔH and ΔS, not ΔS alone, in order to predict the direction of spontaneous change.

(b) The expression $\Delta\bar{G} = \Delta\bar{G}° + 2.303 \cdot RT \cdot \log Q$ allows us to calculate ΔG for any set of activities of reactants and products expressed through the reaction quotient, Q. If $\Delta G < 0$ the reaction is spontaneous for the given conditions; if $\Delta G > 0$, the reaction is nonspontaneous. But in order to use this expression we must have a value of $\Delta\bar{G}°$--the free energy change when reactants and products are in their standard states.

10. We start with the basic expression $\Delta\bar{G} = \Delta\bar{G}° + 2.303 \cdot RT \cdot \log Q$. If a reaction is at equilibrium, $\Delta\bar{G} = 0$ and $Q = K$. This leads to the expression $0 = \Delta\bar{G}° + 2.303 \cdot RT \cdot \log K$ and $\Delta\bar{G}° = -2.303 \cdot RT \cdot \log K$.

11. (a) $C_5H_{10}(l) \rightarrow C_5H_{10}(g)$

$$\Delta\bar{H}°_{vap} = \Delta\bar{H}°_f[C_5H_{10}(g)] - \Delta\bar{H}°_f[C_5H_{10}(l)] = -77.2 - (-105.9) = 28.7 \text{ kJ/mol}$$

To estimate the normal boiling point, assume Trouton's rule:

$$\Delta\bar{S}°_{vap} = 88 \text{ J mol}^{-1} \text{ K}^{-1} \text{ and}$$

$$T = \frac{\Delta\bar{H}°}{\Delta\bar{S}°} = \frac{28.7 \text{ kJ/mol}}{0.088 \text{ kJ mol}^{-1} \text{ K}^{-1}} = 3.3 \times 10^2 \text{ K}$$

(b) $$\Delta\bar{G}°_{vap} = \Delta\bar{H}°_{vap} - T\Delta\bar{S}°_{vap} = 28.7 \text{ kJ/mol} - (298 \text{ K} \times 0.088 \text{ kJ mol}^{-1} \text{ K}^{-1})$$

$$= 28.7 \text{ kJ/mol} - 26 \text{ kJ/mol} = 3 \text{ kJ/mol}$$

(c) The positive sign of $\Delta\bar{G}°$ signifies that the production of $C_5H_{10}(g)$ at 1.00 atm pressure from $C_5H_{10}(l)$ will not proceed spontaneously at 298 K. That is, the equilibrium vapor pressure of C_5H_{10} at 298 K must be less than 1.00 atm.

12. (a) $NH_4NO_3(s) \rightarrow N_2O(g) + 2\ H_2O(l)$

$$\Delta\bar{H}° = \Delta\bar{H}°_f[N_2O(g)] + 2\ \Delta\bar{H}°_f[H_2O(l)] - \Delta\bar{H}°_f[NH_4NO_3(s)]$$

$$= 81.55 + 2 \times (-285.85) - (-365.56) = -124.59 \text{ kJ/mol}$$

The reaction is exothermic.

(b) $$\Delta\bar{S}° = \bar{S}°[N_2O(g)] + 2\ \bar{S}°(H_2O(l)] - \bar{S}°[NH_4NO_3(s)]$$

$$= 219.99 + (2 \times 69.96) - 151.08 = 208.8 \text{ J mol}^{-1} \text{ K}^{-1}$$

$$\Delta\bar{G}° = \Delta\bar{H}° - T\Delta\bar{S}° = -124.59 - (298 \times 0.2088) = -186.81 \text{ kJ/mol}$$

(c) $$\log K = \frac{-\Delta\bar{G}°}{2.303 \cdot RT} = \frac{-(-186.81) \times 10^3 \text{ J/mol}}{2.303 \times 8.314 \text{ J mol}^{-1} \text{ K}^{-1} \times 298 \text{ K}} = 32.7 \qquad K = 5 \times 10^{32}$$

(d) Because $\Delta\bar{H}° < 0$ and $\Delta\bar{S}° > 0$ the reaction is spontaneous at all temperatures (see case 1, Table 16-2).

Chapter 17

Acids and Bases

Review Problems

17-1. Acids and bases are indicated under each species in the equations below.

(a) $HOBr + H_2O \rightleftharpoons H_3O^+ + OBr^-$
acid base acid base

(b) $HSO_4^- + H_2O \rightleftharpoons H_3O^+ + SO_4^{2-}$
acid base acid base

(c) $HS^- + H_2O \rightleftharpoons H_2S + OH^-$
base acid acid base

(d) $C_6H_5NH_3^+ + OH^- \rightleftharpoons C_6H_5NH_2 + H_2O$
acid base base acid

17-2. The Lewis structures are written first. From these structures, determine whether the species is electron deficient or has lone pair electrons available for covalent bond formation.

(a) $[:\ddot{O}-H]^-$ (Available lone pair electrons--a Lewis base.)

(b) H-Ö-B-Ö-H with :O: bonded to B, and H bonded to that O (The B atom has an available orbital to receive an electron pair--a Lewis acid.)

(c) :Cl-Al-Cl: with :Cl: bonded to Al (The situation with the Al atom is the same as with the B atom in part (b)--$AlCl_3$ is a Lewis acid.)

(d) H-C(H)(H)-N̈(H)-H (The N atom has a lone pair of electrons available for sharing--a Lewis base.)

17-3. All the substances are either strong acids or strong bases. Determine the concentration of one ion (H_3O^+ or OH^-) from the molar concentration of acid or base. Determine the other concentration with the K_w expression for water.

(a) 0.0030 M HCl: $[H_3O^+] = 3.0\times10^{-3}$ M

$$[OH^-] = \frac{K_w}{[H_3O^+]} = \frac{1.0\times10^{-14}}{3.0\times10^{-3}} = 3.3\times10^{-12}\ M$$

(b) 0.045 M NaOH: $[OH^-] = 0.045\ M = 4.5\times10^{-2}$ M

$$[H_3O^+] = \frac{K_w}{[OH^-]} = \frac{1.0\times10^{-14}}{4.5\times10^{-2}} = 2.2\times10^{-13}\ M$$

(c) 0.0015 M $Sr(OH)_2$: $[OH^-] = \frac{0.0015 \text{ mol } Sr(OH)_2}{L} \times \frac{2 \text{ mol } OH^-}{1 \text{ mol } Sr(OH)_2} = 3.0\times10^{-3}$ M

$$[H_3O^+] = \frac{K_w}{[OH^-]} = \frac{1.0\times10^{-14}}{3.0\times10^{-3}} = 3.3\times10^{-12}\ M$$

(d) 7.2×10^{-3} M HNO_3: $[H_3O^+] = 7.2\times10^{-3}$ M

$$[OH^-] = \frac{K_w}{[H_3O^+]} = \frac{1.0\times10^{-14}}{7.2\times10^{-3}} = 1.4\times10^{-12}\ M$$

17-4. In a similar fashion to Review Problem 3, determine $[H_3O^+]$ (the solutions are either strong acids or strong bases). Then calculate the pH.

(a) 1.0×10^{-3} M HCl: $[H_3O^+] = 1.0\times10^{-3}$ M $\quad$ pH = $-\log [H_3O^+]$

$$pH = -(\log 1.0\times10^{-3}) = -(-3.00) = 3.00$$

(b) 0.000180 M HBr: $[H_3O^+] = 1.80\times10^{-4}$ M

$$pH = -\log [H_3O^+] = -\log (1.80\times10^{-4}) = 3.745$$

(c) 1.15×10^{-3} M NaOH: $[OH^-] = 1.15\times10^{-3}$ M $\quad [H_3O^+] = \dfrac{1.0\times10^{-14}}{1.15\times10^{-3}} = 8.7\times10^{-12}$ M

$$pH = -\log [H_3O^+] = -\log (8.7\times10^{-12}) = 11.06$$

Alternatively, $pOH = -\log (1.15\times10^{-3}) = 2.939$

$$pH = 14.00 - pOH = 14.00 - 2.94 = 11.06$$

(d) 4.1×10^{-4} M NaOH: $[OH^-] = 4.1\times10^{-4}$ M $\quad pOH = -\log (4.1\times10^{-4}) \doteq 3.39$

$$pH = 14.00 - pOH = 14.00 - 3.39 = 10.61$$

17-5. This problem is similar to Review Problem 4, except that the final result must be expressed as pOH rather than pH.

(a) 1.0×10^{-2} M NaOH: $[OH^-] = 1.0\times10^{-2}$ M; $pOH = -(\log 1.0\times10^{-2}) = 2.00$

(b) 0.0068 M LiOH: $[OH^-] = 6.8\times10^{-3}$ M; $pOH = -(\log 6.8\times10^{-3}) = 2.17$

(c) 0.00520 M $Ba(OH)_2$: $[OH^-] = \dfrac{0.00520 \text{ mol } Ba(OH)_2}{L} \times \dfrac{2 \text{ mol } OH^-}{1 \text{ mol } Ba(OH)_2} = 1.04\times10^{-2}$ M

$$pOH = -\log (1.04\times10^{-2}) = 1.983$$

(d) 3.51×10^{-4} M HCl: $[H_3O^+] = 3.51\times10^{-4}$ M; $pH = -(\log 3.51\times10^{-4}) = 3.455$

$$pOH = 14.00 - pH = 14.00 - 3.455 = 10.54$$

17-6. The calculations here are the reverse of Review Problems 4 and 5--conversion of pH to $[H_3O^+]$. This requires use of antilogarithms.

(a) pH = 6.0: $\log [H_3O^+] = -pH = -6.0$ $\quad [H_3O^+] = 1\times10^{-6}$ M

(b) pH = 3.15: $\log [H_3O^+] = -pH = -3.15$ $\quad [H_3O^+] = 7.1\times10^{-4}$ M

(c) pH = 0.65: $\log [H_3^+] = -pH = -0.65$ $\quad [H_3O^+] = 0.22$ M

(d) pH = 4.10: pH = 14.00 − pOH = 14.00 − 4.10 = 9.90

$$[H_3O^+] = \text{antilog } (-9.90) = 1.3\times10^{-10} \text{ M}$$

(e) pOH = 11.15 $\quad$ pH = 14.00 − 11.15 = 2.85 $\quad [H_3O^+] = \text{antilog } (-2.85) = 1.4\times10^{-3}$ M

17-7. Convert the data about the $Ba(OH)_2(aq)$ to $[OH^-]$.

$$[OH^-] = \frac{1.06 \text{ g } Ba(OH)_2{\cdot}8H_2O \times \dfrac{1 \text{ mol } Ba(OH)_2{\cdot}8H_2O}{316 \text{ g } Ba(OH)_2{\cdot}8H_2O} \times \dfrac{2 \text{ mol } OH^-}{1 \text{ mol } Ba(OH)_2{\cdot}8H_2O}}{0.575 \text{ L}} = 1.17\times10^{-2} \text{ M}$$

$$pOH = -\log [OH^-] = -\log (1.17\times10^{-2}) = 1.932$$

$$pH = 14.00 - 1.932 = 12.07$$

17-8. Determine no. mol H_3O^+ in the acid and no. mol OH^- in the base.

$$\text{no. mol } H_3O^+ = 0.02500 \text{ L} \times \frac{0.150 \text{ mol } HNO_3}{L} \times \frac{1 \text{ mol } H_3O^+}{1 \text{ mol } HNO_3} = 3.75\times10^{-3} \text{ mol } H_3O^+$$

$$\text{no. mol } OH^- = 0.01000 \text{ L} \times \frac{0.412 \text{ mol KOH}}{L} \times \frac{1 \text{ mol } OH^-}{1 \text{ mol KOH}} = 4.12\times10^{-3} \text{ mol } OH^-$$

H_3O^+ and OH^- combine in a 1:1 mole ratio to produce H_2O. H_3O^+ is the limiting reagent and OH^- is in excess.

no. mol OH^- in excess = 4.12×10^{-3} mol OH^- available - 3.75×10^{-3} mol OH^- reacted

$= 0.37\times10^{-3}$ mol $OH^- = 3.7\times10^{-4}$ mol OH^-

In the mixed solution of 25.00 + 10.00 = 35.00 mL,

$$[OH^-] = \frac{3.7\times10^{-4}\text{ mol }OH^-}{0.03500\text{ L}} = 1.1\times10^{-2}\text{ M}$$

$$[H_3O^+] = \frac{1.0\times10^{-14}}{1.1\times10^{-2}} = 9.1\times10^{-13}\text{ M}$$

17-9. The molar concentration of butyric acid placed in solution is

$$[HC_4H_7O_2] = \frac{0.355\text{ mol }HC_4H_7O_2}{0.715\text{ L}} = 0.497\text{ M}$$

In the ionization equilibrium:

	$HC_4H_7O_2$	+	H_2O	$\rightleftharpoons$	H_3O^+	+	$C_4H_7O_2^-$
Place in soln:	0.497 M				----		----
Changes:	-2.73×10^{-3} M				$+2.73\times10^{-3}$ M		$+2.73\times10^{-3}$ M
Equil. Concns:	0.494 M				2.73×10^{-3} M		2.73×10^{-3} M

$$K_a = \frac{[H_3O^+][C_4H_7O_2^-]}{[HC_4H_7O_2]} = \frac{(2.73\times10^{-3})(2.73\times10^{-3})}{0.494} = 1.51\times10^{-5}$$

17-10. (a) In the acetic acid solution $[H_3O^+] = [C_2H_3O_2^-]$. Assume that $[HC_2H_3O_2] = 0.815$ M (i.e., assume that the $HC_2H_3O_2$ is largely nonionized).

$$K_a = \frac{[H_3O^+][C_2H_3O_2^-]}{[HC_2H_3O_2]} = \frac{[C_2H_3O_2^-][C_2H_3O_2^-]}{0.815} = 1.74\times10^{-5}$$

$$[C_2H_3O_2^-] = \sqrt{0.815\times1.74\times10^{-5}} = 3.77\times10^{-3}\text{ M}$$

(b) Proceed as in (a) to calculate $[H_3O^+]$ and then determine the pH.

$$K_a = \frac{[H_3O^+][C_8H_7O_2^-]}{[HC_8H_7O_2]} = \frac{[H_3O^+][H_3O^+]}{0.105} = 4.9\times10^{-5}$$

$$[H_3O^+] = \sqrt{0.105\times4.9\times10^{-5}} = 2.3\times10^{-3}\text{ M}$$

$$pH = -\log[H_3O^+] = -\log(2.3\times10^{-3}) = 2.64$$

(c) If pH = 4.86, $\log[H_3O^+] = -4.86$, and $[H_3O^+] = 1.4\times10^{-5}$ M. In the ionization equilibrium, $[H_3O^+] = [C_6H_4ClO^-]$ and assume that the *o*-chlorophenol is essentially nonionized. That is, substitute and solve for $[HC_6H_4ClO]$ in the expression

$$K_a = \frac{[H_3O^+][C_6H_4ClO^-]}{[HC_6H_4ClO]} = \frac{(1.4\times10^{-5})(1.4\times10^{-5})}{[HC_6H_4ClO]} = 3.2\times10^{-9}$$

$$[HC_6H_4ClO] = \frac{(1.4\times10^{-5})^2}{3.2\times10^{-9}} = 6.1\times10^{-2} = 0.061\text{ M}$$

17-11. The ionization equilibrium can be represented as

	$(CH_3)_3N$	+	H_2O	$\rightleftharpoons$	$(CH_3)_3NH^+$	+	OH^-
Place in Soln:	1.52 M				----		---
Changes:	$-x$ M				$+x$ M		$+x$ M
Equil. Concns:	$(1.52 - x)$M				x M		x M

$$K_b = \frac{[(CH_3)_3NH^+][OH^-]}{[(CH_3)_3N]} = \frac{x \cdot x}{(1.52 - x)} = 6.2\times10^{-5}$$

Assume that $x << 1.52$ and obtain

$x^2 = 1.52 \times 6.2\times10^{-5} = 9.4\times10^{-5}$ $\qquad x = [(CH_3)_3NH^+] = 9.7\times10^{-3}$ M

(The assumption that $9.7\times10^{-3} << 1.52$ is a reasonably good one.)

17-12. Both parts of this question can be answered by determining $[H_3O^+]$.

	$HC_3H_5O_2$	+	H_2O	$\rightleftharpoons$	H_3O^+	+	$C_3H_5O_2^-$
Place in Soln:	0.25 M						
Changes:	$-x$ M				$+x$ M		$+x$ M
Equil. Concns:	$(0.25 - x)$M				x M		x M

$$K_a = \frac{[H_3O^+][C_3H_5O_2^-]}{[HC_3H_5O_2]} = \frac{x \cdot x}{0.25 - x} = 1.35\times10^{-5}$$

Assume that $x << 0.25$ and obtain

$x^2 = 0.25 \times 1.34\times10^{-5} = 3.4\times10^{-6}$

$x = [H_3O^+] = 1.8\times10^{-3}$ M

(a) $\alpha = \dfrac{1.8\times10^{-3} \text{ mol } H_3O^+/L}{0.25 \text{ mol } HC_3H_5O_2/L} = 0.0072$

(b) % dissociation = 100 × 0.0072 = 0.72%

17-13. $[H_3O^+]$ and $[HCO_3^-]$ are calculated from Ka_1 and $[CO_3^{2-}]$ from Ka_2.

	H_2CO_3	+	H_2O	$\rightleftharpoons$	H_3O^+	+	HCO_3^+
Placed in Soln:	0.015 M						
Changes:	$-x$ M				$+x$ M		$+x$ M
Equil. Concns:	$(0.015 - x)$M				x M		x M

Assume that $x << 0.015$ and write

$$Ka_1 = \frac{[H_3O^+][HCO_3^-]}{[H_2CO_3]} = \frac{x \cdot x}{0.015} = 4.2\times10^{-7}$$

$x^2 = 0.015 \times 4.2\times10^{-7} = 6.3\times10^{-9}$ $\qquad x = 7.9\times10^{-5}$

(a) $[H_3O^+] = 7.9\times10^{-5}$ M

(b) $[HCO_3^-] = [H_3O^+] = 7.9\times10^{-5}$ M

(c) $[CO_3^{2-}] = Ka_2 = 5.6\times10^{-11}$ M

17-14. Use the idea that salts of strong acids and strong bases are pH neutral; the salt of a strong acid and a weak base is acidic; and the salt of a weak acid and a strong base is basic.

(a) KCl (strong acid + strong base): neutral

(b) NH_4NO_3 (strong acid + weak base): acidic

(c) $NaNO_2$ (weak acid + strong base): basic

(d) NaI (strong acid + strong base): neutral

(e) $Ca(OCl)_2$ (weak acid + strong base): basic

17-15. The hydrolysis equilibrium is

	NH_4^+	+	H_2O	$\rightleftharpoons$	H_3O^+	+	NH_3
Place in Soln:	0.25 M						
Changes:	$-x$ M				$+x$ M		$+x$ M
Equil. Concns:	$(0.25 - x)$M				x M		x M

$$K_a = \frac{[H_3O^+][NH_3]}{[NH_4^+]} = \frac{x \cdot x}{0.25 - x} = 5.75\times10^{-10}$$

Assume $x<<0.25$ and write

$$x^2 = 0.25 \times 5.57\times10^{-10} = 1.4\times10^{-10}$$

$$x = [H_3O^+] = \sqrt{1.4\times10^{-10}} = 1.2\times10^{-5} \text{ M}$$

$$pH = -\log [H_3O^+] = -\log (1.2\times10^{-5}) = 4.92$$

17-16. What is required in each case is either K_w/K_a or K_w/K_b where K_a and K_b refer to the conjugate acids and bases. Numerical values of K_a and K_b are obtained from Table 17-2.

(a) $K_a(C_5H_5NH^+) = \dfrac{K_w}{K_b(C_5H_5N)} = \dfrac{1.0\times10^{-14}}{2.0\times10^{-9}} = 5.0\times10^{-6}$

(b) $K_b(CHO_2^-) = \dfrac{K_w}{K_a(HCHO_2)} = \dfrac{1.0\times10^{-14}}{1.8\times10^{-4}} = 5.6\times10^{-11}$

(c) $K_b(C_6H_5O^-) = \dfrac{K_w}{K_a(HC_6H_5O)} = \dfrac{1.0\times10^{-14}}{1.6\times10^{-10}} = 6.2\times10^{-5}$

17-17. (a) HI is more acidic than HBr because the H - I bond is weaker than is the H - Br bond.

(b) HOClO is expected to be a stronger acid than HOI for two reasons: The central atom Cl is more electronegative than I and is better able to weaken the O - H bond by withdrawing electrons from it. Also, the Cl atom has one O atom bonded directly to it (i.e., other than -OH). This too, according to the rule stated in Section 17-8, increases the acid strength.

(c) Cl_3CCH_2COOH is a stronger acid than H_3CCH_2COOH because the highly electronegative Cl atoms withdraw electrons from the C - O - H bond (inductive effect).

(d) $H_3CCH_2CF_2COOH$ is a stronger acid than $I_3CCH_2CH_2COOH$ because the F atoms are much more electronegative than I atoms and because they are closer to the C - O - H bond from which electrons are to be withdrawn by the inductive effect.

Exercises

Brønsted-Lowry theory of acids and bases

17-1. (a) HNO_2 is a proton donor; it is acidic.

(b) OCl^- cannot be a proton donor. We can only think of it as accepting a proton to form HOCl; it is basic.

(c) In order to act as an acid (proton donor), NH_2^- would have to form the dinegative ion NH^{2-}. More plausible is that NH_2^- should gain a proton to form NH_3. The ion NH_2^- is basic.

(d) NH_4^+ does not have the capacity to form an additional N - H bond; it cannot act as a base. On the other hand, by losing a proton NH_4^+ is converted to the neutral molecule, NH_3. NH_4^+ is acidic.

(e) Similar to NH_4^+, methyl ammonium ion acts as a proton donor to form methyl amine. $C_6H_5NH_3^+$ is acidic.

17-2. In each case the conjugate acid is the product of adding a proton to the anion listed:

(a) HOH (or H_2O) (b) HCl (c) HOCl (d) HCN

17-3. The conjugate base is the product remaining after the species in question loses a proton.

(a) IO_4^- (b) $C_3H_5O_2^-$ (c) $C_6H_5COO^-$ (d) $C_6H_5NH_2$

17-4. (a) OH^- is not amphiprotic. It can easily gain a proton to form H_2O, but the loss of a proton would produce O^{2-}, which does not exist in aqueous solution.

(b) NH_3 is amphiprotic. Loss of a proton produces NH_2^- and gain of a proton, NH_4^+.

(c) H_2O is amphiprotic. Loss of a proton produces OH^-, and gain of a proton, H_3O^+. Both of these species can exist in aqueous solution.

(d) HS^- is amphiprotic. Loss of a proton produces S^{2-} and gain of a proton, H_2S.

(e) NO_3^- can act as a base in very strongly acidic solutions, but it cannot act as an acid.

(f) HCO_3^- is amphiprotic. Loss of a proton produces CO_3^{2-} and gain of a proton, H_2CO_3.

(g) HSO_4^- is amphiprotic. Loss of a proton produces SO_4^{2-} and gain of a proton (in strongly acidic solutions), H_2SO_4.

(h) HNO_3 is a strong acid. It has no basic properties.

17-5. (a) $C_2H_3O_2^- + H^+$ (from an acid) $\rightarrow HC_2H_3O_2$ (the solvent)

$C_2H_3O_2^-$ is a base.

(b) $H_2O + HC_2H_3O_2 \rightarrow (H_2C_2H_3O_2)^+ + OH^-$ $H_2O + HC_2H_3O_2 \rightarrow H_3O^+ + C_2H_3O_2^-$

H_2O is an acid. H_2O is a base.

(c) $HC_2H_3O_2 + HC_2H_3O_2 \rightarrow (H_2C_2H_3O_2)^+ + C_2H_3O_2^-$

$HC_2H_3O_2$ is both an acid and a base.

(d) $HClO_4 + HC_2H_3O_2 \rightarrow (H_2C_2H_3O_2)^+ + ClO_4^-$

$HClO_4$ is an acid.

Lewis theory of acids and bases

17-6. (a) According to the following Lewis structures, SO_3 appears to be the electron pair acceptor (acid) and H_2O, the electron pair donor (base). Note that an additional sulfur-to-oxygen bond is formed.

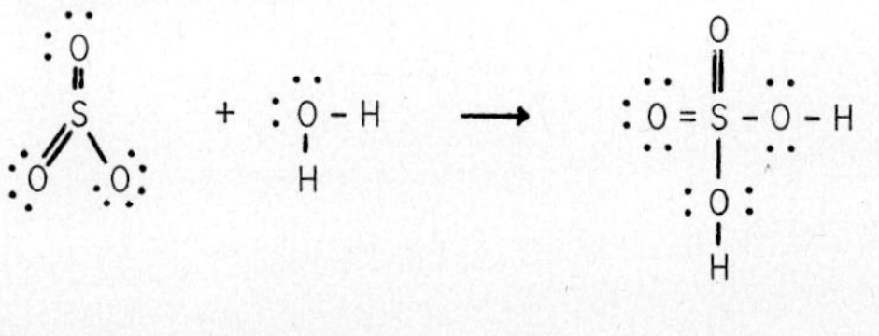

(b) The OH^- has lone pair electrons that it can make available for bond formation with Al^{3+} in $Al(OH)_3$. The result is a complex ion $[Al(OH)_4]^-$. This situation is similar to the formation of $[Zn(NH_3)_4]^{2+}$ described in Example 17-2b. $Al(OH)_3(s)$ is the acid and $OH^-(aq)$ is the base.

17-7. (a) The ion O^{2-} in CaO can be represented as $[:\ddot{O}:]^{2-}$. Because of the presence of available electron pairs, we should expect O^{2-} (and thus CaO) to act as a base. The $SO_2(g)$ must act as an acid. The acid-base reaction is similar to that depicted in Example 17-2(c). One possible representation of the reaction is shown below.

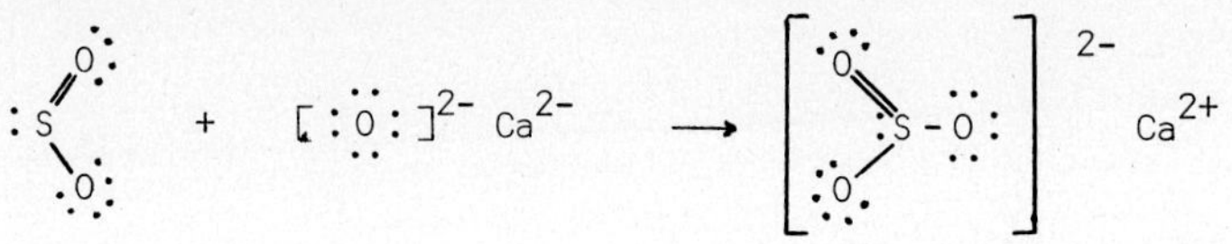

$$SO_2 + [:\ddot{O}:]^{2-}\ Ca^{2-} \longrightarrow [SO_3]^{2-}\ Ca^{2+}$$

(b) Here, LiOH is a base and CO_2, an acid.

$$O{=}C{=}O \; + \; [:\ddot{O}:H]^- Li^+ \longrightarrow \left[O{=}C(-O{:}H){=}O \right]^- Li^+ \longrightarrow \left[O{-}C(-O{:}H){=}O \right]^- Li^+$$

Strong acids, strong bases, and pH

17-8. Proceed in the following steps:

Calculate the number of moles of HCl:

$$PV = nRT; \; n = \frac{PV}{RT} = \frac{(740/760)\text{atm} \times 0.0312 \text{ L}}{0.0821 \text{ L atm mol}^{-1}\text{K}^{-1} \times 303 \text{ K}} = 1.22\times10^{-3} \text{ mol HCl}$$

Calculate the number of moles of H_3O^+ in solution: $HCl + H_2O \rightarrow H_3O^+ + Cl^-$

$$\text{no. mol } H_3O^+ = 1.22\times10^{-3} \text{ mol HCl} \times \frac{1 \text{ mol } H_3O^+}{1 \text{ mol HCl}} = 1.22\times10^{-3} \text{ mol } H_3O^+$$

Calculate $[H_3O^+]$:

$$[H_3O^+] = \frac{1.22\times10^{-3} \text{ mol } H_3O^+}{3.25 \text{ L}} = 3.75\times10^{-3} \text{ M}$$

17-9. $\text{no. mol } H_3O^+ = 0.215 \text{ L} \times \frac{3.00\times10^{-3} \text{ mol HI}}{1 \text{ L}} \times \frac{1 \text{ mol } H_3O^+}{1 \text{ mol HI}} = 6.45\times10^{-4} \text{ mol } H_3O^+$

$$\text{no. mol } H_3O^+ = 0.475 \text{ L} \times \frac{6.40\times10^{-2} \text{ mol HCl}}{1 \text{ L}} \times \frac{1 \text{ mol } H_3O^+}{1 \text{ mol HCl}} = 3.04\times10^{-2} \text{ mol } H_3O^+$$

$$\text{Total no. mol } H_3O^+ = 6.45\times10^{-4} + 3.04\times10^{-2} = 3.10\times10^{-2}$$

$$[H_3O^+] = \frac{3.10\times10^{-2} \text{ mol } H_3O^+}{(0.215 + 0.475)\text{L}} = 4.49\times10^{-2} \text{ M}$$

$$pH = -\log [H_3O^+] = -\log (4.49\times10^{-2}) = 1.348$$

17-10. The simplest approach here is to calculate the number of moles of H_3O^+ in the final dilute solution, and then the volume of the concentrated acid required to provide this much H_3O^+. Required first is a calculation of $[H_3O^+]$ from pH.

$$pH = -\log [H_3O^+] = 1.85; \; \log [H_3O^+] = -1.85; \; [H_3O^+] = 1.4\times10^{-2} \text{ M}$$

$$\text{no. mL conc. acid} = 6.55 \text{ L} \times \frac{1.4\times10^{-2} \text{ mol } H_3O^+}{1 \text{ L}} \times \frac{1 \text{ mol HCl}}{1 \text{ mol } H_3O^+} \times \frac{36.5 \text{ g HCl}}{1 \text{ mol HCl}} \times \frac{100 \text{ g conc. acid}}{36.0 \text{ g HCl}}$$

$$\times \frac{1 \text{ mL conc. acid}}{1.13 \text{ g conc. acid}} = 7.9 \text{ mL conc. acid}$$

17-11. If $pH = 10.53$, $pOH = 14.00 - 10.53 = 3.47$, $[OH^-] = \text{antilog } (-3.47) = 3.4\times10^{-4}$ M

$$\text{no. mol } Mg(OH)_2/L = 3.4\times10^{-4} \frac{\text{mol } OH^-}{L} \times \frac{1 \text{ mol } Mg(OH)_2}{2 \text{ mol } OH^-} = 1.7\times10^{-4} \text{ mol } Mg(OH)_2$$

$$\text{no. mg } Mg(OH)_2/L = \frac{1.7\times10^{-4} \text{ mol } Mg(OH)_2}{L} \times \frac{58.3 \text{ g } Mg(OH)_2}{1 \text{ mol } Mg(OH)_2} \times \frac{1000 \text{ mg } Mg(OH)_2}{1 \text{ g } Mg(OH)_2} = 9.9 \text{ mg } Mg(OH)_2/L$$

17-12. A solution with pH = 8 is basic. There is no way that a basic solution can be prepared by dissolving an *acid* in water. If one prepared a 1×10^{-8} M HCl solution in water, $[H_3O^+] \neq 1\times10^{-8}$ M because in this dilute a solution more H_3O^+ would appear in solution as a result of the self-ionization of water than is derived from the strong acid HCl. Including the self-ionization of water in the calculation of $[H_3O^+]$ would result in $[H_3O^+] > 1\times10^{-7}$ M, which it should be for an acidic solution.

Neutralization

17-13. The neutralization reaction is $HCl(aq) + NaOH(aq) \rightarrow NaCl(aq) + H_2O$.

$$\text{no. mol NaOH} = 0.0251\text{L soln} \times \frac{0.218 \text{ mol NaOH}}{\text{L soln}} = 5.47\times10^{-3} \text{ mol NaOH}$$

no. mol HCl required for the neutralization = 5.47×10^{-3} mol HCl

$$\text{no. mL HCl soln} = 5.47\times10^{-3} \text{ mol HCl} \times \frac{1 \text{ L soln}}{0.151 \text{ mol HCl}} \times \frac{1000 \text{ mL soln}}{1 \text{ L soln}} = 36.2 \text{ mL}$$

17-14. Here the neutralization reaction is $KOH(aq) + HBr(aq) \rightarrow KBr(aq) + H_2O$.

$$\text{no. mol HBr} = 0.02500 \text{ L soln} \times \frac{0.1051 \text{ mol HBr}}{\text{L soln}} = 2.628\times10^{-3} \text{ mol HBr}$$

no. mol KOH required for neutralization reaction = 2.628×10^{-3} mol KOH

$$\text{molar concentration} = \frac{2.628\times10^{-3} \text{ mol KOH}}{0.02122 \text{ L}} = 0.1238 \text{ M KOH}$$

17-15. Convert the information about the two solutions to no. mol H_3O^+ and no. mol OH^-.

pH = 2.52 $[H_3O^+]$ = antilog (-2.52) = 3.0×10^{-3} M

$$\text{no. mol } H_3O^+ = 0.02500 \text{ L} \times \frac{3.0\times10^{-3} \text{ mol } H_3O^+}{\text{L}} = 7.5\times10^{-5} \text{ mol } H_3O^+$$

pH = 12.05 pOH = 14.00 - 12.05 = 1.95 $[OH^-]$ = antilog (-1.95) = 1.1×10^{-2} M

$$\text{no. mol } OH^- = 0.02500 \text{ L} \times \frac{1.1\times10^{-2} \text{ mol } OH^-}{\text{L}} = 2.8\times10^{-4} \text{ mol } OH^-$$

OH^- is in excess; H_3O^+ is the limiting reactant. The final solution is basic.

no. mol OH^- in excess = $2.8\times10^{-4} - 7.5\times10^{-5} = 2.0\times10^{-4}$ mol OH^-

The final solution volume is 25.00 mL + 25.00 mL = 50.00 mL.

$$[OH^-] = \frac{2.0\times10^{-4} \text{ mol } OH^-}{0.0500 \text{ L}} = 4.0\times10^{-3} \text{ M}$$

pOH = -log $[OH^-]$ = -log (4.0×10^{-3}) = 2.40

pH = 14.00 - pOH = 14.00 - 2.40 = 11.60

Weak acids, weak bases, and pH

17-16.

$$\text{no. mol } HC_6H_{11}O_2/\text{L} = \frac{11 \text{ g } HC_6H_{11}O_2 \times \frac{1 \text{ mol } HC_6H_{11}O_2}{116 \text{ g } HC_6H_{11}O_2}}{1.00 \text{ L}} = 0.095 \text{ M}$$

pH = 2.94 log $[H_3O^+]$ = -2.94 = 0.06 - 3.00 $[H_3O^+] = 1.1\times10^{-3}$ M

	$HC_6H_{11}O_2(aq)$	+	H_2O	$\rightleftharpoons$	$H_3O^+(aq)$	+	$C_6H_{11}O_2^-(aq)$
Placed in Soln:	0.095 M				---		---
Changes:	-1.1×10^{-3} M				$+1.1\times10^{-3}$ M		$+1.1\times10^{-3}$ M
Equilibrium:	0.094 M				1.1×10^{-3} M		1.1×10^{-3} M

$$K_a = \frac{[H_3O^+][C_6H_{11}O_2^-]}{[HC_6H_{11}O_2]} = \frac{(1.1\times10^{-3})(1.1\times10^{-3})}{0.094} = 1.3\times10^{-5}$$

17-17. pH = 10.86; pOH = 14.00 - 10.86 = 3.14 = -log $[OH^-]$; log $[OH^-]$ = -3.14; $[OH^-] = 7.2\times10^{-4}$

	$CH_3C_5H_4N$	+	H_2O	$\rightleftharpoons$	$CH_3C_5H_4NH$	+	OH^-
Placed in Soln:	0.250 M				---		---
Changes:	-7.2×10^{-4} M				7.2×10^{-4} M		7.2×10^{-4} M
Equil. Concns:	$(0.250 - 7.2\times10^{-4})$M				7.2×10^{-4} M		7.2×10^{-4} M

$$K_b = \frac{[CH_3C_5H_4NH^+][OH^-]}{[CH_3C_5H_4N]} = \frac{(7.2\times10^{-4})(7.2\times10^{-4})}{0.249} = 2.1\times10^{-6}$$

17-18. pH = 4.53 = -log $[H_3O^+]$ log $[H_3O^+]$ = -4.53 = 0.47 - 5.00 $[H_3O^+] = 3.0\times10^{-5}$ M

	$HC_6H_4NO_3$	+	H_2O	$\rightleftharpoons$	H_3O^+	+	$C_6H_4NO_3^-$; $K_a = 5.9\times10^{-8}$
Dissolve:	x M				---		---
Changes:	-3.0×10^{-5} M				$+3.0\times10^{-5}$ M		$+3.0\times10^{-5}$ M
Equilibrium:	$(x - 3.0\times10^{-5})$M				3.0×10^{-5} M		3.0×10^{-5} M

Assume $x >> 3.0\times10^{-5}$, so that $(x - 3.0\times10^{-5}) \approx x$.

$$K_a = \frac{[H_3O^+][C_6H_4NO_3^-]}{[HC_6H_4NO_3]} = \frac{(3.0\times10^{-5})(3.0\times10^{-5})}{x} = 5.9\times10^{-8}$$

$x = [HC_6H_4NO_3] = 1.5\times10^{-2}$ M--the assumption is valid.

no. g $HC_6H_4NO_3$/L = 1.5×10^{-2} mol $HC_6H_4NO_3$/L x 139 g $HC_6H_4NO_3$/mol $HC_6H_4NO_3$ = 2.1 g $HC_6H_4NO_3$/L

17-19. The number of moles of acetylsalicyclic acid per liter of solution is:

$$\frac{\text{no. mol } HC_9H_7O_4}{L} = 2 \times 0.32 \text{ g } HC_9H_7O_4 \times \frac{1 \text{ mol } HC_9H_7O_4}{180 \text{ g } HC_9H_7O_4} = 1.4\times10^{-2} \text{ mol } HC_9H_7O_4/L$$

In the ionization of acetylsalicylic acid, let $x = [H_3O^+] = [C_9H_7O_4^-]$.

$[HC_9H_7O_4] = 1.4\times10^{-2} - x \approx 1.4\times10^{-2}$.

$$K_a = \frac{[H_3O^+][C_9H_7O_4^-]}{[HC_9H_7O_4]} = \frac{x^2}{1.4\times10^{-2}} = 2.75\times10^{-5};\ x^2 = 3.8\times10^{-7}$$

$x = [H_3O^+] = 6.2\times10^{-4}$ M; pH = -log $[H_3O^+]$ = -log (6.2×10^{-4}) = 3.21

To test the assumption that $1.4\times10^{-2} - x \approx 1.4\times10^{-2}$, substitute $x = 6.2\times10^{-4}$. That is, $1.4\times10^{-2} - 6.2\times10^{-4} = 1.3\times10^{-2} \approx 1.4\times10^{-2}$ (to two significant figures).

17-20. pH = 1.30 = -log $[H_3O^+]$ log $[H_3O^+]$ = -1.30 = 0.70 - 2.00 $[H_3O^+] = 5.0\times10^{-2}$

	$HC_2HCl_2O_2$	+	H_2O	$\rightleftharpoons$	H_3O^+	+	$C_2HCl_2O_2^-$
Placed in Soln:	0.10 M				---		---
Changes:	-0.05 M				+0.05 M		+0.05 M
Equilibrium:	0.05 M				0.05 M		0.05 M

$$K_a = \frac{[H_3O^+][C_2HCl_2O_2^-]}{[HC_2HCl_2O_2]} = \frac{0.05 \times 0.05}{0.05} = 5\times10^{-2}$$

17-21. $pH = 4.62 = -\log[H_3O^+]$ $\quad \log[H_3O^+] = -4.62 = 0.38 - 5.00$ $\quad [H_3O^+] = 2.4\times10^{-5}$ M

	$HC_2H_3O_2$	+	H_2O	$\rightleftharpoons$	H_3O^+	+	$C_2H_3O_2^-$
Placed in Soln:	x M				---		---
Changes:	-2.4×10^{-5} M				$+2.4\times10^{-5}$ M		$+2.4\times10^{-5}$ M
Equilibrium:	$(x - 2.4\times10^{-5})$M				2.4×10^{-5} M		2.4×10^{-5} M

$$K_a = \frac{[H_3O^+][C_2H_3O_2^-]}{[HC_2H_3O_2]} = \frac{(2.4\times10^{-5})(2.4\times10^{-5})}{x - 2.4\times10^{-5}} = 1.74\times10^{-5}$$

Assume the $x >> 2.4\times10^{-5}$, leading to $\frac{(2.4\times10^{-5})^2}{x} = 1.74\times10^{-5}$

$x = 3.3\times10^{-5}$ M. The assumption just made is *not* valid.

$$\frac{(2.4\times10^{-5})^2}{x - 2.4\times10^{-5}} = 1.74\times10^{-5} \qquad 5.8\times10^{-10} = 1.74\times10^{-5}x - 4.2\times10^{-10} \qquad x = 5.7\times10^{-5}\text{ M}$$

$$\text{no. mg } HC_2H_3O_2 = 1.00\text{ L} \times \frac{5.7\times10^{-5}\text{ mol } HC_2H_3O_2}{1.00\text{ L}} \times \frac{60\text{ g } HC_2H_3O_2}{1\text{ mol } HC_2H_3O_2} \times \frac{1000\text{ mg } HC_2H_3O_2}{1.00\text{ g } HC_2H_3O_2} = 3.4\text{ mg } HC_2H_3O_2$$

17-22.

	$HC_2H_2ClO_2$	+	H_2O	$\rightleftharpoons$	H_3O^+	+	$C_2H_2ClO_2^-$
Placed in Soln:	4.5×10^{-3} M				---		---
Changes:	$-x$ M				$+x$ M		$+x$ M
Equilibrium:	$(4.5\times10^{-3} - x)$M				x M		x M

Because this acid is rather dilute and the value of K_a is rather large, we should not expect the usual simplifying assumption to hold.

$$K_a = \frac{[H_3O^+][C_2H_2ClO_2^-]}{[HC_2H_2HClO_2]} = \frac{x\cdot x}{(4.5\times10^{-3} - x)} = 1.35\times10^{-3} \qquad x^2 + 1.35\times10^{-3}x - 6.1\times10^{-6} = 0$$

$$x = [C_2H_2ClO_2^-] = \frac{-1.35\times10^{-3} \pm \sqrt{(1.35\times10^{-3})^2 + 4\times6.1\times10^{-6}}}{2} = 1.9\times10^{-3}\text{ M}$$

17-23.

$$[C_5H_{11}N] = \frac{0.118\text{ g } C_5H_{11}N \times \frac{1\text{ mol } C_5H_{11}N}{85.1\text{ g } C_5H_{11}N}}{0.287} = 4.83\times10^{-3}\text{ mol } C_5H_{11}N/L$$

	$C_5H_{11}N$	+	H_2O	$\rightleftharpoons$	$C_5H_{11}NH^+$	+	OH^-
Placed in Soln:	4.83×10^{-3} M				---		---
Changes:	$-x$ M				$+x$ M		$+x$ M
Equilibrium:	$(4.83\times10^{-3} - x)$M				x M		x M

As in Exercise 22, because we are dealing with a fairly dilute solution of a base with $K_b > 1\times10^{-5}$, we should not expect the usual simplifying assumption to work well. That is, 4.84×10^{-3} is not large with respect to x.

$$K_a = \frac{[C_5H_{11}NH^+][OH^-]}{[C_5H_{11}N]} = \frac{x\cdot x}{(4.83\times10^{-3} - x)} = 1.6\times10^{-3} \qquad x^2 + 1.6\times10^{-3}x - 7.7\times10^{-6} = 0$$

$$x = \frac{-1.6\times10^{-3} \pm \sqrt{(1.6\times10^{-3})^2 + 4\times7.7\times10^{-6}}}{2} = 2.1\times10^{-3}$$

$[OH^-] = 2.1\times10^{-3}$ M $\quad pOH = -\log[OH^-]$ $\quad pOH = -\log(2.1\times10^{-3})$ $\quad pOH = 2.68$

$pH = 14.00 - 2.68 = 11.32$

Degree of ionization

17-24. If an $HC_2H_3O_2(aq)$ solution of molarity x is 1.00% ionized, we can write:

	$HC_2H_3O_2$	+	H_2O	$\rightleftharpoons$	H_3O^+	+	$C_2H_3O_2^-$
Placed in Soln:	x M				---		---
Changes:	$-0.0100x$ M				$+0.0100x$ M		$+0.0100x$ M
Equilibrium:	$(x-0.0100x)$M				$0.0100x$ M		$0.0100x$ M

$$K_a = \frac{[H_3O^+][C_2H_3O_2^-]}{[HC_2H_3O_2]} = \frac{(0.0100\,x)(0.0100\,x)}{0.9900x} = 1.74\times10^{-5} \qquad 1.00\times10^{-4}x = 1.7\times10^{-5}$$

$x = 1.72\times10^{-1}$ M $HC_2H_3O_2$ = 0.172 M $HC_2H_3O_2$

17-25. If we could make the same simplifying assumption as in Example 17-10 [that x << M (the molarity of acid)], we would expect 0.0010 M $HC_2H_3O_2$ to be 13% ionized and 0.00010 M $HC_2H_3O_2$ to be 42% ionized. But this assumption is not valid when the molarity of the acid is very low (below about 0.01 M), therefore, these predictions of the percent ionization of acetic acid are also *invalid*.

17-26. Calculate $[H_3O^+]$ in the trichloroacetic acid solution, and then determine the % ionization.

	$HC_2Cl_3O_2$	+	H_2O	$\rightleftharpoons$	H_3O^+	+	$C_2Cl_3O_2^-$
Placed in Soln:	0.050 M				---		---
Changes:	$-x$ M				$+x$ M		$+x$ M
Equilibrium:	$(0.050 - x)$M				x M		x M

$$K_a = \frac{[H_3O^+][C_2Cl_3O_2^-]}{[HC_2Cl_3O_2]} = \frac{x\cdot x}{0.050 - x} = 2.0\times10^{-1}$$

$$x^2 + 0.20x - 0.010 = 0$$

$$x = \frac{-0.20 \pm \sqrt{(0.20)^2 + 4\times0.010}}{2} = \frac{-0.20 \pm 0.28}{2} = \frac{0.08}{2} = 0.04$$

$$\%\ \text{ionization} = \frac{[H_3O^+]}{\text{initial }[HC_2Cl_3O_2]}\times100 = \frac{x}{[HC_2Cl_3O_2]_{init.}}\times100 = \frac{0.04}{0.050}\times100 = 8\times10^{1}\ \%$$

17-27. Begin with the usual set-up

	$HC_4H_5O_2$	+	H_2O	$\rightleftharpoons$	H_3O^+	+	$C_4H_5O_2^-$
Placed in Soln:	0.0500 M				---		---
Changes:	$-x$ M				$+x$ M		$+x$ M
Equilibrium:	$(0.0500 - x)$M				x M		x M

The total concentration of particles in solution is $(0.0500 - x) + x + x = (0.0500 + x)$M ≈ $(0.0500 + x)$m (Molarity and molality are essentially equal in a dilute aqueous solution). An Alternative expression for molality comes from the freezing point data.

$$\Delta T_f = K_f\cdot m \qquad m = \frac{\Delta T_f}{K_f} = \frac{0.096°C}{1.86°C\ (\text{kg solv})\ \text{mol}^{-1}} = 0.052\ m$$

$0.052\ m = (0.0500 + x)m \qquad x = 0.052 - 0.0500 = 0.002\ m = 0.002$ M

Now set-up the expression for K_a.

$$K_a = \frac{[H_3O^+][C_4H_5O_2^-]}{[HC_4H_5O_2^-]} = \frac{x\cdot x}{0.0500 - x} = \frac{(0.002)(0.002)}{0.048} = 8\times10^{-5}$$

Polyprotic acids

17-28. Because in each of these solutions the second ionization step is so limited, we can assume that $[H_3O^+] = [HS^-]$ in each solution and that $[S^{2-}] = Ka_2 = 1.0\times10^{-14}$ M.

(a)

	H_2S	+	H_2O	$\rightleftharpoons$	H_3O^+	+	HS^-
Placed in Soln:	0.075 M				---		---
Changes:	$-x$ M				$+x$ M		$+x$ M
Equilibrium:	$(0.075 - x)$M				x M		x M

$$Ka_1 = \frac{[H_3O^+][HS^-]}{[H_2S]} = \frac{x \cdot x}{0.075 - x} \qquad \text{Assume } x << 0.075 \qquad \frac{x^2}{0.075} = 1.1\times10^{-7}$$

$x^2 = 8.2\times10^{-9}$ $\quad x = 9.1\times10^{-5}$ $\quad [H_3O^+] = [HS^-] = 9.1\times10^{-5}$ M $\quad [S^{2-}] = 1.0\times10^{-14}$ M

(b) Assume that $x << 0.0050$.

$$Ka_1 = \frac{x \cdot x}{0.0050 - x} = \frac{x^2}{0.0050} = 1.1\times10^{-7} \qquad x^2 = 5.5\times10^{-10} \qquad x = 2.3\times10^{-5}\text{ M}$$

(The assumption is valid.)

$[H_3O^+] = [HS^-] = 2.3\times10^{-5}$ M $\qquad [S^{2-}] = 1.0\times10^{-14}$ M

(c) Assume that $x << 1.0\times10^{-5}$.

$$Ka_1 = \frac{x \cdot x}{(1.0\times10^{-5} - x)} = \frac{x^2}{1.0\times10^{-5}} = 1.1\times10^{-7} \qquad x^2 = 1.1\times10^{-12} \qquad x = 1.0\times10^{-6}$$

$$1.0\times10^{-5} - x = 1.0\times10^{-5} - 1.0\times10^{-6} = 0.9\times10^{-5} \simeq 1.0\times10^{-5}$$

The assumption works fairly well.

$[H_3O^+] = [HS^-] = 1.0\times10^{-6}$ M $\qquad [S^{2-}] = 1.0\times10^{-14}$ M

17-29. (a) For the same reasons given in the text in the discussion of a diprotic acid, for a triprotic acid we should expect $Ka_1 > Ka_2 > Ka_3$.

(b) If $Ka_1 >> Ka_2$, the extent of ionization in the second step is much less than in the first step.

We can assume that $[H_3O^+] = [H_2A^-]$. Then, $Ka_2 = \frac{\cancel{[H_3O^+]}[HA^{2-}]}{\cancel{[H_2A^-]}}$. For this expression to hold true it is also necessary that very little HA^{2-} ionize further. This will be the case if $Ka_2 >> Ka_3$.

(c) We should not expect $[A^{3-}]$ to be equal to Ka_3, because in the following expression $[H_3O^+]$ is not equal to $[HA^{2-}]$.

$$Ka_3 = \frac{[H_3O^+][A^{3-}]}{[HA^{2-}]}$$

17-30. Assume that ionization occurs primarily in the first step. Calculate $[H_3O^+]$, $[H_2PO_4^-]$, and $[H_3PO_4]$ in the usual way.

	H_3PO_4	+	H_2O	$\rightleftharpoons$	H_3O^+	+	$H_2PO_4^{2-}$
Placed in Soln:	0.100 M				---		---
Changes:	$-x$ M				$-x$ M		$-x$ M
Equilibrium:	$(0.100 - x)$M				x M		x M

$$Ka_1 = \frac{[H_3O^+][H_2PO_4^{2-}]}{[H_3PO_4]} = \frac{x \cdot x}{0.100 - x} = 5.9\times10^{-3} \qquad x^2 + 5.9\times10^{-3}x - 5.9\times10^{-4} = 0$$

$$x = \frac{-5.9\times10^{-3} \pm \sqrt{(5.9\times10^{-3})^2 + 4 \times 5.9\times10^{-4}}}{2} = 2.2\times10^{-2}$$

Use the Ka_2 expression to calculate $[HPO_4^{2-}]$: $Ka_2 = \frac{[H_3O^+][HPO_4^{2-}]}{[H_2PO_4^-]} = 6.2\times10^{-8}$

$[HPO_4^{2-}] = 6.2\times10^{-8}$ M

Use the Ka_3 expression to calculate $[PO_4^{3-}]$: $Ka_3 = \frac{[H_3O^+][PO_4^{3-}]}{[HPO_4^{2-}]} = \frac{2.2\times10^{-2}\ [PO_4^{3-}]}{6.2\times10^{-8}} = 4.8\times10^{-13}$

$[PO_4^{3-}] = 1.4\times10^{-18}$ M

Equilibrium concentrations: $[H_3O^+] = [H_2PO_4^-] = 0.022$ M; $[HPO_4^{2-}] = 6.2\times10^{-8}$ M;

$[PO_4^{3-}] = 1.4\times10^{-18}$ M

17-31. (a) $C_{20}H_{24}O_2N_2 + H_2O \rightleftharpoons C_{20}H_{24}O_2N_2H^+ + OH^-$; $Kb_1 = 1.08\times10^{-6}$

$C_{20}H_{24}O_2N_2H^+ + H_2O \rightleftharpoons C_{20}H_{24}O_2N_2H_2^{2+} + OH^-$; $Kb_2 = 1.5\times10^{-10}$

(b) Make the usual assumption that ionization in the first step occurs to a much greater extent than in the second step.

$$Kb_1 = \frac{[C_{20}H_{24}O_2N_2H^+][OH^-]}{[C_{20}H_{24}O_2N_2]} = 1.08\times10^{-6}$$

$[C_{20}H_{24}O_2N_2] = (1.00\text{ g} \times 1\text{ mol}/324\text{ g})/1.900\text{ L} = 0.00162$ mol/L

(In the formulation of the equilibrium constant expression, assume that $0.00162 - x \simeq 0.00162$. Also $[C_{20}H_{24}O_2N_2H_2] = [OH^-]$.)

$$Kb_1 = \frac{[OH^-]^2}{0.00162} = 1.08\times10^{-6} \qquad [OH^-]^2 = 1.75\times10^{-9} \qquad [OH^-] = 4.18\times10^{-5}\text{ M}$$

$pOH = -\log(4.18\times10^{-5}) = 4.379$ $\qquad$ $pH = 14.00 - 4.379 = 9.62$

17-32. (a) In the first ionization step:

	H_2SO_4	+	H_2O	$\longrightarrow$	H_3O^+	+	HSO_4^-
Place in Soln:	0.750 M				---		---
Changes:	−0.750 M				+0.750 M		+0.750 M

In the second ionization step:

	HSO_4^-	+	H_2O	$\rightleftharpoons$	H_3O^+	+	SO_4^{2-}
From 1st Step:	0.750 M				0.750 M		
Changes:	$-x$				$+x$		$+x$
At Equil:	$(0.750 - x)$M				$(0.750 + x)$M		x M

$$Ka_2 = \frac{[H_3O^+][SO_4^{2-}]}{[HSO_4^-]} = \frac{(0.750 + x)x}{(0.750 - x)} = 1.29\times10^{-2} \qquad x \simeq 1.29\times10^{-2}$$

$[H_3O^+] = 0.750 + x = (0.750 + 0.0129) = 0.763$ M

(b) Proceed as in Part (a), substituting 1.10 M for 0.750 M. $[SO_4^{2-}] = x = 1.29\times10^{-2}$ M.

17-33. Proceed as in Exercise 32 but without simplifying assumptions. For the second ionization step the relevant data are

	HSO_4^-	+ H_2O ⇌	H_3O^+	+	SO_4^{2-}
From 1st Step:	0.0100 M		0.0100 M		
Changes:	$-x$ M		$+x$ M		$+x$ M
Equilibrium:	$(0.0100 - x)$M		$(0.0100 + x)$M		x M

$$K_{a_2} = \frac{[H_3O^+][SO_4^{2-}]}{[HSO_4^-]} = \frac{(0.0100 - x)x}{(0.0100 - x)} = 0.0129$$

$$0.0100x + x^2 = 0.000129 - 0.0129\,x$$

$$x^2 + 0.0229x - 0.000129 = 0$$

$$x = \frac{-0.0229 \pm \sqrt{(2.29\times10^{-2})^2 + 4 \times 1.29\times10^{-4}}}{2} = 4.68\times10^{-3}\text{ M}$$

$[H_3O^+] = 0.0100 + x = 0.0100 + 0.00468 = 0.0147$ M

$[HSO_4^-] = 0.0100 - x = 0.0100 - 0.00468 = 0.0053$ M

$[SO_4^{2-}] = x = 4.68\times10^{-3}$ M

17-34. The molar concentration of sodium benzoate is:

$$\text{no. mol } NaC_7H_5O_2/L = \frac{0.10\text{ g } NaC_7H_5O_2 \times \frac{1\text{ mol } NaC_7H_5O_2}{144\text{ g } NaC_7H_5O_2}}{0.100\text{ L}} = 0.0069\text{ M}$$

In the equilibrium expression for the hydrolysis of benzoate ion, $K_b(C_7H_5O_2^-) = K_w/K_a(HC_7H_5O_2)$ $= 1.00\times10^{-14}/6.3\times10^{-5} = 1.6\times10^{-10}$.

	$C_7H_5O_2^-$	+ H_2O ⇌	$HC_7H_5O_2$	+	OH^-
Placed in Soln:	0.0069 M		---		---
Changes:	$-x$ M		$+x$ M		$+x$ M
Equilibrium:	$(0.0069 - x)$M		x M		x M

$$K_a = \frac{[HC_7H_5O_2][OH^-]}{[C_7H_5O_2^-]} = \frac{x \cdot x}{0.0069 - x} = 1.6\times10^{-10}$$

Assume $x << 0.0069$: $\frac{x^2}{0.0069} = 1.6\times10^{-10}$ $\quad x^2 = 1.1\times10^{-12}$ $\quad x = [OH^-] = 1.0\times10^{-6}$

$pOH = -\log[OH^-] = -\log(1.0\times10^{-6}) = 6.00$ $\quad pH = 14.00 - 6.00 = 8.00$

17-35. The salt chosen must be that of a strong base and a weak acid if the aqueous solution has pH > 7. Of those listed, this would be KNO_2.

	NO_2^-	+ H_2O ⇌	HNO_2	+	OH^-
Placed in Soln:	x M		---		---
Changes:	-5.6×10^{-6} M		$+5.6\times10^{-6}$ M		$+5.6\times10^{-6}$ M
Equilibrium:	$(x - 5.6\times10^{-6})$M		5.6×10^{-6} M		5.6×10^{-6} M

The value of $[OH^-]$ listed above comes from pH = 8.75; pOH = 5.25; $\log[OH^-] = 5.25 = 0.75 - 6$; $[OH^-] = 5.6\times10^{-6}$ M. (The value of K_b is from Table 17-4.)

$$K_b = \frac{[HNO_2][OH^-]}{[HNO_2]} = \frac{(5.6\times10^{-6})^2}{x - 5.6\times10^{-6}} = \frac{(5.6\times10^{-6})^2}{x} = 1.95\times10^{-11}$$

$$\frac{(5.6\times10^{-6})^2}{x} = 1.95\times10^{-11} \qquad x = 1.6\text{ M } KNO_2$$

17-36. (a) $C_5H_5NH^+ + H_2O \rightleftharpoons H_3O^+ + C_5H_5N$; $K_a = K_w/K_b = 1.0\times10^{-14}/2.0\times10^{-9} = 5.0\times10^{-6}$

(b) In the usual fashion we can write that

$[H_3O^+] = [C_6H_5N] = x$; $[C_5H_5NH^+] = 0.0482 - x \simeq 0.0482$

$$K_a = \frac{[H_3O^+][C_5H_5N]}{[C_5H_5NH^+]} = \frac{x\cdot x}{0.0482} = 5.0\times10^{-6} \qquad x^2 = 2.4\times10^{-7} \qquad x = [H_3O^+] = 4.9\times10^{-4}\ M$$

$pH = -\log(4.9\times10^{-4}) = 3.31$

17-37. (a) ionization as an acid: $HSO_3^- + H_2O \rightleftharpoons H_3O^+ + SO_3^{2-}$; $K_{a_2} = 6.3\times10^{-8}$

hydrolysis: $HSO_3^- + H_2O \rightleftharpoons H_2SO_3 + OH^-$; $K_b = K_w/K_{a_1} = 1.0\times10^{-14}/1.3\times10^{-2} = 7.7\times10^{-13}$

Because $K_{a_2} >> K_b$, the further ionization of HSO_3^- occurs to a greater extent than its hydrolysis. The solution is acidic.

(b) ionization as an acid: $HS^- + H_2O \rightleftharpoons H_3O^+ + S^{2-}$; $K_{a_2} = 1.0\times10^{-14}$

hydrolysis: $HS^- + H_2O \rightleftharpoons H_2S + OH^-$; $K_b = K_w/K_{a_1} = 1.0\times10^{-14}/1.1\times10^{-7} = 9.1\times10^{-8}$

Because $K_b >> K_{a_2}$, the hydrolysis of HS^- occurs to a greater extent than its further ionization. The solution is basic.

(c) ionization as an acid: $HC_2O_4^- + H_2O \rightleftharpoons H_3O^+ + C_2O_4^{2-}$; $K_{a_2} = 5.4\times10^{-5}$

hydrolysis: $HC_2O_4^- + H_2O \rightleftharpoons H_2C_2O_4 + OH^-$; $K_b = K_w/K_{a_1} = 1.0\times10^{-14}/5.4\times10^{-2} = 1.9\times10^{-13}$

Because $K_{a_2} >> K_b$, ionization of $HC_2O_4^-$ occurs to a greater extent than its hydrolysis and the solution is acidic.

Molecular structure and acid strength

17-38. (a) NH_3 is a stronger base than H_2O (see Table 17-4). This means that $HC_2H_3O_2$ is more able to donate protons in $NH_3(\ell)$ than in $H_2O(\ell)$; it is a stronger acid in $NH_3(\ell)$ than in $H_2O(\ell)$.

(b) The situation here is that acetic acid is a stronger acid than is water. NH_3 accepts protons from $HC_2H_3O_2$ more readily than from H_2O. Therefore, NH_3 is a stronger base in $HC_2H_3O_2$ than in H_2O.

17-39. (a) This is a combination of a very strong acid (HBr) and a very strong base (OH^-). The reaction proceeds in the forward reaction.

(b) NO_3^- is a weaker base than HSO_4^-; it will not take a proton away from HSO_4^-. The reverse reaction predominates.

(c) Methoxide ion, CH_3O^-, is a very strong base. It will extract a proton from $HC_2H_3O_2$ and will favor the reverse reaction shown.

(d) CO_3^{2-} is a stronger base than $C_2H_3O_2^-$. It will extract a proton from $HC_2H_3O_2$. The forward reaction predominates.

(e) ClO_4^- is an exceptionally weak base. It will not extract a proton from HNO_2. The reverse reaction predominates.

(f) Since carbonate ion, CO_3^{2-}, is a stronger base than bicarbonate ion, HCO_3^-, the forward reaction predominates.

17-40. Because the H - I bond is *weaker* than the H - Cl bond, we should expect HI to ionize *more readily* than HCl; HI is a stronger acid than HCl. And because they are both almost completely ionized in aqueous solution, we should expect HCl and HI to be stronger than the acids listed, all of which are weak. The four weak acids are all substituted acetic acids. The more electronegative the substituent atoms the weaker the O - H bond and the stronger the acid. The order of the acid strength increases as follows: (c) < (f) < (d) < (e) < (b) < (a).

17-41. The conjugate acids are $CH_3CH_2CH_2NH_3^+$ and C_6H_5-NH_3^+ (benzene ring)-NH_3^+. Because of the inductive effect, the phenyl group, C_6H_5, withdraws electrons away from (benzene ring)-NH_3^+, making $C_6H_5NH_3^+$ a stronger acid than $CH_3CH_2CH_2NH_3^+$. In turn the conjugate base $C_6H_5NH_2$ (aniline) is a weaker base than is $CH_3CH_2NH_2$ (propylamine). The *stronger* of the two bases is propylamine.

17-42. Consider that a molecule of HOCl loses a proton and that the "extra" electron, and hence the negative charge is centered on the O atom.

$$^{-}:\ddot{O}-\ddot{Cl}:$$

With $HClO_2$, the "extra" electron can be spread between two O atoms. This makes ClO_2^- a weaker base than OCl^-, and $HClO_2$ a stronger acid than HOCl.

$$^{-}:\ddot{O}-\ddot{Cl}(-\ddot{O}:): \longleftrightarrow :\ddot{O}-\ddot{Cl}(-\ddot{O}:^{-}):$$

The possibility for spreading the electronic charge increases as the number of O atoms increases. This makes ClO_3^- and ClO_4^- progressively weaker bases, and the acids, $HClO_3$ and $HClO_4$, progressively stronger acids.

17-43. Because both molecules contain one O atom bonded directly to the central P atom, Ka_1 (H_3PO_3) should be about the same as for H_3PO_4.

```
    H
    O                      OH
    |  ..                  |  ..
HO - P = O:            HO - P = O:
    |                      |
    H                      OH
```

phosphorous acid — phosphoric acid

The Lewis structure of phosphorous acid is also consistent with the fact that H_3PO_3 is a *diprotic* acid.

17-44. (a) Ka_1 (H_3AsO_4):

```
      ..
      :O
       ||
HO - As - OH
       |
       OH
```

$EO_m(OH)_n$ $m = 1$ $Ka_1 \cong 1\times10^{-2}$

(b) If $pKa_1 = 1.1$, the formula of hypophosphorous acid must be of the type, $EO_m(OH)_n$, where m = 1.

```
       OH
  ..   |
 :O = P - H
       |
       H
```

(H_3PO_2, hyposphosphorous acid)

Self-test Questions

1. (*c*) The set up leading to a calculation of the number of moles of OH^- is:

$0.30\ L \times 0.0050\ mol\ Ba(OH)_2/L \times 2\ mol\ OH^-/mol\ Ba(OH)_2 = 0.0030\ mol\ OH^-$.

2. (*a*) If the solution pH = 5, the solution is acidic; its pOH = 14 - 5 = 9, and its $[OH^-] = 1\times10^{-9}$ M.

3. (*b*) Propionic acid is a weak acid. In a 0.10 M solution of the acid, $[H_3O^+] < 0.10$ M since the acid is largely nonionized. Item (d) is incorrect. So is item (c), since in 0.10 M HBr $[H_3O^+] = 0.10$ M. $[H_3O^+]$ in 0.10 M $HC_3H_5O_2$ and in 0.10 M HNO_2 are not expected to be equal unless K_a has exactly the same value for these two acids (which is quite unlikely). Since $[H_3O^+]$ in 0.10 M HI (a strong acid) is 0.10 M, it must also be true that $[H_3O^+]$ in 0.10 M $HC_3H_5O_2$ (a weak acid) is smaller than in 0.10 M HI.

4. (*d*) CH_3NH_2 is a *weak* base. A 0.10 M solution must have $[OH^-] < 0.10$ M. Items (a) and (c) are incorrect because both refer to an acidic solution. The solution in question is basic, but with pH < 13. (pH = 13 corresponds to $[OH^-] = 0.10$ M.)

5. (*d*) Neither NaCl nor KNO_3 hydrolyze in aqueous solution; they have pH = 7. NH_4Cl hydrolyzes but to produce an acidic solution. Acetate ion hydrolyzes (acts as a base) and produces a basic solution.

6. (*a*) Of the solutions listed, NaCl(aq) has pH = 7 because neither Na^+ nor Cl^- hydrolyzes. In $NaC_2H_3O_2$ and Na_2S it is the anions that hydrolyze (i.e. accept a proton from H_2O). These solutions are basic. $NaHSO_4$(aq) is acidic because of the ionization reaction $HSO_4^-(aq) + H_2O \rightleftharpoons H_3O^+(aq) + SO_4^{2-}(aq)$.

7. (*a*) The amphiprotic ion must contain H if it is to be able to donate a proton. This eliminates CO_3^{2-} and Cl^-. HCO_3^- can lose a proton to form CO_3^{2-} or can gain a proton to form H_2CO_3. Although NH_4^+ can lose a proton, it is incapable of gaining an additional proton.

8\. (*b*) Cl^- is a very weak base and $HClO_4$ is a very strong acid. $HC_2H_3O_2$ can transfer protons to H_2O but it is only partially ionized. $HC_2H_3O_2 + H_2O \rightleftharpoons H_3O^+ + C_2H_3O_2^-$. Because NH_3 is a stronger base than H_2O, ionization of $HC_2H_3O_2$ proceeds essentially to completion in NH_3. $HC_2H_3O_2 + NH_3 \rightarrow NH_4^+ + C_2H_3O_2^-$.

9\. The ten solutions can be arranged according to decreasing $[H_3O^+]$ without need for detailed calculations. The highest $[H_3O^+]$ is expected from the diprotic H_2SO_4; this is followed by the monoprotic HNO_3. $HC_2H_3O_2$, a weak acid, comes next. An 0.01 M NH_4ClO_4 solution is slightly acidic because of the hydrolysis of NH_4^+. Because no hydrolysis is possible, an 0.01 M NaCl solution is neutral--pH = 7. The $NaNO_2$ solution is slightly basic because of hydrolysis of NO_2^-. NH_3 is a weak base, and NaOH is a strong base. The highest $[OH^-]$ and lowest $[H_3O^+]$ are found in 0.01 M $Ba(OH)_2$. The overall order of decreasing $[H_3O^+]$ is 0.01 M H_2SO_4 > 0.01 M HNO_3 > 0.01 M $HC_2H_3O_2$ > 0.01 M NH_4ClO_4 > 0.01 M NaCl > 0.01 M $NaNO_2$ > 0.01 M NH_3 > 0.01 M NaOH > 0.01 M $Ba(OH)_2$.

10\. $pH = 2.60 = -\log [H_3O^+]$; $\log [H_3O^+] = -2.60 = 0.40 - 3.00$; $[H_3O^+] = 2.5\times10^{-3}$.

	$HC_7H_5O_2$	+	H_2O	$\rightleftharpoons$	H_3O^+	+	$C_7H_5O_2^-$
Placed in Soln:	x M				---		---
Changes:	-2.5×10^{-3} M				$+2.5\times10^{-3}$ M		$+2.5\times10^{-3}$ M
Equilibrium:	$(x - 2.5\times10^{-3})$M				2.5×10^{-3} M		2.5×10^{-3} M

$$K_a = \frac{[H_3O^+][C_7H_5O_2^-]}{[HC_7H_5O_2]} = \frac{(2.5\times10^{-3})^2}{x - 2.5\times10^{-3}} \simeq \frac{(2.5\times10^{-3})^2}{x} = 6.3\times10^{-5} \qquad x = 9.9\times10^{-2}\text{ M } C_7H_5O_2$$

$$\text{no. g } HC_7H_5O_2 = 0.250\text{ L} \times \frac{9.9\times10^{-2}\text{ mol } C_7H_5O_2}{\text{L}} \times \frac{122\text{ g } C_7H_5O_2}{1\text{ mol } C_7H_5O_2} = 3.0\text{ g } HC_7H_5O_2$$

11\. (a) Compare Lewis structures of the two acids.

```
  ..                    ..
 :O                    :O:
  ||                ..  |  ..
  N                :O - Cl - O - H
 / \ ..             ..  |  ..
O   O - N              :O:
    ..                  ..
```

In HNO_3, there are two O atoms and one -OH group bonded to the central nonmetal atom (N). In $HClO_4$ there are three O atoms and one -OH group. Because of the larger number of O atoms bonded directly to the central atom, we should expect $HClO_4$ to be stronger than HNO_3. Also, the oxidation state of Cl is +7, compared to +5 for N, and the electronegativity of Cl is greater than that of N. These are additional factors contributing to the very great acid strength of $HClO_4$.

(b) Electrons are attracted away from the -OH group by the presence of neighboring F atoms in $HC_2F_3O_2$. This leads to an easier loss of protons and to an increased acid strength.

```
      ..   ..
     :F:   O:
  ..  |    ||  ..
 :F - C -  C - O - H
  ..  |        ..
     :F:
      ..
```

(c) From the structures given for the two molecules, we might reason that the electronegative Cl atom withdraws electrons from the $-NH_2$ group (inductive effect). This makes $C_6H_4ClNH_2$ less able to donate electron pairs (or to receive a proton), and hence a weaker base than $C_6H_5NH_2$. Alternatively, the conjugate acid $C_6H_4ClNH_3^+$ is a stronger acid than $C_6H_5NH_3^+$ because of the withdrawal of electrons from the $-NH_3^+$ group by the presence of Cl in the benzene ring.

12\. (a) A strong acid ionizes completely in water solution. This means that the concentration of ions increases by the same factor as the solution concentration. Thus, if the molarity of a strong acid HX is increased from 0.1 M to 0.2 M (doubled), the $[H_3O^+]$ also increases from 0.1 M to 0.2 M (doubles). For a weak acid solution with [HA] = M, we may write

$$K_{HA} = \frac{[H_3O^+][A^-]}{[HA]} = \frac{[H_3O^+]^2}{M} \; ; \; [H_3O^+]^2 = M \times K_{HA} \; ; \; [H_3O^+] = \sqrt{M \times K_{HA}}$$

For a weak acid solution with [HA] = 2 M,

$$K_{HA} = \frac{[H_3O^+]^2}{2\text{ M}} \; ; \; [H_3O^+]^2 = 2 \times M \times K_{HA} \; ; \; [H_3O^+] = \sqrt{2} \times \sqrt{M \times K_{HA}}$$

(b) One S^{2-} ion and one H_3O^+ ion are produced by the ionization of one HS^- ion. For the HS^- ion to appear in solution, one H_2S molecule must also have ionized, producing one H_3O^+ ion as well. Thus, for every S^{2-} that does appear in solution, two H_3O^+ must be formed as well.

$H_2S + H_2O \rightleftharpoons H_3O^+ + HS^-$ $\qquad$ $HS^- + H_2O \rightleftharpoons H_3O^+ + S^{2-}$

However, the vast majority of HS^- ions produced by the ionization of H_2S do not ionize further to S^{2-}. For each of these HS^- ions in solution there must also exist an H_3O^+ ion. Thus, $[H_3O^+] >> 2 \times [S^{2-}]$.

Chapter 18

Additional Aspects of Acid-Base Equilibrium

Review Problems

18-1. (a) The solution is a mixture of a weak acid and a strong acid. The $[H_3O^+]$ is established by the 0.106 M HI: $[H_3O^+] = 0.106$ M.

(b) The only source of OH^- in this mixture of strong and weak acid is the self ionization of water.

$$K_w = [H_3O^+][OH^-] = 1.0\times10^{-14}$$

$$[OH^-] = \frac{1.0\times10^{-14}}{[H_3O^+]} = \frac{1.0\times10^{-14}}{0.106} = 9.4\times10^{-14}\text{ M}$$

(c) $C_3H_5O_2^-$ is produced in the ionization equilibrium.

	$HC_3H_5O_2$	+ H_2O ⇌	H_3O^+	+	$C_3H_5O_2^-$
Placed in Soln:	0.355 M		0.106 M		---
Changes:	$-x$ M		$+x$ M		$+x$ M
At Equilibrium:	$(0.355 - x)$M		$(0.106 + x)$M		x M

Assume that $x << 0.106$ and also $x << 0.355$.

$$K_a = \frac{[H_3O^+][C_3H_5O_2^-]}{[HC_3H_5O_2]} = \frac{0.106\cdot x}{0.355} = 1.34\times10^{-5}$$

$$x = [C_3H_5O_2^-] = \frac{0.355 \times 1.34\times10^{-5}}{0.106} = 4.49\times10^{-5}\text{ M}$$

(d) I^- is derived from the HI. $[I^-] = [H_3O^+] = 0.106$ M.

18-2. (a) In this solution of a weak base and its salt, NH_4^+ is the common ion. Its pressure represses the ionization of NH_3, reducing $[OH^-]$ to a value below that for the weak base alone.

	NH_3	+ H_2O ⇌	NH_4^+	+	OH^-
Placed in Soln:	0.143 M		0.0875 M		---
Changes:	$-x$ M		$+x$ M		$+x$ M
At Equilibrium:	$(0.143 - x)$M		$(0.0875 + x)$M		x M

Assume that $x << 0.143$ and $x << 0.0875$.

$$K_b = \frac{[NH_4^+][OH^-]}{[NH_3]} = \frac{0.0875\cdot x}{0.143} = 1.74\times10^{-5}$$

$$x = [OH^-] = \frac{0.143 \times 1.74\times10^{-5}}{0.0875} = 2.84\times10^{-5}\text{ M}$$

(b) As noted in part (a), $[NH_4^+] = 0.0875$ M.

(c) The sole source of Cl^- is NH_4Cl: $[Cl^-] = [NH_4^+] = 0.0875$ M.

(d) H_3O^+ is derived from the self ionization of water:

$$[H_3O^+] = \frac{K_w}{[OH^-]} = \frac{1.0\times10^{-14}}{2.84\times10^{-5}} = 3.5\times10^{-10}\text{ M}$$

18-3. (a) with acid: $CHO_2^- + H_3O^+ \rightarrow HCHO_2 + H_2O$

with base: $HCHO_2 + OH^- \rightarrow CHO_2^- + H_2O$

(b) with acid: $C_6H_5NH_2 + H_3O^+ \rightarrow C_6H_5NH^+ + H_2O$

with base: $C_6H_5NH_3^+ + OH^- \rightarrow C_6H_5NH_2 + H_2O$

(c) with acid: $HPO_4^{2-} + H_3O^+ \rightarrow H_2PO_4^- + H_2O$

with base: $H_2PO_4^- + OH^- \rightarrow HPO_4^{2-} + H_2O$

18-4. The two methods of handling buffer solution calculations introduced in the text are illustrated here --one method in (a) and the other in (b).

(a)

	$HC_7H_5O_2$	+	H_2O	$\rightleftharpoons$	H_3O^+	+	$C_7H_5O_2^-$
Placed in Soln:	0.0552 M				---		0.132 M
Changes:	$-x$ M				$+x$ M		$+x$ M
At Equilibrium:	$(0.0552 - x)$M				x M		$(0.132 + x)$M

Assume that $x << 0.0552$ and also $x << 0.132$.

$$K_a = \frac{[H_3O^+][C_7H_5O_2^-]}{[HC_7H_5O_2]} = \frac{x(0.132)}{0.0552} = 6.3\times10^{-5}$$

$$x = [H_3O^+] = \frac{0.0552 \times 6.3\times10^{-5}}{0.132} = 2.6\times10^{-5} \text{ M}$$

$$pH = -\log [H_3O^+] = -\log (2.6\times10^{-5}) = 4.59$$

(b) Use the expression $pOH = pK_b + \log \frac{[NH_4^+]}{[NH_3]}$

$$pK_b = -\log (1.74\times10^{-5}) = 4.76$$

$[NH_4^+] = 0.17$ M and $[NH_3] = 0.085$ M

$$pOH = 4.76 + \log \frac{0.17}{0.085} = 4.76 + \log 2.0 = 4.76 + 0.30 = 5.06$$

$$pH = 14.00 - pOH = 14.00 - 5.06 = 8.94$$

18-5. Here the simplest approach is to use the equation $pH = pK_a + \log \frac{[A^-]}{[HA]}$. Substitute the given data and solve for $[A^-]$. $pK_a = -\log (1.8\times10^{-4}) = 3.74$

$$4.12 = 3.74 + \log \frac{[A^-]}{0.472}$$

$$\log [A^-] = 4.12 - 3.74 + \log 0.472$$
$$= 4.12 - 3.74 - 0.326$$
$$= 0.05$$

$$[CHO_2^-] = [A^-] = \text{antilog } 0.05 = 1.12 \text{ M}$$

18-6. This problem can be solved in a similar fashion to the preceeding one, but with $[NH_3]$ as the unknown. That is,

$$pOH = pK_b + \log \frac{[BH^+]}{[B]}$$

$$pOH = -\log (1.74\times10^{-5}) + \log \frac{0.812}{[NH_3]}$$

$$14.00 - pH = 4.76 + \log 0.812 - \log [NH_3]$$

$$14.00 - 9.15 = 4.76 - 0.0904 - \log [NH_3]$$

$$\log [NH_3] = -14.00 + 9.15 + 4.76 - 0.09 = -0.18$$

$$[NH_3] = \text{antilogarithm } (-0.18) = 0.66 \text{ M}$$

18-7. Since the pH range over which phenol red changes color is essentially at the neutral point (pH 7), in any solution with acidic properties the indicator color should be yellow. In any solution with basic properties the solution should be red.

(a) 0.10 M KOH--a strongly alkaline solution--red.

(b) 0.10 M $HC_2H_3O_2$--a weak acid--yellow.

(c) 0.10 M NH_4NO_3--weakly acidic through hydrolysis of the NH_4^+ ion--yellow.

(d) 0.10 M HBr--a strong acid--yellow.

(e) 0.10 M NaCN--a basic solution as a result of hydrolysis of the CN^- ion--red.

18-8. The no. mol OH^- to be neutralized = 0.02000 L x 0.350 mol OH^-/L = 7.00×10^{-3} mol OH^-.

(a) The no. mol H_3O^+ added at the 15.00 mL point is

no. mol H_3O^+ = 0.01500 L x 0.425 mol H_3O^+/L = 6.38×10^{-3} mol H_3O^+

no. mol OH^- in excess = $(7.00\times10^{-3}) - (6.38\times10^{-3}) = 0.62\times10^{-3}$ mol OH^-

$$[OH^-] = \frac{0.62\times10^{-3}\text{ mol }OH^-}{(0.02000 + 0.01500)L} = 0.018\text{ M}$$

pOH = -log 0.018 = 1.74 pH = 14.00 - 1.74 = 12.26

(b) At the 20.00 mL point

no. mol H_3O^+ = 0.02000 L x 0.425 mol H_3O^+/L = 8.50×10^{-3} mol H_3O^+

no. mol H_3O^+ in excess = $8.50\times10^{-3} - 7.00\times10^{-3} = 1.50\times10^{-3}$ mol H_3O^+

$$[H_3O^+] = \frac{1.50\times10^{-3}\text{ mol }H_3O^+}{(0.02000 + 0.02000)L} = 3.75\times10^{-2}\text{ M}$$

pH = -log $[H_3O^+]$ = -log 3.75×10^{-2} = 1.426

18-9. (a) The number of moles of HNO_2 present initially is

no. mol HNO_2 = 0.02500 L x 0.108 mol HNO_2/L = 2.70×10^{-3} mol HNO_2

The no. mol OH^- added at the 10.00 mL point is

no. mol OH^- = 0.01000 L x 0.162 mol OH^-/L = 1.62×10^{-3} mol OH^-

In a familiar fashion we can write

	HNO_2	+	OH^-	→	NO_2	+	H_2O
Initial:	2.70×10^{-3} mol		---		---		---
Add:	---		1.62×10^{-3} mol		---		---
After Reaction:	1.08×10^{-3} mol		≈0		1.62×10^{-3} mol		---

$$pH = pK_a + \log\frac{[NO_2^-]}{[HNO_2]} \qquad pH = -\log(5.1\times10^{-4}) + \log\frac{1.62\times10^{-3}/0.03500\text{ L}}{1.08\times10^{-3}/0.03500\text{ L}}$$

pH = 3.29 + log 1.500 = 3.29 + 0.176 = 3.47

(b) no. mol OH^- = 0.02000 L x 0.162 mol OH^-/L = 3.24×10^{-3} mol OH^-

	HNO_2	+	OH^-	→	NO_2^-	+	H_2O
Initial:	2.70×10^{-3} mol				---		
Add:			3.24×10^{-3} mol				
After Reaction:	≈0		0.54×10^{-3} mol		2.70×10^{-3} mol		

$$[OH^-] = \frac{0.54\times10^{-3}\ mol\ OH^-}{(0.02500 + 0.02000)L} = 1.2\times10^{-2}\ M$$

$pOH = -\log(1.2\times10^{-2}) = 1.92$ $\qquad$ $pH = 14.00 - 1.92 = 12.08$

18-10. (a) $HClO_4$: eq. wt. = mol. wt. = 100.5

(b) $Mg(OH)_2$: 2 mol OH^- per mol $Mg(OH)_2$. eq. wt. = ½ f. wt. = 29.2

(c) $HC_3H_5O_2$: Only one H atom per molecule is ionizable. eq. wt. = mol. wt. = 74.1

18-11. (a) KOH: Normality = molarity. 0.24 M KOH = 0.24 N KOH

(b) $H_2C_2O_4$: Normality = molarity if neutralization is carried only to $HC_2O_4^-$. That is, 0.15 M $H_2C_2O_4$ = 0.15 N $H_2C_2O_4$. For complete neutralization (to $C_2O_4^{2-}$), normality = 2 x molarity. 0.15 M $H_2C_2O_4$ = 0.30 N $H_2C_2O_4$.

(c) $Ca(OH)_2$: There are two mol OH^- per mol compound. normality = 2 x molarity. 2×10^{-3} M $Ca(OH)_2$ = 4×10^{-3} N $Ca(OH)_2$

18-12. (a) $V_{NaOH} \times N_{NaOH} = V_{H_2SO_4} \times V_{H_2SO_4}$

$V \times 0.1090\ meq/L = 25.00\ ml \times 0.1471\ meq/L$

$$V = \frac{0.1471}{0.1090} \times 25.00 = 33.74\ ml$$

(b) Before using an equation of the type in (a), the molarity of H_2SO_4 must be converted to normality. 0.08511 M H_2SO_4 = 2 x 0.08511 = 0.1702 N H_2SO_4

$V_{NaOH} \times N_{NaOH} = V_{H_2SO_4} \times N_{H_2SO_4}$

$V \times 0.1090\ meq/L = 15.00\ ml \times 0.1702\ meq/L$

$$V = \frac{0.1702}{0.1090} \times 15.00 = 23.42\ ml$$

Exercises

The common-ion effect in acid-base equilibria

18-1. (a) NO_2^- from $NaNO_2$ favors the *reverse* of the ionization reaction.

$HNO_2 + H_2O \rightleftharpoons H_3O^+ + NO_2^-$ $\quad$ This reduces $[H_3O^+]$ and raises the pH.

(b) HNO_3 is a *strong* acid that is completely ionized in aqueous solution. The addition of NO_3^- (from $NaNO_3$) has no effect on its ionization. $[H_3O^+]$ and pH are not affected.

18-2. (a) In this mixture of a strong and a weak acid, we can assume that essentially all of the H_3O^+ is derived from the strong acid. $[H_3O^+]$ = 0.100 M

(b) In a combination of a weak acid and its salt, we can assume that essentially all of the anion is derived from the salt. In 0.100 M $NaNO_2$, $[NO_2^-]$ = 0.100 M.

(c) All of the Cl^- is derived from the HCl, a strong acid. Moreover, Cl^- does not participate in any reaction in solution. $[Cl^-]$ = 0.200 M

(d) $C_2H_3O_2^-$ is produced by ionization of the weak acid.

	$HC_2H_3O_2$	+	H_2O	$\rightleftharpoons$	H_3O^+	+	$C_2H_3O_2^-$
From Weak Acid:	$(0.300 - x)$M				x M		x M
From Strong Acid:	---				0.100 M		---
At Equilibrium:	$(0.300 - x)$M				$(0.100 + x)$M		x M

Assume that $x \ll 0.100$.

$$K_a = \frac{[H_3O^+][C_2H_3O_2^-]}{[HC_2H_3O_2]} = \frac{(0.100)x}{(0.300)} = 1.74\times10^{-5} \qquad x = [C_2H_3O_2^-] = 5.22\times10^{-5}\text{ M}$$

(e) $NH_3 + H_2O \rightleftharpoons NH_4^+ + OH^-$

	NH_3	NH_4^+	OH^-
From Weak Base:	$(0.500 - x)$M	x M	x M
From Salt:	---	(2×0.200)M	---
At Equilibrium:	$(0.500 - x)$M	$(0.400 + x)$M	x M

Assume $x << 0.400$

$$K_b = \frac{[NH_4^+][OH^-]}{[NH_3]} = \frac{(0.400)}{(0.500)} = 1.74\times10^{-5} \qquad x = [OH^-] = 2.18\times10^{-5}\text{ M}$$

18-3. First determine the molar concentration of sodium lactate,

$$\text{no. mol } NaC_3H_5O_3/L = 10.0\text{ g } NaC_3H_5O_3 \times \frac{1\text{ mol } NaC_3H_5O_3}{112\text{ g } NaC_3H_5O_3} \times \frac{1}{0.100\text{ L}} = 0.893\text{ M}$$

and then $[H_3O^+]$: $\log [H_3O^+] = -4.11 = 0.89 - 5$ $\qquad [H_3O^+] = 7.8\times10^{-5}$ M

$HC_3H_5O_3 + H_2O \rightleftharpoons H_3O^+ + C_3H_5O_3^-$

	$HC_3H_5O_3$	H_3O^+	$C_3H_5O_3^-$
From Weak Acid:	$(0.0500 - x)$M	x M	x M
From Salt:	---	---	0.893 M
At Equilibrium:	$(0.0500 - x)$M	x M	$(0.893 + x)$M

But $x = 7.8\times10^{-5}$ M. Furthermore $x << 0.0500$.

$$K_a = \frac{[H_3O^+][C_3H_5O_3^-]}{[HC_3H_5O_3]} = \frac{(7.8\times10^{-5}) \times 0.893}{0.0500} = 1.4\times10^{-3}$$

18-4. Let $x = [OH^-]$ and express the following concentrations in terms of x. $[C_6H_5NH_2] = 0.105 - x = 0.105$ (This assumes that $x << 0.105$.)

$$[C_6H_5NH_3^+] = \left\{1.37\times10^{-3}\text{ g } C_6H_5NH_3Cl \times \frac{1\text{ mol } C_6H_5NH_3Cl}{130\text{ g } C_6H_5NH_3Cl} \times \frac{1\text{ mol } C_6H_5NH_3^+}{1\text{ mol } C_6H_5NH_3Cl} \times \frac{1}{3.25\text{ L}}\right\} + x$$

$$= 3.24\times10^{-6} + x$$

$$K_b = \frac{[C_6H_5NH_3^+][OH^-]}{[C_2H_5NH_2]} = \frac{(3.24\times10^{-6} + x)x}{0.105} = 4.30\times10^{-10} \qquad x^2 + 3.24\times10^{-6}x - 4.52\times10^{-11} = 0$$

$$x = \frac{-3.24\times10^{-6} \pm \sqrt{(3.24\times10^{-6})^2 + 4 \times 4.52\times10^{-11}}}{2}$$

$$x = [OH^-] = \frac{-3.24\times10^{-6} \pm 1.38\times10^{-5}}{2} = 5.28\times10^{-6}$$

$$pOH = -\log [OH^-] = -\log (5.28\times10^{-6}) = 5.277$$

18-5. In each case let the volume, in mL, of indicated solution = x. Write an expression for $[H_3O^+]$ and set this equal to the value calculated from the pH.

(a) Since this will be a fairly acidic solution (pH = 1.00; $[H_3O^+]$ = 0.10 M), assume that the pH is determined solely by the strong acid, HCl. That is, treat the solution as if HCl had been added to pure water.

$$\text{no. mol HCl} = x\text{ mL} \times \frac{1.00\text{ L}}{1000\text{ mL}} \times \frac{1.00\text{ mol HCl}}{1.00\text{ L}} = 0.00100\,x\text{ mol HCl}$$

$$\text{total volume} = (250.0 + x)\text{mL} = [(250.0 + x)/1000]\text{L} \quad [H_3O^+] = \frac{0.00100\,x}{[(250.0 + x)/1000]} = \frac{x}{250.0 + x} = 0.10$$

$$x = 25.00 + 0.10\,x \qquad 0.90\,x = 25.0 \qquad x = 28\text{ mL}$$

(b) $HC_3H_7O_2 + H_2O \rightleftharpoons H_3O^+ + C_3H_7O_2^-$

	$HC_3H_7O_2$	H_3O^+	$C_3H_7O_2^-$
From Weak Acid:	$(0.100 - x)$M	x M	x M
From Salt:	---	---	y M
At Equilibrium:	$(0.100 - x)$M	x M	$(x + y)$M

But if pH = 4.00, $[H_3O^+] = x = 1.0\times10^{-4}$

Also assume $x << y$ and $x << 0.100$.

$$K_a = \frac{[H_3O^+][C_3H_7O_2^-]}{[HC_3H_7O_2]} = \frac{(1.0\times10^{-4}) \times y}{0.100} = 1.35\times10^{-5} \qquad y = 0.014$$

The final solution must be 0.014 M in $NaC_3H_7O_2$. Assume that the final solution volume remains at 250.0 mL.

no. mol $NaC_3H_7O_2$ = .2500 L × 0.014 mol/L = 3.5×10^{-3} mol $NaC_3H_7O_2$

$$\text{no. mL required} = 3.5\times10^{-3} \text{ mol } NaC_3H_7O_2 \times \frac{1.00 \text{ L}}{1.00 \text{ mol } NaC_3H_7O_2} \times \frac{1000 \text{ mL}}{1.00 \text{ L}} = 3.5 \text{ mL}$$

Because the added solution volume (3.5 mL) is much smaller than the initial volume (250 mL), the assumption of a constant solution volume is valid.

(c) One approach is to calculate the pH of 0.100 M $HC_3H_7O_2$(aq); add 0.15 to this value; determine the corresponding $[H_3O^+]$; establish the molarity of an $HC_3H_7O_2$(aq) having this $[H_3O^+]$; and, finally, calculate the volume of water to dilute the 0.100 M $HC_3H_7O_2$(aq) to the new molarity. Another approach, outlined below, does all of this symbolically, without intermediate calculations. Let M_i = initial molarity of the acid solution; pH_i, its initial pH; M_f = final molarity of the acid solution; and pH_f, the final pH. Also, assume that the acid ionizes to such a limited extent that the molar concentration of nonionized acid is equal to the total molarity. [This is the assumption usually expressed in a form such as $(0.100 - x) \cong 0.100$.]

$$\text{Initially: } K_a = \frac{[H_3O^+][A^-]}{[HA]} = \frac{[H_3O^+]_i^2}{M_i} = K_a$$

$$[H_3O^+]_i = \sqrt{K_a \times M_i} \qquad \log [H_3O^+] = 1/2 \log (K_a \times M_i)$$

$$pH_i = -\log [H_3O^+] = -1/2 \log (K_a \times M_i)$$

$$pH_f = pH_i + 0.15$$

$$-1/2 \log (K_a \times M_f) = -1/2 \log (K_a \times M_i) + 0.15$$

$$\cancel{-1/2 \log K_a} - 1/2 \log M_f = \cancel{-1/2 \log K_a} - 1/2 \log M_i + 0.15$$

$$1/2 \log \frac{M_i}{M_f} = 0.15 \qquad \log \frac{M_i}{M_f} = 0.30 \qquad \frac{M_i}{M_f} = 2.0 \quad \text{or} \quad M_f = 1/2\ M_i$$

To raise the pH of the weak acid solution by 0.15 unit requires that the acid solution be diluted to one-half its original molar concentration. In the present case this means adding 250.0 mL H_2O to 250.0 mL 0.100 M $HC_3H_7O_2$(aq).

Buffer solutions

18-6. (a) 0.100 M NaCl is not a buffer solution. Addition of even small quantities of acid or base will change its pH drastically.

(b) 0.100 M NaCl - 0.100 M NH_4Cl is not a buffer solution. The solution has no capacity to neutralize small added quantities of acid. And although NH_4^+ is capable of reacting with added base, the solution pH would change greatly. [A solution of NH_4^+ is acidic by hydrolysis; a NH_3-NH_4^+ solution is basic.]

(c) A 0.100 M CH_3NH_2 - 0.150 M $CH_3NH_3^+Cl^-$ solution contains a weak base and its salt (in approximately equal concentrations). It is a buffer solution.

(d) Following reaction between HCl and NO_2^-, the solution contains a weak acid, HNO_2, and an excess of strong acid, HCl. This is not a buffer solution.

$$H^+ + Cl^- + Na^+ + NO_2^- \rightarrow HNO_2 + Na^+ + Cl^-$$

(e) A reaction occurs between a strong acid and the salt of a weak acid. Because there is an excess of the salt, the final solution is that of a weak acid and its salt. This is a buffer solution.

$$H^+ + Cl^- + Na^+ + \underset{\text{(excess)}}{C_2H_3O_2^-} \rightarrow HC_2H_3O_2 + Na^+ + Cl^-$$

(f) This is a mixture of a weak acid and the salt of a (different) weak acid. It should still function well as a buffer solution. This is,

$$HC_2H_3O_2 + OH^- \rightarrow H_2O + C_2H_3O_2^- \quad \text{and} \quad C_3H_5O_2^- + H_3O^+ \rightarrow HC_3H_5O_2 + H_2O$$

18-7. (a) In the presence of acid: $H_3O^+ + HPO_4^{2-} \rightarrow H_2PO_4^- + H_2O$

In the presence of base: $OH^- + H_2PO_4^- \rightarrow HPO_4^{2-} + H_2O$

(b) $pH = pK_{a_2} + \log \frac{[HPO_4^{2-}]}{[H_2PO_4^-]}$. This buffer will be most effective when $[HPO_4^{2-}] = [H_2PO_4^-]$.

$pH = -(\log K_{a_2}) + \log 1 = -\log (6.2\times10^{-8}) + 0 = 7.21$

(c) $pH = 7.21 + \log \frac{0.150}{0.050} = 7.21 + \log 3.0 = 7.69$

18-8. Our problem is to determine the concentration of NH_4^+ that must be present in 0.100 M NH_3 to yield a solution with a pH of 9.35. ($[OH^-]$ in the solution can be calculated from the pH value.)

$NH_3 + H_2O \rightleftharpoons NH_4^+ + OH^-$; $K_b = 1.74\times10^{-5}$

$pOH = 14.00 - pH = 14.00 - 9.35 = 4.65$; $\log [OH^-] = -pOH = -4.65$; $pOH = 0.35 - 5.00$; $[OH^-] = 2.2\times10^{-5}$

$$K_b = \frac{[NH_4^+][OH^-]}{[NH_3]} = \frac{[NH_4^+] \times 2.2\times10^{-5}}{0.105} = 1.74\times10^{-5};\ [NH_4^+] = 8.3\times10^{-2}$$

$$\text{no. g } (NH_4)_2SO_4 = 32.0 \text{ mL} \times \frac{1 \text{ L}}{1000 \text{ mL}} \times \frac{8.3\times10^{-2} \text{ mol } NH_4^+}{1 \text{ L}} \times \frac{1 \text{ mol } (NH_4)_2SO_4}{2 \text{ mol } NH_4^+} \times \frac{132 \text{ g } (NH_4)_2SO_4}{1 \text{ mol } (NH_4)_2SO_4}$$

$$= 1.8 \text{ g } (NH_4)_2SO_4$$

18-9. Look up K_a values for the three acids in Tables 17-2 and 17-3, establish the corresponding pK_a values, and substitute into an equation similar to (18.9) to obtain:

$pH = 3.74 + \log \frac{[CHO_2^-]}{[HCHO_2]}$ for *formic acid-sodium formate*

$pH = 4.76 + \log \frac{[C_2H_3O_2^-]}{[HC_2H_3O_2]}$ for *acetic acid-sodium acetate*

$pH = 2.23 + \log \frac{[H_2PO_4^-]}{[H_3PO_4]}$ for *phosphoric acid-sodium dihydrogen phosphate*

The capacity of a buffer solution is limited to about one pH unit above and below the pK_a value. This fact suggests that the formic acid-sodium formate buffer is best to use the preparing a buffer solution with pH = 3.50.

Determine the ratio $[CHO_2^-]/[HCHO_2]$ corresponding to pH = 3.50.

$3.50 = 3.74 + \log \frac{[CHO_2^-]}{[HCHO_2]}$ $\qquad$ $\log \frac{[CHO_2^-]}{[HCHO_2]} = 3.50 - 3.74 = -0.24$

$\frac{[CHO_2^-]}{[HCHO_2]} = \text{antilog } (-0.24) = 0.58$ $\qquad$ $[CHO_2^-] = 0.58\ [HCHO_2]$

Since the two buffer components are present in the same solution, the ratio of their amounts, in moles, is the same as their ratio of concentrations, 0.58:1.00. For convenience let us take

1.00 L 0.100 M $HCHO_2$ (which contains 0.10 mol $HC_2H_3O_2$). To obtain the desired buffer solution, we must mix this with *0.58 L 0.100 M* $NaCHO_2$ (which contains 0.058 mol $NaCHO_2$).

18-10. This problem differs in two ways from the preceding exercise: (1) Weak acid solutions are not available and (2) an exact volume of buffer solution is required. We can begin, however, at the point where we left off in Exercise 9--$[CHO_2^-]/[HCHO_2] = 0.58$. The required weak acid in the buffer must be produced by allowing the strong acid (1.00 M HCl) to react with the salt of the weak acid (0.100 M $NaCHO_2$). Let us call the required volumes of these solution, in liters, x and y, respectively. We need to find two equations to relate these unknowns. The first of the equations is $x + y = 1.00$. The other equation is obtained as follows:

	CHO_2^-	+	H_3O^+	⟶	$HCHO_2$	+	H_2O
Initial:	$0.100x$ mol		---		---		
Add:	---		$1.00x$ mol				
Buffer:	$(0.100y - 1.00x)$mol		pH = 3.50		$1.00x$ mol		

Because the total solution volume is 1.00 L, the molar amounts and the molar concentrations are numerically equal. We can write:

$$\text{pH} = \text{p}K_a + \log\frac{[CHO_2^-]}{[HCHO_2]} \qquad 3.50 = 3.74 + \log\frac{(0.100y - 1.00x)}{1.00x}$$

$$\frac{0.100y - 1.00x}{1.00x} = \text{antilog}(-0.24) = 0.58 \qquad 0.100y - 1.00x = 0.58x$$

However, since $x + y = 1.00$; $0.100(1.00 - x) - 1.00x = 0.58x$ $\qquad 1.68x = 0.100$

$x = 0.0595$. $y = 0.94$. Prepare the buffer solution by mixing 59.5 mL 1.00 M HCl with enough 0.100 M $NaHCO_2$ to make 1.00 L of solution.

18-11. The required expression is $\text{pH} = \text{p}K_a + \log\frac{[HCO_3^-]}{[H_2CO_3]} = 6.4 + \log\frac{20}{1} = 6.4 + 1.3 = 7.7$

A buffer having $[C_2H_3O_2^-]/[HC_2H_3O_2] = 20$ would not function well as a general purpose buffer. It would have considerable capacity to react with added acid ($C_2H_3O_2^- + H_3O^+ \rightarrow HC_2H_3O_2 + H_2O$), but only 1/20th of that capacity to react with added base because of the limited quantity of $HC_2H_3O_2$ present ($HC_2H_3O_2 + OH^- \rightarrow C_2H_3O_2^- + H_2O$).

18-12. (a) 0.010 M $HC_2H_3O_2$ - 0.010 M $NaC_2H_3O_2$: $\text{pH} = -\log(1.74\times10^{-5}) + \log\frac{0.010}{0.010} = 4.76 + \log 1 = 4.76$

1.0 M $HC_2H_3O_2$ - 0.50 M $NaC_2H_3O_2$: $\text{pH} = -\log(1.74\times10^{-5}) + \log\frac{0.50}{1.0} = 4.76 + \log 0.50$

$= 4.76 - 0.30 = 4.46$

(b) 0.010 M $HC_2H_3O_2$ - 0.010 M $NaC_2H_3O_2$ has a capacity of neutralizing 0.010 mol/L of either H_3O^+ or OH^- before its buffering action is totally destroyed.

1.0 M $HC_2H_3O_2$ - 0.50 M $NaC_2H_3O_2$ has a capacity of neutralizing 0.50 mol H_3O^+/L or 1.0 mol OH^-/L.

18-13. (a)

	$C_2H_3O_2^-$	+	H_3O^+	⟶	$HC_2H_3O_2$	+	H_2O
Original Soln:	0.050 mol				0.0150 mol		
Add 0.00100 L of 6.00 M HCl:			0.00600 mol				
Changes:	-0.00600 mol		--		+0.00600 mol		
Final Amounts:	0.044 mol		--		0.021 mol		
Final Concns:	$\frac{0.044\text{ mol}}{0.101\text{ L}} = 0.44$ M		?		$\frac{0.021\text{ mol}}{0.101\text{ L}} = 0.21$ M		

$$\text{pH} = -\log(1.74\times10^{-5}) + \log\frac{0.44}{0.21} = 4.76 + \log 2.1 = 4.76 + 0.32 = 5.08$$

(b) If 10.00 mL 6.00 M HCl were added to 100 mL of the buffer this would be equivalent to adding 0.060 mol H_3O^+. All of the $C_2H_3O_2^-$ would be converted to $HC_2H_3O_2$ and there would remain an excess of 0.010 mol H_3O^+.

$[H_3O^+] = \frac{0.010 \text{ mol } H_3O^+}{0.110 \text{ L}} = 0.091 \text{ M}$ $\qquad$ $pH = -\log 0.091 = 1.04$

18-14. $[NH_3] = 1.51 \text{ g } NH_3 \times \frac{1 \text{ mol } NH_3}{17.0 \text{ g } NH_3} \times \frac{1}{0.500 \text{ L}} = 0.178 \text{ M}$

$$[NH_4^+] = 3.85 \text{ g } (NH_4)_2SO_4 \times \frac{1 \text{ mol } (NH_4)_2SO_4}{132 \text{ g } (NH_4)_2SO_4} \times \frac{2 \text{ mol } NH_4^+}{1 \text{ mol } (NH_4)_2SO_4} \times \frac{1}{0.500 \text{ L}} = 0.117 \text{ M}$$

(a) $pOH = -\log(1.74 \times 10^{-5}) + \log \frac{[NH_4^+]}{[NH_3]} = 4.76 + \log \frac{0.117}{0.178} = 4.76 + \log 0.657 = 4.76 - 0.18 = 4.58$

$pH = 14.00 - 4.58 = 9.42$

(b) The addition of 1.00 g NaOH to 0.500 L of the buffer means adding

$$1.00 \text{ g NaOH} \times \frac{1 \text{ mol NaOH}}{40.0 \text{ g NaOH}} \times \frac{1 \text{ mol } OH^-}{1 \text{ mol NaOH}} \times \frac{1}{0.500 \text{ L}} = 0.0500 \text{ mol } OH^-/\text{L}$$

	NH_4^+	+	OH^-	$\longrightarrow$	NH_3	+	H_2O
Original Buffer:	0.117 M				0.178 M		
Add:			0.0500 M				
Changes:	−0.0500 M		−0.0500 M		+0.0500 M		
Equil. Concns:	0.067 M		?		0.228 M		

$pOH = 4.76 + \log \frac{0.067}{0.228} = 4.76 + \log 0.29 = 4.76 - 0.54 = 4.22$

$pH = 14.00 - pOH = 14.00 - 4.22 = 9.78$

(c) Here let us work backwards: $pH = 9.00$, $pOH = 5.00$, $pOH = 4.76 + \log \frac{[NH_4^+]}{[NH_3]} = 5.00$

$\log \frac{[NH_4^+]}{[NH_3]} = 0.24$ $\qquad$ $\frac{[NH_4^+]}{[NH_3]} = \text{antilog } 0.24 = 1.7$

The ratio $[NH_4^+]/[NH_3]$ in the original buffer is 0.117/0.178 = 0.657. Thus, some NH_3 must be converted to NH_4^+. Let this quantity, expressed as mol/L, be x.

$\frac{(0.117 + x)}{(0.178 - x)} = 1.7$ $\qquad$ $0.117 + x = 0.31 - 1.7$ $\qquad$ $2.7\,x = 0.19$ $\qquad$ $x = 0.070$

Enough 12 M HCl must be added to produce 0.070 M H_3O^+ (which then converts 0.070 mol NH_3/L to 0.070 mol NH_4^+/L).

$$\text{no. mL} = 0.500 \text{ L buffer} \times \frac{0.070 \text{ mol } H_3O^+}{\text{L buffer}} \times \frac{1 \text{ mol HCl}}{1 \text{ mol } H_3O^+} \times \frac{1 \text{ L HCl(aq)}}{12 \text{ mol HCl}} \times \frac{1000 \text{ mL HCl(aq)}}{1 \text{ L HCl(aq)}} = 2.9 \text{ mL}$$

18-15. (a) Determine the number of moles of the two reactants, $C_2H_3O_2^-$ and H_3O^+.

$$\text{no. mol } C_2H_3O_2^- = 10.0 \text{ g } NaC_2H_3O_2 \times \frac{1 \text{ mol } NaC_2H_3O_2}{82.0 \text{ g } NaC_2H_3O_2} \times \frac{1 \text{ mol } C_2H_3O_2^-}{1 \text{ mol } NaC_2H_3O_2} = 0.122 \text{ mol } C_2H_3O_2^-$$

$$\text{no. mol } H_3O^+ = 0.300 \text{ L} \times \frac{0.200 \text{ mol HCl}}{1 \text{ L}} \times \frac{1 \text{ mol } H_3O^+}{1 \text{ mol HCl}} = 0.0600 \text{ mol } H_3O^+$$

no. mol $C_2H_3O_2^-$ = 0.122 mol initially = 0.0600 mol consumed = 0.062 mol $C_2H_3O_2^-$

These values are now substituted into the expression

$$pH = pK_a + \log \frac{[C_2H_3O_2^-]}{[HC_2H_3O_2]}$$

$= -\log(1.74\times10^{-5}) + \log \frac{0.062/0.300}{0.060/0.300}$

$= 4.76 + 0.014 = 4.77$

(b) The amount of OH^- added to the buffer solution is calculated first.

$$\text{no. mol } OH^- = 1.00 \text{ g } Ba(OH)_2 \times \frac{1 \text{ mol } Ba(OH)_2}{171 \text{ g } Ba(OH)_2} \times \frac{2 \text{ mol } OH^-}{1 \text{ mol } Ba(OH)_2} = 0.0117 \text{ mol } OH^-$$

In the buffer reaction 0.0117 mol $HC_2H_3O_2$ is converted to 0.0117 mol $C_2H_3O_2^-$.

$$HC_2H_3O_2 + OH^- \rightarrow C_2H_3O_2^- + H_2O$$

In the new equilibrium condition, no. mol $HC_2H_3O_2$ = 0.060 - 0.0117 = 0.048 mol $HC_2H_3O_2$; no. mol $C_2H_3O_2^-$ = 0.062 + 0.0117 = 0.074 mol $C_2H_3O_2^-$.

These new values are now substituted into expression (18.9).

$$pH = 4.76 + \log \frac{0.074/0.300}{0.048/0.300}$$

$= 4.76 + 0.19 = 4.95$

(c) The capacity of the buffer toward $Ba(OH)_2$ is simply the amount of $Ba(OH)_2$ required to react with all of the $HC_2H_3O_2$.

$$\text{no. g } Ba(OH)_2 = 0.060 \text{ mol } HC_2H_3O_2 \times \frac{1 \text{ mol } OH^-}{1 \text{ mol } HC_2H_3O_2} \times \frac{1 \text{ mol } Ba(OH)_2}{2 \text{ mol } OH^-} \times \frac{171 \text{ g } Ba(OH)_2}{1 \text{ mol } Ba(OH)_2} = 5.1 \text{ g } Ba(OH)_2$$

(d) Since the amount of $Ba(OH)_2$ added (5.2 g) exceeds the buffer capacity (5.1 g), all the acetic acid is consumed. The resulting solution contains sodium and barium acetates and 0.1 g of excess barium hydroxide, which establishes the pH of the solution.

$$[OH^-] = 0.1 \text{ g } Ba(OH)_2 \times \frac{1 \text{ mol } Ba(OH)_2}{171 \text{ g } Ba(OH)_2} \times \frac{2 \text{ mol } OH^-}{1 \text{ mol } Ba(OH)_2} \times \frac{1}{0.300 \text{ L}} = 4\times10^{-3} \text{ M}$$

$pOH = -\log(4\times10^{-3}) = 2.4$ $\qquad$ $pH = 14.00 - 2.4 = 11.6$

Acid-base indicators

18-16. (a) In an acid-base titration the pH range where neutralization occurs can be established from an appropriate titration curve. Then an indicator can be selected that changes color in this range. If the pH of an unknown solution can have a value from 0 to 14, several different indicators, each with a pH range of about 2, are required to establish this pH.

(b) An indicator is itself a weak acid and consumes a small quantity of the titrant in being converted to its anion. This volume is treated as though it were part of the volume required for the titration itself. The larger the volume of indicator used, the greater will be the error introduced. (Generally, in laboratory procedures a correction is made for this effect.)

18-17. (a) Take the negative logarithms of the K_a values. This yields values of pK_a. For each indicator $pH = pK_a$ at the midpoint of the indicator color change.

acidic solution: bromphenol blue (pK_a = 3.85); bromcresol green (pK_a = 4.68); 2,4-dinitrophenol (pK_a = 3.90); chlorophenol red (pK_a = 6.00)

approx. neutral: bromthymol blue (pK_a = 7.10)

basic solution: thymolphthalein (pK_a = 10.00)

(b) If bromcresol green assumes a green color it is at about the midpoint of its color change. The pH of the solution must be about equal to pK_a for bromcresol green--about pH = 5. Chlorophenol red assumes an orange color at about pH = 6.

18-18. (a) 0.100 M HCl(aq); pH = 1.000; 2,4-dinitrophenol is colorless.

(b) 1.00 M NaCl(aq); pH = 7.00; chlorophenol red is red.

(c) In 1.00 M $NH_3(aq)$; $K_b = \frac{[NH_4^+][OH^-]}{[NH_3]} = \frac{[OH^-]^2}{1.00} = 1.74\times10^{-5}$ $[OH^-] = 4.17\times10^{-3}$ M

$pOH = -\log(4.17\times10^{-3}) = 2.380$ $pH = 14.00 - 2.38 = 11.62$

At this pH thymolphthalein is blue.

(d) Hydrolysis of NH_4^+ occurs in NH_4NO_3.

$NH_4^+ + H_2O \rightleftharpoons NH_3 + H_3O^+$ $K_a = K_w/K_b = 1.0\times10^{-14}/1.74\times10^{-5} = 5.8\times10^{-10}$

$K_h = \frac{[NH_3][H_3O^+]}{[NH_4^+]} = \frac{[H_3O^+]^2}{1.00} = 5.8\times10^{-10}$ $[H_3O^+] = 2.4\times10^{-5}$ $pH = 4.62$

At this pH bromthymol blue is yellow.

(e) According to Figure 17-3 the pH of seawater is about 8. At this pH bromcresol green is blue.

(f) Considering just the first ionization of H_2CO_3, we can write

$K_{a_1} = \frac{[H_3O^+][HCO_3^-]}{[H_2CO_3]} = \frac{[H_3O^+]^2}{0.034} = 4.2\times10^{-7}$ $[H_3O^+] = 1.2\times10^{-4}$ $pH = 3.92$

At this pH bromphenol blue is at about the middle of its change from yellow to blue; the color should be green.

18-19. (a) First calculate the pH of the buffer solution:

$$pH = pK_a + \log\frac{[C_2H_3O_2^-]}{[HC_2H_3O_2]} = -\log(1.74\times10^{-5}) + \log\frac{0.10}{0.10} = 4.76$$

Now determine the ratio $[In^-]/[HIn]$ at this pH:

$pH = pK_a + \log\frac{[In^-]}{[HIn]} = 4.95 + \log\frac{[In^-]}{[HIn]} = 4.76$ $\log\frac{[In^-]}{[HIn]} = -0.19$ $\frac{[In^-]}{[HIn]} = 0.65$

If we take $[HIn] = 1.00$, $[In^-] = 0.65$, and total indicator concentration = 1.65. This means the percent of the indicator in the anion form is $(0.65/1.65) \times 100 = 39\%$.

(b) When the indicator is at its pK_a (4.95), the ratio $[In^-]/[HIn] = 1.00$. At the midpoint of its color change (about pH = 5.3) the ratio $[In^-]/[HIn]$ is greater than 1.00. Thus, even though $[HIn] < [In^-]$ at the midpoint, the contribution of HIn to establishing the color of the solution is the same as one would normally expect when $[HIn] = [In^-]$. This must mean that HIn (red) is more strongly colored than In^- (yellow).

Neutralization reactions

18-20. no. mol OH^- = $0.01867 L \times \frac{0.07152 \text{ mol HI}}{1.00 L} \times \frac{1 \text{ mol } H_3O^+}{1 \text{ mol } H_2SO_4} \times \frac{1 \text{ mol } OH^-}{1 \text{ mol } H_3O^+} = 1.335\times10^{-3}$ mol OH^-

$[OH^-] = \frac{1.335\times10^{-3} \text{ mol } OH^-}{0.02500 \text{ L}} = 0.05340$ M

18-21. no. mol H_3O^+ = $0.0107 \text{ L} \times \frac{0.1032 \text{ mol HCl}}{L} \times \frac{1 \text{ mol } H_3O^+}{1 \text{ mol HCl}} = 1.10\times10^{-3}$ mol H_3O^+

no. mol $Ca(OH)_2$ neutralized = 1.10×10^{-3} mol $H_3O^+ \times \frac{1 \text{ mol } OH^-}{1 \text{ mol } H_3O^+} \times \frac{1 \text{ mol } Ca(OH)_2}{2 \text{ mol } OH^-} = 5.50\times10^{-4}$ mol $Ca(OH)_2$

no. g $Ca(OH)_2$/L = 5.50×10^{-4} mol $Ca(OH)_2 \times \frac{74.1 \text{ g } Ca(OH)_2}{1 \text{ mol } Ca(OH)_2} \times \frac{1}{0.0500 \text{ L}} = 0.815$ g $Ca(OH)_2$/L

18-22. Calculate the no. mol H_3O^+ in the acid and no. mol OH^- in the base. Determine which is in excess and then establish the final pH.

Solution A: pH = 2.50 $\log [H_3O^+] = -2.50$ $[H_3O^+] = 3.2\times10^{-3}$ M

no. mol H_3O^+ = 0.100 L x 3.2×10^{-3} mol H_3O^+/L = 3.2×10^{-4} mol H_3O^+

Solution B: pH = 11.00 pOH = 3.00 $[OH^-] = 1.0\times10^{-3}$ M

no. mol OH^- = 0.100 L x 1.0×10^{-3} mol/L = 1.0×10^{-4} mol OH^-

The neutralization reaction involves 1.0×10^{-4} mol OH^- reacting with 1.0×10^{-4} mol H_3O^+. The excess reactant is 2.2×10^{-4} mol H_3O^+.

$$[H_3O^+] = \frac{2.2\times10^{-4} \text{ mol } H_3O^+}{0.200 \text{ L}} = 1.1\times10^{-3} \text{ M} \qquad pH = 2.96$$

18-23. (a) In the event that the acid and base in Exercise 22 are weak rather than strong, the final pH will be higher than in the strong acid/strong base case. If the molar concentration of weak base to establish a pH = 11.00 is greater than the molar concentration of weak acid to establish a pH = 2.50, then upon mixing excess base will be present. The pH in this case would certainly be greater than 7. Even if weak acid is left in excess following the neutralization reaction, its ionization will be repressed by the salt formed in the neutralization reaction and the final pH will be above 2.50 (although below 7).

(b) K_a and K_b would have to be given to obtain an exact solution to the problem. That is, if K_a of a weak acid and the pH of its solution are known, its molarity can be calculated. The same is true for a weak base. Once the molarities of the weak acid and weak base are known, the result of the neutralization reaction can be worked out.

18-24. (a) The neutralization reaction for a strong acid and a strong base is the reverse of the self-ionization reaction, and the K value is the reciprocal of K_w.

(1) $H_3O^+ + OH^- \rightleftharpoons 2\ H_2O$ $K = 1/K_w = 1.0\times10^{14}$

The neutralization reaction between NH_3 and a strong acid can be described as the following combination of two reactions:

$NH_3 + H_2O \rightleftharpoons NH_4^+ + OH^-$ $K_b = 1.74\times10^{-5}$

$H_3O^+ + OH^- \rightleftharpoons 2\ H_2O$ $K = 1/K_w = 1.0\times10^{14}$

(2) $NH_3 + H_3O^+ \rightleftharpoons NH_4^+ + H_2O$ $K = 1.74\times10^{-5} \times 1.0\times10^{14} = 1.7\times10^{9}$

(b) For each neutralization reaction the value of K is very large (1.0×10^{14} in one case and 1.74×10^{-9} in the other). These large values are a sufficient indication that the reactions go essentially to completion. Alternatively, convert the K values to large positive $\Delta\bar{G}°$ values to arrive at the same conclusion.

Titration curves

18-25. The equivalence point in a titration is the point where complete neutralization has occurred. For a given volume of acid (say 25.00 mL) of a fixed concentration (say 0.100 M), the number of moles of acid to be titrated is the same, regardless of whether the acid is strong or weak. The volumes of an NaOH solution required to reach the equivalence point in the two cases will also be the same. The major difference in the two cases is that in the strong acid all the available acid is ionized initially, whereas in a weak acid ionization occurs throughout the titration to the equivalence point.

18-26. (a) The solution being titrated is 25.0 mL of a 0.100 M KOH. Initially, $[OH^-]$ = 0.100; pOH = 1.00 and *pH = 13.00*. At the equivalence point the solution contains the neutral salt KI; *pH = 7.00*. The volume of 0.200 M HI required to neutralize the base is *12.5 mL*. Because the pH changes so rapidly at the equivalence point, many different indicators may be used in the titration--all of those in Figure 18-2 except alizarin yellow R.

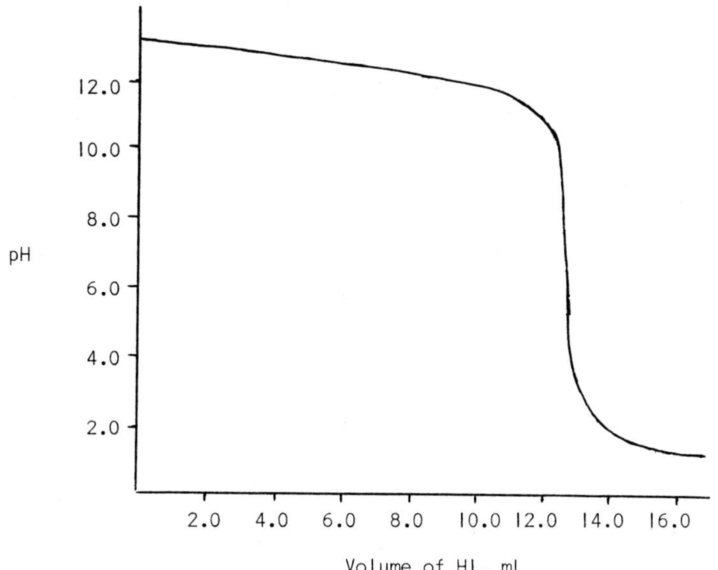

(b) The solution being titrated is 10.0 mL of 1.00 M NH_3.

Initial pH: $K_b = \frac{[NH_4^+][OH^-]}{[NH_3]} = \frac{x^2}{1.00} = 1.74\times10^{-5}$ $\qquad x = [OH^-] = 4.17\times10^{-3}$

pOH = 2.380 $\qquad$ pH = 11.62

Volume of titrant: no. mL acid = $0.0100 \text{ L} \times \frac{1.00 \text{ mol } NH_3}{1 \text{ L}} \times \frac{1 \text{ L acid}}{0.250 \text{ mol HCl}} \times \frac{1000 \text{ mL acid}}{1 \text{ L acid}}$

= 40.0 mL acid

pH at the equivalence point: The volume of 0.250 M HCl required for the titration is 40.0 mL. At this point, 0.0100 mol NH_4Cl is present in 50.00 mL of solution. The molarity of NH_4Cl is 0.0100/0.0500 = 0.200 M NH_4Cl. The pH at the equivalence point is determined by the hydrolysis of NH_4^+.

$$NH_4^+ + H_2O \rightleftharpoons NH_3 + H_3O^+$$

$$K_a = \frac{[H_3O^+][NH_3]}{[NH_4^+]} = \frac{x \cdot x}{0.200} = \frac{K_w}{K_b} = \frac{1.0\times10^{-14}}{1.74\times10^{-5}} = 5.7\times10^{-10}$$

$x^2 = 1.1 \qquad x = [H_3O^+] = 1.0 \qquad$ pH = 5.00

Selection of indicator: The indicator chosen should undergo a color change at pH ≈ 5. Methyl orange and bromcresol green should be satisfactory.

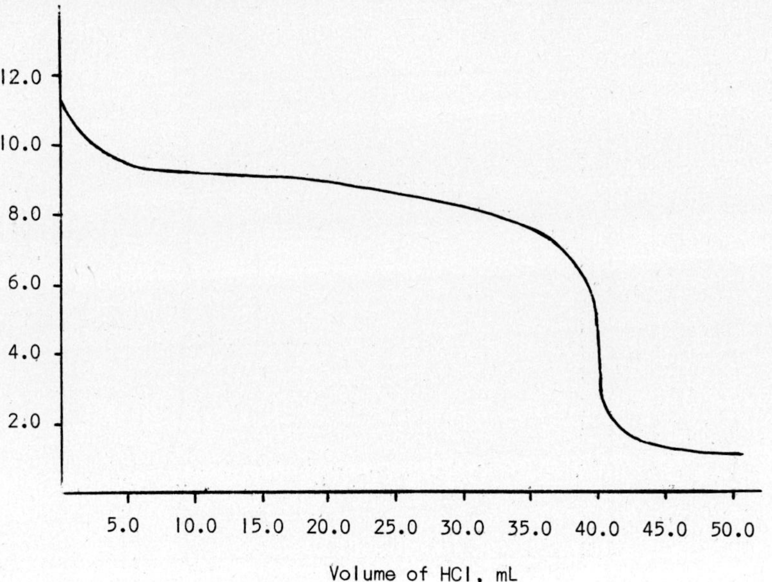

18-27. The initial pH in each case is derived as follows:

$[H_3O^+]^2/0.100 = K_a$ $\qquad [H_3O^+] = (0.100\ K_a)^{\frac{1}{2}}$ $\qquad pH = -\log [H_3O^+]$

The pH at the midpoint of each titration is pH = pK_a.

At the equivalence point the solutions are 0.0500 M in the salts, NaX, NaY, and NaZ, respectively. For the hydrolysis of these salts.

$[OH^-]^2 = 0.0500 \times K_b = 0.0500 \times K_w/K_a$ and $[OH^-] = (0.0500\ K_w/K_a)^{\frac{1}{2}}$

$pOH = -\log [OH^-]$ and $pH = 14.00 - pOH$

0.100 M HX

initial pH = 2.00

midpoint of titration: $pH = pK_a = -\log (1.0\times10^{-3}) = 3.00$

equivalence point: $[OH^-] = (0.0500 \times 1.0\times10^{-14}/1.0\times10^{-3})^{1/2} = 7.1\times10^{-7}$ M

pOH = 6.15 $\qquad$ pH = 14.00 - 6.15 = 7.85

0.100 M HY

initial pH = 3.00

midpoint of titration: $pH = -\log (1.0\times10^{-5}) = 5.00$

equivalence point: $[OH^-] = (0.0500 \times 1.0\times10^{-14}/1.0\times10^{-5})^{1/2} = 7.1\times10^{-6}$ M

pOH = 5.15 $\qquad$ pH = 14.00 - 5.15 = 8.85

0.100 M HZ

initial pH = 4.00

midpoint of titration: $pH = -\log (1.0\times10^{-7}) = 7.00$

equivalence point: $[OH^-] = (0.0500 \times 1.0\times10^{-14}/1.0\times10^{-7})^{1/2} = 7.1\times10^{-5}$ M

pOH = 4.15 $\qquad$ pH = 14.00 - 4.15 = 9.85

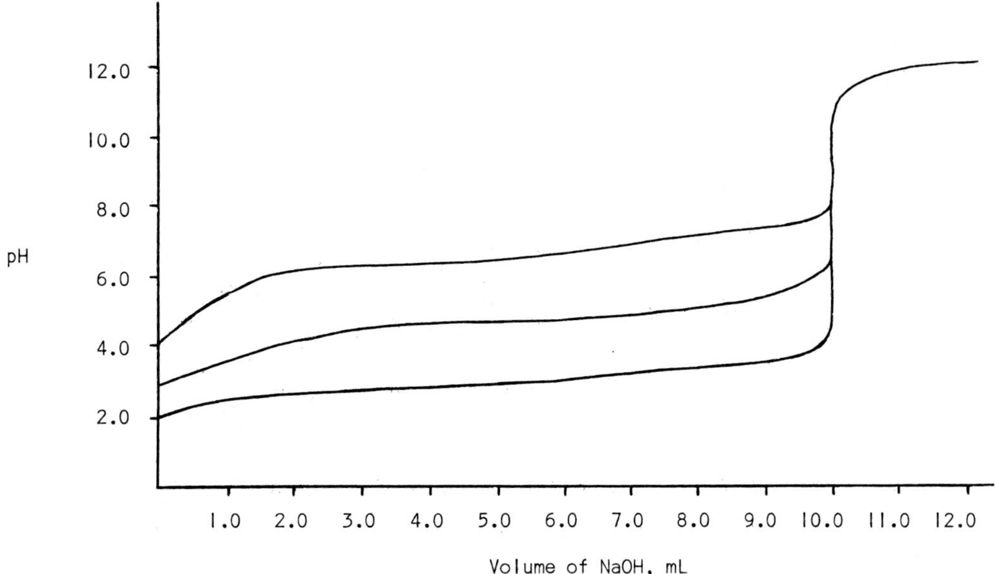

18-28. Refer to Figure 18-5. The first equivalence point corresponds to the conversion of H_3PO_4 to NaH_2PO_4. The volume of 0.0200 M NaOH required is

$$\text{no. mL} = 0.0100 \text{ L} \times \frac{0.0400 \text{ mol } H_3PO_4}{\text{L}} \times \frac{1 \text{ mol NaOH}}{1 \text{ mol } H_3PO_4} \times \frac{1 \text{ L}}{0.0200 \text{ mol NaOH}} \times \frac{1000 \text{ mL}}{1 \text{ L}} = 20.0 \text{ mL}$$

Present in the solution at the first equivalence point is 4.00×10^{-4} mol NaH_2PO_4 produced by neutralizing the H_3PO_4, *plus* 1.50×10^{-4} mol NaH_2PO_4 present initially. Thus, to reach the second equivalence point, 5.50×10^{-4} mol $H_2PO_4^-$ must be converted to HPO_4^{2-}. The required volume of NaOH, which will be greater than that for the first equivalence point is,

$$\text{no. mL} = 5.50\times10^{-4} \text{ mol } H_2PO_4^- \times \frac{1 \text{ mol NaOH}}{1 \text{ mol } H_2PO_4^-} \times \frac{1 \text{ L}}{0.0200 \text{ mol NaOH}} \times \frac{1000 \text{ mL}}{1 \text{ L}} = 27.5 \text{ mL}$$

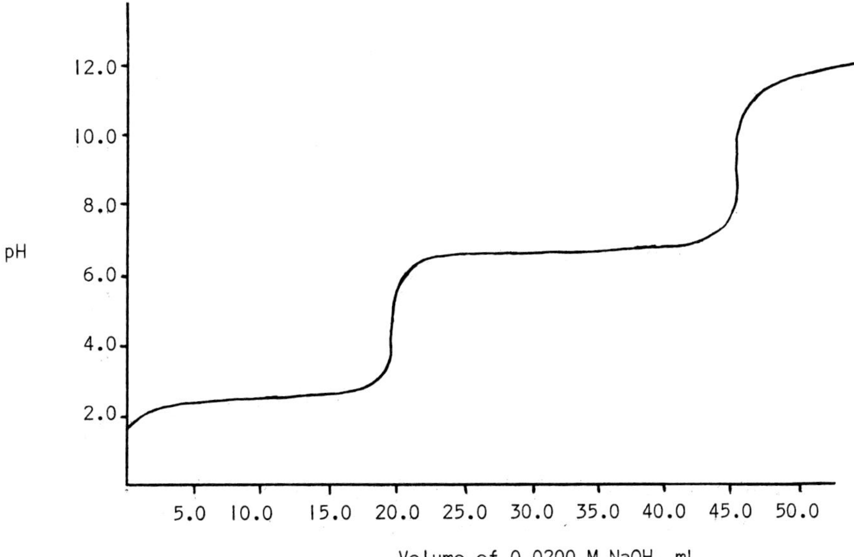

18-29. (a) tot. no. mol H_3O^+ = $0.02500\ L \times \frac{0.1000\ mol\ HCl}{1.00\ L} \times \frac{1\ mol\ H_3O^+}{1\ mol\ HCl} = 2.500\times10^{-3}\ mol\ H_3O^+$

When the acid is 90% neutralized, 10% of the acid is unreacted.

no. mol H_3O^+ unreacted = $0.10 \times 2.500\times10^{-3}\ mol\ H_3O^+ = 2.5\times10^{-4}\ mol\ H_3O^+$

The volume of 0.1000 M NaOH added at this point = 0.90 × 25.00 mL = 22.50 mL.

$$[H_3O^+] = \frac{2.5\times10^{-4}\ mol\ H_3O^+}{[(25.00 + 22.50)/1000]L} = 5.3\times10^{-3}\ M \qquad pH = 2.28$$

(b) At the point of 90% neutralization, 90% of the original $HC_2H_3O_2$ has been converted to $C_2H_3O_2^-$. The ratio $[C_2H_3O_2^-]/[HC_2H_3O_2]$ is 90:10 = 9.0. For a mixture of a weak acid and its salt,

$$pH = pK_a + \log\frac{[C_2H_3O_2^-]}{[HC_2H_3O_2]} = 4.76 + \log 9.0 = 5.71$$

18-30. (a) There would not be a sharp color change at the indicator end point. As indicated in Figure 18-2, thymol blue would have been completely converted to its yellow color before the sharp rise in pH at the equivalence point.

(b) Let the no. mL 0.1000 M NaOH added to reach pH = 2 be x. The number of moles of H_3O^+ unreacted at this point would be 0.02500 × 0.1000 mol H_3O^+ (initially) - (x/1000) × 0.1000 mol H_3O^+ (reacted) = $2.500\times10^{-3} - (1.000\times10^{-4})x$.

The volume of solution, in liters, would be $(25.00 + x)/1000$. If pH = 2.00,

$[H_3O^+] = 1.00\times10^{-2}$ M. Therefore, $[H_3O^+] = \frac{2.500\times10^{-3} - (1.000\times10^{-4})x}{(25.00 + x)/1000} = 0.0100$

$2.500 - 0.1000x = 0.2500 + 0.0100x \qquad 0.1100x = 2.250 \qquad x = 20.46$ mL 0.1000 M NaOH

For complete neutralization of the HCl, 25.00 mL 0.1000 M NaOH is required. At the point where 20.46 mL has been added, 20.46/25.00 = 0.8184 (81.84%) of the acid has been neutralized. The percent unneutralized = 100.00 - 81.84 = 18.16%.

18-31. (a) Review the solution to part (b) of Exercise 30. In a similar fashion let x = no. mL 0.1000 M NaOH required to reach a pH = 3.00.

$$[H_3O^+] = \frac{2.500\times10^{-3} - (1.000\times10^{-4})x}{(25.00 + x)/1000} = 0.0010$$

$2.500 - 0.1000x = 0.0250 + 0.0010 \qquad 0.1010x = 2.475 \qquad x = 24.50$ mL 0.1000 M NaOH

(b) Use the equation $pH = pK_a + \log\frac{[C_2H_3O_2^-]}{[HC_2H_3O_2]}$ and solve for the ratio $[C_2H_3O_2^-]/[HC_2H_3O_2]$.

$pH = 4.76 + \log\frac{[C_2H_3O_2^-]}{[HC_2H_3O_2]} = 5.25 \qquad \log\frac{[C_2H_3O_2^-]}{[HC_2H_3O_2]} = 5.25 - 4.76 = 0.49 \qquad \frac{[C_2H_3O_2^-]}{[HC_2H_3O_2]} = 3.1$

If at this point $[HC_2H_3O_2] = x$, $[C_2H_3O_2^-]$ is $3.1x$ and the total acetate concentration = $4.1x$.

$$\text{Fraction neutralized} = \frac{[C_2H_3O_2^-]}{\text{total acetate}} = \frac{3.1x}{4.1x} = 0.76$$

For complete neutralization, 25.00 mL 0.1000 M NaOH is required. To reach a pH = 5.25, 0.76 × 25.00 = 19 mL 0.1000 M NaOH is required.

(c) When the pH of the solution is 7.50, the titration is that of NaH_2PO_4 being converted to Na_2HPO_4. This occurs in the range from 10.00 to 20.00 mL 0.1000 M NaOH. As in part (b), determine the ratio $[HPO_4^{2-}]/[H_2PO_4^-]$.

$$pH = pK_{a_2} + \log\frac{[HPO_4^{2-}]}{[H_2PO_4^-]} = -\log(6.2\times10^{-8}) + \log\frac{[HPO_4^{2-}]}{[H_2PO_4^-]} = 7.50$$

$\log \frac{[HPO_4^{2-}]}{[H_2PO_4^-]} = 7.50 - 7.21 = 0.29$ $\qquad$ $[HPO_4^{2-}]/[H_2PO_4^-] = 1.9$

If at this point $[H_2PO_4^-] = x$, $[HPO_4^{2-}] = 1.9$ and the total phosphate concentration is $2.95x$.

$$\text{Fraction neutralized} = \frac{[HPO_4^{2-}]}{\text{total phosphate}} = \frac{1.9}{2.9} = 0.661$$

From complete conversion of $H_2PO_4^-$ to HPO_4^{2-}, 10.00 mL 0.1000 M NaOH is required. To reach a pH = 7.50, 0.66 x 10.00 = 6.6 mL 0.1000 M NaOH is required, *in addition to* the 10.00 m required to reach the first equivalence point. Total volume is 16.6 mL.

Hydrolysis of salts of polyprotic acids

18-32. An $Na_2S(aq)$ solution is likely to be rather strongly basic because of the reaction

$S^{2-} + H_2O \rightleftharpoons HS^- + OH^-$ $\qquad$ $K_b = K_w/K_{a_2} = 1.0\times10^{-14}/1.0\times10^{-14} = 1.0$

The base ionization constant for S^{2-} (K_b = 1.0) is quite large for a weak base.

18-33. (a) For 1.0 M $Na_2CO_3(aq)$ we can write

	CO_3^{2-}	+	H_2O	$\rightleftharpoons$	HCO_3^-	+	OH^-
Placed in Soln:	1.0 M				---		---
Changes:	$-x$ M				$+x$ M		$+x$ M
At Equilibrium:	$(1.0 - x)$M				x M		x M

$K_b = K_w/K_{a_2} = 1.0\times10^{-14}/5.6\times10^{-11} = 1.8\times10^{-4}$

Assume $x << 1.0$,

$$K_b = \frac{[HCO_3^-][OH^-]}{[CO_3^{2-}]} = \frac{x \cdot x}{1.0} = 1.8\times10^{-4}$$

$x^2 = 1.8\times10^{-4}$ $\qquad$ $x = [OH^-] = 1.3\times10^{-2}$ M

$pOH = -\log 1.3\times10^{-2} = 1.89$ $\qquad$ $pH = 14.00 - 1.89 = 12.11$

(b) Proceed as in part (a), but with 0.010 M substituting for 1.0 M.

$$K_b = \frac{[HCO_3^-][OH^-]}{[CO_3^{2-}]} = \frac{x \cdot x}{0.010 - x} = 1.8\times10^{-4}$$

If we assume that $(0.010 - x) \cong 0.010$

$\frac{x^2}{0.010} = 1.8\times10^{-4}$ $\qquad$ $x^2 = 1.8\times10^{-6}$ $\qquad$ $x = 1.3\times10^{-3}$

Although $0.010 - x = 0.010 - 0.0013 = 0.009 \approx 0.010$, the assumption still works reasonably well since we are carrying only two significant figures.

$pOH = -\log(1.3\times10^{-3}) = 2.89$ $\qquad$ $pH = 14.00 - 2.89 = 11.11$

18-34. The solutions under comparison have the same molar concentrations. The means by which the substances produce H_3O^+ (or OH^-) are

$C_2H_3O_2^- + H_2O \rightleftharpoons HC_2H_3O_2 + OH^-$ $K_b = 5.75\times10^{-10}$

$HC_2O_4^- + H_2O \rightleftharpoons H_3O^+ + C_2O_4^{2-}$ $K_{a_2} = 5.4\times10^{-5}$

$HSO_4^- + H_2O \rightleftharpoons H_3O^+ + SO_4^{2-}$ $K_{a_2} = 1.29\times10^{-2}$

$H_2PO_4^- + H_2O \rightleftharpoons H_3O^+ + HPO_4^{2-}$ $K_{a_2} = 6.2\times10^{-8}$

The three substances that produce H_3O^+ are all more acidic than $C_2H_3O_2^-$ (which acts as a base). The three acids can be arranged in order of decreasing K values, and $C_2H_3O_2^-$ comes last.

$HSO_4^- > HC_2O_4^- > H_2PO_4^- > C_2H_3O_2^-$

Equivalent weight and normality concentration

18-35. From the information given about the reaction of CO_3^{2-} and H^+ we see that for Na_2CO_3 or $Na_2CO_3\cdot10\ H_2O$, equiv. wt. = 1/2 ft. wt.; equiv. wt. $Na_2CO_3\cdot10\ H_2O$ = 1/2(286.2) = 143.1

$$\text{no. equiv.} = 2.00\ L \times \frac{0.175\ \text{equiv. } Na_2CO_3}{1.00\ L} \times \frac{1\ \text{equiv. } Na_2CO_3\cdot10\ H_2O}{1\ \text{equiv. } Na_2CO_3} \times \frac{143.1\ g\ Na_2CO_3\cdot10\ H_2O}{1\ \text{equiv. } Na_2CO_3\cdot10\ H_2O}$$

$$= 50.1\ g\ Na_2CO_3\cdot10\ H_2O$$

18-36. Assume that the solution density is approximately 1.0 g/mL, so that the solubility can be expressed as 3.9 g $Ba(OH)_2$ per 100 mL or 39 g $Ba(OH)_2$/L.

$$\text{molar concentration} = \frac{39\ g\ Ba(OH)_2}{1.00\ L} \times \frac{1\ mol\ Ba(OH)_2}{171\ g\ Ba(OH)_2} = 0.23\ M$$

For $Ba(OH)_2$, normality = 2 x molarity = 0.46 N

18-37. The volume of NaOH given is that required for titration to the second equivalence point. To the first equivalence point, the volume required is 31.15/2 = 15.58 mL.

For use in reaction (18-32), $N_{H_3PO_4} = \dfrac{15.58\ ml \times 0.242\ mequiv/mL}{25.00\ mL} = 0.151\ mequiv/mL = 0.151\ N$

For use in reaction (18-33), $N_{H_3PO_4} = 2 \times 0.151 = 0.302\ N$

For use in reaction (18-34), $N_{H_3PO_4} = 3 \times 0.151 = 0.453\ N$

18-38. $\text{no. g } H_2SO_4 \text{ in sample titrated} = 0.03240\ L \times \dfrac{0.0100\ equiv\ Ba(OH)_2}{1.00\ L} \times \dfrac{1\ equiv\ H_2SO_4}{1\ equiv\ Ba(OH)_2}$

$$\times \frac{49.0\ g\ H_2SO_4}{1\ equiv\ H_2SO_4} = 0.0159\ g\ H_2SO_4$$

The sample titrated (10.00 mL) is only a portion of the total solution prepared (250.0 mL).

$$\text{total g } H_2SO_4 = \frac{0.0159\ g\ H_2SO_4}{10.00\ mL} \times 250.0\ mL = 0.398\ g\ H_2SO_4$$

This mass of H_2SO_4 was derived from 1.239 g battery acid.

$$\%\ H_2SO_4\text{, by mass} = \frac{0.398\ g\ H_2SO_4}{1.239\ g\ \text{battery acid}} \times 100 = 32.1\%\ H_2SO_4$$

Self-test Questions

1. (*c*) Of the substances listed only $NaCHO_2$ has a common ion with $HCHO_2$. NaCl and $NaNO_3$ are expected to have very little effect on the ionization of $HCHO_2$. NaOH, by neutralizing $HCHO_2$, enhances rather than represses the ionization of $HCHO_2$.

2. (*d*) $NaCHO_2$ represses the ionization of $HCHO_2$; so does H_2SO_4. NaCl would have little effect on the ionization of $HCHO_2$. $NaHCO_3$ increases the ionization of $HCHO_2$ through the reaction $HCHO_2 + HCO_3^- \rightarrow CHO_2^- + H_2O + CO_2(g)$.

3. (*c*) The amount of HCl present is 0.50 mol. Adding a weak acid ($HC_2H_3O_2$) to a strong acid (HCl) will not affect the pH of the strong acid. Adding only 0.40 mol NaOH will also not effect the pH significantly, since this will leave 0.10 mol of the acid as free acid. Addition of NaCl will have very little effect on the pH of HCl. The addition of 0.60 mol $NaC_2H_3O_2$ converts the strong acid to 0.50 mol $HC_2H_3O_2$ and leaves an excess of 0.10 mol of $NaC_2H_3O_2$. The pH of this $HC_2H_3O_2$ - $NaC_2H_3O_2$ buffer solution will be several units higher (about 3-4) than that of 0.50 M HCl.

4. (*b*) KNO_3 would have essentially no effect on the NH_3-NH_4^+ equilibrium. Neither would $BaSO_4(s)$ since it is insoluble. The addition of H_3O^+ would convert NH_3 to NH_4^+. To convert NH_4^+ to NH_3 add OH^- ($NH_4^+ + OH^- \rightarrow NH_3 + H_2O$). This can be accomplished by raising the pH.

5. (*a*) The NH_3-NH_4Cl solution is a buffer. It can accommodate a small quantity of KOH (0.001 mol) with very little change in pH. The change that does occur is a very slight increase in pH.

6. (*b*) Of the four salts listed, $NaHSO_4$ is rather strongly acidic: $HSO_4^- + H_2O \rightleftharpoons H_3O^+ + SO_4^{2-}$. The other three salts are all basic (either slightly or strongly) through the hydrolysis of an anion--S^{2-}, HCO_3^- and HPO_4^{2-}.

7. (*b*) Indicators are weak acids that function in a pH range of about two units with the center of the range being at pH = pK_a (indicator). If K_a for the indicator is 1×10^{-9}, $pK_a = 9$.

8. (*d*) In the neutralization of a weak acid by a strong base, a buffer solution is produced, with $pH = pK_a + \log \frac{[A^-]}{[HA]}$. If the acid is one-half neutralized, $[A^-] = [HA]$, $\log [A^-]/[HA] = 0$, and $pH = pK_a$.

9. The indicator end point is the point in a titration at which the indicator changes color. The equivalence point of a titration is the point at which the amounts of the acid and base are exactly equivalent--both are consumed and neither is in excess. The object in a titration is to select an indicator whose end point matches the equivalence point.

10. (a) The titration is $HCO_3^- + OH^- \rightarrow H_2O + CO_3^{2-}$. At the equivalence point the solution is $Na_2CO_3(aq)$. Hydrolysis of CO_3^{2-} produces a basic solution (pH > 7). $CO_3^{2-} + H_2O \rightleftharpoons HCO_3^- + OH^-$.

    (b) The titration is $HCl + NH_3 \rightarrow NH_4^+ + Cl^-$. At the equivalence point the solution is $NH_4Cl(aq)$. Hydrolysis of NH_4^+ produces an acidic solution (pH < 7). $NH_4^+ + H_2O \rightleftharpoons H_3O^+ + NH_3$.

    (c) The titration is $KOH + HI \rightarrow H_2O + KI$. At the equivalence point the solution is KI(aq). Neither $K^+(aq)$ nor $I^-(aq)$ hydrolyzes. pH = 7.

11. (a) Use the equation $pH = pK_a + \log \frac{[CHO_2^-]}{[HCHO_2]}$ and solve for $[CHO_2^-]$.

$$3.90 = -\log(1.8\times10^{-4}) + \log \frac{[CHO_2^-]}{0.650} \qquad 3.90 = 3.74 + \log \frac{[CHO_2^-]}{0.650}$$

$$\log \frac{[CHO_2^-]}{0.650} = 3.90 - 3.74 = 0.16 \qquad \frac{[CHO_2^-]}{0.650} = 1.4 \qquad [CHO_2^-] = 0.91\ M$$

$$\text{no. g } NaCHO_2 = 0.500\ L \times \frac{0.91\ \text{mol } NaCHO_2}{1.00\ L} \times \frac{68.0\ \text{g } NaCHO_2}{1\ \text{mol } NaCHO_2} = 31\ \text{g } NaCHO_2$$

    (b) The concentration of OH^- being added to the buffer solution is:

$$[OH^-] = \frac{0.20\ \text{g NaOH} \times \frac{1\ \text{mol NaOH}}{40.0\ \text{g NaOH}} \times \frac{1\ \text{mol } OH^-}{1\ \text{mol NaOH}}}{0.500\ L} = 0.010\ M$$

	$HCHO_2$	+	OH^-	⟶	CHO_2^-	+	H_2O
Original:	0.650 M				0.91 M		
Add:			0.010 M				
Changes:	-0.010 M		-0.010 M		+0.010 M		
Equilibrium:	0.640 M		≈0		0.92 M		

$$pH = 3.74 + \log \frac{0.92}{0.640} = 3.74 + \log 1.4 = 3.74 + 0.15 = 3.89$$

12. (a) $K_a = \frac{[H_3O^+][C_7H_5O_2^-]}{[HC_7H_5O_2]} = \frac{[H_3O^+]^2}{0.0100} = 6.3\times10^{-5}$ $\qquad [H_3O^+]^2 = 6.3\times10^{-7}$

$[H_3O^+] = 7.9\times10^{-4}$ M $\qquad$ pH = 3.10

(b) no. mol $HC_7H_5O_2$ = 0.0250 L x 0.0100 mol/L = 2.50×10^{-4} mol

no. mol OH^- added = 0.00625 L x 0.0200 mol OH^-/L = 1.25×10^{-4} mol OH^-

no. mol $HC_7H_5O_2$ reacted = 1.25×10^{-4} mol

no. mol $HC_7H_5O_2$ unreacted = 2.50×10^{-4} mol - 1.25×10^{-4} mol = 1.25×10^{-4} mol

$[HC_7H_5O_2] = 1.25\times10^{-4}$ mol/(0.0250 + 0.00625)L = 4.00×10^{-3} M

$[C_7H_5O_2^-] = [HC_7H_5O_2]$ (since the acid is half neutralized).

$$pH = -\log (6.3\times10^{-5}) + \log \frac{[C_7H_5O_2^-]}{[HC_7H_5O_2]} = 4.20 + \log 1.0 = 4.20$$

(c) At the equivalence point there is present 2.50×10^{-4} mol $NaC_7H_5O_2$ in (25.0 + 12.5) = 37.5 mL.

$$[C_7H_5O_2^-] = \frac{2.50\times10^{-4}\ \text{mol}}{0.0375\ \text{L}} = 6.67\times10^{-3}\ \text{M}$$

In the hydrolysis reaction $C_7H_5O_2^- + H_2O \rightleftharpoons HC_7H_5O_2 + OH^-$,

$$K_b = \frac{[HC_7H_5O_2][OH^-]}{[C_7H_5O_2^-]} = \frac{[OH^-]^2}{6.67\times10^{-3}} = \frac{1.0\times10^{-14}}{6.3\times10^{-5}} = 1.6\times10^{-10} \qquad [OH^-]^2 = 1.1\times10^{-12}$$

$[OH^-] = 1.0\times10^{-6}$ M $\qquad$ pOH = 6.00 $\qquad$ pH = 8.00

(d) The addition of 15.00 mL 0.0100 M $Ba(OH)_2$ introduces no. mol OH^- = 0.01500 L x 0.0200 mol OH^-/L = 3.00×10^{-4} mol OH^-; excess mol OH^- = $(3.00\times10^{-4}) - (2.50\times10^{-4}) = 0.50\times10^{-4}$ mol OH^-; soln. volume = 40.00 mL = 0.04000L.

$$[OH^-] = \frac{0.50\times10^{-4}\ \text{mol OH}^-}{0.04000\text{L}} = 1.2\times10^{-3} \qquad pOH = 2.92 \qquad pH = 11.08$$

Chapter 19

Solubility and Complex Ion Equilibria

Review Problems

19-1. (a) $K_{sp} = [Ag^+]^2[SO_4^{2-}]$

(b) $K_{sp} = [Ra^{2+}][IO_3^-]^2$

(c) $K_{sp} = [Ni^{2+}]^3[PO_4^{2-}]^2$

(d) $K_{sp} = [Hg_2^{2+}][C_2O_4^{2-}]$

(e) $K_{sp} = [PuO_2^{2+}][CO_3^{2-}]$

19-2. (a) $Fe(OH)_3(s) \rightleftharpoons Fe^{3+}(aq) + 3\ OH^-(aq)$

(b) $BiOOH(s) \rightleftharpoons BiO^+(aq) + OH^-(aq)$

(c) $Hg_2I_2(s) \rightleftharpoons Hg_2^{2+}(aq) + 2\ I^-(aq)$

(d) $Pb_3(AsO_4)_2(s) \rightleftharpoons 3\ Pb^{2+}(aq) + 2\ AsO_4^{3-}(aq)$

(e) $Cu_2[Fe(CN)_6](s) \rightleftharpoons 2\ Cu^{2+}(aq) + [Fe(CN)_6]^{4-}(aq)$

(f) $MgNH_4PO_4(s) \rightleftharpoons Mg^{2+}(aq) + NH_4^+(aq) + PO_4^{3-}(aq)$

19-3. In each case denote the molar solubility by S and ion concentrations in terms of S. Substitute ion concentrations into the K_{sp} expression and solve for S.

(a) $BaCrO_4$: $[Ba^{2+}] = S$ and $[CrO_4^{2-}] = S$.

$$K_{sp} = [Ba^{2+}][CrO_4^{2-}] = (S)(S) = S^2 = 1.2\times10^{-10}$$

$$S = \text{molar solubility of } BaCrO_4 = \sqrt{1.2\times10^{-10}} = 1.1\times10^{-5} \text{ mol } BaCrO_4/L$$

(b) $PbBr_2$: $[Pb^{2+}] = S$ and $[Br^-] = 2\ S$

$$K_{sp} = [Pb^{2+}][Br^-]^2 = S(2S)^2 = 4\ S^3 = 4.0\times10^{-5}$$

$$S = (1.0\times10^{-5})^{1/3} = 2.2\times10^{-2} \text{ mol } PbBr_2/L$$

(c) CeF_3: $[Ce^{3+}] = S$ and $[F^-] = 3\ S$

$$K_{sp} = [Ce^{3+}][F^-]^3 = S(3S)^3 = 27\ S^4 = 8\times10^{-16}$$

$$S = \left(\frac{8\times10^{-16}}{27}\right)^{\frac{1}{4}} = 7\times10^{-5} \text{ mol } CeF_3/L$$

(d) $Mg_3(AsO_4)_2$: $[Mg^{2+}] = 3\,S$ and $AsO_4^{3-} = 2\,S$

$K_{sp} = [Mg^{2+}]^3[AsO_4^{3-}]^2 = (3\,S)^3 \times (2\,S)^2 = 27\,S^3 \times 4\,S^2 = 108\,S^5 = 2.1\times10^{-20}$

$S = \left(\frac{2.1\times10^{-20}}{108}\right)^{1/5} = 4.5\times10^{-5}$ mol $Mg_3(AsO_4)_2$/L

19-4. (a) $CsMnO_4(s) \rightleftharpoons Cs^+(aq) + MnO_4^-(aq)$

If the molar solubility = S, then $[Cs^+] = [MnO_4^-] = S = 3.8\times10^{-3}$

$K_{sp} = [Cs^+][MnO_4^-] = S \times S = (3.8\times10^{-3})(3.8\times10^{-3}) = 1.4\times10^{-5}$

(b) $Pb(ClO_2)_2(s) \rightleftharpoons Pb^{2+}(aq) + 2\,ClO_2^-(aq)$

If the molar solubility = S, then $[Pb^{2+}] = S$ and $[ClO_2^-] = S$

$K_{sp} = [Pb^{2+}][ClO_2^-]^2 = S\,(2\,S)^2 = 4\,S^3 = 4(2.8\times10^{-3})^3 = 8.8\times10^{-8}$

(c) $Li_3PO_4(s) \rightleftharpoons 3\,Li^+(aq) + PO_4^{3-}(aq)$

If the molar solubility = S, then $[Li^+] = 3\,S$ and $[PO_4^{3-}] = S$

$K_{sp} = [Li^+]^3[PO_4^{3-}] = (3\,S)^3 \times S = 27\,S^4$; $S = 2.9\times10^{-3}$ M

$K_{sp} = 27\,S^4 = 27 \times (2.9\times10^{-3})^4 = 1.9\times10^{-9}$

19-5. (a) $Mg(OH)_2(s) \rightleftharpoons Mg^{2+}(aq) + 2\,OH^-(aq)$

Let molar solubility = S; $[Mg^{2+}] = S$ and $[OH^-] = 2\,S$

$K_{sp} = [Mg^{2+}][OH^-]^2 = S \times (2\,S)^2 = 4\,S^3 = 1.8\times10^{-11}$

$S^3 = 4.5\times10^{-12}$ $S = (4.5\times10^{-12})^{1/3} = 1.7\times10^{-4}$ mol $Mg(OH)_2$/L

(b)

	$Mg(OH)_2(s) \rightleftharpoons$ $Mg^{2+}(aq)$	+	$2\,OH^-(aq)$
From $Mg(OH)_2$:	S		$2\,S$
From 0.015 M $MgCl_2$:	0.015 M		-
At equilibrium	$(0.015 + S)$ M		$2S$ M

$K_{sp} = [Mg^{2+}][OH^-]^2 = (0.015 + S)(2\,S)^2 = 1.8\times10^{-11}$

Assume that $S << 0.015$ and write

$0.015(2\,S)^2 = 1.8\times10^{-11}$ $S = \left(\frac{1.8\times10^{-11}}{0.015 \times 4}\right)^{\frac{1}{2}} = 1.7\times10^{-5}$ mol $Mg(OH)_2$/L

(c) In a set up similar to the one in part (b) we would write

$[Mg^{2+}] = S$ and $[OH^-] = (0.217 + 2\ S)$. Assume that $S << 0.217$.

$$K_{sp} = [Mg^{2+}][OH^-]^2 = S(0.217)^2 = 1.8\times10^{-11}$$

$$S = \frac{1.8\times10^{-11}}{(0.217)^2} = 3.8\times10^{-10} \text{ mol Mg(OH)}_2\text{/L}$$

19-6. (a) $MgCO_3(s) \rightleftharpoons Mg^{2+}(aq) + CO_3^{2-}(aq)$

$$Q = [Mg^{2+}][CO_3^{2-}] = (0.015)(0.0072) = 1.1\times10^{-4} > K_{sp}(3.5\times10^{-8})$$

$MgCO_3(s)$ will precipitate.

(b) $Ag_2SO_4(s) \rightleftharpoons 2\ Ag^+(aq) + SO_4^{3-}(aq)$

$$Q = [Ag^+]^2[SO_4^{2-}] = (0.0038)^2(0.0105) = 1.5\times10^{-7} < K_{sp}(1.4\times10^{-5})$$

$Ag_2SO_4(s)$ will not precipitate.

(c) $Cr(OH)_3(s) \rightleftharpoons Cr^{3+}(aq) + 3\ OH^-(aq)$

$$[OH^-] = \frac{K_w}{[H_3O^+]} = \frac{1.0\times10^{-14}}{0.0016} = 6.2\times10^{-12}$$

$$Q = [Cr^{3+}][OH^-]^3 = 0.041(6.2\times10^{-12})^3 = 9.8\times10^{-36} < K_{sp}(6.3\times10^{-31})$$

$Cr(OH)_3$ will not precipitate.

19-7. (a) The precipitate to form first is the one that will begin to precipitate at the lowest $[I^-]$.

To precipitate PbI_2: $[Pb^{2+}][I^-]^2 = K_{sp} = 7.1\times10^{-9}$

$$(0.10)[I^-]^2 = 7.1\times10^{-9}$$

$$[I^-]^2 = 7.1\times10^{-8} \quad [I^-] = 2.7\times10^{-4}\ M$$

To precipitate AgI: $[Ag^+][I^-] = K_{sp} = 8.5\times10^{-17}$

$$(0.10)[I^-] = 8.5\times10^{-17}$$

$$[I^-] = 8.5\times10^{-16}\ M$$

Since AgI can precipitate at a lower $[I^-]$ than PbI_2, AgI will precipitate first.

(b) The second cation to precipitate is Pb^{2+}. The $[I^-]$ required to bring this about is

$$[I^-]^2 = \frac{K_{sp}}{[Pb^{2+}]} = \frac{7.1\times10^{-9}}{0.10} = 7.1\times10^{-8}$$

$$[I^-] = 2.7\times10^{-4}\ M$$

(c) What is $[Ag^+]$ in saturated AgI(aq) when $[I^-] = 2.7\times10^{-4}$ M?

$$[Ag^+] = \frac{K_{sp}}{[I^-]} = \frac{8.5\times10^{-17}}{2.7\times10^{-4}} = 3.1\times10^{-13}\ M$$

(d) Before Pb^{2+} begins to precipitate as $PbI_2(s)$, the concentration of Ag^+ would have been reduced from 0.10 M to 3.1×10^{-13} M. The separation of Pb^{2+} and Ag^+ by selective precipitation of AgI should be very effective.

19-8. (a) All combinations of Na^+, Zn^{2+}, I^- and SO_4^{2-} yield soluble ionic compounds. No reaction occurs.

$NaI(aq) + ZnSO_4(aq) \rightarrow$ no reaction

(b) Insoluble $CuCO_3(s)$ is produced in this reaction.

$Cu^{2+}(aq) + CO_3^{2-}(aq) \rightarrow CuCO_3(s)$

(c) Insoluble AgCl(s) is formed.

$Ag^+(aq) + Cl^-(aq) \rightarrow AgCl(s)$

(d) Both CuS and $BaSO_4$ are insoluble.

$Ba^{2+}(aq) + S^{2-}(aq) + Cu^{2+}(aq) + SO_4^{2-}(aq) \rightarrow BaSO_4(s) + CuS(s)$

(e) OH^- from $Al(OH)_3$ combines with H^+ from HCl to form H_2O. This means that $Al^{3+}(aq)$ appears in solution.

$Al(OH)_3(s) + 3\ H^+(aq) \rightarrow Al^{3+}(aq) + 3\ H_2O$

(f) $C_2O_4^{2-}$ from slightly soluble CaC_2O_4 combines with H^+ from HCl to form the weak acid oxalic acid, $H_2C_2O_4$.

$CaC_2O_4(s) + 2\ H^+(aq) \rightarrow Ca^{2+}(aq) + H_2C_2O_4(aq)$.

(g) In Example 19-20 we learned that CdS will precipitate from H_2S (satd. aq) with $[H_3O^+] = 0.10$ M (or greater). $HC_2H_3O_2$, a weak acid, does not produce $[H_3O^+]$ at 0.10 M or greater. CdS(s) does not dissolve in $HC_2H_3O_2(aq)$.

$CdS(s) + HC_2H_3O_2(aq) \rightarrow$ no reaction

19-9. Calculate $[OH^-]$ in the buffer solution and then compare the ion products Q with K_{sp} values.

$$pH = pK_a + \log\frac{[C_2H_3O_2^-]}{[HC_2H_3O_2]} = 4.76 + \log\frac{0.25}{0.50} = 4.76 - 0.30 = 4.46$$

$pOH = 14.00 - 4.46 = 9.54 \quad [OH^-] = \text{antilog}(-9.54) = 2.9\times10^{-10}$ M

(a) $Ca(OH)_2$: $Q = [Ca^{2+}][OH^-]^2 = 0.10\ (2.9\times10^{-10})^2 < K_{sp}\ (5.5\times10^{-6})$

$Ca(OH)_2(s)$ does not precipitate. $[Ca^{2+}]$ can be maintained at 0.10 M or higher.

(b) $Al(OH)_3$: $Q = [Al^{3+}][OH^-]^3 = 0.10(2.9\times10^{-10})^3 = 2.4\times10^{-30} > K_{sp}\ (1.3\times10^{-33})$

$[Al^{3+}]$ cannot be maintained at 0.10 M; $Al(OH)_3(s)$ precipitates.

(c) $Cr(OH)_3$: $Q = [Cr^{3+}][OH^-]^3 = 0.10(2.9\times10^{-6})^3 = 2.4\times10^{-30} > K_{sp}\ (6.3\times10^{-31})$

$[Cr^{3+}]$ cannot be maintained at 0.10 M; $Cr(OH)_3(s)$ precipitates.

19-10. (a) First, determine $[NH_3]$ and $[Mg^{2+}]$ in the mixed solution (0.350 L + 0.150 L = 0.500 L).

$$[Mg^{2+}] = 0.150 \text{ L} \times \frac{0.100 \text{ mol } Mg^{2+}}{\text{L}} \times \frac{1}{0.500 \text{ L}} = 0.0300 \text{ M}$$

$$[NH_3] = 0.350 \text{ L} \times \frac{0.100 \text{ mol } NH_3}{\text{L}} \times \frac{1}{0.500 \text{ L}} = 0.0700 \text{ M}$$

Now calculate $[OH^-]$ in 0.0700 M NH_3. That is,

$$\underset{(0.0700 - x)\text{ M}}{NH_3} + H_2O \rightleftharpoons \underset{x\text{ M}}{NH_4^+} + \underset{x\text{ M}}{OH^-}$$

Assume that $(0.0700 - x) \cong 0.0700$

$$K_b = \frac{[NH_4^+][OH^-]}{[NH_3]} = \frac{x \cdot x}{0.0700} = 1.74\text{x}10^{-5}. \quad x = [OH^-] = 1.10\text{x}10^{-3} \text{ M}$$

For $Mg(OH)_2$: $Q = [Mg^{2+}][OH^-]^2 = (0.0300)(1.10\text{x}10^{-3})^2 = 3.6\text{x}10^{-8} > K_{sp}(1.8\text{x}10^{-11})$

$Mg(OH)_2(s)$ will precipitate from 0.0700 M $NH_3(aq)$.

(b) To prevent the precipitation of $Mg(OH)_2(s)$ requires that

$$[OH^-]^2 = \frac{K_{sp}}{[Mg^{2+}]} = \frac{1.8\text{x}10^{-11}}{0.0300} = 6.0\text{x}10^{-10} \qquad [OH^-] = 2.4\text{x}10^{-5} \text{ and pOH} = 4.62$$

Now determine $[NH_4^+]$ required to maintain a pOH = 4.62 in a solution with $[NH_3]$ = 0.0700 M.

$$pOH = pK_b + \log\frac{[NH_4^+]}{[NH_3]} \qquad 4.62 = 4.76 + \log\frac{[NH_4^+]}{0.0700}$$

$$\log[NH_4^+] = 4.62 - 4.76 + \log(0.0700) = 4.62 - 4.76 - 1.155 = -1.29$$

$$[NH_4^+] = \text{antilog}(-1.29) = 5.1\text{x}10^{-2} \text{ M}$$

$$\text{no. g } (NH_4)_2SO_4 = 0.500 \text{ L} \times \frac{5.1\text{x}10^{-2} \text{ mol } NH_4^+}{\text{L}} \times \frac{1 \text{ mol } (NH_4)_2SO_4}{2 \text{ mol } NH_4^+} \times \frac{132 \text{ g } (NH_4)_2SO_4}{1 \text{ mol } (NH_4)_2SO_4}$$

$$= 1.7 \text{ g } (NH_4)_2SO_4$$

19-11. In each case assume that practically all of the metal cation is in the form of the complex ion.

(a) $Ag^+(aq) + 2\ CN^-(aq) \rightleftharpoons [Ag(CN)_2]^-(aq) \qquad K_f = 5.6\text{x}10^{18}$

$$K_f = \frac{[[Ag(CN)_2]^-]}{[Ag^+][CN^-]^2} = \frac{0.012}{[Ag^+](1.05)^2} = 5.6\text{x}10^{18}$$

$$[Ag^+] = \frac{0.012}{(1.05)^2 \times 5.6\text{x}10^{18}} = 1.9\text{x}10^{-21} \text{ M}$$

Compare $Q = [Ag^+][I^-]$ with K_{sp} for AgI.

$$Q = (1.9\times10^{-21})(2.0) = 3.8\times10^{-21} < K_{sp}(8.5\times10^{-17})$$

AgI(s) does not precipitate.

(b) $Ag^+(aq) + 2\ S_2O_3^{2-}(aq) \rightleftharpoons [Ag(S_2O_3)_2^{3-}](aq) \qquad K_f = 1.7\times10^{13}$

$$K_f = \frac{[[Ag(S_2O_3)_2]^{3-}]}{[Ag^+][S_2O_3^{2-}]^2} = \frac{0.012}{[Ag^+](1.05)^2} = 1.7\times10^{13}$$

$$[Ag^+] = \frac{0.012}{(1.05)^2 \times 1.7\times10^{13}} = 6.4\times10^{-16}\ M$$

Now compare Q and K_{sp} for AgI, that is,

$$Q = [Ag^+][I^-] = 6.4\times10^{-16} \times 2.0 = 1.3 \times 10^{-15} > K_{sp}(8.5\times10^{-17})$$

AgI(s) should precipitate from this solution.

(c) $Cu^{2+}(aq) + 4\ NH_3(aq) \rightleftharpoons [Cu(NH_3)_4]^{2+}(aq) \qquad K_f = 1.1\times10^{13}$

$$K_f = \frac{[[Cu(NH_3)_4]^{2+}]}{[Cu^{2+}][NH_3]^4} = \frac{0.055}{[Cu^{2+}](1.8)^4} = 1.1\times10^{13}$$

$$[Cu^{2+}] = \frac{0.055}{(1.8)^4 \times 1.1\times10^{13}} = 4.8\times10^{-16}$$

Compare Q and K_{sp} for CuS.

$$Q = [Cu^{2+}][S^{2-}] = (4.8\times10^{-16})(3.2\times10^{-5}) = 1.5\times10^{-20} > K_{sp}(6.3\times10^{-36})$$

A precipitate of CuS(s) will form under these conditions.

19-12. If AgCl is not to precipitate from a solution in which $[Cl^-] = 0.327$ M,

$[Ag^+][Cl^-] \leq K_{sp} \qquad [Ag^+] \times 0.327 \leq 1.6\times10^{-10} \qquad [Ag^+] \leq 4.9\times10^{-10}$ M

CN^- converts Ag^+ to the complex ion $[Ag(CN)_2]^-$.

$Ag^+(aq) + 2\ CN^-(aq) \rightleftharpoons [Ag(CN)_2]^-(aq) \qquad K_f = 5.6\times10^{18}$

Essentially all the Ag^+ from $AgNO_3$ must be converted to the complex ion, leaving $[Ag^+] = 4.9\times10^{-10}$ M and producing $[[Ag(CN)_2]^-] = 1.8$ M. Solve the K_f expression for $[CN^-]$.

$$K_f = \frac{[[Ag(CN)_2]^-]}{[Ag^+][CN^-]^2} = \frac{1.8}{4.9\times10^{-10}[CN^-]^2} = 5.6\times10^{18}$$

$$[CN^-]^2 = \frac{1.8}{4.9\times10^{-10} \times 5.6\times10^{18}} = 6.6\times10^{-10} \qquad [CN^-] = 2.6\times10^{-5}\ M$$

19-13. (a) If a solution is 0.01 M $H_2S(aq)$, with no added acid or base, $[S^{2-}] = K_{a_2} = 1.0\times10^{-14}$ M. Compare ion products, Q, with K_{sp} values.

For CdS; $Q = [Cd^{2+}][S^{2-}] = 0.10 \times 1.0\times10^{-14} = 1.0\times10^{15} > K_{sp}(8.0\times10^{-27})$

For CuS: $Q = [Cu^{2+}][S^{2-}] = 0.10 \times 1.0\times10^{-14} = 1.0\times10^{-15} > K_{sp}(6.3\times10^{-36})$

For FeS: $Q = [Fe^{2+}][S^{2-}] = 0.10 \times 1.0\times10^{-14} = 1.0\times10^{-15} > K_{sp}(6.3\times10^{-18})$

All three metal sulfides precipitate.

(b) First determine $[S^{2-}]$ in the solution. Use the expression

$$K_{a_1} \times K_{a_2} = \frac{[H_3O^+]^2[S^{2-}]}{[H_2S]} = 1.1\times10^{-21}$$

Substitute the data given, that is,

$$\frac{(0.010)^2[S^{2-}]}{(0.010)} = 1.1\times10^{-21} \qquad [S^{2-}] = 1.1\times10^{-19} \text{ M}$$

Now compare Q and K_{sp} values of CdS, CuS and CdS as in part (b). Q for each sulfide is $0.10 \times 1.1\times10^{-19} = 1.1\times10^{-20}$

CdS and CuS still precipitate from solution; FeS does not.

19-14. Determine $[S^{2-}]$ with the expression.

$$K_{a_1} \times K_{a_2} = \frac{[H_3O^+]^2[S^{2-}]}{[H_2S]} = 1.1\times10^{-21} \quad \text{and} \quad [S^{2-}] = \frac{1.1\times10^{-21}[H_2S]}{[H_3O^+]^2}$$

$[H_3O^+]$ = antilog(-3.55) = 2.8×10^{-4} M

$$[S^{2-}] = \frac{0.10 \times 1.1\times10^{-21}}{(2.8\times10^{-4})^2} = 1.4\times10^{-15} \text{ M}$$

For FeS: $Q = [Fe^{2+}][S^{2-}] = 0.0022 \times 1.4\times10^{-15} = 3.1\times10^{-18} < K_{sp}(6.3\times10^{-18})$

FeS(s) should not precipitate from this solution.

19-15. If FeS(s) is to precipitate, $[Fe^{2+}][S^{2-}] \geq K_{sp}$ and

$$[S^{2-}] \geq \frac{K_{sp}}{[Fe^{2+}]} = \frac{6.3\times10^{-18}}{0.015} = 4.2\times10^{-16} \text{ M}$$

Calculate the value of $[H_3O^+]$ corresponding to the given conditions by using the expression

$$K_{a_1} \times K_{a_2} = \frac{[H_3O^+]^2[S^{2-}]}{[H_2S]} = \frac{[H_3O^+]^2 \times 4.2\times10^{-16}}{0.10} = 1.1\times10^{-21}$$

$$[H_3O^+]^2 = \frac{0.10 \times 1.1\times10^{-21}}{4.2\times10^{-16}} = 2.6\times10^{-7} \qquad [H_3O^+] = 5.1\times10^{-4} \text{ M}$$

pH = $-\log(5.1\times10^{-4})$ = 3.29

If FeS(s) is to precipitate, $[S^{2-}] \geq 4.2\times10^{-16}$ M and $[H_3O^+] \leq 5.1\times10^{-4}$. The pH must be 3.29 or higher.

Exercises

K_{sp} *and solubility*

19-1. The molar solubility of a solute is S mol/L. If the solute is of the type XY the ionic concentrations in solution are $[X^{n+}] = S$ and $[Y^{n-}] = S$, and $K_{sp} = S \times S = S^2$. If the solute is of the type X_2Y, $[X^{n+}] = 2S$; $[Y^{2n-}] = S$; and $K_{sp} = (2S)^2(S) = 4S^3$. No matter what the solute, K_{sp} will be equal to S^2 or some higher power of S. (If the solubility of a solute of the type XY were S = 1.00 mol/L one might argue that $K_{sp} = S \times S = 1.00$. Recall, however, that the solubility product expression loses its meaning in solutions as concentrated as this.)

19-2. Refer to Table 19-1 and write a solubility product constant expression for each solute in terms of its solubility S (in mol/L). Solve each expression for S; relate $[Mg^{2+}]$ to S; and find the largest of the three.

(a) $[Mg^{2+}][CO_3^{2-}] = (S)(S) = 3.5\times10^{-8}$ $\qquad S = 1.9\times10^{-4}$

$[Mg^{2+}] = S = 1.9\times10^{-4}$ M

(b) $[Mg^{2+}][F^-]^2 = (S)(2S)^2 = 4S^3 = 3.7\times10^{-8}$ $\qquad S = 2.1\times10^{-3}$

$[Mg^{2+}] = S = 2.1\times10^{-3}$ M

(c) $Mg_3(PO_4)_2(s) \rightleftharpoons 3\ Mg^{2+}(aq) + 2\ PO_4^{3-}(aq)$

S $\longrightarrow$ 3S 2S

$[Mg^{2+}][PO_4^{3-}]^2 = (3S)^3(2S)^2 = 108S^5 = 1\times10^{-25}$

$S = 4\times10^{-6}$ $\qquad [Mg^{2+}] = 3S = 3 \times 4\times10^{-6} = 1.2\times10^{-5}\text{ M} = 1\times10^{-5}$ M

The highest concentration of Mg^{2+} is found in saturated MgF_2.

19-3. $BaC_2O_4(s) \rightleftharpoons Ba^{2+}(aq) + C_2O_4^{2-}(aq)$

$$\text{no. mol } BaC_2O_4/L = \frac{9\text{ mg }BaC_2O_4}{0.100\text{ L}} \times \frac{1\text{ g }BaC_2O_4}{1000\text{ mg }BaC_2O_4} \times \frac{1\text{ mol }BaC_2O_4}{225\text{ g }BaC_2O_4} = 4\times10^{-4}\text{ mol }BaC_2O_4/L$$

$[Ba^{2+}] = [C_2O_4^{2-}] = 4\times10^{-4}$ M

$K_{sp} = [Ba^{2+}][C_2O_4^{2-}] = (4\times10^{-4})^2 = 1.6\times10^{-7} = 2\times10^{-7}$

19-4. $CaF_2(s) \rightleftharpoons Ca^{2+}(aq) + 2\ F^-(aq)$

at equil: S mol/L 2 S mol/L

$[Ca^{2+}][F^-]^2 = S \times (2S)^2 = 4S^3 = 2.7\times10^{-11}$ $\qquad S^3 = 6.8\times10^{-12}$

$S = 1.9\times10^{-4}$ $\qquad [F^-] = 2S = 3.8\times10^{-14}\text{ mol }F^-/L$

Assume that 1.00 L of the saturated solution has a mass of 1000 g. The volume of solution weighing 1×10^6 g would be 1000 L. Determine the number of grams of F^- in 1000 L of saturated solution.

$$\text{no. g }F^- = 1000\text{ L} \times \frac{3.8\times10^{-4}\text{ mol }F^-}{\text{L}} \times \frac{19.0\text{ g }F^-}{1\text{ mol }F^-} = 7.2\text{ g }F^-$$

The saturated solution contains 7.2 ppm F^-.

19-5. Let us base the entire calculation on 1.00 L of the water sample.

$$\text{no. mol } CaSO_4 = 1.00 \text{ L} \times \frac{1000 \text{ cm}^3}{1 \text{ L}} \times \frac{1.00 \text{ g water}}{1.00 \text{ cm}^3} \times \frac{131 \text{ g } CaSO_4}{1\times10^6 \text{ g water}} \times \frac{1 \text{ mol } CaSO_4}{136 \text{ g } CaSO_4}$$

$$= 9.63\times10^{-4} \text{ mol } CaSO_4$$

Next, let us determine the molar solubility of $CaSO_4$ in water.

$$CaSO_4(s) \rightleftharpoons Ca^{2+}(aq) \quad + \quad SO_4^{2-}(aq)$$

S mol/L   S mol/L

$$K_{sp} = [Ca^{2+}][SO_4^{2-}] = S \times S = 9.1\times10^{-6} \qquad S = 3.0\times10^{-3} \text{ mol } CaSO_4/L$$

The liter of water at the outset is not saturated. It contains 9.63×10^{-4} mol $CaSO_4$ but could tolerate 3.0×10^{-3} mol $CaSO_4$. Let us find the volume of a 3.0×10^{-3} M $CaSO_4$ solution that contains 9.63×10^{-4} mol $CaSO_4$.

$$\text{no. L} = 9.63\times10^{-4} \text{ mol } CaSO_4 \times \frac{1.00 \text{ L}}{3.0 \times 10^{-3} \text{ mol } CaSO_4} = 0.32 \text{ L}$$

To reduce the original 1.00 L of solution to a volume of 0.32 L (at which point $CaSO_4$ begins to precipitate) requires that 0.68 L be evaporated. The fraction of the water that must be evaporated, then, is 0.68 (or 68%).

19-6. First, determine the molarity of saturated PbI_2. (Again, refer to this as S.)

$$PbI_2(s) \rightleftharpoons Pb^{2+}(aq) \quad + \quad 2\ I^-(aq)$$

S mol/L   2 S mol/L

$$K_{sp} = [Pb^{2+}][I^-]^2 = (S)(2S)^2 = 4S^3 = 7.1\times10^{-9} \qquad S = 1.2\times10^{-3} \text{ mol } PbI_2/L$$

The titration reaction is $I^-(aq) + Ag^+(aq) \rightarrow AgI(s)$.

$$\text{no. mol } AgNO_3 = 25.00 \text{ mL} \times \frac{1.00 \text{ L}}{1000 \text{ mL}} \times \frac{1.2\times10^{-3} \text{ mol } PbI_2}{\text{L}} \times \frac{2 \text{ mol } I^-}{1 \text{ mol } PbI_2} \times \frac{1 \text{ mol } Ag^+}{1 \text{ mol } I^-} \times \frac{1 \text{ mol } AgNO_3}{1 \text{ mol } Ag^+}$$

$$= 6.0\times10^{-5} \text{ mol } AgNO_3$$

This amount of $AgNO_3$ was found in 13.3 mL of solution. The molar concentration of this solution is:

$$\frac{6.0\times10^{-5} \text{ mol } AgNO_3}{0.0133 \text{ L}} = 4.5\times10^{-3} \text{ M } AgNO_3$$

19-7. Use the titration data to determine $[C_2O_4^{2-}]$ in the saturated solution. (Note also that $[Ca^{2+}] = [C_2O_4^{2-}]$ in this solution.)

$$\text{no. mol } C_2O_4^{2-} = 4.9 \text{ mL} \times \frac{1.00 \text{ L}}{1000 \text{ mL}} \times \frac{0.00131 \text{ mol } MnO_4^-}{1.00 \text{ L}} \times \frac{5 \text{ mol } C_2O_4^{2-}}{2 \text{ mol } MnO_4^-} = 1.6\times10^{-5} \text{ mol } C_2O_4^{2-}$$

$$[C_2O_4^{2-}] = \frac{1.6\times10^{-5} \text{ mol } C_2O_4^{2-}}{0.2500 \text{ L}} = 6.4\times10^{-5} \text{ M}$$

$$K_{sp} = [Ca^{2+}][C_2O_4^{2-}] = (6.4\times10^{-5})^2 = 4.1\times10^{-9}$$

19-8. First, we must establish $[Ag^+]$ in saturated $AgBrO_3(aq)$.

$$2\ Ag^+(aq) + H_2S(g) \rightarrow Ag_2S + 2\ H^+$$

$$\text{no. mol } H_2S = n = \frac{PV}{RT} = \frac{(748/760) \times 0.0304}{0.0821 \times 296} = 1.23\times10^{-3} \text{ mol } H_2S$$

$$\text{no. mol } Ag^+ = 1.23\times10^{-3} \text{ mol } H_2S \times \frac{2 \text{ mol } Ag^+}{1 \text{ mol } H_2S} = 2.46\times10^{-3} \text{ mol } Ag^+$$

$$[Ag^+] \text{ in saturated } AgBrO_3(aq) = \frac{2.46\times10^{-3} \text{ mol } Ag^+}{0.338 \text{ L}} = 7.28\times10^{-3} \text{ M}$$

$$[BrO_3^-] = [Ag^+] = 7.28\times10^{-3} \text{ M}$$

$$K_{sp} = [Ag^+][BrO_3^-] = (7.28\times10^{-3})^2 = 5.30\times10^{-5}$$

19-9. (a) $Mg(OH)_2(s) \rightleftharpoons Mg^{2+}(aq) + 2\ OH^-(aq)$

$$S \longrightarrow S \qquad 2S$$

$$K_{sp} = [Mg^{2+}][OH^-]^2 = (S)(2S)^2 = 4S^3$$

$$S^3 = \frac{K_{sp}}{4} \qquad S = \left(\frac{K_{sp}}{4}\right)^{1/3} \qquad S = \left(\frac{1.8\times10^{-11}}{4}\right)^{1/3} = 1.7\times10^{-4} \text{ mol } Mg(OH)_2/L$$

(b) $Ag_2CO_3(s) \rightleftharpoons 2\ Ag^+(aq) + CO_3^{2-}(aq)$

$$S \longrightarrow 2S \qquad S$$

$$K_{sp} = [Ag^+]^2[CO_3^{2-}] = (2S)^2(S) = 4S^3 \qquad S^3 = \frac{K_{sp}}{4}$$

$$S = \left(\frac{K_{sp}}{4}\right)^{1/3} = \left(\frac{8.1\times10^{-12}}{4}\right)^{1/3} = 1.3\times10^{-4} \text{ mol } Ag_2CO_3/L$$

(c) $Al(OH)_3(s) \rightleftharpoons Al^{3+}(aq) + 3\ OH^-(aq)$

$$S \longrightarrow S \qquad 3S$$

$$K_{sp} = [Al^{3+}][OH^-]^3 = (S)(3S)^3 = 27S^4 \qquad S^4 = \frac{K_{sp}}{27}$$

$$S = \left(\frac{K_{sp}}{27}\right)^{1/4} = \left(\frac{1.3\times10^{-33}}{27}\right)^{1/4} = 2.6\times10^{-9} \text{ mol } Al(OH)_3/L$$

(d) $Li_3PO_4(s) \rightleftharpoons 3\ Li^+(aq) + PO_4^{3-}(aq)$

$$S \longrightarrow 3S \qquad S$$

$$K_{sp} = [Li^+]^3[PO_4^{3-}] = (3S)^3(S) = 27S^4 \qquad S^4 = \frac{K_{sp}}{27}$$

$$S = \left(\frac{K_{sp}}{27}\right)^{1/4} = \left(\frac{3.2\times10^{-9}}{27}\right)^{1/4} = 3.3\times10^{-3} \text{ mol } Li_3PO_4/L$$

(e) $Bi_2S_3(s) \rightleftharpoons 2\ Bi^{3+}(aq) + 3\ S^{2-}(aq)$

$$S \longrightarrow \quad 2S \qquad 3S$$

$$K_{sp} = [Bi^{3+}]^2[S^{2-}]^3 = (2S)^2(3S)^3 = (4S^2)(27S^3) = 108S^5 \qquad S^5 = \frac{K_{sp}}{108} \qquad S = \left(\frac{K_{sp}}{108}\right)^{1/5}$$

$$S = \left(\frac{1\times10^{-97}}{108}\right)^{1/5} = 2\times10^{-20} \text{ mol } Bi_2S_3/L$$

The common ion effect

19-10. Through the common ion I^-, KI displaces the solubility equilibrium of AgI to the left, that is, toward its precipitation. Thus AgI(s) is less soluble in KI(aq) than in water. KNO_3, on the other hand, reduces the activities of the Ag^+ and I^- ions in solution. Additional AgI must dissolve to maintain equilibrium. The solubility of AgI(s) is increased somewhat by the presence of K^+ and NO_3^- ions (the salt effect.)

19-11. $Ag_2SO_4(s) \rightleftharpoons 2\ Ag^+(aq) + SO_4^{2-}(aq)$

$$K_{sp} = [Ag^+]^2[SO_4^{2-}] = (9.2\times10^{-3})^2(0.200) = 1.7\times10^{-5}$$

19-12. The method employed in Example 19-4 was

$Ag_2CrO_4(s)$	$\rightleftharpoons$	$2\ Ag^+(aq)$	+	$CrO_4^{2-}(aq)$
from Ag_2CrO_4:		2S mol/L		S mol/L
from 0.10 M $AgNO_3$:		0.10 mol/L		--
equil. cocn:		(0.10 + 2S)mol/L		S mol/L

$$K_{sp} = [Ag^+]^2[CrO_4^{2-}] = (0.10 + 2S)^2(S) = 2.4\times10^{-12}$$

Assume S << 0.10: $(0.10)^2 \times S = 2.4\times10^{-12}$ $\qquad S = 2.4\times10^{-10}$ mol Ag_2CrO_4/L

in 0.10 M $AgNO_3$: $S = 2.4\times10^{-10}$ mol Ag_2CrO_4/L

in 0.10 M K_2CrO_4: $S = 2.4\times10^{-6}$ mol Ag_2CrO_4/L

in pure water: $S = 8.4\times10^{-5}$ mol Ag_2CrO_4/L

The common ion effect due to Ag^+ is seen to be much more pronounced than that of CrO_4^{2-}.

19-13. Let S = no. mol $Pb(IO_3)_2/L$ and M = molarity of KIO_3. The calculation of some data points follows:

$$Pb(IO_3)_2(s) \rightleftharpoons Pb^{2+}(aq) + 2\ IO_3^-(aq); \quad K_{sp} = 3.2\times10^{-13}$$

$$K_{sp} = [Pb^{2+}][IO_3^-]^2 = S(2S + M)^2 = 3.2\times10^{-13}$$

Assume S << M: $S\cdot M^2 = 32.\times10^{-13}$ $\qquad S = \dfrac{3.2\times10^{-13}}{M^2}$

in 0.10 M KIO_3: $S = \frac{3.2\times10^{-13}}{(0.10)^2} = 3.2\times10^{-11}$ mol $Pb(IO_3)_2$/L log S = -10.49

in 0.050 M KIO_3: $S = \frac{3.2\times10^{-13}}{(0.05)^2} = 1.3\times10^{-10}$ mol $Pb(IO_3)_2$/L log S = -9.89

in 0.020 M KIO_3: $S = \frac{3.2\times10^{-13}}{(0.020)^2} = 8.0\times10^{-10}$ mol $Pb(IO_3)_2$/L log S = -9.10

in 0.010 M KIO_3: $S = \frac{3.2\times10^{-13}}{(0.010)^2} = 3.2\times10^{-9}$ mol $Pb(IO_3)_2$/L log S = -8.49

in 0.0050 M KIO_3: $S = \frac{3.2\times10^{-13}}{(0.0050)^2} = 1.3\times10^{-8}$ mol $Pb(IO_3)_2$/L log S = -7.89

in pure water: $4S^3 = 3.2\times10^{-13}$ $S = 4.3\times10^{-5}$ mol $Pb(IO_3)_2$/L log S = -4.37

Because of the extreme range of solubilities, a logarithmic plot is necessary.

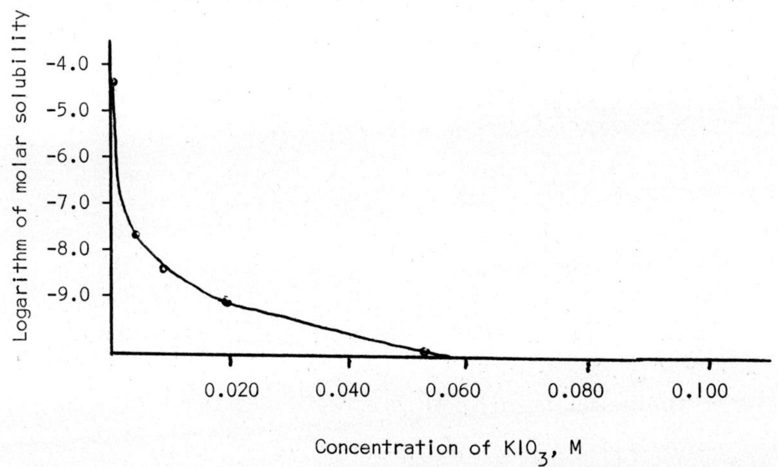

19-14. (a) The solution volume is increased from 500.0 mL to 1000.0 mL. Additional $Mg(OH)_2$ dissolves but the concentration of the saturated solution is unchanged.

$$Mg(OH)_2(s) \rightleftharpoons Mg^{2+} + 2\ OH^-(aq);\ K_{sp} = 1.8\times10^{-11}$$

S mol/L 2 S mol/L

$$K_{sp} = [Mg^{2+}][OH^-]^2 = S \times (2S)^2 = 4S^3 = 1.8\times10^{-11}$$

$$[Mg^{2+}] = S = 1.7\times10^{-4}\ M$$

(b) This involves a dilution of the solution described above.

$$[Mg^{2+}] = \frac{0.1000\ L \times \frac{1.7\times10^{-4}\ mol\ Mg^{2+}}{L}}{0.6000\ L} = 2.8\times10^{-5}\ M$$

(c) The addition of $MgCl_2$ to the saturated solution of $Mg(OH)_2$ would, through the commonion effect, cause the precipitation of some $Mg(OH)_2(s)$. However, the amount of $Mg(OH)_2(s)$ precipitated in this way would be very small. At most, it would correspond to one half the number of moles of OH^- present in the saturated solution.

max. no. mol $Mg(OH)_2$ precipitating = max. no. mol Mg^{2+} removed by pptn

$= 0.02500 \text{ L} \times 1.7\times10^{-4} \text{ mol } Mg^{2+}/\text{L} = 4.2\times10^{-6} \text{ mol } Mg^{2+}$.

The final solution, then, contains all the Mg^{2+} ion from the 250.0 mL 0.065 M $MgCl_2$ and some of the Mg^{2+} from the 25.00 mL saturated $Mg(OH)_2(aq)$. And the first source is far more important than the second.

no. mol Mg^{2+} = $0.2500 \text{ L} \times 0.065 \text{ mol } Mg^{2+}/\text{L} = 0.016 \text{ mol } Mg^{2+}$

$$[Mg^{2+}] = \frac{0.016 \text{ mol } Mg^{2+}}{0.2750 \text{ L}} = 0.058 \text{ M}$$

(d) The 0.150 M KOH solution is diluted by the addition of 50.00 ml saturated $Mg(OH)_2(aq)$, and again some slight precipitation of $Mg(OH)_2(s)$ occurs because of the common ion effect. In the mixed solution:

$$[OH^-] = \frac{0.150 \text{ L} \times 0.150 \text{ mol } OH^-/\text{L}}{0.200 \text{ L}} = 0.112 \text{ M}$$

$$[Mg^+][OH^-]^2 = [Mg^{2+}](0.112)^2 = 1.8\times10^{-11} \qquad [Mg^{2+}] = 1.4\times10^{-9} \text{ M}$$

19-15. Convert the Ca^{2+} content to molarity. Then use the K_{sp} expression to solve for $[F^-]$.

$$[Ca^{2+}] = \frac{115 \text{ g } Ca^{2+}}{1\times10^6 \text{ g water}} \times \frac{1\times10^3 \text{ g water}}{1 \text{ L water}} \times \frac{1 \text{ mol } Ca^{2+}}{40.1 \text{ g } Ca^{2+}} = 2.87\times10^{-3} \text{ M}$$

$$K_{sp} = [Ca^{2+}][F^-]^2 = (2.87\times10^{-3})[F^-]^2 = 2.7\times10^{-11} \qquad [F^-]^2 = 9.4\times10^{-9} \qquad [F^-] = 9.7\times10^{-5} \text{ M}$$

Now convert $[F^-]$ to ppm F^-.

$$\frac{9.7\times10^{-5} \text{ mol } F^-}{\text{L water}} \times \frac{19.0 \text{ g } F^-}{1 \text{ mol } F^-} \times \frac{1 \text{ L water}}{1000 \text{ g water}} = \frac{1.8\times10^{-6} \text{ g } F^-}{\text{g water}} \times \frac{10^6}{10^6} = \frac{1.8 \text{ g } F^-}{10^6 \text{ g water}} = 1.8 \text{ ppm } F^-$$

19-16. Display the data for this exercise in the usual form:

$MgF_2(s)$ $\rightleftharpoons$	$Mg^{2+}(aq)$	+	$2\,F^-(aq)$
From MgF_2:	S mol/L		2 S mol/L
From $MgCl_2(aq)$:	5.50×10^{-4} mol/L		--
Equil. concn:	$(S + 5.50\times10^{-4})$mol/L		2 S mol/L

$$K_{sp} = [Mg^{2+}][F^-]^2 = (S + 5.50 \times 10^{-4})(2S)^2$$

If we make the usual assumption, it would be $S << 5.50\times10^{-4}$; $5.50\times10^{-4}(2S)^2 = 3.7\times10^{-8}$; $S = 4.1\times10^{-3}$. But 4.1×10^{-3} is not greatly smaller than 5.50×10^{-4}; it is larger. The assumption is not valid.

$$K_{sp} = 4S^2(S + 5.50\times10^{-4}) = 4S^3 + 2.2\times10^{-2}S^2 = 3.7\times10^{-8}$$

$$S^3 + 5.5\times10^{-4}S^2 - 9.2\times10^{-9} = 0$$

Try $S = 1.0\times10^{-3}$: $1.0\times10^{-9} + 5.5\times10^{-10} - 9.2\times10^{-9} = -7.6\times10^{-9} < 0$

Try $S = 2.0\times10^{-3}$: $8.0\times10^{-9} + 2.2\times10^{-9} - 9.2\times10^{-9} = +1.0\times10^{-9} > 0$

Try $S = 1.8\times10^{-3}$: $5.8\times10^{-9} + 1.8\times10^{-9} - 9.2\times10^{-9} = -1.6\times10^{-9} < 0$

Try $S = 1.9\times10^{-3}$: $6.9\times10^{-9} + 2.0\times10^{-9} - 9.2\times10^{-9} = -0.3\times10^{-9} < 0$

To two significant figures the molar solubility of MgF_2 in 5.50×10^{-4} M $MgCl_2$(aq) is 1.9×10^{-3} mol MgF_2/L.

Criterion for precipitation from solution

19-17. First determine $[Mg^{2+}]$ in the solution in question.

$$[Mg^{2+}] = \frac{17.5 \text{ mg } MgCl_2 \cdot 6\,H_2O \times \dfrac{1 \text{ g } MgCl_2 \cdot 6\,H_2O}{1000 \text{ mg } MgCl_2 \cdot 6\,H_2O} \times \dfrac{1 \text{ mol } MgCl_2 \cdot 6\,H_2O}{203 \text{ g } MgCl_2 \cdot 6\,H_2O} \times \dfrac{1 \text{ mol } Mg^{2+}}{1 \text{ mol } MgCl_2}}{0.325 \text{ L}}$$

$[Mg^{2+}] = 2.65\times10^{-4}$ M

$Q = [Mg^{2+}][F^-]^2 = (2.65\times10^{-4})(0.045)^2 = 5.4\times10^{-7} > 3.7\times10^{-8} (K_{sp})$

Precipitation of MgF_2(s) should occur.

19-18. Here each ion concentration after mixing must first be determined.

$$[Pb^{2+}] = \frac{0.245 \text{ L} \times 0.175 \text{ mol } Pb(NO_3)_2/\text{L} \times 1 \text{ mol } Pb^{2+}/\text{mol } Pb(NO_3)_2}{(0.245 + 0.155)\text{L}}$$

$[Pb^{2+}] = 0.107$ M

$$[Cl^-] = \frac{0.155 \text{ L} \times 0.016 \text{ mol KCl/L} \times 1 \text{ mol } Cl^-/\text{mol KCl}}{(0.245 + 0.155)\text{L}} = 6.2\times10^{-3} \text{ M}$$

$Q = [Pb^{2+}][Cl^-]^2 = (0.107)(6.2\times10^{-3})^2 = 4.1\times10^{-6} < 1.6\times10^{-5} (K_{sp})$

Precipitation of $PbCl_2$(s) will not occur.

19-19. If precipitation is just to occur, $Q = [Fe^{3+}][OH^-]^3 = (0.17)[OH^-]^3 = K_{sp} = 4\times10^{-38}$

$[OH^-]^3 = 2\times10^{-37}$ $[OH^-] = 6\times10^{-13}$ M pOH = 12.2 pH = 1.8

19-20. (a)
$$[Cl^-] = \frac{1.0 \text{ mg NaCl} \times \dfrac{1.00 \text{ g NaCl}}{1000 \text{ mg NaCl}} \times \dfrac{1 \text{ mol NaCl}}{58.5 \text{ g NaCl}} \times \dfrac{1 \text{ mol } Na^+}{1 \text{ mol NaCl}}}{1.00 \text{ L}} = 1.7\times10^{-5} \text{ M}$$

$Q = [Ag^+][Cl^-] = (0.10)(1.7\times10^{-5}) = 1.7\times10^{-6} > 1.6\times10^{-10} (K_{sp})$

Precipitation of AgCl(s) should occur.

(b)
$$[Br^-] = \frac{0.05 \text{ mL} \times \dfrac{1 \text{ L}}{1000 \text{ mL}} \times \dfrac{0.20 \text{ mol KBr}}{1.00 \text{ L}} \times \dfrac{1 \text{ mol } Br^-}{1 \text{ mol KBr}}}{0.200 \text{ L}} = 5\times10^{-5} \text{ M}$$

In saturated AgCl(aq), $[Ag^+][Cl^-] = K_{sp} = 1.6\times10^{-10}$; $[Ag^+]^2 = 1.6\times10^{-10}$; $[Ag^+] = 1.3\times10^{-5}$ M. (Note $[Ag^+] = [Cl^-]$.)

To test for precipitation:

$$Q = [Ag^+][Br^-] = (1.3\times10^{-5})(5 \times 10^{-5}) = 6\times10^{-10} > 5.0\times10^{-13} (K_{sp})$$

Precipitation of AgBr(s) should occur.

(c) $$[Mg^{2+}] = \frac{2.0 \text{ mg } Mg^{2+} \times \frac{1.00 \text{ g } Mg^{2+}}{1000 \text{ mg } Mg^{2+}} \times \frac{1 \text{ mol } Mg^{2+}}{24.3 \text{ g } Mg^{2+}}}{1.00 \text{ L}} = 8.2\times10^{-5} \text{ M}$$

$$[OH^-] = \frac{0.05 \text{ mL} \times \frac{1.00 \text{ L}}{1000 \text{ mL}} \times \frac{0.0150 \text{ mol NaOH}}{1.00 \text{ L}} \times \frac{1 \text{ mol } OH^-}{1 \text{ mol NaOH}}}{5.0 \text{ L}} = 2\times10^{-7} \text{ M}$$

$$Q = [Mg^{2+}][OH^-]^2 = (8.2\times10^{-5})\ (2\times10^{-7})^2 = 3\times10^{-18} < 1.8\times10^{-11} (K_{so})$$

Precipitation of $Mg(OH)_2$(s) will not occur.

19-21. First determine $[Ag^+]$ in saturated Ag_2CO_3(aq). Let the molar solubility be S.

$$Ag_2CO_3(s) \rightleftharpoons 2\ Ag^+(aq) + CO_3^{2-}(aq)$$

$$S \longrightarrow 2S \qquad S$$

$$K_{sp} = [Ag^+]^2[CO_3^{2-}] = (2S)^2(S) = 4S^3 = 8.1\times10^{-12} \qquad S = 1.3\times10^{-4} \text{ M}$$

$$[Ag^+] = 2S = 2 \times 1.3\times10^{-4} = 2.6\times10^{-4} \text{ M}$$

$$[Cl^-] = \frac{0.025 \text{ g KCl} \times \frac{1 \text{ mol KCl}}{74.6 \text{ g KCl}} \times \frac{1 \text{ mol } Cl^-}{1 \text{ mol KCl}}}{0.75 \text{ L}} = 4.5\times10^{-4} \text{ M}$$

$$Q = [Ag^+][Cl^-] = (2.6\times10^{-4})(4.5\times10^{-4}) = 1.2\times10^{-7} > 1.6\times10^{-10} (K_{sp})$$

Precipitation of AgCl(s) should occur.

Completeness of precipitation

19-22. (a) The mixing of the two solutions causes each ion concentration to be diluted to one half of its original value.

$$[Ag^+] = \frac{0.200 \text{ L} \times 0.100 \text{ mol } Ag^+/\text{L}}{0.400 \text{ L}} = 0.0500 \text{ M} \qquad [CrO_4^{2-}] = \tfrac{1}{2} \times 0.350 = 0.175 \text{ M}$$

$$Q = [Ag^+]^2[CrO_4^{2-}] = (0.0500)^2(0.175) = 4.4\times10^{-4} > 2.4\times10^{-12} (K_{sp})$$

Precipitation of Ag_2CrO_4 should occur.

(b) Write down the information in the manner shown below:

$Ag_2CrO_4(s)$ $\rightleftharpoons$	$2\ Ag^+(aq)$	+	$CrO_4^{2-}(aq)$
After mixing:	0.0500 M		0.175 M
Consumed in pptn:	$0.0500 - x$		$\tfrac{1}{2}(0.0500 - x)$
Equil. concn:	x		$0.175 - \tfrac{1}{2}(0.0500 - x)$

$$[CrO_4^{2-}] = 0.175 - 0.0250 + x/2$$

$$= 0.150 + x/2$$

$$K_{sp} = [Ag^+]^2[CrO_4^{2-}] = (x)^2(0.150 + x/2) = 2.4\times10^{-12}$$

Assume $x << 0.150$: $x^2 = \frac{2.4\times10^{-12}}{0.150} = 1.6\times10^{-11}$ $\quad x = [Ag^+] = 4.0\times10^{-6}$ M

19-23. (a) Calculate $[OH^-]$ in a saturated $Mg(OH)_2(aq)$ solution having $[Mg^{2+}] = 1\times10^{-6}$ M.

$$K_{sp} = [Mg^{2+}][OH^-]^2 = (1\times10^{-6})[OH^-]^2 = 1.8\times10^{-11} \qquad [OH^-]^2 = 1.8\times10^{-5} \qquad [OH^-] = 4.2\times10^{-3}\text{ M}$$

(b) To effect the removal of 90% of the Mg^{2+} means that $[Mg^{2+}]$ must be reduced to 10% of its original value, that is, $[Mg^{2+}] = 0.10\times1.0\times10^{-2} = 1.0\times10^{-3}$ M. Again, determine $[OH^-]$ in saturated $Mg(OH)_2(aq)$ in which $[Mg^{2+}] = 1.0\times10^{-3}$ M.

$$K_{sp} = [Mg^{2+}][OH^-]^2 = (1.0\times10^{-3})[OH^-]^2 = 1.8\times10^{-11} \qquad [OH^-]^2 = 1.8\times10^{-8}$$

$$[OH^-] = 1.3\times10^{-4}\text{ M}$$

19-24. The ionic concentrations immediately after mixing the solutions are one half of their original values, since equal volumes of solutions are mixed.

$BaCO_3(s) \rightleftharpoons$	$Ba^{2+}(aq)$	+	$CO_3^{2-}(aq)$
After mixing:	5.0×10^{-4} M		1.0×10^{-3} M
Consumed in pptn:	$(5.0\times10^{-4} - x)$M		$(5.0\times10^{-4} - x)$M
Equil. concn:	x M		$1.0\times10^{-3} - (5.0\times10^{-4} - x)$
			$= (5.0\times10^{-4} + x)$M

$$K_{sp} = [Ba^{2+}][CO_3^{2-}] = x(5.0\times10^{-4} + x) = 5.1\times10^{-9}$$

Assume that $x << 5.0\times10^{-4}$: $(5.0\times10^{-4})x = 5.1\times10^{-9}$ $\quad x = [Ba^{2+}] = 1.0\times10^{-5}$ M

$$\%\ Ba^{2+}\text{ precipitated} = \frac{(5.0\times10^{-4} - 1.0\times10^{-5})\text{M}}{5.0\times10^{-4}\text{ M}} \times 100 = 98\%$$

19-25. First, calculate the ion concentrations immediately after mixing the solutions.

$$[Pb^{2+}] = \frac{0.135\text{ L} \times 0.12\text{ mol } Pb^{2+}/\text{L}}{(0.135 + 0.225)\text{L}} = 0.045\text{ M} \qquad [Cl^-] = \frac{0.225\text{ L} \times 0.15\text{ mol } K^+/\text{L}}{(0.135 + 0.225)\text{L}} = 0.094\text{ M}$$

Now, write down relevant information in the manner of Example 19-7b.

$PbCl_2(s) \rightleftharpoons$	$Pb^{2+}(aq)$	+	$2\ Cl^-(aq)$
After mixing:	0.045 M		0.094 M
Consumed in pptn:	$(0.045 - x)$M		$2 \times (0.045 - x)$M
Equil. conc:	x M		$0.094 - 2 \times (0.045 - x)$
			$= 0.094 - 0.090 + 2x$
			$= (0.004 + 2x)$M

$$K_{sp} = [Pb^{2+}][Cl^-]^2 = x(0.004 + 2x)^2 = 1.6\times10^{-5}$$

Here try the assumption that $0.004 << 2x$: $4x^3 = 1.6\times10^{-5}$ $\qquad x^3 = 4.0\times10^{-6}$

$x = [Pb^{2+}] = 1.6\times10^{-2}$ M

The assumption is not completely valid, but 0.004 is a relatively small quantity compared to $2x = 3.2\times10^{-2} = 0.032$.

Fractional precipitation

19-26. (a) Determine $[Ca^{2+}]$ in the seawater sample.

$$[Ca^{2+}] = \frac{440 \text{ g } Ca^{2+}}{1000 \text{ kg seawater}} \times \frac{1.00 \text{ kg seawater}}{1000 \text{ g seawater}} \times \frac{1.03 \text{ g seawater}}{1.00 \text{ cm}^3 \text{ seawater}} \times \frac{1000 \text{ cm}^3 \text{ seawater}}{1.00 \text{ L seawater}}$$

$$\times \frac{1 \text{ mol } Ca^{2+}}{40.1 \text{ g } Ca^{2+}} = 1.1\times10^{-2} \text{ M}$$

$$Q = [Ca^{2+}][OH^-]^2 = (1.1\times10^{-2})(2.0\times10^{-3})^2 = 4.4\times10^{-8} < 5.5\times10^{-6} (K_{sp})$$

$Ca(OH)_2$ will not precipitate under these conditions.

(b) In Example 19-6 it was established that precipitation of $Mg(OH)_2(s)$ is complete. In part (a) of this exercise it was established that $Ca(OH)_2(s)$ does not precipitate while the precipitation of $Mg(OH)_2(s)$ is being carried to completion. Therefore the separation of Ca^{2+} and Mg^{2+} by fractional precipitation from seawater is feasible.

19-27. (a) The precipitate that should form first is the one with the smaller value of K_{sp}--AgBr.

(b) Determine the $[Ag^+]$ required to just begin the precipitation of AgCl.

$$K_{sp} = [Ag^+][Cl^-] = [Ag^+](0.250) = 1.6\times10^{-10} \qquad [Ag^+] = 6.4\times10^{-10} \text{ M}$$

Now determine $[Br^-]$ remaining in solution at this value of $[Ag^+]$.

$$K_{sp} = [Ag^+][Br^-] = 5.0\times10^{-13} \qquad [Br^-] = \frac{5.0\times10^{-13}}{6.4\times10^{-10}} = 7.8\times10^{-4} \text{ M}$$

$$\% \; Br^- \text{ remaining} = \frac{7.8\times10^{-4} \text{ mol } Br^-/L}{2.2\text{-}10^{-3} \text{ mol } Br^-/L} \times 100 = 35\%$$

The precipitation of AgBr is not complete at the point where AgCl begins to precipitate. Fractional precipitation does not work here.

19-28. NaCl cannot be used since both $BaCl_2$ and $CaCl_2$ are water soluble. Even though K_{sp} values are listed in Table 16-1 for $Ba(OH)_2$ and $Ca(OH)_2$, an examination of the ion concentrations involved shows that neither K_{sp} value is exceeded and, therefore, neither precipitates. Of the two remaining possibilities--the sulfates and the carbonates--compare the differences in K_{sp} values between the calcium and barrium compound [i.e., compare the difference between $K_{sp}(BaSO_4)$ and $K_{sp}(CaSO_4)$ to the difference between $K_{sp}(BaCO_3)$ and $K_{sp}(CaCO_3)$.] The greatest difference is found for the sulfates. The best reagent to use is 0.50 M Na_2SO_4(aq).

19-29. It is likely that an appreciable quantity of white AgCl would have to be formed before it could be detected in the presence of a much larger quantity of yellow AgI. The two ions could not be effectively separated in this case.

19-30. (a) A mixture of 1.05×10^{-3} M HI and 1.05×10^{-3} M NaI is a mixture of a strong acid and its salt. It has $[I^-] = 2.10\times10^{-3}$ M. $[Pb^{2+}] = 1.1\times10^{-3}$ M

$Q = [Pb^{2+}][I^-]^2 = (1.1\times10^{-3})(2.10\times10^{-3})^2 = 4.9\times10^{-9} < 7.1\times10^{-9}\ (K_{sp})$

Precipitation of $PbI_2(s)$ should not occur.

(b) Addition of 1 drop (0.05 ml) 1.00 M NH_3 to 2.50 L of an aqueous solution yields

$$[NH_3] = \frac{0.05\text{ mL} \times \frac{1.00\text{ L}}{1000\text{ mL}} \times \frac{1.00\text{ mol } NH_3}{1.00\text{ L}}}{2.50\text{ L}} = 2\times10^{-5}\text{ M}$$

Next, determine $[OH^-]$ in this weak base solution

	NH_3	+	$H_2O \rightleftharpoons$	NH_4^+	+	OH^-
Dissolve:	2×10^{-5} M					
Changes:	$-x$ M			$+x$ M		$+x$ M
Equil. conc:	$(2\times10^{-5} - x)$M			x M		x M

$$K_b = \frac{[NH_4^+][OH^-]}{[NH_3]} = \frac{x \cdot x}{2\times10^{-5} - x} = 1.74\times10^{-5}$$

Because the solution is so dilute in NH_3, the assumption that $x << 2\times10^{-5}$ is probably not valid.

$$x^2 + (1.74\times10^{-5})x - 3\times10^{-10} = 0$$

$$x = \frac{-1.74\times10^{-5} \pm \sqrt{(1.74\times10^{-5})^2 + 4 \times 3 \times 10^{-10}}}{2} = 1\times10^{-5}\text{ M} = [OH^-]$$

$Q = [Mg^{2+}][OH^-]^2 = (0.0150)\ (1\times10^{-5})^2 = 2\times10^{-12} < 1.8\times10^{-11}\ (K_{sp})$

$Mg(OH)_2(s)$ will not precipitate under these conditions.

(c) Determine the pH of the $HC_2H_3O_2$ - $NaC_2H_3O_2$ buffer solution.

$$pH = pK_a + \log\frac{[C_2H_3O_2^-]}{[HC_2H_3O_2]} = 4.76 + \log\frac{0.010}{0.010} = 4.76$$

Then determine $[OH^-]$ in the solution. pOH = 14.00 - pH = 14.00 - 4.76 = 9.24

$\log [OH^-] = -9.24 = 0.76 - 10.00 \qquad [OH^-] = 5.8\times10^{-10}$ M

$Q = [Al^{3+}][OH^-]^3 = (1.0\times10^{-2})(5.8\times10^{-10})^3 = 2.0\times10^{-30} > 1.3\times10^{-33}\ (K_{sp})$

$Al(OH)_3(s)$ should precipitate.

19-31. In the saturated solution $[Mg^{2+}][OH^-]^2 = K_{sp} = 1.8\times10^{-11}$

From the solubility value given:

$$[Mg^{2+}] = \frac{0.95 \text{ g } Mg(OH)_2 \times \dfrac{1 \text{ mol } Mg(OH)_2}{58.3 \text{ g } Mg(OH)_2} \times \dfrac{1 \text{ mol } Mg^{2+}}{1 \text{ mol } Mg(OH)_2}}{1.00 \text{ L}} = 0.016 \text{ M}$$

$(0.016)[OH^-]^2 = 1.8\times10^{-11}$ $[OH^-] = 3.4\times10^{-5}$ pOH = 4.47 pH = 14.00 - 4.47 = 9.53

19-32. Combine the two equilibrium expressions and obtain a value of K

$$Mg(OH)_2(s) \rightleftharpoons Mg^{2+}(aq) + 2\ OH^-(aq);\quad K_{sp} = 1.8\times10^{-11}$$

$$2\ NH_4^+(aq) + 2\ OH^-(aq) \rightleftharpoons 2\ NH_3(aq) + 2\ H_2O;\quad 1/K_b^2 = 1/(1.74\times10^{-5})^2$$

$$Mg(OH)_2(s) + 2\ NH_4^+(aq) \rightleftharpoons Mg^{2+}(aq) + 2\ NH_3(aq) + 2\ H_2O$$

$$K = \frac{1.8\times10^{-11}}{(1.74\times10^{-5})^2} = 5.9\times10^{-2}$$

Let the molar solubility of $Mg(OH)_2$ = S

	$Mg(OH)_2(s)$ +	$2\ NH_4^+(aq)$ ⇌	$Mg^{2+}(aq)$ +	$2\ NH_3(aq)$ +	$2\ H_2O$
Initial:		1.00 M			
Changes:		$-2S$ M	$+S$ M	$+2S$ M	
Equil:		$(1.00 - 2S)$M	S M	$2S$ M	

$$K = \frac{[Mg^{2+}][NH_3]^2}{[NH_4^+]^2} = \frac{S \times (2S)^2}{(1.00 - 2S)^2} = 5.9\times10^{-2}$$

$4S^3 = 5.9\times10^{-2}\,(1.00 - 4S + 4S^2)$ $S^3 - 0.059S^2 + 0.059S - 1.5\times10^{-2} = 0$

Solve by successive approximations:

Try S = 0.20: $8.0\times10^{-3} - 2.4\times10^{-3} + 1.2\times10^{-2} - 1.5\times10^{-2} = 2.6\times10^{-3} > 0$

Try S = 0.19: $6.9\times10^{-3} - 2.1\times10^{-3} + 1.1\times10^{-2} - 1.5\times10^{-2} = 8\times10^{-4} > 0$

Try S = 0.18: $5.8\times10^{-3} - 1.9\times10^{-3} + 1.1\times10^{-2} - 1.5\times10^{-2} = 1\times10^{-4} < 0$

The molar solubility is 0.18 mol $Mg(OH)_2$/L.

19-33. From the measured solubility determine $[Mg^{2+}]$ and $[F^-]$ in the saturated solution.

$$[Mg^{2+}] = \frac{0.049 \text{ g } MgF_2}{L} \times \frac{1 \text{ mol } MgF_2}{62.3 \text{ g } MgF_2} \times \frac{1 \text{ mol } Mg^{2+}}{1 \text{ mol } MgF_2} = 7.9\times10^{-4} \text{ M}$$

$[F^-] = 2 \times [Mg^{2+}] = 2 \times 7.9\times10^{-4} = 1.6\times10^{-3}$ M

The two equilibria to consider are

$$MgF_2(s) \rightleftharpoons Mg^{2+}(aq) + 2\ F^-(aq) \qquad K_{sp} = 3.7\times10^{-8}$$

$$2\ H_3O^+(aq) + 2\ F^-(aq) \rightleftharpoons 2\ HF(aq) + 2\ H_2O \qquad K' = 1/(K_a)^2 = \frac{1}{(6.7\times10^{-4})^2}$$

$$MgF_2(s) + 2\ H_3O^+(aq) \rightleftharpoons Mg^{2+}(aq) + 2\ HF(aq) + 2\ H_2O$$

$$K = K_{sp}/(K_a)^2 = 3.7\times10^{-8}/(6.7\times10^{-4})^2 = 8.2\times10^{-2}$$

A summary of the dissolving reaction (with $x = [H_3O^+]$) is given below.

	$MgF_2(s) + 2\ H_3O^+$	$\rightleftharpoons$	Mg^{2+}	+	$2\ HF$	+	$2\ H_2O$
Initially:	x M						
At Equil:	x M		7.9×10^{-4} M		1.6×10^{-3} M		

$$K = \frac{[Mg^{2+}][HF]^2}{[H_3O^+]^2} = \frac{(7.9\times10^{-4})(1.6\times10^{-3})^2}{x^2} = 8.2\times10^{-2}$$

$$x^2 = \frac{(7.9\times10^{-4})(1.6\times10^{-3})^2}{8.2\times10^{-2}} = 2.5\times10^{-8}$$

$$x = [H_3O^+] = \sqrt{2.5\times10^{-8}} = 1.6\times10^{-4}\ M$$

$$pH = -\log [H_3O^+] = -\log (1.6\times10^{-4}) = 3.80$$

19-34. (a)

	$CaHPO_4(s)$	$\rightleftharpoons$	$Ca^{2+}(aq)$	+	$HPO_4^{2-}(aq)$
Dissolve	S mol/L	$\longrightarrow$	S mol/L		S mol/L

$$K_{sp} = [Ca^{2+}][HPO_4^{2-}] = (S)(S) = 1\times10^{-7} \qquad S = 3\times10^{-4}\ \text{mol}\ CaHPO_4/L$$

Expressed as g $CaHPO_4 \cdot 2\ H_2O$/L, the calculated solubility is:

$$\text{no. g}\ CaHPO_4 \cdot 2\ H_2O/L = \frac{3\times10^{-4}\ \text{mol}\ CaHPO_4}{1.00\ L} \times \frac{1\ \text{mol}\ CaHPO_4 \cdot 2\ H_2O}{1\ \text{mol}\ CaHPO_4} \times \frac{172\ g\ CaHPO_4 \cdot 2\ H_2O}{1\ \text{mol}\ CaHPO_4 \cdot 2\ H_2O}$$

$$= 0.05\ g\ CaHPO_4 \cdot 2\ H_2O/L$$

These data do not appear to be consistent with one another. The solubility based on K_{sp} is 0.05 g/L and the actual observed solubility is 0.32 g/L.

(b) Several factors may be involved in accounting for this discrepancy. The K_{sp} calculation assumes that HPO_4^{2-} is the only phosphate species, when, in fact, some HPO_4^{2-} hydrolyzes to $H_2PO_4^-$ (and a very small quantity ionizes further to PO_4^{3-}). Also, some ion pair formation might occur $(Ca^{2+})(HPO_4^{2-})$. All of these factors would lead to a greater observed solubility than that calculated from K_{sp}.

Complex ion equilibria

19-35. In HCl(aq) the following equilibria occur:

$$PbCl_2(s) \rightleftharpoons Pb^{2+}(aq) + 2\ Cl^-(aq)$$

$$Pb^{2+}(aq) + 3\ Cl^-(aq) \rightleftharpoons [PbCl_3]^-(aq)$$

$$PbCl_2(s) + Cl^-(aq) \rightleftharpoons [PbCl_3]^-(aq)$$

Thus a high $[Cl^-]$, such as found in HCl(aq), promotes the dissolution of $PbCl_2(s)$ through formation of the complex ion $[PbCl_3]^-$. Pb^{2+} does not form a complex ion with NO_3^-, and the solubility of $PbCl_2(s)$ is not enhanced in $HNO_3(aq)$.

19-36. with HCl(aq): This is an acid-base reaction involving H^+ from the acid and OH^- from $Zn(OH)_2$.

$$Zn(OH)_2(s) + 2\ H^+(aq) \rightarrow Zn^{2+}(aq) + 2\ H_2O$$

with $HC_2H_3O_2(aq)$: Again this is an acid-base reaction, this time with H^+ donated by the weak acid $HC_2H_3O_2$.

$$Zn(OH)_2(s) + 2\ HC_2H_3O_2(aq) \rightarrow Zn^{2+}(aq) + 2\ C_2H_3O_2^-(aq) + 2\ H_2O$$

with $NH_3(aq)$: Dissolving occurs because of the formation of the stable complex ion $[Zn(NH_3)_4]^{2+}$.

$$Zn(OH)_2(s) + 4\ NH_3(aq) \rightarrow [Zn(NH_3)_4]^{2+}(aq) + 2\ OH^-(aq)$$

with NaOH(aq): Again complex ion formation is involved; this time the complex ion is $[Zn(OH)_4]^{2-}$

$$Zn(OH)_2(s) + 2\ OH^-(aq) \rightarrow [Zn(OH)_4]^{2-}$$

19-37. First, calculate $[Cu^{2+}]$ from K_f for $[Cu(NH_3)_4]^{2+}$

$$K_f = \frac{[[Cu(NH_3)_4]^{2+}]}{[Cu^{2+}][NH_3]^4} = \frac{0.15}{[Cu^{2+}](0.10)^4} = 1.1\times10^{-13}$$

$$[Cu^{2+}] = \frac{0.15}{1.1\times10^{13}(0.10)^4} = 1.4\times10^{-10}$$

Next, determine pOH and $[OH^-]$ in the NH_3-NH_4^+ buffer.

$$pOH = pK_a + \log\frac{[NH_4^+]}{[NH_3]} \qquad pOH = 4.76 + \log\frac{0.10}{0.10} = 4.76$$

$$[OH^-] = \text{antilog}(-4.76) = 1.7\times10^{-5}$$

Now compare Q and K_{sp} for $Cu(OH)_2$.

$$Q = [Cu^{2+}][OH^-]^2 = (1.4\times10^{-10})(1.7\times10^{-5})^2 = 4.0\times10^{-20} < 1.6\times10^{-19}(K_{sp})$$

$Cu(OH)_2(s)$ will not precipitate from this solution.

19-38. The $[Ag^+]$ in Example 19-15 was found to be 9.8×10^{-9} M. Saturated AgI(aq) with $[Ag^+] = 9.8\times10^{-9}$ M has

$$[I^-] = \frac{K_{sp}}{[Ag^+]} = \frac{8.5\times10^{-17}}{9.8\times10^{-9}} = 8.7\times10^{-9}\ M$$

In terms of a quantity of KI,

$$\text{no. g KI} = 1.00\ L \times \frac{8.7\times10^{-9}\ \text{mol } I^-}{L} \times \frac{1\ \text{mol KI}}{1\ \text{mol } I^-} \times \frac{166\ \text{g KI}}{1\ \text{mol KI}} = 1.4\times10^{-6}\ \text{g KI}$$

19-39. Combine the equilibrium $AgBr(s) \rightleftharpoons Ag^+(aq) + Br^-(aq)$ $K_{sp} = 5.0\times10^{-13}$ with the appropriate complex ion formation equilibrium for each part. Let the molar solutility = x and convert this to a solubility in g/L.

(a) $AgBr(s) \rightleftharpoons Ag^+(aq) + Br^-(aq)$ $K_{sp} = 5.0\times10^{-13}$

$Ag^+(aq) + 2\ NH_3(aq) \rightleftharpoons [Ag(NH_3)_2]^+(aq)$ $K_f = 1.6\times10^7$

$AgBr(s) + 2\ NH_3(aq) \rightleftharpoons [Ag(NH_3)_2]^+(aq) + Br^-(aq)$ $K = K_{sp} \times K_f$

In the familiar fashion write

	$AgBr(s)$ +	$2\ NH_3$ $\rightleftharpoons$	$[Ag(NH_3)_2]^+$ +	Br^-
Initially:		1.50 M	--	--
Changes:		$-2x$ M	$+x$ M	$+x$ M
At Equil:		$(1.50-2x)$M	x M	x M

$$K = \frac{[[Ag(NH_3)_2]^+][Br^-]}{[NH_3]^2} = \frac{x \cdot x}{1.50 - 2} = K_{sp} \times K_f = 5.0\times10^{-13} \times 1.6\times10^7 = 8.0\times10^{-6}$$

Assume that $x << 1.50$ and write

$$\frac{x^2}{1.50} = 8.0\times10^{-6} \qquad x^2 = 1.2\times10^{-5} \qquad x = 3.5\times10^{-3}\ M$$

$$\text{no. g AgBr} = \frac{3.5\times10^{-3}\ \text{mol AgBr}}{L} \times \frac{188\ \text{g AgBr}}{1\ \text{mol AgBr}} = 0.66\ \text{g AgBr}$$

(b)

	$AgBr(s)$ +	$2\ CN^-$ $\rightleftharpoons$	$[Ag(CN)_2]^-$ +	Br^-
Initially:		0.10 M	-	-
Changes:		$-2x$ M	$+x$ M	$+x$ M
At equil:		$(0.10 - 2x)$M	x M	x M

$$K = \frac{[[Ag(CN)_2]^-][Br^-]}{[CN^-]^2} \times \frac{x \cdot x}{(0.10 - 2x)^2} = K_{sp} \times K_f = 5.0\times10^{-13} \times 5.6\times10^{18} = 2.8\times10^6$$

The assumption $(0.10 - 2x) \simeq 0.10$ will not be valid because of the large value of K.

$$x^2 = 2.8\times10^6(0.10 - 2x)^2 = 2.8\times10^6(0.01 - 0.40x + 4x^2)$$

$$x^2 = 2.8\times10^4 - 1.1\times10^6 x + 1.1\times10^7 x^2$$

$$1.1\times10^7 x^2 - 1.1\times10^6 x + 2.8\times10^4 = 0$$

$$x^2 - 0.10x + 2.5\times10^{-3} = 0$$

$$x = \frac{0.10 \pm \sqrt{(0.10)^2 - 4 \times 2.5 \times 10^{-3}}}{2} = 0.050\,M$$

$$\text{no. g AgBr} = \frac{0.050\text{ mol AgBr}}{L} \times \frac{188\text{ g AgBr}}{1\text{ mol AgBr}} = 9.4\text{ g AgBr}$$

(Note: is actually slightly less than 0.050M; otherwise the denominator in the K expression would be zero.)

(c)

	AgBr(s)	+	$2\ S_2O_3^{2-}$	$\rightleftharpoons$	$[Ag(S_2O_3)_2]^{3-}$	+	Br^-
Initially:			0.50 M		--		--
Changes:			$-2x$ M		$+x$ M		$+x$ M
At equil:			$(0.50 - 2x)$M		x M		x M

$$K = \frac{[[Ag(S_2O_3)_2]^{3-}][Br^-]}{[S_2O_3^{2-}]^2} = \frac{x \cdot x}{(0.50 - 2x)^2} = K_{sp} \times K_f = 5.0\times10^{-13} \times 1.7\times10^{13} = 8.5$$

Again, the simplifying assumption, that $(0.50 - 2x) \simeq 0.50$, is not expected to work.

$$x^2 = 8.5(0.50 - 2x)^2 = 8.5(0.25 - 2x + 4x^2)$$

$$x^2 = 2.1 - 17x + 34x^2$$

$$33x^2 - 17x + 2.1 = 0 \qquad x = \frac{17 \pm \sqrt{(17)^2 - 4 \times 33 \times 2.1}}{2 \times 33} = 0.21\ M$$

$$\text{no. g AgBr} = \frac{0.21\text{ mol AgBr}}{L} \times \frac{188\text{ g AgBr}}{1\text{ mol AgBr}} = 39\text{ g AgBr}$$

The other root of the quadratic equation $x = 0.30$ M is rejected because $2x$ cannot be greater than 0.50 M.

Precipitation and solubilities of metal sulfides

19-40. (a) Calculate the pH and then $[H_3O^+]$ in the buffer.

$$pH = pK_a + \log\frac{[C_2H_3O_2^-]}{[HC_2H_3O_2]} = 4.76 + \log\frac{0.15}{0.25} = 4.76 - 0.22 = 4.54$$

$$[H_3O^+] = \text{antilog}(-4.54) = 2.9\times10^{-5}$$

Now use equation (19.30) with $[H_2S] = 0.10$ M

$$\frac{[H_3O^+]^2[S^{2-}]}{[H_2S]} = \frac{(2.9\times10^{-5})^2[S^{2-}]}{0.10} = 1.1\times10^{-21}$$

$$[S^{2-}] = \frac{0.10 \times 1.1 \times 10^{-21}}{(2.9\times10^{-5})^2} = 1.3\times10^{-13}$$

Now compare Q and K_{sp} for MnS.

$$Q = [Mn^{2+}][S^{2-}] = 0.015 \times 1.3 \times 10^{-13} = 2.0\times10^{-15} < 2.5\times10^{-3} (K_{sp})$$

MnS(s) will not precipitate from this solution.

(b) If precipitation of MnS is to begin, $[S^{2-}]$ must increase, $[H_3O^+]$ must decrease, and the anion concentration $[C_2H_3O_2^-]$ must be increased. The required $[S^{2-}]$ for precipitation to begin is

$$[S^{2-}] = K_{sp}/[Mn^{2+}] = 2.5\times10^{-13}/0.015 = 1.7\times10^{-11} \text{ M}$$

The required $[H_3O^+]$ is

$$[H_3O^+]^2 = 1.1\times10^{-21} \times \frac{[H_2S]}{[S^{2-}]} = \frac{1.1\times10^{-21} \times 0.10}{1.7\times10^{-11}} = 6.5\times10^{-12}$$

$$[H_3O^+] = 2.5\times10^{-6} \qquad pH = -\log(2.5\times10^{-6}) = 5.60$$

The required $[C_2H_3O_2^-]$ is obtained through the expression

$$pH = pK_a + \log\frac{[C_2H_3O_2^-]}{[HC_2N_3O_2]} \qquad 5.60 = 4.76 + \log\frac{[C_2H_3O_2^-]}{0.25}$$

$$\log[C_2H_3O_2^-] = 5.60 - 4.76 + \log 0.25 = 5.60 - 4.76 - 0.60 = 0.24$$

$$[C_2H_3O_2^-] = \text{antilog}(0.24) = 1.7 \text{ M}$$

19-41. The first step in this solution is to calculate $[Pb^{2+}]$ in saturated $PbCl_2(aq)$. Let $[Pb^{2+}]$ in this solution = x; $[Cl^-] = 2x$.

$$[Pb^{2+}][Cl^-]^2 = (x)(2x)^2 = 4x^3 = 1.6\times10^{-5} \qquad x^3 = 4.0\times10^{-6} \qquad x = [Pb^{2+}] = 1.6\times10^{-2}$$

Next, calculate $[S^{2-}]$. Since the solution is saturated in H_2S, use equation (19.30) with $[H_2S] = 0.10$ M.

$$[H_3O^+][S^{2-}] = 1.1\times10^{-22}$$

The value of $[H_3O^+]$ is determined from the pH.

$$pH = 0.5 \qquad \log[H_3O^+] = -0.5 = 0.5 - 1.00 \qquad [H_3O^+] = 3\times10^{-1}$$

Now we are in a position to calculate $[S^{2-}]$ in the solution.

$$(3\times10^{-1})^2 \times [S^{2-}] = 1.1\times10^{-22} \qquad [S^{2-}] = 1\times10^{-21}$$

Finally, compare the ion product expression and K_{sp} for PbS.

$$[Pb^{2+}][S^{2-}] = (1.6\times10^{-2}) \times (1\times10^{-21}) = 2\times10^{-23} > 8.0\times10^{-28} (K_{so})$$

PbS *does* precipitate under the conditions described here.

19-42. Following the hint given in the exercise, first solve for pH and $[H_3O^+]$ in the buffer solution.

$$pH = pK_a + \log \frac{[C_2H_3O_2^-]}{[HC_2H_3O_2]} = 4.76 + \log \frac{0.250}{0.500} = 4.76 - 0.30 = 4.46$$

$$\log [H_3O^+] = -4.46 = 0.54 - 5.00 \qquad [H_3O^+] = 3.5\times10^{-5}\ M$$

$[H_3O^+]$ is assumed to remain constant in the following reaction:

$$FeS(s) + 2\ H_3O^+(aq) \rightleftharpoons Fe^{2+}(aq) + H_2S(aq) + 2\ H_2O$$

To obtain K for this reaction proceed in the same fashion as was used to establish equation (19.33) in the text. The result obtained is:

$$K = \frac{K_{sp}}{K_{a_1} \times K_{a_2}} = \frac{6.3\times10^{-18}}{1.1\times10^{-21}} = 5.7\times10^3$$

Let the molar solubility of FeS(s) = $[Fe^{2+}] = [H_2S(aq)] = S$

$$K = \frac{[Fe^{2+}][H_2S]}{[H_3O^+]^2} = \frac{S \times S}{(3.5\times10^{-5})^2} = 5.7\times10^3 \qquad S^2 = 7.0\times10^{-6} \qquad S = 2.6\times10^{-3}\ M$$

$$\text{no. g FeS/L} = \frac{2.6\times10^{-3}\ \text{mol FeS}}{1.00\ \text{L}} \times \frac{87.9\ \text{g FeS}}{1\ \text{mol FeS}} = 0.23\ \text{g FeS/L}$$

19-43. The three equilibrium expressions referred to in the exercise are

$$CoS(s) \rightleftharpoons Co^{2+}(aq) + \cancel{S^{2-}(aq)}; \quad K_{sp} = 4\times10^{-21}$$

$$\cancel{S^{2-}(aq)} + \cancel{2\ H_3O^+(aq)} \rightleftharpoons H_2S(aq) + \cancel{2\ H_2O}; \quad K = 1/(K_{a_1} \times K_{a_2}) = 1/(1.1\times10^{-21})$$

$$2\ HC_2H_3O_2(aq) + \cancel{2\ H_2O} \rightleftharpoons \cancel{2\ H_3O^+(aq)} + 2\ C_2H_3O_2^-(aq); \quad K = K_a^2 = (1.74\times10^{-5})^2$$

$$CoS(s) + 2\ HC_2H_3O_2(aq) \rightleftharpoons Co^{2+}(aq) + 2\ C_2H_3O_2^-(aq) + H_2S(aq)$$

$$K = \frac{K_{sp}\,(K_a)^2}{K_{a_1} \times K_{a_2}} = \frac{4\times10^{-21} \times (1.74\times10^{-5})^2}{1.1\times10^{-21}} = 1.1\times10^{-9} = 1\times10^{-9}$$

Let the molar solubility of CoS = $[Co^{2+}] = S$. Assume $[HC_2H_3O_2]$ remains constant at 0.100 M.

$$K = \frac{[Co^{2+}][C_2H_3O_2^-]^2[H_2S]}{[HC_2H_3O_2]^2} = \frac{S \times (2S)^2 \times S}{(0.100)^2} = \frac{4S^4}{(0.100)^2} = 1\times10^{-9}$$

$$S^4 = 2\times10^{-12} \qquad S^2 = 1\times10^{-6} \qquad S = 1\times10^{-3}\ M$$

The molar solubility of CoS in 0.100 M $HC_2H_3O_2$ is 1.3×10^{-3} mol CoS/L.

19-44. (a) There must be at least one ion present from group I (Ag^+, Hg_2^{2+}, or Pb^{2+}). However, the failure of the group I filtrate to yield a sulfide precipitate seems to eliminate Pb^{2+}, since it is expected to precipitate both in group I and in group 2. Thus we would predict the presence of Ag^+ and/or Hg_2^{2+}. Statement (a) is accurate.

(b) This is not an accurate statement. No conclusion can be drawn about either the possible presence or possible absence of Mg^{2+} from the observations noted.

(c) This is an accurate statement, Pb^{2+} is probably not present if the group I filtrate fails to yield a precipitate with H_2S.

(d) This is not an accurate statement, since Fe^{2+} does not precipitate in either group I or group 2. No conclusion can be drawn about the presence or absence of Fe^{2+}.

19-45. In group I, Pb^{2+} precipitates as the chloride ($K_{sp} = 1.6\times10^{-5}$) and in group 2 as the sulfide ($K_{sp} = 8.0\times10^{-28}$). The $[Pb^{2+}]$ remaining in a saturated solution of $PbCl_2$, is fairly high, certainly high enough so that the product $[Pb^{2+}][S^{2-}]$ exceeds K_{sp}(PbS) when the solution is later treated with H_2S. No other common cation is first precipitated as a moderately soluble precipitate, followed by treatment with a reagent that yields a much more insoluble precipitate.

19-46. (a) Precipitate the Ba^{2+} as the sulfate or carbonate. That is, use Na_2SO_4(aq) or Na_2CO_3(aq) as the reagent.

(b) Na_2CO_3 is water soluble and $MgCO_3$ is not. Water can be used as the reagent.

(c) Use a chloride ion solution, such as KCl(aq). The $AgNO_3$(s) is converted to AgCl(s) and the KNO_3(s) dissolves.

(d) $Pb(NO_3)_2$ is water soluble and $PbSO_4$ is not. Use water as the reagent to separate them.

19-47. Let the molar solubility of $PbCl_2$ = S. In a saturated solution, $[Pb^{2+}] = S$ and $[Cl^-] = 2S$.

$$K_{sp} = [Pb^{2+}][Cl^-]^2 = (S)(2S)^2 = 4S^3$$

at 25°C: $4S^3 = 1.6\times10^{-5}$ $\quad S = 1.6\times10^{-2} = 0.016$ M

at 80°C: $4S^3 = 3.3\times10^{-3}$ $\quad S = 9.4\times10^{-2} = 0.094$ M

$PbCl_2$ is moderately soluble at both temperatures. The reason that precipitation of $PbCl_2$ can be carried essentially to completion at 25°C is that an excess of Cl^- (a common ion) is maintained in the precipitating solution.

19-48. (a) In general, a test for one cation cannot be made until other interfering ions have been removed. That is, the precipitating reagents of the qualitative analysis scheme are not specific for individual ions. If one attempted to identify Ba^{2+} by precipitation with $(NH_4)_2CO_3$ there would be interferences from Sr^{2+}, Ca^{2+}, Cu^{2+}, Zn^{2+}, and so on.

(b) The tests used for ions within a group require that there be no ions present from a preceding group. For example, if the group 2 reagent were used before group I, any Ag^+ present in solution would precipitate as the sulfide, complicating the sulfide separations of group 2. And the test used for Ag^+ in group I, based on AgCl, would not work for Ag_2S.

Self-Test Questions

1. (*d*) In saturated PbI_2, if $[Pb^{2+}] = S$, $[I^-] = 2S$, $K_{sp} = 4S^3$ and $S = (K_{sp}/4)^{1/3}$. Item (a) is incorrect because $[Pb^{2+}] \neq [I^-]$. Items (b) and (c) are incorrect because $[Pb^{2+}] = S = (K_{sp}/4)^{1/3}$, not K_{sp} and not $\sqrt{K_{sp}}$. Since $[I^-] = 2[Pb^{2+}]$, then $[Pb^{2+}] = [I^-]/2$.

2. (*a*) The common ion effect results from adding Na_2SO_4 to saturated $BaSO_4$(aq). The solubility of $BaSO_4$ is *reduced*. This means that $[Ba^{2+}]$ is *reduced*. ($[SO_4^{2-}]$ is increased because of the added Na_2SO_4.)

3. (*c*) K_2CrO_4 and $AgNO_3$ produce common ions to the saturated Ag_2CrO_4(aq). They cause a decrease in the solubility of Ag_2CrO_4. KNO_3, which has no ions in common with Ag_2CrO_4, produces the salt effect. Ag_2CrO_4 is most soluble in KNO_3(aq).

4. (*b*) Both ions form an insoluble sulfide and neither forms an insoluble nitrate. Use H_2SO_4(aq). $PbSO_4$(s) precipitates and $CuSO_4$ is soluble.

5. (*c*) Ba^{2+}, Ca^{2+}, and Pb^{2+} all form insoluble carbonates. All ammonium salts, including $(NH_4)_2CO_3$ are soluble.

6. (*a*) Because of the large excess of undissolved solute, the addition of pure water to saturated MgF_2(aq) just results in a larger volume of saturated solution. However, since $[Mg^{2+}]$ in saturated MgF_2(aq) is independent of the volume of solution, $[Mg^{2+}]$ does not change in this process.

7. (*a*) The solubility of $CaCO_3$(s) can be increased by increasing H_3O^+ in a solution. This occurs through reactions such as $CO_3^{2-} + H_3O^+ \rightarrow HCO_3^- + H_2O$ and $HCO_3^- + H_2O^+ \rightarrow H_2CO_3 + H_2O$ and $H_2CO_3 \rightarrow H_2O + CO(g)$. The only one of the species listed that has acidic properties is $NaHSO_4$.

8. (*c*) $[H_2S]$ in a solution cannot be raised beyond its saturation limit--0.10 M. Heating a solution will simply expel H_2S(g) and not promote more complete precipitation. The need is to increase $[S^{2-}]$. $H_2S + H_2O \rightleftharpoons H_3O^+ + HS^-$ and $HS^- + H_2O \quad H_3O^+ + S^{2-}$. By adding OH^- both of these equilibria are shifted to the right. $[S^{2-}]$ increases and the metal ion concentration decreases, ensuring more complete precipitation. The pH must be raised.

9.

	$PbI_2(s) \rightleftharpoons$	$Pb^{2+}(aq)$	+	$2\ I^-(aq)$
From PbI_2:		x M		$2x$ M
From 0.065 M KI:		--		0.065 M
At equilibrium:		x M		$(0.065 + 2x)$M

Assume x << 0.065 and write

$$K_{sp} = [P_b^{2+}][I^-]^2 = x(0.065)^2 = 7.1\times10^{-9} \qquad x = 1.7\times10^{-6}\ M$$

$$\text{no. mg } Pb^{2+}/mL = \frac{1.7\times10^{-6}\ mol\ Pb^{2+}}{L} \times \frac{1.00\ L}{1000\ mL} \times \frac{207\ g\ Pb^{2+}}{1\ mol\ Pb^{2+}} \times \frac{1000\ mg\ Pb^{2+}}{1\ g\ Pb^{2+}} = 3.5\times10^{-4}\ mg\ Pb^{2+}/mL$$

10. If the solid has acidic properties, it will be more soluble in basic solution. Solids with basic properties (e.g., carbonates, hydroxides, and sulfides) are more soluble in acidic solution.

more soluble in acidic solution: $MgCO_3$, CdS, $Ca(OH)_2$

more soluble in basic solution: $H_2C_2O_4$

solubility independent of pH: KCl, $NaNO_3$

11. (a) The first to precipitate should be the compound with the lower K_{sp} (since the two are of similar type). This should be $PbCrO_4$.

(b) At the point where $PbSO_4$ beings to precipitate $K_{sp} = [Pb^{2+}][SO_4^{2-}] = [Pb^{2+}](0.010) = 1.6\times10^{-8}$
$[Pb^{2+}] = 1.6\times10^{-6}$ M

(c) $[CrO_4^{2-}]$ remaining in solution at point where $PbSO_4$ beings to precipitate:

$$K_{sp} = [Pb^{2+}][CrO_4^{2-}] = (1.6\times10^{-6})[CrO_4^{2-}] = 2.8\times10^{-13} \qquad [CrO_4^{2-}] = 1.8\times10^{-7}\ M$$

$$\%\ CrO_4^{2-}\ \text{unprecipitated} = \frac{1.8\times10^{-7}\ mol/L}{1.0\times10^{-2}\ mol/L} \times 100 = 1.8\times10^{-3}\%$$

The precipitation of CrO_4^{2-} is essentially complete before precipitation of $PbSO_4$ beings. Separation by fractional precipitation is effective.

12. Combine equilibrium data into a single expression to describe the dissolving of a copper compound with complex ion formation. If K for this combined expression is large, dissolving should occur to an appreciable extent. If K is very small, very little dissolving will occur.

(a) $CuS(s) \rightleftharpoons Cu^{2+}(aq) + S^{2-}(aq) \qquad K_{sp} = 6.3\times10^{-136}$

$Cu^{2+}(aq) + 4\ NH_3(aq) \rightleftharpoons [Cu(NH_3)_4]^{2+}(aq) \qquad K_f = 1.1\times10^{13}$

$CuS(s) + 4\ NH_3(aq) \rightleftharpoons [Cu(NH_3)_4]^{2+}(aq) + S^{2-}(aq)$

$K = K_{sp} \times K_f = 6.3\times10^{-36} \times 1.1\times10^{13} = 6.9\times10^{-23}$

The very small value of K indicates that no appreciable dissolution of CuS occurs.

(b) $CuCO_3(s) + 4\ NH_3(aq) \rightleftharpoons [Cu(NH_3)_4]^{2+}(aq) + CO_3^{2-}(aq)$

$K = K_{sp} \times K_f = 1.4\times10^{-10} \times 1.1\times10^{13} = 1.5\times10^{3}$

The large value of K suggests that $CuCO_3(s)$ will dissolve to an app-eciable extent in $NH_3(aq)$.

Chapter 20

Oxidation-Reduction and Electrochemistry

Review Problems

20-1. (a) reduction: $S_2O_8^{2-} + 2\,e^- \rightarrow 2\,SO_4^{2-}$

(b) $2\,HNO_3 \rightarrow N_2O$

$2\,HNO_3 \rightarrow N_2O + 5\,H_2O$

$2\,HNO_3 + 8\,H^+ \rightarrow N_2O + 5\,H_2O$

reduction: $2\,HNO_3 + 8\,H^+ + 8\,e^- \rightarrow N_2O + 5\,H_2O$

(c) $CH_4 \rightarrow CO_2$

$CH_4 + 2\,H_2O \rightarrow CO_2 + 8\,H^+$

oxidation: $CH_4 + 2\,H_2O \rightarrow CO_2 + 8\,H^+ + 8\,e^-$

(d) $Br^- \rightarrow BrO_3^-$

$Br^- + 6\,OH^- \rightarrow BrO_3^- + 3\,H_2O$

oxidation: $Br^- + 6\,OH^- \rightarrow BrO_3^- + 3\,H_2O + 6\,e^-$

(e) $NO_3^- \rightarrow NH_3$

$NO_3^- + 3\,H_2O \rightarrow NH_3 + 6\,OH^-$

reduction: $NO_3^- + 6\,H_2O + 8\,e^- \rightarrow NH_3 + 9\,OH^-$

20-2. (a) oxid: $3\{Cu(s) \rightarrow Cu^{2+}(aq) + 2\,e^-\}$

red: $2\{NO_3^- + 4\,H^+ + 3\,e^- \rightarrow NO(g) + 2\,H_2O\}$

---

net: $3\,Cu(s) + 8\,H^+ + 2\,NO_3^- \rightarrow 3\,Cu^{2+} + 2\,NO(g) + 4\,H_2O$

(b) oxid: $4\{Zn(s) \rightarrow Zn^{2+}(aq) + 2\,e^-\}$

red: $NO_3^- + 10\,H^+ + 8\,e^- \rightarrow NH_4^+ + 3\,H_2O$

---

net: $4\,Zn(s) + 10\,H^+ + NO_3^- \rightarrow 4\,Zn^{2+} + NH_4^+ + 3\,H_2O$

(c) oxid: $5\{H_2O_2 \rightarrow O_2(g) + 2\,H^+ + 2\,e^-\}$

red: $2\{MnO_4^- + 8\,H^+ + 5\,e^- \rightarrow Mn^{2+} + 4\,H_2O\}$

---

net: $5\,H_2O_2 + 2\,MnO_4^- + 6\,H^+ \rightarrow 2\,Mn^{2+} + 8\,H_2O + 5\,O_2(g)$

(d) oxid: $S_2O_3^{2-} + 5\ H_2O \rightarrow 2\ HSO_4^- + 8\ H^+ + 8\ e^-$

red: $4\{Cl_2(g) + 2\ e^- \rightarrow 2\ Cl^-\}$

net: $S_2O_3^{2-} + 4\ Cl_2(g) + 5\ H_2O \rightarrow 2\ HSO_4^- + 8\ Cl^- + 8\ H^+$

(e) oxid: $3\{P(s) + 4\ H_2O \rightarrow H_2PO_4^- + 6\ H^+ + 5\ e^-\}$

red: $5\{NO_3^- + 4\ H^+ + 3\ e^- \rightarrow NO(g) + 2\ H_2O\}$

net: $3\ P(s) + 2\ H_2O + 2\ H^+ + 5\ NO_3^- \rightarrow 3\ H_2PO_4^- + 5\ NO(g)$

20-3. (a) oxid: $3\{CN^- + 2\ OH^- \rightarrow CNO^- + H_2O + 2\ e^-\}$

red: $2\{MnO_4^- + 2\ H_2O + 3\ e^- \rightarrow MnO_2(s) + 4\ OH^-\}$

net: $3\ CN^- + 2\ MnO_4^- + H_2O \rightarrow 2\ MnO_2 + 3\ CNO^- + 2\ OH^-$

(b) oxid: $N_2H_4(g) + 4\ OH^- \rightarrow N_2(g) + 4\ H_2O + 4\ e^-$

red: $4\{[Fe(CN)_6]^{3-} + e^- \rightarrow [Fe(CN)_6]^{4-}\}$

net: $4[Fe(CN)_6]^{3-} + N_2H_4(g) + 4\ OH^- \rightarrow 4[Fe(CN)_6]^{4-} + N_2(g) + 4\ H_2O$

(c) oxid: $4\{Fe(OH)_2(s) + OH^- \rightarrow Fe(OH)_3(s) + e^-\}$

red: $O_2(g) + 2\ H_2O + 4\ e^- \rightarrow 4\ OH^-$

net: $4\ Fe(OH)_2(s) + O_2(g) + 2\ H_2O \rightarrow 4\ Fe(OH)_3(s)$

(d) oxid: $3\{C_2H_5OH + 5\ OH^- \rightarrow C_2H_3O_2^- + 4\ H_2O + 4\ e^-\}$

red: $4\{MnO_4^- + 2\ H_2O + 3\ e^- \rightarrow MnO_2 + 4\ OH^-\}$

net: $3\ C_2H_5OH + 4\ MnO_4^- \rightarrow 3\ C_2H_3O_2^- + 4\ MnO_2 + 4\ H_2O + OH^-$

20-4. (a) oxid: $Cl_2(g) + 2\ H_2O \rightarrow 2\ HClO + 2\ H^+ + 2\ e^-$

red: $Cl_2(g) + 2\ e^- \rightarrow 2\ Cl^-$

net: $2\ Cl_2(g) + 2\ H_2O \rightarrow 2\ HClO + 2\ H^+ + 2\ Cl^-$

or $Cl_2(g) + H_2O \rightarrow HClO + H^+ + Cl^-$

(b) oxid: $Br_2(l) + 12\ OH^- \rightarrow 2\ BrO_3^- + 6\ H_2O + 10\ e^-$

red: $5\{Br_2(l) + 2\ e^- \rightarrow 2\ Br^-\}$

net: $6\ Br_2(l) + 12\ OH^- \rightarrow 10\ Br^- + 2\ BrO_3^- + 6\ H_2O$

or $3\ Br_2(l) + 6\ OH^- \rightarrow 5\ Br^- + BrO_3^- + 3\ H_2O$

(c) oxid: $S_2O_4^{2-} + 2\ H_2O \rightarrow 2\ HSO_3^- + 2\ H^+ + 2\ e^-$

red: $S_2O_4^{2-} + 2\ H^+ + 2\ e^- \rightarrow S_2O_3^{2-} + H_2O$

---

net: $2\ S_2O_4^{2-} + H_2O \rightarrow S_2O_3^{2-} + 2\ HSO_3^-$

(d) oxid: $2\{MnO_4^{2-} \rightarrow MnO_4^- + e^-\}$

red: $MnO_4^{2-} + 2\ H_2O + 2\ e^- \rightarrow MnO_2(s) + 4\ OH^-$

---

net: $3\ MnO_4^{2-} + 2\ H_2O \rightarrow MnO_2(s) + 2\ MnO_4^- + 4\ OH^-$

20-5. (a) oxid: $Zn(s) \rightarrow Zn^{2+}(aq) + 2\ e^-$ $E^\circ_{ox} = -(-0.763) = +0.763\ V$

red: $Sn^{2+}(aq) + 2\ e^- \rightarrow Sn(s)$ $E^\circ_{red} = -0.136\ V$

---

net: $Zn(s) + Sn^{2+}(aq) \rightarrow Zn^{2+}(aq) + Sn(s)$ $E^\circ_{cell} = +0.627\ V$

(b) oxid: $2\{Fe^{2+}(aq) \rightarrow Fe^{3+}(aq) + e^-\}$ $E^\circ_{ox} = -(+0.771) = -0.771\ V$

red: $Sn^{4+}(aq) + 2\ e^- \rightarrow Sn^{2+}(aq)$ $E^\circ_{red} = +0.154\ V$

---

net: $2\ Fe^{2+}(aq) + Sn^{4+}(aq) \rightarrow 2\ Fe^{3+}(aq) + Sn^{2+}(aq)$ $E^\circ_{cell} = -0.617\ V$

(c) oxid: $Cu(s) \rightarrow Cu^{2+}(aq) + 2\ e^-$ $E^\circ_{ox} = -(+0.337) = -0.337\ V$

red: $Cl_2(g) + 2\ e^- \rightarrow 2\ Cl^-(aq)$ $E^\circ_{red} = +1.360\ V$

---

net: $Cu(s) + Cl_2(g) \rightarrow Cu^{2+}(aq) + 2\ Cl^-(aq)$ $E^\circ_{cell} = +1.023\ V$

20-6. (a) oxid: $2\ Cl^-(aq) \rightarrow Cl_2(g) + 2\ e^-$ $E^\circ_{ox} = -(+1.360) = -1.360\ V$

red: $PbO_2(s) + 4\ H^+(aq) + 2\ e^- \rightarrow Pb^{2+}(aq) + 2\ H_2O$ $E^\circ_{red} = +1.455\ V$

---

net: $PbO_2(s) + 2\ Cl^-(aq) + 4\ H^+(aq) \rightarrow Pb^{2+}(aq) + Cl_2(g) + 2\ H_2O$ $E^\circ_{cell} = +0.095\ V$

(b) oxid: $3\{Mg(s) \rightarrow Mg^{2+}(aq) + 2\ e^-\}$ $E^\circ_{ox} = -(-2.375) = +2.375\ V$

red: $2\{Sc^{3+}(aq) + 3\ e^- \rightarrow Sc(s)\}$ $E^\circ_{red} = ?$

---

net: $3\ Mg(s) + 2\ Sc^{3+}(aq) \rightarrow 3\ Mg^{2+}(aq) + 2\ Sc(s)$ $E^\circ_{cell} = +0.35\ V$

$E^\circ_{cell} = E^\circ_{ox} + E^\circ_{red} = 2.38 + E^\circ_{red} = 0.35$ $E^\circ_{red} = -2.03\ V$

(c) oxid: $Cu^+(aq) \rightarrow Cu^{2+}(aq) + e^-$ $E^\circ_{ox} = ?$

red: $Ag^+(aq) + e^- \rightarrow Ag(s)$ $E^\circ_{red} = +0.800$ V

net: $Cu^+(aq) + Ag^+(aq) \rightarrow Cu^{2+}(aq) + Ag(s)$ $E^\circ_{cell} = +0.647$ V

$E^\circ_{cell} = E^\circ_{ox} + E^\circ_{red} = E^\circ_{ox} + 0.800 = 0.647$ $E^\circ_{ox} = -0.153$ V

$E^\circ_{red}(Cu^{2+}/Cu^+) = -E^\circ_{ox} = -(-0.153\text{ V}) = +0.153$ V

20-7. (a) oxid: $Sn(s) \rightarrow Sn^{2+}(aq) + 2\,e^-$ $E^\circ_{ox} = -(-0.136) = +0.136$ V

red: $Zn^{2+}(aq) + 2\,e^- \rightarrow Zn(s)$ $E^\circ_{red} = -0.763$ V

net: $Sn(s) + Zn^{2+}(aq) \rightarrow Sn^{2+}(aq) + Zn(s)$ $E^\circ_{cell} = -0.627$ V

The forward reaction is nonspontaneous.

(b) oxid: $2\,I^-(aq) \rightarrow I_2(s) + 2\,e^-$ $E^\circ_{ox} = -(+0.535) = -0.535$ V

red: $2\{Fe^{3+}(aq) + e^- \rightarrow Fe^{2+}(aq)\}$ $E^\circ_{red} = +0.771$ V

net: $2\,I^-(aq) + 2\,Fe^{3+}(aq) \rightarrow I_2(s) + 2\,Fe^{2+}(aq)$ $E^\circ_{cell} = +0.236$ V

The forward reaction is spontaneous.

(c) oxid: $3\{2\,H_2O \rightarrow O_2(g) + 4\,H^+(aq) + 4\,e^-\}$ $E^\circ_{ox} = -(+1.229) = -1.229$ V

red: $4\{NO_3^-(aq) + 4\,H^+(aq) + 3\,e^- \rightarrow NO(g) + 2\,H_2O\}$ $E^\circ_{red} = +0.96$ V

net: $4\,NO_3^-(aq) + 4\,H^+(aq) \rightarrow 4\,NO(g) + 3\,O_2(g) + 2\,H_2O$ $E^\circ_{cell} = -0.27$ V

The forward reaction is nonspontaneous.

(d) oxid: $2\{Cl^-(aq) + 2\,OH^-(aq) \rightarrow ClO^-(aq) + H_2O + 2\,e^-\}$ $E^\circ_{ox} = -(+0.89) = -0.89$ V

red: $O_2(g) + 2\,H_2O + 4\,e^- \rightarrow 4\,OH^-(aq)$ $E^\circ_{red} = +0.401$ V

net: $2\,Cl^-(aq) + O_2(g) \rightarrow 2\,ClO^-$ $E^\circ_{cell} = -0.49$ V

The forward reaction is nonspontaneous.

(e) oxid: $H_2O_2(aq) \rightarrow O_2(g) + 2\,H^+(aq) + 2\,e^-$ $E^\circ_{ox} = -(+0.682) = -0.682$ V

red: $H_2O_2(aq) + 2\,H^+(aq) + 2\,e^- \rightarrow 2\,H_2O$ $E^\circ_{red} = +1.77$ V

net: $2\,H_2O_2(aq) \rightarrow 2\,H_2O + O_2(g)$ $E^\circ_{cell} = +1.09$ V

The forward reaction is spontaneous.

20-8. This problem is similar to Review Problem 7 but in each case it is necessary to write an equation from the verbal description of the process.

(a) oxid: $Mg(s) \rightarrow Mg^{2+}(aq) + 2\ e^-$ $\quad E^\circ_{ox} = -(-2.375) = +2.375$ V

red: $Sn^{2+}(aq) + 2\ e^- \rightarrow Sn(s)$ $\quad E^\circ_{red} = -0.136$ V

net: $Mg(s) + Sn^{2+}(aq) \rightarrow Mg^{2+}(aq) + Sn(s)$ $\quad E^\circ_{cell} = +2.239$ V

The displacement of $Sn^{2+}(aq)$ by Mg(s) should go essentially to completion (because of the large positive value of E°_{cell}).

(b) oxid: $Pb(s) \rightarrow Pb^{2+}(aq) + 2\ e^-$ $\quad E^\circ_{ox} = -(-0.126) = +0.126$ V

red: $2\ H^+(aq) + 2\ e^- \rightarrow H_2(g)$ $\quad E^\circ_{red} = 0.0000$ V

net: $Pb(s) + 2\ H^+(aq) \rightarrow Pb^{2+}(aq) + H_2(g)$ $\quad E^\circ_{cell} = +0.126$ V

Lead metal should dissolve in HCl(aq) and liberate $H_2(g)$. [Of course the precipitation of $PbCl_2(s)$ and formation of the complex ion $[PbCl_3]^-$ would also occur to some extent.]

(c) oxid: $2\{Fe^{2+}(aq) \rightarrow Fe^{3+}(aq) + e^-\}$ $\quad E^\circ_{ox} = -(+0.771) = -0.771$ V

red: $SO_4^{2-}(aq) + 4\ H^+(aq) + 2\ e^- \rightarrow SO_2(g) + 2\ H_2O$ $\quad E^\circ_{red} = +0.17$ V

net: $2\ Fe^{2+}(aq) + 4\ H^+(aq) + SO_4^{2-}(aq) \rightarrow 2\ Fe^{3+}(aq) + SO_2(g) + 2\ H_2O$ $\quad E^\circ_{cell} = -0.60$ V

Because of the large negative value of E°_{cell} we conclude that the oxidation of $Fe^{2+}(aq)$ by $SO_4^{2-}(aq)$ in acidic solution will not occur to a significant extent.

(d) oxid: $2\{Fe^{2+}(aq) \rightarrow Fe^{3+}(aq) + e^-\}$ $\quad E^\circ_{ox} = -(+0.771) = -0.771$ V

red: $SO_4^{2-}(aq) + H_2O + 2\ e^- \rightarrow SO_3^{2-}(aq) + 2\ OH^-(aq)$ $\quad E^\circ_{red} = -0.93$ V

net: $2\ Fe^{2+}(aq) + SO_4^{2-}(aq) + H_2O \rightarrow 2\ Fe^{3+}(aq) + SO_3^{2-}(aq) + 2\ OH^-(aq)$ $\quad E^\circ_{cell} = -1.70$ V

The oxidation of $Fe^{2+}(aq)$ by $SO_4^{3-}(aq)$ occurs to an even lesser extent than in part (c).

(e) oxid: $2\ I^-(aq) \rightarrow I_2 + 2\ e^-$ $\quad E^\circ_{ox} = -(+0.535) = -0.535$ V

red: $Cl_2(g) + 2\ e^- \rightarrow 2\ Cl^-(aq)$ $\quad E^\circ_{red} = +1.360$ V

net: $Cl_2(g) + 2\ I^-(aq) \rightarrow I_2 + 2\ Cl^-(aq)$ $\quad E^\circ_{cell} = +0.825$ V

The large positive value of E°_{cell} signifies that I_2 will displace $Cl^-(aq)$ from solution.

20-9. Separate each reaction into half-reactions, for which E°_{ox} and E°_{red} can be written. Combine the half-equations into a net equation and E°_{ox} and E°_{red} into E°_{cell}. Determine K from the expression:

$2.303\ RT \log K = n\mathcal{F}E^\circ_{cell}$, or $E^\circ_{cell} = \frac{0.0592}{n} \log K$

(a) oxid: $Ag(s) \rightarrow Ag^{+}(aq) + e^{-}$ $\quad E^{\circ}_{ox} = -(+0.800) = -0.800$ V

red: $Fe^{3+}(aq) + e^{-} \rightarrow Fe^{2+}(aq)$ $\quad E^{\circ}_{red} = +0.771$ V

net: $Fe^{3+}(aq) + Ag(s) \rightarrow Fe^{2+}(aq) + Ag^{+}(aq)$ $\quad E^{\circ}_{cell} = -0.029$ V

$$\log K = \frac{n \times E^{\circ}_{cell}}{0.0592} = \frac{-1 \times 0.029}{0.0592} = -0.49 \qquad K = 0.32$$

$$K = \frac{[Fe^{2+}][Ag^{+}]}{[Fe^{3+}]} = 0.32$$

(b) oxid: $2\ Cl^{-}(aq) \rightarrow Cl_2(g) + 2\ e^{-}$ $\quad E^{\circ}_{ox} = -(+1.360) = -1.360$ V

red: $MnO_2(s) + 4\ H^{+} + 2\ e^{-} \rightarrow Mn^{2+}(aq) + 2\ H_2O$ $\quad E^{\circ}_{red} = +1.23$ V

net: $MnO_2(s) + 2\ Cl^{-}(aq) + 4\ H^{+}(aq) \rightarrow Mn^{2+}(aq) + Cl_2(g) + 2\ H_2O$ $\quad E^{\circ}_{cell} = -0.13$ V

$$\log K = \frac{n \times E^{\circ}_{cell}}{0.0592} = \frac{-2 \times 0.13}{0.0592} = -4.4 \qquad K = 4\times10^{-5}$$

$$K = \frac{[Mn^{2+}]P_{Cl_2}}{[Cl^{-}]^2[H^{+}]^4} = 4\times10^{-5}$$

(c) oxid: $4\ OH^{-}(aq) \rightarrow O_2(g) + 2\ H_2O + 4\ e^{-}$ $\quad E^{\circ}_{ox} = (+0.401) = -0.401$ V

red: $2\{ClO^{-}(aq) + H_2O + 2\ e^{-} \rightarrow Cl^{-}(aq) + 2\ OH^{-}(aq)\}$ $\quad E^{\circ}_{red} = +0.89$ V

net: $2\ ClO^{-}(aq) \rightarrow 2\ Cl^{-}(aq) + O_2(g)$ $\quad E^{\circ}_{cell} = +0.49$ V

$$\log K = \frac{n \times E^{\circ}_{cell}}{0.0592} = \frac{4 \times 0.49}{0.0592} = 33 \qquad K = 1\times10^{33}$$

$$K = \frac{[Cl^{-}]^2 P_{O_2}}{[ClO^{-}]^2} = 1\times10^{33}$$

20-10. (a) $$E_{cell} = E^{\circ}_{cell} - \frac{0.0592}{n} \log \frac{[Cu^{2+}]}{[Ag^{+}]^2}$$

$E^{\circ}_{cell} = E^{\circ}_{ox} + E^{\circ}_{red} = -(+0.337) + 0.800 = +0.463$ V $\qquad n = 2$

$$E_{cell} = 0.463 - \frac{0.0592}{2} \log \frac{[Cu^{2+}]}{[Ag^{+}]^2} = 0.463 - 0.0296 \log \frac{[Cu^{2+}]}{[Ag^{+}]^2}$$

(b) $E_{cell} = E°_{cell} - \frac{0.0592}{n} \log \frac{[Al^{3+}]^2}{[Cu^{2+}]^3}$

$E°_{cell} = E°_{ox} + E°_{red} = -(-1.66) + 0.337 = 2.00$ V   n = 6

$E_{cell} = 2.00 - \frac{0.0592}{6} \log \frac{[Al^{3+}]^2}{[Cu^{2+}]^3} = 2.00 - 0.00987 \log \frac{[Al^{3+}]^2}{[Cu^{2+}]^3}$

(c) $E_{cell} = E°_{cell} - \frac{0.0592}{n} \log \frac{[Fe^{3+}]^5[Mn^{2+}]}{[Fe^{2+}]^5[MnO_4^-][H^+]^8}$

$E°_{cell} = E°_{ox} + E°_{red} = -(+0.771) + 1.51 = +0.74$ V   n = 5

$E_{cell} = 0.74 - \frac{0.0592}{5} \log \frac{[Fe^{3+}]^5[Mn^{2+}]}{[Fe^{2+}]^5[MnO_4^-][H^+]^8} = 0.74 - 0.0118 \log \frac{[Fe^{3+}]^5[Mn^{2+}]}{[Fe^{2+}]^5[MnO_4^-][H^+]^8}$

20-11. For the cell pictured in Figure 20-9, the relationship between E_{cell} and pH is given by equation (20.37): $E_{cell} = 0.0592$ pH

(a) If pH = 5.12, $E_{cell} = 0.0592 \times 5.12 = 0.303$ V

(b) In 0.00185 M HCl, $[H_3O^+] = 1.85 \times 10^{-3}$ and pH = $-\log [H_3O^+] = -\log (1.85 \times 10^{-3}) = 2.732$

$E_{cell} = 0.0592 \times 2.732 = 0.162$ V

(c) Let $[H_3O^+]$ in 0.357 M $HC_2H_3O_2$ = x. $[H_3O^+] = [C_2H_3O_2^-] = x$. $[HC_2H_3O_2] = 0.357 - x \simeq 0.357$ M.

$K_a = \frac{[H_3O^+][H_3O_2^-]}{[HC_2H_3O_2]} = \frac{x \cdot x}{0.357} = 1.74 \times 10^{-5}$ $x = [H_3O^+] = 2.49 \times 10^{-3}$ M.

pH = $-\log [H_3O^+] = -\log (2.49 \times 10^{-3}) = +2.604$

$E_{cell} = 0.0592 \times 2.604 = 0.154$ V

20-12. In general, choose the combination of oxidation and reduction half-reactions that can be produced with the least negative $E°_{cell}$.

(a) oxid: $2\ Cl^-(aq) \rightarrow Cl_2(g) + 2\ e^-$ $E°_{ox} = -1.360$ V

red: $Cu^{2+}(aq) + 2\ e^- \rightarrow Cu(s)$ $E°_{red} = +0.337$ V

net: $Cu^{2+}(aq) + 2\ Cl^-(aq) \rightarrow Cu(s) + Cl_2(g)$ $E°_{cell} = -1.023$ V

The probable products are Cu(s) at the cathode and Cl_2(g) at the anode.

(b) oxid: $2\ Cl^-(aq) \rightarrow Cl_2(g) + 2\ e^-$ $E°_{ox} = -1.360$ V

red: $2\ H^+(aq) + 2\ e^- \rightarrow H_2(g)$ $E°_{red} = 0.0000$ V

net: $2\ H^+(aq) + 2\ Cl^-(aq) \rightarrow H_2(g) + Cl_2(g)$ $E°_{cell} = -1.360$ V

The probable products are H_2(g) at the cathode and Cl_2(g) at the anode.

(c) oxid: $2\ H_2O \rightarrow O_2(g) + 4\ H^+(aq) + 4\ e^-$ $E°_{ox} = -1.229$ V

red: $2\{2\ H_2O + 2\ e^- \rightarrow 2\ OH^-(aq) + H_2(g)\}$ $E°_{red} = -0.828$ V

net: $6\ H_2O \rightarrow O_2(g) + \underbrace{4\ H^+(aq) + 4\ OH^-(aq)}_{4\ H_2O} + 2\ H_2(g)$

or $2\ H_2O \rightarrow O_2(g) + 2\ H_2(g)$ $E°_{cell} = -2.057$ V

The probable products are $H_2(g)$ at the cathode and $O_2(g)$ at the anode. (The oxidation of SO_4^{2-} to $S_2O_8^{2-}$ and the reduction of Na^+ to Na are both much more difficult to accomplish than the half-reactions listed here.)

(d) oxid: $2\ Cl^- \rightarrow Cl_2(g)$

red: $Ba^{2+} + 2\ e^- \rightarrow Ba(l)$

net: $Ba^{2+} + 2\ Cl^- \rightarrow Ba(l) + Cl_2(g)$

The products are $Ba(l)$ at the cathode and $Cl_2(g)$ at the anode. These are the only possible products from molten $BaCl_2$.

(e) oxid: $2\ I^-(aq) \rightarrow I_2 + 2\ e^-$ $E°_{ox} = -(+0.535) = -0.535$ V

red: $2\ H_2O + 2\ e^- \rightarrow 2\ OH^-(aq) + H_2(g)$ $E°_{red} = -0.828$ V

net: $2\ I^-(aq) + 2\ H_2O \rightarrow I_2 + 2\ OH^-(aq) + H_2(g)$ $E°_{cell} = -1.363$ V

The probable products are $H_2(g)$ and $OH^-(aq)$ at the cathode and I_2 at the anode.

(f) oxid: $2\ H_2O \rightarrow O_2(g) + 4\ H^+(aq) + 4\ e^-$ $E°_{ox} = -1.229$ V

red: $2\{2\ H_2O + 2\ e^- \rightarrow H_2(g) + 2\ OH^-(aq)\}$ $E°_{red} = -0.828$ V

net: $6\ H_2O \rightarrow O_2(g) + 2\ H_2(g) + \underbrace{4\ H^+ + 4\ OH^-}_{4\ H_2O}$ $E°_{cell} = -2.057$ V

or $2\ H_2O \rightarrow O_2(g) + 2\ H_2(g)$

The substance most easily oxidized and reduced is H_2O. The probable product at the cathode is $H_2(g)$; at the anode, $O_2(g)$.

20-13. First determine the amount of electric charge involved in these electrolyses.

$$\text{no. mol } e^- = 2.25\ \text{h} \times \frac{60\ \text{min.}}{1\ \text{h}} \times \frac{60\ \text{s}}{1\ \text{min}} \times \frac{1.56\ \text{C}}{\text{s}} \times \frac{1\ \text{mol}e^-}{96.500\ \text{C}} = 0.131\ \text{mol}e^-$$

(a) $Zn^{2+}(aq) + 2\ e^- \rightarrow Zn(s)$

$$\text{no. g Zn} = 0.131\ \text{mol}e^- \times \frac{1\ \text{mol Zn}}{2\ \text{mol } e^-} \times \frac{65.38\ \text{g Zn}}{1\ \text{mol Zn}} = 4.28\ \text{g Zn}$$

(b) $Al^{3+}(aq) + 3\ e^- \rightarrow Al(s)$

$$\text{no. g Al} = 0.131\ \text{mol}\ e^- \times \frac{1\ \text{mol Al}}{3\ \text{mol}\ e^-} \times \frac{26.98\ \text{g Al}}{1\ \text{mol Al}} = 1.18\ \text{g Al}$$

(c) $Ag^+(aq) + e^- \rightarrow Ag(s)$

$$\text{no. g Ag} = 0.131\ \text{mol}\ e^- \times \frac{1\ \text{mol Ag}}{1\ \text{mol}\ e^-} \times \frac{107.87\ \text{g Ag}}{1\ \text{mol Ag}} = 14.1\ \text{g Ag}$$

(d) $Ni^{2+}(aq) + 2\ e^- \rightarrow Ni(s)$

$$\text{no. g Ni} = 0.131\ \text{mol}\ e^- \times \frac{1\ \text{mol Ni}}{2\ \text{mol}\ e^-} \times \frac{58.71\ \text{g Ni}}{1\ \text{mol Ni}} = 3.85\ \text{g Ni}$$

20-14. (a) The oxidation of SO_3^{2-} in reaction (20.6) is

$SO_3^{2-} + H_2O \rightarrow SO_4^{2-} + 2\ H^+ + 2\ e^-$

$1\ \text{mol}\ Na_2SO_3 \Leftrightarrow 2\ \text{mol}\ e^-$

$\frac{1}{2}\ \text{mol}\ Na_2SO_3 \Leftrightarrow 1\ \text{mol}\ e^-$ eq. wt. = $\frac{1}{2} \times 126.06 = 63.0$

(b) The oxidation of Al in reaction (20.27) is

$Al \rightarrow Al^{3+} + 3\ e^-$

$1\ \text{mol Al} \Leftrightarrow 3\ \text{mol}\ e^-$ $\frac{1}{3}\ \text{mol Al} \Leftrightarrow 1\ \text{mol}\ e^-$ eq. wt. = $\frac{1}{3} \times 26.98 = 8.99$

(c) The reduction of PbO_2 in reaction (20.40) is

$PbO_2 + 4\ H^+ + SO_4^{2-} + 2\ e^- \rightarrow PbSO_4 + 2\ H_2O$

$1\ \text{mol}\ PbO_2 \Leftrightarrow 2\ \text{mol}\ e^-$

$\frac{1}{2}\ \text{mol}\ PbO_2 \Leftrightarrow 1\ \text{mol}\ e^-$ eq. wt. = $\frac{1}{2} \times 239.2 = 119.6$

(d) The reduction of $O_2(g)$ in reaction (20.41) is

$O_2(g) + 2\ H_2O + 4\ e^- \rightarrow 4\ OH^-(aq)$

$1\ \text{mol}\ O_2 \Leftrightarrow 4\ \text{mol}\ e^-$

$\frac{1}{4}\ \text{mol}\ O_2 \Leftrightarrow 1\ \text{mol}\ e^-$ eq. wt. = $\frac{1}{4} \times 32.00 = 8.00$

20-15. $$\text{no. eq Fe} = 0.02513\ \text{L} \times \frac{0.2821\ \text{eq}\ MnO_4^-}{\text{L}} \times \frac{1\ \text{eq}\ Fe^{2+}}{1\ \text{eq}\ MnO_4^-} \times \frac{1\ \text{eq Fe}}{1\ \text{eq}\ Fe^{2+}} = 7.089\times10^{-5}\ \text{eq Fe}$$

$$\text{no. g Fe} = 7.089\times10^{-3}\ \text{eq Fe} \times \frac{55.85\ \text{g Fe}}{1\ \text{eq Fe}} = 0.3959\ \text{g Fe}$$

$$\%\ \text{Fe} = \frac{0.3959\ \text{g Fe}}{0.8312\ \text{g ore}} \times 100 = 47.63\%\ \text{Fe}$$

Exercises

Definitions and terminology

20-1. (a) In an oxidation process some element undergoes an increase in oxidation state as a result of a substance losing electrons. In a reduction process a substance gains electrons and an element undergoes a decrease in its oxidation state.

(b) An oxidizing agent gains the electrons lost in an oxidation process; it is reduced. A reducing agent loses electrons and is itself oxidized, thereby making possible a reduction process.

(c) A half-reaction refers either to an oxidation or reduction process. When oxidation and reduction half-reactions are combined, a net reaction results.

(d) A voltaic (galvanic) cell produces electricity as a result of a spontaneous oxidation-reduction reaction occurring in an electrochemical cell. In an electrolytic cell a nonspontaneous reaction is produced by the use of electricity.

(e) An anode is an electrode in an electrochemical cell at which oxidation occurs. Reduction occurs at a cathode.

(f) E_{cell} is the electromotive force associated with an oxidation-reduction reaction occurring in an electrochemical cell. If the reactants and products are in their standard states, the value is $E°_{cell}$.

Oxidation-reduction reactions

20-2. (a) oxid: $2\{Fe_2S_3(s) + 6\ H_2O \rightarrow 2\ Fe(OH)_3(s) + 3\ S(s) + 6\ H^+ + 6\ e^-\}$

red: $3\{O_2(g) + 4\ H^+ + 4\ e^- \rightarrow 2\ H_2O\}$

net: $2\ Fe_2S_3(s) + 6\ H_2O + 3\ O_2(g) \rightarrow 4\ Fe(OH)_3(s) + 6\ S(s)$

(b) oxid: $3\{IBr + 3\ H_2O \rightarrow IO_3^- + Br^- + 6\ H^+ + 4\ e^-\}$

red: $2\{BrO_3^- + 6\ H^+ + 6\ e^- \rightarrow Br^- + 3\ H_2O\}$

net: $3\ IBr + 2\ BrO_3^- + 3\ H_2O \rightarrow 3\ IO_3^- + 5\ Br^- + 6\ H^+$

(c) oxid: $As_2S_3 + 40\ OH^- \rightarrow 2\ AsO_4^{3-} + 3\ SO_4^{2-} + 20\ H_2O + 28\ e^-$

red: $14\{H_2O_2 + 2\ e^- \rightarrow 2\ OH^-\}$

net: $As_2S_3(s) + 14\ H_2O_2 + 12\ OH^- \rightarrow 2\ AsO_4^{3-} + 3\ SO_4^{2-} + 20\ H_2O$

(d) oxid: $2\{CrI_3(s) + 32\ OH^- \rightarrow CrO_4^{2-} + 3\ IO_4^- + 16\ H_2O + 27\ e^-\}$

red: $27\{H_2O_2 + 2\ e^- \rightarrow 2\ OH^-\}$

net: $2\ CrI_3(s) + 27\ H_2O_2 + 10\ OH^- \rightarrow 2\ CrO_4^{2-} + 6\ IO_4^- + 32\ H_2O$

(e) oxid: $4\ OH^- \rightarrow O_2(g) + 2\ H_2O + 4\ e^-$

red: $2\{F_5SeOF + 6\ OH^- + 2\ e^- \rightarrow SeO_4^{2-} + 6\ F^- + 3\ H_2O\}$

net: $2\ F_5SeOF + 16\ OH^- \rightarrow 2\ SeO_4^{2-} + O_2(g) + 8\ H_2O + 12\ F^-$

(f) oxid: $4\{Ag(s) + 2\ CN^- \rightarrow [Ag(CN)_2]^- + e^-\}$

red: $O_2(g) + 2\ H_2O + 4\ e^- \rightarrow 4\ OH-$

net: $4\ Ag(s) + 8\ CN^- + O_2(g) + 2\ H_2O \rightarrow 4[Ag(CN)_2]^- + 4\ OH^-$

(g) oxid: $B_2Cl_4 + 8\ OH^- \rightarrow 2\ BO_2^- + 4\ Cl^- + 4\ H_2O + 2\ e^-$

red: $2\ H_2O + 2\ e^- \rightarrow H_2(g) + 2\ OH^-$

net: $B_2Cl_4 + 6\ OH^- \rightarrow 2\ BO_2^- + 2\ H_2O + 4\ Cl^- + H_2(g)$

(h) oxid: $3\{Sn \rightarrow Sn^{2+} + 2\ e^-\}$

red: $C_2H_5NO_3 + 6\ H^+ + 6\ e^- \rightarrow C_2H_5OH + NH_2OH + H_2O$

net: $C_2H_5NO_3 + 3\ Sn + 6\ H^+ \rightarrow 3\ Sn^{2+} + C_2H_5OH + NH_2OH + H_2O$

20-3. The power of an oxidizing agent is measured by the magnitude of the standard reduction potential. For example, when comparing $F_2(g) + 2\ e^- \rightarrow 2\ F^-(aq)$, $E^\circ_{red} = +2.87$ V; with $I_2(s) + 2\ e^- \rightarrow 2\ I^-(aq)$, $E^\circ_{red} = +0.535$ V. We conclude that $F_2(g)$ is a more powerful oxidizing agent that $I_2(s)$. For the species listed the order is

$Na^+(aq) < Zn^{2+}(aq) < I_2(s) < IO_3^-(aq) < PbO_2(s) < F_2(g)$.

20-4. Substances that have their central atom in its highest oxidation state can only be reduced; they can act only as oxidizing agents. If the central atom is in its lowest oxidation state, its oxidation state can only increase, that is, it can only be oxidized and thereby function as a reducing agent. Only if the oxidation state has an intermediate value can the substance act either as an oxidizing or reducing agent.

oxidizing agents only: Al^{3+}, $Cr_2O_7^{2-}$

reducing agents only: Zn, S^{2-}

either an oxidizing or a reducing agent: Fe^{2+}, I_2, S

Standard electrode potentials

Combine the known electrode potential for the reduction of dichromate ion with the unknown potential for the oxidation of palladium to obtain the measured cell potential for the oxidation-reduction reaction.

$3\ Pd \rightarrow 3\ Pd^{2+} + 6\ e^-$ $\quad E^\circ_{ox} = -E^\circ_{red}$

$Cr_2O_7^{2-} + 14\ H^+ + 6\ e^- \rightarrow 2\ Cr^{3+} + 7\ H_2O$ $\quad E^\circ_{red} = +1.33$ V

$3\ Pd + Cr_2O_7^{2-} + 14\ H^+ \rightarrow 3\ Pd^{2+} + 2\ Cr^{3+} + 7\ H_2O$ $\quad E^\circ_{cell} = +0.34$ V

$-E^\circ_{red} + 1.33\ V = +0.34\ V$ $\quad E^\circ_{red} = 1.33\ V - 0.34\ V = +0.99\ V$

20-5. (a) Because the metal does not dissolve in HCl(aq), its standard reduction potential must be positive. The metal displaces silver ion, so the reduction potential must be less than +0.800 V. It does not displace copper ion, so its reduction potential must be greater than +0.337 V. For the reduction process, $M^{2+}(aq) + 2\,e^- \rightarrow M(s)$, $+0.337 < E° < +0.800$ V.

(b) Because the metal dissolves in HCl(aq), its standard reduction potential must be negative. The fact that it displaces neither zinc ion (E° = -0.763 V) nor ferrous ion (E° = -0.440 V) means that $-0.440 < E° < 0.000$ V.

20-6. Arrange the available metal ion solutions in decreasing order of standard electrode potentials. Determine which metals indium displaces from solutions of their ions and which it does not. This will enable you to bracket the standard electrode potential between thw known values of two other metals. For example, indium is found to displace Sn^{2+} but not Fe^{2+}. This suggests that for the reduction process, $In^{3+}(aq) + 3\,e^- \rightarrow In(s)$, $-0.440 < E° < -0.136$ V. (The tabulated value is -0.345 V.)

20-7. Consider the second reaction first. Express it as the sum of two half-reactions.

oxid: $V^{3+} + H_2O \rightarrow VO^{2+} + 2\,H^+ + e^-$ $\quad E°_{ox} = -E°_{red} = ?$

red: $Ag^+ + e^- \rightarrow Ag(s)$ $\quad E°_{red} = +0.800$ V

net: $V^{3+} + Ag^+ + H_2O \rightarrow VO^{2+} + 2\,H^+ + Ag(s)$ $\quad E°_{cell} = +0.439$ V

$E°_{cell} = 0.799\text{ V} - E°_{red} = +0.439\text{ V} \qquad E°_{red} = 0.800 - 0.439 = +0.361\text{ V}$

Now write the first reaction as the sum of two half-reactions.

oxid: $V^{2+} \rightarrow V^{3+} + e^-$ $\quad E°_{ox} = -E°_{red} = ?$

red: $VO^{2+} + 2\,H^+ + e^- \rightarrow V^{3+} + H_2O$ $\quad E°_{red} = +0.361$ V

net: $V^{2+} + VO^{2+} + 2\,H^+ \rightarrow 2\,V^{3+} + H_2O$ $\quad E°_{cell} = +0.616$ V

$E°_{cell} = 0.361\text{ V} - E°_{red} = 0.616\text{ V} \qquad E°_{red} = 0.361 - 0.616 = -0.255\text{ V}$

The standard electrode potential for the reduction process, $V^{3+} + e^- \rightarrow V^{2+}$ is -0.255 V.

Predicting oxidation-reduction reactions

20-8. A reduction that can occur in aqueous solution, and which does occur more readily than the reduction of Mg^{2+}, is: $2\,H_2O + 2\,e^- \rightarrow H_2(g) + 2\,OH^-$. Or, considered in another way, the favored reaction below is the one with the most positive value of $E°_{cell}$.

(1) $2\,Na(s) + Mg^{2+}(aq) \rightarrow Mg(s) + 2\,Na^+(aq)$ $\quad E°_{cell} = +0.339$ V

(2) $2\,Na(s) + 2\,H_2O \rightarrow 2\,Na^+(aq) + 2\,OH^-(aq) + H_2(g)$ $\quad E°_{cell} = +1.886$ V

20-9. (a) To dissove in HCl(aq), Cu(s) would have to liberate $H_2(g)$. This is a reaction with $E°_{cell} < 0$. It is *nonspontaneous*.

$Cu(s) + 2\,H^+(aq) \rightarrow Cu^{2+}(aq) + H_2(g)$ $\quad E°_{cell} = -0.337$ V

(b) The reduction half-reaction when Cu(s) is dissolved in $HNO_3(aq)$ is not the reduction of $H^+(aq)$ to $H_2(g)$, but the reduction of $NO_3^-(aq)$ to NO(g). The resulting oxidation-reduction has $E°_{cell} > 0$. It is spontaneous.

oxid. $3\{Cu(s) \rightarrow Cu^{2+}(aq) + 2\ e^-\}$ $E^\circ_{ox} = -(+0.34) = -0.34$ V

red: $2\{NO_3^-(aq) + 4\ H^+ + 3\ e^- \rightarrow 2\ H_2O + NO(g)\}$ $E^\circ_{red} = +0.96$ V

net: $3\ Cu(s) + 2\ NO_3^- + 8\ H^+ \rightarrow 3\ Cu^{2+} + 4\ H_2O + 2\ NO(g)$ $E^\circ_{cell} = +0.62$ V

20-10. (a) If the strip of copper metal is weighed before and after the reaction, its mass is found to remain unchanged. Copper does not dissolve in HCl(aq).

(b) The reaction is $Zn(s) + 2\ H^+(aq) \rightarrow Zn^{2+}(aq) + H_2(g)$.

(c) Electrons lost in the oxidation of zinc [$Zn(s) \rightarrow Zn^{2+}(aq) + 2\ e^-$] are conducted to the copper metal, where hydrogen gas is evolved [$2\ H^+(aq) + 2\ e^- \rightarrow H_2(g)$].

20-11. (a) No reaction would occur since neither copper nor silver can displace H_2 from an acid solution.

(b) Both metals can dissolve in $HNO_3(aq)$. The oxidizing agent is nitrate ion. Oxides of nitrogen (NO, NO_2), but no hydrogen, are produced.

$3\ Cu(s) + 8\ H^+ + 2\ NO_3^- \rightarrow 3\ Cu^{2+}(aq) + 4\ H_2O + 2\ NO(g)$ $E^\circ_{cell} = +0.62$ V

$3\ Ag(s) + 4\ H^+ + NO_3^- \rightarrow 3\ Ag^+(aq) + 2\ H_2O + NO(g)$ $E^\circ_{cell} = +0.16$ V

We should expect the dissolving of Cu to proceed more readily. Electrons from the copper are transferred to the Ag, where the reduction of NO_3^- occurs.

20-12. oxid: $H_2O_2(aq) \rightarrow O_2(g) + 2\ H^+(aq) + 2\ e^-$

red: $H_2O_2(aq) + 2\ H^+(aq) + 2\ e^- \rightarrow 2\ H_2O$

net: $2\ H_2O_2(aq) \rightarrow 2\ H_2O + O_2(g)$

Voltaic (galvanic) cells

20-13. (anode) $\xrightarrow{\text{electron flow}}$ (cathode)

(a) $Cu(s)\ |\ Cu^{2+}(aq)\ ||\ Fe^{3+}(aq),\ Fe^{2+}(aq)\ |\ Pt(s)$

oxid: $Cu(s) \rightarrow Cu^{2+}(aq) + 2\ e^-$ $E^\circ_{ox} = -0.337$ V

red: $2\{Fe^{3+}(aq) + e^- \rightarrow Fe^{2+}(aq)\}$ $E^\circ_{red} = +0.771$ V

net: $Cu(s) + 2\ Fe^{3+}(aq) \rightarrow Cu^{2+}(aq) + 2\ Fe^{2+}(aq)$ $E^\circ_{cell} = +0.434$ V

(anode) $\xrightarrow{\text{electron flow}}$ (cathode)

(b) $Pt\ |\ Sn^{2+}(aq),\ Sn^{4+}(aq)\ ||\ Cr_2O_7^{2-}(aq),\ Cr^{3+}(aq)\ |\ Pt$

oxid: $3\{Sn^{2+}(aq) \rightarrow Sn^{4+}(aq) + 2\ e^-\}$ $E^\circ_{ox} = -0.154$ V

red: $Cr_2O_7^{2-}(aq) + 14\ H^+ + 6\ e^- \rightarrow 2\ Cr^{3+}(aq) + 7\ H_2O$ $E^\circ_{red} = +1.33$ V

net: $3\ Sn^{2+} + Cr_2O_7^{2-} + 14\ H^+ \rightarrow 3\ Sn^{4+} + 2\ Cr^{3+} + 7\ H_2O$ $E^\circ_{cell} = 1.18$ V

(anode) —— electron flow ——→ (cathode)

(c) Pt, $O_2(g)$ | H_2O, H^+ || $Cl^-(aq)$ | $Cl_2(g)$, Pt

oxid: $2\ H_2O \rightarrow O_2(g) + 4\ H^+ + 4\ e^-$ $E^\circ_{ox} = -1.229$ V

red: $2\{Cl_2(g) + 2\ e^- \rightarrow 2\ Cl^-(aq)\}$ $E^\circ_{red} = +1.360$ V

---

net: $2\ Cl_2(g) + 2\ H_2O \rightarrow 4\ H^+ + 4\ Cl^- + O_2(g)$ $E^\circ_{cell} = +0.131$ V

20-14. (a) the result we wish to achieve here is:

oxid: $Fe(s) \rightarrow Fe^{2+} + 2\ e^-$ $E^\circ_{ox} = +0.440$ V

red: $Cl_2(g) + 2\ e^- \rightarrow 2\ Cl^-$ $E^\circ_{red} = +1.360$ V

---

net: $Fe(s) + Cl_2(g) \rightarrow Fe^{2+} + 2\ Cl^-$ $E^\circ_{cell} = 1.800$ V

The cell in which the reaction occurs is: Fe(s) | $Fe^{2+}(aq)$ || $Cl^-(aq)$ | $Cl_2(g)$, Pt(s)

(b) If Zn^{2+} is to be reduced to Zn(s), for which $E^\circ_{red} = -0.763$ V, we must select an oxidation couple for which $E^\circ_{ox} > +0.763$ V. For example:

oxid: $Mg(s) \rightarrow Mg^{2+}(aq) + 2\ e^-$ $E^\circ_{ox} = +2.375$ V

red: $Zn^{2+}(aq) + 2\ e^- \rightarrow Zn(s)$ $E^\circ_{red} = -0.763$ V

---

net: $Mg(s) + Zn^{2+}(aq) \rightarrow Mg^{2+}(aq) + Zn(s)$ $E^\circ_{cell} = +1.612$ V

Mg(s) | $Mg^{2+}(aq)$ || $Zn^{2+}(aq)$ | Zn(s)

(c) The required half-reactions are:

oxid: $Cu^+(aq) \rightarrow Cu^{2+}(aq) + e^-$

red: $Cu^+(aq) + e^- \rightarrow Cu(s)$

---

net: $2\ Cu^+(aq) \rightarrow Cu^{2+}(aq) + Cu(s)$

Pt | $Cu^+(aq)$, $Cu^{2+}(aq)$ || $Cu^+(aq)$ | Cu(s)

Note that whereas the cathode can be made of Cu(s), the anode cannot. Cu(s) is not involved in the oxidation half-reaction. Since E° for the couple Cu^{2+}/Cu^+ is not listed in Table 20-2, an E°_{cell} value cannot be calculated.

$\Delta\bar{G}$, E°_{cell}, and K

20-15. (a) oxid: $2\{Al(s) \rightarrow Al^{3+} + 3\ e^-\}$ $E^\circ_{ox} = +1.66$ V

red: $3\{Zn^{2+} + 2\ e^- \rightarrow Zn(s)\}$ $E^\circ_{red} = -0.763$ V

---

net: $2\ Al(s) + 3\ Zn^{2+} \rightarrow 2\ Al^{3+} + 3\ Zn(s)$ $E^\circ_{cell} = 0.90$ V

$\Delta\bar{G}^\circ = -n\mathcal{F}E^\circ_{cell} = -6 \times 96{,}500\ C/mol \times 0.90\ V = -5.2\times10^5\ J/mol = -5.2\times10^2\ kJ/mol$

(b) oxid: $5\{Pb^{2+} + 2\ H_2O \rightarrow PbO_2(s) + 4\ H^+ + 2\ e^-\}$ $E^\circ_{ox} = -1.455$ V

red: $2\{MnO_4^- + 8\ H^+ + 5\ e^- \rightarrow Mn^{2+} + 4\ H_2O\}$ $E^\circ_{red} = +1.51$ V

net: $5\ Pb^{2+} + 2\ MnO_4^- + 2\ H_2O \rightarrow 5\ PbO_2(s) + 2\ Mn^{2+} + 4\ H^+$ $E^\circ_{cell} = 0.05$ V

$\Delta\bar{G}^\circ = -n\mathcal{F}E^\circ_{cell} = -10 \times 96{,}500$ C/mol $\times\ 0.05$ V $= -5\times10^4$ J/mol $= -5\times10^1$ kJ/mol

(c) oxid: $2\ Cl^- \rightarrow Cl_2(g) + 2\ e^-$ $E^\circ_{ox} = -1.36$ V

red: $MnO_2(s) + 4\ H^+ + 2\ e^- \rightarrow Mn^{2+} + 2\ H_2O$ $E^\circ_{red} = +1.23$ V

net: $MnO_2(s) + 4\ H^+ + 2\ Cl^- \rightarrow Mn^{2+} + 2\ H_2O + Cl_2(g)$ $E^\circ_{cell} = -0.13$ V

$\Delta\bar{G}^\circ = -n\mathcal{F}E^\circ_{cell} = -2 \times 96{,}500$ C/mol $\times (-0.13)$V $= +2.5\times10^4$ J/mol $= +25$ kJ/mol

20-16. $\Delta\bar{G}^\circ = -2.303\cdot RT\cdot\log K = -5.2\times10^5$ J/mol

$$\log K = \frac{-5.2\times10^5 \text{ J/mol}}{-2.303 \times 8.314 \text{ K mol}^{-1}\text{ K}^{-1} \times 298 \text{ K}} = 91 \qquad K = 1\times10^{91}$$

	2 Al(s)	+	$3\ Zn^{2+}$	$\rightleftharpoons$	$2\ Al^{3+}$	+	3 Zn(s)
Initial:			1.0 M		--		
Changes:			$(-1.0 - x)$M		$+2/3(1.0 - x)$M		
Equil:			x M		$2/3(1.0 - x)$M		
Assume x << 1.0:			x M		0.67 M		

$$K = \frac{[Al^{3+}]^2}{[Zn^{2+}]^3} = \frac{(0.67)^2}{x^3} = 1\times10^{91} \qquad x^3 = 4\times10^{-92} \qquad x = [Zn^{2+}] = 3\times10^{-31}\text{ M}$$

The displacement of Zn^{2+} by Al(s) goes to completion.

20-17. oxid: $Fe(s) \rightarrow Fe^{2+} + 2\ e^-$ $E^\circ_{ox} = +0.440$ V

red: $2\{Cr^{3+} + e^- \rightarrow Cr^{2+}\}$ $E^\circ_{red} = -0.407$ V

net: $Fe(s) + 2\ Cr^{3+} \rightarrow Fe^{2+} + 2\ Cr^{2+}$ $E^\circ_{cell} = +0.033$ V

$\Delta\bar{G}^\circ = -n\mathcal{F}E^\circ_{cell} = -2.303\ RT\ \log K$

$$\log K = \frac{n\mathcal{F}E^\circ_{cell}}{2.303\ RT} = \frac{2 \times 96{,}500 \text{ C/mol} \times 0.033 \text{ V}}{2.303 \times 8.314 \text{ J mol}^{-1}\text{ K}^{-1} \times 298} = 1.1 \qquad K = 13$$

$$Fe(s) \quad + \quad 2\ Cr^{3+} \rightleftharpoons Fe^{2+} \quad + \quad 2\ Cr^{2+}$$

Initial:		1.00 M	--	
Changes:		$-2x$ M	$+x$ M	$+2x$ M
Equil.		$(1.00 - 2x)$M	x M	$2x$ M

$$K = \frac{[Fe^{2+}][Cr^{2+}]^2}{[Cr^{3+}]^2} = \frac{x\ (2x)^2}{(1.00 - 2x)^2} = 13$$

We are seeking a value of x that will make the above expression equal to 13. Note that x must be less than 0.50 (which corresponds to $[Cr^{3+}] = 1.00 - 2x = 0$).

Try $x = 0.40$: $\dfrac{0.40(0.80)^2}{(1.00 - 0.80)^2} = 6.4 < 13$ Try $x = 0.41$: $\dfrac{0.41(0.82)^2}{(1.00 - 0.82)^2} = 8.5 < 13$

Try $x = 0.42$: $\dfrac{0.42(0.84)^2}{(1.00 - 0.84)^2} = 11.6 < 13$ Try $x = 0.43$: $\dfrac{9.43(0.86)^2}{(1.00 - 0.86)^2} = 16.2 > 13$

Our result is $x = [Fe^{2+}] = 0.42$ M

Concentration dependence of E_{cell}--the Nernst equation

20-18. (a) oxid: $Fe(s) \rightarrow Fe^{2+}(aq) + 2\ e^-$ $E^\circ_{ox} = +0.440$ V

red: $Sn^{2+}(aq) + 2\ e^- \rightarrow Sn(s)$ $E^\circ_{red} = -0.136$ V

net: $Fe(s) + Sn^{2+}(aq) \rightarrow Fe^{2+}(aq) + Sn(s)$ $E^\circ_{cell} = +0.304$ V

$$E_{cell} = E^\circ_{cell} - \frac{0.0592}{n} \log \frac{[Fe^{2+}]}{[Sn^{2+}]}$$ where $n = 2$; $[Fe^{2+}] = 0.20$ M; $[Sn^{2+}] = 0.0050$ M

$$E_{cell} = 0.304 - \frac{0.0592}{2} \log \frac{0.20}{0.050} = 0.304 - 0.0296 \log 4.0 = 0.304 - 0.018 = +0.286 \text{ V}$$

(b) oxid: $2\ Cl^-(aq) \rightarrow Cl_2(g) + 2\ e^-$ $E^\circ_{ox} = -1.360$ V

red: $2\{Ag^+(aq) + e^- \rightarrow Ag(s)\}$ $E^\circ_{red} = +0.800$ V

net: $2\ Ag^+(aq) + 2\ Cl^-(aq) \rightarrow 2\ Ag(s) + Cl_2(g)$ $E^\circ_{cell} = -0.560$ V

$$E_{cell} = E^\circ_{cell} - \frac{0.0592}{n} \log \frac{P_{Cl_2}}{[Ag^+]^2[Cl^-]^2}$$ where $n = 2$; $P_{Cl_2} = 0.50$ atm; $[Ag^+] = 0.35$ M;

$[Cl^-] = 1.2$ M

$$E_{cell} = -0.560 - \frac{0.0592}{2} \log \frac{0.50}{(0.35)^2(1.2)^2} = -0.560 - 0.0296 \log 2.8 = -0.360 - 0.013 = -0.573 \text{ V}$$

(c) oxid: $Mg(s) \rightarrow Mg^{2+}(aq) + 2\ e^-$ $E^\circ_{ox} = +2.375$ V

red: $2\ H_2O + 2\ e^- \rightarrow H_2(g) + 2\ OH^-$ $E^\circ_{red} = -0.828$ V

net: $Mg(s) + 2\ H_2O \rightarrow Mg^{2+}(aq) + 2\ OH^-(aq) + H_2(g)$ $E^\circ_{cell} = +1.547$ V

$$E_{cell} = 1.547 - \frac{0.0592}{2} \log [Mg^{2+}][OH^-]^2\ P_{H_2} = 1.547 - 0.0296 \log (0.012)(0.75)^2(0.50)$$

$$E_{cell} = 1.547 - 0.0296 \log(3.4\times10^{-3}) = 1.547 + 0.0731 = +1.620 \text{ V}$$

20-19. oxid: $2\{Mn^{2+} + 2\ H_2O \rightarrow MnO_2(s) + 4\ H^+ + 2\ e^-\}$ $E^\circ_{ox} = -1.23$ V

red: $O_2(g) + 4\ H^+ + 4\ e^- \rightarrow 2\ H_2O$ $E^\circ_{red} = +1.229$ V

net: $2\ Mn^{2+} + O_2(g) + 2\ H_2O \rightarrow 2\ MnO_2(s) + 4\ H^+$ $E^\circ_{cell} = 0.00$ V

The reaction will be essentially at equilibrium if all reactants and products are in their standard states.

(a) $[H^+] = 5.5$ M. Increasing $[H^+]$ above its standard state value favors the reverse of the reaction shown. The forward reaction would be nonspontaneous.

(b) $[H^+] = 1.0$ M. This is the condition under which the reaction is essentially in a state of equilibrium, with all reactants and products in their standard states and $E_{cell} = E^\circ_{cell} = 0$.

(c) $[H^+] = 0.050$ M. This condition, according to Le Chatelier's principle, should favor the forward reaction.

(d) pH = 9.13; $[H_3O^+] = 7.4\times10^{-10}$. Following the reasoning in part (c), this condition should very much favor the forward reaction.

20-20. oxid: $3\{2\ Cl^- \rightarrow Cl_2(g) + 2\ e^-\}$ $E^\circ_{ox} = -1.36$ V

red: $Cr_2O_7^{2-} + 14\ H^+ + 6\ e^- \rightarrow 2\ Cr^{3+} + 7\ H_2O$ $E^\circ_{red} = +1.33$ V

net: $Cr_2O_7^{2-} + 6\ Cl^- + 14\ H^+ \rightarrow 2\ Cr^{3+} + 3\ Cl_2(g) + 7\ H_2O$ $E^\circ_{cell} = -0.03$ V

Because $E^\circ_{cell} < 0$, the reaction is not spontaneous if all reactants and products are in their standard states. The reason that this reaction can be used as a laboratory preparation of $Cl_2(g)$ is that as the gas escapes (especially if the reaction mixture is heated) the forward reaction must proceed to replace the lost $Cl_2(g)$. The reaction never reaches a condition of equilibrium.

20-21. (a) The power of an oxidizing agent is pH dependent for every reduction half-reaction that involves H^+. Of the species listed, this includes $O_2(g)$, $MnO_4^-(aq)$ and $H_2O_2(aq)$.

(b) the oxidizing power is independent of pH if H^+ does not appear in the reduction half-reaction at all. This is the case with $Cl_2(g)$ and $F_2(g)$.

(c) If the reduction half-equation has H^+ appearing on the left side, the power of the oxidizing agent increases in acidic solution. This is the case with the following:

$O_2(g) + 4\ H^+(aq) + 4\ e^- \rightarrow 2\ H_2O$

$MnO_4^-(aq) + 8\ H^+(aq) + 5\ e^- \rightarrow Mn^{2+}(aq) + 4\ H_2O$

$H_2O_2(aq) + 2\ H^+(aq) + 2\ e^- \rightarrow 2\ H_2O$

In order for the oxidizing agent to be more powerful in basic solution, the reduction half-reaction would have to have H^+ appearing on the right side of the half-equation. This is a situation that we would not frequently expect to encounter.

20-22. The reaction of interest in this exercise is:

$$Zn(s) + Cu^{2+}(aq) \rightarrow Cu(s) + Zn^{2+}(aq) \qquad E^\circ_{cell} = 1.10 \text{ V}$$

The dependence of the cell potential on ion concentrations is given by the Nernst equation.

$$E_{cell} = 1.10 - \frac{0.0592}{2} \log \frac{[Zn^{2+}]}{[Cu^{2+}]}$$

(a) If $[Zn^{2+}] = 1.0$ M, the smallest concentration of Cu^{2+} for which the reaction is still spontaneous is obtained by solving the expression:

$$0.00 = 1.10 - \frac{0.0592}{2} \log \frac{1}{[Cu^{2+}]} \qquad -1.10 = \frac{0.0592}{2} \log [Cu^{2+}]$$

$$\log [Cu^{2+}] = \frac{-2 \times 1.10}{0.0592} = -37.2 \qquad [Cu^{2+}] = 6\times10^{-38}$$

(b) The concentration of Cu^{2+} calculated in part (a) is immeasurably small. The displacement of copper metal from an aqueous solution of copper ion by zinc metal is a reaction that goes to completion.

20-23. The spontaneous cell reaction is the sum of these two half-reactions:

oxid: $H_2(g, 1 \text{ atm}) \rightarrow 2\ H^+(\text{in } 0.85 \text{ M KOH}) + 2\ e^-$

red: $2\ H^+(1.0 \text{ M}) + 2\ e^- \rightarrow H_2(g, 1 \text{ atm})$

net: $2\ H^+(1.0 \text{ M}) \rightarrow 2\ H^+(\text{in } 0.85 \text{ M KOH})$

$$E_{cell} = E^\circ_{cell} - \frac{0.0592}{2} \log \frac{[H^+]^2}{(1.0)^2} = 0.00 - 0.0592 \log [H^+]$$

(a) In 0.85 M KOH, $[OH^-] = 0.85$ and $[H^+] = 1.2\times10^{-14}$

$$E_{cell} = -0.0592 \log (1.2\times10^{-14}) = 13.92 \times 0.0592 = +0.824 \text{ V}$$

(b) The value of E° (see Table 20-2) for the reduction half-reaction, $2\ H_2O + 2\ e^- \rightarrow H_2(g) + 2\ OH^-$, is -0.828 V. This is for the case where $[OH^-]$ is 1 M, $[H^+] = 10^{-14}$ M and $P_{H_2} = 1$ atm. The calculation in (a) was for the oxidation of $H_2(g)$ in a very similar solution; the value obtained is almost exactly the negative of the standard electrode potential. (It is slightly smaller than +0.828 V because $[H_3O^+]$ in 0.85 M KOH is slightly larger than in 1.00 M KOH.)

20-24. Substitution of 0.85 M NH_3 for 0.85 M KOH in Exercise 20-23 means that the cell reaction is: $2\ H^+(1.0 \text{ M}) \rightarrow 2\ H^+(\text{in } 0.85 \text{ M } NH_3)$. The Nernst expression for the cell reaction remains unchanged.

$$E_{cell} = E^\circ_{cell} - \frac{0.0592}{2} \log \frac{[H^+]^2}{(1.0)^2} = 0.00 - 0.0592 \log [H^+]$$

(a) Since $[H^+]$ in 0.85 M NH_3 is *larger* than in 0.85 M KOH, E_{cell} is *lower*.

(b) To determine the value of E_{cell} requires first a calculation of $[H^+]$ in 0.85 M NH_3.

$$\frac{[NH_4^+][OH^-]}{[NH_3]} = \frac{[OH^-]^2}{0.85} = 1.74\times10^{-5} \qquad [OH^-]^2 = 1.5\times10^{-5}$$

$$[OH^-] = 3.9\times10^{-3} \qquad [H^+] = \frac{1.0\times10^{-14}}{3.9\times10^{-3}} = 2.6\times10^{-12}$$

$$E_{cell} = -0.0592\cdot\log 2.6\times10^{-12} = 0.686\text{ V}$$

20-25. oxid: $Ag(s) \rightarrow Ag^+(\text{satd. } Ag_2SO_4(aq)) + e^-$

red: $Ag^+(0.125\text{ M}) + e^- \rightarrow Ag(s)$

---

net: $Ag^+(0.125\text{ M}) \rightarrow Ag^+(\text{satd. } Ag_2SO_4(aq))$

$$K_{sp}(Ag_2SO_4) = [Ag^+]^2[SO_4^{2-}] = 1.4\times10^{-5}$$

let $x = [SO_4^{2-}]$ and $2x = [Ag^+]$ $\qquad (2x)^2(x) = 4x^3 = 1.4\times10^{-5}$

$x = 1.5\times10^{-2}$ $\qquad [Ag^+] = 2x = 3.0\times10^{-2}$ M

net: $Ag^+(0.125\text{ M}) \rightarrow Ag^+(0.030\text{ M})$

$$E_{cell} = E^\circ_{cell} - \frac{0.0592}{n}\log\frac{0.030}{0.125} = 0.00 - \frac{0.0592}{1}\log 0.24 = +0.037\text{ V}$$

20-26. First determine E°_{cell}

oxid: $Sn(s) \rightarrow Sn^{2+}(aq) + 2\,e^-$ $\qquad E^\circ_{ox} = +0.136$ V

red: $Pb^{2+}(aq) + 2\,e^- \rightarrow Pb(s)$ $\qquad E^\circ_{red} = -0.126$ V

---

net: $Sn(s) + Pb^{2+}(aq) \rightarrow Sn^{2+}(aq) + Pb(s)$ $\qquad E^\circ_{cell} = +0.010$ V

(a) $E_{cell} = E^\circ_{cell} - \dfrac{0.0592}{n}\log\dfrac{[Sn^{2+}]}{[Pb^{2+}]}$ where n = 2; $[Sn^{2+}] = 0.150$ M; $[Pb^{2+}] = 0.550$ M

$$E_{cell} = +0.010 - \frac{0.0592}{2}\log\frac{0.150}{0.550} = +0.010 + 0.017 = +0.027\text{ V}$$

(b) As the cell operates spontaneously (forward direction), $[Sn^{2+}]$ increases and $[Pb^{2+}]$ decreases; log $[Sn^{2+}]/[Pb^{2+}]$ increases; and E_{cell} decreases. [According to Le Chatelier's principle the reaction should become less spontaneous (E_{cell} decreases) as the concentrations on the left side of the equation decrease and those on the right increase.]

(c) When $[Pb^{2+}]$ falls from 0.550 M to 0.500 M, $[Sn^{2+}]$ increases from 0.150 M to 0.200 M.

$$E_{cell} = +0.010 - \frac{0.0592}{2}\log\frac{0.200}{0.500} = +0.010 + 0.012 = +0.022\text{ V}$$

(d) Let the no mol/L of Pb^{2+} replaced by Sn^{2+} = x. At the point in question $[Pb^{2+}]$ = $(0.550 - x)$M and $[Sn^{2+}] = (0.150 + x)$M.

$$E_{cell} = 0.020 = 0.010 - \frac{0.0592}{2}\log\left(\frac{0.150 + x}{0.500 - x}\right)$$

$$\log\left(\frac{0.150 + x}{0.550 - x}\right) = \frac{-2 \times 0.010}{0.0592} = -0.34 \qquad \frac{0.150 + x}{0.550 - x} = 0.46$$

$$0.150 + x = 0.25 - 0.46x \qquad 1.46x = 0.10 \qquad x = 0.068 \text{ M}$$

$$[Sn^{2+}] = (0.150 + x) = 0.150 + 0.068 = 0.218 \text{ M}$$

(e) Use a set up similar to (d) but set $E_{cell} = 0$

According to the rules for significant figures, the underlined digits are not significant. However, if they are dropped, E_{cell} comes far from equalling zero.

$$0.00 = 0.010 - \frac{0.0592}{2} \log\left(\frac{0.150 + x}{0.550 - x}\right)$$

$$\log\left(\frac{0.150 + x}{0.550 - x}\right) = \frac{2 \times 0.010}{0.0592} = 0.34 \qquad \frac{0.150 + x}{0.550 - x} = 2.2$$

$$0.150 + x = 1.2\underline{1} - 2.2x \qquad 3.2x = 1.0\underline{6} \qquad x = 0.3\underline{3}$$

$$[Sn^{2+}] = 0.150 + x = 0.4\underline{8} \text{ M}; \qquad [Pb^{2+}] = 0.550 - x = 0.2\underline{2} \text{ M}$$

20-27. oxid: $Ag(s) \rightarrow Ag^+(\text{satd. AgI(aq)}) + e^-$

red: $Ag^+(\text{satd. AgCl}, x \text{ M } Cl^-) + e^- \rightarrow Ag(s)$

net: $Ag^+(\text{satd. AgCl}, x \text{ M } Cl^-) \rightarrow Ag^+(\text{satd. AgI(aq)})$

$$K_{sp}(AgI) = [Ag^+][I^-] = [Ag^+]^2 = 8.5\times10^{-17} \qquad [Ag^+] = 9.2\times10^{-9} \text{ M}$$

$$E_{cell} = E°_{cell} - \frac{0.0592}{1} \log \frac{9.2\times10^{-9}}{[Ag^+]}$$ ← This is $[Ag^+]$ in saturated AgCl, x M Cl^-

$$0.110 = 0.00 - 0.0592 (\log 9.2\times10^{-9} - \log[Ag^+])$$

$$0.110 = -0.0592 \times (-8.04) + 0.0592 \log [Ag^+]$$

$$\log [Ag^+] = \frac{0.110 - 0.476}{0.0592} = -6.18 \qquad [Ag^+] = 6.6\times10^{-7}$$

$$K_{sp}(AgCl) = [Ag^+][Cl^-] = (6.6\times10^{-7})[Cl^-] = 1.6\times10^{-10}$$

$$[Cl^-] = 2.4\times10^{-4} \text{ M}$$

Batteries and fuel cells

20-28. *Anode (oxid)* $Cd(s) + 2\ OH^-(aq) \rightarrow Cd(OH)_2(s) + 2\ e^-$

Cathode (red) $2\{Ni(OH)_3(s) + e^- \rightarrow Ni(OH)_2(s) + OH^-(aq)\}$

Discharge reaction $Cd(s) + 2\ Ni(OH)_3(s) \rightarrow Cd(OH)_2(s) + 2\ Ni(OH)_2(s)$

20-29. *Anode (oxid)* $CH_4(g) + 4\ O^{2-} \rightarrow CO_2(g) + 2\ H_2O(g) + 8\ e^-$

Cathode (red) $2\ \{O_2(g) + 4\ e^- \rightarrow 2\ O^{2-}\}$

Net reaction $CH_4(g) + 2\ O_2(g) \rightarrow CO_2(g) + 2\ H_2O(g)$

Electrochemical mechanism of corrosion

20-30. In a galvanic cell electrons lost at the anode in an oxidation half-reaction are conveyed through an external electrical circuit to the cathode, where a reduction half-reaction occurs. In a corroding metal an oxidation half-reaction occurs in one region (the anodic reaction) and electrons are transferred to another region, where a reduction half-reaction occurs (the cathodic reaction). The two processes are quite similar. The essential difference is that the galvanic cell reaction is a desirable one that produces useful work whereas the corrosion reaction is undesirable and does no useful work.

20-31. (a) Oxidation continues to occur at the exposed ends of the iron nail, yielding a blue precipitate. The cathodic (reduction) half-reaction occurs on the copper wire. The wire develops a pink color, signaling the presence of OH^-.

(c) The zinc metal oxidizes (yielding a precipitate of zinc ferricyanide). The reduction half-reaction occurs on the iron nail, detectable by the characteristic color of phenolphtalein in a basic medium. No corrosion of the iron occurs; no precipitate appears on the iron.

20-32. The combination of the transmission pipe and the inert electrode functions as an electrolysis cell. The inert electrode is made to be the anode (positive electrode). Whatever oxidation half-reaction can occur does so on this electrode. The pipe is made the cathode (negative electrode). Here a reduction half-reaction occurs, but this does not affect the metal. (A metal corrodes by oxidation, not reduction.)

Electrolysis reactions

20-33. In an electrolysis reaction we generally expect that whatever combination of oxidation and reduction half-reactions can proceed with the smallest applied potential difference will occur. In the electrolysis of $MgCl_2$ the anode (oxidation) reaction is the same whether in concentrated aqueous solution or in the pure molten salt. But the reduction of water in aqueous solutions proceeds more readily than that of Mg^{2+}.

Aqueous soln. $Mg^{2+} + 2\ Cl^- + 2\ H_2O \rightarrow Mg^{2+} + 2\ OH^- + H_2(g) + Cl_2(g)$

$E^\circ_{cell} = -2.19\ V$

Molten state $Mg^{2+} + 2\ Cl^- \rightarrow Mg(l) + Cl_2(g)$

2-34. (a) The gases produced by overcharging a lead storage battery are $O_2(g)$ and $H_2(g)$.

(b) *Anode (oxid)* $2\ H_2O \rightarrow O_2(g) + 4\ H^+ + 4\ e^-$

Cathode (red) $2\{2\ H^+ + 2\ e^- \rightarrow H_2(g)\}$

Net reaction $2\ H_2O \rightarrow O_2(g) + 2\ H_2(g)$

20-35. The key to complete separation of the ions by electrolysis is a large difference in their E° values.

(a) The reduction $Cu^{2+} + 2\ e^- \rightarrow Cu(s)$ (E° = +0.337 V) occurs readily. The reduction $K^+ + e^-$ K(s) (E° = -2.925 V) occurs with great difficulty. In fact, in aqueous solutions H_2O is reduced, not K^+. As a result, separation of Cu^{2+} and K^+ can easily be achieved by electrolysis.

(b) Here the comparative E° values for reduction of the cations are +0.800 V for Ag^+ and +0.337 V for Cu^{2+}. This is a larger difference, with Ag^+ being reduced more easily than Cu^{2+}. From a solution containing both Ag^+ and Cu^{2+} we should expect Ag^+ to deposit more readily than Cu^{2+} and for the deposition of Ag^+ to go essentially to completion before Cu^{2+} deposits.

(c) Here the reduction potentials are nearly alike in magnitude. Although Pb^{2+} is reduced a little more easily than Sn^{2+}, as the $[Pb^{2+}]$ decreases the potential for its reduction becomes smaller. It soon reaches that for the reduction of Sn^{2+} and the two metals deposit together at this point. A complete separation of the cations is not possible by electrolysis.

20-36. (a) $Zn^{2+} + 2\ e^- \rightarrow Zn(s)$

$$\text{no. g Zn} = \frac{1.05\ \text{C}}{\text{s}} \times 756\ \text{s} \times \frac{1\ \text{mol}\ e^-}{96.500\ \text{C}} \times \frac{1\ \text{mol Zn}}{2\ \text{mol}\ e^-} \times \frac{65.38\ \text{g Zn}}{1\ \text{mol Zn}} = 0.269\ \text{g Zn}$$

(b) $$\text{no. mol}\ e^- = 2.18\ \text{g}\ I_2 \times \frac{1\ \text{mol}\ I_2}{253.8\ \text{g}\ I_2} \times \frac{2\ \text{mol}\ e^-}{1\ \text{mol}\ I_2} = 0.0172\ \text{mol}\ e^-$$

$$\text{no. C} = 0.0172\ \text{mol}\ e^- \times \frac{96{,}500\ \text{C}}{1\ \text{mol}\ e^-} = 1.66 \times 10^3\ \text{C}$$

$$\text{no. s} = 1.66 \times 10^3\ \text{C} \times \frac{1\ \text{s}}{4.28\ \text{C}} = 388\ \text{s}$$

(c) First determine the no. mol Cu deposited

$$\text{no. mol Cu} = 235\ \text{s} \times \frac{2.17\ \text{C}}{\text{s}} \times \frac{1\ \text{mol}\ e^-}{96{,}500\ \text{C}} \times \frac{1\ \text{mol Cu}}{2\ \text{mol}\ e^-} = 2.64 \times 10^{-3}\ \text{mol Cu}$$

Then determine the no. mol Cu^{2+} present initially.

$$\text{no. mol}\ Cu^{2+} = 0.335\ \text{L} \times \frac{0.215\ \text{mol}\ Cu^{2+}}{\text{L}} = 7.20 \times 10^{-2}\ \text{mol}\ Cu^{2+}$$

$$\text{no. mol}\ Cu^{2+}\ \text{remaining} = 7.20 \times 10^{-2} - 2.64 \times 10^{-3} = 6.94 \times 10^{-2}\ \text{mol}\ Cu^{2+}$$

$$[Cu^{2+}] = \frac{6.94 \times 10^{-2}\ \text{mol}\ Cu^{2+}}{0.335\ \text{L}} = 0.207\ \text{M.}$$

(d) $$\text{no. mol}\ Ag^+\ \text{initially} = 0.415\ \text{L} \times \frac{0.185\ \text{mol}\ Ag^+}{\text{L}} = 0.0768\ \text{mol}\ Ag^+$$

$$\text{find no. mol}\ Ag^+ = 0.415\ \text{L} \times \frac{0.175\ \text{mol}\ Ag^+}{\text{L}} = 0.0726\ \text{mol}\ Ag^+$$

$$\text{no. mol Ag to be deposited} = 0.0768 - 0.0726 = 0.0042\ \text{mol Ag}$$

$$\text{no. C} = 0.0042\ \text{mol Ag} \times \frac{1\ \text{mol}\ e^-}{1\ \text{mol Ag}} \times \frac{96{,}500\ \text{C}}{1\ \text{mol}\ e^-} = 4.1 \times 10^2$$

$$\text{no. s} = 4.1 \times 10^2\ \text{C} \times \frac{1\ \text{s}}{3.12\ \text{C}} = 1.3 \times 10^2\ \text{s}$$

20-37. The oxidation half-reaction, in which $O_2(g)$ is produced, is $2\ H_2O \rightarrow O_2(g) + 4\ H^+ + 4\ e^-$. Four moles of electrons are transferred for every mole of $O_2(g)$ produced.

(a) This calculation is performed as if the water solution being electrolyzed had no vapor pressure. The $O_2(g)$ is assumed to be dry.

$$\text{no. C} = 1.32\ \text{h} \times \frac{60\ \text{min}}{1\ \text{hr}} \times \frac{60\ \text{s}}{1\ \text{min}} \times \frac{1.75\ \text{C}}{1\ \text{s}} = 8.32 \times 10^3\ \text{C}$$

$$\text{no. mol}\ O_2 = 8.32 \times 10^3\ \text{C} \times \frac{1\ \text{mol}\ e^-}{96{,}500\ \text{C}} \times \frac{1\ \text{mol}\ O_2}{4\ \text{mol}\ e^-} = 2.16 \times 10^{-2}\ \text{mol}\ O_2$$

$PV = nRT$; $V = nRT/P$

$$V = \frac{2.16\times10^{-2}\ \text{mol} \times 0.0821\ \text{L atm mol}^{-1}\ \text{K}^{-1} \times 298\ \text{K}}{(747/760)\text{atm}} = 0.538\ \text{L}$$

(b) When the gas produced is saturated with water vapor, the partial pressure of $O_2(g)$ is $P_{bar} - P_{H_2O} = 747 - 23.8 = 723$ mmHg. The calculation of part (a) can be repeated with a simple substitution of 723 mmHg for 747 mmHg as the pressure of the gas, or, according to Boyle's law, the 0.538 L of dry gas would expand by the factor, 747/723, if it were saturated with water vapor.

$$V = 0.538\ \text{L} \times \frac{747\ \text{mmHg}}{723\ \text{mmHg}} = 0.556\ \text{L}\ O_2(g)$$

20-38. The electrodeposition reaction on which the following calculations are made is: $Ag^{+} + e^{-} \rightarrow Ag(s)$. In part (b) the definition of an ampere is used: 1 A = 1 C/s.

(a) $$\text{no. C} = 1.96\ \text{g Ag} \times \frac{1\ \text{mol Ag}}{108\ \text{g Ag}} \times \frac{1\ \text{mol}\ e^-}{1\ \text{mol Ag}} \times \frac{96{,}500\ \text{C}}{1\ \text{mol}\ e^-} = 1.75\times10^{3}\ \text{C}$$

(b) $$\text{no. A} = \frac{1.75\times10^{3}\ \text{C}}{787\ \text{s}} = 2.22\ \text{C/s} = 2.22\ \text{A}$$

20-39. First determine the amount of Cu^{2+} present initially.

$$\text{no. mol}\ Cu^{2+} = 0.2500\ \text{L} \times \frac{0.1000\ \text{mol}\ Cu^{2+}}{\text{L}} = 2.500\times10^{-2}\ \text{mol}\ Cu^{2+}$$

Next determine the amount of Cu^{2+} that is electrodeposited as Cu.

$$\text{no. mol}\ Cu^{2+} = 1368\ \text{s} \times \frac{3.512\ \text{C}}{\text{s}} \times \frac{1\ \text{mol}\ e^-}{96{,}487\ \text{C}} \times \frac{1\ \text{mol}\ Cu^{2+}}{2\ \text{mol}\ e^-} = 2.490\times10^{-2}\ \text{mol}\ Cu^{2+}$$

The difference between these two quantities represents the amount of Cu^{2+} remaining in solution.

$$\text{no. mol}\ Cu^{2+} = 2.500\times10^{-2}\ \text{mol}\ Cu^{2+} - 2.490\times10^{-2}\ \text{mol}\ Cu^{2+} = 1.0\times10^{-4}\ \text{mol}\ Cu^{2+}$$

The total concentration of Cu^{2+} would be

$$\frac{1.0\times10^{-4}\ \text{mol}\ Cu^{2+}}{0.2500\ \text{L}} = 4.0\times10^{-4}\ \text{M}$$

Part of this Cu^{2+} is present as free Cu^{2+}, i.e., $[Cu^{2+}] = x$, and part is present as $[Cu(NH_3)_4]^{2+}$, i.e., $(4.0\times10^{-4} - x)$.

$$K_f = \frac{[[Cu(NH_3)_4]^{2+}]}{[Cu^{2+}][NH_3]^4} = \frac{(4.0\times10^{-4} - x)}{x(0.10)^4} = 1.1\times10^{13}$$

$$1.1\times10^{9}\ x = 4.0\times10^{-4} - x$$

$$x = \frac{4.0\times10^{-4}}{1.1\times10^{9}} = 4.0\times10^{-13}\ \text{M} = [Cu^{2+}]$$

$$[Cu(NH_3)_4]^{2+} = 4.0\times10^{-4} - 3.6\times10^{-13} = 4.0\times10^{-4}\ \text{M}$$

Since this concentration is considerably larger than the detectable amount of $[Cu(NH_3)_4]^{2+}$, the blue color should form.

20-40. (a) $$\text{no. mol } K_2Cr_2O_7 = 0.2050 \text{ g sample} \times \frac{99.72 \text{ g } FeSO_4}{100.0 \text{ sample}} \times \frac{1 \text{ mol } FeSO_4}{151.9 \text{ g } FeSO_4} \times \frac{1 \text{ mol } Fe^{2+}}{1 \text{ mol } FeSO_4}$$

$$\times \frac{1 \text{ mol } Cr_2O_7^{2-}}{6 \text{ mol } Fe^{2+}} \times \frac{1 \text{ mol } K_2Cr_2O_7}{1 \text{ mol } Cr_2O_7^{2-}} = 2.243\times10^{-4} \text{ mol } K_2Cr_2O_7$$

$$\text{molarity of } K_2Cr_2O_7(aq) = \frac{2.243\times10^{-4} \text{ mol } K_2Cr_2O_7}{0.04010 \text{ L}} = 5.594\times10^{-3} \text{ M}$$

(b) The equation on which the calculation in part (a) is based involves the transfer of 6 mol of electrons, that is, 6 equiv of Fe^{2+} and 6 equiv of $Cr_2O_7^{2-}$.

$$\text{normality of } K_2Cr_2O_7(aq) = 6 \times \text{molarity} = 6 \times 5.594\times10^{-3} = 3.356\times10^{-2} \text{ N}$$

20-41. The reaction that occurs is

oxid: $3\{H_2O_2 \rightarrow O_2(g) + 2\ H^+ + 2\ e^-\}$

red: $Cr_2O_7^{2-} + 14\ H^+ + 6\ e^- \rightarrow 2\ Cr^{3+} + 7\ H_2O$

net: $Cr_2O_7^{2-} + 3\ H_2O_2 + 8\ H^+ \rightarrow 2\ Cr^{3+} + 7\ H_2O + 3\ O_2(g)$

$$\text{no. mol } H_2O_2 = 0.02500 \text{ L} \times \frac{0.0101 \text{ mol } H_2O_2}{\text{L}} = 2.52\times10^{-4} \text{ mol } H_2O_2$$

$$\text{no. mol } Cr_2O_7^{2-} \text{ available} = 0.02500 \text{ L} \times \frac{5.594\times10^{-3} \text{ mol } Cr_2O_7^{2-}}{\text{L}} = 1.398\times10^{-4} \text{ mol } Cr_2O_7^{2-}$$

$$\text{no. mol } Cr_2O_7^{2-} \text{ that reacts} = 2.52\times10^{-4} \text{ mol } H_2O_2 \times \frac{1 \text{ mol } Cr_2O_7^{2-}}{3 \text{ mol } H_2O_2} = 8.40\times10^{-5} \text{ mol } Cr_2O_7^{2-}$$

$$\text{no. mol } Cr_2O_7^{2-} \text{ that remains} = 1.398\times10^{-4} - 0.840\times10^{-4} = 5.58\times10^{-5} \text{ mol } Cr_2O_7^{2-}$$

$$[Cr_2O_7^{2-}] \text{ that remains} = \frac{5.58\times10^{-5} \text{ mol } Cr_2O_7^{2-}}{0.05000 \text{ L}} = 1.12\times10^{-3} \text{ M}$$

20-42. The $KMnO_4(aq)$ in Example 20-14 is 0.1070 N for a reaction involving the transfer of 5 mol of electrons. Its molarity is (0.1070/5) = 0.02140 M. In the reaction of MnO_4^- and $S_2O_3^{2-}$.

oxid: $3\{S_2O_3^{2-} + 5\ H_2O \rightarrow 2\ SO_4^{2-} + 10\ H^+ + 8\ e^-\}$

red: $8\{MnO_4^- + 4\ H^+ + 3\ e^- \rightarrow MnO_2(s) + 2\ H_2O\}$

net: $3\ S_2O_3^{2-} + 8\ MnO_4^- + 2\ H^+ \rightarrow 6\ SO_4^{2-} + 8\ MnO_2(s) + H_2O$

The equivalent of MnO_4^- is based on a gain of 3 electrons.

normality = 3 x molarity = 3 x 0.02140 = 0.06420 N

eq wt $Na_2S_2O_3$ = 1/8 f wt = 1/8(158.1) = 19.76 g $Na_2S_2O_3$/eq

$$\text{no. meq } KMnO_4 = 0.0417 \text{ g } Na_2S_2O_3 \times \frac{1 \text{ eq } Na_2S_2O_3}{19.76 \text{ g } Na_2S_2O_3} \times \frac{1 \text{ eq } KMnO_4}{1 \text{ eq } Na_2S_2O_3} \times \frac{1000 \text{ meq } KMnO_4}{1 \text{ eq } KMnO_4}$$

$$= 2.11 \text{ meq } KMnO_4$$

$$\text{no. mL } KMnO_4(aq) = 2.11 \text{ meq } KMnO_4 \times \frac{1 \text{ mL } KMnO_4(aq)}{0.06420 \text{ meq } KMnO_4} = 32.9 \text{ mL}$$

Self-Test Questions

1. (*d*) This reaction involves the oxidation of Cu to Cu^{2+} by H_2SO_4. H_2SO_4 is reduced to $SO_2(g)$.

2. (*c*) The oxidation state of Np decreases from +5 in NpO_2^+ to +4 in Np^{4+}. This is a reduction process.

3. (*b*) Since E°_{red} for Hg_2^{2+} is a positive quantity, Hg_2^{2+} is more readily reduced than H^+. Also, E°_{cell} for the reactions of Hg(l) with HCl(aq) and with $Zn^{2+}(aq)$ would both be negative (nonspontaneous).

4. (*b*) The reaction in question is nonspontaneous. This reaction cannot be made to produce electricity but it can be carried out in an electrolysis cell.

5. (*b*) For this reaction, the Nernst equation is $E_{cell} = E^\circ_{cell} - \frac{0.0592}{2} \log \frac{[Zn^{2+}]}{[Pb^{2+}]}$. Since $[Zn^{2+}] = [Pb^{2+}]$, $\log [Zn^{2+}]/[Pb^{2+}] = \log 1 = 0$. $E_{cell} = E^\circ_{cell} = +0.66$ V.

6. (*c*) Because $E^\circ_{cell} > 0$, we should expect a spontaneous reaction in the forward direction. However, because of the small magnitude of E°_{cell} (+0.03 V), we should not expect the reaction to go to completion.

7. (*a*) Oxidation occurs at the anode. SO_4^{2-} is oxidized to $S_2O_8^{2-}$ only with great difficulty. K^+ cannot be oxidized. Only H_2O can be oxidized, to $O_2(g)$.

8. (*a*) The reduction of Al^{3+} to Al requires 3 mol of electrons per mol Al (27.0 g). To deposit 4.5 g Al (0.167 mol) requires 3 x 0.167 = 0.50 mol electrons. In turn 0.50 mol electrons will produce 0.25 mol (5.6 L) $H_2(g)$: $2\ H^+(aq) + 2\ e^- \rightarrow H_2(g)$.

9. oxid: $3\{PbO(s) + 2\ OH^- \rightarrow PbO_2(s) + H_2O + 2\ e^-\}$

   red: $2\{MnO_4^- + 2\ H_2O + 3\ e^- \rightarrow MnO_2(s) + 4\ OH^-\}$

   net: $3\ PbO(s) + 2\ MnO_4^- + H_2O \rightarrow 3\ PbO_2(s) + 2\ MnO_2(s) + 2\ OH^-$

10. oxid: $3\{Zn(s) \rightarrow Zn^{2+} + 2\ e^-\}$ $\quad E^\circ_{ox} = +0.763$ V

    red: $2\{NO_3^- + 4\ H^+ + 3\ e^- \rightarrow NO(g) + 2\ H_2O\}$ $\quad E^\circ_{red} = +0.96$ V

    net: $3\ Zn(s) + 2\ NO_3^- + 3\ e^- \rightarrow 3\ Zn^{2+} + 2\ NO(g) + 4\ H_2O$ $\quad E^\circ_{cell} = 1.72$ V (anode) (cathode)

$Zn(s)\ |\ Zn^{2+}(aq)\ ||\ H^{+}(aq),\ NO_3^{-}(aq)\ |\ NO(g),\ Pt(s)$

11. oxid: $H_2(g,\ 1\ atm) \rightarrow 2\ H^{+}(x\ M) + 2\ e^{-}$

red: $2\ H^{+}(0.10\ M) + 2\ e^{-} \rightarrow H_2(g,\ 1\ atm)$

net: $2\ H^{+}(0.10\ M) \rightarrow 2\ H^{+}(x\ M)$

$$E_{cell} = E°_{cell} - \frac{0.0592}{2} \log \frac{x^2}{(0.10)^2} = 0.00 - \frac{0.0592}{2} \times 2 \times \log \frac{x}{0.10} = +0.108$$

$$\log x - \log 0.10 = \frac{0.108}{-0.0592} = -1.82 \qquad \log x = -1.82 + \log 0.10 = -1.82 - 1.00 = -2.82$$

$$pH = -\log [H^{+}] = -\log x = -(-2.82) = 2.82$$

12. Determine K from the value given for $E°_{cell}$.

$$E°_{cell} = \frac{0.0592}{n} \log K \text{ and } \log K = \frac{n\ E°_{cell}}{0.0592}$$

$$\log K = \frac{2}{0.0592} \times (-0.0050) = -0.17$$

K = antilog (-0.17) = 0.68

Based on the initial concentrations given, $Q = \frac{[Cu^{2+}]^2[Sn^{2+}]}{[Cu^{+}]^2[Sn^{4+}]} = 1.00 > 0.68.$

The reverse reaction must occur.

	$2\ Cu^{+}$	+ Sn^{4+} $\rightleftharpoons$	$2\ Cu^{2+}$	+ Sn^{2+}
Initial:	1.00 M	1.00 M	1.00 M	1.00 M
Changes:	$+2x$ M	$+x$ M	$-2x$ M	$-x$ M
Equil:	$(1.00 + 2x)$M	$(1.00 + x)$M	$(1.00 - 2x)$M	$(1.00 - x)$M

$$K = \frac{[Cu^{2+}]^2[Sn^{2+}]}{[Cu^{+}]^2[Sn^{4+}]} = \frac{(1.00 - 2x)^2(1.00 - x)}{(1.00 + 2x)^2(1.00 + x)} = 0.68$$

Use a method of successive approximations:

Try $x = 0.10$: $\frac{(1.00 - 0.20)^2(1.00 - 0.10)}{(1.00 + 0.20)^2(1.00 + 0.10)} = 0.36 < 0.68$

Try $x = 0.010$: $\frac{(1.00 - 0.020)^2(1.00 - 0.010)}{(1.00 + 0.020)^2(1.00 + 0.010)} = 0.90 > 0.68$

Try $x = 0.050$: $\frac{(1.00 - 0.10)^2(1.00 - 0.050)}{(1.00 + 0.10)^2(1.00 + 0.050)} = 0.61 < 0.68$

Try $= 0.040$: $\dfrac{(1.00 - 0.080)^2(1.00 - 0.040)}{(1.00 + 0.080)^2(1.00 + 0.040)} = 0.67 \simeq 0.68$

Equilibrium concentrations: $[Cu^+] = 1.08$ M; $[Sn^{4+}] = 1.04$ M; $[Cu^{2+}] = 0.92$ M; $[Sn^{2+}] = 0.96$ M.

Chapter 21

Chemistry of the Representative Elements I: Nonmetals

Review Problems

21-1. (a) $KBrO_3$ = potassium bromate

(b) ClF_3 = chlorine trifluoride

(c) sodium hypoiodite = NaOI

(d) trisilane = Si_3H_8

(e) KOCN = potassium cyanate

(f) sodium dithionate = $Na_2S_2O_6$

(g) silver azide = AgN_3

(h) NaH_2PO_4 = sodium dihydrogen phosphate

21-2. (a) $CaCl_2(s) + H_2SO_4(\text{conc. aq}) \xrightarrow{\Delta} CaSO_4(s) + 2\ HCl(g)$

(b) $I_2(s) + Cl^-(aq) \rightarrow$ no reaction

(c) $PbS(s) + HCl(aq) \rightarrow$ no reaction

(d) $NH_3(aq) + HClO_4(aq) \rightarrow NH_4ClO_4(aq)$

(e) $2\ NO(g) + O_2(g) \rightarrow 2\ NO_2(g)$

(f) $2\ Hg(NO_3)_2(s) \xrightarrow{\Delta} 2\ HgO(s) + 4\ NO_2(g) + O_2(g)$

followed by: $2\ HgO(s) \xrightarrow{\Delta} 2\ Hg(l) + O_2(g)$

21-3. (a) Decomposition of $KClO_3(s)$ (with MnO_2 as a catalyst):

$2\ KClO_3(s) \xrightarrow{\Delta} 2\ KCl(s) + 3\ O_2(g)$

(b) Action of $H_2SO_4(\text{conc. aq})$ on NaCl(s):

$NaCl(s) + H_2SO_4(\text{conc. aq}) \xrightarrow{\Delta} NaHSO_4(s) + HCl(g)$

followed by: $NaCl(s) + NaHSO_4(s) \xrightarrow{\Delta} Na_2SO_4(s) + HCl(g)$

(c) Decomposition of NH_4NO_3:

$NH_4NO_3(s) \xrightarrow{\Delta} N_2O(g) + 2\ H_2O(l)$

(d) $SO_3^{2-}(aq) + Ba^{2+}(aq) \rightarrow BaSO_3(s)$

21-4. (a) $Cl_2(g) + 2\ OH^-(aq) \rightarrow OCl^-(aq) + Cl^-(aq) + H_2O$

(b) $2\ NaI(s) + 3\ H_2SO_4(\text{conc. aq}) \xrightarrow{\Delta} 2\ NaHSO_4(s) + 2\ H_2O + SO_2(g) + I_2(g)$

(c) $Cl_2(g) + 2\ Br^-(aq) \rightarrow 2\ Cl^-(aq) + Br_2(aq)$

(d) $3\ CdS(s) + 8\ H^+ + 2\ NO_3^- \rightarrow 3\ Cd^{2+} + 3\ S(s) + 4\ H_2O + 2\ NO(g)$

(e) $2\ Fe(s) + 3\ N_2O(g) \xrightarrow{\Delta} Fe_2O_3(s) + 3\ N_2(g)$

(f) $3\ Pb(s) + 8\ H^+ + 2\ NO_3^- \rightarrow 3\ Pb^{2+} + 4\ H_2O + 2\ NO(g)$

21-5. $H_5IO_6 + H^+ + 2\ e^- \rightarrow IO_3^- + 3\ H_2O$ $\quad E° = +1.60$ V

$4\ H_3PO_4 + 4\ H^+ + 4\ e^- \rightarrow P_4 + 8\ H_2O$ $\quad E° = -0.51$ V

$Sb_2O_5 + 6\ H^+ + 4\ e^- \rightarrow 2\ SbO^+ + 3\ H_2O$ $\quad E° = +0.58$ V

$OCl^- + H_2O + 2\ e^- \rightarrow Cl^- + 2\ OH^-$ $\quad E° = +0.88$ V

$B_2H_6 + 2\ H_2O + 4\ e^- \rightarrow 2\ BH_4^- + 2\ OH^-$ $\quad E° = +0.78$ V

$HXeO_4^- + 3\ H_2O + 6\ e^- \rightarrow Xe + 7\ OH^-$ $\quad E° = +1.24$ V

21-6. (a) *oxid:* $HClO_2 + H_2O \rightarrow 3\ H^+ + ClO_3^- + 2\ e^-$ $\quad E°_{ox} = -1.21$ V

red: $H_2O_2 + 2\ H^+ + 2\ e^- \rightarrow 2\ H_2O$ $\quad E°_{red} = +1.77$ V

net: $HClO_2 + H_2O_2 \rightarrow H^+ + ClO_3^- + H_2O$ $\quad E°_{cell} = +0.56$ V

The forward reaction should occur to a significant extent.

(b) *oxid:* $2\ Cl^- \rightarrow Cl_2(g) + 2\ e^-$ $\quad E°_{ox} = -1.36$ V

red: $2\{NO_3^- + 2\ H^+ + e^- \rightarrow NO_2(g) + H_2O\}$ $\quad E°_{red} = +0.81$ V

net: $2\ Cl^- + 2\ NO_3^- + 4\ H^+ \rightarrow 2\ H_2O + 2\ NO(g) + Cl_2(g)$ $\quad E°_{cell} = -0.55$ V

The forward reaction does not occur to a significant extent.

(c) *oxid:* $S^{2-} \rightarrow S(s) + 2\ e^-$ $\quad E°_{ox} = +0.48$ V

red: $Cl_2(g) + 2\ e^- \rightarrow 2\ Cl^-$ $\quad E°_{red} = +1.36$ V

net: $Cl_2(g) + S^{2-} \rightarrow 2\ Cl^- + S(s)$ $\quad E°_{cell} = +1.84$ V

The forward reaction should occur to a significant extent (essentially to completion).

(d) *oxid:* $OCl^- + 2\ OH^- \rightarrow ClO_2^- + H_2O + 2\ e^-$ $\quad E°_{ox} = -0.65$ V

red: $2\ OCl^- + 2\ H_2O + 2\ e^- \rightarrow Cl_2(g) + 4\ OH^-$ $\quad E°_{red} = +0.40$ V

net: $3\ OCl^- + H_2O \rightarrow ClO_2^- + Cl_2(g) + 2\ OH^-$ $\quad E°_{cell} = -0.25$ V

The forward reaction will not occur to any significant extent.

21-7. (a) $4\ SO_2 + 4\ H^+ + 6\ e^- \rightarrow S_4O_6^{2-} + 2\ H_2O$ $\quad -\Delta\bar{G}° = 6\mathfrak{F}(0.51)$

$S_4O_6^{2-} + 2\ e^- \rightarrow 2\ S_2O_3^{2-}$ $\quad -\Delta\bar{G}° = 2\mathfrak{F}(0.08)$

$4\ SO_2 + 4\ H^+ + 8\ e^- \rightarrow 2\ S_2O_3^{2-} + 2\ H_2O$ $\quad -\Delta\bar{G}° = 6\mathfrak{F}(0.51) + 2\mathfrak{F}(0.08)$

Also, $\Delta\bar{G}° = -n\mathcal{F}E°$ where n = 8.

$8\mathcal{F}E° = 6\mathcal{F}(0.51) + 2\mathcal{F}(0.08)$

$$E° = \frac{(6 \times 0.51) + (2 \times 0.08)}{8} = 0.40\ V$$

(b) $4\ SO_3^{2-} + 6\ H_2O + 6\ e^- \rightarrow S_4O_6^{2-} + 12\ OH^-$ $\qquad -\Delta\bar{G}° = 6\mathcal{F}(-0.79)$

$S_4O_6^{2-} + 2\ e^- \rightarrow 2\ S_2O_3^{2-}$ $\qquad -\Delta\bar{G}° = 2\mathcal{F}(0.08)$

$4\ SO_3^{2-} + 6\ H_2O + 8\ e^- \rightarrow 2\ S_2O_3^{2-} + 12\ OH^-$ $\qquad -\Delta G° = 6\mathcal{F}(-0.79) + 2\mathcal{F}(0.08)$

Also, $\Delta\bar{G}° = -n\mathcal{F}E°$ where n = 8

$8\mathcal{F}E° = 6\mathcal{F}(-0.79) + 2\mathcal{F}(0.08)$

$$E° = \frac{-(6 \times 0.79) + (2 \times 0.08)}{8} = -0.57\ V$$

21-8. The required oxidation half-reaction is

$H_2O_2 \rightarrow O_2(g) + 2\ H^+ + 2\ e^-$ $\qquad E°_{ox} = -0.682\ V$

The oxidizing agents suitable for this oxidation must have $E°_{red} > 0.682$ V. From Table 20-2 the oxidizing agents that meet this requirement are (a) $Cl_2(g)$; (c) $Cr_2O_7^{2-}(aq)$; (d) $MnO_2(s)$.

21-9. (a) Since $I^-(aq)$ is easier to oxidize than H_2O

oxid: $2\ I^-(aq) \rightarrow I_2 + 2\ e^-$ $\qquad E°_{ox} = -0.535\ V$

oxid: $2\ H_2O \rightarrow O_2(g) + 4\ H^+ + 4\ e^-$ $\qquad E°_{ox} = -1.229\ V$

We should expect $Cl_2(g)$ to oxidize I^- to I_2.

(b) NH_4^+ has no ability to act as an oxidizing agent (it cannot be reduced). This means that the H_2O_2 must act as an oxidizing agent (i.e., by oxidizing NH_4^+ to NH_3OH^+ or some higher oxidization state). If H_2O_2 is an oxidizing agent, it must be reduced, to H_2O.

(c) For the oxidation of Ag

$Ag(s) \rightarrow Ag^+(aq) + e^-$ $\qquad E°_{ox} = -0.800\ V$

The oxidizing agent capable of oxidizing Ag must have $E°_{red} > 0.800$ V. The reduction of NO_3^- to NO(g) in acidic solution has $E°_{red} = +0.96$ V, and this is the only possible reduction with a large enough E° value. The gaseous product should be NO(g).

21-10. Write an equation for the formation of one mole of each substance. Assess the total energy required to break old bonds and to form new ones, and hence $\bar{H}°_f$ of the substance.

(a) $\frac{1}{2}\ Cl_2(g) + \frac{1}{2}\ F_2(g) \rightarrow ClF(g)$

bond breakage: $\frac{1}{2}(243) + \frac{1}{2}(159) = 122 + 80 = 202$ kJ/mol

bond formation: -251 kJ/mol

$\Delta\bar{H}°_f = 202 - 251 = -49$ kJ/mol

(b) $\frac{1}{2}\ O_2(g) + F_2(g) \rightarrow OF_2(g)$

bond breakage: $\frac{1}{2}(499) + 159 = 409$ kJ/mol

bond formation: $2 \times (-213) = -426$ kJ/mol $\Delta\bar{H}^\circ_f = 409 - 426 = -17$ kJ/mol

(c) $\frac{1}{2}\ O_2(g) + Cl_2(g) \rightarrow OCl_2(g)$

bond breakage: $\frac{1}{2}(499) + 243 = 493$ kJ/mol

bond formation: $2 \times (-205) = -410$ kJ/mol

$\Delta\bar{H}^\circ_f = 493 - 410 = 83$ kJ/mol

(d) $\frac{1}{2}\ N_2(g) + \frac{3}{2}\ F_2(g) \rightarrow NH_3(g)$

bond breakage: $\frac{1}{2}(946) + \frac{3}{2}(159) = 711$ kJ/mol

bond formation: $3 \times (-280) = -840$ kJ/mol

$\Delta\bar{H}^\circ_f = 711 - 840 = -129$ kJ/mol

21-11. (a) calcium orthophosphate = $Ca_3(PO_4)_2$ (orthophosphoric acid = H_3PO_4)

(b) potassium pyrophosphate = $K_4P_2O_7$ (pyrophosphoric acid = $H_4P_2O_7$)

(c) $NaSbO_2$ = sodium metaantimonite (Sb is in the oxidation state +3, and the acid $HSbO_2$ is a "meta" acid.)

(d) $NaBiO_3$ = sodium metabismuthate (Bi is in the oxidation state +5, and the acid $HBiO_3$ is a "meta" acid.)

(e) Na_3BiO_4 = sodium orthobismuthate (The sodium salt of an "ortho" acid -- H_3BiO_4.)

21-12. Use the method of Example 9-12.

(a) XeO_3: no. valence electron pairs $= \dfrac{8 + (3 \times 6)}{2} = 13$

no. bond pairs = 4 - 1 = 3

no. central pairs = 13 - (3 x 3) = 4

no. lone pairs = 4 - 3 = 1

distribution of central pairs: tetrahedral

geometric shape: AX_3E, trigonal pyramidal

(b) XeO_4: no. valence pairs $= \dfrac{8 \times (4 \times 6)}{2} = 16$

no. bond pairs = 5 - 1 = 4

no. central pairs - 16 - (3 x 4) = 4

no. lone pairs = 4 - 4 = 0

distribution of central pairs: tetrahedral

geometric shape: AX_4, tetrahedral

(c) $OXeF_4$: no. valence pairs = $\frac{6 + 8 + (4 \times 7)}{2}$ = 21

no. bond pairs = 6 - 1 = 5

no. central pairs = 21 - (3 × 5) = 6

no. lone pairs = 6 - 5 = 1

distribution of central pairs = octahedral

geometric shape: AX_5E, square-based pyramidal

Exercises

The Halogens

21-1. The more direct approach is to calculate E°_{cell} for each of the following reactions. Because E°_{cell} > 0 for the first reaction, this one occurs spontaneously; the others do not.

$Cl_2 + 2\ I^- \rightarrow 2\ Cl^- + I_2$ $\quad E^\circ_{cell} = +1.36 - 0.54 = +0.82$ V

$I_2 + 2\ Br^- \rightarrow 2\ I^- + Br_2$ $\quad E^\circ_{cell} = +0.535 - 1.065 = -0.530$ V

$Br_2 + 2\ Cl^- \rightarrow 2\ Br^- + Cl_2$ $\quad E^\circ_{cell} = +1.065 - 1.36 = -0.30$ V

A principle that can be derived from the results of the above calculations is that a given halogen element will displace from solution any halide ions *below* it in Group VIIA of the periodic table.

Thus, Cl_2 will displace I^-, but I_2 will not displace Cl^-.

21-2. (a) In principle, because $F_2(g)$ has the greatest tendency of all substances to undergo reduction (to F^-) it should be able to displace all three halide ions, Cl^-, Br^-, and I^-.

(b) The difficulty in attempting to displace Cl^-, Br^- or I^- from aqueous solution with $F_2(g)$ is that $F_2(g)$ oxidizes H_2O to O_2 instead of oxidizing the halide ions.

(c) There is no common reagent that is capable of oxidizing $F^-(aq)$ to $F_2(g)$. If one could be found that had a sufficiently high E°_{red} to do so, it would probably oxidize H_2O to $O_2(g)$ more easily than it could oxidize $F^-(aq)$.

21-3. The following reaction is displaced to the right in alkaline solution (a solution containing a high $[OH^-]$) and should occur spontaneously.

$3\ Cl_2(g) + 3\ H_2O \rightarrow 5\ Cl^- + ClO_3^- + 6\ H^+$ $\quad E^\circ_{cell} = -0.11$ V

Another method of arriving at this same conclusion is to write the Nernst expression for the reaction, substitute $[H^+] \simeq 10^{-14}$ and unit activities for all other terms, and calculate E_{cell}. The value obtained is positive.

$$E_{cell} = -0.11 - \frac{0.0592}{5} \log [H^+]^6 = (-0.11) - \frac{6 \times 0.0592 \times \log (10^{-14})}{5} = -0.11 + \frac{6 \times 14 \times 0.0592}{5}$$

$$= +0.88 \text{ V}$$

21-4. (a) $2\ HClO + 2\ H^+ + 2\ e^- \rightarrow Cl_2 + 2\ H_2O$ $\quad \Delta\bar{G}^\circ = -2\mathcal{F}\cdot(1.62)$

$Cl_2 + 2\ e^- \rightarrow 2\ Cl^-$ $\quad \Delta\bar{G}^\circ = -2\cdot\mathcal{F}\cdot(1.36)$

$2\ HClO + 2\ H^+ + 4\ e^- \rightarrow 2\ Cl^- + 2\ H_2O$ $\quad \Delta G^\circ = -4\cdot\mathcal{F}\cdot(E^\circ)$

$-4\cdot\mathcal{F}\cdot(E^\circ) = -2\cdot\mathcal{F}\cdot(1.62 + 1.36)$ $\quad E^\circ = (1.62 + 1.36)/2 = +1.49$ V

(b) Use the result of part (a) in writing the reduction half-equation below:

oxid: $HClO + H_2O \quad HClO_2 + 2\ H^+ + 2\ e^-$ $\qquad E^\circ_{ox} = -(1.63)\ V$

red: $HClO + H^+ + 2\ e^- \quad Cl^- + H_2O$ $\qquad E^\circ_{red} = +1.49\ V$

---

net: $2\ HClO \quad HClO_2 + H^+ + Cl^-$ $\qquad E^\circ_{cell} = -0.14\ V$

This reaction *will not* occur spontaneously as written.

21-5. (a) $HCl(g) \rightarrow \frac{1}{2}\ H_2(g) + \frac{1}{2}\ Cl_2(g) \qquad \Delta\bar{G}^\circ = -\Delta\bar{G}^\circ_f = -(-95.27) = +95.27\ kJ/mol$

$$\log K_p = \frac{-\Delta\bar{G}^\circ}{2.303\ RT} = \frac{-95.27\times10^3\ J\ mok^{-1}}{2.303 \times 8.314\ J\ mol^{-1}\ K^{-1} \times 298\ K} = -16.7$$

$K_p = \text{antilog}(-16.7) = 2\times10^{-17}$

(b) For this reaction $K_c = K_p$. Start with 1.00 mol HCl in a volume V.

	$HCl(g)$ $\rightleftharpoons$	$\frac{1}{2}\ H_2(g)$	+	$\frac{1}{2}\ Cl_2(g)$
Initial amount:	1.00 mol	-		-
Changes:	$-x$ mol	$+\frac{x}{2}$ mol		$+\frac{x}{2}$ mol
Equil. amounts:	$(1.00 - x)$ mol	$(x/2)$ mol		$(x/2)$ mol
Equil. concns:	$(1.00 - x)/V$	$(x/2)/V$		$(x/2)/V$

$$K_c = \frac{[H_2]^{\frac{1}{2}}[Cl_2]^{\frac{1}{2}}}{[HCl]} = \frac{(x/2)^{\frac{1}{2}}(x/2)^{\frac{1}{2}}}{(V^{\frac{1}{2}})(V^{\frac{1}{2}})\left(\frac{1.00 - x}{V}\right)} = \frac{x}{2(1.00 - x)} = 2\times10^{-17}$$

$x = 4\times10^{-17} - 4\times10^{-17}x \qquad x = 4\times10^{-17}$

$$\%\ \text{dissoc.} = \frac{4\times10^{-17}\ mol}{1.00\ mol} \times 100 = 4\times10^{-15}\%$$

21-6. (a) $CaCl_2$ can be most easily prepared through the reaction of CaO(s) with HCl(aq): $CaO(s) + 2\ HCl(aq) \rightarrow CaCl_2(aq) + H_2O$. The HCl(aq), in turn, can be prepared by the reaction of NaCl(s) with H_2SO_4(conc. aq).

$NaCl(s) + H_2SO_4(\text{conc. aq}) \rightarrow NaHSO_4(s) + HCl(g)$

$HCl(g) \xrightarrow{\text{water}} HCl(aq)$

(b) KBr results from the neutralization of KOH(aq) with HBr(aq): $KOH(aq) + HBr(aq) \rightarrow KBr(aq) + H_2O$. HBr can be prepared in a manner similiar to that given in part (a) for HCl, except that a nonoxidizing acid must be used. $H_3PO_4(aq)$ serves this purpose well.

$NaBr(s) + H_3PO_4(\text{conc. aq}) \rightarrow NaH_2PO_4(s) + HBr(g)$

$HBr(g) \xrightarrow{\text{water}} HBr(aq)$

(c) A bromate salt can be prepared in a manner similar to a chlorate, that is, by the disproportionation of Br_2 in a warm alkaline solution (KOH).

$3\ Br_2 + 6\ OH^- \rightarrow BrO_3^- + 5\ Br^- + 3\ H_2O$

The KBr and $KBrO_3$ are separated by fractional crystallization.

The Br_2 can be produced by heating NaBr(s) with the oxidizing acid H_2SO_4.

$$2\ NaBr(s) + 3\ H_2SO_4(conc.\ aq) \rightarrow 2\ NaHSO_4(s) + 2\ H_2O + SO_2(g) + Br_2(g)$$

21-7. (a) BrF_3: Total no. valence electron pairs $= \frac{7 + (3 \times 7)}{2} = 14$

no. bond pairs = 4 - 1 = 3

no. central pairs = 14 - (3 x 3) = 5

no. lone pairs = 5 - 3 = 2

VSEPR geometry: AX_3E_2 = T-shaped

(b) IF_5: Total no. valence electron pairs $= \frac{7 + (5 \times 7)}{2} = 21$

no. bond pairs = 6 - 1 = 5

no. central pairs = 21 - (3 x 5) = 6

no. lone pairs = 6 - 5 = 1

VSEPR geometry: AX_5E = square-based pyramidal

(c) ICl_2^-: Total no. valence electron pairs $= \frac{7 + (2 \times 7) + 1}{2} = 11$

no. bond pairs = 3 - 1 = 2

no. central pairs = 11 - (2 x 3) = 5

no. lone pairs = 5 - 2 = 3

VSEPR geometry: AX_2E_5 = linear

(d) Cl_3IF^-: Total no. valence electron pairs $= \frac{7 + (4 \times 7) + 1}{2} = 18$

no. bond pairs = 5 - 1 = 4

no. central pairs = 18 - (4 x 3) = 6

no. lone pairs = 6 - 4 = 2

VSEPR geometry: AX_4E_2 = square planar

21-8. For the IO_3^-/I^- couple,

oxid: $3\{H_2SO_3 + H_2O \rightarrow SO_4^{2-} + 4\ H^+ + 2\ e^-\}$ $\qquad E^\circ_{ox} = -0.17\ V$

red: $IO_3^- + 6\ H^+ + 6\ e^- \rightarrow I^- + 3\ H_2O$ $\qquad E^\circ_{red} = ?$

net: $3\ H_2SO_3 + IO_3^- \rightarrow 3\ SO_4^{2-} + I^- + 6\ H^+$ $\qquad E^\circ_{cell} = E^\circ_{ox} + E^\circ_{red} = 0.92\ V$

$E^\circ_{red} = 0.92 - (-0.17) = +1.09\ V$

For the HIO/I_2 couple,

$$2\ I^- \rightarrow I_2(s) + 2\ e^- \qquad E°_{ox} = -0.54\ V$$

$$2\ HIO + 2\ H^+ + 2\ e^- \rightarrow I_2(s) + 2\ H_2O \qquad E°_{red} = ?$$

$$HIO + H^+ + I^- \rightarrow I_2 + H_2O \qquad E°_{cell} = E°_{ox} + E°_{red} = 0.91\ V$$

$$E°_{red} = 0.91 - (-0.54) = +1.45\ V$$

For the IO_3^-/HIO couple,

$$IO_3^- + 5\ H^+ + 4\ e^- \rightarrow HIO + 2\ H_2O \qquad \Delta\bar{G}° = -n\ E° = -4\ E°$$

$$2\ HIO + 2\ H^+ + 2\ e^- \rightarrow I_2 + 2\ H_2O \qquad \Delta\bar{G}° = -2 \times \mathcal{F} \times (1.45)$$

$$I_2 + 2\ e^- \rightarrow 2\ I^- \qquad \Delta\bar{G}° = -2 \times \mathcal{F} \times (0.54)$$

Rewrite these equations as follows:

$$IO_3^- + 5\ H^+ + 4\ e^- \rightarrow \cancel{HIO} + 2\ H_2O \qquad \Delta\bar{G} = -4\mathcal{F}E°$$

$$\cancel{HIO} + H^+ + e^- \rightarrow \cancel{\tfrac{1}{2}\ I_2} + H_2O \qquad \Delta\bar{G}° = -\mathcal{F} \times (1.45)$$

$$\cancel{\tfrac{1}{2}\ I_2} + e^- \rightarrow I^- \qquad \Delta\bar{G}° = -\mathcal{F} \times (0.54)$$

$$IO_3^- + 6\ H^+ + 6\ e^- \rightarrow I^- + 3\ H_2O \qquad \Delta\bar{G}° = -\mathcal{F}\{4\ E° + 1.45 + 0.54\}$$

Also, $\Delta\bar{G}° = -6 \times \mathcal{F} \times (1.09)$ $\qquad -6 \times \mathcal{F} \times (1.09) = -\mathcal{F}\{4\ E° + 1.45 + 0.54\}$

$$E° = \frac{(6 \times 1.09) - 1.45 - 0.54}{4} = 1.14\ V$$

Oxygen

21-9. (a) $2\ HgO(s) \xrightarrow{\Delta} 2\ Hg(l) + O_2(g)$

(b) $KClO_4(s) \xrightarrow{\Delta} KCl(s) + 2\ O_2(g)$

(c) $2\ Hg(NO_3)_2 \xrightarrow{\Delta} 2\ HgO + 4\ NO_2(g) + O_2(g)$

followed by: $2\ HgO \xrightarrow{\Delta} 2\ Hg(l) + O_2(g)$

(d) $2\ H_2O_2(l) \xrightarrow{\Delta} 2\ H_2O + O_2(g)$

21-10. (a) *oxid:* $H_2O_2 \rightarrow O_2(g) + 2\ H^+ + 2\ e^-$

red: $H_2O_2 + 2\ H^+ + 2\ e^- \rightarrow 2\ H_2O$

net: $2\ H_2O_2 \rightarrow O_2(g) + 2\ H_2O$

(b) *oxid:* $H_2O_2 + 2\ OH^- \rightarrow O_2(g) + 2\ H_2O + 2\ e^-$

red: $H_2O_2 + 2\ e^- \rightarrow 2\ OH^-$

net: $2\ H_2O_2 \rightarrow 2\ H_2O + O_2(g)$

21-11. (a) $3\ O_2(g) \rightarrow 2\ O_3(g)$

bond breakage: $3 \times 499 = 1497$ kJ/mol

bond formation: $-\ 4x$

$\Delta\bar{H}^\circ_f = 1497 - 4x = 285$

$$x = \frac{1497 - 285}{4} = 303 \text{ kJ/mol}$$

(b) According to the structure on page 391 the oxygen-to-oxygen bond in O_3 is $1\frac{1}{2}$ bond. We might assess the energy of this bond as the average of the single and double bond energies $(499 + 142)/2 = 320$ kJ/mol. The agreement between this result and part (a) is good.

21-12. Determine the number of bonding electrons in each species. Bond distances decrease and bond energies increase with number of bonding electrons O_2: 4 bonding electrons; O_2^+: 5 bonding electrons; O_2^-: 3 bonding electrons; O_2^{3-}: 2 bonding electrons.

(a) Increasing bond distance: $O_2^+ < O_2 < O_2 < O_2^{2-}$

(b) Increasing bond energy: $O_2^{2-} < O_2^- < O_2 < O_2^+$

Sulfur

21-13. (a) ZnS = zinc sulfide

(b) $KHSO_3$ = potassium hydrogen sulfite

(c) S_4^{2-} = tetrasulfide ion

(d) $K_2S_4O_6$ = potassium tetrathionate

(e) OSF_2 = thionyl fluoride

21-14. (a) $MnS(s) + 2\ HCl(aq) \rightarrow MnCl_2(aq) + H_2S(aq)$

(b) $Na_2SO_4 + 2\ H^+(aq) \rightarrow 2\ Na^+(aq) + H_2O + SO_2(g)$

(c) *oxid:* $SO_2(aq) + 2\ H_2O \rightarrow SO_4^{2-}(aq) + 4\ H^+ + 4\ e^-$

red: $MnO_2(s) + 4\ H^+ + 4\ e^- \rightarrow Mn^{2+}(aq) + 2\ H_2O$

net: $MnO_2(s) + SO_2(aq) \rightarrow Mn^{2+}(aq) + SO_4^{2-}(aq)$

(d) *oxid:* $2\{S^{2-} \rightarrow S + 2\ e^-\}$

red: $O_2(g) + 2\ H_2O + 4\ e^- \rightarrow 4\ OH^-$

net: $O_2(g) + 2\ S^{2-} + 2\ H_2O \rightarrow 2\ S + 4\ OH^-$

21-15. (a) Allow the Na and H_2O to react to form NaOH(aq)

$2\ Na + 2\ H_2O \rightarrow 2\ NaOH(aq) + H_2(g)$

Burn S in air to form $SO_2(g)$: $S(s) + O_2(g) \rightarrow SO_2(g)$. Dissolve the $SO_2(g)$ in NaOH(aq):

$SO_2(g) + H_2O \rightarrow H_2SO_3(aq)$

$H_2SO_3(aq) + 2\ NaOH(aq) \rightarrow Na_2SO_3(aq) + 2\ H_2O$

(b) $Na_2SO_4(aq)$ is formed by the oxidation of $Na_2SO_3(aq)$. Use the $Na_2SO_3(aq)$ from part (a) and $Cl_2(g)$ as an oxidizing agent.

$SO_3^{2-} + Cl_2(g) + H_2O \rightarrow 2\ Cl^-(aq) + SO_4^{2-} + 2\ H^+$

(c) Boil $Na_2SO_3(aq)$ from part (a) with sulfur.

$Na_2SO_3(aq) + S(s) \rightarrow Na_2S_2O_3(aq)$

21-16. The structure must feature an -O-O- bond. Start with H_2SO_4.

```
  O                                                O
  ‖                                                ‖
O=S-OH     and introduce the peroxo group    O=S-O-O-H
  |                                                |
  OH                                               OH
```

The formula is H_2SO_5.

21-17. One must compare not the values of K_{sp} but the actual molar solubilities to determine which is the least soluble of the two sulfides. If we let x equal the solubility in moles per liter of HgS and y, that of Bi_2S_3, we obtain the following expressions:

HgS

$$[Hg^{2+}][S^{2-}] = K_{sp}$$

$$(x)(x) = 3 \times 10^{-52}$$

$$x \simeq 2 \times 10^{-26}\ M$$

Bi_2S_3

$$[Bi^{3+}]^2[S^{2-}]^3 = K_{sp}$$

$$(2y)^2(3y)^3 = 1 \times 10^{-96}$$

$$4y^2 \times 27y^3 = 1 \times 10^{-96}$$

$$108y^5 = 1 \times 10^{-96}$$

$$y \simeq 2 \times 10^{-20}\ M$$

21-18. The sodium sulfite in acidic solution reduces permanganate ion to manganese(II) ion and is itself oxidized to sulfate ion.

$$5\ SO_3^{2-} + 2\ MnO_4^- + 6\ H^+ \rightarrow 5\ SO_4^{2-} + 2\ Mn^{2+} + 3\ H_2O$$

$$\text{no. g } Na_2SO_3 = 0.02650\ L \times \frac{0.0510\ mol\ MnO_4^-}{1\ L} \times \frac{5\ mol\ SO_3^{2-}}{2\ mol\ MnO_4^-} \times \frac{1\ mol\ Na_2SO_3}{1\ mol\ SO_3^{2-}} \times \frac{126\ g\ Na_2SO_3}{1\ mol\ Na_2SO_3}$$

$$= 0.426\ g\ Na_2SO_3$$

21-19. no. g Cu = $0.01212 \text{ L} \times \frac{0.1000 \text{ mol } S_2O_3^{2-}}{\text{L}} \times \frac{1 \text{ mol } I_2}{2 \text{ mol } S_2O_3^{2-}} \times \frac{2 \text{ mol } Cu^{2+}}{1 \text{ mol } I_2} \times \frac{1 \text{ mol Cu}}{1 \text{ mol } Cu^{2+}} \times \frac{63.55 \text{ g Cu}}{1 \text{ mol Cu}}$

$= 7.702\times10^{-2}$ g Cu

$\% \text{ Cu} = \frac{7.702\times10^{-2} \text{ g Cu}}{1.100 \text{ g sample}} \times 1000 = 7.002\% \text{ Cu}$

Nitrogen

21-20. (a) $2\ NO_2(g) \rightleftharpoons N_2O_4(g)$

(b) $2\ NaNO_3(s) \xrightarrow{\Delta} 2\ NaNO_2(s) + O_2(g)$

(c) $2\ NH_3(aq) + H_2SO_4(aq) \rightarrow (NH_4)_2SO_4(aq)$

(d) $3\ Ag(s) + 4\ H^+ + NO_3^- \rightarrow 3\ Ag^+ + 2\ H_2O + NO(g)$

(e) $(CH_3)_2NNH_2 + 4\ O_2 \rightarrow 2\ CO_2(g) + 4\ H_2O(\ell) + N_2(g)$

21-21. (a) $NH_2OH + H_2O \rightleftharpoons NH_3OH^+ + OH^-$ $K_b = 9.1\times10^{-9}$

let $[NH_3OH^+] = [OH^-] = x$ $[NH_2OH] = 0.032 - x \simeq 0.032$ M

$\frac{[NH_3OH^+][OH^-]}{[NH_2OH]} = \frac{x \cdot x}{0.032} = 9.1\times10^{-9}$ $x^2 = 2.9\times10^{-10}$ $x = 1.7\times10^{-5}$ M

$pOH = -\log(1.7\times10^{-5}) = 4.77$ $pH = 14.00 - 4.77 = 9.23$

(b) $NH_3OH^+ + H_2O \rightleftharpoons H_3O^+ + NH_2OH$ $K_a = K_w/K_b = 1.0\times10^{-14}/9.1\times10^{-9} = 1.1\times10^{-6}$

let $[H_3O^+] = [NH_2OH] = x$ $[NH_3OH^+] = 0.018 - x \simeq 0.018$ M

$\frac{[NH_2OH][H_3O^+]}{[NH_3OH^+]} = \frac{x \cdot x}{0.018} = 1.1\times10^{-6}$ $x^2 = 2.0\times10^{-8}$ $x = 1.4\times10^{-4}$ M

$pH = -\log(1.4\times10^{-4}) = 3.85$

21-22. (a) $NH_3OH^+ + 2\ H^+ + 2\ e^- \rightarrow NH_4^+ + H_2O$

(b) This calculation requires a combination of two half-reactions in the manner illustrated in Example 21-2.

$2\ NH_3OH^+ + H^+ + 2\ e^- \rightarrow \cancel{N_2H_5^+} + 2\ H_2O$ $\Delta\bar{G}° = 2\mathcal{F}(1.46)$

$\cancel{N_2H_5^+} + 3\ H^+ + 2\ e^- \rightarrow 2\ NH_4^+$ $\Delta\bar{G}° = -2\mathcal{F}(1.24)$

$2\ NH_3OH^+ + 4\ H^+ + 4\ e^- \rightarrow 2\ NH_4^+ + 2\ H_2O$ $\Delta\bar{G}° = -4\mathcal{F}E° = -2\mathcal{F}[1.46 + 1.24]$

$E° = \frac{-2\mathcal{F}(2.70)}{-4\mathcal{F}} = +1.35$ V

(c) $2\{Fe^{2+} \rightarrow Fe^{3+} + e^-\}$ $E°_{ox} = -(0.771)$ V

$NH_3OH^+ + 2\ H^+ + 2\ e^- \rightarrow NH_4^+ + H_2O$ $E°_{red} = +1.35$ V

$NH_3OH^+ + 2\ Fe^{2+} + 2\ H^+ \rightarrow NH_4^+ + 2\ Fe^{3+} + H_2O$ $E°_{cell} = +0.58$ V

This reaction *will* occur as written.

21-23. Use data from Appendix D, together with the value given for $\Delta\bar{H}^\circ_f[N_2H_4(l)]$, to determine $\bar{H}°$ for the reaction.

$$N_2H_4(l) + O_2(g) \rightarrow N_2(g) + 2\ H_2O(l)$$

$$\Delta\bar{H}° = 2\ \Delta\bar{H}^\circ_f[H_2O(l)] - \Delta\bar{H}^\circ_f[N_2H_4(l)] = 2 \times (-285.85) - (50.63) = -622.33 \text{ kJ/mol}$$

21-24. (a) N_2O_3 has 28 valence electrons or 18 pairs. A plausible Lewis structure begins with

|O̲-N̄-N-O̲| (with |O̲| bonded below N) which becomes |O̲=N̄-N⊕=O̲| (with ⊖|O̲| bonded below N) ⇌ |O̲=N̄-N⊕⊖-O̲| (with O̲| double-bonded below N)

(b) N_2O_5 has 40 valence electrons or 20 pairs. A plausible Lewis structure begins with

|O̲-N-O̲-N̄-O̲| (with |O̲| bonded below each N) which becomes |O̲=N⊕-O̲-N⊕=O̲| (with ⊖|O̲| and |O̲|⊖ bonded below each N) or other structures contributing to a resonance hybrid.

21-25. $N_2O_4(s) \xrightarrow{-11°C} N_2O_4(l) \xrightarrow{21°C} N_2O_4(g) \xrightarrow[\text{to } 150°C]{21°C} 2\ NO_2(g) \xrightarrow{600\text{-}700°C} 2\ NO(g) + O_2(g)$

(recall marginal note on page 653)

21-26. We are given $E°_{cell}$ for reaction (21.51). We need to establish the value of $E°_{red}$ for NH_2OH/NH_3. Then we can calculate the value of E° for $Fe(OH)_3/Fe(OH)_2$. From Figure 21-11

$2\ NH_2OH + 2\ e^- \rightarrow N_2H_4 + 2\ OH^-$ $-\Delta\bar{G}° = 2\mathfrak{F}(0.74)$

$N_2H_4 + 2\ H_2O + 2\ e^- \rightarrow 2\ NH_3 + 2\ OH^-$ $-\Delta\bar{G}° = 2\mathfrak{F}(0.10)$

$2\ NH_2OH + 2\ H_2O + 4\ e^- \rightarrow 2\ NH_3 + 4\ OH^-$ $-\Delta G° = 2\mathfrak{F}(0.74 + 0.10)$

Also, $-\Delta G° = n\mathfrak{F}E°$, where n = 4

$$E° = \frac{-\Delta G°}{n\mathfrak{F}} = \frac{2\mathfrak{F}(0.74 + 0.10)}{4\mathfrak{F}} = +0.42 \text{ V}$$

Now return to reaction (21.51)

oxid: $2\{Fe(OH)_2 + OH^- \rightarrow Fe(OH)_3 + e^-\}$ $E°_{ox} = ?$

red: $NH_2OH + H_2O + 2\ e^- \rightarrow NH_3 + 2\ OH^-$ $E°_{red} = +0.42$ V

---

net: $2\ Fe(OH)_2 + NH_2OH + H_2O \rightarrow 2\ Fe(OH)_3 + NH_3$

$E^\circ_{cell} = E^\circ_{ox} + E^\circ_{red} = E^\circ_{ox} + 0.42 = +0.98$ V

$E^\circ_{ox} = 0.98 - 0.42 = 0.56$ V

E°_{red} for $Fe(OH)_3/Fe(OH)_2 = -0.56$ V

Phosphorus

21-27. (a) The hydrolysis of PO_4^{3-} is

$$PO_4^{3-} + H_2O \rightleftharpoons HPO_4^{2-} + OH^- \qquad K_b = K_w/K_{a_3}$$

$$K_b = 1.0\times10^{-14}/4.8\times10^{-13} = 2.1\times10^{-2}$$

The large value of K_b indicates that the forward reaction proceeds to a considerable extent.

(b) At the first equivalence point the solution is $Na_2HPO_4(aq)$. As we learned in Chapter 18 for such a solution the further ionization of $H_2PO_4^-$ occurs to a greater extent than the hydrolysis of $H_2PO_4^-$ and the solution is acidic.

Also, $pH = \frac{1}{2}(pK_{a_1} + pK_{a_2}) = \frac{2.23 + 7.21}{2} = 4.72.$

21-28. (a) HPO_4^{2-} = (mono)hydrogen phosphate ion

(b) $Ca_2P_2O_7$ = calcium pyrophosphate

(c) $H_6P_4O_{13}$ = tetrapolyphosphoric acid

(d) $(NaPO_3)_4$ = sodium tetrametaphosphate

21-29. The key is first to produce the salts NaH_2PO_4 and Na_2HPO_4, by neutralizing $H_3PO_4(aq)$ to the first and second equivalence points, respectively.

$$NaOH(aq) + H_3PO_4(aq) \rightarrow NaH_2PO_4(aq) + H_2O$$

$$2\ NaOH(aq) + H_3PO_4(aq) \rightarrow Na_2HPO_4(aq) + 2\ H_2O$$

The solid salts are recrystallized from solution. Fuse a mixture of two parts Na_2HPO_4 and one part NaH_2PO_4 to obtain $Na_5P_3O_{10}$.

$$2\ Na_2HPO_4 + NaH_2PO_4 \xrightarrow{\Delta} Na_5P_3O_{10}(s) + 2\ H_2O(g)$$

Fuse NaH_2PO_4 to obtain $(NaPO_3)_n$.

$$n\ NaH_2PO_4 \rightarrow (NaPO_3)_n + n\ H_2O$$

21-30. The name sodium metaphosphate is not as specific in identifying a chemical substance as are most chemical names because the compound $(NaPO_3)_n$ is a mixture of species. That is, the name metaphosphate in itself does not signify the value of n. A more specific name would be sodium hexametaphosphate for $(NaPO_3)_6$.

21-31. Reaction (13.65) is used to produce $P_4(g)$. The $P_4(g)$ is oxidized to P_4O_{10}, which then reacts with water to produce H_3PO_4. This is followed by reaction (21.67).

$$2\ Ca_3(PO_4)_2(s) + 10\ C(s) + 6\ SiO_2(s) \rightarrow 6\ CaSiO_3(l) + 10\ CO(g) + P_4(g)$$

$$P_4(g) + 5\ O_2(g) \rightarrow P_4O_{10}(s)$$

$$P_4O_{10}(s) + 6\ H_2O(l) \rightarrow 4\ H_3PO_4(aq)$$

$$[3\ Ca_3(PO_4)_2 \cdot CaF_3] + 14\ H_3PO_4 + 10\ H_2O \rightarrow \underbrace{10\ [Ca(H_2PO_4)_2 \cdot H_2O]}_{\text{triple superphosphate}} + 2\ HF$$

Carbon, Silicon

21-32. (a) $KCN(aq) + AgNO_3(aq) \rightarrow AgCN(s) + KNO_3(aq)$

(b) $Al_4C_3(s) + 12\ H_2O \rightarrow 4\ Al(OH)_3(s) + 3\ CH_4(g)$

(c) $Si_3H_8 + 5\ O_2(g) \rightarrow 3\ SiO_2(s) + 4\ H_2O(l)$

21-33. Silanes are silicon-hydrogen compounds of the type SiH_4, Si_2H_6, ... (up to six Si atoms in a chain). A silanol is a silicon based molecule in which one or more H atoms of a silane is replaced by an OH group. Silanes must be converted to silanols for the preparation of silicones.

21-34. (a)

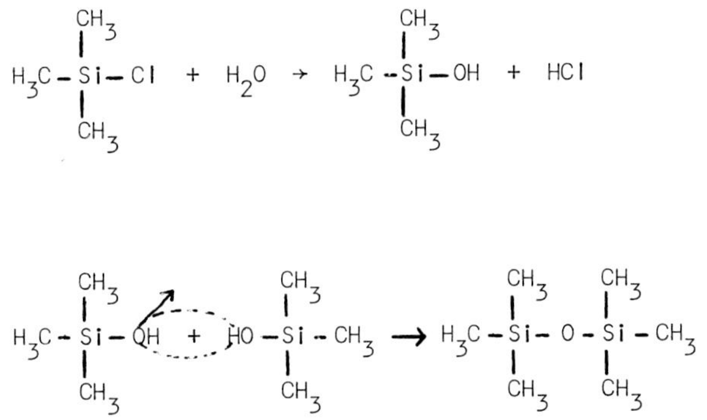

$$H_3C-\overset{CH_3}{\underset{CH_3}{Si}}-Cl + H_2O \rightarrow H_3C-\overset{CH_3}{\underset{CH_3}{Si}}-OH + HCl$$

$$H_3C-\overset{CH_3}{\underset{CH_3}{Si}}-OH + HO-\overset{CH_3}{\underset{CH_3}{Si}}-CH_3 \longrightarrow H_3C-\overset{CH_3}{\underset{CH_3}{Si}}-O-\overset{CH_3}{\underset{CH_3}{Si}}-CH_3$$

(b) A silicone polymer does not form. At best two of the $(CH_3)_3SiOH$ molecules (monomers) join to form a dimer. The chains cannot grow longer than two units in length.

(c) $CH_3SiCl_3 + 3\ H_2O \rightarrow H_3C-Si(OH)_3 + 3\ HCl$

A silicone polymer can be produced from $CH_3Si(OH)_3$ because each monomer unit is able to link up with three other units in a three-dimensional structure.

21-35. (a) Each B atom has four available orbitals and each hydrogen has one. This makes a total of 26 atomic orbitals available for bonding. In a normal compound (such as C_4H_{10}) these 26 orbitals and 26 electrons would produce a molecule with 13 bonds. The total number of valence electrons in B_4H_{10} is only 22. The compound appears to be lacking *four* electrons. In the hypothetical structure below these deficiencies are noted by open circles.

$$\begin{array}{ccccccccccc} & & H & & H & & H & & H & & \\ & & \cdot\times & & \cdot\times & & \cdot\times & & \cdot\times & & \\ H & \overset{\bullet}{\times} & B & \overset{\times}{\circ} & B & \overset{\times}{\circ} & B & \overset{\times}{\circ} & B & \overset{\bullet}{\circ} & H \\ & & \cdot\times & & \cdot\times & & \cdot\times & & \cdot\times & & \\ & & H & & H & & H & & H & & \end{array}$$

(b) If we reason by analogy to B_2H_6 pictured in Figure 21-15 we might write the following structure in which the electrons are numbered.

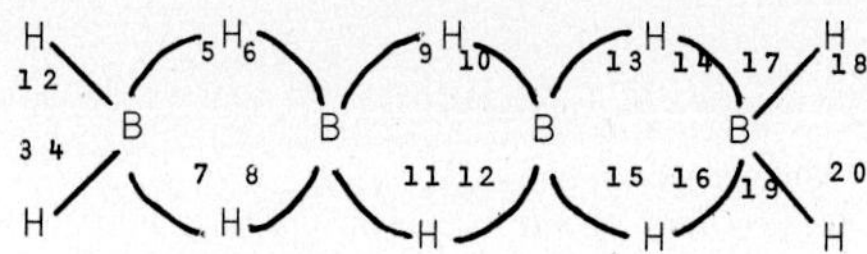

This structure has six B-H-B bridge bonds, but it utilizes a total of only 20 electrons. The number of electrons available is 22. It should be possible to write a structure in which the additional two electrons are used and the number of B-H-B bridge bonds reduced to four, as indicated below.

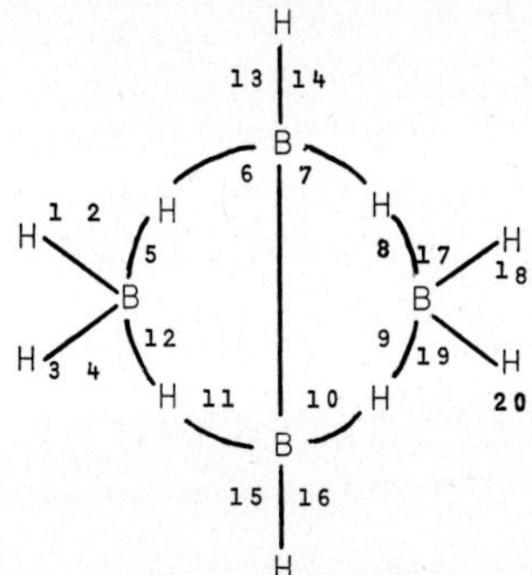

(c) The carbon analog referred to in part (a) is simply

$$\begin{array}{ccccccccccc} & & H & & H & & H & & H & & \\ & & \cdot\times & & \cdot\times & & \cdot\times & & \cdot\times & & \\ H & \overset{\bullet}{\times} & C & : & C & : & C & : & C & \overset{\bullet}{\times} & H \\ & & \cdot\times & & \cdot\times & & \cdot\times & & \cdot\times & & \\ & & H & & H & & H & & H & & \end{array}$$

21-36. (a) $2\ BBr_3 + 3\ H_2(g) \rightarrow 2\ B + 6\ HBr(g)$

(b) $B_2O_3 + 3\ CaF_2 + 3\ H_2SO_4 \rightarrow 2\ BF_3 + 3\ CaSO_4 + 3\ H_2O$

(c) $2\ B + 3\ N_2O \xrightarrow{\Delta} B_2O_3 + 3\ N_2$

(At high temperatures N_2O decomposes to $N_2(g)$ and $O_2(g)$; the O_2 combines with B.)

21-37. (a) O_2XeF_2: total electron pairs: $\frac{(2 \times 6) + 8 + (2 \times 7)}{2} = 17$

no. bond pairs = 5 - 1 = 4

no. central pairs = 17 - (3 x 4) = 5

no. lone pairs = 5 - 4 = 1

distribution of electron pairs: trigonal bipyramid

geometric shape: AX_4E, irregular tetrahedral

(b) O_3XeF_2: total electron pairs = $\frac{(3 \times 6) + 8 + (2 \times 7)}{2} = 20$

no. bond pairs = 6 - 1 = 5

no. central pairs = 20 - (3 x 5) = 5

no. lone pairs = 5 - 5 = 0

distribution of electron pairs: trigonal bipyramidal

geometric shape: trigonal bipyramidal

(c) O_2XeF_4: total electron pairs = $\frac{(2 \times 6) + 8 + (4 \times 7)}{2} = 24$

no. bond pairs = 7 - 1 = 6

no. central pairs = 24 - (3 x 6) = 6

no. lone pairs = 6 - 6 = 0

distribution of electron pairs: octahedral

geometric shape: octahedral

(d) XeF_5^+: total electron pairs = $\frac{8 + (5 \times 7) - 1}{2} = 21$

no. bond pairs = 6 - 1 = 5

no. central pairs = 21 - (3 x 5) = 6

no. lone pairs = 6 - 5 = 1

distribution of electron pairs = octahedral

geometric shape: AX_5E, square-based pyramidal

21-38. XeF_6: total electron pairs $= \frac{8 + (6 \times 7)}{2} = 25$

no. bond pairs = 7 - 1 = 6

no. central pairs = 25 - (6 × 3) = 7

no. lone pairs = 7 - 6 = 1

distribution of electron pairs: pentagonal bipyramidal

geometric shape: pentagonal-based pyramidal

The difficulty in applying the valence bond method is in devising a hybridization scheme for the bonding (it would seem to be sp^3d^3) and in determining into which orbital the lone pair electrons should be placed.

21-39. Because F is more electronegative than Cl, we should expect the Xe-F bond energy to be greater than that of Xe-Cl. Now consider the approximate enthalpies of formation of XeF_2 and $XeCl_2$:

$Xe(g) + F_2(g) \rightarrow XeF_2(g)$

bond breakage = +155 kJ/mol

bond formation = -2(Xe-F) bond energy

$\Delta\bar{H}_f^\circ[XeF_2] = 155 - 2(\text{Xe-F})$

$Xe(g) + Cl_2(g) \rightarrow XeCl_2(g)$

bond breakage = +243 kJ/mol

bond formation = -2(Xe-Cl) bond energy

$\Delta\bar{H}_f^\circ[XeCl_2] = 243 - 2(\text{Xe-Cl})$

Both because the bond energy of F_2 is smaller than that of Cl_2 and the expected bond energy Xe-F is greater than Xe-Cl, $\Delta\bar{H}_f^\circ$ should be greater (more negative) for XeF_2 than for $XeCl_2$. Having a greater $\Delta\bar{H}_f^\circ$, we should expect XeF_2 to be more stable than $XeCl_2$.

21-40. *oxid.* $XeF_4 + 3\ H_2O \rightarrow XeO_3 + 4\ F^- + 6\ H^+ + 2\ e^-$

oxid. $\frac{3}{2}\{2\ H_2O \rightarrow O_2 + 4\ H^+ + 4\ e^-\}$

red. $2\{XeF_4 + 4\ e^- \rightarrow Xe + 4\ F^-\}$

net: $3\ XeF_4 + 6\ H_2O \rightarrow 2\ Xe + XeO_3 + \underbrace{12\ H^+ + 12\ F^-}_{12\ HF} + \frac{3}{2}\ O_2(g)$

The statement that both XeO_3 and O_2 are produced suggests two oxidation half-reactions--the oxidation of XeF_4 and the oxidation of H_2O. The statement that Xe and XeO_3 are formed in a 2:1 ratio provides the clue on how the half-equations are to be combined: the reduction half-reaction is multiplied by 2 and the half-equation for the oxidation of XeF_4 to XeO_3, by 1. This requires that the half-equation for the oxidation of H_2O be multiplied by 3/2.

Self-Test Questions

1. (*b*) The displacement of Br_2 from a solution of Br^- is an oxidation process. The oxidizing agent employed must be a stronger one than Br_2. $Cl_2(aq)$ meets this requirement, but not $I_2(aq)$ nor $I_3^-(aq)$. $Cl^-(aq)$ can act as a reducing agent but not as an oxidizing agent.

2. (*d*) $CaCO_3$ yields CaO(s) and CO_2(g) when strongly heated $KClO_3$ yields KCl and O_2; HgO yields Hg and O_2; and N_2O yields N_2 and O_2.

3. (*a*) Cu is not an active enough metal to displace H_2 from an acidic solution. A strong oxidizing agent (e.g., HNO_3) is required to dissolve it. From concentrated HNO_3(aq) reduction occurs only to NO_2(g).

4. (*d*) The basic properties of N_2H_4, NH_2OH, and NH_3 are all described in the chapter. Also described is the ionization of hydrozoic acid as a weak acid.

$$HN_3 + H_2O \rightleftharpoons H_3O^+ + N_3^-$$

5. (*c*) HgS is the most insoluble of all metal sulfides. In order to dissolve HgS, it is necessary to oxidize S^{2-} to S (the function of HNO_3) and complex Hg^{2+} as $HgCl_4^{2-}$ (the function of HCl). HgS will only dissolve in aqua regia--a mixture of HNO_3 and HCl.

6. (*a*) Cl^- is a poor reducing agent (difficult to oxidize). O_3 and SO_4^{2-} are oxidizing agents, not reducing agents. H_2S is rather easily oxidized and is therefore a good reducing agent.

7. (*b*) One possibility is to apply the VSEPR method to each of these species to discover the one that is not tetrahedral in shape. However, two of these species have been identified as tetrahedral on more than one occasion: ClO_4^- and SO_4^{2-}. If VSEPR theory is applied to the two Xe compounds it is seen that in XeF_4 there is an octahedral distribution of six electron pairs and a structure corresponding to AX_4E_2. This is a square planar structure.

8. (*d*) The first three compounds are sodium thiosulfate, sodium thiocarbonate, and sodium thiocyanate. Na_2S_3 is a polysulfide.

9. (a) $2\ Pb(NO_3)_2 \xrightarrow{\Delta} 2\ PbO + 4\ NO_2(g) + O_2(g)$

(b) $Cl_2(g) + 2\ OH^- \rightarrow ClO^- + Cl^- + H_2O$

(c) $H_3PO_4(aq) + 2\ KOH(aq) \rightarrow K_2HPO_4(aq) + 2\ H_2O$

(d) $2\ KBr(s) + 2\ H_2SO_4 \xrightarrow{\Delta} K_2SO_4 + SO_2(g) + 2\ H_2O + Br_2$

(e) $2\ H_3PO_4 \xrightarrow{\Delta} H_4P_2O_7 + H_2O$

10. (a) Ag(s) cannot displace H^+(aq); it is not active enough a metal. Thus Ag(s) does not dissolve in the mineral acid HCl. In HNO_3(aq) it is the NO_3^- (not H^+) that oxidizes Ag to Ag^+.

(b) I_2 and I^- form the polyhalide ion I_3^- (triiodide ion): $I_2 + I^- \rightleftharpoons I_3^-$. The greater the concentration of I^-, the more I_2 that can be dissolved and converted to I_3^-. In pure water polyhalide ion formation does not occur and the solubility of I_2 is quite limited. In KI(aq), I_3^- formation is significant and the solubility of I_2 is high.

(c) H_2S(g) has a molecular weight of 34; and for species of this molecular weight, in the absence of strong intermolecular forces, we expect the gaseous state at room temperature. H_2O, with a molecular weight of 18, would also be a gas if it were not for the unusually large intermolecular forces attributed to hydrogen bond formation. Thus H_2S behaves "normally" and H_2O is the unusual substance.

11. (a) $$\text{no. g } F_2 = 1\ \text{km}^3\ \text{seawater} \times \frac{(1000)^3\ \text{cm}^3}{1\ \text{km}^3} \times \frac{(100)^3\ \text{cm}^3}{1\ \text{m}^3} \times \frac{1.03\ \text{g seawater}}{1\ \text{cm}^3} \times \frac{116\ \text{seawater}}{454\ \text{g seawater}}$$

$$\times \frac{1\ \text{ton seawater}}{2000\ \text{lb seawater}} \times \frac{1\ \text{g } F^-}{1\ \text{ton seawater}} \times \frac{1\ \text{mol } F^-}{19.0\ \text{g } F^-} \times \frac{1\ \text{mol } F_2}{2\ \text{mol } F^-} \times \frac{38.0\ \text{g } F_2}{1\ \text{mol } F_2}$$

$$= 1\times10^9\ \text{g } F_2 = 1\times10^6\ \text{kg } F_2$$

[Note that the last three terms in the set up are not required (they are equivalent to 1). That is, the no. g F_2 that can be extracted should be the same as the no. g F^-, since F^- is 100% fluorine.]

(b) The process for extracting fluorine from seawater cannot be expected to resemble that used for bromine. First, there is no oxidizing agent strong enough to oxidize $F^-(aq)$ to $F_2(g)$. And $F_2(g)$ would oxidize H_2O and be reduced back to $F^-(aq)$. The process would have to involve crystallizing a fluoride salt from the water and oxidation of F^- in the molten salt by electrolysis.

12. $H_2SeO_3 + 4\,H^+ + 4\,e^- \rightarrow Se + 3\,H_2O$ $\qquad \Delta\bar{G}° = -4\mathcal{F} \times (0.74)V$

$Se + 2\,H^+ + 2\,e^- \rightarrow H_2Se$ $\qquad \Delta\bar{G}° = -2\mathcal{F}\ (-0.35)V$

$H_2SeO_3 + 6\,H^+ + 6\,e^- \rightarrow H_2Se + 3\,H_2O$ $\qquad \Delta\bar{G}° = -4\mathcal{F}(0.74) + 2\mathcal{F}(0.35)\ V$

$$\Delta\bar{G}° = -n\mathcal{F}E°$$

$$E° = \frac{-\Delta\bar{G}°}{n\mathcal{F}} = \frac{-\{-4\mathcal{F}(0.74) + 2\mathcal{F}(0.35)\}V}{6\mathcal{F}} = \frac{2.96 - 0.70}{6} = 0.38\ V$$

Chapter 22

Chemistry of the Representative Elements II: Metals

Review Problems

22-1. (a) $Pb(CH_3COO)_2$ = lead acetate

(b) magnesium hydrogen carbonate = $Mg(HCO_3)_2$

(c) Hg_2Br_2 = mercury(I) bromide (mercurous bromide)

(d) $CaCl_2 \cdot 6\ H_2O$ = calcium chloride hexahydrate

(e) aluminum ammonium alum = $NH_4Al(SO_4)_2 \cdot 12\ H_2O$

(f) zincate ion = $[Zn(OH)_4]^{2-}$

(g) $NaAl(OH)_4$ = sodium aluminate

(h) $K_2Sn(OH)_6$ = potassium stannate

22-2. (a) $MgCO_3(s) \xrightarrow{\Delta} MgO(s) + CO_2(g)$

(b) $CaO(s) + 2\ HCl(aq) \rightarrow CaCl_2(aq) + H_2O$

(c) $2\ Al(s) + 2\ KOH(aq) + 6\ H_2O \rightarrow 2\ KAl(OH)_4(aq) + 3\ H_2(g)$

(d) $3\ Pb(s) + 8\ HNO_3(aq) \rightarrow 3\ Pb(NO_3)_2(aq) + 4\ H_2O + 2\ NO(g)$

(e) $Hg(l) + HCl(dil.\ aq) \rightarrow$ no reaction

(f) $ZnO(s) + CO(g) \xrightarrow{\Delta} Zn(g) + CO_2(g)$

22-3. (a) Dissolve $MgCO_3(s)$ in HCl(aq) and recrystallize $MgCl_2 \cdot 6\ H_2O$ from the aqueous solution.

$MgCO_3(s) + 2\ HCl(aq) \rightarrow MgCl_2(aq) + H_2O + CO_2(g)$

(b) Dissolve Na(s) in H_2O to produce NaOH(aq).

$2\ Na(s) + 2\ H_2O \rightarrow 2\ NaOH(aq) + H_2(g)$

Then dissolve Al in NaOH(aq) to obtain $NaAl(OH)_4(aq)$

$2\ Al + 2\ NaOH(aq) + 6\ H_2O \rightarrow 2\ NaAl(OH)_4 + 3\ H_2(g)$

(c) Dissolve ZnO(s) in HCl(aq): $ZnO(s) + 2\ HCl(aq) \rightarrow ZnCl_2(aq) + H_2O$

Then pass $H_2S(g)$ into the $ZnCl_2(aq)$: $Zn^{2+}(aq) + H_2S(g) \rightarrow ZnS(s) + 2\ H^+(aq)$

(d) HgS(s) will form to some extent by heating together Hg(l) and S(s), but a more effective synthesis would probably be to dissolve Hg(l) in an oxidizing acid (HNO_3)

$3\ Hg(l) + 8\ H^+ + 2\ NO_3^- \rightarrow 3\ Hg^{2+}(aq) + 4\ H_2O + 2\ NO(g)$

or

$Hg(l) + 4\ H^+ + 2\ NO_3^- \rightarrow Hg^{2+}(aq) + 2\ H_2O + 2\ NO_2(g)$

Then pass $H_2S(g)$ into the $Hg^{2+}(aq)$.

$Hg^{2+}(aq) + H_2S(g) \rightarrow HgS(s) + 2\ H^+(aq)$

22-4. (a) $Ba^{2+}(aq) + CO_3^{2-}(aq) \rightarrow BaCO_3(s)$

(b) $Mg(HCO_3)_2(aq) \xrightarrow{\Delta} MgCO_3(s) + H_2O + CO_2(g)$

(c) $SnO(s) + C(s) \rightarrow Sn(l) + CO(g)$

(d) $Sn(s) + 4\ HNO_3(\text{conc. aq}) \rightarrow SnO_2(s) + 2\ H_2O + 4\ NO_2(g)$

(e) $CdO(s) + H_2SO_4(aq) \rightarrow CdSO_4(aq) + H_2O$

(f) $PbO_2(s) + 4\ H^+ + 4\ I^- \rightarrow PbI_2(s) + 2\ H_2O + I_2$

22-5. In each case determine a value of K for the overall reaction by combining K values of two or more equilibria. Conclusions about the direction of spontaneous change are based on the magnitude of K.

(a) $Ba(OH)_2(s) \rightleftharpoons Ba^{2+}(aq) + 2\ OH^-(aq) \qquad K_{sp} = 5\times10^{-3}$

$Ba^{2+}(aq) + SO_4^{2-}(aq) \rightleftharpoons BaSO_4(s) \qquad 1/K_{sp} = 1/1.1\times10^{-10}$

$Ba(OH)_2(s) + SO_4^{2-}(aq) \rightleftharpoons BaSO_4(s) + 2\ OH^-(aq) \qquad K = \frac{5\times10^{-3}}{1.1\times10^{-10}} = 5\times10^{7}$

The large value of K suggests that equilibrium is displaced far to the right.

(b) $Mg(OH)_2(s) \rightleftharpoons Mg^{2+}(aq) + 2\ OH^-(aq) \qquad K_{sp} = 1.8\times10^{-11}$

$Mg^{2+}(aq) + CO_3^{2-}(aq) \rightleftharpoons MgCO_3(s) \qquad 1/K_{sp} = 1/3.5\times10^{-8}$

$Mg(OH)_2(s) + CO_3^{2-}(aq) \rightleftharpoons MgCO_3(s) + 2\ OH^-(aq) \qquad K = \frac{1.8\times10^{-11}}{3.5\times10^{-8}} = 5.1\times10^{-4}$

Here, the small value of K signifies that equilibrium is displaced to the left.

(c) $Ca(OH)_2(s) \rightleftharpoons Ca^{2+}(aq) + 2\ OH^- \qquad K_{sp} = 5.5\times10^{-6}$

$Ca^{2+}(aq) + 2\ F^-(aq) \rightleftharpoons CaF_2(s) \qquad 1/K_{sp} = 1/2.7\times10^{-11}$

$Ca(OH)_2(s) + 2\ F^-(aq) \rightleftharpoons CaF_2(s) + 2\ OH^-(aq) \qquad K = \frac{5.5\times10^{-6}}{2.7\times10^{-11}} = 2.0\times10^{5}$

Equilibrium is displaced to the right. [CaF_2 is considerably more insoluble than $Ca(OH)_2$.]

22-6. In some cases we may reason from Le Chatelier's principle. In other cases we made need to combine equilibrium constant expressions as in Review Problem 5.

(a) The key here is that $CO_2(g)$ is produced and escapes from the reaction mixture. Also the H_3O^+ required for the conversions $CO_3^{2-} \rightarrow HCO_3^- \rightarrow H_2CO_3 \rightarrow H_2O + CO_2$ comes from $HC_2H_3O_2$, whose ionization is promoted by the removal of H_3O^+ in the conversion of CO_3^{2-} to $CO_2(g)$. All of these factors cause equilibria to shift in the direction in which $SrCO_3$ is dissolved (to the right).

(b) The hydrolysis of NH_4^+ produces an acidic solution which should promote the dissolving of $Ba(OH)_2$, but we need to combine equilibrium expressions to see how significant this may be.

$$Ba(OH)_2(s) \rightleftharpoons Ba^{2+}(aq) + \cancel{2\ OH^-(aq)} \qquad K_{sp} = 5\times10^{-3}$$

$$2\ NH_4^+ + 2\ H_2O \rightleftharpoons 2\ NH_3 + \cancel{2\ H_3O^+} \qquad (K_a)^2 = \frac{(K_w)^2}{(K_b)^2} = \frac{(1.0\times10^{-14})^2}{(1.74\times10^{-5})^2}$$

$$\cancel{2\ H_3O^+} + \cancel{2\ OH^-} \rightleftharpoons 4\ H_2O \qquad 1/(K_w)^2 = 1/(1.0\times10^{-14})^2$$

$$Ba(OH)_2(s) + 2\ NH_4^+ \rightleftharpoons Ba^{2+} + 2\ NH_3 + 2\ H_2O$$

$$K = \frac{5\times10^{-3}(1.0\times10^{-14})^2}{(1.74\times10^{-5})^2(1.0\times10^{-14})^2} = 2\times10^7$$

The large value of K suggests that the dissolution of $Ba(OH)_2(s)$ goes to completion.

(c) $$ZnS(s) \rightleftharpoons Zn^{2+}(aq) + \cancel{S^{2-}(aq)} \qquad K_{sp} = 1.0\times10^{-21}$$

$$\cancel{S^{2-}(aq)} + H_3O^+ \rightleftharpoons \cancel{HS^-(aq)} + H_2O \qquad 1/K_{a_2} = 1/1.0\times10^{-14}$$

$$\cancel{HS^-(aq)} + H_3O^+ \rightleftharpoons H_2S(aq) + H_2O \qquad 1/K_{a_1} = 1/1.1\times10^{-7}$$

$$ZnS(s) + 2\ H_3O^+(aq) \rightleftharpoons Zn^{2+}(aq) + H_2S(aq) + 2\ H_2O$$

$$K = \frac{1.0\times10^{-21}}{(1.0\times10^{-14})(1.1\times10^{-7})} = 0.91$$

The value of K is large enough to suggest that some dissolving of ZnS should occur. Considering as well that $[H_3O^+]$ is maintained at a high value and that $[H_2S]$ cannot exceed 0.10 M (the saturation limit of H_2S in water), we should expect dissolution of ZnS to be complete.

22-7. The ion exchange reaction is $Na_2R + Ca^{2+} \rightarrow CaR + 2\ Na^+$

$$[Ca^{2+}] = \frac{185\ g\ Ca^{2+}}{1.0\times10^6\ g\ H_2O} \times \frac{1.00\times10^3\ g\ H_2O}{1.00\ L\ H_2O} \times \frac{1\ mol\ Ca^{2+}}{40.08\ g\ Ca^{2+}} = 4.6\times10^{-3}\ M$$

According to the ion exchange equation

$$[Na^+] = \frac{4.6\times10^{-3}\ mol\ Ca^{2+}}{L} \times \frac{2\ mol\ Na^+}{1\ mol\ Ca^{2+}} = 9.2\times10^{-3}\ M$$

22-8. The ion exchange reaction is $H_2R + Ca^{2+} \rightarrow CaR + 2\ H^+$

$$[Ca^{2+}] = \frac{77.5\ g\ Ca^{2+}}{1.0\times10^6\ g\ H_2O} \times \frac{1.00\times10^3\ g\ H_2O}{1.00\ L\ H_2O} \times \frac{1\ mol\ Ca^{2+}}{40.08\ g\ Ca^{2+}} = 1.9\times10^{-3}\ M$$

$$[H^+] = \frac{1.9\times10^{-3}\ mol\ Ca^{2+}}{L} \times \frac{2\ mol\ H^+}{1\ mol\ Ca^{2+}} = 3.8\times10^{-3}\ M$$

$$pH = -\log[H^+] = -\log(3.8\times10^{-3}) = -(-2.42) = 2.42$$

22-9. (a) $CdCO_3(s) \xrightarrow{\Delta} CdO(s) + CO_2(g)$

(b) $MgO(s) \xrightarrow{\Delta}$ no reaction

(c) $SnO_2(s) + CO(g) \xrightarrow{\Delta} SnO(s) + CO_2(g)$

followed by

$SnO(s) + CO(g) \xrightarrow{\Delta} Sn(l) + CO_2(g)$

(d) *oxid:* $2\ H_2O \rightarrow 4\ H^+ + O_2(g) + 4\ e^-$

red: $2\{Cd^{2+} + 2\ e^- \rightarrow Cd(s)\}$

---

net: $2\ Cd^{2+} + 2\ H_2O \xrightarrow{\text{electrolysis}} 2\ Cd(s) + 4\ H^+(aq) + O_2(g)$

(e) $2\ HgO(s) \xrightarrow{\Delta} 2\ Hg(l) + O_2(g)$

(f) $MgO(s) + Zn(s) \xrightarrow{\Delta}$ no reaction (Mg is a better reducing agent than Zn. That is, Mg will reduce ZnO to Zn.)

22-10. Combine two reactions to yield the desired result. Obtain approximate $\Delta\bar{G}°$ values of the two reactions from the appropriate figures. Then establish $\Delta\bar{G}°$ for the net reaction. If $\Delta\bar{G}° < 0$ we would expect the forward reaction to occur to a significant extent. If $\Delta\bar{G}° > 0$ the forward reaction would not occur to a significant extent.

(a) $2\ C(s) + O_2(g) \rightarrow 2\ CO(g)$ $\Delta\bar{G}° \simeq -275$ kJ/mol

$2\ HgO(s) \rightarrow 2\ Hg(l) + O_2(g)$ $\Delta\bar{G}° \simeq +100$ kJ/mol

---

$2\ HgO(s) + 2\ C(s) \rightarrow 2\ Hg(l) + 2\ CO(g)$ $\Delta\bar{G}° \simeq -175$ kJ/mol

$HgO(s) + C(s) \rightarrow Hg(l) + CO(g)$ $\Delta\bar{G}° \simeq -90$ kJ/mol

This reduction should occur at 100°C.

(b) $2\ Hg(l) + O_2(g) \rightarrow 2\ HgO(s)$ $\Delta\bar{G}° \simeq -75$ kJ/mol

$2\ ZnO(s) \rightarrow 2\ Zn(s) + O_2(g)$ $\Delta\bar{G}° \simeq +600$ kJ/mol

---

$2\ ZnO(s) + 2\ Hg(l) \rightarrow 2\ Zn(s) + 2\ HgO(s)$ $\Delta\bar{G}° \simeq +500$ kJ/mol

$ZnO(s) + Hg(l) \rightarrow Zn(s) + HgO(s)$ $\Delta\bar{G}° \simeq +250$ kJ/mol

The very large positive value of $\Delta\bar{G}°$ indicates that the reduction reaction will not occur to any significant extent.

(c) $2\ Mg(s) + O_2(g) \rightarrow 2\ MgO(s)$ $\Delta\bar{G}° \simeq -1100$ kJ/mol

$2\ ZnO(s) \rightarrow 2\ Zn(s) + O_2(g)$ $\Delta\bar{G}° \simeq +600$ kJ/mol

---

$2\ ZnO(s) + 2\ Mg(s) \rightarrow 2\ Zn(s) + 2\ MgO(s)$ $\Delta\bar{G}° \simeq -500$ kJ/mol

$ZnO(s) + Mg(s) \rightarrow Zn(s) + MgO(s)$ $\Delta\bar{G}° \simeq -250$ kJ/mol

The very large negative value of $\Delta\bar{G}°$ suggests that the reduction should go essentially to completion.

Exercises

Alkali (IA) Metals

22-1. $3\ Cl_2 + 6\ OH^-(aq) \xrightarrow{\Delta} ClO_3^-(aq) + 5\ Cl^-(aq)$

If NaOH(aq) is used the salts obtained are $NaClO_3$ and NaCl, which can be separated by fractional crystallization.

$Cl_2 + 2\ OH^-(aq) \rightarrow OCl^- + Cl^- + H_2O$

If NaOH(aq) is used, the hypochlorite solution is NaOCl(aq).

Dissolve $SO_2(g)$ in H_2O: $H_2O + SO_2(g) \rightarrow H_2SO_3(aq)$. Now allow NaOH(aq) and $H_2SO_3(aq)$ to react:

$2\ NaOH(aq) + H_2SO_3(aq) \rightarrow Na_2SO_3(aq) + 2\ H_2O$

Prepare $HNO_3(aq)$ by the reaction of $NO_2(g)$ with H_2O. $H_2O + 3\ NO_2 \rightarrow 2\ HNO_3(aq) + NO(g)$
Follow this by the reaction of $HNO_3(aq)$ and NaOH(aq). $HNO_3(aq) + NaOH(aq) \rightarrow NaNO_3(aq) + H_2O$.

Prepare NaH_2PO_4 and Na_2HPO_4 by neutralizing H_3PO_4 with NaOH(aq).

$H_3PO_4(aq) + NaOH(aq) \rightarrow NaH_2PO_4(aq) + H_2O$

$H_3PO_4(aq) + 2\ NaOH(aq) \rightarrow Na_2HPO_4(aq) + 2\ H_2O$.
Crystallize the sodium phosphates from solution and obtain polyphosphates by fusing them. For example, when fused NaH_2PO_4 yields $(NaPO_3)_n$ and a mixture of 2 mol Na_2HPO_4 and 1 mol NaH_2PO_4 yields $Na_5P_3O_{10}$.

To obtain $Na_2S_2O_3$ from Na_2SO_3, boil $Na_2SO_3(aq)$ with S(s) in an alkaline solution:
$Na_2SO_3(aq) + S(s) \rightarrow Na_2S_2O_3(aq)$.

To obtain $NaNO_2$ from $NaNO_3(aq)$: $2\ NaNO_3(s) \xrightarrow{\Delta} 2\ NaNO_2(s) + O_2(g)$

22-2. The reaction in question is

	$Ca(OH)_2(s)$	+	$CO_3^{2-}(aq) \rightleftharpoons$	$CaCO_3(s)$	+	$2\ OH^-(aq)$; $K = 2.0\times10^3$
Initially:			1.00 M			
Changes:			$-x$ M			$+2x$ M
At equil:			$(1.00 - x)$M			$2x$ M

$$K = \frac{[OH^-]^2}{[CO_3^{2-}]} = \frac{(2\ )^2}{(1.00 - x)} = 2.0\times10^3 \qquad 4x^2 + 2.0\times10^3x - 2.0\times10^3 = 0$$

$$x = \frac{-2.0\times10^3 \pm \sqrt{(2.0\times10^3)^2 + (4 \times 4 \times 2.0\times10^3)}}{2 \times 4} = 0.998$$

equilibrium concentrations: $[CO_3^{2-}] = 1.000 - x = 0.002 \approx 2\times10^{-3}$ M

$[OH^-] = 2x = 1.996\text{ M} \approx 2.00\text{ M}$

(Additional significant figures beyond what is normally allowed must be carried if a meaningful answer is to be obtained.)

22-3. The reaction in question here is

$$Ca(OH)_2(s) + SO_4^{2-}(aq) \rightleftharpoons CaSO_4(s) + 2\ OH^-(aq)$$

(a) Compare K_{sp} of $CaCO_3(2.8\times10^{-9})$ with K_{sp} of $CaSO_4(9.1\times10^{-6})$. The lower solubility of $CaCO_3$ means that reaction (22.4) should go more toward completion than the one written above. That is, the above reaction is not expected to go to completion.

(b) This calculation is similar to that of Exercise 2. First we need a value of K for the reaction written above.

$$Ca(OH)_2(s) \rightleftharpoons Ca^{2+}(aq) + 2\ OH^-(aq) \qquad K_{sp} = 5.5\times10^{-6}$$

$$Ca^{2+}(aq) + SO_4^{2-}(aq) \rightleftharpoons CaSO_4(s) \qquad 1/K_{sp} = 1/9.1\times10^{-6}$$

$$Ca(OH)_2(s) + SO_4^{2-}(aq) \rightleftharpoons CaSO_4(s) + 2\ OH^-(aq); \quad K = \frac{5.5\times10^{-6}}{9.1\times10^{-6}} = 0.60$$

	SO_4^{2-}	OH^-
Initial:	1.00 M	
Changes:	$-x$ M	$+2x$ M
At equil:	$(1.00 - x)$M	$2x$ M

$$K = \frac{[OH^-]^2}{[SO_4^{2-}]} = \frac{(2x)^2}{(1.00 - x)} = 0.60 \qquad 4x^2 + 0.60x - 0.60 = 0$$

$$x = \frac{-0.60 \pm \sqrt{(0.60)^2 + (4 \times 4 \times 0.60)}}{2 \times 4} = \frac{-0.60 \pm 3.2}{8} = 0.32$$

equilibrium concentrations: $[SO_4^{2-}] = 1.00 - 0.32 = 0.68$ M

$[OH^-] = 2x = 2 \times 0.32 = 0.64$ M

22-4. (a) The net reaction for the conversion of NaCl to $NaHCO_3$ is

$$NaCl(aq) + NH_3(g) + CO_2(g) + H_2O \rightarrow NaHCO_3(s) + NH_4Cl(aq)$$

We must calculate the theoretical yield of $NaHCO_3(s)$ and compare it with the actual yield. Note that $NaHCO_3$ and NaCl must be involved in the same ratio as their formula weights.

$$\text{no. t } NaHCO_3 = 1.00 \text{ t NaCl} \times \frac{84.0 \text{ t } NaHCO_3}{58.5 \text{ t NaCl}} = 1.44 \text{ t } NaHCO_3$$

(theoretical)

$$\%\ \text{efficiency} = \frac{1.03 \text{ t, actual}}{1.44 \text{ t, theoretical}} \times 100 = 71.5\%$$

(b) NH_3 is recycled in the Solvay process. Theoretically none should be consumed in an operating plant; however, a small loss of material is experienced in the recycling and must be replaced.

22-5. The reduction half reaction in the electrolysis is $2\ H_2O + 2\ e^- \rightarrow 2\ OH^- + H_2(g)$.

$$\text{no. mol } OH^- = 137 \text{ s} \times \frac{1.08 \text{ C}}{1 \text{ s}} \times \frac{1 \text{ mol } e^-}{96{,}500 \text{ C}} \times \frac{2 \text{ mol } OH^-}{2 \text{ mol } e^-} = 1.53\times10^{-3} \text{ mol } OH^-$$

$$[OH^-] = \frac{1.53\times10^{-3}\ \text{mol } OH^-}{0.445\ \text{L}} = 3.44\times10^{-3}\ \text{M}$$

$$pOH = -\log[OH^-] = -\log(3.44\times10^{-3}) = 2.463$$

$$pH = 14.00 - pH = 14.00 - 2.463 = 11.54$$

Magnesium

22-6. $$\text{no. t Mg} = 4\ \text{km}^3\ \text{seawater} \times \frac{(1000)^3\ \text{m}^3\ \text{seawater}}{1\ \text{km}^3\ \text{seawater}} \times \frac{(100)^3\ \text{cm}^3\ \text{seawater}}{1\ \text{m}^3\ \text{seawater}} \times \frac{1.03\ \text{g seawater}}{1.00\ \text{cm}^3\ \text{seawater}}$$

$$\times \frac{1\ \text{lb seawater}}{454\ \text{g seawater}} \times \frac{1\ \text{t seawater}}{2000\ \text{lb seawater}} \times \frac{1{,}272\ \text{g Mg}}{1\ \text{t seawater}} \times \frac{1\ \text{lb Mg}}{454\ \text{g Mg}} \times \frac{1\ \text{t Mg}}{200\ \text{lb Mg}} = 6\times10^6\ \text{t Mg}$$

22-7. First use the ideal gas equation to calculate the number of moles of $H_2(g)$. Next, calculate the number of moles of OH^- produced along with the $H_2(g)$ by the electrolysis reaction. Then, calculate $[OH^-]$ that would be present in solution if no precipitation occurred. Finally, use the mass-action expression to determine if precipitation will occur.

$$n = \frac{PV}{RT} = \frac{[(748 - 21)/760]\ \text{atm} \times 0.104\ \text{L}}{0.0821\ \text{L atm mol}^{-1}\ \text{K}^{-1} \times 296\ \text{K}} = 4.09\times10^{-3}\ \text{mol } H_2(g)$$

The reduction half-reaction is: $2\ H_2O + 2\ e^- \rightarrow 2\ OH^- + H_2(g)$.

$$\text{no. mol } OH^- = 4.09\times10^{-3}\ \text{mol } H_2 \times \frac{2\ \text{mol } OH^-}{1\ \text{mol } H_2} = 8.18\times10^{-3}\ \text{mol } OH^-$$

$$[OH^-] = \frac{8.18\times10^{-3}\ \text{mol } OH^-}{0.250\ \text{L}} = 3.27\times10^{-2}\ \text{M}$$

$$Q = [Mg^{2+}][OH^-]^2 = (2.20\times10^{-1}) \times (3.27\times10^{-2})^2 = 2.35\times10^{-4} > K_{sp} = 1.8\times10^{-11}$$

$Mg(OH)_2$ does precipitate from the solution.

22-8. The dissolving of $MgCO_3(s)$ in a solution of $NH_4^+(aq)$ occurs because the hydrolysis of $NH_4^+(aq)$ produces H_3O^+, which promotes the conversions $CO_3^{2-} \rightarrow HCO_3^- \rightarrow H_2CO_3 \rightarrow H_2O + CO_2$. Thus the more acidic the solution the more soluble the $MgCO_3(s)$.

(a) We must arrange the solutions in order of *decreasing* acidity. The most acidic of the three solutions is 1.00 M NH_4Cl. The solution 1.00 M NH_3 - 1.00 M NH_4Cl is a basic buffer solution. The solution 0.100 M NH_3 - 1.00 M NH_4Cl is also a buffer but not quite so basic as the preceding solution. The order we are seeking is 1.00 M NH_4Cl > 0.1000 M NH_3 - 1.00 M NH_4Cl > 1.00 NH_3 - 1.00 M NH_4Cl.

(b) The following equilibria must be combined.

$MgCO_3(s) \rightleftharpoons Mg^{2+}(aq) + \cancel{CO_3^{2-}}(aq)$ $\qquad K_{sp} = 3.5\times10^{-8}$

$NH_4^+ + \cancel{OH^-} \rightleftharpoons NH_3 + \cancel{H_2O}$ $\qquad K = 1/K_b = 1/1.74\times10^{-5}$

$\cancel{CO_3^{2-}} + \cancel{H_3O^+} \rightleftharpoons HCO_3^- + \cancel{H_2O}$ $\qquad K = 1/K_{a_2} = 1/5.6\times10^{-11}$

$\cancel{2\ H_2O} \rightleftharpoons \cancel{H_3O^+} + \cancel{OH^-}$ $\qquad K_w = 1.0\times10^{-14}$

$MgCO_3(s) + NH_4^+ \rightleftharpoons Mg^{2+} + HCO_3^- + NH_3$ $\qquad K = 3.6\times10^{-7}$

In 1.00 M NH_4Cl:

Initial:	1.00 M	-	-	-
Changes:	$-x$ M	$+x$ M	$+x$ M	$+x$ M
At equil:	$(1.00 - x)$M	x M	x M	x M

$$K = \frac{[Mg^{2+}][HCO_3^-][NH_3]}{[NH_4^+]} = \frac{x \cdot x \cdot x}{1.00 - x} = 3.6\times10^{-7}$$

Assume $x \ll 1.00$ $\quad x^3 = 3.6\times10^{-7} \quad x = [Mg^{2+}] = 7.1\times10^{-3}$ M

In 0.100 M NH_3 - 1.00 M NH_4Cl;

Initial:	1.00 M	-	-	0.100 M
Changes:	$-x$ M	$+x$ M	$+x$ M	$+x$ M
At equil:	$(1.00 - x)$M	x M	x M	$(1.00 + x)$M

$$K = \frac{x \cdot x \cdot (0.100 + x)}{(1.00 - x)} = \frac{x \cdot x(0.100)}{1.00} = 3.6\times10^{-7} \qquad x = [Mg^{2+}] = 1.9\times10^{-3} \text{ M}$$

In 1.00 M NH_3 - 1.00 M NH_4Cl:

Initial:	1.00 M	-	-	1.00 M
Changes:	$-x$ M	$+x$ M	$+x$ M	$+x$ M
At equil:	$(1.00 - x)$M	x M	x M	$(1.00 + x)$M

$$K = \frac{x \cdot x \cdot (1.00 + x)}{(1.00 - x)} = 3.6\times10^{-7} \qquad x = [Mg^{2+}] = 6.0\times10^{-4} \text{ M}$$

Hard Water

22-9. Initial precipitation of $CaCO_3(s)$: $Ca^{2+}(aq) + 2\ OH^-(aq) + CO_2(aq) \rightarrow CaCO_3(s) + H_2O$

Redissolving of $CaCO_3(s)$: $CaCO_3(s) + H_2O + CO_2(aq) \rightarrow Ca^{2+}(aq) + 2\ HCO_3^-(aq)$

Precipitation of $CaCO_3(s)$ by heating: $Ca^{2+}(aq) + 2\ HCO_3^-(aq) \rightarrow CaCO_3(s) + H_2O + CO_2(g)$

22-10. We start by writing the equations

$CaO + H_2O \rightarrow Ca(OH)_2 \qquad Ca(OH)_2 \rightarrow Ca^{2+} + 2\ OH^-$

followed by equations (22.16) and (22.17). For every mole of Cal consumed, 2 mol HCO_3^- is removed from solution. The HCO_3^- content of the hard water is taken to be 180.0 g HCO_3^- per 1×10^6 g H_2O.

$$\text{no. kg CaO} = 1.00\times10^6 \text{ gal} \times \frac{3.78 \text{ L}}{1 \text{ gal}} \times \frac{1000 \text{ cm}^3}{1 \text{ L}} \times \frac{1.00 \text{ g } H_2O}{1 \text{ cm}^3} \times \frac{180.0 \text{ g } HCO_3^-}{10^6 \text{ g } H_2O} \times \frac{1 \text{ mol } HCO_3^-}{61.0 \text{ g } HCO_3^-}$$

$$\times \frac{1 \text{ mol } OH^-}{1 \text{ mol } HCO_3^-} \times \frac{1 \text{ mol } Ca(OH)_2}{2 \text{ mol } OH^-} \times \frac{1 \text{ mol CaO}}{1 \text{ mol } Ca(OH)_2} \times \frac{56.1 \text{ g CaO}}{1 \text{ mol CaO}} \times \frac{1 \text{ kg CaO}}{1000 \text{ g CaO}} = 313 \text{ kg CaO}$$

22-11. (a) $$\text{no. g } CaCO_3 = 1.00\times10^6 \text{ gal} \times \frac{3.78 \text{ L}}{1 \text{ gal}} \times \frac{1000 \text{ cm}^3}{1 \text{ L}} \times \frac{1.00 \text{ g } H_2O}{1 \text{ cm}^3} \times \frac{180.0 \text{ g } HCO_3^-}{10^6 \text{ g } H_2O} \times \frac{1 \text{ mol } HCO_3^-}{61.0 \text{ g } HCO_3^-}$$

$$\times \frac{1 \text{ mol } CaCO_3}{1 \text{ mol } HCO_3^-} \times \frac{100 \text{ g } CaCO_3}{1 \text{ mol } CaCO_3} = 1.12\times10^6 \text{ g } CaCO_3 = 1.12\times10^3 \text{ kg } CaCO_3$$

(b) The water softening reactions are shown below, the calcium ion associated with the water itself as in italic.

$$CaO + H_2O \rightarrow Ca^{2+} + 2\ OH^-$$

$$\mathit{Ca}^{2+} + 2\ HCO_3^- + Ca^{2+} + 2\ OH^- \rightarrow \mathit{CaCO_3} + CaCO_3 + 2\ H_2O$$

As these equations indicate, half the calcium ion in the precipitated $CaCO_3$ comes from the lime and half from the water itself.

22-12. (a) $2\ Na^+ + CO_3^{2-} + Ca^{2+} + SO_4^{2-} \rightarrow CaCO_3(s) + 2\ Na^+ + SO_4^{2-}$

(b) $$\text{no. g } Na_2CO_3 = 385 \text{ L} \times \frac{1000 \text{ cm}^3}{1 \text{ L}} \times \frac{1 \text{ g } H_2O}{1 \text{ cm}^3} \times \frac{131 \text{ g } SO_4^{2-}}{10^6 \text{ g } H_2O} \times \frac{1 \text{ mol } SO_4^{2-}}{96.1 \text{ g } SO_4^{2-}} \times \frac{1 \text{ mol } Na_2CO_3}{1 \text{ mol } SO_4^{2-}}$$

$$\times \frac{106 \text{ g } Na_2CO_3}{1 \text{ mol } Na_2CO_3} = 55.6 \text{ g } Na_2CO_3$$

22-13. First pass the sample of hard water through a cation exchange resin that has H^+ as its counterions.

$H_2R + Mg^{2+} \rightarrow MgR + 2\ H^+$; $H_2R + Ca^{2+} \rightarrow CaR + 2\ H^+$; and so on.

The sample now contains H^+ as the only cation and HCO_3^- and SO_4^{2-}, as its anions. Pass this sample through an anion exchange resin having OH^- as its counterions.

$ROH + HCO_3^- \rightarrow RHCO_3 + OH^-$; $2\ ROH + SO_4^{2-} \rightarrow R_2SO_4 + 2\ OH^-$

When present in the same sample, H^+ and OH^- ions combine to form water. Because all other ions were removed in the two ion exchange processes, the sample of water is deionized.

Aluminum

22-14. In the text it is noted that Al is soluble (displaces H_2) both in acidic and basic solutions. Thus, we would expect Al to corrode readily at low and high pH values. It is at intermediate pH values (4.5 to 8.5) that it is most corrosion resistant. Or, viewed in another way, the protective coating on aluminum metal, i.e., Al_2O_3, is amphoteric. It is soluble at low and high pH values and insoluble at intermediate pH values.

22-15. There are many ways in which one may distinguish between Aluminum 2S (99.2% Al) and Magnalium (70% Al + 30% Mg) by simple chemical tests. For example, one might calculate the volume of hydrogen gas anticipated if a sample of the metal is dissolved in HCl(aq). The volume of hydrogen obtained per gram of metal is about 9% greater for Aluminum 2S than for Magnalium.

A much simpler test is to dissolve the metal in HCl(aq), followed by neutralization with NaOH(aq) and treatment with excess NaOH(aq). If the metal is Aluminum 2S, the $Al(OH)_3$ first precipitated will redissolve, because $Al(OH)_3$ is amphoteric. (The aluminum is obtained as $[Al(OH)_4]^-(aq)$). If the alloy is Magnalium, some $Mg(OH)_2$ will form and this will not redissolve; $Mg(OH)_2$ is not amphoteric.

22-16. $[Al(OH)_4]^-$ may be converted back to $Al(OH)_3(s)$ either by dilution of $[Al(OH)_4]^-(aq)$ to a large volume or by adding a weak acid. In both cases the purpose is to reduce the pH of the solution to about pH = 7. Water solutions of H_2CO_3 are weakly acidic. If acidification is attempted with HCl(aq), a strong acid, the precipitated $Al(OH)_3$ redissolves and appears in solution as Al^{3+}. [Can you establish that this redissolving would occur at about pH < 3?]

22-17. Concerning the compound $NaAl(SO_4)_2 \cdot 12\ H_2O$

(a) "Sodium alum" is not a sufficient name because although the M(I) ion is identified--Na^+--the M(III) ion is not. That is, it could be Al^{3+}, but it could also be Fe^{3+} or Cr^{3+}, etc.

(b) A more appropriate name would be sodium aluminum alum.

(c) An aqueous solution of this alum would be acidic. Neither Na^+ nor SO_4^{2-} hydrolyzes, but $[Al(H_2O)_6]^{3+}$ ionizes as an acid.

$$[Al(H_2O)_6]^{3+} + H_2O \rightleftharpoons [Al(H_2O)_5OH]^{2+} + H_3O^+$$

22-18. As described in the previous exercise $[Al(H_2O)_6]^{3+}$ produces an acidic solution. H_3O^+ from this ionization would react with either HCO_3^- or CO_3^{2-}, producing H_2O and $CO_2(g)$. Thus, if either compound, $Al(HCO_3)_3$ or $Al_2(CO_3)_3$, were produced, dissociation would quickly follow.

Extractive Metallurgy

22-19. (a)

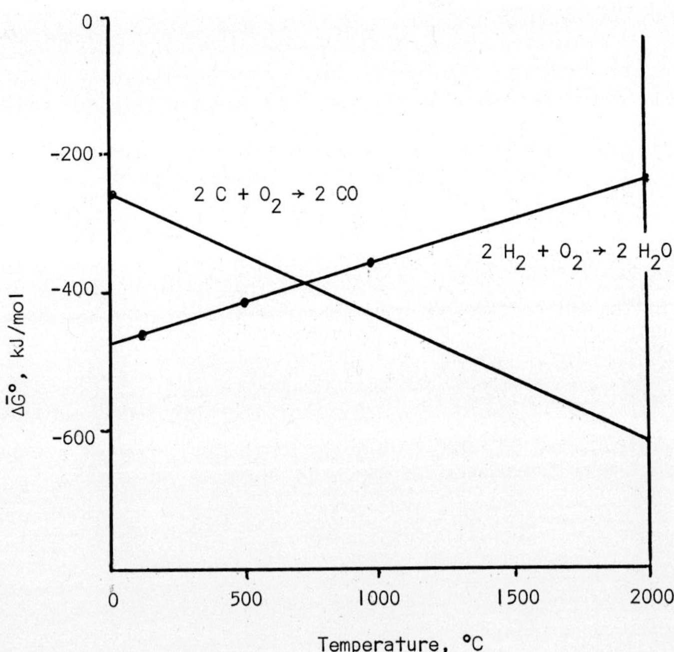

(b) For the water gas reaction

$C + \frac{1}{2}\ O_2 \rightarrow CO$	$\Delta\bar{G}° = \frac{1}{2}$ value listed in graph
$H_2O \rightarrow H_2 + \frac{1}{2}\ O_2$	$\Delta\bar{G}° = -\frac{1}{2}$ value listed in graph
$C + H_2O \rightarrow CO + H_2$	$\Delta\bar{G}° = \frac{1}{2}$ value listed for oxidation of C $-\frac{1}{2}$ value listed for oxidation of H_2

At temperatures below about 750°C the sum of the two terms noted above is positive and the reaction is not expected to occur to a significant extent in the forward direction. Above about 750°C the sum of terms (and hence $\Delta\bar{G}°$) becomes increasingly more negative, and the forward reaction becomes increasingly more spontaneous.

22-20.

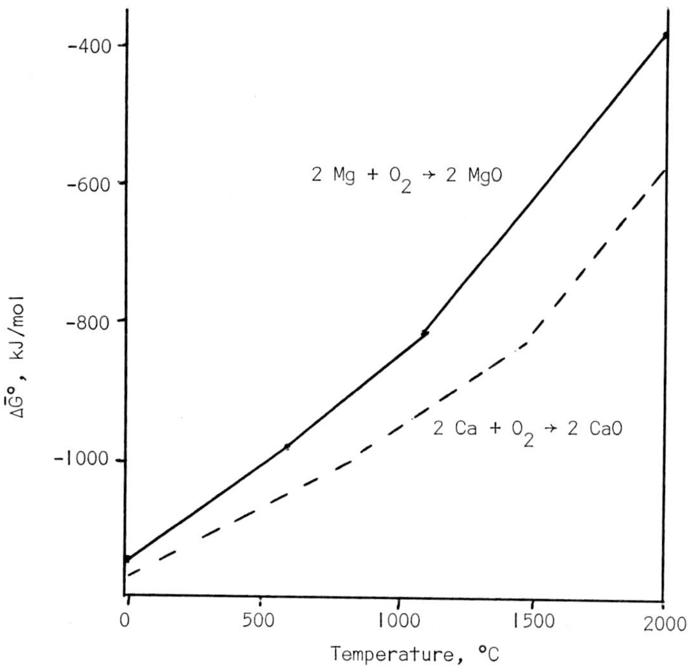

The reaction in question is

$2\ MgO \rightarrow 2\ Mg + O_2$

$2\ Ca + O_2 \rightarrow 2\ CaO$

$2\ Ca + 2\ MgO \rightarrow 2\ Mg + 2\ CaO$

If this reaction is spontaneous at all temperatures from 0° to 2000°C, then the graph for the oxidation of Ca must lie below the graph for the oxidation of Mg. (Recall that the sign of $\Delta\bar{G}°$ must be changed from that of the graph for $2\ Mg + O_2 \rightarrow 2\ MgO$ in the above summation.) Breaks are indicated in the graph at the melting and boiling points of Ca. The hypothetical graph for $2\ Ca + O_2 \rightarrow 2\ CaO$ is plotted with a broken line.

22-21. Assume that $\Delta\bar{H}°$ is practically independent of temperature and assess how $\Delta\bar{G}°$ changes with temperature, with special attention to the $T\Delta S$ term in $\Delta\bar{G}° = \Delta\bar{H}° - T\Delta\bar{S}°$.

(a) In the reaction $C(s) + O_2(g) \rightarrow CO_2(g)$ we would expect $\Delta\bar{S}°$ to be rather small (no change in number of moles of gas). The $T\Delta\bar{S}°$ should similarly be of minor importance and the variation of $\Delta\bar{G}°$ with temperature should be about the same as that of $\Delta\bar{H}°$. In short, for this reaction we would not expect much of a variation of $\Delta\bar{G}°$ with temperature.

(b) $2\ CO(g) + O_2(g) \rightarrow 2\ CO_2(g)$. Here there is a decrease in number of moles of gas; $\Delta\bar{S}° < 0$. The term $-T\Delta\bar{S}° > 0$, and the higher the temperature the larger this term becomes. Thus, regardless of the value of $\Delta\bar{H}°$, we should expect $\Delta\bar{G}°$ for this reaction to increase with temperature.

22-22. (a) Breaks occur at the melting and boiling points of the metals in Figure 22-12 because at these points the reaction conditions change, e.g., from $2\ M(s) + O_2(g) \rightarrow 2\ MO(s)$ to $2\ M(l) + O_2(g) \rightarrow 2\ MO(s)$. Because the physical condition of the metal M differs (solid in one case and liquid in the other), we should expect a different $\Delta\bar{G}°$.

(b) Consider that the second of the two equations written in (a) can be represented by the following sum.

$2\ M(s) + O_2(g) \rightarrow 2\ MO(s)$ $\Delta\bar{G}°$ = a negative quantity

$2\ M(l) \rightarrow 2\ M(s)$ $\Delta\bar{G}° = ?$

$2\ M(l) + O_2(g) \rightarrow 2\ MO(s)$ $\Delta\bar{G}° = ?$

At the melting point, for the transition $l \rightarrow s$, $\Delta\bar{G}° = 0$. Above the melting point $\Delta\bar{G}° > 0$ for this transition (the freezing of a liquid is nonspontaneous at temperatures above the melting point). The sum of the two expressions is not as negative as it would be in the absence of melting. The $\Delta\bar{G}°$ value lies above the straight-line extrapolation of the low-temperature line, and thus the slope of the new line is more positive.

(c) The break at the boiling point is sharper than at the melting point because for the transition, gas → liquid, $\Delta\bar{S}°$ is much larger (more negative) than for liquid → solid. This causes $\Delta\bar{G}°$ for the transition; gas → liquid, to increase more rapidly with increased temperature than for liquid → solid.

Tin and Lead

22-23. The name dilead(II)lead(IV) oxide is quite specific about the composition of the compound: two Pb atoms in the oxidation state +2 and one Pb atm in the oxidation state +4. Each Pb(II) atom must have one O atom (oxid. state -2) associated with it, and the Pb(IV) atom, two O atoms. This means a formula unit with three Pb atoms and four O atoms--2 (PbO)PbO_2 or Pb_3O_4.

If we propose a hypothetical orthoplumbic acid it would be $Pb(OH)_4$ or H_4PbO_4. Salts of this acid could be called orthoplumbates and the lead(II) salt would be Pb_2PbO_4. This is equivalent to writing Pb_3O_4.

22-24. (a) $Pb(s) + 4\ H^+ + 2\ NO_3^- \rightarrow Pb^{2+} + 2\ H_2O + 2\ NO_2(g)$

(b) $4\ H^+ + 2\ Cl^- + PbO_2(s) \rightarrow Pb^{2+} + 2\ H_2O + Cl_2(g)$

(c) $2\ Fe^{3+} + Sn^{2+} \rightarrow 2\ Fe^{2+} + Sn^{4+}$

(d) $3\ PbO(s) + H_2O + 2\ CO_2 \rightarrow 2\ PbCO_3{\cdot}Pb(OH)_2(s)$

22-25. (a) The reaction of $Pb_3O_4(s)$ with $HNO_3(aq)$ yields $Pb(NO_3)_2(aq)$, which can then be combined with $K_2CrO_4(aq)$ to yield $PbCrO_4(s)$.

$Pb_3O_4(s) + 4\ HNO_3(aq) \rightarrow 2\ Pb(NO_3)_2(aq) + PbO_2(s) + 2\ H_2O$ (22.57)

$Pb(NO_3)_2(aq) + K_2CrO_4(aq) \rightarrow PbCrO_4(s) + 2\ KNO_3(aq)$

(b) PbS(s) is soluble in $HNO_3(aq)$.

$3\ PbS(s) + 8\ H^+ + 2\ NO_3^- \rightarrow 3\ Pb^{2+}(aq) + 4\ H_2O + 2\ NO(g) + 3\ S(s)$

$Pb^{2+}(aq)$ can be precipitated by the addition of HCl(aq).

$Pb^{2+}(aq) + 2\ HCl(aq) \rightarrow PbCl_2(s) + 2\ H^+(aq)$

22-26. The dissolution of Sn in NaOH(aq) is represented by the equation

$Sn + OH^-(aq) + 2\ H_2O \rightarrow Sn(OH)_3^- + H_2(g)$

The reduction of $Sn(OH)_3^-$ by electrolysis can be represented by the reduction half-equation $Sn(OH)_3^- + 2\ e^- \rightarrow Sn(s) + 3\ OH^-$

22-27. Pb(IV) is a powerful oxidizing agent, capable of oxidizing I^- to I_2 and Br^- to Br_2. Thus Pb(IV) and I^- and Br^- cannot be maintained in contact; a spontaneous reaction would occur between Pb(IV) and the halide ion, as indicated by the following electrode potential data.

(a) *red.* $Pb^{4+} + 2\ e^- \rightarrow Pb^{2+}$ $\qquad E^\circ_{red} = +1.5\ V$

(b) *oxid.* $2\ I^- \rightarrow I_2 + 2\ e^-$ $\qquad E^\circ_{ox} = -0.535\ V$

(c) *oxid.* $2\ Br^- \rightarrow Br_2 + 2\ e^-$ $\qquad E^\circ_{ox} = -1.065\ V$

The combination of (a) with either (b) or (c) leads to $E^\circ_{cell} > 0$ and suggests a spontaneous forward reaction. That is Pb(IV) is expected to oxidize I^- to I_2 and Br^- to Br_2.

22-28. If metallic tin is able to reduce Sn^{4+} to Sn^{2+}, then it should be successful in maintaining tin in its lower oxidation state.

oxid. $Sn \rightarrow Sn^{2+} + 2\ e^-$ $\qquad E^\circ_{ox} = -(-0.136) = +0.136\ V$

red. $Sn^{4+} + 2\ e^- \rightarrow Sn^{2+}$ $\qquad E^\circ_{red} = 0.15\ V$

---

net: $Sn + Sn^{4+} \rightarrow 2\ Sn^{2+}$ $\qquad E^\circ_{cell} = 0.29\ V$

Thus any Sn^{4+} that is produced by the air oxidation of Sn^{2+} is easily reduced back to Sn^{2+} with Sn as the reducing agent.

Group IIB metals

22-29. (a) $Cd(s) + 2\ H_2SO_4(conc.\ aq) \rightarrow CdSO_4(aq) + 2\ H_2O + SO_2(g)$

(b) $3\ Hg + 8\ H^+ + 2\ NO_3^- \rightarrow 3\ Hg^{2+} + 4\ H_2O + 2\ NO(g)$

(c) $ZnO(s) + 2\ HC_2H_3O_2(aq) \rightarrow Zn(C_2H_3O_2)_2(aq) + H_2O$

(d) $Zn^{2+}(aq) + OH^-(aq) \rightarrow Zn(OH)_2(s)$

followed by $Zn(OH)_2(s) + 2\ OH^-(aq) \rightarrow [Zn(OH)_4]^{2-}(aq)$

22-30. Combine the following half cells and $\Delta\bar{G}^\circ$ values:

$2\ Hg^{2+} + 2\ e^- \rightarrow Hg_2^{2+}$ $\qquad -\Delta\bar{G}^\circ = 2\mathcal{F}E^\circ$

$Hg_2^{2+} + 2\ e^- \rightarrow 2\ Hg$ $\qquad -\Delta\bar{G}^\circ = 2\mathcal{F}(0.796)$

$2\ Hg^{2+} + 4\ e^- \rightarrow 2\ Hg$ $\qquad -\Delta\bar{G}^\circ = 4\mathcal{F}(0.851) = 2\mathcal{F}E^\circ + 2\ (0.796)$

$2\ E^\circ = 4 \times 0.851 - 2 \times 0.796$ $\qquad E^\circ = (2 \times 0.851) - 0.796 = +0.90\ V$

22-31. First determine the vapor pressure of Hg at 25°C with the equation provided.

$$\log p\ (\text{mmHg}) = \frac{-(0.05223 \times 61{,}960)}{298.15\ \text{K}} + 8.118$$

$$= -10.85 \quad + 8.118 = -2.73$$

$$p\ (\text{mmHg}) = \text{antilog}(-2.73) \quad = 0.0019\ \text{mmHg}$$

Now use the ideal gas equation to calculate the mass of Hg in 1 m^3 = 1000 L.

$$PV = \frac{m}{M}RT \qquad m = \frac{MPV}{RT} = \frac{200.6\ \text{g mol}^{-1} \times (0.0019/760)\ \text{atm} \times 1000\ \text{L}}{0.0821\ \text{L atm mol}^{-1}\ \text{K}^{-1} \times 298\ \text{K}} = 0.020\ \text{g Hg}$$

This quantity of Hg greatly exceeds the maximum permissible level of 0.05 mg/m^3.

22-32. In the presence of acidic foods, e.g., canned tomatoes, zinc might dissolve to some extent over a period of time and produce aqueous Zn^{2+} at higher concentrations than can be tolerated. On the other hand, tin is so much less active than Zn that its dissolution is negligible. Another problem with using Zn is that it is almost always contaminated with small amounts of Cd, and Cd^{2+} is extremely deleterious to human health.

Self-Test Questions

1. (*b*) Aluminum is the most difficult of the listed metals to reduce from its combined forms. If this reduction were to be accomplished with C, it would require a very high temperature to do so.

2. (*c*) To soften temporary hard water requires that HCO_3^- be converted to CO_3^{2-}, by the reaction $HCO_3^- + OH^- \rightarrow H_2O + CO_3^{2-}$. The substance used must have basic properties. Of these listed all do except NH_4Cl (which is slightly acidic by hydrolysis).

3. (*a*) If one assumes that each of the sulfides produces the metal oxide on roasting, the question becomes one of which element oxide is easily decomposed to the free metal and oxygen at moderate temperatures. The oxide that decomposes in this way is HgO, so it is HgS that we expect to yield the free metal upon roasting.

4. (*c*) To be soluble in NaOH(aq) an oxide must be either acidic or amphoteric. ZnO, Al_2O_3 and SnO_2 are all amphoteric. By contrast Fe_2O_3 is a basic oxide. Fe_2O_3 dissolves in an acid but not in a basic medium.

5. (*a*) Hg is less active than H_2(g). This means that Hg^{2+} is not difficult to reduce to Hg(l), i.e., the reduction potential is positive. For the other three metals, E°_{red} is negative. These metals will not displace H_2(g) from H^+(aq). Of the three, however, E°_{red} is most negative for Zn^{2+}/Zn.

6. (*a*) Na^+ and K^+ do not hydrolyze; neither does SO_4^{2-}. The aluminate in $[Al(OH)_4]^-$ can only exist in a basic medium. Hydrated zinc ion produces an acidic solution by the ionization

$$[Zn(H_2O)_6]^{2+} + H_2O \rightleftharpoons [Zn(H_2O)_5OH]^+ + H_3O^+$$

7. (*b*) The +4 oxidation state of Sn (e.g., SnO_2) and the +2 oxidation state of Hg (e.g. HgO) are quite stable. The +2 oxidation state of Mg(e.g., MgO) is very stable (difficult to reduce). The +4 oxidation state of Pb is one that is difficult to obtain and maintain. PbO_2 is a very strong oxidizing agent [easily reduced to Pb(II)].

8. (*d*) To be used in a low melting alloy a metal must usually have a low melting point. The melting points of Sn, Pb and Cd are quite low, but the melting point of Mg is high enough that it is not used in low melting solids. The low melting points of Sn, Pb, and Cd were all encountered in one or more contexts in Chapter 22.

9. (a) $Sn(s) + 4\ HNO_3(\text{conc. aq}) \rightarrow SnO_2(s) + 2\ H_2O + 4\ NO_2(g)$

(b) $MgCO_3(s) + H_2O + CO_2 \rightarrow Mg^{2+} + 2\ HCO_3^-$

(c) $Zn(s) + 2\ OH^- + 2\ H_2O \rightarrow [Zn(OH)_4]^{2-} + H_2(g)$

(d) $2\ CdS(s) + 3\ O_2(g) \rightarrow 2\ CdO(s) + 2\ SO_2(g)$

followed by $CdO(s) + C(s) \rightarrow Cd + CO(g)$

10. (a) MgO and BaO are of a similar type ($M^{2+}O^{2-}$), but because of the smaller size of Mg^{2+} than Ba^{2+} the lattice energy of MgO is greater than that of BaO. A higher temperature is required to destroy the crystalline lattice of MgO than of BaO.

(b) Again this is a question of the compound with the higher lattice energy being more difficult to dissolve (i.e., being less soluble). Because of the smaller size of F^- compared to Cl^-, MgF_2 has a higher lattice energy than $MgCl_2$.

11. A particularly simple method is to treat the metal with concentrated $HNO_3(aq)$. The Pb-Cd alloy should dissolve completely, but the Pb-Sn alloy will leave a residue (SnO_2).

12. The exchange reaction can be represented as $R(OH)_2 + SO_4^{2-} \quad RSO_4 + 2\ OH^-$.

Determine the no. mol OH^- displaced by the SO_4^{2-}.

$$\text{no. mol } OH^- = 0.02158\ L \times \frac{1.00\times10^{-3}\ \text{mol } H_2SO_4}{L} \times \frac{2\ \text{mol } OH^-}{1\ \text{mol } H_2SO_4} = 4.32\times10^{-5}\ \text{mol OH}$$

$$\text{no. mol } SO_4^{2-} = 4.32\times10^{-5}\ \text{mol } OH^- \times \frac{1\ \text{mol } SO_4^{2-}}{2\ \text{mol } OH^-} = 2.16\times10^{-5}\ \text{mol } SO_4^{2-}$$

$$\text{no. mol } Ca^{2+} = 2.16\times10^{-5}\ \text{mol } SO_4^{2-} \times \frac{1\ \text{mol } Ca^{2+}}{1\ \text{mol } SO_4^{2-}} = 2.16\times10^{-5}\ \text{mol } Ca^{2+}$$

$$\text{no. g } Ca^{2+} \text{ in } 25.00\ \text{mL} = \frac{2.16\times10^{-5}\ \text{mol } Ca^{2+}}{25.00\ \text{g } H_2O} \times \frac{40.08\ \text{g } Ca^{2+}}{1\ \text{mol } Ca^{2+}} = 3.46\times10^{-5}\ \text{g } Ca^{2+}/\text{g } H_2O$$

$$\text{per million g } H_2O\text{:} \quad \frac{3.46\times10^{-5}\ \text{g } Ca^{2+}}{\text{g } H_2O} \times \frac{1.0\times10^{6}}{1.0\times10^{6}} = \frac{3.46\times10^{1}\ \text{g } Ca^{2+}}{1.0\times10^{6}\ \text{g } H_2O} = 34.6\ \text{ppm } Ca^{2+}$$

Chapter 23

Chemistry of the Transition Elements

Review Problems

23-1. (a) barium dichromate = $BaCr_2O_7$ (b) $Sc(OH)_3$ = scandium hydroxide

(c) chromium trioxide = CrO_3 (d) MnO = manganese(II) oxide

(e) iron(II) silicate = $FeSiO_3$ (f) $Fe(CO)_5$ = iron(penta)carbonyl

23-2. (a) ferromangagese = Fe + Mn (b) cast iron = Fe with more than 1% C

(c) chromite ore = $Fe(CrO_2)_3$ (d) chrome alum = $KCr(SO_4)_2 \cdot 12\ H_2O$

(e) aqua regia = 1 part HNO_3 + 3 parts HCl (f) Prussian blue = $Fe_4[Fe(CN)_6]_3$

23-3. (a) $TiCl_4(g) + 4\ Na(l) \xrightarrow{\Delta} 4\ NaCl(l) + Ti(s)$ (Since Mg will reduce $TiCl_4$, we should expect Na to do so also.)

(b) $Cr_2O_3(s) + 2\ Al(s) \xrightarrow{\Delta} Al_2O_3(s) + 2\ Cr(l)$ (The thermite reaction)

(c) $Ag(s) + HCl(aq) \rightarrow$ no reaction (Ag is less active than H_2.)

(d) $K_2Cr_2O_7(aq) + 2\ KOH(aq) \rightarrow 2\ K_2CrO_4 + H_2O$ (This is the chromate-dichromate equilibrium, which favors CrO_4^{2-} in basic solution.)

(e) $MnO_2(s) + 2\ C(s) \xrightarrow{\Delta} Mn + 2\ CO(g)$ (This is similar to the reaction for the preparation of ferromanganese.)

(f) $Fe(OH)_3(s) + NaOH(aq) \rightarrow$ no reaction [$Fe(OH)_3$ is a basic compound with no amphoteric properties; it will not dissolve in excess NaOH(aq).]

23-4. (a) *oxid.* $2\{Fe_2S_3(s) + 6\ OH^- + 6\ e^- \rightarrow 2\ Fe(OH)_3(s) + 3\ S(s)\}$

red. $3\{O_2(g) + 2\ H_2O + 4\ e^- \rightarrow 4\ OH^-\}$

---

net: $2\ Fe_2S_3(s) + 6\ H_2O + 3\ O_2(g) \rightarrow 4\ Fe(OH)_3(s) + 6\ S(s)$

(b) *oxid.* $2\{Mn^{2+} + 4\ H_2O \rightarrow MnO_4^- + 8\ H^+ + 5\ e^-\}$

red. $5\{S_2O_8^{2-} + 2\ e^- \rightarrow 2\ SO_4^{2-}\}$

---

net: $2\ Mn^{2+} + 5\ S_2O_8^{2-} + 8\ H_2O \rightarrow 2\ MnO_4^- + 10\ SO_4^{2-} + 16\ H^+$

(c) *oxid.* $4\{Ag(s) + 2\ CN^- \rightarrow [Ag(CN)_2]^- + e^-\}$

red. $O_2(g) + 2\ H_2O + 4\ e^- \rightarrow 4\ OH^-$

---

net: $4\ Ag(s) + 8\ CN^- + O_2(g) + 2\ H_2O \rightarrow 4[Ag(CN)_2]^- + 4\ OH^-$

23-5. (a) Acidify $Na_2CrO_4(aq)$ and recrystallize $Na_2Cr_2O_7(s)$.

$2\ CrO_4^{2-}(aq) + 2\ H^+(aq) \rightarrow Cr_2O_7^{2-}(aq) + H_2O$

(b) Several possibilities exist for obtaining $Cr_2O_3(s)$ from Na_2CrO_4. One involves converting Na_2CrO_4 to $Na_2Cr_2O_7$, as in part (a). Then, if favorable conditions (of temperature) can be found, separate NaCl and $(NH_4)_2Cr_2O_7$ from a solution of $Na_2Cr_2O_7$ and NH_4Cl by fractional crystallization. Cr_2O_3 is produced by strongly heating $(NH_4)_2Cr_2O_7$.

$$(NH_4)_2Cr_2O_7 \xrightarrow{\Delta} Cr_2O_3(s) + 4\ H_2O(g) + N_2(g) \qquad (23.12)$$

Alternatively, reduce $Cr_2O_7^{2-}$ to Cr^{3+}. e.g.,

$$Cr_2O_7^{2-} + 8\ H^+ + 3\ SO_3^{2-} \rightarrow 2\ Cr^{3+} + 3\ SO_4^{2-} + 4\ H_2O$$

Precipitate Cr^{3+} as the hydroxide $Cr(OH)_3$,

$$Cr^{3+} + 3\ OH^- \rightarrow Cr(OH)_3(s)$$

and decompose $Cr(OH)_3(s)$ by heating.

$$2\ Cr(OH)_3 \xrightarrow{\Delta} Cr_2O_3(s) + 3\ H_2O(g)$$

(c) Once Cr_2O_3 has been obtained in (b), dissolve it in HCl(aq) and recrystallize $CrCl_3$ from aqueous solution.

$$Cr_2O_3(s) + 6\ H^+ + 6\ Cl^- \rightarrow \underbrace{2\ Cr^{3+} + 6\ Cl^-}_{\text{recrystallize as } CrCl_3} + 3\ H_2O$$

(d) Freshly prepared $Cr(OH)_3$, as in part (b), dissolves in excess OH^- to yield $[Cr(OH)_4]^-(aq)$. Also Cr_2O_3 can be dissolved in excess NaOH(aq).

$$Cr(OH)_3(s) + Na^+(aq) + OH^-(aq) \rightarrow Na^+(aq) + [Cr(OH)_4]^-(aq)$$

23-6. (a) An oxidizing acid would be one containing an anion that is a good oxidizing agent, e.g. $HNO_3(aq)$

$$3\ Cu + 8\ H^+ + 2\ NO_3^- \rightarrow 3\ Cu^{2+} + 2\ NO(g) + 4\ H_2O$$

(b) The transition metal oxide must have acidic properties to dissolve in NaOH(aq). Cr_2O_3 is a good example of an amphoteric transition metal oxide.

$$Cr_2O_3(s) + 2\ OH^- + 3\ H_2O \rightarrow 2[Cr(OH)_4]^-$$

(c) This must be a sulfide of the ammonium sulfide group (cation group 3), of which MnS is most soluble in acid. ZnS and FeS are also soluble.

$$FeS(s) + 2\ H_3O^+(aq) \rightarrow Fe^{2+}(aq) + 2\ H_2O + H_2S(g)$$

(Recall that this reaction can be used for the laboratory preparation of H_2S.)

(d) The iron triad is Fe, Co, and Ni. In the qualitative analysis scheme $NH_3(aq)$ is used to precipitate $Fe(OH)_3(s)$, while converting Co^{3+} and Ni^{2+} to the complex ions $[Co(NH_3)_6]^{3+}$ and $[Ni(NH_3)_6]^{2+}$. Our example *cannot* be $Fe(OH)_3$, but $Ni(OH)_2$ works fine.

$$Ni(OH)_2(s) + 6\ NH_3(aq) \rightarrow [Ni(NH_3)_6]^{2+}(aq) + 2\ OH^-(aq)$$

(e) Any of the inner transition metals can be chosen for this example, or simply use the general symbol Ln.

$$2\ Ln(s) + 6\ H_3O^+(aq) \rightarrow 2\ Ln^{3+}(aq) + 6\ H_2O + 3\ H_2(g)$$

23-7. (a) For the diagram written below, the reduction potentials given are from Table 20-2, Table 23-1, Table 23-2, and expression (23.15)

$$Cr_2O_7^{2-} \xrightarrow{+1.33\ V} Cr^{3+} \xrightarrow{-0.41\ V} Cr^{2+} \xrightarrow{-0.91\ V} Cr$$

Cr^{3+} to Cr: -0.744 V

(b) For the species in Example 23-1

$Zn \rightarrow Zn^{2+} + 2\ e^-$; $E^\circ_{ox} = +0.763$ V

$Sn^{2+} \rightarrow Sn^{4+} + 2\ e^-$; $E^\circ_{ox} = -0.154$ V

$2\ I^- \rightarrow I_2(s) + 2\ e^-$; $E^\circ_{ox} = -0.535$ V

Because of the large positive value of E°_{ox} for the couple $Cr_2O_7^{2-}/Cr^{3+}$, any of the three--Zn, Sn^{2+}, or I^---is capable of reducing $Cr_2O_7^{2-}$ to Cr^{3+}. However, for the reduction of Cr^{3+}, for which $E^\circ_{red} = -0.41$ V only a reducing agent whose $E^\circ_{ox} > 0.41$ V will work. This means only Zn.

23-8. Use equation (23.19) in parts (a) and (b). First convert the pH value to $[H^+]$.

(a) At pH = 5.0, $[H^+] = 1.0\times10^{-5}$ M

$$\frac{[Cr_2O_7^{2-}]}{[CrO_4^{2-}]^2} = 3.2\times10^{14}\ [H^+]^2 = 32.\times10^{14}(1.0\times10^{-5})^2 = 3.2\times10^4$$

For example, if $[CrO_4^{2-}] = 0.010$ M, $[Cr_2O_7^{2-}] = 3.2$ M

(b) At pH = 9.12, $[H^+] = \text{antilog}(-9.12) = 7.6\times10^{-10}$ M

$$\frac{[Cr_2O_7^{2-}]}{[CrO_4^{2-}]^2} = 3.2\times10^{14}[H^+]^2 = 3.2\times10^{14}(7.6\times10^{-10})^2 = 1.8\times10^{-4}$$

For example, if $[CrO_4^{2-}] = 1.0$ M, $[Cr_2O_7^{2-}] = 1.8\times10^{-4}$ M

23-9. For this problem we use expression (23.46)

$$K = \frac{[Cu^{2+}]}{[Cu^+]^2} = 1.8\times10^6$$

(a) If $[Cu^{2+}] = [Cu^+]$

$$\frac{[Cu^{2+}]}{[Cu^+]^2} = \frac{[Cu^{2+}]}{[Cu^{2+}]^2} = \frac{1}{[Cu^{2+}]} = 1.8\times10^6$$

$[Cu^+] = [Cu^{2+}] = 1/1.8\times10^6 = 5.6\times10^{-7}$ M

(b) Here, $[Cu^{2+}] + [Cu^+] = 1.00\times10^{-4}$ and $[Cu^{2+}] = (1.00\times10^{-4} - [Cu^+])$

$$\frac{(1.00\times10^{-4} - [Cu^+])}{[Cu^+]^2} = 1.8\times10^6$$ let $[Cu^+] = x$ and solve the

quadratic equation $1.00\times10^{-4} - x = 1.8\times10^6\ x^2$

Rearrange the equation to $x^2 + 5.6\times10^{-7}x - 5.6\times10^{-11} = 0$

$$x = \frac{-5.6\times10^{-7} \pm \sqrt{(5.6\times10^{-7})^2 + 4 \times 5.6\times10^{-11}}}{2} = \frac{(-5.6\times10^{-7}) + (1.5\times10^{-5})}{2} = 7.0\times10^{-6}$$

$[Cu^+] = 7.0\times10^{-6}$ M

$$\% \; Cu^+ = \frac{7.0\times10^{-6} \text{ M}}{1.00\times10^{-4} \text{ M}} \times 100 = 7.0\% \; Cu^+$$

23-10. (a) H_2O: NaOH is highly water soluble and $Fe(OH)_3$ is insoluble.

(b) NH_3: Ni^{2+} forms to complex ion $[Ni(NH_3)_6]^{2+}$ and Fe^{3+} does not form a complex ion with NH_3. This is the procedure used in qualitative analysis cation group 3 to separate Ni^{2+} from Fe^{3+}.

(c) NaOH: $Cr_2O_3(s)$ is amphoteric and will dissolve in $OH^-(aq)$ to form $[Cr(OH)_4]^-$. Fe_2O_3 is a basic oxide. Having no acidic properties (not being amphoteric) it is insoluble in NaOH(aq).

(d) HCl: MnS is quite soluble in HCl(aq). CoS requires aqua regia for its dissolution. HCl(aq) is used in qualitative analysis cation group 3 to separate these two sulfides.

Exercises

Properties of the transition elements

23-1. Transition metal atoms display several oxidation states. Representative metal atoms, such as those in group IIA, display but one oxidation state. Group IIA metal atoms and ions are diamagnetic. Group IA metal atoms are paramagnetic but their ions are diamagnetic. Most transition metal atoms and ions are paramagnetic. Most transition metal ions are colored; representative metal ions are not. Representative metal ions form few complexes, whereas transition metal ions form many.

23-2. In a series of transition elements, additional electrons go into an inner electron shell whereas the number of electrons in the outer shell tends to remain constant. Thus, the outershell electrons experience an essentially unvarying force of attraction. Atomic sizes change very little in a transition series. The atomic radius decreases from the alkali metal atom of group IA to the alkaline earth metal atom of group IIA and successively throughout a period of elements, as described on pages 204-205 of the text.

23-3. Transition elements are characterized by the involvement of d orbitals in bond formation; this includes the metallic bonds between atoms in a metallic crystal. Cr has the maximum number of half-filled $3d$ orbitals--five. Its potential for forming metallic bonds would seem to be greatest. For the transition elements preceding Cr, the number of available $3d$ electrons is less than the number of $3d$ orbitals, and for the transition elements following Cr the number of half-filled $3d$ orbitals falls steadily. In Zn all the $3d$ orbitals are filled and metallic bonding involves only $4s$ and $4p$ orbitals. This leads to weaker metallic bonding and a lower melting point than for the transition metals.

23-4. (a) The difference among lanthanoid elements is mostly in the electron configuration of the f subshell. This subshell is so "buried" within an atom that its occupancy has little effect on the behavior of outer shell electrons, e.g. on the ease with which these electrons are lost or gained. The occupancy of $3d$ subshell has a greater effect on electrons of the valence shell, and hence on reduction potentials. Moreover, with the elements Cr and Cu the $3d$ subshell is directly involved in the reduction process $M^{2+} + 2\,e^- \rightarrow M(s)$.

(b) The lanthanoids, for the reason stated above about the relative noninvolvement of the f subshell in determining properties, would be expected to be about as active as the group of elements that precedes them in the periodic table--group IIA.

Oxidation-reduction

23-5. (a) $VO_2^+ + 2\,H^+ + e^- \rightarrow V^{3+} + H_2O$

(b) $Cr^{2+} \rightarrow Cr^{3+} + e^-$

(c) $Fe(OH)_3 + 5\ OH^- \rightarrow FeO_4^{2-} + 4\ H_2O + 3\ e^-$

(d) $[Ag(CN)_2]^- + e^- \rightarrow Ag(s) + 2\ CN^-$

23-6. Separate each net equation into an oxidation and a reduction half-reaction. Combine the half-cell potentials to evaluate $E°_{cell}$.

(a) *oxid.* $3\{2\ Br^- \rightarrow Br_2 + 2\ e^-\}$ $\qquad E°_{ox} = -1.065$ V

red. $2\{VO_2^+ + 4\ H^+ + 3\ e^- \rightarrow V^{2+} + 2\ H_2O\}$ $\qquad E°_{red} = +0.37$ V

net: $2\ VO_2^+ + 6\ Br^- + 8\ H^+ \rightarrow V^{2+} + 3\ Br_2 + 4\ H_2O$ $\qquad E°_{cell} = -0.69$ V

The large negative value of $E°_{cell}$ suggests that this reaction will not occur to any significant extent in the forward direction.

To obtain $E°_{red}$ use the method of Example 21-2:

$VO_2^+ + 2\ H^+ + e^- \rightarrow VO^{2+} + H_2O$ $\qquad -\Delta\bar{G}° = \mathcal{F}(1.00)$

$VO^{2+} + 2\ H^+ + e^- \rightarrow V^{3+} + H_2O$ $\qquad -\Delta\bar{G}° = \mathcal{F}(0.361)$

$V^{3+} + e^- \rightarrow V^{2+}$ $\qquad -\Delta\bar{G}° = \mathcal{F}(-0.255)$

$VO_2^+ + 4\ H^+ + 3\ e^- \rightarrow V^{2+} + 2\ H_2O$

$$-\Delta\bar{G}° = -3\mathcal{F}(E°_{red}) = \mathcal{F}(1.00 + 0.361 - 0.255)$$

$$E°_{red} = \frac{1.00 + 0.361 - 0.255}{3} = 0.367$$

An alternative way to consider this question is in terms of a stepwise reduction of VO_2^+ to V^{2+} with data from Figure 23-1. With its $E°_{ox}$ of -1.065 V we would not even expect the Br^-/Br_2 even to be very effective in reducing VO_2^+ to VO^{2+}.

(b) *oxid.* $Fe^{2+} \rightarrow Fe^{3+} + e^-$ $\qquad E°_{ox} = -0.771$ V

red. $VO_2^+ + 2\ H^+ + e^- \rightarrow VO^{2+} + H_2O$ $\qquad E°_{red} = +1.00$ V

net: $Fe^{2+} + VO_2^+ + 2\ H^+ \rightarrow Fe^{3+} + VO_2^{2+} + H_2O$ $\qquad E°_{cell} = +0.23$ V

The positive value of $E°_{cell}$ suggests that equilibrium in this reaction is displaced to the right, that is, that the forward reaction occurs to a significant extent.

(c) *oxid.* $5\{H_2O_2 \rightarrow O_2(g) + 2\ H^+ + 2\ e^-\}$ $\qquad E°_{ox} = -0.682$ V

red. $2\{MnO_4^- + 8\ H^+ + 5\ e^- \rightarrow Mn^{2+} + 4\ H_2O\}$ $\qquad E°_{red} = +1.51$ V

net: $5\ H_2O_2 + 2\ MnO_4^- + 6\ H^+ \rightarrow 2\ Mn^{2+} + 5\ O_2(g) + 8\ H_2O$ $\qquad E°_{cell} = +0.83$ V

Again, a large positive value of $E°_{cell}$ signifies that the forward reaction occurs to a significant extent (essentially going to completion).

23-7. $Cr_2O_7^{2-} + 3\ Zn(s) + 14\ H^+ \rightarrow 2\ Cr^{3+} + 7\ H_2O + 3\ Zn^{2+}$

(orange) (green)

$2\ Cr^{3+} + Zn(s) \rightarrow 2\ Cr^{2+} + Zn^{2+}$

(green) (blue)

$4\ Cr^{2+} + O_2 + 4\ H^+ \rightarrow 4\ Cr^{3+} + 2\ H_2O$

(blue) (green)

23-8. Use the method of Example 21-2 as follows:

$MnO_4^- + 4\ H^+ + 3\ e^- \rightarrow \cancel{MnO_2(s)} + 2\ H_2O \qquad \Delta\bar{G}° = -3\mathcal{F}(1.70)$

$\cancel{MnO_2(s)} + 4\ H^+ + 2\ e^- \rightarrow Mn^{2+} + 2\ H_2O \qquad \Delta\bar{G}° = -2\mathcal{F}(1.23)$

$MnO_4^- + 8\ H^+ + 5\ e^- \rightarrow Mn^{2+} + 4\ H_2O \qquad \Delta\bar{G}° = -5\mathcal{F}E°_{red}$

$-5\mathcal{F}E°_{red} = -3\mathcal{F}(1.70) - 2\mathcal{F}(1.23) \qquad E°_{red} = \frac{(3 \times 1.70) + (2 \times 1.23)}{5} = 1.51\ V$

23-9. For the first reduction pick a reducing agent for which $E°_{ox}$ is between -0.361 V and -1.00 V. This reducing agent will carry the reduction of VO_2^+ to VO^{2+} but no further. For example, with $H_2O_2(aq)$

oxid. $H_2O_2(aq) \rightarrow O_2(g) + 2\ H^+ + 2\ e^- \qquad E°_{ox} = -0.682\ V$

red. $2\{VO_2^+ + 2\ H^+ + e^- \rightarrow VO^{2+} + H_2O\} \qquad E°_{red} = 1.00\ V$

net: $2\ VO_2^+ + H_2O_2 + 2\ H^+ \rightarrow 2\ VO^{2+} + 2\ H_2O + O_2(g) \qquad E°_{cell} = +0.32\ V$

This means that H_2O_2 will reduce VO_2^+ to VO^{2+}, but now for the further reduction to V^{3+},

oxid. $H_2O_2(aq) \rightarrow O_2(g) + 2\ H^+ + 2\ e^- \qquad E°_{ox} = -0.682\ V$

red. $2\{VO^{2+} + 2\ H^+ + e^- \rightarrow V^{3+} + H_2O\} \qquad E°_{red} = +0.361\ V$

net: $2\ VO^{2+} + H_2O_2 + 2\ H^+ \rightarrow 2\ V^{3+} + 2\ H_2O + O_2(g) \qquad E°_{cell} = -0.321\ V$

This reaction is not expected to proceed to any significant extent in the forward direction.

Now, seek a reducing agent with $E°_{ox}$ between -0.321 V and +0.255 V. This will reduce VO^{2+} to V^{3+}, but will not reduce V^{3+} to V^{2+}. A suitable reducing agent is $Sn^{2+}(aq)$.

The net reducing agent must have $+0.255\ V < E°_{ox} < +1.18\ V$. A suitable choice is Zn(s).

The final reducing agent must have $E°_{ox} > 1.18\ V$, e.g., Al(s).

23-10. The two half-equations listed must be combined in the manner of Example 21-2.

$$Cr^{2+}(aq) + 2\,e^- \rightarrow Cr(s) \qquad \Delta\bar{G}° = -n\mathcal{F}E° = +2\mathcal{F}(0.91)$$

$$Cr^{3+}(aq) + e^- \rightarrow Cr^{2+}(aq) \qquad \Delta\bar{G}° = -n\mathcal{F}E° = +\mathcal{F}(0.41)$$

$$Cr^{3+}(aq) + 3\,e^- \rightarrow Cr(s) \qquad \Delta\bar{G}° = -n\mathcal{F}E° = -3\mathcal{F}E° = 2.2\,\mathcal{F}$$

$E° = -2.2/3 = -0.73$ V Note how well this result agrees with the value listed in Table 23-2.

23-11. The large positive reduction potential for the half-reaction

$$Cr_2O_7^{2-} + 14\,H^+ + 6\,e^- \rightarrow 2\,Cr^{3+} + 7\,H_2O \qquad E°_{red} = +1.33\ V$$

means that many substances can be oxidized by $Cr_2O_7^{2-}$ in acidic medium, e.g., Br^-, Fe^{2+}, I^-, etc. By contrast the reduction potential for CrO_4^{2-} in basic solution is small (slightly negative).

$$CrO_4^{2-}(aq) + 4\,H_2O + 3\,e^- \rightarrow Cr(OH)_3(s) + 5\,OH^- \qquad E°_{red} = -0.13\ V$$

$CrO_4^{2-}(aq)$ is not capable of oxidizing Br^- to Br_2; I^- to I_2; Fe^{2+} to Fe^{3+}, etc.

Because so many cations form insoluble chromates (e.g., Ba^{2+}, Pb^{2+}, Ag^+, ...) $CrO_4^{2-}(aq)$ is a useful precipitating agent.

23-12. The $CO_2(g)$ produces an acidic reaction through the equilibria

$$CO_2(g) + H_2O \rightleftharpoons H_2CO_3(g)$$

$$H_2CO_3(aq) + H_2O \rightleftharpoons H_3O^+(aq) + HCO_3^-(aq)$$

The H_3O^+ formed in this way then causes a shift in the dichromate-chromate equilibrium toward the dichromate side.

$$2\,CrO_4^- + 2\,H^+ \rightleftharpoons Cr_2O_7^{2-} + H_2O$$

23-13. According to the text, the equilibrium between CrO_4^{2-} and $Cr_2O_7^{2-}$ can be thought of in terms of the following two reactions:

$$2\,H^+ + 2\,CrO_4^{2-} \rightleftharpoons 2\,HCrO_4^- \qquad K = (1/3.2\times10^{-7})^2$$

$$2\,HCrO_4^- \rightleftharpoons Cr_2O_7^{2-} + H_2O \qquad K = ?$$

$$2\,H^+ + 2\,CrO_4^{2-} \rightleftharpoons Cr_2O_7^{2-} + H_2O \qquad K = 3.2\times10^{14}$$

But also, K for the net reaction, (3.2×10^{14}) is equal to the product of the other K values, that

is, $3.2\times10^{14} = K \times \dfrac{1}{(3.2\times10^{-7})^3}$ $K = 33$

23-14. The total chromium concentration must be 0.100 M, partly as CrO_4^{2-} and partly as $Cr_2O_7^{2-}$. (In the expression below $[Cr_2O_7^{2-}]$ must be multiplied by two since two mol/L of Cr would appear in solution for every mol/L of $Cr_2O_7^{2-}$.)

$$[CrO_4^{2-}] + 2[Cr_2O_7^{2-}] = 0.100$$

From equation (23.19), at pH 7.00

$$\frac{[Cr_2O_7^{2-}]}{[CrO_4^{2-}]^2} = 32.\times10^{14}(1.0\times10^{-7})^2 = 3.2$$

Let $[CrO_4^{2-}] = x$ and $[Cr_2O_7^{2-}] = (0.100 - x)/2$

$$\frac{(0.100 - x)}{2\,x^2} = 3.2 \qquad 0.100 - x = 6.4\,x^2 \qquad x^2 + 0.16x - 0.016 = 0$$

$$x = \frac{-0.16 \pm \sqrt{(0.16)^2 + 0.064}}{2} = \frac{-0.16 \pm 0.30}{2} = 0.070\text{ M}$$

$[CrO_4^{2-}] = x = 0.070$ M

$[Cr_2O_7^{2-}] = (0.100 - x)/2 = (0.100 - 0.070)/2 = 0.015$ M

23-15. First use equation (18.10) to determine $[H_3O^+]$ in the acetic acid-ammonium acetate buffer.

$$pH = 4.76 + \log\frac{[C_2H_3O_2^-]}{[HC_2H_3O_2]} = 4.76 + \log\frac{1.0}{1.0} = 4.76 \qquad [H_3O^+] = 1.7\times10^{-5}\text{ M}$$

Now use equation (23.19) to calculate $[CrO_4^{2-}]$ in a buffer solution with $[H_3O^+] = 1.74\times10^{-5}$ M.

$$\frac{[Cr_2O_7^{2-}]}{[CrO_4^{2-}]^2} = 3.2\times10^{14} \times (1.7\times10^{-5})^2 = 9.2\times10^4$$

Since the ratio $[Cr_2O_7^{2-}]/[CrO_4^{2-}]^2$ is so large--9.2×10^4--it is safe to assume that most of the chromium remains as $Cr_2O_7^{2-}$ at this pH. That is, assume that $[Cr_2O_7^{2-}] = 1.0\times10^{-3}$ M and that

$$[CrO_4^{2-}]^2 = \frac{[Cr_2O_7^{2-}]}{9.2\times10^4} = \frac{1.0\times10^{-3}}{9.2\times10^4} = 1.1\times10^{-8} \qquad [CrO_4^{2-}] = 1.0\times10^{-4}\text{ M}$$

Only the solubility product constant of $BaCrO_4$ is exceeded.

$$[Ba^{2+}][CrO_4^{2-}] = (0.10)(1.0\times10^{-4}) = 1.0\times10^{-5} > 1.2\times10^{-10}$$

$$[Sr^{2+}][CrO_4^{2-}] = (0.10)(1.0\times10^{-4}) = 1.0\times10^{-5} < 2.2\times10^{-5}$$

$$[Ca^{2+}][CrO_4^{2-}] = (0.10)(1.0\times10^{-4}) = 1.0\times10^{-5} < 7.1\times10^{-4}$$

The separation of Ba^{2+} from Sr^{2+} and Ca^{2+} does not occur under the stated conditions.

23-16. The chrome plating bath contains chromium as $Cr_2O_7^{2-}$. The electrolytic reduction is

$Cr(VI) + 6\,e^- \rightarrow Cr$

no. coloumbs required: $22.7\text{ cm}^2 \times 0.0010\text{ mm} \times \frac{1\text{ cm}}{10\text{ mm}} \times \frac{7.14\text{ g Cr}}{1\text{ cm}^3} \times \frac{1\text{ mol Cr}}{52.0\text{ g Cr}} \times \frac{6\text{ mol e}^-}{1\text{ mol Cr}}$

$$\times \frac{96{,}500\text{ C}}{1\text{ mol e}^-} = 1.8\times10^2$$

no. s $= 1.8\times10^2\text{ C s}\,\frac{1\text{ s}}{4.2\text{ A}} = 43$ s

The iron triad

23-17. Differences in the electron configurations of Fe, Co, and Ni are limited to the $3d$ subshell (that is, $3d^6$, $3d^7$ and $3d^8$, respectively). Each atom has two electrons in the $4s$ subshell. The atomic sizes are very similar and the way in which the electron configurations become altered through ionization and compound formation are also similar. As a result, these three elements resemble one another in physical and chemical properties.

23-18. Paramagnetism requires simply that unpaired electrons be present in individual atoms, ions or molecules. Ferromagnetism requires that, in addition to being paramagnetic, ions cluster into magnetic domains. When these domains are oriented in a magnetic field, an object displays ferromagnetism. Magnetic domains can only be produced by species having ionic radii corresponding to those of iron, cobalt, and nickel.

23-19. The reduction half-reaction is $O_2 + 4\ H^+ + 4\ e^- \rightarrow 2\ H_2O$ $\qquad E° = +1.229$ V.

We should expect Fe^{2+} to be oxidized to Fe^{3+}, as suggested by the half-reaction:

$Fe^{2+} \rightarrow Fe^{3+} + e^-$ $\qquad E°_{ox} = -(0.771)$ V [from Table 20-2]. But we should not expect Co^{2+} to be oxidized to Co^{3+}: $Co^{2+} \rightarrow Co^{3+} + e^-$ $\qquad E°_{ox} = -1.82$ V (equation 23.31).

For the reason stated on page 722 of the text, we should expect Ni^{2+} to behave in a similar manner to Co^{2+}. We should not expect it to be oxidized to Ni^{3+} by $O_2(g)$ in acidic solution.

Carbonyls

23-20. (a) Mo has an atomic number of 42. To bring the total number of electrons in its carbonyl to that of the noble gas at the end of the fifth period, Xe (Z = 54), requires that 12 electrons or 6 electron pairs be made available through CO molecules. This means six CO molecules and a carbonyl with the formula $Mo(CO)_6$ and an octahedral structure.

(b) Five pairs of electrons brought to the carbonyl structure by CO molecules produces a noble gas electron configuration, that of Rn (Z = 86). The formula is $Os(CO)_5$ and the structure is trigonal bipyramidal.

(c) Rhenium carbonyl must form a binuclear carbonyl. The basic unit, $Re(CO)_5$, requires an extra electron from a second unit to produce the electron configuration of Rn (Z = 86). The formula of the carbonyl is $Re_2(CO)_{10}$.

23-21. Nickel and iron carbonyls are $Ni(CO)_4$ and $Fe(CO)_5$, respectively, whereas cobalt forms a *binuclear* carbonyl, $Co_2(CO)_9$. Because the binuclear carbonyl has a considerably higher molecular weight than the other two, we should expect it to exist as a solid at a higher temperature than would be the case for the others.

23-22. The total number of electrons associated with the V atom in $V(CO)_6$ is 35. We might expect this covalent carbonyl molecule to acquire one extra electron, say from a sodium atom. The result is a compound consisting of the ions Na^+ and $[V(CO)_6]^-$.

The coinage metals

23-23. In the parting process, Ag dissolves in the concentrated acid and Au remains undissolved.

$HNO_3(aq)$: $Ag(s) + 2\ H^+ + NO_3^- \rightarrow Ag^+ + H_2O + NO_2(g)$

$H_2SO_4(aq)$: $2\ Ag(s) + 4\ H^+ + SO_4^{2-} \rightarrow 2\ Ag^+ + 2\ H_2O + SO_2(g)$

Some $Ag_2SO_4(s)$ might form in the treatment of Ag with H_2SO_4, but this can be dissolved in a large volume of water (Ag_2SO_4 is moderately soluble).

At a minimum Ag^+ would have to be displaced from solution to obtain pure Ag(s). [Zn(s) might be the reducing agent.]. It might also be necessary to redissolve the gold (e.g., by cyanidation or with aqua regia) and displace Au(s) from solution.

23-24. (a) *Dissolution:* $AgO + 2\ H_3O^+ \rightarrow Ag^{2+} + 3\ H_2O$

Oxidation-reduction:

oxid. $2\ H_2O \rightarrow 4\ H^+ + O_2(g) + 4\ e^-$ $\quad E^\circ_{ox} = -1.229$ V

red. $4\{Ag^{2+} + e^- \rightarrow Ag^+\}$ $\quad E^\circ_{red} = ?$

net: $4\ Ag^{2+} + 2\ H_2O \rightarrow 4\ Ag^+ + 4\ H^+ + O_2(g)$ $\quad E^\circ_{cell} = ?$

(b) We must determine E°_{red} and E°_{cell} for the reaction in (a). The method of Example 21-2 requires

$Ag^{2+} + e^- \rightarrow Ag^+$ $\quad -\Delta\bar{G}^\circ = \mathcal{F}(E^\circ_{red})$

$Ag^+ + e^- \rightarrow Ag$ $\quad -\Delta\bar{G}^\circ = \mathcal{F}(0.800)$

$Ag^{2+} + 2\ e^- \rightarrow Ag$ $\quad -\Delta\bar{G}^\circ = 2\mathcal{F}(1.39)$

$$-\Delta\bar{G}^\circ = 2\ (1.39) = \mathcal{F}(E^\circ_{red}) + \mathcal{F}(0.800)$$

$$E^\circ_{red} = (2 \times 1.39) - 0.800 = +1.98\ V$$

(The value 1.39 V is taken from Table 23-10).

Now for the reaction in part (a) we can write

$$E^\circ_{cell} = E^\circ_{ox} + E^\circ_{red} = -1.229\ V + 1.98\ V = +0.75\ V$$

The large positive value of E°_{cell} suggests that the liberation of $O_2(g)$ from H_2O by Ag^{2+} occurs to a significant extent (essentially going to completion).

23-25. The relationship between K and E°_{cell} is

$$E^\circ_{cell} = \frac{0.0592}{n} \log K$$

Reaction (23.47): $n = 1$ and $-1.88 = \frac{0.0592}{1} \log K$

$\log K = -19.9 \quad K = 1\times10^{-20}$

Reaction (23.48): $n = 2$ and $0.39 = \frac{0.0592}{2} \log K$

$\log K = 13 \quad K = 1\times10^{13}$

23-26. For the stated reaction

oxid. $2\ I^- \rightarrow I_2 + 2\ e^-$ $\quad E^\circ_{ox} = -0.535$ V

red. $2\{Cu^{2+} + e^- \rightarrow Cu^+\}$ $\quad E^\circ_{red} = +0.152$ V

net: $2\ Cu^{2+} + 2\ I^- \rightarrow 2\ Cu^+ + I_2$ $\quad E^\circ_{cell} = -0.383$ V

We would not expect this reaction to occur to any appreciable extent if reactants and products are in their standard states. However, if Cu^+ is to precipitate as CuI(s), $[Cu^+]$ will be considerably less than 1.00 M. Reducing $[Cu^+]$ and maintaining a high $[I^-]$ do tend to favor the forward reaction, perhaps enough so to make it spontaneous. To determine $[Cu^+]$ in saturated CuI(aq) in the presence of I^- as a common ion we proceed in a familiar fashion. Suppose $[I^-] = 1.00$ M, then $[Cu^+] = K_{sp}/[I^-] = 1.1\times10^{-12}/1.00 = 1.1\times10^{-12}$ M.

According to the Nernst equation

$$E_{cell} = E^\circ_{cell} - \frac{0.0592}{n} \log \frac{[Cu^+]^2}{[Cu^{2+}]^2} \quad \text{when } n = 2$$

$$E_{cell} = -0.383 - 0.0592 \log [Cu^+]/[Cu^{2+}]$$

Suppose $[Cu^{2+}]$ is also 1.00 M. Then $E_{cell} = -0.383 - 0.0592 \log(1.1\times10^{-12}) = -0.383 + 0.708 = 0.325$ V

Thus, a spontaneous reaction should occur with large values of $[Cu^{2+}]$ and $[I^-]$ and a low value of $[Cu^+]$ (because of the formation of a precipitate).

23-27. If the disproportionation reaction were *unimportant* then $[Au^{3+}] << [Au^+]$, where $[Au^+]$ is established by the solubility of AuCl. Let $[Au^+] = x = [Cl^-]$ and

$$K_{sp} = [Au^+][Cl^-] = x^2 = 2.0\times10^{-13} \qquad x = [Au^+] = 4.5\times10^{-7} \text{ M}$$

Now let us set up the Nernst equation for reaction (23.48), substitute this value for $[Au^+]$ and see what conclusion we reach.

$$E_{cell} = E^\circ_{cell} - \frac{0.0592}{2} \log \frac{[Au^{3+}]}{[Au^+]^3}$$

Let us ask this question: What is $[Au^{3+}]$ when reaction (23.48) reaches equilibrium and $[Au^+] = 4.5\times10^{-7}$ M? If $[Au^{3+}]$ proves to be negligibly small, then little decomposition of AuCl occurs. If $[Au^{3+}]$ is large relative to $[Au^+]$ then decomposition must occur. At equilibrium $E_{cell} = 0.000$.

$$0.00 = 0.39 - \frac{0.0592}{2} \log \frac{[Au^{3+}]}{(4.5\times10^{-7})^3}$$

$$0.00 = 0.39 - 0.0296 \log[Au^{3+}] + 0.0296 \log(4.5\times10^{-7})^3$$

$$0.00 = 0.39 - 0.0296 \log[Au^{3+}] - 0.564$$

$$\log [Au^{3+}] = \frac{0.39 - 0.564}{0.0296} = -5.7 \qquad [Au^{3+}] = 2\times10^{-6} \text{ M}$$

The calculated value of $[Au^{3+}]$ is larger than that of $[Au^+]$. Disproportionation of Au^+ must occur. In the final equilibrium $[Au^+] < 45.\times10^{-7}$ M and $[Au^{3+}]$ will be significant compared to $[Au^+]$. (An exact solution would require substituting the expression $[Au^+] + [Au^{3+}] + 2\,[Au^{3+}] = 4.5\times10^{-7}$ M into the Nernst equation. The term "2 $[Au^{3+}]$" represents the concentration of gold that deposits as the pure metal.)

Qualitation Analysis

23-28. (a) $FeS(s) + 2\,H^+ \rightarrow Fe^{2+} + H_2S(g)$

(b) $CoS + 4\,H^+ + 2\,NO_3^- + 4\,Cl^- \rightarrow [CoCl_4]^{2-} + S + 2\,H_2O + 2\,NO_2(g)$

(c) $2\,Cr^{3+} + 3\,H_2O_2 + 10\,OH^- \rightarrow 2\,CrO_4^{2-} + 8\,H_2O$

(d) $Ni(OH)_2(s) + 6\,NH_3(aq) \rightarrow [Ni(NH_3)_6]^{2+}(aq) + 2\,OH^-(aq)$

(e) $Fe^{2+} + K_3[Fe(CN)_6] \rightarrow$ Turnbull's blue

(f) $MnO_2(s) + 4\ H^+ + 2\ Cl^- \rightarrow Mn^{2+} + 2\ H_2O + Cl_2(g)$

23-29. (a) In the absence of NH_4Cl, the hydroxide ion concentration would be great enough that $Mg(OH)_2$ would precipitate with the ammonium sulfide group.

(b) Without the addition of H_2O_2, Cr^{3+} would not be oxidized to CrO_4^{2-} nor would cobalt be oxidized to the oxidation state +3 [in $Co(OH)_3$].

(c) The hydroxides of Fe, Co, and Ni would all precipitate if NaOH were used as the reagent instead of NH_3. The desired separation of Co^{3+} and Ni^{3+} from Fe^{3+} through the formation of amine complex ions would not occur.

(d) Dissolving of CoS and NiS would occur along with MnS; a separation of Co^{2+}, Ni^{2+} and Mn^{2+} would not be achieved.

(e) Traces of Fe^{3+} would fail to be complexed and would interfere with the test for cobalt if NaF were omitted from the reagent in step 7.

23-30. Dissolve the sample in HCl(aq). The expected ions that would appear in solution are Fe^{2+} and some combination of Mn^{2+}, Ni^{2+} and Cr^{3+}. Treat this solution with NaOH and H_2O_2. If chromium is present, a yellow color (CrO_4^{2-}) appears in solution. The precipitate that forms is $Fe(OH)_3$ and either $Ni(OH)_2$ or MnO_2 or a mixture of the two. Treat the precipitate with HCl(aq). Dissolving should be complete. Now add $NH_3(aq)$. The precipitate that forms is $Fe(OH)_3$. The filtrate contains Mn^{2+}, $[Ni(NH_3)_6]^{2+}$, both or neither. Add HCl and H_2S. If precipitation occurs, the precipitate is NiS. Add a strong oxidizing agent, such as $(NH_4)_2S_2O_8$, to the remaining solution. If Mn^{2+} is present it will be oxidized to the strongly colored purple MnO_4^-.

23-31. The absence of a precipitate in the basic buffer solution signifies the *absence* of Fe^{3+}, Cr^{3+}, and Al^{3+}. The *absence* of Ni^{2+} and Co^{3+} is indicated by the lack of color in the sulfide precipitate and by the fact that the precipitate is completely soluble in HCl(aq). (CoS and NiS only dissolve in aqua regia.) ZnS and MnS are both light-colored and soluble in HCl(aq). Very likely one or both Fe^{2+} and Mn^{2+} is (are) present. FeS is a dark-colored precipitate, but if present in small quantities mixed with ZnS and/or MnS, its presence may go unnoticed. Further tests would be required for Fe^{2+}, Mn^{2+} and Zn^{2+}.

23-32. If the initial group 3 precipitate is partly soluble in NaOH(aq) then either Cr or Al must be present, since they would be formed in the group 3 precipitate as hydroxides and their hydroxides are amphoteric. ZnS, if present, would not be expected to dissolve in NaOH(aq). Following dissolution in aqua regia and treatment with NaOH(aq) and $H_2O_2(aq)$, the possible precipitates are $Fe(OH)_3$, $Co(OH)_3$, $Ni(OH)_2$ and MnO_2. If any Zn was present in the original sample it would now be found in the NaOH(aq) as $[Zn(OH)_4]^-$. The fact that the precipitate is completely soluble in $NH_3(aq)$, eliminates $Fe(OH)_3$ and MnO_2 as possibly being present. The precipitate must be either $Co(OH)_3$ or $Ni(OH)_2$.

Thus, no metal is definitely proved to be present. Fe and Mn are absent. At least one of the following must be present: Al or Cr; and at least one of these: Ni or Co. The presence of Zn is uncertain.

Quantitative analysis

23-33. (a) Formula of nickel dimethylglyoximate: $NiC_8H_{14}N_4O_4$

(b) $\text{no. g Ni} = 0.259\text{ g } NiC_8H_{14}N_4O_4 \times \frac{1\text{ mol Ni}}{289\text{ g } NiC_8H_{14}N_4O_4} \times \frac{58.71\text{ g Ni}}{1\text{ mol Ni}} = 0.0526\text{ g Ni}$

$\%\ Ni = \frac{0.0526\text{ g Ni}}{1.502\text{ g steel}} \times 100 = 3.50\%$

23-34. no. mol $H_2C_2O_4$ available = $1.651\ g\ H_2C_2O_4 \cdot 2\ H_2O \times \frac{1\ mol\ H_2C_2O_4}{126.0\ g\ H_2C_2O_4 \cdot 2\ H_2O} = 0.01310\ mol\ H_2C_2O_4$

no. mol $H_2C_2O_4$ reacting with MnO_4^- = $0.03006\ L \times \frac{0.1000\ mol\ MnO_4^-}{1\ L} \times \frac{5\ mol\ H_2C_2O_4}{2\ mol\ MnO_4^-} = 0.007515\ mol\ H_2C_2O_4$

no. mol $H_2C_2O_4$ reacting with MnO_2 = 0.01310 - 0.00752 = 0.00558 mol $H_2C_2O_4$

Self-Test Questions

1. (*c*) Strong metallic bonding in the transition elements leads to high melting points. The ionization energies, though not as low as for IA and IIA metals, are not particularly high. Most of the transition elements are more metallic than hydrogen and have negative standard reduction potentials. In general, the transition elements do have a variety of oxidation states.

2. (*d*) The ion Sc^{3+} has a noble gas electron configuration ([Ar]) and therefore is diamagnetic. That the other ions are paramagnetic can be established by writing out their electron configurations.

3. (*a*) The maximum oxidation state of Ti, a member of group IVB is +4. Ti is not expected to display an oxidation state of +6; the other elements listed do.

4. (*b*) $Cl_2(g)$ is produced from HCl(aq) through an oxidation half-reaction; this requires an oxidizing agent. Of the substances listed, only MnO_2 is a strong oxidizing agent.

5. (*d*) In order to disproportionate the ion must yield another ion of higher charge and either the metal or a cation of lower charge. Fe^{3+} is not easily oxidized; it does not disproportionate. Cr^{2+} is easily oxidized but is not simultaneously reduced. Ag^+ does not disproportionate; it is very difficult to oxidize to Ag^{2+}. Cu^+ disproportionates regularly.

6. (*c*) All of the ions are oxidizing agents in that they are rather easily reduced (they have positive reduction potentials). Ag^{2+} is the most powerful oxidizing agent of the group, in that the +2 oxidation state is difficult to achieve.

7. (*a*) By shifting the chromate-dichromate equilibrium to $Cr_2O_7^{2-}$ the solubility of $BaCrO_4(s)$ can be increased ($BaCr_2O_7$ is water soluble). This shift in equilibrium requires an acidic solution. Of the solutions listed, only $NH_4Cl(aq)$ is acidic (because of hydrolysis of NH_4^+). Furthermore Na_2CrO_4 and $BaCl_2$ would lower the solubility of $BaCrO_4$ through the common ion effect.

8. (*b*) Wrought iron is a low carbon-content iron and steels generally have a maximum of 1.5% C. By contrast, cast iron may have several percent C.

9. Conversion of Fe^{2+} ([Ar]$3d^6$) to Fe^{3+} ([Ar]$3d^5$) leads to an electron configuration in which the $3d$ subshell is half-filled. This is an energetically favorable electron configuration and is rather easily attainable (e.g., through air oxidation of Fe^{2+}). Loss of an addition electron by Ni^{2+} or Co^{2+} does not lead to a half-filled $3d$ subshell and so this loss occurs with a great deal more difficulty.

10. With a limited amount of OH^-, both Mg^{2+} and Cr^{3+} yield hydroxide precipitates.

$Mg^{2+} + 2\ OH^- \rightarrow Mg(OH)_2(s)$ and $Cr^{3+} + 3\ OH^- \rightarrow Cr(OH)_3(s)$

With an excess of NaOH, the $Cr(OH)_3(s)$ would redissolve whereas the $Mg(OH)_2(s)$ would not. This is simply a case of $Cr(OH)_3$ being an amphoteric hydroxide whereas $Mg(OH)_2$ has only basic properties.

Mg^{2+} + excess $OH^- \rightarrow Mg(OH)_2(s)$ and Cr^{3+} + excess $OH^- \rightarrow [Cr(OH)_4]^-$

11. (a) $CuSO_4$ will dissolve in H_2O and HCl, but with NaOH(aq) it yields a precipitate, $Cu(OH)_2(s)$, that does not dissolve in excess NaOH(aq).

(b) $Ni(OH)_2$ is insoluble in water, but soluble in HCl(aq). Because it is not amphoteric. $Ni(OH)_2$ will not dissolve in NaOH(aq).

(c) $AgNO_3$ is soluble in water but not in NaOH, where a hydroxide precipitate would form, nor in HCl, from which the chloride would deposit as a precipitate.

(d) FeS is soluble in HCl(aq) [recall Table 21-5] but not in H_2O or NaOH(aq).

(e) $Cr(OH)_3$ is an amphoteric hydroxo compound. It is soluble both in HCl(aq) and in NaOH(aq), but not in water.

12. The formation of a precipitate with NaOH(aq) and H_2O_2 establishes that Fe^{3+} and/or Ni^{2+} is present. That the solution obtained is colorless suggests that Cr^{3+} is absent (that is, no yellow CrO_4^{2-} is formed). Zn^{2+} may or may not be present. That the precipitate [$Fe(OH_3)$ and/or $Ni(OH)_2$] redissolves in HCl(aq) but does *not* reform upon addition of NH_3(aq), establishes that the initial precipitate was $Ni(OH)_2$ only. (If Fe^{3+} had been present it would have reprecipitated as $Fe(OH)_3$ in the presence of NH_3(aq); Ni^{2+} remains in solution as the complex ion $[Ni(NH_3)_6]^{2+}$.) Ions present: Ni^{2+}; ions absent: Cr^{3+}, Fe^{3+}; further tests needs: Zn^{2+}.

Chapter 24

Complex Ions and Coordination Compounds

Review Problems

24-1. (a) $[Ni(NH_3)_6]^{2+}$ Coordination no. = 6; oxidation state = +2

(b) $[AlF_6]^{3-}$ Coordination no. = 6. Each F^- carries a charge of -1; the total for six F^- is -6; the net charge on the complex ion is -3; the oxidation state of Al must be +3.

(c) $[Cu(CN)_4]^{2-}$ Coordination no. = 4. Each CN^- carries a charge of -1; the total for four CN^- is -4; the net charge on the complex ion is -2; the oxidation state of Cu must be +2.

(d) $[Cr(NH_3)_3Br_3]$ The coordination no. = 6. NH_3 molecules are neutral; each Br^- carries a charge of -1; the total for three Br^- is -3; the complex is neutral; the oxidation state of Cr must be +3.

(e) $[Fe(ox)_3]^{3-}$ Oxalate ions are bidentate ligands carrying a charge of 2-. The coordination number is 3 x 2 = 6. The total charge associated with three oxalate ions is -6 and the net charge on the complex ion is -3. The oxidation state of Fe must be +3.

(f) $[Cr(EDTA)]^-$ EDTA is a hexadentate ligand carrying a charge of -4. The coordination number is 6. To account for a net charge of -1 on the complex ion, the Cr must be in the oxidation state +3.

24-2. (a) $[Ag(NH_3)_2]^+$ = diamminesilver(I) ion

(b) $[Fe(H_2O)_5OH]^{2+}$ = pentaaquahydroxoiron(III) ion

(c) $[ZnCl_4]^{2-}$ = tetrachlorozincate(II) ion

(d) $[Pt(en)_2]^{2+}$ = bis(ethylenediamine)platinum(II) ion

(e) $[Co(NH_3)_4(NO_2)Cl]^+$ = tetraamminechloronitrocobalt(III) ion

24-3. (a) $[Co(NH_3)_5Br]SO_4$ = pentaamminebromocobalt(III) sulfate

(b) $[Co(NH_3)_5SO_4]Br$ = pentaamminesulfatocobalt(III) bromide

(c) $[Cr(NH_3)_6][Co(CN)_6]$ = hexaamminechromium(III) hexacyanocobaltate(III)

(d) $Na_3[Co(NO_2)_6]$ = sodium hexanitrocobaltate(III)

(e) $[Co(en)_3]Cl_3$ = tris(ethylenediammine)cobalt(III) chloride

24-4. (a) dicyanosilver(I) ion = $[Ag(CN)_2]^-$

(b) diamminetetrachloronickelate(II) ion = $[Ni(NH_3)_2Cl_4]^{2-}$

(c) tris(ethylenediamine)copper(II) sulfate = $[Cu(en)_3]SO_4$

(d) sodium diaquatetrahydroxoaluminate(III) = $Na[Al(H_2O)_2(OH)_4]$

24-5. (a) iron(III) chloride hexahydrate = $FeCl_3 \cdot 6\ H_2O$

(b) cobalt(II) hexachloroplatinate(IV) hexahydrate = $Co[PtCl_6] \cdot 6\ H_2O$

24-6. *trans*-$[Cr(NH_3)_4ClOH]^+$

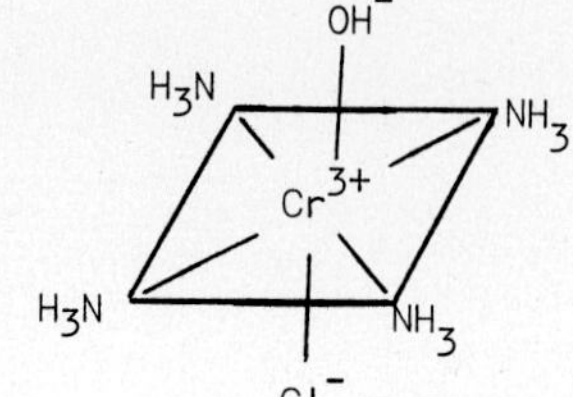

24-7. (a) $[PtCl_4]^{2-}$

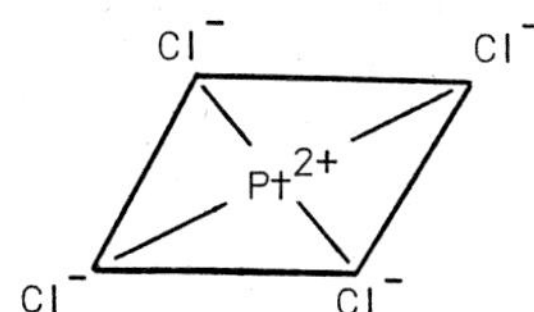

(b) $[Fe(en)Cl_4]^-$

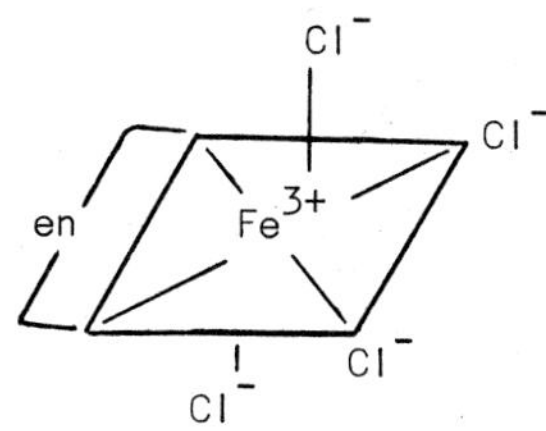

(c) *cis*-$[Fe(en)(ox)Cl_2]^-$

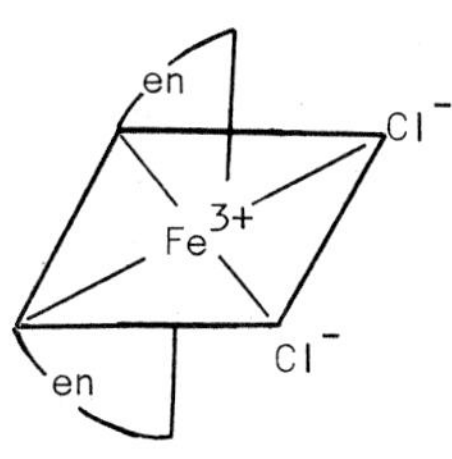

24-8. (a) $[Co(NH_3)_5H_2O]^{3+}$ There is only one possible structure.

(b) $[Co(NH_3)_4(H_2O)_2]^{3+}$ The two H_2O molecules may be either *cis* or *trans*. Two structures are possible.

(c) $[Co(NH_3)_3(H_2O)_3]^{3+}$ Consider either the three NH_3 or the three H_2O. They can all be situated on the same face of an octahedron (*cis* or *fac*) or around the perimeter of the octahedron (*trans* or *mer*). Two structures are possible. (Recall Example 24-3.)

(d) $[Co(NH_3)_2(H_2O)_4]^{3+}$ This situation is exactly the same as (b) but with NH_3 and H_2O interchanged. There are two possible structures.

24-9. (a) $[Zn(NH_3)_4][CuCl_4]$ Coordination isomerism is possible. That is, the possible isomer is $[Cu(NH_3)_4][ZnCl_4]$.

(b) $[Fe(CN)_5SCN]^{4-}$ Linkage isomerism is possible. That is, the possible isomer is $[Fe(CN)_5NCS]^{4-}$. The isomers are a thiocyanato and an isothiocyanato.

(c) $[Ni(NH_3)_5Cl]^+$ No isomerism is found here.

(d) $[Pt(py)Cl_3]^-$ No isomerism is found here. (The py attaches to one corner of a square and the three Cl^- to the other three corners.)

(e) $[Cr(NH_3)_3(OH)_3]$ Two isomers exist, a *cis* (or *fac*) and a *trans* (or *mer*). This is similar to the case encountered in Example 24-3.

24-10. (a)

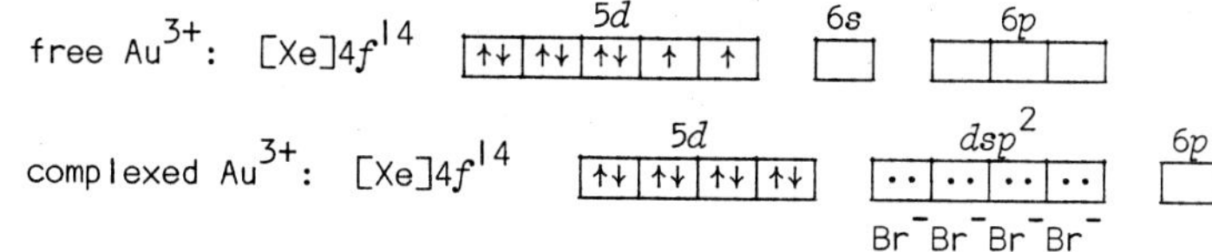

structure:

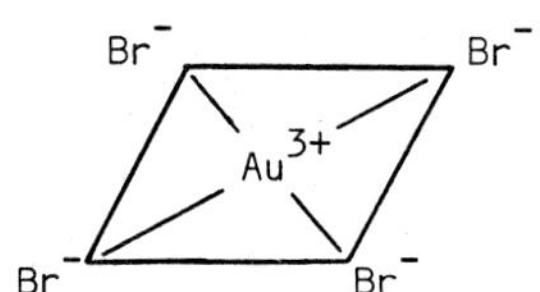

(b) From the orbital diagram shown for the complexed Au^{3+} we conclude that the complex ion is diamagnetic (no unpaired electrons). (This same conclusion would be reached by using the *d* level energy diagram of Figure 24-13.)

Exercises

Definitions and terminology

24-1. (a) Octahedral geometry refers to the structure of a complex ion in which the central ion displays a coordination number of six. Bonds between the central ion and the ligands are directed to the six corners of an octahedron.

(b) In dsp^2 hybrid bonding one *d* orbital, one *s* orbital, and two *p* orbitals are hybridized into a set of four orbitals directed from the central metal ion to the four corners of a square.

(c) An aqua complex has H_2O molecules as its ligands, as in $[Al(H_2O)_6]^{3+}$.

24-2. (a) The coordination number of a metal ion establishes the number of ligands attached to the ion by secondary valence forces (e.g., coordinate covalent bonds). The oxidation number is related to the primary valence of the central ion, that is, signifying the number of electrons lost by a metal atom in forming a metal ion. Thus, in the formulation $[Co(NH_3)_6]^{3+}$ the oxidation number is +3 and the coordination number is 6.

(b) A unidentate ligand has one pair of electrons available for donation to a central ion, as in $:NH_3$. In a multidentate ligand two or more electron pairs are available and attachment of the ligand occurs at two or more points, as in ethylenediamine

```
 H      H
:NCH2CH2N:
 H      H
```

(c) Structural isomerism is based on the same number and types of atoms but bonded differently, as in bonding through N (as in $[Co(NH_3)_5NO_2]^{2+}$) or through O (as in $[Co(NH_3)_5ONO]^{2+}$). In stereoisomerism the bonds in the isomers are identical but the orientation of certain groups in space differs, as in whether two groups are bonded on adjacent corners (*cis*) or opposite corners (*trans*) of a square planar complex.

(d) One type of *cis-trans* isomerism, that of a square planar complex, was used as an example in part (c). Another example is whether two distinctive groups are attached along the same edge (*cis*) or on opposite vertices (*trans*) of an octahedral complex.

(e) *d* and *l* isomers are identical in all respects, except that their structures are nonsuperimposable (mirror images) and that they rotate the plane of polarized light in opposing directions.

(f) In a nitro complex the ligand NO_2^- is attached to a central ion through the N atom. In a nitrito complex bonding occurs through an O atom (see also, part c).

(g) In the formation of a complex ion a *d* energy level of the central ion is split into two (or more) groups. If the *d* electrons of the central ion remain paired in the lowest energy states the complex is low spin. If the electron remain unpaired to the greatest extent possible (by occupying higher *d* energy levels in the set) the complex is high spin. Whether a complex is low or high-spin depends on the energy differences within the *d* level compared to electron spin pairing energy.

24-3. A chelate complex is based on multidentate ligands that, by attaching to two or more points in the coordination sphere of a metal ion, produce five- or six-membered ring structures with the central ion. In an ordinary complex a unidentate ligand attaches at a single point and no ring formation occurs.

Nomenclature

24-4. (a) $[Co(NH_3)_4(H_2O)(OH)]^{2+}$ = tetraammineaquahydroxocobalt(III) ion

(b) $[Co(NH_3)_3(-NO_2)_3]$ = triamminetrinitrocobalt(III) (a neutral complex)

(c) $[Pt(NH_3)_4][PtCl_6]$ = tetraammineplatinum(II) hexachloroplatinate(IV)

(d) $[Fe(ox)_2(H_2O)_2]^-$ = diaquabis(oxalato)ferrate(III) ion

(e) $[Fe(py)(CN)_5]^{3-}$ = pentacyanopyridineferrate(II) ion

(f) $Ag_2[HgI_4]$ = silver tetraiodomercurate(II)

24-5. (a) potassium hexacyanoferrate(II) = $K_4[Fe(CN)_6]$

(b) bis(ethylenediamine)copper(II) ion = $[Cu(en)_2]^{2+}$

(c) tetraaquadihydroxoaluminum(III) chloride = $[Al(H_2O)_4(OH)_2]Cl$

(d) amminechlorobis(ethylenediamine)chromium(III) sulfate = $[Cr(en)_2(NH_3)Cl]SO_4$

(e) tris(ethylenediamine)iron(II) hexacyanoferrate(III) = $[Fe(en)_3]_3[Fe(CN)_6]_2$

Bonding and structure in complex ions

24-6. The only orbital bonding scheme that accounts for the coordination number 2 is *sp* hybrid bonding. The two *sp* hybrid orbitals are directed at a 180° angle about the central metal ion. This means that the structure is linear.

24-7. The hybridization scheme that accounts for tetrahedral geometry is sp^3.

		3*d*	4*s*	4*p*
Fe atom:	[Ar]	↑↓ ↑ ↑ ↑ ↑	↑↓	☐☐☐
free Fe^{3+} ion:		↑ ↑ ↑ ↑ ↑	☐	☐☐☐
complexed Fe^{3+}:		↑ ↑ ↑ ↑ ↑	sp^3: ·· ·· ·· ··	
			Cl^- Cl^- Cl^- Cl^-	

The complex ion has five unpaired electrons.

24-8. The electron configuration of Co is $[Ar]3d^7 4s^2$ and for Co(II), $[Ar]3d^7$. Since $[Co(H_2O)_6]^{2+}$ is described as a high spin complex, this implies a maximum number of unpaired electrons, an "outer-orbital" complex, and sp^3d^2 hybridization.

		3*d*	$5p^3d^2$	4*d*
Co^{2+}	[Ar]	↑↓ ↑↓ ↑ ↑ ↑	·· ·· ·· ·· ·· ··	☐☐☐
			(H_2O)	

$[Co(NH_3)]^{3+}$ is a $3d^6$ low-spin, inner-orbital complex.

24-9. (a) Pt(II) complexes are square planar (see Figures 24-2 and 24-4).

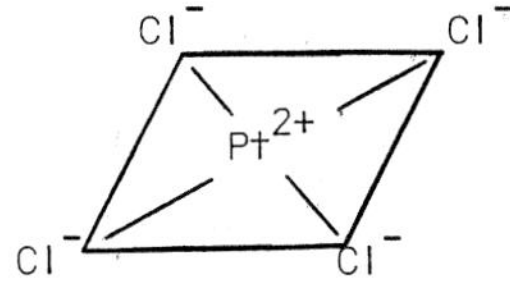

(b) In Figure 24-2 $[Zn(NH_3)_4]^{2+}$ is shown to have a tetrahedral structure. We might rewrite that structure, but with OH^- substituting for three of the NH_3 molecules.

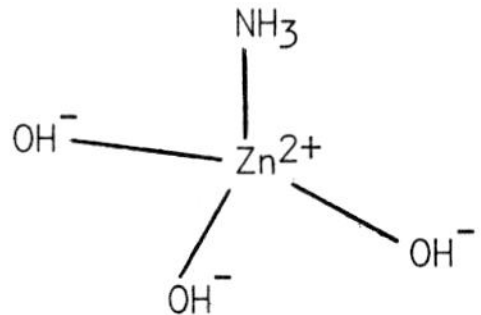

(c) This is an octahedral complex of Cr(III).

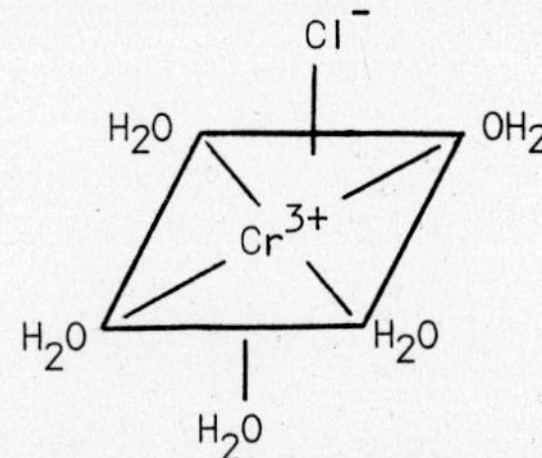

24-10. (a) Pt(II) complexes as indicated in Figures 24-2 and 24-4 are square planar. Oxalate ion is a bidentate ligand.

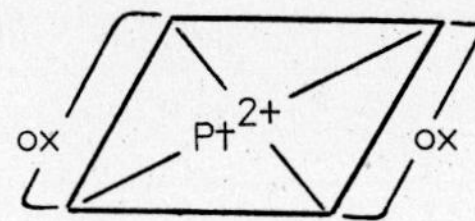

(b) As in part (a), oxalate is a bidentate ligand. The coordination number of Cr(III) is 6; the structure is octahedral.

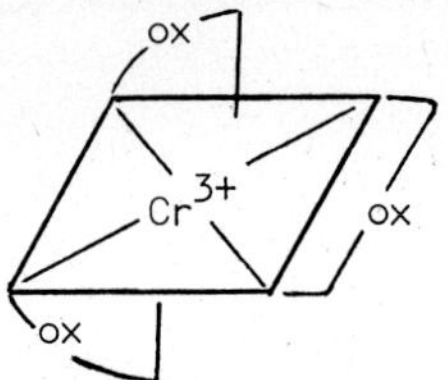

or

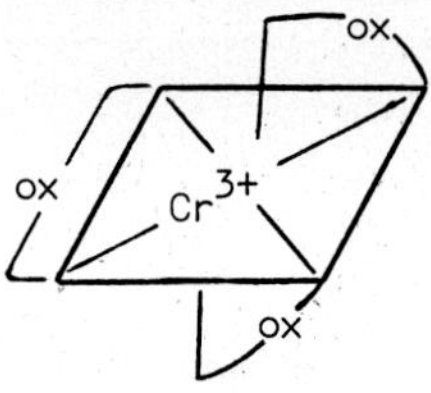

(Optical isomerism occurs here.)

(c) $[Fe(EDTA)]^{2-}$. The EDTA attaches to all six points in the coordination sphere (see Table 24-4 and Figure 24-16).

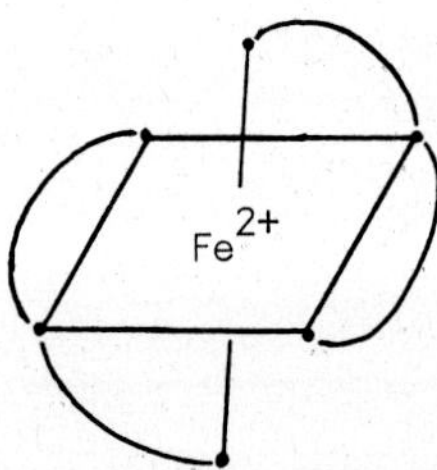

Isomerism

24-11. (a) *Cis-trans* isomerism does not exist in complex ions with a tetrahedral structure because all positions that ligands occupy are equivalent.

(b) There are two basic orientations that a pair of ligands may have with respect to one another in a square planar structure, either along the same edge or at opposite corners of the square. For this reason, square planar complexes do exhibit *cis-trans* isomerism.

(c) The two positions that ligands may occupy in a linear complex ion are identical. Whether the two ligands are the same or different, there is still only one possibility for each complex ion.

24-12. Structure (24.6a):

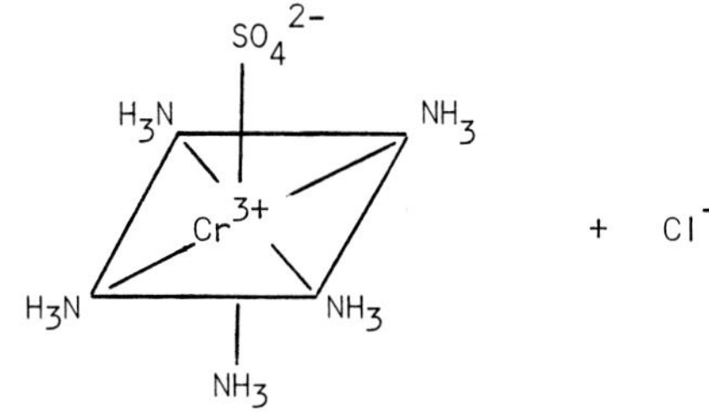

Structure (24.6b):

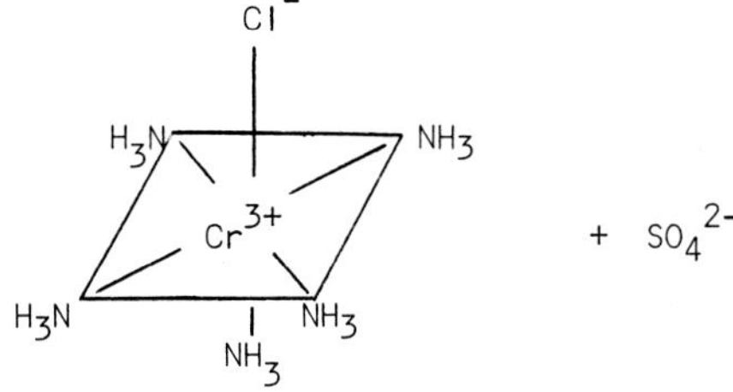

Structure (24.7a):
(An optical isomer is possible for each complex ion.)

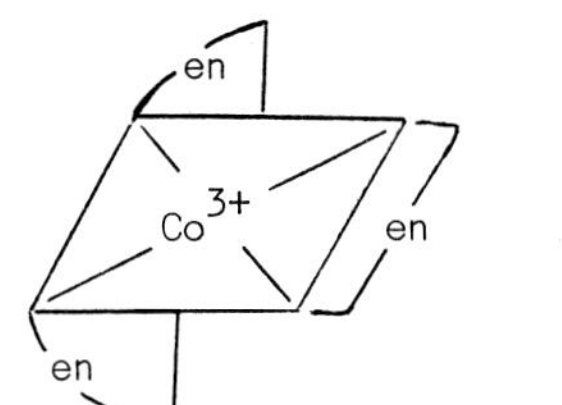

+

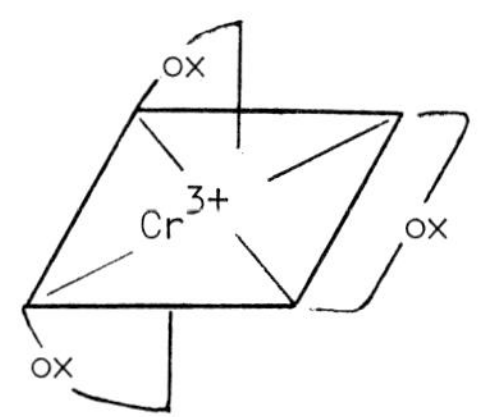

Structure (24.7b):
(An optical isomer is possible for each complex ion.)

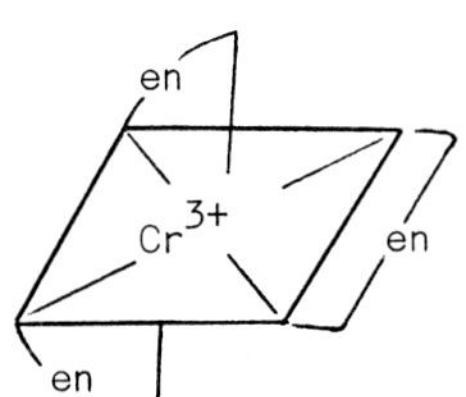

+

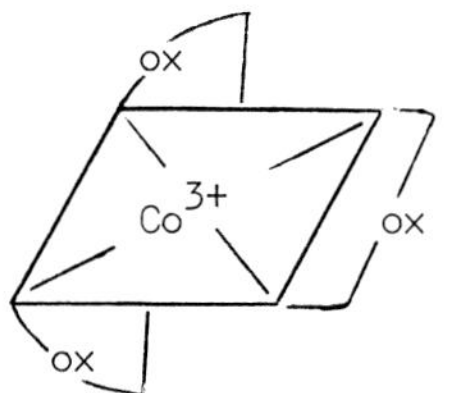

Structure (24.9a):

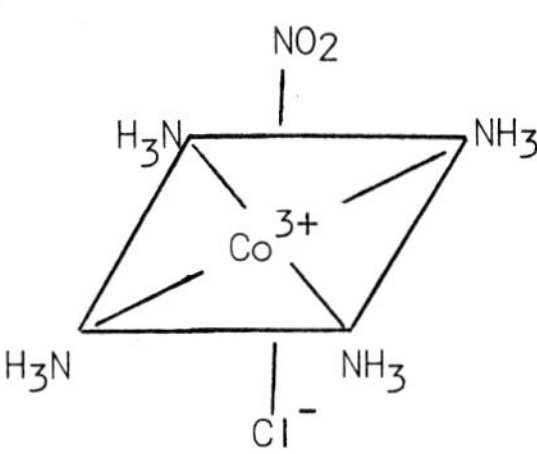

Structure (24.9b):

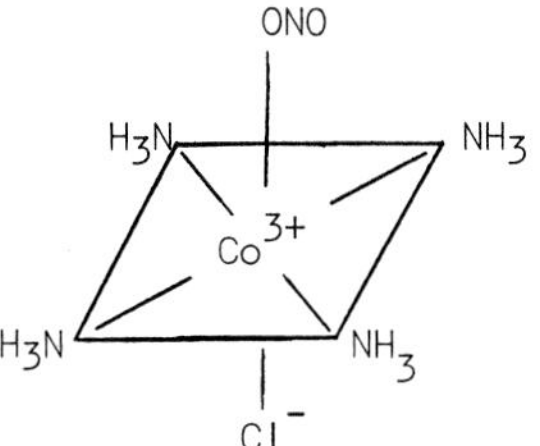

(*Cis-trans* isomerism is also possible in each structure. Only the *trans* forms are shown here.)

24-13. (a) Since in this complex ion all positions that the OH^- ion might assume are equivalent, no geometric isomerism is expected.

(b) Different structures are possible. Two of these possibilities are:

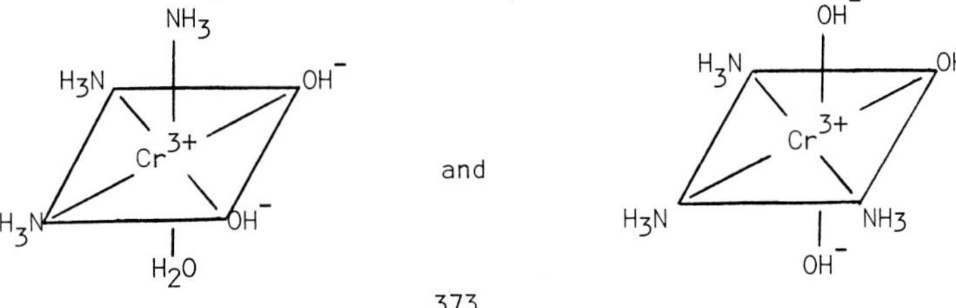

(c) The (en) groups must be in *cis* positions, but the Cl^- can be either *cis* or *trans*.

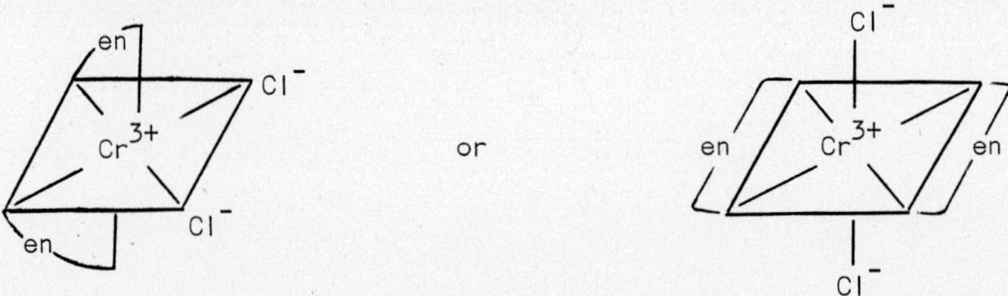

(The *cis* isomer can also exist in two optically active forms.)

(d) Again the (en) group must be in a *cis* orientation. This time there is only one structural possibility and no isomers.

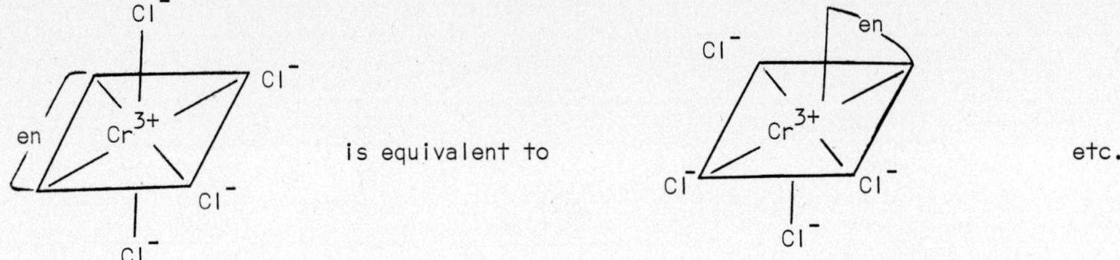

(e) Here the geometrical arrangment of the (en) groups is the same but two isomers exist because the structures are nonsuperimposable. However, these are optical, *not* geometric, isomers.

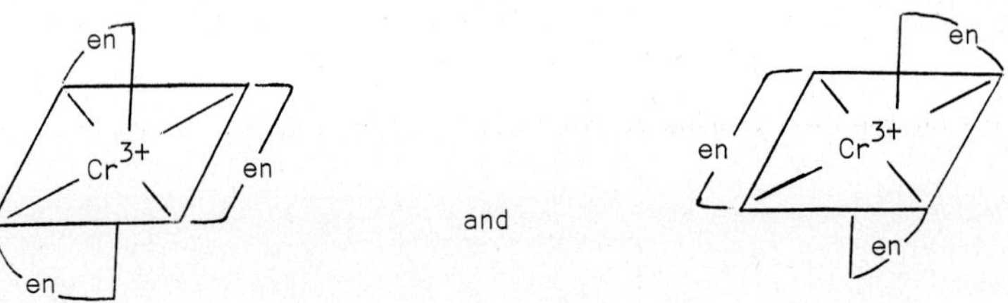

24-14. *cis*-dichlorobis(ethylenediamine)cobalt(III) ion:

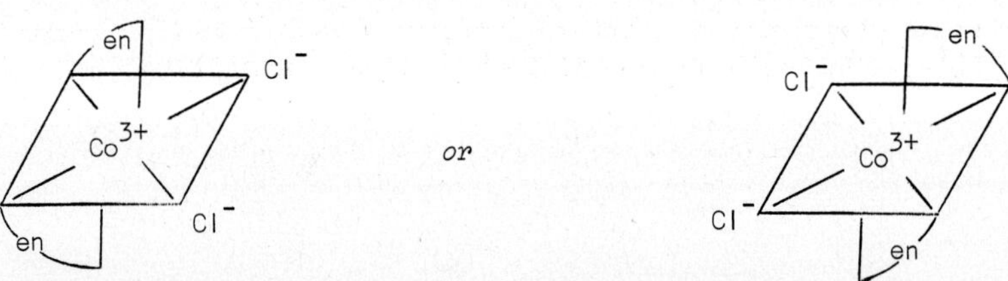

The *cis* isomer forms two optical isomers as well. It is optically active. The *trans* isomer is not optically active because it and its mirror image are superimposable.

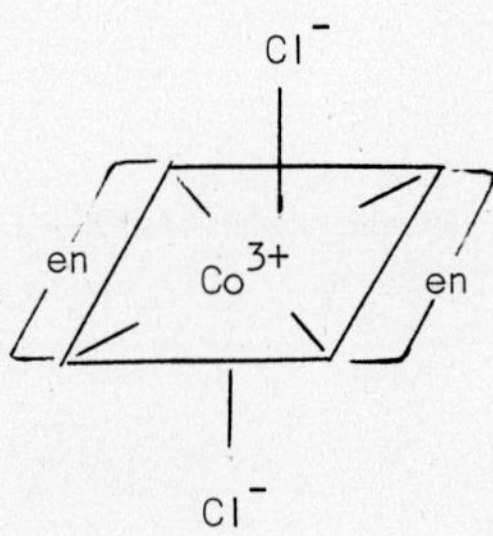

24-15. (a) Draw three structures in which ligand A has, alternately, B, C, and D as the *trans* ligand. Any other structure is equivalent to one of these three, that is there are three isomers.

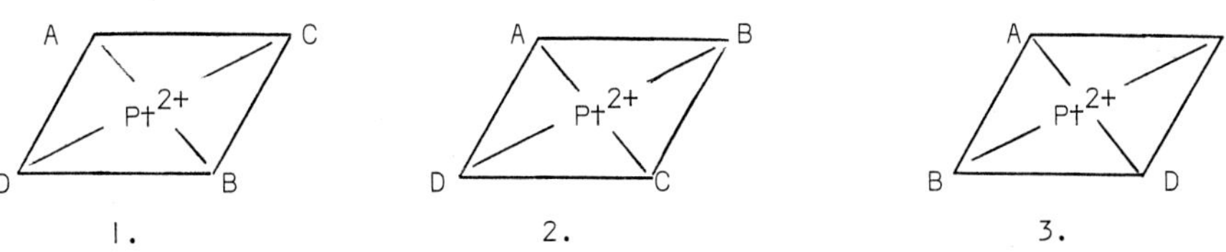

Note that

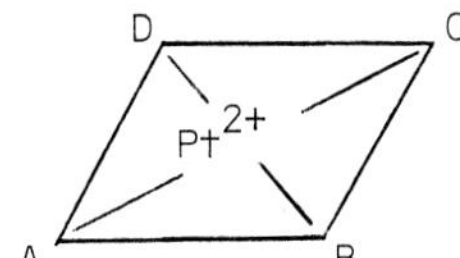

can be flipped over and it is equivalent to structure 2, etc.

(b) Draw a structure $[ZnABCD]^{2+}$ and its mirror image. They are not superimposable and so must exhibit optical isomerism.

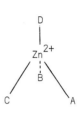

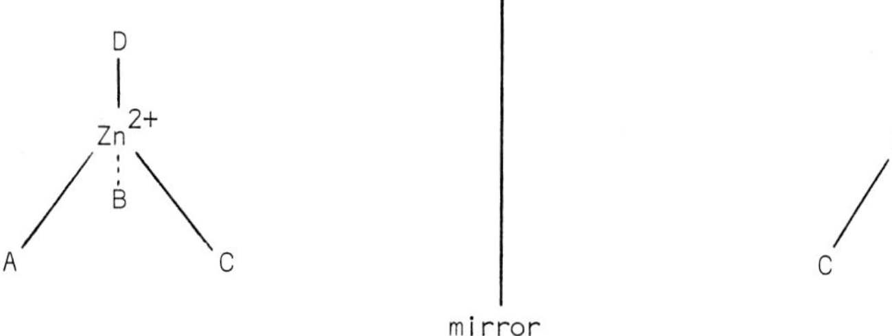

The structure on the left can be rotated 180° and the ligands A, C, and D made to superimpose. However, ligand B will project in front of the plane of the page instead of behind it.

Crystal field theory

24-16. Ligands attached to a transition metal ion produce a splitting of the d energy level of the ion. The magnitude of this splitting (Δ) corresponds roughly to the energy content of visible light. When white light is passed through a solution of the complex ion, that component of the light with energy equal to Δ is absorbed. The transmitted light is of a complementary color to the component that is absorbed.

24-17. Both ions have an octahedral structure and the d-level splitting shown below. Ethylenediamine is a strong-field ligand and Cl^- is a weak-field ligand.

(a) Weak field

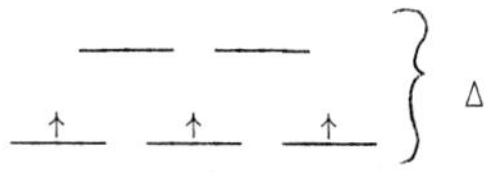

Δ

$[MoCl_6]^{3-}$

paramagnetic

(b) Strong field

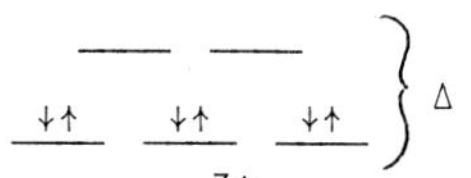

Δ

$[Co(en)_3]^{3+}$

diamagnetic

24-18. The d-level splitting in octahedral complexes of Cr^{2+} and the placement of the *four 3d* electrons follows the pattern indicated below and corresponds to the observations stated in the exercise.

Weak field

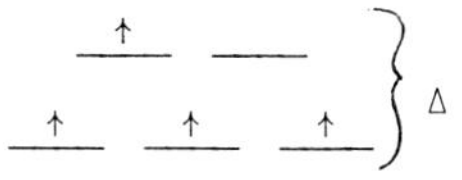

Δ

Strong field

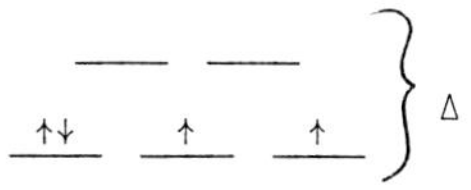

Δ

The weak field, high spin complex should be an outer-orbital complex, that is, with the hybridization sp^3d^2.

Co^{2+} [Ar] 3d [↑|↑|↑|↑|] sp^3d^2 [••|••|••|••|••|••] 4d [| |]

The strong field, low spin complex should be an inner-orbital complex, that is, with the hybridization d^2sp^3.

Co^{2+} [Ar] 3d [↑↓|↑|↑] d^2sp^3 [••|••|••|••|••|••] 4d [| | | |]

24-19. The situation with Cr^{3+} corresponding to that of Cr^{2+} in the preceding exercise is

Weak field *Strong field*

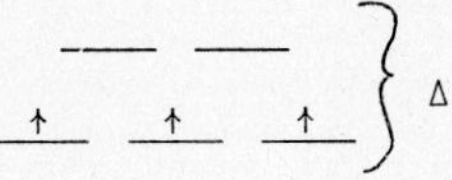

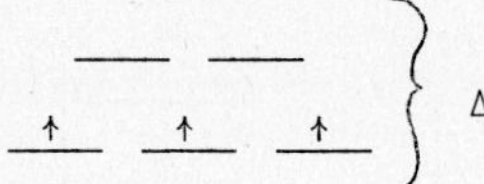

Because Cr^{3+} has *three* $3d$ electrons, all of them remain unpaired, whether in a weak-field or a strong-field complex.

24-20. CN^- is a strong-field ligand and H_2O is weak; d-level splitting is greater in cyano complexes than in aqua complexes. To produce an electronic transition from the lower to higher energy levels in the cyano complexes generally requires an absorption of more energy than is the case for aqua complexes. The highest energy content of the visible colors is associated with blue light. Absorption of blue light means that yellow is the color of the transmitted light. Many solutions of cyano complexes are yellow. In contrast, the absorption of red or yellow light, which have lower energies than blue light, results in solutions that are green and blue, respectively.

24-21. (a) The Co(II) in $[CoCl_4]^{2-}$ has the electron configuration $[Ar]3d^7$. Seven electrons must be distributed among the d orbitals indicated below. Because Cl^- is a weak field ligand, and because the energy separation between the two groups of d orbitals is much less for a tetrahedral complex than for an octahedral or square planar complex, we should expect a high spin comples ion.

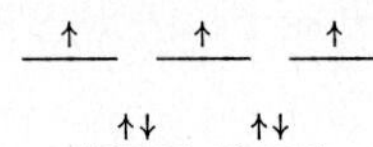

The number of unpaired electrons is 3.

(b) For a square planar complex we use the d-level diagram of Figure 24-13 and distribute *nine* electrons, corresponding to the configuration of Cu^{2+}: $[Ar]3d^9$. There will be one unpaired electron, and the ion should be paramagnetic.

↑

↑↓

↑↓

↑↓ ↑↓

(c) The electron configurations of the central ions are Mn(III): $[Ar]3d^4$ and Fe(III): $[Ar]3d^5$.

$[FeCl_4]^-$ has the greater number of unpaired electrons.

24-22. If the complex ion is $[Ni(NH_3)_6]^{2+}$, octahedral d-level splitting occurs, if $[Ni(NH_3)_4]^{2+}$, tetrahedral. The electron configuration of Ni(II) is $[Ar]3d^8$.

octahedral ↑ ↑ / ↑↓ ↑↓ ↑↓

tetrahedral ↑↓ ↑ ↑ / ↑↓ ↑↓

In either case there would be two unpaired electrons. We would not be able to distinguish between the two possibilities from magnetic data alone.

Complex ion equilibria

24-23. (a) $Mg(OH)_2(s) + NH_3(aq) \rightarrow$ no reaction

$Zn(OH)_2(s) + 4\ NH_3(aq) \rightarrow [Zn(NH_3)_4]^{2+}(aq) + 2\ OH^-(aq)$

(b) $Cu^{2+} + \cancel{SO_4^{2-}} + \cancel{2\ Na^+} + 2\ OH^- \rightarrow Cu(OH)_2(s) + \cancel{2\ Na^+} + \cancel{SO_4^{2-}}$

$Cu(OH)_2(s) + 4\ NH_3(aq) \rightarrow [Cu(NH_3)_4]^{2+} + 2\ OH^-$

$[Cu(NH_3)_4]^{2+} + 4\ H_3O^+ \rightarrow [Cu(H_2O)_4]^{2+} + 4\ NH_4^+$

(Cu^{2+} exists primarily as the light blue aqua complex, $[Cu(H_2O)_4]^{2+}$.)

(c) $CuCl\ (s) + \cancel{2\ H^+} + 2\ Cl^- \rightarrow [CuCl_4]^{2-} + \cancel{2\ H^+}$
yellow

$[CuCl_4]^{2-} + 4\ H_2O \rightarrow [Cu(H_2O)_4]^{2+} + 4\ Cl^-$
yellow blue

The equilibrium mixture of the yellow and blue complex ions accounts for the green color of the solution. With a large excess of water, equilibrium is shifted far to the right and the solution color becomes pale blue.

24-24. $[Fe(H_2O)_6]^{3+} + en \rightleftharpoons [Fe(en)(H_2O)_4]^{3+} + 2\ H_2O \qquad \log K_1 = 4.34$

$[Fe(en)(H_2O)_4]^{3+} + en \rightleftharpoons [Fe(en)_2(H_2O)_2]^{3+} + 2\ H_2O \qquad \log K_2 = 3.31$

$[Fe(en)_2(H_2O)_2]^{3+} + en \rightleftharpoons [Fe(en)_3]^{3+} + 2\ H_2O \qquad \log K_3 = 2.05$

For the overall reaction:

$[Fe(H_2O)_6]^{3+} + 3\ en \rightleftharpoons [Fe(en)_3]^{3+} + 6\ H_2O \qquad K_f = K_1 \times K_2 \times K_3$

$\log K_f = \log K_1 + \log K_2 + \log K_3 = 4.34 + 3.31 + 2.05 = 9.70$

$K_f = \text{antilog}\ (9.70) = 5.0\times10^9$

24-25. The reactions in question are:

$[Cu(H_2O)_4]^{2+} + NH_3 \rightleftharpoons [Cu(H_2O)_3NH_3]^{2+} + H_2O \qquad K_1 = 1.9\times10^4$

$[Cu(H_2O)_3NH_3]^{2+} + NH_3 \rightleftharpoons [Cu(H_2O)_2(NH_3)_2]^{2+} + H_2O \qquad K_2 = 3.9\times10^3$

$[Cu(H_2O)_2(NH_3)_2]^{2+} + NH_3 \rightleftharpoons [Cu(H_2O)(NH_3)_3]^{2+} + H_2O \qquad K_3 = 1.0\times10^3$

$[Cu(H_2O)(NH_3)_3]^{2+} + NH_3 \rightleftharpoons [Cu(NH_3)_4]^{2+} + H_2O \qquad K_4 = 1.5\times10^2$

net $[Cu(H_2O)_4]^{2+} + 4\ NH_3 \rightleftharpoons [Cu(NH_3)_4]^{2+} + 4\ H_2O \qquad K_f = 1.1\times10^{13}$

Each K value is large and the total concentration of NH_3 is maintained at a much higher value than necessary to convert all the $[Cu(H_2O)_4]^{2+}$ to $[Cu(NH_3)_4]^{2+}$. These are conditions that should favor essentially complete conversion to $[Cu(NH_3)_4]^{2+}$.

Another demonstration is to calculate $[Cu(H_2O)_4]^{2+}$ in equilibrium with $[Cu(NH_3)_4]^{2+}$, assuming complete conversion.

$$K_f = \frac{[Cu(NH_3)_4]^{2+}}{[Cu(H_2O)_4]^{2+}[NH_3]^4} = \frac{0.10}{[Cu(H_2O)_4]^{2+}(0.60)^4} = K_f = 1.1\times10^{13}$$

$[Cu(H_2O)_4]^{2+} = 7.0\times10^{-14}$ M

Now use this value of $[Cu(H_2O)_4]^{2+}$ to estimate a value of $[Cu(H_2O)_3NH_3]^{2+}$

$$K_1 = \frac{[Cu(H_2O)_3NH_3]^{2+}}{[Cu(H_2O)_4]^{2+}[NH_3]} = \frac{[Cu(H_2O)_3NH_3]^{2+}}{7.0\times10^{-14}\times 0.60} = K_1 = 1.9\times10^4$$

$[Cu(H_2O)_3NH_3]^{2+} = 7.0\times10^{-14}\times0.60 \times 1.9\times10^4 = 8.0\times10^{-10}$ M

Similarly $[Cu(H_2O)_2(NH_3)_2]^{2+} = 8.0\times10^{-10}\times0.60 \times 3.9\times10^3 = 1.9\times10^{-6}$ M

and $[Cu(H_2O)(NH_3)_3]^{2+} = 1.9\times10^{-6} \times 0.60 \times 1.0\times10^3 = 1.1\times10^{-3}$ M

and $[Cu(NH_3)_4]^{2+} = 1.1\times10^{-3} \times 0.60 \times 1.5\times10^2 = 9.9\times10^{-2}$ M

All the ion concentrations except $[Cu(NH_3)_4]^{2+}$ are rather small and the total of all the ion concentrations is about 0.10 M. These results are consistent with the assumption made, and they would not have been if the assumption were invalid.

If the amount of free NH_3 present is very small, or if there is insufficient NH_3 present to convert all the $[Cu(H_2O)_4]^{2+}$ to $[Cu(NH_3)_4]^{2+}$, then the concentration of intermediate species may become appreciable. For example, repeat the series of calculations sketched above but substitute for the concentrations of *free* NH_3 0.01 M instead of 0.60 M.

24-26. Consider $[[Ca(EDTA)]^{2-}] = 0.10$ M and $[EDTA] = 0.10$ M in the expression

$$K_f = \frac{[[Ca(EDTA)]^{2-}]}{[Ca^{2+}][EDTA]} = 4\times10^{10}$$

This leads to $[Ca^{2+}] = 1/4\times10^{10} = 2\times10^{-11}$. If $[Ca^{2+}]$ is this low then K_{sp} is not exceeded for the common substances listed in Table 19-1. For Mg^{2+}, $[Mg^{2+}] \approx 2\times10^{-9}$, and here too we would not expect many K_{sp} values to be exceeded [e.g., $K_{sp}(MgCO_3) = 3.5\times10^{-8}$; $K_{sp}(CaCO_3) = 2.8\times10^{-9}$].

Acid-base properties

24-27. (a) as an acid: $[Cr(H_2O)_5OH]^{2+} + H_2O \rightleftharpoons [Cr(H_2O)_4(OH)_2]^+ + H_3O^+$

as a base: $[Cr(H_2O)_5OH]^{2+} + H_3O^+ \rightleftharpoons [Cr(H_2O)_6]^{3+} + H_2O$

24-28. (a) For reaction (24.25)

$$Ka_1 = \frac{[Fe(H_2O)_5OH]^{2+}[H_3O^+]}{[Fe(H_2O)_6]^{3+}} = 9\times10^{-4}$$

Let $[[Fe(H_2O)_5OH]^{2+}] = [H_3O^+] = x$

$[[Fe(H_2O)_6]^{3+}] = 0.100 - x \approx 0.100$

$\frac{x\cdot x}{0.100} = 9\times10^{-4}$ $\quad x^2 = 9\times10^{-5}$ $\quad x = [H_3O^+] = 9\times10^{-3}$ M

$pH = -\log(9\times10^{-3}) = 2.0$

(b) Here $[H_3O^+] = 0.100$ M (This is a mixture of strong and weak acid.)

$\frac{[[Fe(H_2O)_5OH]^{2+}] \times 0.100}{0.100} = 9\times10^{-4}$ $\quad [[Fe(H_2O)_5OH]^{2+}] = 9\times10^{-4}$ M

(c) If $[[Fe(H_2O)_6]^{3+}] = 0.100$ M and $[[Fe(H_2O)_5OH]^{2+}]$ is to be held to 1×10^{-6} M, then the required $[H_3O^+]$ is

$\frac{1\times10^{-6} \times [H_3O^+]}{0.100} = 9\times10^{-4}$ $\quad [H_3O^+] = 9\times10^1$ M

It is impossible to obtain $[H_3O^+] = 9\times10^1$ M and therefore, impossible to maintain $[[Fe(H_2O)_5OH]^{2+}]$ as low as 1×10^{-6} M.

Applications

24-29. (a) We must combine the following equilibrium expressions and their K values.

$$AgBr(s) \rightleftharpoons Ag^+(aq) + Br^-(aq) \qquad K_{sp} = 5.0\times10^{-13}$$

$$Ag^+(aq) + 2\ S_2O_3^{2-}(aq) \rightleftharpoons [Ag(S_2O_3)_2]^{3-} \qquad K_f = 1.7\times10^{13}$$

$$AgBr(s) + 2\ S_2O_3^{2-}(aq) \rightleftharpoons [Ag(S_2O_3)_2]^{3-} + Br^-$$

$$K = K_{sp} \times K_f = 5.0\times10^{-13} \times 1.7\times10^{13} = 8.5$$

This value of K is large enough that in the presence of excess $S_2O_3^{2-}$ significant reaction occurs in the forward direction.

(b) A similar set up for the dissolution of AgBr in NH_3 would yield

$$AgBr(s) + 2\ NH_3(aq) \rightleftharpoons [Ag(NH_3)_2]^+ + Br^-$$

$$K = K_{sp} \times K_f = 5.0\times10^{-13} \times 1.6\times10^{7} = 8.0\times10^{-6}$$

The small value of K suggests that little dissolution of AgBr(s) occurs in $NH_3(aq)$.

24-30. Use data from Section 24-13 and Figure 21-7.

oxid. $2\ \{[Co(NH_3)_6]^{2+} \rightarrow [Co(NH_3)_6]^{3+} + e^-\} \qquad E^\circ_{ox} = -0.10\ V$

red. $HO_2^- + H_2O + 2\ e^- \rightarrow 3\ OH^- \qquad E^\circ_{red} = +0.878\ V$

net: $2\ [Co(NH_3)_6]^{2+} + HO_2^- + H_2O \rightarrow 2[Co(NH_3)_6]^{3+} + 3\ OH^- \qquad E^\circ_{cell} = 0.78\ V$

24-31. (a) $[Cu(H_2O)_4]^{2+} + 2\ e^- \rightarrow Cu(s) + 4\ H_2O$

$$\text{no. g Cu} = 0.347\ h \times \frac{3600\ s}{1\ h} \times \frac{2.13\ C}{1\ s} \times \frac{1\ mol\ e^-}{96{,}500\ C} \times \frac{1\ mol\ Cu}{2\ mol\ e^-} \times \frac{63.55\ g\ Cu}{1\ mol\ Cu} = 0.876\ g\ Cu$$

(b) $[Cu(CN)_4]^{3-} + e^- \rightarrow Cu(s) + 4\ CN^-$

In this reduction there is 1 mol Cu produced per mol e^- instead of 1/2 mol Cu per mol e^-. The mass of Cu produced should be twice that of part (a): no. g Cu = 2 × 0.876 = 1.75 g Cu.

24-32. The basic calculations required are for $[Cd^{2+}]$, $[Cu^+]$, and $[S^{2-}]$. Then the ion products (Q) must be compared to K_{sp} values. In a solution that has $[[Cd(CN)_4]^{2-}] = [[Cu(CN)_4]^{3-}] = 0.10$ M and a *free* $[CN^-] = 0.10$ M.

$$\frac{[[Cd(CN)_4]^{2-}]}{[Cd^{2+}][CN^-]^4} = K_f = 7.1\times10^{18} = \frac{0.10}{[Cd^{2+}](0.10)^4} \qquad [Cd^{2+}] = 1.4\times10^{-16}\ M$$

$$\frac{[[Cu(CN)_4]^{3-}]}{[Cu^+][CN^-]^4} = K_f = 1\times10^{28} = \frac{0.10}{[Cu^+](0.10)^4} \qquad [Cu^+] = 1\times10^{-25}\ M$$

To determine $[S^{2-}]$ in saturated $H_2S(aq)$ with pH 7 use the expression

$$\frac{[H_3O^+]^2[S^{2-}]}{[H_2S]} = Ka_1 \times Ka_2 = 1.1\times10^{-7} \times 1.0\times10^{-14} = 1.1\times10^{-21}$$

$$[S^{2-}] = \frac{0.10 \times 1.1\times10^{-21}}{[H_3O^+]^2} = \frac{0.10 \times 1.1\times10^{-21}}{(1\times10^{-7})^2} = 1.1\times10^{-8}\ M$$

Finally, we make the comparisons:

$$[Cd^{2+}][S^{2-}] = 1.4\times10^{-16} \times 1.1\times10^{-8} > K_{sp}\ (= 8.0\times10^{-27})$$

$$[Cu^+]^2[S^{2-}] = (1\times10^{-25})^2 \times 1.1\times10^{-8} < K_{sp}\ (= 1.2\times10^{-49})$$

Self-test Questions

1. (*c*) The ligands bring a charge of -5 to the complexion (four CN^- and one I^-). The net charge on the complex ion is -3. The charge (oxidation state) on the central ion is +2.

2. (*d*) The (en) groups are bidentate. Two of these groups attach at four parts in the coordination sphere; the two Cl^- attach at two more points. The coordination number is 6.

3. (*b*) There is no isomerism in $[Ag(NH_3)_2]^+$. The structure in (b) does exhibit isomerism because attachment of NO_2 can occur either through the N atom ($-NO_2$) or an O atom (-ONO). The complex $[Pt(en)Cl_2]$ does not exhibit isomerism because the (en) group must occupy *cis* positions. The Cl^- ions must also occupy *cis* position; they cannot be *trans*. With just one Cl^- and five NH_3 ligands, the complex ion (d) does not exhibit isomerism.

4. (*a*) The ions (b) and (c) each exhibit *cis-trans* isomerism. The isomer *cis*-$[Co(en)_2Cl_2]^+$ does display optical isomerism.

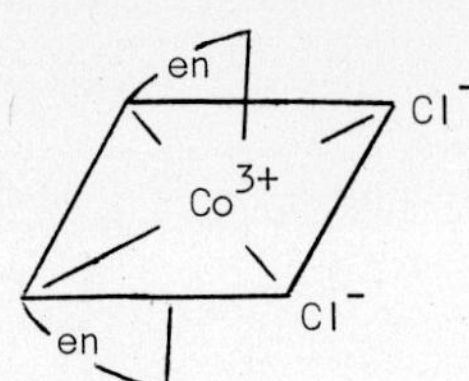

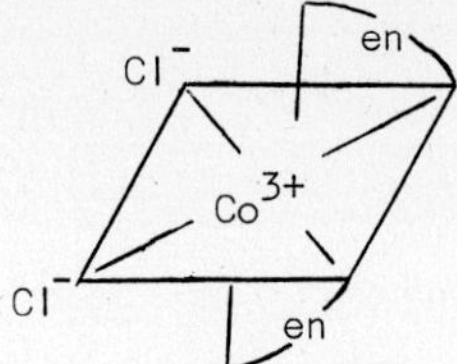

The *trans* isomer and its mirror image are identical.

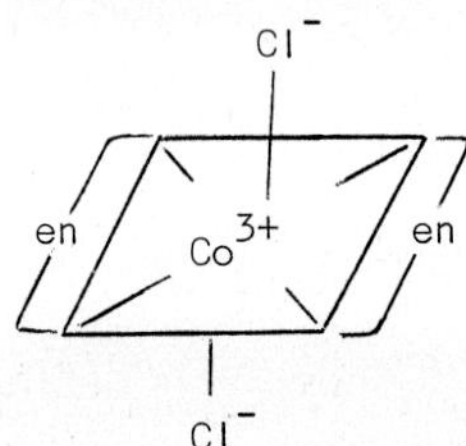

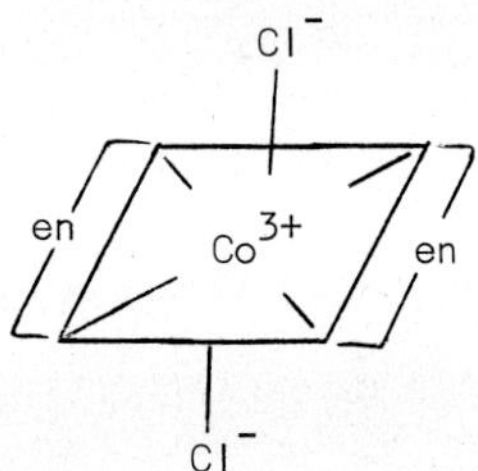

5. (*a*) Cr^{2+} has the electron configuration $[Ar]4d^4$. The filling of *d* orbitals in a strong-field octahedral complex would be

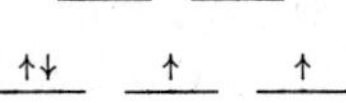

Thus, there are 2 unpaired electrons.

6. (*c*) The Brønsted-Lowry acid must have ligands capable of donating a proton. This cannot be the case if the ligands are OH^- or Cl^-, and NH_3 would be an extremely weak acid in aqueous solution. H_2O molecules in aqua complexes are often able to donate protons; $[Fe(H_2O)_6]^{3+}$ is the acid.

7. (*d*) Because of the chelate effect we should expect the most stable complex ion and the largest value of K_f in the complex ion with multidentate ligands: $[Co(en)_3]^{3+}$.

8. (*b*) $Cu(OH)_2(s)$ dissolves in $NH_3(aq)$ through the formation of the complex ion $[Cu(NH_3)_4]^{2+}$. There is no equivalent mechanism to promote the solubility of the others, e.g., $MgCO_3(s)$ is soluble in acidic solution and SiO_2 is slightly soluble in *strongly* alkaline media.

9. (*a*) $[Cr(NH_3)_4(OH)_2]Br$: *cis-* or *trans*-tetraamminedihydroxochromium(III) bromide

   (*b*) $K_3[Co(-NO_2)_6]$: potassium hexanitrocobaltate(III)

   (*c*) $[Fe(en)_2(H_2O)_2]^{2+}$: *cis-* or *trans*-diaquabis(ethylenediamine)iron(II) ion

   (*d*) $[Pt(en)_2Cl_2]SO_4$: *cis-* or *trans*-dichlorobis(ethylenediamine)platinum(IV) sulfate

10. (a) Structure of $[Co(NH_3)_5Cl]^{2+}$

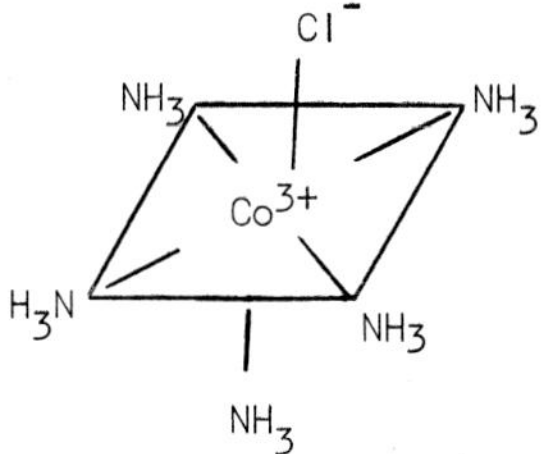

(b) Structure of $[Cr(en)_2(ox)]^+$

optical isomer:

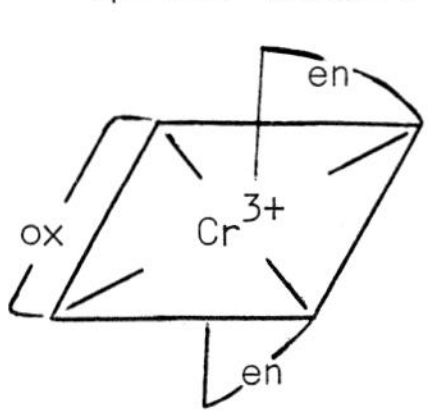

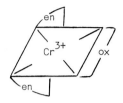

(c) *cis*-diamminedinitroplatinum(II)

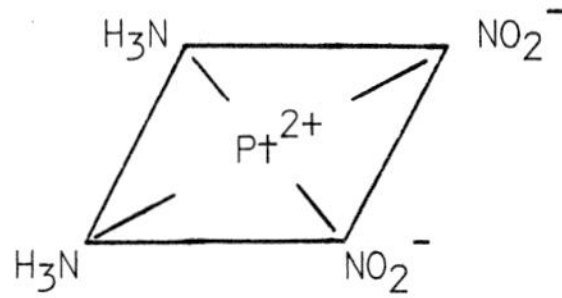

(d) *trans*-triamminetrichlorocobalt(III) [In this *trans* (*mer*) isomer three ligands of the same type are situated around a perimeter of the octahedron.]

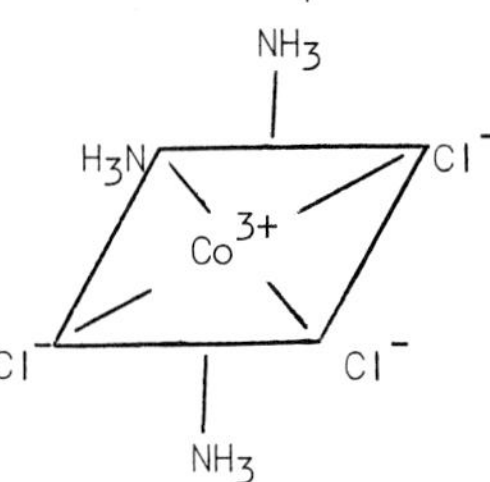

11. (a) Aluminum forms a hydroxo complex, such as $[Al(H_2O)_2(OH)_4]^-$ but not an ammine complex. As a result $Al(OH)_3$ is soluble in NaOH(aq) but not NH_3(aq).

(b) The dissolving of $ZnCO_3$(s) in NH_3 occurs through the formation of the complex ion $[Zn(NH_3)_4]^{2+}$. With ZnS(s) the concentration of Zn^{2+}(aq) is so low that the complex ion will not form. In either case the equilibrium constant for the dissolution reaction is $K = K_{sp} \times K_f$. The substance that should be more soluble is the one with the larger value of K_{sp} (and this is $ZnCO_3$).

(c) The reduction of Co^{3+} to Co^{2+} in aqueous solution can be presented if $[Co^{3+}]$ is kept to a very low value. The complex ion $[Co(NH_3)_6]^{3+}$ is very stable and the free $[Co^{3+}]$ is kept low enough to prevent its reduction to Co^{2+}.

12. CN^- is a strong-field ligand and H_2O is weak-field. The ion Fe^{2+} has the electron configuration $[Ar]3d^6$, and the distribution of the d electrons in the two cases is

Strong field		*Weak field*
___ ___		↑ ↑
↑↓ ↑↓ ↑↓		↑↓ ↑ ↑
$[Fe(CN)_6]^{4-}$		$[Fe(H_2O)_6]^{2+}$
diamagnetic		paramagnetic

Chapter 25

Nuclear Chemistry

Review Problems

25-1. (a) Gamma radiation has a much higher penetrating power through matter than do either α or β particles.

(b) Because of their large mass relative to other particles of radiation, α particles have the highest ionizing power.

(c) Gamma radiation is not deflected in a magnetic field; α and β particles are. Of these two, β particles experience the strongest deflections, because of their very small mass (and low inertia).

25-2. (a) The total of the superscripts on the right must be 32. Since the mass number of a β^- particle is zero, the mass number of Cl must be 32.

$$^{32}_{16}S \rightarrow ^{32}_{17}Cl + ^{0}_{-1}e$$

(b) Here we must supply both the mass number and atomic number of one of the species on the right. The mass number must be 0 and the atomic number, +1. The particle is a positron, β^+.

$$^{14}_{8}O \rightarrow ^{14}_{7}N + ^{0}_{+1}e$$

(c) In this problem we must supply an atomic number for U, and this can only be 92. The element thorium has an atomic number of 90. This means that the other species on the right has an atomic number of 2; it is an α particle. The mass number of an α particle is 4; and this requires the mass number of Th to be 235-4 = 231.

$$^{235}_{92}U \rightarrow ^{231}_{90}Th + ^{4}_{2}He$$

(d) Po has an atomic number of 84. The nuclide on the left must have an atomic number of 84-1 = 83. It is Bi; its mass number is 214.

$$^{214}_{83}Bi \rightarrow ^{214}_{84}Po + ^{0}_{-1}e$$

25-3. (a) $^{23}_{11}Na + ^{2}_{1}H \rightarrow ^{24}_{11}Na + ^{1}_{1}H$

(b) $^{59}_{27}Co + ^{1}_{0}n \rightarrow ^{56}_{25}Mn + ^{4}_{2}He$

(c) $^{238}_{92}U + ^{2}_{1}H \rightarrow ^{240}_{94}Pu + ^{0}_{-1}e$

(d) $^{246}_{96}Cm + ^{13}_{6}C \rightarrow ^{254}_{102}No + 5\ ^{1}_{0}n$

(e) $^{238}_{92}U + ^{14}_{7}N \rightarrow ^{246}_{99}Es + 6\ ^{1}_{0}n$

25-4. (a) $^{2}_{1}H + ^{2}_{1}H \rightarrow ^{3}_{2}He + ^{1}_{0}n$

(b) $^{241}_{95}Am + ^{4}_{2}He \rightarrow ^{243}_{97}Bk + 2\ ^{1}_{0}n$

(c) $^{121}_{51}Sb + ^{4}_{2}He \rightarrow ^{124}_{53}I + ^{1}_{0}n$

25-5. (a) The radioisotope with the *largest* decay constant is the one with the *shortest* half-life, since $\lambda = 0.693/t_{\frac{1}{2}}$. This is $^{214}_{84}Po$.

(b) If the radioactivity is to be reduced by 75% in two days, this means that the activity has fallen to 25% of its original value in two days. This corresponds to two half-life periods. We are looking for the nuclide with $t_{\frac{1}{2}} \simeq 1d$ or 24 h. It is ${}^{28}_{12}Mg$.

(c) To lose 99% of its radioactivity, that is, to have its activity reduced to 1% (1/100th) of some original value, requires that a nuclide pass through 7 half-life periods: 1.00 → 0.50 → 0.25 → 0.125 → 0.0625 → 0.03125 → 0.015625 → 0.008. We are looking for the nuclides whose half-life are less than about 4 days. They are ${}^{13}_{8}O$, ${}^{28}_{12}Mg$, ${}^{80}_{35}Br$, ${}^{214}_{84}Po$, ${}^{222}_{86}Rn$.

25-6. First determine λ: $\lambda = 0.693/87.9d = 7.88\times10^{-3}d^{-1}$. Then use equation (25.18), solving for t in each case.

(a) $N_0 = 1.00\times10^3$; $N_t = 115$: $\log \frac{115}{1000} = \frac{-7.88\times10^{-3}\ t}{2.303} = -0.939$ $\qquad t = 2.74\times10^2$ d

(b) $N_0 = 1.00\times10^3$; $N_t = 86$: $\log \frac{86}{1000} = \frac{-7.88\times10^{-3}\ t}{2.303} = -1.1$ $\qquad t = 3.2\times10^2$ d

(c) $N_0 = 1.00\times10^3$; $N_t = 43$: $\log \frac{43}{1000} = \frac{-7.88\times10^{-3}\ t}{2.303} = -1.4$ $\qquad t = 4.1\times10^2$ d

[Note also that since $43 = \frac{1}{2} \times 86$, the answer in (c) should be 3.2×10^2 d + $t_{\frac{1}{2}}$ = 3.2×10^2 d + 87.9d = 4.1×10^2 d.]

25-7. Use the method of Example 25-4 with $N_0 \propto 15/\lambda$ and $N_t \propto 12/\lambda$.

$$\log N - \log N_0 = \frac{-\lambda t}{2.303}$$

$$\log (12/\lambda) - \log (15/\lambda) = \frac{-1.21\times10^{-4}y^{-1} \times t}{2.303}$$

$$\log 12 - \log \lambda - \log 15 + \log \lambda = -5.25\times10^{-5}y^{-1} \times t$$

$$1.08 - 1.18 = -5.25\times10^{-5}y^{-1} \times t$$

$$t = \frac{1.08 - 1.18}{-5.25\times10^{-5}} = \frac{-0.10}{-5.25\times10^{-5}} = 1.9\times10^3 \text{ y}$$

The item in question is less than 2000 years old. It can not have been an artifact of the ancient Egyptians.

25-8. (a) Use the expression $E = mc^2$, with mass in kg and c in m/s.

$$E = (1.05\times10^{-23} \text{ g} \times \frac{1 \text{ kg}}{1000 \text{ g}}) \times (3.00\times10^8)^2 \text{ m}^2/\text{s}^2 = 9.45\times10^{-10} \text{ J}$$

(b) The mass of an alpha particle is given in Figure 25.5. Use this together with the conversion factor (25.25).

$$\text{no. MeV} = 4.0015 \text{ u} \times \frac{931.2 \text{ MeV}}{1 \text{ u}} = 3.726\times10^3 \text{ MeV}$$

(c) The mass of a neutron is 1.0087 u. Again, expression (25.25) can be used.

$$\text{no. neutrons} = 1.50\times10^6 \text{ MeV} \times \frac{1 \text{ u}}{931.2 \text{ MeV}} \times \frac{1 \text{ neutron}}{1.0087 \text{ u}} = 1.60\times10^3 \text{ neutrons}$$

25-9. Determine the sum of the masses of the fundamental particles in an atom of Ne-20. The difference between this sum and the measured nuclidic mass is the quantity of mass that is converted to nuclear binding energy.

10p = 10 × 1.0073 u	=	10.073 u
10n = 10 × 1.0087	=	10.087
10e = 10 × 0.00055	=	0.006
calculated mass	=	20.166 u
measured mass	=	19.992 u
mass defect	=	0.174 u

$$\text{Total nuclear binding energy} = 0.174 \text{ u} \times \frac{931.2 \text{ MeV}}{1 \text{ u}} = 162 \text{ MeV}$$

Binding energy per nucleon = 162 MeV/20 = 8.10 MeV

25-10. The nuclides with both the number of protons and neutrons as odd numbers are least likely to occur naturally, however, deuterium $^{2}_{1}H$ does exist.

(a) $^{2}_{1}H$ (1p, 1n) occurs naturally

(b) ^{32}S (16p, 16n) occurs naturally

(c) ^{80}Br (35p, 45n) does not occur naturally

(d) ^{132}Cs (55p, 77n) does not occur naturally

(e) ^{184}W (74p, 110n) occurs naturally

Exercises

Definitions and terminology

25-1. (a) The neutron-to-proton ratio is the number of neutrons in the nucleus of an atom divided by the number of protons. Only when the ratio n:p falls within certain limits is the nucleus stable. If these limits are exceeded, the nucleus either cannot exist at all or it is radioactive.

(b) A nucleon is one of two fundamental nuclear particles, a proton or a neutron.

(c) The mass-energy relationship is the expression, $E = mc^2$, which indicates the quantity of energy that is liberated when a given quantity of mass is destroyed.

(d) Background radiation refers to a constant presence of readioactivity from natural sources, like cosmic rays and emanations from rock containing radioactive elements (such as uranium and thorium).

(e) A radioactive decay series describes a sequence of nuclear disintegrations in which one radioactive nucleus (a parent) gives rise to another (a daughter), which gives rise to still another (a daughter), and so on, until eventually a stable nucleus is formed.

(f) A nuclear accelerator is a device used to accelerate charged particles, increasing their energies to the point where these particles can penetrate atomic nuclei and produce nuclear reactions.

25-2. (a) A naturally occurring radioisotope is a radioactive species which, though sometimes rare, can be found in natural sources. An artificial radioisotope cannot be found in natural sources; it can only be produced by bombarding stable atomic nuclei with appropriate particles, that is, by a nuclear reaction.

(b) An electron is the basic unit of negative electric charge found in all atoms. A positron is a particle that is produced as a result of certain nuclear reactions. It has the same mass as an electron and the same magnitude charge, but *positive*. A positron is a positive electron, so to speak.

(c) Primary ionization refers to the loss of electrons by atoms (and the consequent production of ions) as a result of a direct impact by radiation, such as α, β, and γ rays. Energetic electrons ionized from atoms by primary ionization may themselves strike atoms and dislodge additional electrons; this is the process of secondary ionization.

(d) The transuranium elements are those having an atomic number greater than uranium ($Z > 92$). The actinoids are a group of 14 elements with electron configurations featuring the filling of the $5f$ subshell. The transactinoids are the elements following the actinoids. They have $Z > 103$.

(e) Nuclear fission is a process in which a heavy nucleus disintegrates into smaller nuclei and some free neutrons. Energy is released in the process. Nuclear fusion is a process in which small nuclei coalesce into a larger nucleus. Again, the destruction of a small quantity of matter is accompanied by the release of energy.

25-3. (a) The symbol α designates an alpha particle.

(b) Gamma rays are represented by the symbol γ.

(c) The half-life of a radioisotope is designated by the symbol $t_{\frac{1}{2}}$. This is the time required for one-half of the nuclei in a radioactive sample to disintegrate.

(d) The disintegration of a radioactive sample is a first-order process. The rate constant for this process, known as the radioactive decay constant, is denoted by the symbol λ. The decay constant is related to the half-life through the familiar equation: $\lambda = 0.693\ t_{\frac{1}{2}}$.

(e) A β^+ particle is a positron--a positive electron.

25-4. (a) ${}^{4}_{2}He$ represents an alpha particle, α.

(b) A beta particle, an electron, can be represented either as β^- or ${}^{0}_{-1}e$.

(c) A neutron has the symbol ${}^{1}_{0}n$.

(d) The symbol ${}^{1}_{1}H$ stands for a proton.

(e) The symbol ${}^{0}_{+1}e$ represents a unit of positive electric charge with the same mass as an electron--a positron. At times the positron is also represented as β^+.

(f) Tritium is a nuclide of hydrogen with mass number 3, that is, ${}^{3}_{1}H$.

Radioactive processes

25-5. (a) ${}^{234}_{94}Pu \rightarrow {}^{230}_{92}U + {}^{4}_{2}He$

(b) ${}^{248}_{97}Bk \rightarrow {}^{248}_{98}Cf + {}^{0}_{-1}e$

(c) ${}^{196}Pb \xrightarrow{E.C.} {}^{196}Tl \xrightarrow{E.C.} {}^{196}Hg$

(d) ${}^{214}_{82}Pb \rightarrow {}^{214}_{84}Po + 2\ {}^{0}_{-1}e$

(e) ${}^{226}_{88}Ra \rightarrow {}^{214}_{82}Pb + 3\ {}^{4}_{2}He$

(f) ${}^{69}_{33}As \rightarrow {}^{69}_{32}Ge + {}^{0}_{+1}e$

25-6. All naturally occurring radioisotopes of high atomic number have n:p ratios greater than allowable for stable nuclei. Conversion of neutrons to protons within their nuclei and the emission of β^- particles is a means by which these ratios can be lowered. Some artificially produced radioisotopes have n:p ratios above and some below the range for stable nuclei. Thus, some of the nuclei emit β^- and some β^+ particles.

Radioactive decay series

25-7.

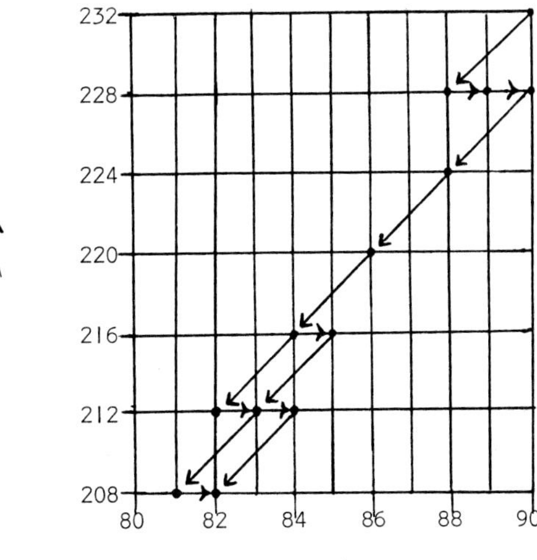

25-8. Start with the first member of the series: ${}^{238}_{92}U$.

$A = 4n + 2 = 238;\ 4n = 236;\ n = 59$

Now note that th only mass numbers that appear in Figure 25-3 are 238, 234, 230, and so on. Since n is an integer when the "4n + 2" formula is applied to A = 238, it must also be an integer as A is decreased four units at a time. All members of the U-238 decay series follow the "4n + 2" formula.

Nuclear reactions

25-9. (a) ${}^{7}_{3}Li + {}^{1}_{1}H \rightarrow {}^{8}_{4}Be + \gamma$ (b) ${}^{33}_{16}S + {}^{1}_{0}n \rightarrow {}^{33}_{15}P + {}^{1}_{1}H$

(c) ${}^{239}_{94}Pu + {}^{4}_{2}He \rightarrow {}^{242}_{96}Cm + {}^{1}_{0}n$ (d) ${}^{238}_{92}U + {}^{4}_{2}He \rightarrow {}^{239}_{94}Pu + 3\ {}^{1}_{0}n$

25-10. ${}^{238}_{92}U \rightarrow {}^{234}_{90}Th + {}^{4}_{2}He$; ${}^{234}_{90}Th \rightarrow {}^{234}_{91}Pa + {}^{0}_{-1}e$; ${}^{234}_{91}Pa \rightarrow {}^{234}_{91}U + {}^{0}_{-1}e$; ${}^{234}_{92}U \rightarrow {}^{230}_{90}Th + {}^{4}_{2}He$; ${}^{230}_{90}Th \rightarrow {}^{226}_{88}Ra + {}^{4}_{2}He$;

${}^{226}_{88}Ra \rightarrow {}^{222}_{86}Rn + {}^{4}_{2}He$; ${}^{222}_{86}Rn \rightarrow {}^{218}_{84}Po + {}^{4}_{2}He$; $\{$ (a) ${}^{218}_{84}Po \rightarrow {}^{214}_{82}Pb + {}^{4}_{2}He$; ${}^{214}_{82}Pb \rightarrow {}^{214}_{83}Bi + {}^{0}_{-1}e$ *or*

(b) ${}^{218}_{84}Po \rightarrow {}^{218}_{85}At + {}^{0}_{-1}e$; ${}^{218}_{85}At \rightarrow {}^{214}_{83}Bi + {}^{4}_{2}He$ $\}$ $\{$ (a) ${}^{214}_{83}Bi \rightarrow {}^{210}_{81}Tl + {}^{4}_{2}He$; ${}^{210}_{81}Tl \rightarrow {}^{210}_{82}Pb + {}^{0}_{-1}e$ *or*

(b) ${}^{214}_{83}Bi \rightarrow {}^{214}_{83}Po + {}^{0}_{-1}e$; ${}^{214}_{84}Po \rightarrow {}^{210}_{82}Pb + {}^{4}_{2}He$ $\}$ ${}^{210}_{82}Pb \rightarrow {}^{210}_{83}Bi + {}^{0}_{-1}e$; $\{$ (a) ${}^{210}_{83}Bi \rightarrow {}^{206}_{81}Tl + {}^{4}_{2}He$;

${}^{206}_{81}Tl \rightarrow {}^{206}_{82}Pb + {}^{0}_{-1}e$ *or* (b) ${}^{210}_{83}Bi \rightarrow {}^{210}_{84}Po + {}^{0}_{-1}e$; ${}^{210}_{84}Pb \rightarrow {}^{206}_{82}Pb + {}^{4}_{2}He$ $\}$

Rate of radioactive decay

25-11. The equation required here is (25.17): rate of decay = λN. The rate of decay is given. The value of λ is obtained from the half-life; however, this must be expressed in the unit min^{-1}.

$$\lambda = \frac{0.693}{t_{1/2}} = \frac{0.693}{5.2\ y} \times \frac{1\ y}{365\ d} \times \frac{1\ d}{24\ h} \times \frac{1\ h}{60\ min} = 2.5\times10^{-7}\ min^{-1}$$

$$N = \frac{\text{rate of decay}}{\lambda} = \frac{185\ \text{atom min}^{-1}}{2.5\times10^{-7}\ min^{-1}} = 7.4\times10^{8}\ \text{atoms}$$

25-12. Use the method of Review Problem 6 with $N_0 \propto 185$ and $N_t \propto 101$. For λ, substitute $(0.693/5.2)\ y^{-1}$.

$$\log\frac{101}{185} = \frac{-\lambda t}{2.303} = \frac{-(0.693/5.2)\ y^{-1} \times t}{2.303} = -0.263$$

$$t = \frac{2.303 \times 0.263 \times 5.2}{0.693} = 4.5\ y$$

25-13. Again, the situation is similar to those encountered in preceding exercises. The key in this case is that the initial number of atoms is 1000 times greater than at the limit of detection. For example, take 1.00×10^{3} and $N_t = 1.00$. For λ substitute $(0.693/14.2)\ day^{-1}$.

$$\log\frac{1.00}{1.00\times10^{3}} = \frac{-\lambda t}{2.303} = \frac{-(0.693/14.2)\ day^{-1} \times t}{2.303} = -3.000$$

$$t = \frac{14.2 \times 2.303 \times (-3.000)}{-0.693}\ day = 142\ day$$

25-14. The simplest approach is to use equation (25.18) with two representative data points, say $t = 0$, 1000 cpm and $t = 100$ h, 452 cpm. Note that as in previous cases the decay rate is proportional to the number of atoms in a radioactive sample. This means that we can take $N_0 = 1000$ and $N_t = 452$, with $t = 100$ h.

$$\log \frac{452}{1000} = \frac{-\lambda t}{2.303} = \frac{-\lambda \times 100 \text{ h}}{2.303} = -0.345$$

$$\lambda = \frac{2.303 \times 0.345}{100 \text{ h}} = 7.95\times10^{-3} \text{ h}^{-1}$$

$$t_{1/2} = \frac{0.693}{\lambda} = \frac{0.693}{7.95\times10^{-3}} \text{ h} = 87.2 \text{ h} = 3.63 \text{ d}$$

This result may be checked by (a) substituting other pairs of data points into equation (25.18), (b) plotting N_t versus time and estimating the time required for N_t to decrease to one-half of some former value, or (d) plotting log N_t versus t and determining λ from the slope of this plot.

25-15. First convert the half-life from the unit, y, to the unit, s. Then use equation (25.19) to convert half-life to decay constant, λ. This is followed by a straightforward application of equation (25.17) to determine the number of atoms of Ra-226 required to produce the measured activity--1 millicurie = 3.7×10^7 dis s^{-1}. Finally, conversion from number of atoms to mass follows in a familiar fashion.

$$t_{1/2} = 1602 \text{ y} \times \frac{365 \text{ d}}{1 \text{ y}} \times \frac{24 \text{ h}}{1 \text{ d}} \times \frac{60 \text{ min}}{1 \text{ h}} \times \frac{60 \text{ s}}{1 \text{ min}} = 5.05\times10^{10} \text{ s}$$

$$\lambda = \frac{0.693}{t_{1/2}} = \frac{0.693}{5.05\times10^{10} \text{ s}} = 1.37\times10^{-11} \text{ s}^{-1}$$

$$\text{Rate of decay} = \lambda N; \; N = \frac{\text{rate of decay}}{\lambda} = \frac{3.7\times10^{7} \text{ atom s}^{-1}}{1.37\times10^{-11} \text{ s}^{-1}} = 2.7\times10^{18} \text{ atoms}$$

$$\text{no. g Ra-226} = 2.7\times10^{18} \text{ atoms Ra-226} \times \frac{226 \text{ g Ra-226}}{6.02\times10^{23} \text{ atoms Ra-226}} = 1.0\times10^{-3} \text{ g Ra-226}$$

25-16. Consider starting with 1 mol ^{232}Th and calculate the amount remaining after 4.5×10^9 y. Then determine the amount of ^{208}Pb that must have been formed. Convert these amounts to a gram basis and establish the mass ratio.

$$\lambda = \frac{0.693}{t_{1/2}} = \frac{0.693}{1.39\times10^{10} \text{ y}} = 4.99\times10^{-11} \text{ y}^{-1}$$

$$\log \frac{N_t}{N_0} = \log \frac{N_t}{1.00} = \frac{-(4.99\times10^{-11} \text{ y}^{-1}) \times (4.5\times10^{9} \text{ y})}{2.303} = -0.098$$

$$\frac{N_t}{1.00} = 0.80 \qquad N_t = 0.80 \text{ mol } ^{232}\text{Th remaining}$$

$$\text{no. mol } ^{208}\text{Pb produced} = 1.000 - 0.80 = 0.20 \text{ mol } ^{208}\text{Pb}.$$

$$\text{mass ratio} = \frac{0.20 \text{ mol} \times \frac{208 \text{ g Pb}}{1 \text{ mol}}}{0.80 \text{ mol} \times \frac{232 \text{ g Th}}{1 \text{ mol}}} = \frac{42 \text{ g } ^{208}\text{Pb}}{1.9\times10^{2} \; ^{232}\text{Th}} = 0.22 \text{ g } ^{208}\text{Pb}/1.00 \text{ g } ^{232}\text{Th}$$

25-17. This problem is simplified if we note that the initial nuclide (^{87}Rb) and the final nuclide (^{87}Sr) have the same mass numbers (i.e., the same atomic weights). In this case the mass ratio (0.004:1.00) and the mol ratio are identical. If we start with 1.00 mol of ^{87}Rb, disintegration proceeds to the point where 0.004 mol ^{87}Sr has formed and 0.996 mol ^{87}Rb remains. That is, we can use N_0 = 1.00 mol and N_t = 0.996 mol.

$$\lambda = \frac{0.693}{t_{1/2}} = \frac{0.693}{5\times10^{11} \text{ y}} = 1\times10^{-12} \text{ y}^{-1}$$

$$\log \frac{N_t}{N_0} = \log \frac{0.996}{1.00} = -1.74\times10^{-3} = \frac{(1\times10^{-12} \text{ y}^{-1}) \times t}{2.303}$$

$$t = \frac{2.303 \times 1.74\times10^{-3}}{1\times10^{-12}} = 4\times10^{9} \text{ y}$$

Radiocarbon dating

25-18. Since the ratio of disintegration are proportional to the numbers of atoms, use N_0 = 15 and solve for N_t at t = 1100 + 1985 = 3.1x10^3 y (the elapsed time since 1100 B.C.).

$$\log \frac{N_t}{15} = \frac{-(0.693/5730\ y) \times 3.1\times10^3\ y}{2.303} = -0.16$$

$$\frac{N_t}{15} = \text{antilog}\ (-0.16) = 0.69$$

$$N_t = 0.69 \times 15 = 1.0\times10^1$$

That is, the decay rate is 1.0x10^1 dis/mm per g C.

25-19. Use the same basic equation as in Exercise 18 but substitute $N_t \propto 0.03/\lambda$.

$$\log (0.03/\lambda) - \log (15/\lambda) = \log \frac{0.03}{15} = \frac{-1.21\times10^{-4}\ y^{-1} \times t}{2.303} = -2.7$$

$$-5.25\times10^{-5}\ y^{-1} \times t = -2.7$$

$$t = 2.7/5.25\times10^{-5} = 5.1\times10^4\ y$$

25-20. The assumption of a constant rate of production of C-14 in the atmosphere may be invalidated in the future because of the possible large amounts of C-14 that were produced in the 1950's and early 1960's as a result of the atmospheric testing of nuclear weapons. The failure of this assumption might be particularly significant in tests made generations from now on objects that were in equilibrium with C-14 during the period mentioned.

Energetics of nuclear reactions

25-21. First determine the mass change in the nuclear reaction and then convert this to energy.

Mass Change = 13.00335 + 1.00783 - 10.01294 - 4.00260 = -0.00436 u

$$\text{Energy (release)} = -0.00436\ u \times \frac{931.2\ \text{MeV}}{1\ u} = -4.06\ \text{MeV}$$

25-22. This exercise is similar to Exercise 21.

Mass Change = 4.00260 + 3.01604 - 6.01513 - 1.008665 = -0.00515 u

$$\text{Energy (release)} = -0.00515\ u \times \frac{931.2\ \text{MeV}}{1\ u} = -4.80\ \text{MeV}$$

Nuclear stability

25-23. (a) $^{20}_{10}Ne$: For nuclides of low atomic number the most stable (and most abundant) nuclei are expected to be those that have equal numbers of protons and neutrons.

(b) $^{18}_{8}O$: A nuclide in which the number of protons and number of neutrons are both even is generally more stable than a nuclide in which one number is odd and the other even (or both numbers are odd).

(c) $^{7}_{3}Li$: A nuclide in which the number of protons and the number of neutrons are both odd is generally less stable (and less common) than one in which one or both of these quantities are even.

25-24. β^- emission results in neutrons being converted to protons and β^+, in protons being converted to neutrons. Thus, β^- emission is expected for a nuclide with a high n:p ratio and β^+ for one with a low n:p ratio.

(a) β^- emission: $^{33}_{15}P$; β^+ emission: $^{29}_{15}P$ (b) β^- emission: $^{134}_{53}I$; β^+ emission: $^{120}_{53}I$

25-25. If the rounded off atomic weight corresponds to a nuclide having a number of protons and number of neutrons both even or one even and one odd, that nuclide is likely to exist naturally. Such is the case with $^{39}_{19}K$, $^{85}_{37}Rb$, and $^{88}_{38}Sr$. If the result of the rounding off is an odd number both of protons and neutrons, the nuclide is expected to be radioactive and not to occur naturally. Such is the case with $^{64}_{29}Cu$.

25-26. A doubly magic nuclide is one having a magic number both for protons and neutrons. Thus, $^{40}_{20}Ca$ is a doubly magic nuclide. Another would be an atom with 82 p and 126 n, that is, $^{208}_{82}Pb$.

Fission and fusion

25-27. (a) A nuclear burner is the type of nuclear reactor in use today. The isotope U-235 is separated from uranium ore. This isotope is the basic nuclear fuel which undergoes fission into lighter fragments and releases energy in the reactor.

(b) In a breeder nuclear reactor energetic neutrons convert U-238, the principal isotope of uranium, to Np-239 and Pu-239. The isotope Pu-239 undergoes fission and releases energy in the reactor. The basic principle of a breeder reactor, then, is that an abudant non-fissionable nuclide, U-238, is converted to a fissionable nuclide, Pu-239.

(c) Thermonuclear reactors, which do not exist presently, are conceived on the principle of nuclear fusion, not fission. For example, in one design deuterium and tritium atoms would be fused into helium atoms. This is a process which is accompanied by a loss of mass, compensated for by the release of an equivalent quantity of energy.

25-28. In both fission and fusion, the new nuclei that are produced have a higher binding energy per nucleon than the nuclei from which they are formed. An inspection of Figure 25-6 suggests, however, that in the fusion of light nuclei this increase is two- or three-fold per nucleon, whereas for the fission process it is only about 10%. We should expect the energy release to be greater with fusion than with fission.

25-29. $$\text{no. m ton coal} = 1.00 \text{ kg U-235} \times \frac{1000 \text{ g U-235}}{1.00 \text{ kg U-235}} \times \frac{8.20\times10^{7} \text{ kJ}}{1.00 \text{ g U-235}} \times \frac{1.00 \text{ g C}}{32.8 \text{ kJ}} \times \frac{1.00 \text{ kg C}}{100 \text{ g C}} \times \frac{1 \text{ m ton C}}{1000 \text{ kg C}}$$

$$\times \frac{100 \text{ m ton coal}}{85 \text{ m ton C}} = 2.9 \quad 10^{3} \text{ m ton coal}$$

To obtain the factor 1.00 g C/32.8 kJ in the above setup, note that $\Delta\overline{H}°$ for the reaction $C(s) + O_2(g) \rightarrow CO_2(g)$ is $\Delta\overline{H}^\circ_f\,[CO_2(g)] = -393.51$ kJ/mol. Thus, for 1.00 g C,

$$\Delta H° = -393.51 \text{ kJ/mol C} \times \frac{1 \text{ mol C}}{12.0 \text{ g C}} = -32.8 \text{ kJ/g C}$$

Effect of radiation on matter

25-30. The rad is based simply on the quantity of energy deposited in matter (1×10^{-2} J/kg). The rem takes into account the fact that the biological damage depends both on the energy deposited and the kind of radiation involved.

25-31. Gamma rays have a lower ionizing power than α and β rays and they penetrate matter mucn more readily. In a cloud chamber a particle is detected through the trail of ions that it leaves behind. If very few ions are formed, the opportunity to detect radiation is limited. In a Geiger-Muller counter, on the other hand, even a single ionizing event gives rise to a pulse of electric current, which can be counted. This fact makes the G-M tube a more sensitive detector, especially for γ rays, than a cloud chamber.

25-32. There are at least two basic difficulties in establishing the effects of low dosages of radiation. If this is to be done by inference from observations made at high dosage levels, where the effects are easily measured, extrapolation of data must be made over a very long interval. Such long extrapolations are usually not reliable. If the effects of low radiation levels are to be made by direct measurement, the problem is that these effects become apparent only after many years. Experiments conducted over very long periods of time are difficult to perform. In both cases heavy reliance must be made on sophisticated statistical tests.

25-33. ^{90}Sr is a particularly hazardous radioactive material for several reasons. Of foremost concern is that if this material is taken into the body (ingested) it tends to follow the metabolic pathways of calcium, an element closely related to strontium. This means that Sr-90 tends to deposit in the bones if it is ingested. A particularly pernicious source of Sr-90 during the period in which atmospheric nuclear tests were being performed was in cow's milk. Another cause for concern about Sr-90 is that its half-life is of such a magnitude (28 y) that it tends to persist in the environment for a long time (say 1000 y) before its radioactivity drops to a safe level.

Application of radioisotopes

25-34. The hydrogen isotope, tritium, ${}^{3}_{1}H$, is radioactive. If a trace quantity of tritium is introduced with the $H_2(g)$ in an ammonia synthesis plant, radioactivity will appear wherever the leak occurs in the hydrogen supply line. The radioactivity can be detected with a Geiger-Mueller tube or other suitable radiation detector.

25-35. If an attempt is made to determine trace amounts of materials by precipitation, extremely small quantities of precipitate are obtained, perhaps too small even to be detected on an analytical balance. If titration is used instead, small volumes of very dilute solutions must be employed. The sources of error are again numerous. Neutron activation analysis is an ideal method because radioactivity can be detected in minute amounts.

25-36. When radioactive NaCl is added to an aqueous solution it dissociates. In this respect the radioactive material is no different from ordinary nonradioactive NaCl. When $NaNO_3$ is later recrystallized from the $NaNO_3$-NaCl solution, Na-23 and Na-24 ions appear in the precipitated material in the same ratio as they appeared in solution. Thus, even though radioactivity was introduced only through the NaCl, it will appear in the recrystallized $NaNO_3$.

25-37. Once a hydrogen ion attaches itself to the N atom of NH_3 through a coordinate covalent bond to form NH_4^+, that H atom becomes indistinguishable from the other three. In particular, when a proton is later donated by NH_4^+ to OH^- to produce $NH_3(g)$ and H_2O, the donated H^+ will be a tritium atom in some cases and not in others; the process is random. Thus, some of the H_2O molecules formed will be radioactive and some will not. Our conclusion must be that the original radioactivity of the HCl appears both in the H_2O and the $NH_3(g)$.

Self-test Questions

1\. (*c*) Beta rays are negatively charged particles (electrons). Neutrons are neutral particles and x rays and γ rays are forms of electromagnetic radiation.

2\. (*b*) A one unit increase in atomic number results from the process $n \rightarrow p + e^-$, that is, β^- emission. Electron capture results in a decrease of one unit in atomic number and α emission, two units. The emission of γ rays does not affect the atomic number.

3\. (*d*) The combination most likely to lead to radioactivity is an odd number of protons and an odd number of neutrons. This condition is found in ${}^{108}Ag$--47 protons and 61 neutrons.

4\. (*c*) Here we look for the combination of an even number of protons and an even number of neutrons. Au and Br have odd numbers of protons (odd atomic numbers). All isotopes of Ra are radioactive. Cd (with atomic number 48) is the only possible answer.

5\. (*a*) Positron emission occurs when the n:p ratio is too low. The most likely of the isotopes to decay by positron emission is the one with the smallest number of neutrons--${}^{59}Cu$.

6\. (*c*) Recall the general shape of Figure 25-6 which plots binding energy per nucleon against atomic number. It has a maximum value at about Z = 26--iron.

7\. (*d*) The rate at which a radioactive isotope decays is not directly related to numbers of protons and neutrons (or to mass numbers). The half-life and radioactive decay constant are related to the rate of decay. (Recall that rate of decay = $\lambda \cdot N$.) However, $t_{\frac{1}{2}}$ is inversely related--the larger the value of $t_{\frac{1}{2}}$ the slower the decay. The most radioactive isotope is the one with the largest value of the decay constant, λ.

8\. (*d*) The nuclide will pass through three half-life periods in three hours. The fraction of radioactive atoms remaining will be $1.00 \rightarrow 0.50 \rightarrow 0.25 \rightarrow 0.12$ Since there were 1000 atoms/s disintegrating initially, after three half-life periods the decay rate will fall to $0.12 \times 1000 = 1.2 \times 10^2$ atoms/s.

9\. (a) ${}^{230}_{90}Th \rightarrow {}^{226}_{88}Ra + {}^{4}_{2}He$ (b) ${}^{54}_{27}Co \rightarrow {}^{54}_{26}Fe + {}^{0}_{+1}e$

(c) ${}^{232}_{90}Th + {}^{4}_{2}He \rightarrow {}^{232}_{92}U + 4\,{}^{1}_{0}n$

10. From the half-life we determine the radioactive decay constant.

$$\lambda = \frac{0.693}{t_{1/2}} = \frac{0.693}{1.7\times10^{7}\ y} = 4.1\times10^{-8}\ y^{-1}$$

The disintegration rate is given by the expression: decay rate = λN. To use this expression we must express N as the number of ^{129}I atoms.

$$\text{decay rate} = 1.00\ mg\ ^{129}I \times \frac{1.00\ g\ ^{129}I}{1000\ mg\ ^{129}I} \times \frac{1\ mol\ ^{129}I}{129\ g\ ^{129}I} \times \frac{6.02\times10^{23}\ ^{129}I\ atoms}{1\ mol\ ^{129}I} \times \frac{4.1\times10^{-8}}{y}$$

$$\times \frac{1\ y}{365\ d} \times \frac{1\ d}{24\ h} \times \frac{1\ h}{60\ min} \times \frac{1\ min}{60\ sec} = 6.1\times10^{3}\ dis/s$$

11. The radioactive decay constant $\lambda = 0.693/t_{1/2} = 0.693/11.4\ d = 6.08\times10^{-2}\ d^{-1}$. Since the decay rate is proportional to the number of atoms, when the decay rate has decreased to 1% of its original value, $N_t = 0.01\ N_0$.

$$\log \frac{N_t}{N_0} = \log \frac{0.01\ N_0}{N_0} = \log 0.01 = -2.0 = \frac{-(6.08\times10^{-2})\ d^{-1} \times t}{2.303}$$

$$t = \frac{2.303 \times 2.0}{6.08\times10^{-2}\ d^{-1}} = 76\ d$$

12. (a) Radioactive isotopes with extremely short half-lives are completely disintegrated before having a chance to do much damage. Those with very long half-lives persist for a long time but their rates of decay are so low as to be of little harm. Isotopes with intermediate half-lives have both a high level of activity and a fairly long persistence in the environment. This makes them potentially quite hazardous.

(b) Gamma emitters are generally hazardous even at long distances because γ rays are so highly penetrating through matter. Alpha emitters, on the other hand, because of the low penetrating power of α rays are generally not very harmful at a distance. They can be extremely hazardous when ingested (taken internally) because this puts them very close to biological tissue. (Recall that α particles have a very high ionizing power.)

(c) Argon is produced from radioactive potassium (^{40}K) and has thus accumulated in the atmosphere as ^{40}K has undergone radioactive decay over several billion years.

(d) Although Fr is an alkali metal and resembles them, it does not occur mixed with them. It is only encountered as one of the transient products in a radioactive decay series. Thus, it is likely to be found mixed, in trace quantities, with other radioactive elements rather than with the nonradioactive alkali metals.

(e) To bring positively charged nuclei in close enough proximity to promote nuclear fusion requires that very large repulsive forces be overcome. For this to occur the nuclei must have high thermal energies, which in turn requires extremely high temperatures.

Chapter 26

Organic Chemistry

Review Problems

26-1. (a)

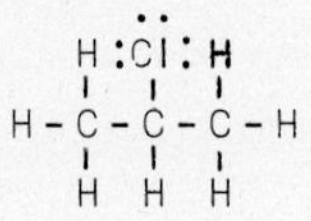

```
     ..
   H:Cl:H
   |  |  |
 H-C- C -C-H
   |  |  |
   H  H  H
```

(b)

```
       H  H
   ..  |  |  ..
 H-O - C- C - O-H
   ..  |  |  ..
       H  H
```

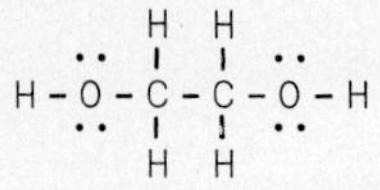

(c)

```
   H  H
   |  |  ..
 H-C- C=O:
   |
   H
```

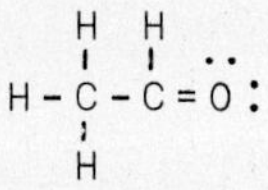

26-2. (a) There are three isomers of pentane. Only the carbon-atom skeletons are shown below:

```
(1) C-C-C-C-C        (2)     C          (3)   C
                             |                |
                       C-C-C-C              C-C-C
                                              |
                                              C
```

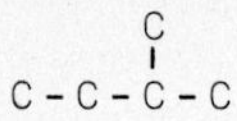

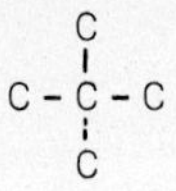

(b) There are nine isomers of heptane; their carbon-atom skeletons are as follows:

```
(1) C-C-C-C-C-C-C

(2)           C    (3)       C      (4) C C        (5)  C     C
              |              |          | |             |     |
    C-C-C-C-C-C-C    C-C-C-C-C-C-C    C-C-C-C-C       C-C-C-C-C

(6)         C      (7)     C        (8)  C C       (9)      C
            |              |             | |                |
    C-C-C-C-C-C        C-C-C-C-C       C-C-C-C              C
            |              |             |                  |
            C              C             C              C-C-C-C-C
```

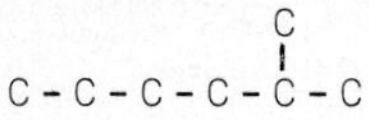

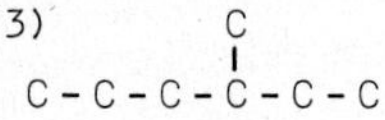

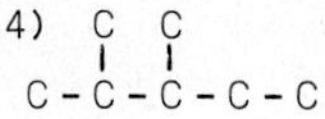

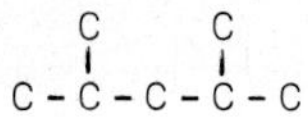

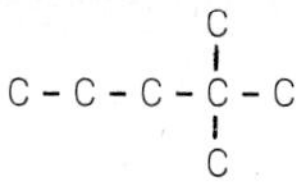

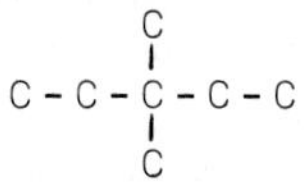

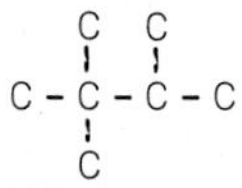

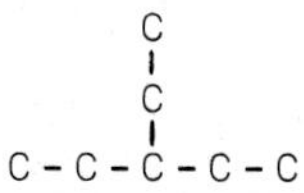

26-3. (a) The compounds have different formulas, C_4H_{10} and C_4H_8. They cannot be isomers.

(b) The compounds have the same formula, C_9H_{20}. They are skeletal isomers.

(c) The structures shown are *identical*. There is a single compound, not isomers.

(d) Again, the structures are *identical* (rotate either structure by 180° to obtain the other).

(e) The two structures are *identical*. There is no isomerism.

(f) The different positions on the benzene ring indicate ortho-para isomerism.

26-4. (a) A three-carbon chain has no skeletal isomerism. The different isomers all result from different positions for the three Cl atoms on a three-carbon chain.

```
(1)   H  H  Cl         (2)   H  Cl Cl          (3)   Cl H  Cl
      |  |  |                |  |  |                 |  |  |
    H-C- C- C-Cl           H-C- C- C-Cl            H-C- C- C-Cl
      |  |  |                |  |  |                 |  |  |
      H  H  Cl               H  H  H                 H  H  H

(4)   Cl Cl Cl         (5)   Cl Cl H
      |  |  |                |  |  |
    H-C- C- C-H            H-C- C- C-H
      |  |  |                |  |  |
      H  H  H                H  Cl H
```

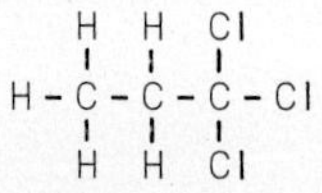

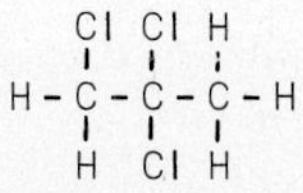

Any other structure that might be written is identical to one of these five.

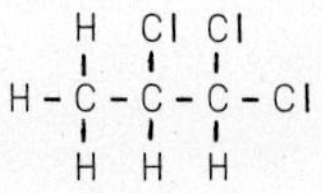

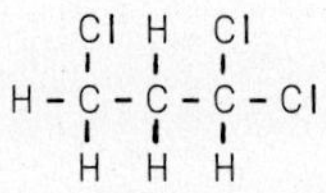

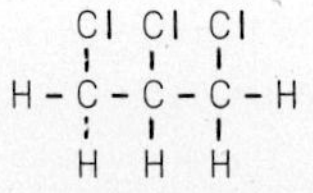

(b) First write all the possibilities based on *n*-butane.

```
(1)   H  H  H  H        (2)   H  H  Cl H        (3)   H  H  Cl Cl
      |  |  |  |              |  |  |  |              |  |  |  |
    H-C- C- C- C-Cl         H-C- C- C- C-H          H-C- C- C- C-H
      |  |  |  |              |  |  |  |              |  |  |  |
      H  H  H  Cl             H  H  Cl H              H  H  H  H

(4)    H  H  H  H       (5)   H  Cl Cl H        (6)   H  Cl H  Cl
       |  |  |  |             |  |  |  |              |  |  |  |
   Cl- C- C- C- C-Cl        H-C- C- C- C-H          H-C- C- C- C-H
       |  |  |  |             |  |  |  |              |  |  |  |
       H  H  H  H             H  H  H  H              H  H  H  H
```

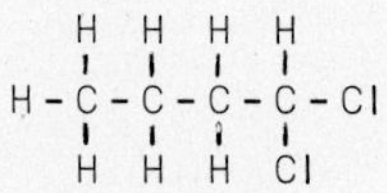

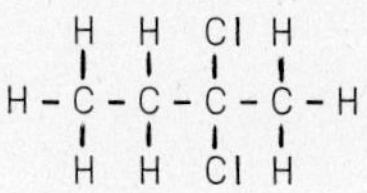

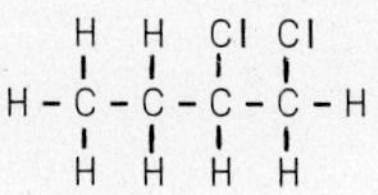

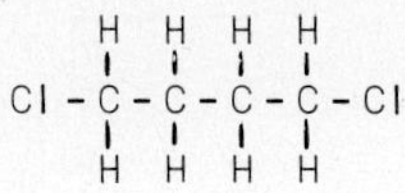

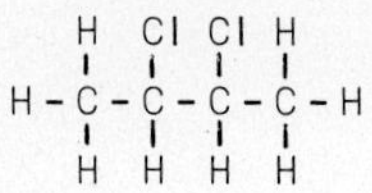

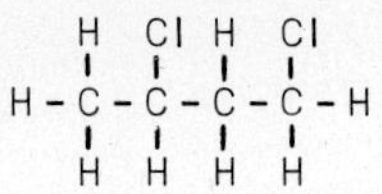

Then write the structures based on *iso*-butane.

(7)
```
    H   H   Cl
    |   |   |
H - C - C - C - Cl
    |   |   |
    H   |   H
        |
    H - C - H
        |
        H
```

(8)
```
    H   Cl  Cl
    |   |   |
H - C - C - C - H
    |   |   |
    H   |   H
        |
    H - C - H
        |
        H
```

(9)
```
     H   H   H
     |   |   |
Cl - C - C - C - Cl
     |   |   |
     H   |   H
         |
     H - C - H
         |
         H
```

26-5. (a) halide (bromide); (b) carboxylic acid; (c) aldehyde; (d) ether; (e) ketone; (f) amine; (g) phenyl; (h) ester.

26-6. (a) This is an eight-carbon chain with two methyl groups at the third C atom: 3,3-dimethyloctane.

(b) This is a three-carbon chain with two methyl groups at the second C atom: 2,2-dimethylpropane.

(c) The longest carbon chain has seven C atoms. The substituents are on carbon atoms 2, 3 and 5 (rather than 3, 5 and 6 if the chain were numbered in the opposite direction) 2,3-dichloro-5-ethylheptane.

(d) Maximum chain branching is achieved if we consider one ethyl and two methyl groups as substituents (rather than one *t*-butyl): 2,2-dimethyl-3-ethylpentane.

(e) Place the double bond at the lowest possible number (3): 5-chloro-3-heptene.

26-7. (a) 3-bromo-2-methylpentane:
```
    H   H   H   H   H
    |   |   |   |   |
H - C - C - C - C - C - H
    |   |   |   |   |
    H   H   Br  |   H
                |
            H - C - H
                |
                H
```

(b) 3-isopropyloctane:
```
    H   H   H   H   H   H   H   H
    |   |   |   |   |   |   |   |
H - C - C - C - C - C - C - C - C - H
    |   |   |   |   |   |   |   |
    H   H   H   H   H   |   H   H
                        |
                    H   |   H
                    |   |   |
                H - C - C - C - H
                    |   |   |
                    H   H   H
```

(c) 2-pentene:
```
    H   H           H
    |   |           |
H - C - C - C = C - C - H
    |   |   |   |   |
    H   H   H   H   H
```

(d) ethyl *n*-propyl ether:
```
    H   H       H   H   H
    |   |       |   |   |
H - C - C - O - C - C - C - H
    |   |       |   |   |
    H   H       H   H   H
```

26-8. (a) isopropyl alcohol: $CH_3CH(OH)CH_3$

(b) tetraethyllead: $Pb(C_2H_5)_4$

(c) 1,1,1-chlorodifluoroethane: $ClCF_2CH_3$

(d) 2-methyl-1,3-butadiene: $CH_2{=}C(CH_3)CH{=}CH_2$

(e) 2-butenal: $CH_3CH{=}CHCHO$

26-9. (a) *p*-dibromobenzene or 1,4-dibromobenzene

(b) *o*-methylaniline or *o*-aminotoluene

(c) 1,3,5-trimethylbenzene:

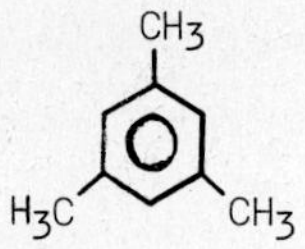

(d) *p*-nitrophenol:

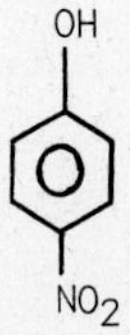

(e) 3-amino-2,5-dichlorobenzoic acid:

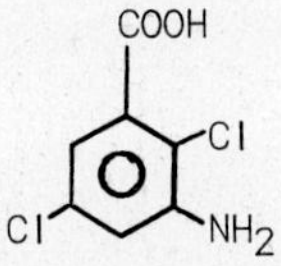

26-10. (a) $CH_3CH_3 + Cl_2 \longrightarrow CH_3CH_2Cl$ (plus some polysubstituted chlorides)

(b) $CH_3CH_2CH{=}CH_2 + H_2 \xrightarrow{Pt} CH_3CH_2CH_2CH_3$

(c) $CH_3CH_2CH(OH)CH_2CH_3 \xrightarrow[H^+]{Cr_2O_7^{2-}} CH_3CH_2\overset{O}{\overset{\|}{C}}CH_2CH_3$ (oxidation of a secondary alcohol produces a ketone)

(d) $CH_3CH_2CH{=}CH_2 + H_2O \xrightarrow[H_2SO_4(aq)]{10\%} CH_3CH_2CH(OH)CH_3$

(e) $CH_3CH_2OH + CH_3\overset{O}{\overset{\|}{C}}{-}OH \rightarrow CH_3CH_2O\overset{O}{\overset{\|}{C}}CH_3$

(f) $C_6H_5{-}\overset{O}{\overset{\|}{C}}{-}OH$ + KOH(aq) → $C_6H_5{-}\overset{O}{\overset{\|}{C}}{-}O^-K^+$

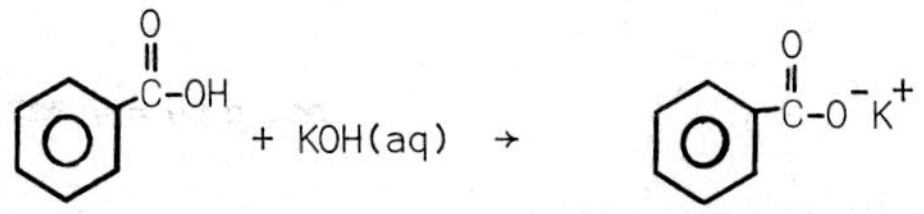

Exercises

Definitions and terminology

26-1. (a) An organic compound is one containing carbon in combination with hydrogen and often other elements, such as nitrogen and/or oxygen.

(b) An alkane hydrocarbon is a carbon-hydrogen compound in which all C-C and C-H bonds are single covalent bonds. Each carbon atom in alkane molecule is bonded simultaneously to four other atoms.

(c) An aromatic hydrocarbon is a carbon-hydrogen compound whose carbon atoms are joined into ring-like structures. The basic structural unit is a hexagonal ring with each carbon atom joined to two other carbon atoms and a third atom which may be carbon or hydrogen. The bonds within the rings possess some multiple bond character.

26-2. (a) Aliphatic molecules have straight or branched chain carbon skeletons. Alicyclic molecules feature carbon atoms joined into a ring. An alicyclic molecule can be thought of as resulting from the elimination of an H atom from each end of an aliphatic chain, followed by ring closure.

(b) An aliphatic compound has a carbon atom skeleton that is either a straight or branched chain. In an aromatic compound the carbon atom skeleton consists of a hexagonal ring of carbon atoms or of two or more rings fused together. The ring system possesses multiple bond character.

(c) Paraffins and olefins are both aliphatics, that is, the carbon atom skeletons are straight or branched chains. The paraffins (alkanes), however, contain only single bonds between carbon atoms whereas the olefins (alkenes) have some double bonds.

(d) An alkane is a hydrocarbon molecule with a formula, C_nH_{2n+2}. An alkyl group has one fewer hydrocarbon atom than the corresponding alkane molecule. It is not a stable species, but it can be bonded to another atom or group to form a molecule. For example.

CH_3CH_3	CH_3CH_2-	CH_3CH_2Cl
ethane	ethyl	ethyl chloride

(e) A normal molecule or group has all carbon atoms in a straight chain. An *iso* group has a branched chain.

$$CH_3CH_2CH_2CH_2- \qquad CH_3-\underset{\displaystyle}{\overset{\overset{\displaystyle CH_3}{|}}{CH}}-CH_2-$$

n-butyl　　　isobutyl

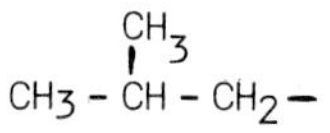

(f) Primary, secondary and tertiary are terms used to indicate the placement of a substituent group on a hydrocarbon chain. If the group *X* is attached to the end of a chain the compound is referred to as primary. If the carbon atom to which the group *X* is attached is bonded to two other carbon atoms, the compound is called secondary; and if the carbon atom to which *X* is attached is bonded to three other carbon atoms, the compound is tertiary.

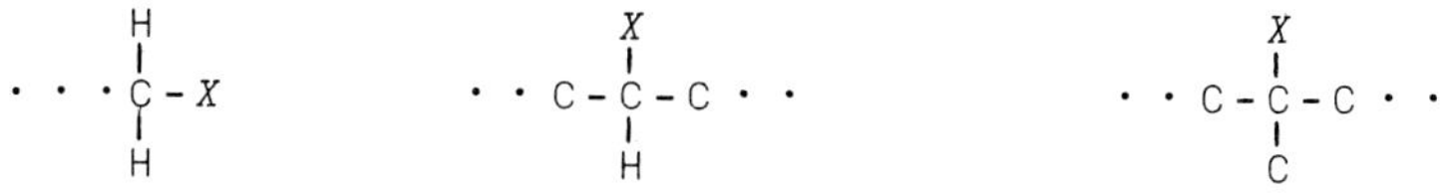

$$\cdots\underset{\underset{\displaystyle H}{|}}{\overset{\overset{\displaystyle H}{|}}{C}}-X \qquad \cdots C-\underset{\underset{\displaystyle H}{|}}{\overset{\overset{\displaystyle X}{|}}{C}}-C\cdots \qquad \cdots C-\underset{\underset{\displaystyle C}{|}}{\overset{\overset{\displaystyle X}{|}}{C}}-C\cdots$$

primary　　　secondary　　　tertiary

(g) Axial and equatorial refer to the placement of substituent groups on the chair conformation of the cyclohexane ring structure (see Figure 26-4). Equatorial groups extend outward from the ring, and axial groups are directed above and below the ring.

26-3. (a) A condensed formula indicates both the total number of atoms of all types in a molecule and the manner in which these atoms are bonded together (that is, single or multiple bonds, straight or branched chains, and so on). The condensed formula of 2,3-dimethylbutane is $CH(CH_3)_2CH(CH_3)_2$ and stands for the structure:

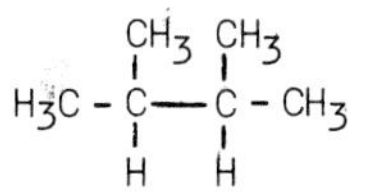

$$H_3C-\underset{\underset{\displaystyle H}{|}}{\overset{\overset{\displaystyle CH_3}{|}}{C}}-\underset{\underset{\displaystyle H}{|}}{\overset{\overset{\displaystyle CH_3}{|}}{C}}-CH_3$$

(b) Compounds in a homologous series differ from one another by some constant unit in their formulas. Each succeeding number of the alkane series of hydrocarbons has an additional $-CH_2$ unit, for example, *n*-butane is $CH_3CH_2CH_2CH_3$ and *n*-pentane is $CH_3CH_2CH_2CH_2CH_3$.

(c) An olefin is an alkene hydrocarbon. The molecule contains a double bond between carbon atoms, as in propene, $CH_3CH{=}CH_2$.

(d) A free radical is a grouping of atoms that may participate in chemical reactions but that cannot be isolated as a stable substance. One example is the methyl radical, $\cdot CH_3$, which participates in chain reactions such as (24.5).

(e) The positions on a benzene ring with respect to the location of a substituent group *X* are denoted as ortho, meta, and para.

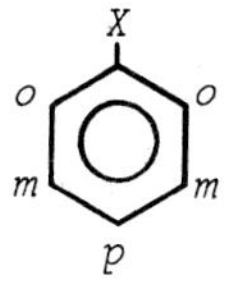

X
o　o
m　m
p

When additional substituents *Y* are added to the ring by appropriate chemical reactions, the group *X* tends to direct the *Y* groups to preferred positions on the ring. If these positions are *o*- and *p*-, the substituent *X* is said to be an ortho/para director (for example, -OH). If the preferred position for the second group *Y* is a meta position, *X* is said to be a meta director (for example, $-NO_2$).

Organic structures

62-4. (a)

```
    H   H   H   H   H
    |   |   |   |   |
H - C - C - C - C - C - H
    |   |   |   |   |
    H   H   H   Br  H
```

(b)

```
      CH3 H   H  CH3 H
       |  |   |   |  |
H3C - C - C - C - C - C - CH3
      |   |   |   |   |
      H   H   H   H   H
```

(c)

```
      CH3 H  CH3 H   H
       |  |   |  |   |
H3C - C - C - C - C - C - CH3
      |   |   |   |   |
      CH3 H   H   H   H
```

(d)

```
      H   CH3     H
      |   |       |
H3C - C - C - C = C
      |   |   |   |
      H   H   |   H
              |
          H - C - H
              |
          H - C - H
              |
              H
```

26-5. All C-H bonds involve the overlap, $(2sp^3, 1s)$ unless otherwise indicated.

(a) Ethane: C–C bond $(2sp^3, 2sp^3)$

(b) H₂C=CHCl: C=C bonds $(2sp^2, 2sp^2)$ and $(2p, 2p)$; C–H bonds $(1s, 2sp^2)$ and $(2sp^2, 1s)$; C–Cl bond $(2sp^2, 3s)$

(c) $H_3C-C\equiv C-H$: $(2sp, 1s)$; $(2p, 2p)$; $(2p, 2p)$; $(2sp, 2sp)$; $(2sp^3, 2sp)$

(d) $H_3C-CO-CH_3$: $(2p, 2sp^2)$; $(2sp^2, 2p)$; $(2sp^2, 2sp^3)$

(e) $H_3C-CH_2-NH_2$: $(2sp^3, 2sp^3)$; $(2sp^3, 1s)$; $(2sp^3, 2sp^3)$

Isomers

26-6. Skeletal isomers differ in the basic carbon atom framework of their molecules, such as straight and branched chains. Positional isomers differ in the placement of substituent groups on the carbon atom framework. Isomers that exist because of restricted rotation about a double bond are called geometric isomers.

(a) positional; (b) skeletal; (c) positional; (d) positional (ortho and meta); (e) geometric.

26-7. Draw the five skeletal structures of C_6H_{14} established in Example 26-1 and determine the number of possibilities of substituting a single Cl atom into each of these.

```
                                        Cl                    Cl
                                        |                     |
(1) C-C-C-C-C-C  leads to Cl-C-C-C-C-C-C; C-C-C-C-C-C; and C-C-C-C-C-C (three)

      C                     C           C             C             C
      |                     |           |             |             |
(2) C-C-C-C-C  leads to Cl-C-C-C-C-C; C-C-C-C-C; C-C-C-C-C; C-C-C-C-C;
                                        |               |               |
                                        Cl              Cl              Cl
      C
      |
    C-C-C-C-C-Cl  (five)

        C                       C         C             C           C-Cl
        |                       |         |             |           |
(3) C-C-C-C-C leads to Cl-C-C-C-C-C; C-C-C-C-C; C-C-C-C-C; C-C-C-C-C  (four)
                                       |               |
                                       Cl              Cl
```

```
         C   C                      C   C          C   C
         |   |                      |   |          |   |
(4)  C - C - C - C   leads to  Cl - C - C - C - C;  C - C - C - C   (two)
                                                       |
                                                       Cl

             C                          C           Cl  C                  C
             |                          |           |   |                  |
(5)  C - C - C - C   leads to  Cl - C - C - C - C;  C - C - C - C;  C - C - C - C - Cl   (three)
             |                          |               |                  |
             C                          C               C                  C
```

Functional groups

26-8. (a) Both carbonyl and carboxyl groups are based on the unit $R-\overset{O}{\overset{\|}{C}}-R$ but in a carboxylic acid one R group is replaced by OH. (In an aldehyde one R group is replaced by an H atom.)

(b) An amide has the $-NH_2$ group attached to a carbonyl group, that is:

$$R-\overset{O}{\overset{\|}{C}}-NH_2$$

When the $-NH_2$ group is attached directly to an organic residue R, the compound is an amine, $R-NH_2$.

(c) A carboxylic acid has the functional group -COOH, that is, R-COOH. Elimination of a water molecule from two acid molecules leads to an acid anhydride.

$$R\text{-}COOH \quad HOOC\text{-}R \rightarrow (RCO)_2O$$

(d) Both aldehydes and ketones contain the carbonyl group, $>C=O$. If the two groups attached to the carbonyl are organic residues, R, the compound is a ketone; if one is an R group and the other an H atom, the molecule is an aldehyde.

26-9. (a) $R-NO_2$, where R is $-CH_3$, $-CH_2CH_3$, $-CH_2CH_2CH_3$, $-CH(CH_3)_2$, and so on.

(b) $R-NH_2$. R groups might be (phenyl) or (naphthyl) or (methylphenyl, CH_3) and so on.

(c) A chlorophenol has one or more Cl atoms on the ring structure of a phenol, such as

OH Cl

(d) An aliphatic diol has two OH groups as substituents on a straight or branched chain hydrocarbon, such as

```
H3CCHCH2OH
   |
   OH
```

(e) An unsaturated aliphatic alcohol is a straight or branched chain hydrocarbon that contains one or more multiple bonds and an OH group as a substituent on the chain, such as

$$H_3CCH=CHCH_2OH$$

(f) An alicyclic ketone has the carbonyl group incorporated into a saturated ring hydrocarbon, such as cyclohexanone.

(cyclohexane ring) = O

(g) A halogenated alkane has one or more halogen atoms X (X = F, Cl, Br or I) substituted on an alkane hydrocarbon chain, such as

$$H_3CCCl_2CH_2CH_3$$

(h) An aromatic dicarboxylic acid has two carboxyl groups as substituents on an aromatic hydrocarbon.

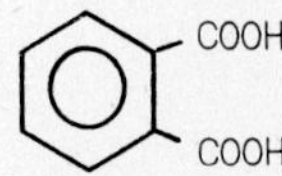

Nomenclature and formulas

26-10. (a) 2,2-dimethylbutane; (b) 2-methyl-1-propene (or simply methylpropene); (c) methylcyclopropane; (d) 4-methyl-2-pentyne; (e) 3,4-dimethylhexane; (f) 3,4-dimethyl-2-*n*-propyl-1-pentene

26-11. (a)

$$\begin{array}{c} CH_3 \\ | \\ H_3C-C-CH_2-CH_3 \\ | \\ H \end{array}$$

(b)

(c) $H_3C-CH_2-C\equiv C-CH_2-CH_3$

(d)

$$\begin{array}{l} H_3C-CH_2-CH-CH_3 \\ \qquad\qquad\quad | \\ \qquad\qquad\quad OH \end{array}$$

(e)

$$\begin{array}{l} H_3C-CH-O-CH_3 \\ \qquad\quad | \\ \qquad\quad CH_3 \end{array}$$

(f)

$$\begin{array}{r} O \\ \| \\ H_3C-CH_2-C-H \end{array}$$

(g)

$$\begin{array}{c} Cl \\ | \\ H_3C-C-CH_3 \\ | \\ CH_3 \end{array}$$

(h)

$$\begin{array}{c} CH_3 \\ | \\ H_3C-CH_2-N-CH_2-CH_3 \end{array}$$

(i)

$$\begin{array}{l} \qquad\qquad\quad O \\ \qquad\qquad\quad \| \\ H_3C-CH-C-OH \\ \qquad\quad | \\ \qquad\quad CH_3 \end{array}$$

(j)

$$\begin{array}{l} \qquad\qquad\qquad\qquad\quad O \\ \qquad\qquad\qquad\qquad\quad \| \\ H_3C-CH-CH_2-O-C-CH_2-CH_3 \\ \qquad\quad | \\ \qquad\quad CH_3 \end{array}$$

26-12. (a) No. The double bond in pentene could be located either at the first or second carbon atom. The compound must be named in such a way as to distinguish between these two possibilities, that is, either 1-pentene or 2-pentene.

(b) Yes. The carbonyl group must be either at the second or third position in the four-carbon chain of butane, but these two positions are identical.

(c) No. The alcohol could be primary, secondary, or tertiary butyl alcohol.

(d) No. Even if it were understood that the $-NH_2$ group is attached to the benzene ring and not directly to the N atom (that is, not N-methylaniline), there are still three posibilities for the $-CH_3$ group with respect to $-NH_2$--ortho, meta, and para.

(e) Yes. All positions on the cyclopentane ring where a methyl group might be substituted are equivalent.

26-13. (a) No. The double bond at the number "3" carbon atom is really at "2" (counting from the other end of the chain).

$H_3C-CH=CH-CH_2-CH_3$ 2-pentene

(b) No. There are two double bonds, but more than one way in which these double bonds may be placed in the molecule, for example

$H_2C=CH-CH=CH-CH_3$ or $H_2C=CH-CH_2-CH=CH_2$

(1,3-pentadiene) (1,4-pentadiene)

(c) No. A ketone must have an organic residue R on either side of the carbonyl group. If the carbonyl group is at the first position on the chain, the compound is an aldehyde.

$$\begin{array}{r} O \\ \| \\ H_3C-CH_2-C-H \end{array} \text{ propanal; } \begin{array}{c} O \\ \| \\ H_3C-C-CH_3 \end{array} \text{ propanone}$$

(d) No. The bromine atom can be either on the first or second carbon atom, that is, 1-bromopropane or 2-bromopropane.

(e) No. The compound in question is *m*-dichlorobenzene. If numbers are to be used in its name they should be the lowest numbers possible, 1,3-dichlorobenzene.

(f) No. The compound in question is the following:

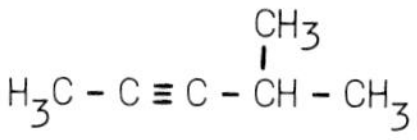

```
            CH3
            |
H3C - C ≡ C - CH - CH3
```

It should be named so that the triple bond appears at the lowest number possible, 4-methyl-2-pentyne.

(g) No. The compound suggested by the name given is

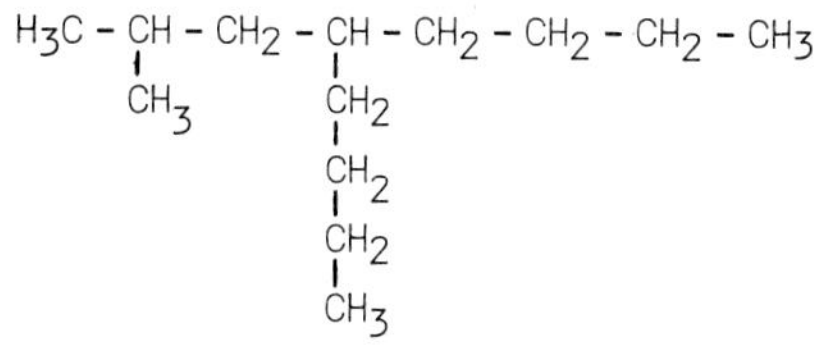

```
H3C - CH - CH2 - CH - CH2 - CH2 - CH2 - CH3
      |          |
      CH3        CH2
                 |
                 CH2
                 |
                 CH2
                 |
                 CH3
```

It should be named 5-isobutylnonane.
(The carbon chain is *nine* atoms long.)

(h) No. The compound indicated is

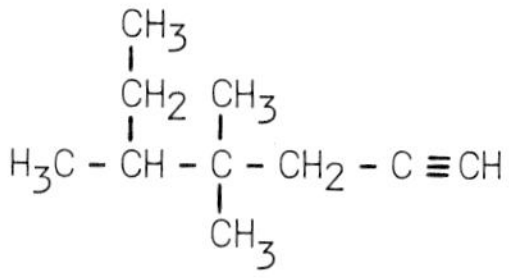

```
      CH3
      |
      CH2 CH3
      |   |
H3C - CH - C - CH2 - C ≡ CH
           |
           CH3
```

The name should be based on a *seven* carbon chain. The correct name is 4,4,5-trimethyl-1-heptyne.

(i) Correct.

(j) Correct.

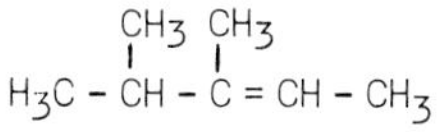

```
      CH3 CH3
      |   |
H3C - CH - C = CH - CH3
```

26-14. (a) $(CH_3)_3CCH_2CH(CH_3)_2$

(b)

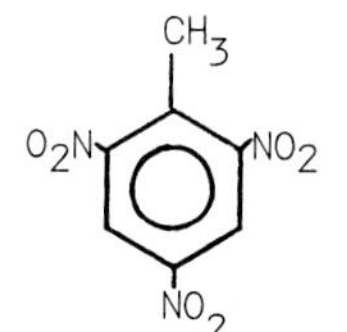

(c) (ring with OH and $\overset{O}{\overset{\|}{C}}$–$OCH_3$ substituents)

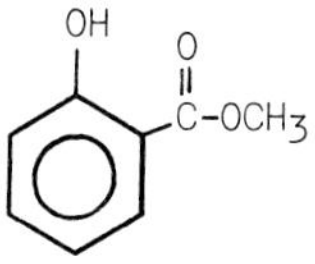

(d)

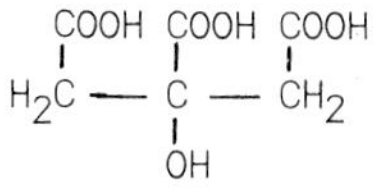

```
COOH COOH COOH
|    |    |
H2C — C — CH2
      |
      OH
```

(e) (cyclooctadiene ring)

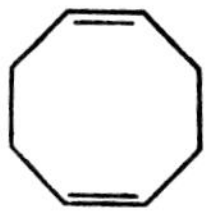

(f) (ring with OH and $C(CH_3)_2$ substituents)

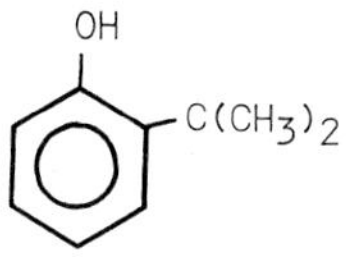

(g) (ring)–$CH_2CH(NH_2)CH_3$

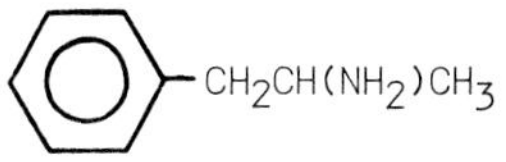

(h) $(CH_3)_2CH(C_{15}H_{31})$

(i) $(CH_3)_2C{=}CH(CH_2)_2C(CH_3){=}CH(CH_2)_2C(CH_3){=}CHCH_2OH$

(j)

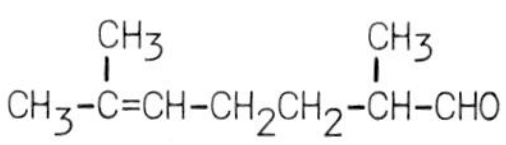

```
     CH3                CH3
     |                  |
CH3-C=CH-CH2CH2-CH-CHO
```

or $(CH_3)_2C{=}CH(CH_2)_2CH(CH_3)CHO$

Experimental determination of formulas

26-15. Recall the methods used in Chapter 3. That is, determine the number of moles of C in the hydrocarbon from the given mass of CO_2 and the number of moles of H from the mass of H_2O. An empirical formula can be written directly at this point.

$$\text{no. mol C} = 0.577 \text{ g } CO_2 \times \frac{1 \text{ mol C}}{44.0 \text{ g } CO_2} = 0.0131 \text{ mol C}$$

$$\text{no. mol H} = 0.236 \text{ g } H_2O \times \frac{2 \text{ mol H}}{18.0 \text{ g } H_2O} = 0.0262 \text{ mol H}$$

Empirical formula: $C_{0.0131}H_{0.0262} = CH_2$

Molecular formula (based on a molecular weight of 56): C_4H_8

Alkanes

26-16. Isobutane has a more compact structure than *n*-butane and a lower boiling point.

$H_3C - CH_2 - CH_2 - CH_3$

n-butane (b.pt. -0.5°C)

$H_3C - CH(CH_3) - CH_3$

isobutane (b.pt. -11.7°C)

The relative boiling points of the pentane isomers also reflect the compactness of the molecular structures.

$H_3C - CH_2 - CH_2 - CH_2 - CH_3$

n-pentane (b.pt. 36.1°C)

$H_3C - CH(CH_3) - CH_2 - CH_3$

isopentane (b.pt. 27.9°C)

$H_3C - C(CH_3)_2 - CH_3$

neopentane (b.pt. 9.5°C)

There are five hexane isomers whose boiling points decrease in the order listed below.

$H_3C - CH_2 - CH_2 - CH_2 - CH_2 - CH_3$

n-hexane (b.pt. 68.7°C)

$H_3C - CH_2 - CH(CH_3) - CH_2 - CH_3$

3-methylpentance (b.pt. 63.3°C)

$H_3C - CH(CH_3) - CH_2 - CH_2 - CH_3$

isohexane (b.pt. 60.3°C)

$H_3C - CH(CH_3) - CH(CH_3) - CH_3$

2,3-dimethylbutane (b.pt. 58.0°C)

$H_3C - C(CH_3)_2 - CH_2 - CH_3$

2,2-dimethylbutane (b.pt. 49.7°C)

26-17. The most stable conformation of t-butylcyclohexane is the chair form with the t-butyl group substituting for a hydrogen atom in an equatorial position (that is, extending outward from the ring).

26-18. (a) If the alkane hydrocarbon has a molecular weight of 44, it must have the formula C_3H_8. The structure is $H_3C - CH_2 - CH_3$ and the monochlorination products are $H_3C - CH_2 - CH - Cl$ and $H_3C - CHCl - CH_3$.

(b) The molecular weight of 58 corresponds to the alkane C_4H_{10}. The structure can be either n-butane or isobutane; each would form two monobromides.

$H_3C - CH_2 - CH_2 - CH_3$ or $H_3C - CH(CH_3) - CH_3$

26-19. (a) $CH_3CH_2CH{=}CH_2 + H_2 \rightarrow CH_3CH_2CH_2CH_3$
n-butane

(b) $2\ H_3CCH_2CH_2Br + 2\ Na \rightarrow H_3CCH_2CH_2{-}CH_2CH_2CH_3 + 2\ NaBr$
n-hexane

(c) $H_3CCH_2CH_2COONa + NaOH \rightarrow H_3CCH_2CH_3 + Na_2CO_3$
n-propane

(d) $H_3CCH_2CH_3 + Cl_2 \xrightarrow{h\nu} H_3CCH_2CH_2Cl + H_3CCHClCH_3$
1-chloropropane 2-chloropropane

26-20. One possibility for the chain termination reaction is the combination of two methyl radicals to form a molecule of ethane. Ethane molecules so formed may themselves undergo chlorination to ethyl chloride, C_2H_5Cl.

Alkenes

26-21. Only one position is available for the double bond in ethene and propene. In butene two nonequivalent positions are available for the double bond. The position of this double bond must be indicated in the name used.

$H_2C{=}CHCH_2CH_3$ (1-butene) $H_3CCH{=}CHCH_3$ (2-butene)

26-22. Alkenes and alicyclic hydrocarbons both can be thought of as resulting from the elimination of two hydrogen atoms and the formation of an additional carbon-to-carbon bond. In alkenes these hydrogen atoms come from *adjacent* carbon atoms, and the introduction of a new bond leads to a *double* bond between the carbon atoms. In alicyclic hydrocarbons the hydrogen atoms are derived from *opposite* ends of an alkane chain. A new carbon-to-carbon *single* bond is formed and the chain is converted to a ring.

26-23. (a) $H_2C{=}CHCH_3 + H_2 \xrightarrow{Pt,\Delta} H_3C{-}CH_2{-}CH_3$

(b) $H_3C{-}CH(OH){-}CH_2{-}CH_3 \xrightarrow[H_2SO_4]{\Delta} H_2C{=}CH{-}CH_2{-}CH_3 + H_3C{-}CH{=}CH{-}CH_3 + H_2O$

(c) $H_3C{-}C(Br)(CH_3){-}CH_3 + Na^{+}\ {}^{-}C{\equiv}CH \rightarrow H_3C{-}C(CH_3)_2{-}C{\equiv}CH + NaBr$

26-24. (a) $H_3C{-}CCl_2{-}CH_3$ (b) $H_3C{-}C(CN){=}CH_2$ (c) $H_3C{-}CH_2{-}C(CH_3)_2{-}Cl$ (d) $C_6H_5{-}CH(Br){-}CH_3$

Aromatic compounds

26-25. (a) *m*-dinitrobenzene (or 1,3-dinitrobenzene); (b) 1,3,5-trihydroxybenzene (or 2,4-dihydroxyphenol); (c) 3,5-dihydroxybenzoic acid; (d) N,N-diethylaniline;

(e) 

(f) 

(g) CH_3, OH, $CH_3{-}CH{-}CH_3$ substituted benzene ring

26-26. (a) 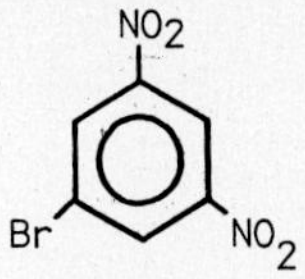 [structure: 1-bromo-3,5-dinitrobenzene] (b) 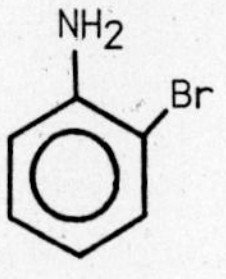 [structures: o-bromoaniline] and [p-bromoaniline]

(c) 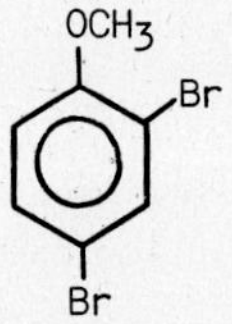 [structure: 2,4-dibromoanisole]

26-27. If the outcome of the substitution were determined by the -CHO group, the result should be 3-methoxy-5-nitrobenzaldehyde.

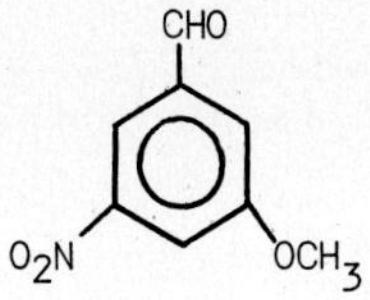

The fact that this compound is not produced suggests that the ortho-para director is stronger than the meta director. The expected isomers are

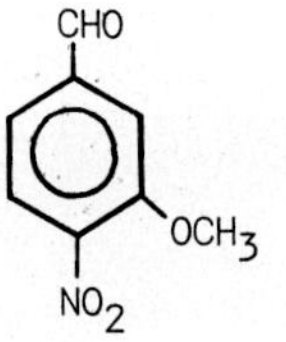

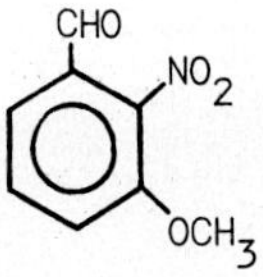

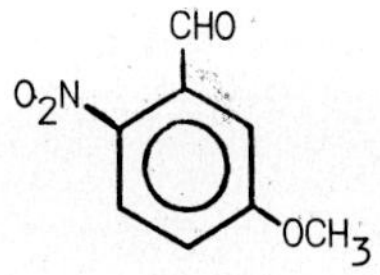

26-28. (a) 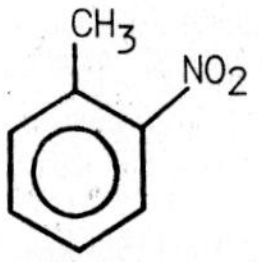 [structures: o-nitrotoluene] and [p-nitrotoluene] (b) 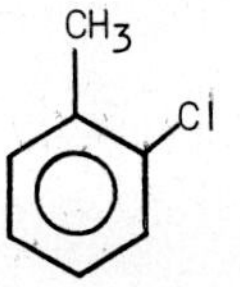 [structures: o-chlorotoluene] and [p-chlorotoluene]

(c) 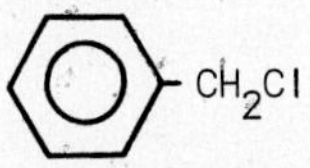 [structure: benzyl chloride, C_6H_5-CH_2Cl]

26-29. Cyclohexatriene is an unsaturated alicyclic hydrocarbon with localized double bonds between alternate pairs of carbon atoms. Benzene, although it has the same empirical formula, is a different substance. It is an aromatic compound, with the three pairs of electrons associated with its double bonds delocalized and spread out over the ring structure.

26-30. Replace the molecular orbital representation by Kekule structures.

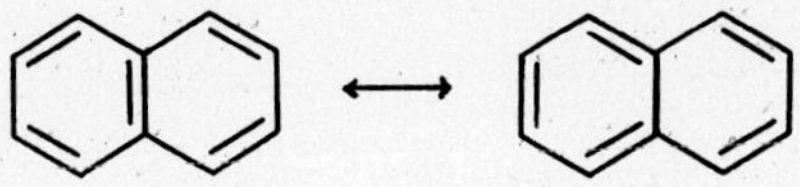

There are 10 electrons associated with the five double bonds in the Kekule structures. Thus, the two circles in the molecular orbital representation must represent a total of 10 electrons in the π bonding system of $C_{10}H_8$.

26-31. (a) The substitution of another group for an H atom on an alkane molecule is an aliphatic substitution reaction, such as reaction (26.5).

(b) In an aromatic substitution reaction a hydrogen atom(s) on a benzene ring or a condensed ring system is(are) replaced by another group. Four examples are given in Figure 26-6.

(c) In an addition reaction, such as those invovling alkenes and alkynes, a small molecule dissociates and the fragments are added to the carbon atoms at the site of a multiple bond [see equation (26.17)].

(d) Atoms or small groups from adjacent carbon atoms are removed and joined into a small molecule in an elimination reaction. A double bond is produced between the carbon atoms where the elimination occurs. Examples are given in equations (26.7) and (26.8).

26-32. (a)

$$CH_3CH_2CH_2CH_2CH_2OH \xrightarrow[H^+]{Cr_2O_7^{2-}} H_3C\text{-}CH_2\text{-}CH_2\text{-}CH_2\text{-}\overset{O}{\overset{\|}{C}}\text{-}OH$$

(b) $$CH_3CH_2CH_2COOH + HOCH_2CH_3 \xrightarrow[H^+]{\Delta} H_3C\text{-}CH_2\text{-}CH_2\text{-}\overset{O}{\overset{\|}{C}}\text{-}O\text{-}CH_2\text{-}CH_3$$

(c) $$o\text{-}NO_2C_6H_4\text{-}OH + NaOH \longrightarrow o\text{-}NO_2C_6H_4\text{-}O^-Na^+ + H_2O$$

(d) $$CH_3CH_2C(CH_3)\text{=}CH \xrightarrow[H_2SO_4]{H_2O} H_3C\text{-}CH_2\text{-}\underset{OH}{\overset{CH_3}{C}}\text{-}CH_3$$

26-33. (a) *oxid:* $2\{Fe \rightarrow Fe^{3+} + 3\ e^-\}$

red: $C_6H_5\text{-}NO_2 + 7\ H^+ + 6\ e^- \longrightarrow C_6H_5\text{-}NH_3^+ + 2\ H_2O$

net: $C_6H_5\text{-}NO_2 + 2\ Fe + 7\ H^+ \longrightarrow C_6H_5\text{-}NH_3^+ + 2\ Fe^{3+} + 2\ H_2O$

(b) *oxid:* $3\{CH_3CH\text{=}CH_2 + 2\ OH^- \rightarrow CH_3CHOHCH_2OH + 2\ e^-\}$

red: $2\{MnO_4^- + 2\ H_2O + 3\ e^- \rightarrow MnO_2 + 4\ OH^-\}$

net: $3\ CH_3CH\text{=}CH_2 + 2\ MnO_4^- + 4\ H_2O \rightarrow 3\ CH_3CHOHCH_2OH + 2\ MnO_2 + 2\ OH^-$

(c) *oxid:* $C_6H_{12}O_2 \rightarrow C_6H_{10}O_2 + 2\ H^+ + 2\ e^-$

red: $Pb(C_2H_3O_2)_4 + 2\ e^- \rightarrow Pb(C_2H_3O_2)_2 + 2\ C_2H_3O_2^-$

net: $C_6H_{12}O_2 + Pb(C_2H_3O_2)_4 \rightarrow C_6H_{10}O_2 + Pb(C_2H_3O_2)_2 + 2\ HC_2H_3O_2$

(d) *oxid:* $H_3C\text{-}C_6H_4\text{-}CH_3 + 4\ H_2O \rightarrow HOOC\text{-}C_6H_4\text{-}COOH + 12\ H^+ + 12\ e^-$

red: $2\{Cr_2O_7^{2-} + 14\ H^+ + 6\ e^- \rightarrow 2\ Cr^{3+} + 7\ H_2O\}$

net: $H_3C\text{-}C_6H_4\text{-}CH_3 + 16\ H^+ + 2\ Cr_2O_7^{2-} \rightarrow HOOC\text{-}C_6H_4\text{-}COOH + 4\ Cr^{3+} + 10\ H_2O$

26-34. First write a balanced equation for this oxidation-reduction reaction.

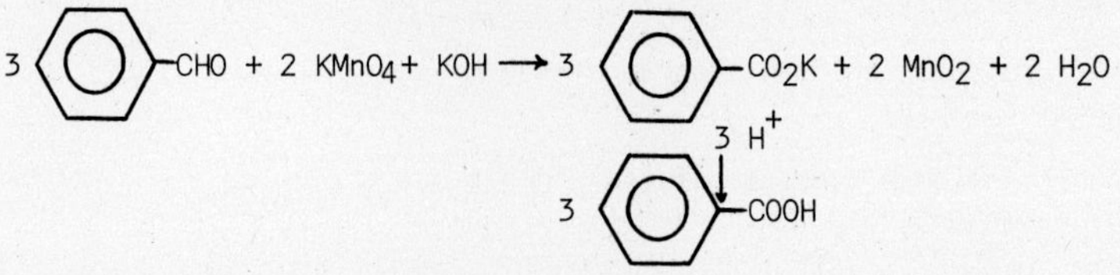

$$3\,C_6H_5CHO + 2\,KMnO_4 + KOH \rightarrow 3\,C_6H_5CO_2K + 2\,MnO_2 + 2\,H_2O$$

$$3\,C_6H_5CO_2K \xrightarrow{3\,H^+} 3\,C_6H_5COOH$$

Next, determine the limiting reagent.

$$\text{no. mol benzaldehyde} = 10.6\ g \times \frac{1\ mol}{106\ g} = 0.100\ mol\ \text{benzaldehyde}$$

$$\text{no. mol } KMnO_4 = 5.9\ g \times \frac{1\ mol}{158\ g} = 0.037\ mol\ KMnO_4$$

All of the $KMnO_4$ is consumed; it is the limiting reagent.

$$\text{no. g benzoic acid} = 0.037\ mol\ KMnO_4 \times \frac{3\ mol\ \text{benzoic acid}}{2\ mol\ KMnO_4} \times \frac{122\ g\ \text{benzoic acid}}{1\ mol\ \text{benzoic acid}} = 6.8\ g\ \text{benzoic acid}$$

$$\%\ \text{Yield} = \frac{6.1\ g\ \text{actual}}{6.8\ g\ \text{theoretical}} \times 100 = 9.0 \times 10^1\ \%$$

Organic synthesis

26-35. (a)

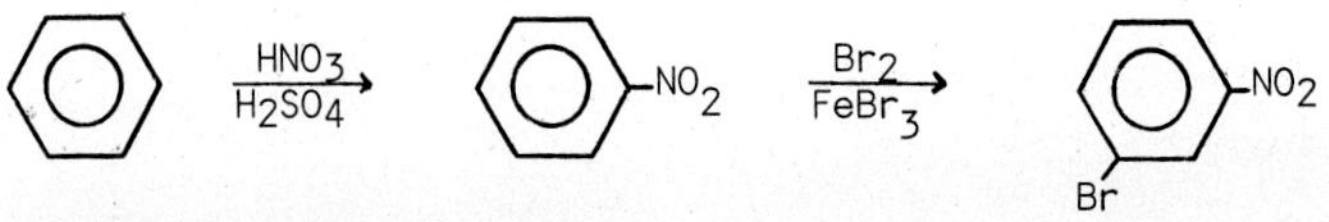

$$C_6H_6 \xrightarrow[H_2SO_4]{HNO_3} C_6H_5NO_2 \xrightarrow[FeBr_3]{Br_2} m\text{-}BrC_6H_4NO_2$$

(b)

$$C_6H_6 \xrightarrow[AlCl_3]{CH_3Cl} C_6H_5CH_3 \xrightarrow[H_2SO_4]{HNO_3} p\text{-}O_2NC_6H_4CH_3 + o\text{-}O_2NC_6H_4CH_3$$

$$p\text{-}O_2NC_6H_4CH_3 \xrightarrow[HCl]{Fe} Cl^-\,{}^+H_3N\text{-}C_6H_4\text{-}CH_3 \xrightarrow{NaOH} H_2N\text{-}C_6H_4\text{-}CH_3$$

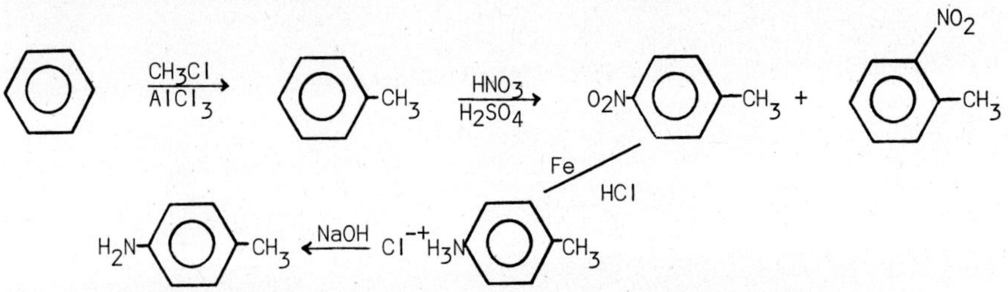

26-36. (a) $HC{\equiv}CH \xrightarrow[H_2]{Pt} H_2C{=}CH_2 \xrightarrow[H_2SO_4]{H_2O} CH_3CH_2OH$ $\quad CH_3CH_2OH \xrightarrow[H^+]{Cr_2O_7^{2-}} CH_3CHO$

(b) $HC{\equiv}CH \xrightarrow{Br_2} CHBr{=}CHBr \xrightarrow{Br_2} CHBr_2CHBr_2$

(c) $HC{\equiv}CH \xrightarrow[H_2]{Pt} H_2C{=}CH_2 \xrightarrow[H_2SO_4]{H_2O} CH_3CH_2OH$ $\quad CH_3CH_2OH \xrightarrow[H^+]{K_2Cr_2O_7} CH_3CO_2H$

$CH_3COOH + NH_3 \rightarrow CH_3COO^- + NH_4^+$

Decompose ammonium acetate as in reactions (26.40) and (26.41).

(d) Prepare isopropyl alcohol by the following series of reactions:

$HC{\equiv}CH \xrightarrow{NaNH_2} HC{\equiv}C^-Na^+ + NH_3$ $\quad HC{\equiv}C^-Na^+ + CH_3Cl \rightarrow CH_3C{\equiv}CH + NaCl$

$CH_3C{\equiv}CH \xrightarrow{Pt/H_2} CH_3CH{=}CH_2 \xrightarrow[H_2SO_4]{H_2O} CH_3CHOHCH_3$

Prepare acetic acid as in part (c) and allow it to react with isopropyl alcohol.

$CH_3CO_2H + (CH_3)_2CHOH \rightarrow CH_3CO_2CH(CH_3)_2 + H_2O$

isopropyl acetate

To prepare the CH_3Cl required in the synthesis of isopropyl alcohol,

$$CH_3CO_2H + NaOH \rightarrow CH_3CO_2^-Na^+ + H_2O$$

$$CH_3CO_2^-Na^+ + NaOH \xrightarrow{\Delta} Na_2CO_3 + CH_4 \qquad CH_4 + Cl_2 \rightarrow CH_3Cl + HCl$$

Polymerization reactions

26-37. An ester is formed by the condensation of a carboxylic acid and an alcohol (that is, by the elimination of a water molecule between them). Dacron is formed by the condensation of a *di*carboxylic acid with a *di*ol. Thus, it contains ester linkages. Because these ester linkages join large numbers of molecules, it is appropriate to call the polymer a *polyester*.

To determine the percent oxygen in Dacron, refer to the basic unit shown in Figure 26-19. It has the formula $C_{10}H_8O_4$.

$$\%O = \frac{(4 \times 16.0)\text{g O}}{192 \text{ g polymer}} \times 100 = 33.3\%$$

26-38. The polymerization of 1,6-hexanediamine with sebacyl chloride proceeds in the following manner:

$$H-\overset{H}{\overset{|}{N}}-(CH_2)_6-\overset{H}{\overset{|}{N}}-H + Cl-\overset{O}{\overset{\|}{C}}-(CH_2)_8-\overset{O}{\overset{\|}{C}}-Cl \quad H-\overset{H}{\overset{|}{N}}-(CH_2)_6-\overset{H}{\overset{|}{N}}-H \rightarrow \left(\overset{O}{\overset{\|}{C}}-(CH_2)_8-\overset{O}{\overset{\|}{C}}-\overset{H}{\overset{|}{N}}-(CH_2)_6-\overset{H}{\overset{|}{N}}\right)_x$$

26-39. To form long-chain molecules every monomer must have at least two functional groups, one on each end of the molecule. Ethyl alcohol has only one functional group (-OH). It cannot participate in a polymerization reaction with dimethyl terephthalate.

26-40. Because it has only two functional groups, we should expect ethylene glycol ($HOCH_2CH_2OH$) to produce polymer chains without crosslinking in its polymerization with phthalic acid. This is the polymer that is soft and tacky. Glycerol has three functional groups ($HOCH_2CHOHCH_2OH$). Not only can it form polymer chains by using two of its functional groups, but the third can be used to join chains together, that is, to crosslink chains. The resulting polymer is hard and brittle.

26-41. (a) $H_2C{=}CHCl + H_2C{=}CCl_2 \longrightarrow \left(H_2C-\overset{Cl}{\overset{|}{C}H}-CH_2-\underset{Cl}{\underset{|}{\overset{Cl}{\overset{|}{C}}}}\right)_x$

vinyl chloride — vinylidene chloride — saran

(b) $C_6H_5CH{=}CH_2 + CH_2{=}CHCN \longrightarrow \left(\underset{C_6H_5}{\underset{|}{CH}}-CH_2-CH_2-\underset{CN}{\underset{|}{\overset{H}{\overset{|}{C}}}}\right)_x$

styrene — acrylonitrile

26-42. In a simple molecular substance like benzene, all molecules are identical (C_6H_6). Whatever number of these molecules is taken in one sample, any other sample with the same number of molecules has the same mass. The molar mass--the mass in grams of one mole of molecules--is a unique quantity. Because the number of monomer units in a polymer chain is quite variable, individual polymer molecules may differ considerably in mass. Thus, the mass of a mole of these molecules must also be variable. We can only speak of this mass, this molecular mass, (and hence the molecular weight) on an average basis.

26-43. In free-radical addition polymerization, the number of radical chains that propagate in the reaction mixture is small, being determined by the amount of free radical initiator that is used. In condensation polymerization, all molecules are equally able to undergo condensation with neighboring molecules. As a result, many initial sites of polymer chain production may arise. This means shorter chains than in free-radical addition polymerization.

Self-test Questions

1. (*b*) The formula C_7H_{14} (a) describes *all* the heptane isomers, not isoheptane uniquely. The formula $CH_3(CH_2)_5CH_3$ (c) is that of *normal* heptane, the straight-chain isomer. Although the compound $C_6H_5CH_3$ (d) has seven carbon atoms, it is not an alkane. It is the aromatic hydrocarbon toluene (note the presence of the C_6H_5 group). Compound (b) has seven carbon atoms, the formula C_7H_{14}, and the necessary chain branching implied by the term "iso".

```
CH3-CHCH2CH2CH2CH3
    |
    CH3
```

2. (*d*) Propane, C_3H_8, has no isomers and butane, C_4H_{10}, has two: $CH_3CH_2CH_2CH_3$ and $(CH_3)_3CH$. Benzene, C_6H_6, may be thought of as a resonance hybrid of two contributing forms but these are not isomers because they cannot be isolated.

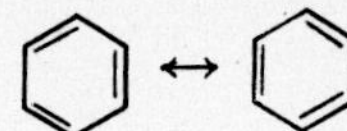

C_5H_{12} (pentane) must be the compound with three isomers: $CH_3(CH_2)_3CH_3$ and $(CH_3)_2CHCH_2CH_3$ and $C(CH_3)_4$.

3. (*b*) The compound C_4H_{10} (a) is butane. Cyclobutane is obtained from it by the elmination of two H atoms and conversion of the straight chain to a ring. Cyclobutane is C_4H_8. 2-Butene (b) has the formula C_4H_8 and, therefore, the same C:H ratio as cyclobutane. 2-Butyne (c) has the formula C_4H_6. The C:H ratio in benzene, C_6H_6, is 1:1, not 1:2 as in cyclobutane.

4. (*a*) The distinct forms of 1,2-dichloroethylene are

```
H       H         H       Cl
 \     /           \     /
  C=C               C=C
 /     \           /     \
Cl      Cl        Cl      H
   cis               trans
```

In 1,1-dichloroethylene (b), since both Cl atoms are on the same C atom, there is no isomerism. That is, the following structures are identical.

```
H       Cl              Cl      H
 \     /                 \     /
  C=C          and        C=C
 /     \                 /     \
H       Cl              Cl      H
```

(Flip one over and the other is obtained.)

Cis-trans isomerism does not occur in saturated hydrocarbons. There is essentially free rotation about the C-C single bond in $ClCH_2CH_2Cl$ (c). In 1,1,2,2-tetrachloroethylene, $Cl_2C{=}CCl_2$ (d), all Cl atoms are equivalent and again there is no isomerism.

5. (*c*) Look for a chain of four carbon atoms with appropriate substituents on the chain. Eliminate (a) because it is only a three-carbon chain. Eliminate (b) because it does not have OH on the first carbon atom. Compound (c) has OH on the first carbon, Cl on the second and $-CH_3$ on the third. Compound (d) has the Cl and $-CH_3$ interchanged.

6. (*d*) The name of this compound can be based either on the $-CH_3$ group (toluene) or the $-NH_2$ group (aniline), but the groups are in meta positions to one another. This eliminates (a) and (b). Compound (c) is eliminated since it does not have an $-NH_2$ group. (Also the name m-methylbenzene is meaningless. The methyl group would be meta to what?) If we use a ring numbering system, starting with $-NH_2$ at the number one C atom, the name is 3-methylaniline.

7. (*b*) Aniline (a) is a weak base. Phenol (c) is a very weak acid, and acetaldehyde has no significant acidic properties. Benzoic acid (b) is a common carboxylic acid--a weak acid.

8. (*c*) Oxidation of 2-propanol would produce dimethyl ketone. Oxidation of 1-butanol would produce the aldehyde $CH_3CH_2CH_2CHO$. Oxidation of 2-butanol produces methyl ethyl ketone. A carbonyl group cannot be present on a *t*-butyl group (five bonds to C).

9. (a) dichlorodifluoromethane:

```
    F
    |
F - C - Cl
    |
    Cl
```

(b) *p*-bromophenol: Br–C_6H_4–OH

(c) 3-hydroxy-2-butanone:

```
       OH   O
       |    ||
H3C - CH - C - CH3
```

(d) methyl *t*-butyl ether:

```
            CH3
            |
H3C - O - C - CH3
            |
            CH3
```

10. The following structures simply represent the carbon-atom chain and the placement of C atoms on the chain. H atoms have been omitted.

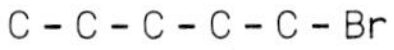

```
                           Br               Br              C
                           |                |               |
C - C - C - C - C - Br     C - C - C - C - C     C - C - C - C - C     C - C - C - C - Br
 1-bromopentane            2-bromopentane        3-bromopentane        1-bromo-3-methylbutane

    C  Br              C                       C                          C
    |  |               |                       |                          |
C - C - C - C      C - C - C - C       Br - C - C - C - C             C - C - C - Br
                       |                                                  |
2-bromo-3-methylbutane Br              1-bromo-2-methylbutane             C

                 2-bromo-2-methylbutane                          1-bromo-2,2-dimethylpropane
```

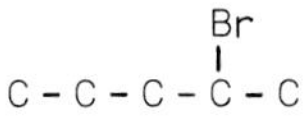

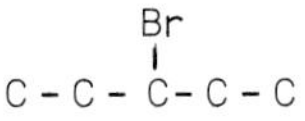

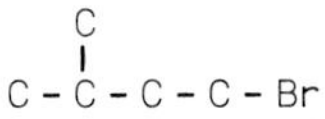

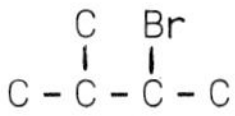

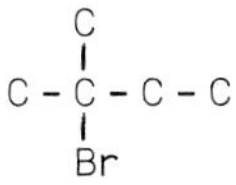

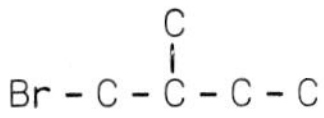

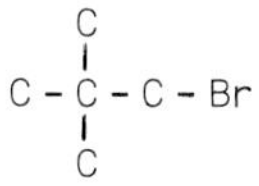

11. (a)

$$CH_3CH_2CH{=}CH_2 + H_2O \xrightarrow{H_2SO_4} CH_3CH_2\overset{OH}{\underset{H}{C}} - CH_3$$

(b) $CH_3CH_2CH_3 + Cl_2 \xrightarrow{h\nu} CH_3CH_2CH_2Cl$ and $CH_3CHClCH_3$

(c) $CH_3CHOHCH_3 + C_6H_5COOH \rightarrow C_6H_5COOCH(CH_3)_2$

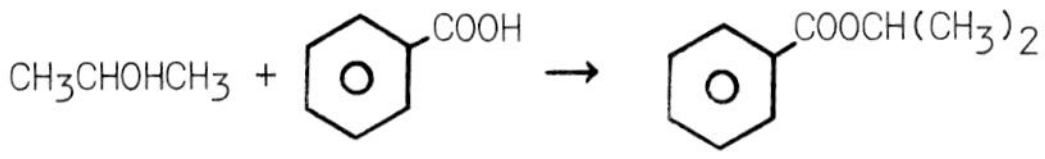

(d) $CH_3CHOHCH_2CH_3 \xrightarrow[H^+]{Cr_2O_7^{2-}} CH_3\underset{\underset{O}{\|}}{C}CH_2CH_3$

12. (a) The physical appearance alone should be sufficient with its low molecular weight, C_2H_6 is a gas; and with its considerably higher molecular weight, C_8H_{18} is a liquid.

(b) C_2H_6 is a saturated hydrocarbon and C_2H_4 is unsaturated. The unsaturated hydrocarbon will react with (and decolorize) MnO_4^-. Recall reaction (26.21).

(c) Low molecular weight alcohols, because of hydrogen bonding are completely imiscible with water. Higher molecular weight alcohols have very limited water solubility.

(d) Prepare a dilute aqueous solution of the substance. C_6H_5COOH will produce a pH < 7 because of its acidic properties. C_6H_5CHO has no acidic properties.

Chapter 27

Chemistry of the Living State

Review Problems

27-1. (a) glyceryl laurooleostearate; (b) glyceryl trilinoleate; (c) sodium myristate

27-2. (a)

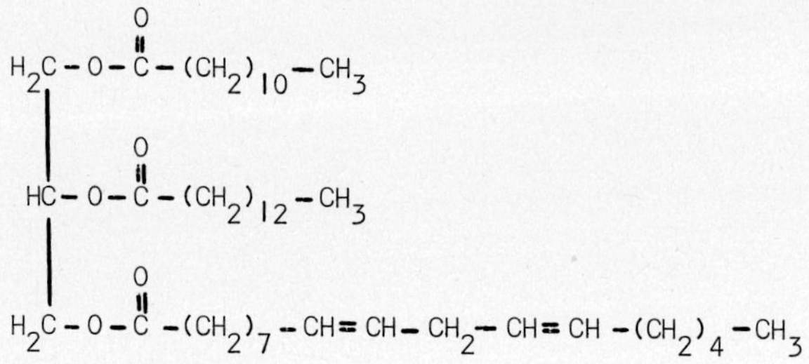

$$\begin{array}{l} H_2C-O-\overset{O}{\overset{\|}{C}}-(CH_2)_{10}-CH_3 \\ | \\ HC-O-\overset{O}{\overset{\|}{C}}-(CH_2)_{12}-CH_3 \\ | \\ H_2C-O-\overset{O}{\overset{\|}{C}}-(CH_2)_7-CH=CH-CH_2-CH=CH-(CH_2)_4-CH_3 \end{array}$$

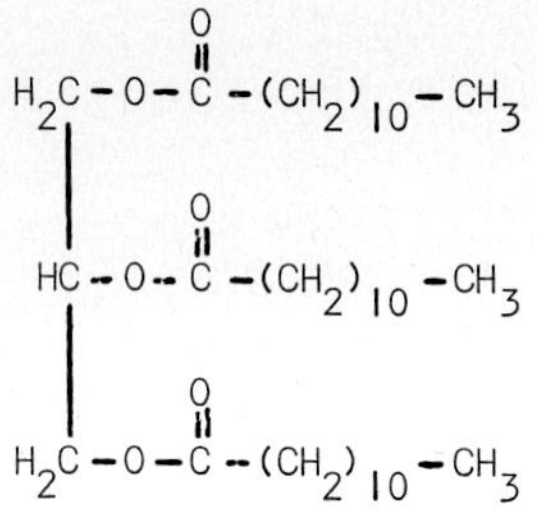

(b)

$$\begin{array}{l} H_2C-O-\overset{O}{\overset{\|}{C}}-(CH_2)_{10}-CH_3 \\ | \\ HC-O-\overset{O}{\overset{\|}{C}}-(CH_2)_{10}-CH_3 \\ | \\ H_2C-O-\overset{O}{\overset{\|}{C}}-(CH_2)_{10}-CH_3 \end{array}$$

(c) $K^+\ ^-O-\overset{O}{\overset{\|}{C}}-(CH_2)_{14}-CH_3$

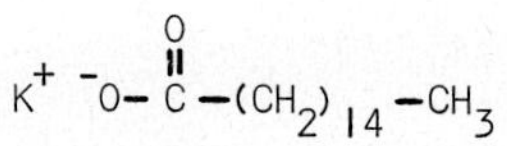

(d) 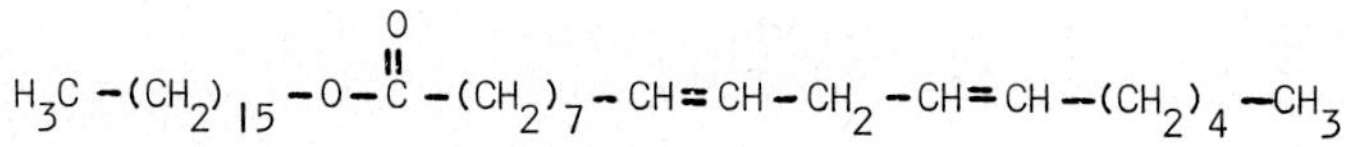 $H_3C-(CH_2)_{15}-O-\overset{O}{\overset{\|}{C}}-(CH_2)_7-CH=CH-CH_2-CH=CH-(CH_2)_4-CH_3$

27-3. The structure formula of glyceryl butyropalmitooelate is given in Example 27-1. First we must determine its molecular weight. The glycerol backbone, $C_3H_5O_3$ formula weight of $(3 \times 12.0) + (5 \times 1.0) + (3 \times 16) = 89$; the butyro group, $CO(CH_2)_2CH_3$: $(4 \times 12.0) + (7 \times 1.0) + 16 = 71$; the palmito group, $CO(CH_2)_{14}CH_3$: $(16 \times 12.0) + (31 \times 1.0) + 16 = 239$; the oleo group, $CO(CH_2)_7CH{=}CH(CH_2)_7CH_3$: $(18 \times 12.0) + (33 \times 1.0) + 16 = 265$. The molecular weight = 89 + 71 + 239 + 265 = 664.

(a) $$\text{no. mg KOH} = 1.00 \text{ g triglyc.} \times \frac{1 \text{ mol triglyc.}}{664 \text{ g triglyc.}} \times \frac{3 \text{ mol KOH}}{1 \text{ mol triglyc.}} \times \frac{56.1 \text{ g KOH}}{1 \text{ mol KOH}} \times \frac{1000 \text{ mg KOH}}{1 \text{ g KOH}}$$

$$= 253 \text{ mg KOH}$$

Saponification value = 253

(b) Only one of the acid residues (oleo) contains a double bond.

$$\text{no. g } I_2 = 1.00 \times 10^2 \text{ triglyc.} \times \frac{1 \text{ mol triglyc.}}{664 \text{ g triglyc.}} \times \frac{1 \text{ mol } I_2}{1 \text{ mol triglyc.}} \times \frac{253.8 \text{ g } I_2}{1 \text{ mol } I_2} = 38.2 \text{ g } I_2$$

Iodine number = 38.2

27-4. The saponification value, as seen from the set up in Example 27-2, is inversely proportional to the molecular weight of the triglyceride. Since the measured saponification value (209) is less than that of tristearin (189), the triglyceride must have a molecular weight that is less than that of tristearin (890). In fact the molecular weight is $890 \times \frac{189}{209} = 805$. Since the triglyceride is a simple glyceride, all three acid residues are the same. The contribution to the molecular weight by the glycerol backbone, $C_3H_5O_3$, is 89. The contributions of the three acid residues is 805 - 89 = 716; or per acid residue, 716/3 = 239. The acyl group (RCO-) in Table 27-2 that has a formula weight of 239 is palmito. The simple glyceride is tripalmitin.

27-5. The structure of D-(−)-arabinose, the enantiomer of L-(+)-arabinose, is shown below.

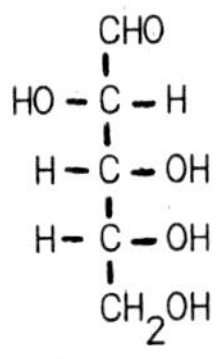

```
      CHO
       |
  HO - C - H
       |
   H - C - OH
       |
   H - C - OH
       |
      CH2OH
```

D-(−)-arabinose

One diastereomer of L-(+)-arabinose is

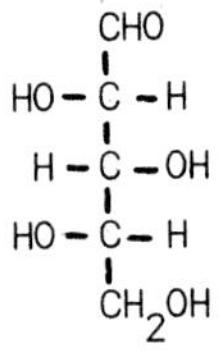

```
      CHO
       |
  HO - C - H
       |
   H - C - OH
       |
  HO - C - H
       |
      CH2OH
```

and there are several others.

27-6. (a) 1.0 M HCl:

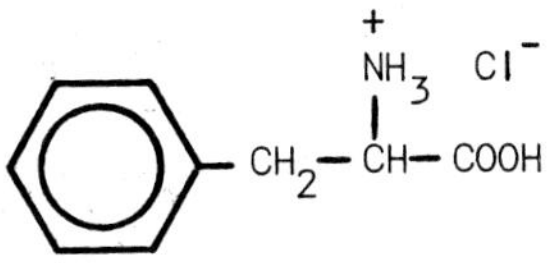

$C_6H_5-CH_2-CH(NH_3^+)-COOH \quad Cl^-$

(b) 1.0 M NaOH:

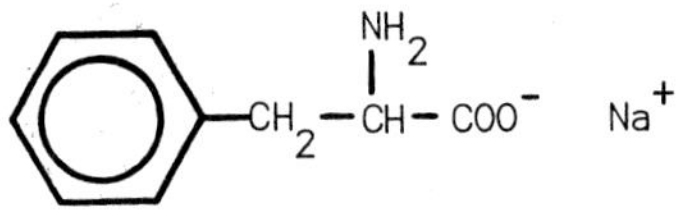

$C_6H_5-CH_2-CH(NH_2)-COO^- \quad Na^+$

(c) pH = 5.5:

$C_6H_5-CH_2-CH(NH_3^+)-COO^-$

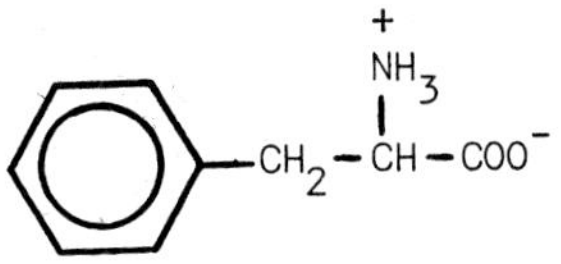

27-7. (a)

```
          O           O
          ‖           ‖
H2  C  –  C – NH – CH – C – OH
    |             |
    NH2           CH2OH
```

glycylmethionine

(b)

```
                NH2 O              O              O
                 |  ‖              ‖              ‖
CH3CH2CH – CH – C – NH – CH – C – NH – CH – C – OH
      |                   |              |
      CH3                 CH2            CH2
                          |              |
                   H3C – CH              OH
                          |
                          CH3
```

isoleucylleucylserine

27-8. (a)

```
               O           O           O            O
               ‖           ‖           ‖            ‖
H2N – CH2 –  C – NH – CH – C – NH – CH – C – NH – CH – C – OH
                      |            |              |
                      CH3          CH2            CHOH
                                   |              |
                                   OH             CH3
```

(b) glycylalanylserylthreonine

27-9. The chain pictured is RNA. This can be inferred in two ways: (a) ribose sugar groups and (b) uracil as a constituent rather than thymine. The chain components are the ribose sugar units, linked by phosphate groups and, as bases, adenine, uracil, guanine, and cytosine (from top to bottom).

27-10. We must make use of the genetic code in Table 25-5 to determine the amino acids involved and their sequence.

Code (mRNA)	ACC	CAU	CCC	UUG	GCG	AGU	GGU	AUG	UAA
Amino Acids	Thr	His	Pro	Leu	Ala	Ser	Gly	Met	(nonsense)

The polypeptide chain is Thr-His-Pro-Leu-Ala-Ser-Gly-Met.

Exercises

Structure and composition of the cell

27-1. First determine the cell volume and then the volume of water in the cell.

$$V = \pi r^2 h = \pi \left\{0.5\ \mu m \times \frac{1\times10^{-4}\ cm}{1\ \mu m}\right\}^2 \times \left\{2\ \mu m \times \frac{1\times10^{-4}\ cm}{1\ \mu m}\right\}^2 = 2\times10^{-12}\ cm^3$$

$$\text{Volume of cell water} = 0.80 \times 2 \times 10^{-12}\ cm^3 = 2\times10^{-12}\ cm^3$$

(a) $pH = -\log[H_3O^+] = 6.4 \qquad [H_3O^+] = 4\times10^{-7}$

$$\text{no. } H^+ \text{ ions} = 2\times10^{-12}\ cm^3 \times \frac{1\ L}{1000\ cm^3} \times \frac{4\times10^{-7}\ mol\ H^+}{1\ L} \times \frac{6.02\times10^{23}\ H^+\ ions}{1\ mol\ H^+} = 5\times10^2$$

(b) $$\text{no. } K^+ \text{ ions} = 2\times10^{-12}\ cm^3 \times \frac{1\ L}{1000\ cm^3} \times \frac{1.5\times10^{-4}\ mol\ K^+}{1\ L} \times \frac{6.02\times10^{23}\ K^+\ ions}{1\ mol\ K^+} = 2\times10^5\ K^+\ ions$$

27-2. Volume of a ribosome $= \frac{4}{3}\pi r^3 = \frac{4}{3}\pi \left\{9 \text{ nm} \times \frac{1\times10^{-7} \text{ cm}}{1 \text{ nm}}\right\}^3 = 3\times10^{-18} \text{ cm}^3$

% Volume occupied by ribosomes $= \dfrac{1.5\times10^4 \text{ ribosomes} \times \dfrac{3\times10^{-18} \text{ cm}^3}{1 \text{ ribosome}}}{2\times10^{-12} \text{ cm}^3} \times 100 = 2\%$

27-3. no. lipid molecules $= (0.02 \times 2 \times 10^{-12})\text{g lipid} \times \frac{1 \text{ mol lipid}}{700 \text{ g lipid}} \times \frac{6.02\times10^{23} \text{ lipid molecules}}{1 \text{ mol lipid}}$

$= 3\times10^7$ lipid molecules

27-4. no. protein molecules $= (0.15 \times 0.90 \times 2 \times 10^{-12})\text{g protein} \times \frac{1 \text{ mol protein}}{3\times10^4 \text{ g protein}}$

$\times \frac{6.02\times10^{23} \text{ protein molecules}}{1 \text{ mol protein}} = 5\times10^6$ protein molecules

27-5. Hypothetical length $= \frac{4.5\times10^6 \text{ units}}{1 \text{ molecule}} \times \frac{450 \text{ pm}}{1 \text{ unit}} \times \frac{1\times10^{-12} \text{ m}}{1 \text{ pm}} \times \frac{1 \text{ μm}}{1\times10^{-6} \text{ m}} = 2.0\times10^3 \text{ μm}$

This length is many times greater than the length of the cell itself. This fact suggests that the DNA molecule must be coiled and folded to be maintained within the cell nucleus.

Lipids

27-6. (a) A lipid is a cell constituent that is soluble in nonpolar solvents.

(b) A triglyceride is an ester of glycerol with three, long-chain monocarboxylic acids (fatty acids).

(c) In a simple glyceride all three fatty acid molecules esterified with glycerol are of the same type, e.g., tristearin.

(d) In a mixed glyceride at least two and often all three fatty acid molecules esterified with glycerol are different.

(e) A fatty acid is a long-chain (C_{12} - C_{18}) monocarboxylic acid.

(f) A soap is a metal salt of a fatty acid. The metal ion is usually Na^+ or K^+ and the fatty acid is derived from a triglyceride by hydrolysis in alkaline solution.

27-7. (a) A lipid is any cell constituent that is soluble in a nonpolar solvent. A fat is a type of lipid; it is a triglyceride in which the fatty acid residues (R groups) are mostly saturated.

(b) Fats and oils are both triglycerides and differ only in the composition of their fatty acid groups. In fats these are mostly saturated, and in oils there is unsaturation (carbon-to-carbon double bonds) in most of the fatty acid groups.

(c) A fat is an ester of the trihydric alcohol, glycerol. A wax is an ester of a fatty acid with a long-chain *monohydric* alcohol.

(d) Butter is a lipid derived from natural sources (milk) in which saturated fatty acids predominate. Margarine is a synthetic, butter-like material that is derived from oils by hydrogenation. Thus, in the manufacture of margarine unsaturated fatty acids are converted to saturated fatty acids.

27-8. Simple and mixed glycerides have only nonpolar components (long-chain fatty acid residues). As a result they tend to dissolve in nonpolar solvents, such as CCl_4, but not in a polar solvent like water. Phospholipids have a polar portion of the molecule (the phosphate ester group) as well as nonpolar portions. As a result their water solubility is greater than that of simple and mixed glycerides.

27-9. Start with the structure of the triglyceride. This is necessary to determine its molecular weight. Three moles of KOH are required per mole of glyceride for the saponification. Express the quantity of KOH consumed as mg KOH per gram of glyceride. The saponification value is 216.

$$\begin{array}{l} H_2C-OOC-C_{11}H_{23} \\ \quad | \\ HC-OOC-C_{15}H_{31} \\ \quad | \\ H_2C-OOC-C_{17}H_{35} \end{array} \qquad \text{molecular weight} = 779$$

$$\text{no. mg KOH} = 1.00 \text{ g glyceride} \times \frac{1 \text{ mol glyceride}}{779 \text{ g glyceride}} \times \frac{3 \text{ mol KOH}}{1 \text{ mol glyceride}} \times \frac{56.1 \text{ g KOH}}{1 \text{ mol KOH}} \times \frac{1000 \text{ mg KOH}}{1 \text{ g KOH}} = 216$$

27-10. The glyceride has all three acid residues the same (it is a simple glyceride). Moreover, the fatty acid residues must be unsaturated, otherwise there would be no iodine number. Finally, the fatty acid must be listed in Table 27-2. All of the possible triglycerides have a molecular weight of about 880.

The four possibilities derived from Table 27-2 all have about the same saponification value but their iodine numbers would be different because of the different degrees of unsaturation in the fatty acids. Triolein would require three moles of iodine per mole of glyceride; trilinolein would require six; and trilinolenin and trieleostearin would require nine. As shown below, the experimentally determined iodine number leads to 6 mol I_2 per mole of glyceride. The glyceride is trilinolein.

$$\frac{\text{no. mol } I_2}{\text{mol glyceride}} = \frac{174 \text{ g } I_2}{1.00 \times 10^2 \text{ g glyceride}} \times \frac{880 \text{ g glyceride}}{1 \text{ mol glyceride}} \times \frac{1 \text{ mol } I_2}{254 \text{ g } I_2} = 6.03$$

27-11. Polyunsaturated acids have more than one double bond per fatty acid chain. Stearic acid is a saturated fatty acid, but eleostearic acid is polyunsaturated. Safflower oil is recommended in dietary programs because it has a much higher proportion of linoleic acid groups in its triglycerides than do most edible oils; it is polyunsaturated.

27-12. The saponification value of a glyceride is related to its molecular weight but not to its degree of unsaturation. The higher the iodine number of a lipid the more unsaturated its fatty acid components and the more desirable it is as a food. The "best" of the lipids listed in Table 27-3 from this standpoint is safflower oil.

Carbohydrates

27-13. (a) A monosaccharide is a simple sugar, for example, glucose. It is a polyhydroxy aldehyde or ketone.

(b) A disaccharide is formed by the joining together (through acetal formation) of two monosaccharide units. Sucrose is a common disaccharide.

(c) An oligosaccharide is a molecule formed by the linking of from two to ten monosaccharide units.

(d) A polysaccharide is formed by the joining together of a large number of monosaccharide units (more than ten). Starch and cellulose are polysaccharides.

(e) Sugar is a term used to denote any monosaccharide or an oligosaccharide.

(f) Glycose is a term used to refer to any carbohydrate, whether a sugar or a polysaccharide.

(g) An aldose is a polyhydroxy aldehyde.

(h) A ketose is a polyhydroxy ketone.

(i) A pentose is a five carbon atom chain (or ring) monosaccharide. It may be either a polyhydroxy aldehyde or a polyhydroxy ketone. Ribose and deoxyribose are examples considered in this chapter.

(j) A hexose is a six carbon atom chain (or ring) monosaccharide. Glucose is a hexose.

27-14. (a) A dextrorotatory compound rotates the plane of polarized light to the right.

(b) A levorotatory compound rotates the plane of polarized light to the left.

(c) A racemic mixture contains equal proportions of the dextrorotatory and levorotatory isomers of an optically active compound. As a result, the racemic mixture does not rotate the plane of polarized light at all. The racemic mixture is denoted as *dl*.

(d) Diastereomers are isomers of a compound that are optically active but are not mirror images.

(e) The symbol (+) is used to indicate a dextrorotatory isomer.

(f) The symbol (-) represents a levorotatory isomer.

(g) The designation D (meaning small capital d) refers to a particular configuration in space of the groups attached to an asymmetric carbon atom. This configuration is described on page 846 of the textbook.

(h) The symbol *l*, like (-), represents a levorotatory isomer.

(i) The symbol *d*, as does (+), designates a dextrorotatory isomer.

27-15. The D configuration of glucose is shown in the text to be (+). The L configuration must be (-), that is, levorotatory.

```
      CHO
       |
 HO - C - H
       |
 H  - C - OH
       |
 HO - C - H
       |
 HO - C - H
       |
      CH2OH
```

27-16. The α and β forms of glucose are diastereomers. The magnitudes of their rotation of plane polarized light are different, and they are both dextrorotatory. If they were enantiomers the magnitudes would be the same and the directions would be opposite, that it one + and the one -. The α and β structures are not mirror images (see Figure 27-6).

27-17. Let the fraction of the α form be x and that of the β form, $1 - x$.

$112x + 18.7(1 - x) = 52.7$

$112x + 18.7 - 18.7x = 52.7$

$93x = 34.0; x = 0.37; 1.00 - x = 0.63$

37% α; 63% β

27-18. The monosaccharide fructose (and the monosaccharide glucose as well) is in equilibrium with the straight chain form of the sugar. This is the form that can undergo reaction with Cu^{2+}. In sucrose both monosaccharide units are tied up (glucose as an acetal and fructose as a ketal). No OH groups remain on the C atoms where ring closure occurs; no straight chain form exists, and so no reduction of Cu^{2+} is possible.

Amino acids, polypeptides, and proteins

27-19. (a) An alpha amino acid is a carboxylic acid that has an amino group ($-NH_2$) attached to the first carbon atom beyond the carboxyl group. By this definition both of the following are α-amino acids.

H_2NCH_2COOH (glycine)

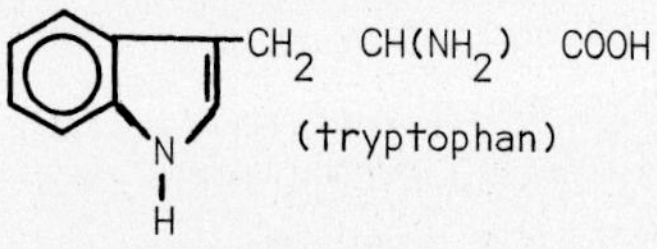

$-CH_2\ CH(NH_2)\ COOH$ (tryptophan)

(b) A zwitterion is a dipolar ion (an ion with both a positive and negative charge center) formed by the transfer of a proton from the carboxyl to the amino group of an α—amino acid. The zwitterion of glycine is depicted below.

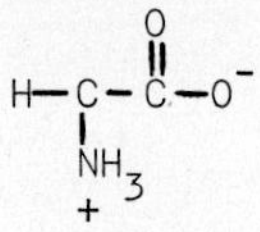

$$H-\underset{\underset{+}{NH_3}}{\underset{|}{C}}-\overset{\overset{O}{\|}}{C}-O^-$$

(c) The isoelectric point of an amino acid is the pH at which the acid exists primarily as the zwitterion. Under these conditions the amino acid unit has no net electrical charge and does not migrate in an electric field.

(d) A peptide bond joins two amino acid units through the elimination of water; the H atom originates from the $-NH_2$ group of one amino acid and the OH from the carboxyl group of the other.

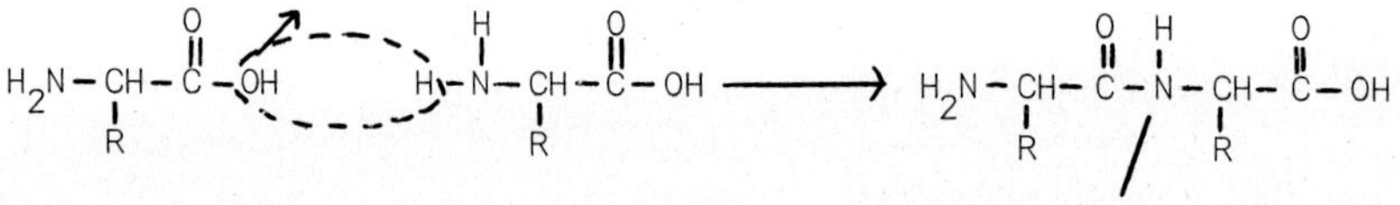

$$H_2N-\underset{R}{\underset{|}{CH}}-\overset{\overset{O}{\|}}{C}-OH\quad H-\overset{\overset{H}{|}}{N}-\underset{R}{\underset{|}{CH}}-\overset{\overset{O}{\|}}{C}-OH \longrightarrow H_2N-\underset{R}{\underset{|}{CH}}-\overset{\overset{O}{\|}}{C}-\overset{\overset{H}{|}}{N}-\underset{R}{\underset{|}{CH}}-\overset{\overset{O}{\|}}{C}-OH$$

peptide bond

(e) A polypeptide is a chain of several α-amino acids joined by peptide bonds.

(f) If the number of α-amino acid units in a polypeptide exceeds about 50-75, the polypeptide is usually called a protein.

(g) In a polypeptide chain the last amino acid unit at one end of the chain has a free $-NH_2$ group. This is called the N-terminal amino acid. (The other end of the chain has a free carboxyl group and is called C-terminal.)

(h) In many proteins the polypeptide chains are coiled, with bonding between successive turns of the coil occurring through hydrogen bonds. The alpha helix is the right-handed spiral structure characteristic of many polypeptide chains (see Figure 27-12).

(i) Denaturation refers to any process by which the secondary, tertiary, and/or quaternary structures of proteins are disrupted. Usually this means a process in which hydrogen bonds, salt (ionic) linkages, or disulfide bonds are broken--by heating, addition of heavy metal ions, treatment with acids and bases, and so forth.

27-20. In a buffer solution with pH = 6.3, amino acids that carry a negative charge (pI < 6.3) migrate to the anode (positive electrode). Those with a positive charge (pI > 6.3) migrate to the cathode. Those with a pI = 6.3 do not migrate at all.

Lysine, cathode; proline, no migration; glutamic acid, anode.

27-21. (a) The six possibilities are Ala-Ser-Lys; Ala-Lys-Ser; Ser-Lys-Ala; Ser-Ala-Lys; Lys-Ala-Ser; Lys-Ser-Ala.

(b) The six possibilities are Ser-Ser-Ala-Ala; Ser-Ala-Ser-Ala; Ser-Ala-Ala-Ser; Ala-Ala-Ser-Ser; Ala-Ser-Ala-Ser; Ala-Ser-Ser-Ala.

27-22. (a) Proceed as in Example 27-5. Arrange the fragments as follows:

```
                Ser - Gly - Val
                            Val - Thr
          Ala - Ser
                            Val - Thr - Leu
                      Gly - Val - Thr
-----------------------------------------------
Sequence: Ala - Ser - Gly - Val - Thr - Leu
```

(b) alanylserylglycylvalylthreonylleucine

27-23. The primary structure of a protein refers to the sequence of amino acid units in the polypeptide chain. The bonds involved are amide bonds (-CO-NH-). The secondary structure of a protein describes the actual structure of the polypeptide chain, that is, whether a spiral, a pleated sheet, or simply random coiling. The bonds responsible for secondary structure are hydrogen bonds between the groups >NH and O=C<. Tertiary structure refers to additional structure beyond the secondary, for example the folding and intertwining of a helical coil (as depicted in Figure 27-15). Several different types of linkages lead to this tertiary structure (illustrated through Figure 27-14). In cases where the protein consists of more than a single polypeptide chain, an additional structural feature is the relationship of these chains to one another. This is called the quaternary structure of the protein (see Figure 27-16, for example). If the protein consists of a single polypeptide chain, it has no quaternary structure.

27-24. If it is assumed that there is a single Ag^+ associated with each enzyme molecule (one active site per molecule), the molar mass of the protein is simply the mass of enzyme required to bind one mole of Ag^+.

$$\text{no. g enzyme} = 1 \text{ mol } Ag^+ \times \frac{1.0 \text{ mg enzyme}}{0.346\times10^{-6} \text{ mol } Ag^+} \times \frac{1 \text{ g enzyme}}{1000 \text{ mg enzyme}} = 2.9\times10^3 \text{ g enzyme}$$

The value obtained is a minimum molecular weight. For example, if we had assumed two active sites per molecule, the molar mass would have been the mass of enzyme required to bind two moles of Ag^+-- twice the value determined above.

27-25. The origin of sicle-cell anemia is in the substitution of valine for glutamic acid at one site in two of the four polypeptide chains in a hemoglobin molecule. That is, there is a very slight difference in the primary structures of normal hemoglobin and sickle-cell hemoglobin. Since this change in molecular structure produces a diseased condition, it is appropriate to refer to the disease as a "molecular disease".

Biochemical reactivity

27-26. (a) A metabolite is a chemical substance involved in a metabolic process.

(b) Anabolism is a metabolic process in which substances are synthesized from smaller molecules.

(c) Catabolism refers to the degradation of large into smaller molecules in a metabolic process.

(d) An endergonic biochemical reaction is one that is accompanied by an increase in free energy.

(e) ADP is the symbol for adenosine diphosphate, an intermediate in metabolic processes.

(f) ATP is the symbol for adenosine triphosphate, also an intermediate in metabolism. The addition of inorganic phosphate and absorption of energy converts ADP to ATP. Thus, ATP is a molecule in which metabolic energy is stored.

27-27. If we assume that 33.5 kJ of energy is stored for every mole of ADP that is converted to ATP, the total quantity of energy stored is 15 x 33.5 = 502 kJ, of a possible 837 kJ.

$$\% \text{ efficiency} = \frac{502 \text{ kJ}}{837 \text{ kJ}} \times 100 = 60.0\%$$

27-28. Use equation (16.25), $\Delta\bar{G}° = -2.303 \cdot RT \cdot \log K$.

$$\log K = \frac{-13.8 \times 10^3 \text{ J/mol}}{-2.303 \times 8.314 \text{ J mol}^{-1} \text{ K}^{-1} \times 298 \text{ K}} = 2.42 \qquad K = 2.6 \times 10^2$$

27-29. An enzyme is a protein. Any process that denatures a protein (disrupts secondary, tertiary and/or quarternary structures) would be expected seriously to inhibit or even to destroy enzyme action. The linkages responsible for higher level structure in proteins (hydrogen bonds and salt linkages, for example) are altered as the pH of the environment of the enzyme changes.

Nucleic acids

27-30. The two types of nucleic acid are ribonucleic acid (RNA) and deoxyribonucleic acid (DNA). Their principal constituents are listed below.

DNA: deoxyribose; the purine bases adenine and guanine; the pyrimidine bases thymine and cytosine; phosphate groups.

RNA: ribose; the purine bases adenine and guanine; the pyrimidine bases cytosine and uracil; phosphate groups.

27-31. The text comments on the manner in which DNA replication occurs and on how the synthesis of proteins is directed by DNA. From this standpoint it would certainly appear that DNA is the basic chemical substance of the living state. Given further, the shape of the molecule--long, coiled, and thread-like--the term "thread of life" seems quite appropriate.

27-32. (a) DNA serves as the template upon which a molecule of messenger RNA (mRNA) is synthesized. The sequence of amino acids in a protein can be traced back ultimately to the sequence of bases on a DNA strand.

(b) mRNA is the molecule that carries the genetic information of DNA into the cell cytoplasm where protein synthesis occurs. This information is coded in the form of the sequence of bases on the mRNA strand.

(c) tRNA is a smaller RNA molecule, formed in the cytoplasm, that carries an amino acid unit to the site on a ribosome where protein synthesis occurs. The tRNA can interact with the mRNA and transfer its amino acid to the protein chain only if the anti-codon of the tRNA matches the codon of the mRNA.

27-33. The ribosome attaches at the site of protein synthesis only the tRNA molecule carrying the amino acid unit called for by the mRNA code. In this sense, the ribosome must "read" the code to make the proper selection of tRNA.

27-34. Here we must work in the reverse fashion from Review Problem 10. That is, given the amino acid sequence in the polypeptide we must determine a corresponding code on the mRNA; but since there is more than one codon for some amino acids, several codes are possible. We will write only one. The next step is to determine the sequence of bases on DNA that is complementary to the code on the mRNA. This requires using the scheme for hydrogen bonding presented in Figure 27-22.

amino acid sequence Ser-Gly-Val-Ala

possible mRNA code UCU GGU GUU GCU

RNA base sequence AGA CCA CAA CGA

27-35. The triplet TCG on one DNA strand is complementary to the triplet AGC on the opposite strand. Hydrogen bonding between the two strands is represented below.

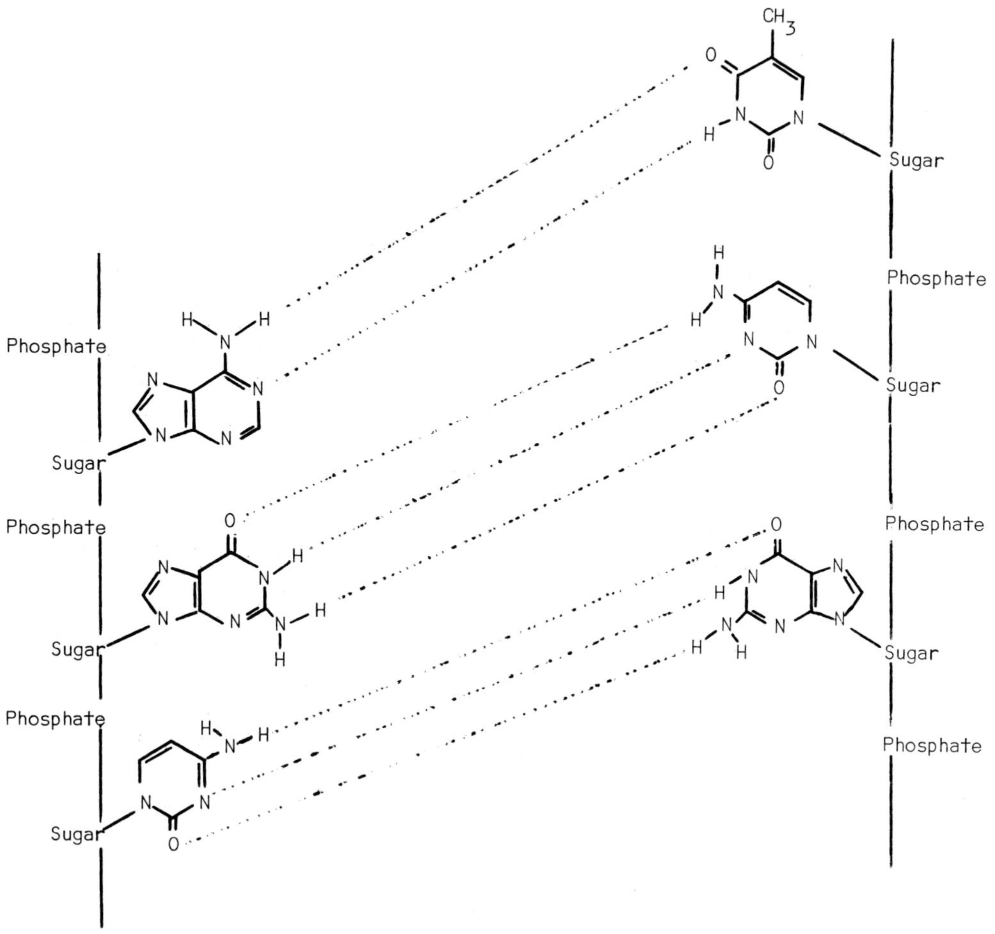

CH_3

O N H N O Sugar

H N H N H N O Phosphate Sugar

H N H Phosphate N N N N Sugar

Phosphate O N H N N N H N H H

O H N H N H N N N N Phosphate Sugar

Phosphate H N H N N O Sugar

Phosphate

Self-Test Questions

1. (*b*) From the name that is given we see that the substance is a simple triglyceride. Thus it must be either a fat or an oil. If a 17-carbon chain is saturated it would correspond to the group $-C_{17}H_{35}$ (based on the hydrocarbon C_nH_{2n+2} or $C_{17}H_{36}$). The group in this case is $-C_{17}H_{31}$. It contains two double bonds and is therefore unsaturated (loss of two H atoms per double bond). The substance is an oil.

2. (*c*) The hydrocarbon group $-C_{17}H_{35}$ is saturated and $-C_{17}H_{33}$ is unsaturated (one double bond). The two triglycerides would differ by only 6 H atoms or 6 units in a molecular weight of about 900. Saponification values, which depend only on molecular weight, would be almost the same for the two triglycerides. On the other hand, only the triglyceride with unsaturation would have an iodine number.

3. (*d*) The *d* isomer rotates the plane of polarized light in one direction, and the *l* isomer, to the same extent in the opposite direction. As a result, the racemic mixture would not rotate the plane of polarized light at all.

4. (*a*) The terms α and β are used to describe the orientation of the OH group on a carbon atom at which ring closure occurs in the transformation of a straight-chain to a ring structure of a sugar. Thus, its appearance in the name of a sugar signifies the ring form.

5. (*d*) Egg white is a protein. Coagulation of a protein is a denaturation process. Hydrolysis of a protein refers to breaking down the polypeptide chain into amino acids. Saponification refers to the hydrolysis of a triglyceride in alkaline solution.

6. (*a*) Glucose is a metabolite and releases energy as it is broken down into H_2O and CO_2. This energy is stored by the conversion of ADP to ATP.

7. (*c*) Purine bases, phosphate groups, and pentose sugars are all constituents of nucleic acids. Glycerol is a fundamental constituent of all triglycerides.

8. (*b*) The structure of DNA is a double helix. The other structures listed are for various proteins.

9.

$$\begin{array}{l} CH_2OOC(CH_2)_{12}CH_3 \\ | \\ CHOOC(CH_2)_{14}CH_3 \\ | \\ CH_2OOC(CH_2)_{14}CH_3 \end{array} + 3\ NaOH \longrightarrow \begin{array}{l} CH_2OH \\ | \\ CHOH \\ | \\ CH_2OH \end{array} + \underset{\text{soap}}{3\ CH_3(CH_2)_{14}COONa}$$

(MW = 807)

$$\text{no. g soap} = 125 \text{ g triglyc.} \times \frac{1 \text{ mol triglyc.}}{807 \text{ g triglyc.}} \times \frac{3 \text{ mol soap}}{1 \text{ mol triglyc.}} \times \frac{278 \text{ g soap}}{1 \text{ mol soap}} = 129 \text{ g soap}$$

10.

Val - Phe

Cys - Val - Phe

Gly - Cys

Phe - Tyr

Sequence: Gly - Cys - Val - Phe - Tyr

11. Enzymes are proteins that catalyze biochemical reactions in regions of the protein molecule called active sites. If the active sites are blocked by a foreign substance, such as a metal ion, the enzyme loses its catalytic activity. If structural changes occur in the protein, this can distort the active sites, leading to a loss of catalytic activity. Such structural changes occur when a protein is subjected to increased temperatures or to changes in pH outside of a fairly narrow range. A precise match must exist between the structure of the substrate and the active site on an enzyme (lock-and-key model). The enzyme is very specific in the reaction it catalyzes.

12. The primary structure of a protein refers to the amino acid sequence in the polypeptide chain. Secondary structure describes the shape of the chain, e.g., whether coiled into a helix or arranged into a pleated sheet. The ternary structure of a protein describes its overall three-dimensional shape, such as the folding and compression of a helical chain into a globular or spherical form.